Water Resources and Hydraulics

This exciting new textbook introduces the concepts and tools essential for upper-level undergraduate study in water resources and hydraulics. Tailored specifically to fit the length of a typical one-semester course, it will prove a valuable resource to students in civil engineering, water resources engineering, and environmental engineering. It will also serve as a reference textbook for researchers, practicing water engineers, consultants, and managers. The book facilitates students' understanding of both hydrologic analysis and hydraulic design. Example problems are carefully selected and solved clearly in a step-by-step manner, allowing students to follow along and gain mastery of relevant principles and concepts. These examples are comparable in terms of difficulty level and content with the end-of-chapter student exercises, so students will become well equipped to handle relevant problems on their own. Physical phenomena are visualized in engaging photos, annotated equations, graphical illustrations, flowcharts, and tables.

Xixi Wang is a professor in the Civil and Environmental Engineering Department at Old Dominion University, where he is Director of the Hydraulics/Water Resources Engineering Laboratory. He has extensive teaching and research experience in hydrology, hydraulics, and water resources. He has published in many high-quality journals and has served as an Editor-in-Chief of the *International Journal of Water Sciences*, and associate editor for the *Journal of Spatial Hydrology*, *Transactions of the American Society of Agricultural and Biological Engineers*, and *Applied Engineering in Agriculture.* In 2018, he won the Most Inspiring Faculty Award at Old Dominion University, and was awarded the National First Award of Science and Technology Research Excellent Achievement in a Higher Institution of China by the Chinese Education Ministry.

"This is an excellent textbook for surface water and groundwater hydraulics, as well as water resources engineering. The systematic and detailed presentation of design examples and the high-quality problem sets at the end of each chapter are major strengths and should help the student grasp complicated concepts and design procedures."

Krishnanand Maillacheruvu, Bradley University

"The book presents a broad range of fundamental topics necessary for design of water resources systems, and illustrates their utility with practical modern-day applications . . . The examples are well laid out with detailed explanations, which makes understanding of the material easy. The appendices provide a quick introduction to advanced spreadsheet analysis and mathematical models, which are essential for modern-day engineering practice."

Venkatesh Uddameri, Texas Tech University

"This book is a much needed update to classical textbooks on water resources. Students often clamor for more examples and this text does an outstanding job of presenting numerous example problems, and it even integrates examples of how to use common engineering software to help solve these problems. I also appreciate the simple use of color to clarify the meaning of each variable in equations . . . designed with specific suggestions for use that will make the book easy to implement in my own course. Overall, this is a much needed, easy to use, and modern undergraduate textbook."

Bradley Striebig, James Madison University

"The textbook *Water Resources and Hydraulics* integrates hydrologic and hydraulic principles. Numerous solved examples within the text illustrate basic concepts and procedures, and the end-of-chapter problems are very instructive for students and course instructors. This will be a very useful textbook for students interested in water resources."

Vijay P. Singh, Texas A&M University

Water Resources and Hydraulics

Xixi Wang
Old Dominion University

CAMBRIDGE
UNIVERSITY PRESS

CAMBRIDGE
UNIVERSITY PRESS

University Printing House, Cambridge CB2 8BS, United Kingdom

One Liberty Plaza, 20th Floor, New York, NY 10006, USA

477 Williamstown Road, Port Melbourne, VIC 3207, Australia

314–321, 3rd Floor, Plot 3, Splendor Forum, Jasola District Centre, New Delhi – 110025, India

79 Anson Road, #06–04/06, Singapore 079906

Cambridge University Press is part of the University of Cambridge.

It furthers the University's mission by disseminating knowledge in the pursuit of education, learning and research at the highest international levels of excellence.

www.cambridge.org
Information on this title: www.cambridge.org/9781108492478
DOI: 10.1017/9781108591768

First published 2021

Printed in Singapore by Markono Print Media Pte Ltd

A catalogue record for this publication is available from the British Library.

ISBN 978-1-108-49247-8 Hardback

Additional resources for this publication at www.cambridge.org/waterresources

Contents

Preface

Water Resources and Hydraulics is written for an upper-level, one-semester course for students majoring in civil, environmental, hydraulic, and/or water resources engineering. Ideally readers will have had previous instruction on hydromechanics or fluid mechanics, but may or may not have exposure to hydrology. This textbook can also be used as a continuing education or self-study reference for practical engineers, consultants, and water resources managers.

This textbook presents fundamental materials, practical examples, and problems in depth to maximize the teaching–learning results. It is devised for students to efficiently gain an elemental understanding of the fundamentals of hydrologic analysis (e.g., estimation of design storm, runoff volume, and peak discharge) and hydraulic design (e.g., selection of pump and turbine; sizing of channel, conduit, bridge opening, and culvert; and computation of water surface profile). Furthermore, this textbook covers the basics of groundwater hydraulics and unsteady-state flow.

Unlike most of the competing books that are currently available, this textbook does not attempt to present the contents that should be covered in other subject-specific books on hydrology, hydromechanics, fluid mechanics, sediment and pollutant transport, and hydrologic and hydraulic modeling. Those detailed subject matters can be redundant, diluting and distracting students from the topics that are imperative for building a basic capability of hydrologic analysis and hydraulic design. The primary mission of a *water resources and hydraulics* class is ultimately to impart the necessary topics for building said capability. Though such a class may have a different name, such as Water Resources Engineering, Hydraulic Engineering, Water Resources and Hydraulics, or Applied Hydrologic Design, depending on the university and/or college, the class is usually taught for one semester as a required junior- and/or senior-level course for all civil and environmental engineering undergraduate students. This textbook intends to best serve the above mission by overcoming the weaknesses of the competing books.

Xixi Wang

This textbook is unique for four reasons. First, its topics are carefully selected so that the cores of hydrologic and hydraulic engineering are introduced with sufficient detail for students to gain and retain knowledge in these fundamentals. Second, example problems, which are carefully selected and solved step by step, clearly demonstrate the relevant principles, concepts, and calculations. They are also comparable, in terms of difficulty level and content, with corresponding end-of-chapter problems. Third, annotated equations, flowcharts, graphical illustrations, monographs, photos, and tables are used to visualize the physical phenomena and facilitate understanding of the related principles. Finally, this textbook provides sufficient detail for all selected topics and can operate as a one-stop information source. Therefore, the instructors who use this textbook will not need additional reference books.

Acknowledgments

I would like to thank Dr. Matt Lloyd, Editor in Earth and Environmental Sciences and Publishing Director in Science, Technology & Medicine, and Americas at Cambridge University Press, for his credence in publishing this textbook. My sincere thanks are also extended to Mr. Dan Kaveney, former Executive Editor at Oxford University Press, who encouraged me to develop the first version of the textbook proposal and coordinated the first-round external reviews. In addition, I highly appreciate the proofreading comments of Mr. Spencer Cotkin, Freelance Development Editor, Mr. David Hemsley, Freelance Copy Editor, and Ms. Lou Attwood, Freelance Proofreader, on the manuscript chapters, which were very helpful for revising the textbook. Further, Ms. Lisa Pinto, Development Team Lead in Higher Education Publishing, Ms. Maggie Jeffers, Associate Development Editor, and Ms. Rachel Norridge, Content Manager in Higher Education Publishing, all at Cambridge University Press, tirelessly offered assistance in acquiring permission for third-party material access and tackling various problems as well as in coordinating both external and internal reviews, proofreading, and production of the textbook. The typesetters at Cambridge University Press did an excellent job in actually producing this textbook. Moreover, I want to express heartfelt gratitude to my parents, Fuying Wang and Erheiyan Zhou, for their selfless support, and to my siblings, Limei Wang, Lixia Wang, Lifang Wang, and Jun Wang, for their encouragement, sacrifice, and support in my growing process. Finally, thank you to my wife, Peilian Cui, and my daughter, Yueying Wang, for their unconditional love and support, without which this textbook would not be possible.

How to Use This Textbook

This textbook is organized into ten chapters and contains five appendixes. **Chapter 1** presents an overview of the practices and principles of water resources engineering. **Chapter 2** reviews the basics of fluid mechanics, which are needed background for subsequent chapters. **Chapter 3** discusses hydrologic processes and analysis methods related to design and management of water resources engineering structures. **Chapter 4** discusses methods for formulating synthetic hydroclimatic extremes for hydrologic engineering design purposes through statistical analysis of observed data on rainfall and streamflow and conceptual empirical formulas. **Chapter 5** discusses the characteristics and selection of two commonly used hydraulic machines, namely pumps and turbines, from a hydraulic engineering perspective. **Chapter 6** describes water surface profile classification and computation, flow measuring, and channel design. **Chapter 7** introduces the application of the continuity equation and/or the energy equation in formulating the governing equations for a single pipe, a pipeline, and a pipe network. **Chapter 8** introduces the principles of hydraulics as they relate to common structures, including culverts, bridges, risers, storm sewers, spillways, and stilling basins. **Chapter 9** discusses the basics of groundwater hydraulics. **Chapter 10** discusses three types of unsteady-state flows. The appendixes present the relevant constants, unit conversion factors, basic calculus, and commonly used hydrology and hydraulic models. There is also a demonstration on how to use Microsoft Excel® Solver.

This textbook is written with a presumption that students already have the prerequisites of calculus, physics, dynamics, fluid mechanics, and probability and statistics. For a standard semester of 18 weeks with classes of three hours per week, the times can be allocated to the chapters in reference to Table 1. The first test covers the material in Chapter 1 through 4, and the second test covers the material in Chapter 5 through 7. The final exam can either be comprehensive or cover only the material in the last three chapters. For schools with a separate hydrology course, the times allocated to Chapter 2 through 4 may be reduced by 0.5 to 1.0 hours, which can be reallocated to the chapters in terms of the instructor's judgement. In addition, if groundwater hydraulics is taught in an independent required course, the 1.5 hours for Chapter 9 can be reallocated to other chapters. Furthermore, some instructors may emphasize steady-state flow, so the 1.5 hours for Chapter 10 can also be reallocated to other chapters. The possible time allocations for these different class setups are presented in Tables 2 through 4.

Table 1 **The suggested time allocations for a normal course setup.**

Allotted weeks	Hours	Class content
1.0	3.0	Chapter 1 and Appendix III
1.5	4.5	Chapter 2
1.5	4.5	Chapter 3
1.5	4.5	Chapter 4
0.5	**1.5**	**First test**
2.0	6.0	Chapter 5
2.0	6.0	Chapter 6
1.5	4.5	Chapter 7
0.5	**1.5**	**Second test**
2.0	6.0	Chapter 8
1.5	4.5	Chapter 9
1.5	4.5	Chapter 10 and Appendix IV
1.0	**3.0**	**Final exam**

Table 2 **Suggested time allocations with a prerequisite of hydrology.**

Allotted weeks	Hours	Class content
1.0	3.0	Chapter 1 and Appendix III
1.5	4.5	Chapter 2
1.0	3.0	Chapter 3
1.0	3.0	Chapter 4
0.5	**1.5**	**First test**
2.0	6.0	Chapter 5
2.5	7.5	Chapter 6
2.0	6.0	Chapter 7
0.5	**1.5**	**Second test**
2.0	6.0	Chapter 8
1.5	4.5	Chapter 9
1.5	4.5	Chapter 10 and Appendix IV
1.0	**3.0**	**Final exam**

Table 3 **Suggested time allocations with a separate groundwater class.**

Allotted weeks	Hours	Class content
1.0	3.0	Chapter 1 and Appendix III
1.5	4.5	Chapter 2
1.5	4.5	Chapter 3
1.5	4.5	Chapter 4
0.5	**1.5**	**First test**
2.0	6.0	Chapter 5

Allotted weeks	Hours	Class content
2.5	7.5	Chapter 6
2.0	6.0	Chapter 7
0.5	**1.5**	**Second test**
2.5	7.5	Chapter 8
1.5	4.5	Chapter 10 and Appendix IV
1.0	**3.0**	**Final exam**

Table 4 **Suggested time allocations with steady-state flow only.**

Allotted weeks	Hours	Class content
1.0	3.0	Chapter 1 and Appendix III
1.5	4.5	Chapter 2
2.0	6.0	Chapter 3
2.0	6.0	Chapter 4
0.5	**1.5**	**First test**
2.0	6.0	Chapter 5
2.5	7.5	Chapter 6
2.0	6.0	Chapter 7
0.5	**1.5**	**Second test**
2.5	7.5	Chapter 8
1.5	4.5	Chapter 9
1.0	**3.0**	**Final exam**

Each chapter has a number of problems closely related to its content and worked examples. Besides a complete solution manual and the relevant Excel spreadsheets, the answers to selected problems are provided online as a stand-alone supplementary sheet. The instructor may assign these and/or other problems for homework and decide whether or not to make the answers available for students before the homework is due. Depending on the difficulty level and number of assigned problems, each homework may take one or two weeks. It is ideal to assign a problem halfway through lecturing on a topic to maximize teaching effectiveness and provide students with quick feedback on their comprehension. In addition, the instructor may use some of the problems as out-of-class exercises or in-class quizzes. Finally, the instructor may relate the problems with the real world to stimulate students' learning interests.

1 Introduction

Since the beginning of time, humans have endeavored to cope with various water-related issues, such as flooding and drought. The successful practices led to the establishment and subsequently the advancement of engineering hydrology and hydraulics, which are branches of science and technology concerned with design of water resources engineering infrastructure, such as channels, pipes, and reservoirs. This chapter presents an overview of the basic concepts and practices used to utilize and protect precious water resources. After highlighting some of the essential aspects of such practices, this chapter discusses the roles of hydrology and hydraulics in water resources engineering design.

1.1 Advances in Water Resources Engineering

Water is critical to human existence and is thus one of the most important resources. In ancient times, hunter-gatherers moved from one location to another in search of fresh water for themselves as well as for their animals. With the advent of agriculture, the first farmers planted crops close to rivers and streams. They passively relied on naturally available water. If the water at a settlement location became insufficient and/or difficult to access, the people moved to a new location where sufficient water could be easily utilized. In dry years, bloody conflicts could break out between tribes competing for limited water, whereas in wet years, flood waters could devastate communities, causing life and property losses. Throughout history people have had to contend with problems resulting from either too little (drought) or too much (flooding) water. Our ancestors' observations and persistent efforts led to some empirical knowledges of water phenomena and successful projects to mitigating drought and flooding. For instance, the Dujiangyan Irrigation System, located in Dujiangyan City, Sichuan Province, southwest China, was originally constructed around 256 BC by the State of Qin as an irrigation and flood control project (https://en.wikipedia.org/wiki/Dujiangyan). It is still in use today and irrigates more than

5300 km^2 of agricultural fields. Prior to construction of this system, the Minjiang River, which is the longest tributary of the Yangtze River, would rush down from the Min Mountains before slowing abruptly in the Chengdu Plain, filling the watercourse with silt and thus making the nearby areas extremely prone to flooding. The Dujiangyan Irrigation System was constructed to harness the river by channeling and dividing the watercourse rather than simply damming it. There are many other examples of ancient hydraulic engineering projects, such as irrigation canals in Egypt and Mesopotamia, diversion channels of the Euphrates River, and ceramic conduits for water supply in Pakistan.

In modern times, from the accumulated experiences gained from those ancient projects, new experimental results, and applications of basic sciences such as physics and calculus, two scientific disciplines, namely hydrology and hydraulics, emerged. These disciplines were established, advanced, and increasingly applied to the design of water resources engineering structures such as channels, culverts, and reservoirs. Hydrology studies the occurrence, storage, and movement of the Earth's water, whereas hydraulics deals with the conveyance of water through pipes and channels. In engineering, hydrologic principles are used to determine the design peak discharge and/or flow hydrograph, which in turn are used in the subsequent hydraulic analyses to determine the size of a structure of interest.

Averaged globally, the Earth receives 1028 mm precipitation annually, about 66% of which is lost to evapotranspiration and infiltration, generating 161 mm runoff as the surface water (Table 1.1). The gross amount of water resources, including groundwater and surface water, is about 10,633,450 km^3 (USGS, 2017), most of which is inaccessible to humans. However, both precipitation and runoff have large spatial variations. South America receives the most precipitation, about 43.8% of which is converted into runoff, whereas Antarctica receives the least precipitation, about 90.9% of which is converted into runoff. In comparison with Asia, Africa receives more precipitation but has a smaller runoff coefficient and thus is drier. North America and Europe receive almost the same amount of precipitation, but Europe has much less runoff because it has a lower runoff coefficient. In Australia and Oceania, only 9.1% of the precipitation (440 mm a^{-1}), which is relatively low, is converted into runoff, making this geographic region both climatically and hydrologically driest among the continents except for Antarctica. The oceans receive the second most precipitation, about 89.5% of which is evaporated back into the atmosphere.

Besides the spatial heterogeneity discussed above, for a given location, the precipitation can greatly vary from season to season in a year and from day to day in a season, resulting in too much water at one time but too little water at another. Such temporal variations have caused major floods and droughts all over the world. A major flood inundates extensive rural and/or urban areas, likely isolating properties and towns and closing major traffic routes. In contrast, a major drought is a prolonged period of abnormally dry weather that can be sufficiently long that the lack of water causes serious problems, including crop damage and/or water supply shortage (NWS, 2019). From 2000 to 2009, there were 175 major floods worldwide, about 70 of which occurred in Asia (Sohoulande-Djebou and Singh, 2016). Previous studies (e.g., Marengo and Espinoza, 2016) have shown that as a result of climate change, larger floods and severer droughts tend to occur more often in more geographic regions. It is interesting that Africa incurred 40 major floods even though its overall runoff was only 110 mm a^{-1}. Major floods also occurred on other continents 5 to 38

Table 1.1 Runoff and number of major floods, droughts, and dams by continents.

Region	Area (km^2)[1]	Precipitation ($mm\ a^{-1}$)[2]	Runoff ($mm\ a^{-1}$)[2]	Runoff coefficient (%)[2]	Number of major floods in the 2000s[3]	Number of major droughts since 1900[4]	Number of major dams by 2019[5]
Africa	30,380,560	740	110	14.9	40	291	1269
Antarctica	13,998,885	110	100	90.9	–	–	–
Asia	44,573,695	650	240	36.9	70	153	34,076
Australia and Oceania	7,692,265	440	40	9.1	5	22	577
Europe	10,181,245	820	230	28	22	42	5480
North and Central America	24,708,485	800	330	41.3	38	134	10,635
South America	17,839,840	1600	700	43.8			4963
Oceans	361,044,345	1140	−120	−10.5	–	–	–
Global	510,419,320	1028	161	34.4	175	642	57,000

[1] *Source*: https://en.wikipedia.org/wiki/Continent, accessed on April 12, 2019.

[2] *Source*: www-das.uwyo.edu/~geerts/cwx/notes/chap10/continents.html, accessed on April 12, 2019.

[3] *Source*: Sohoulande-Djebou and Singh (2016).

[4] *Source*: Masih *et al.* (2014).

[5] *Source*: https://akuinginhijau.files.wordpress.com/2008/11/number-of-dams-country.pdf, accessed on April 12, 2019.

times. In contrast, since 1900, a total of 642 major droughts have occurred worldwide, most of which occurred in Africa (Masih *et al.*, 2014). Countries in Asia and America have also incurred a number of major droughts, while European countries and Australia and Oceania have, so far, incurred relatively fewer major droughts.

To cope with floods and droughts, myriad dams have been constructed, forming reservoirs to regulate uneven runoff flows (Graf, 1999). A reservoir is usually operated to attenuate peak discharges as well as to store extra water during a wet season to be used during the following dry seasons. By doing this, the reservoir has proven to be a successful practice in mitigating flooding and drought issues. Globally, more than 57,000 major (56,700 large and 300 giant) dams have been constructed as of 2019. The height of a large dam is 15 m or more, whereas the height of a giant dam is at least 150 m; the latter forms a reservoir with a storage capacity of more than 3×10^6 m^3. Asia has the greatest number of major dams, followed by North and Central America. Europe has 517 more major dams than South America, while Africa has twice as many major dams than Australia and Oceania combined. In practice, a major dam is always designed to have multiple functions, such as flood control, navigation, power generation, recreation, and water supply. Herein, although the prioritization of these functions may be different from one dam to another, flood control is always the most important function for all dams.

In addition to one or more dams, an actual water engineering system may also include other hydraulic structures (Figure 1.1), such as channels, pipelines, pumps, turbines, bridges, culverts, spillways, and stilling basins, to name a few. At the road crossing, the bridge openings and/or culvert barrels must be large enough to convey the design peak discharge without overtopping the road deck. For the safety of the dam, during large floods, water is released through the spillway downstream into the stilling basin. After the extra energy is dissipated through hydraulic jumps in the basin, the flood water is discharged back into the river. Further, a pipeline conveys high-energy flows from the reservoir into the hydropower plant to drive turbines to generate electricity. After passing through the turbines, water from the plant is released back into the river through a channel. Moreover, a pumping station may be needed to supply water from the reservoir to locations at a higher altitude.

In the past decades, urbanization has rapidly increased throughout the world. This increase could significantly alter natural hydrologic processes by reducing infiltration and increasing runoff (Wang *et al.*, 2017). As illustrated in Figure 1.2, for a given location, as a result of urbanization, the runoff volume and peak discharge will increase, while the baseflow (i.e., groundwater recession flow) will decrease. Also, the peak discharge will appear at an earlier time. The possible consequences are more frequent floods with a larger magnitude and more contaminated stormwater. To minimize such adverse impacts of urbanization, various low-impact development (LID) (Figure 1.3) devices can be installed and/or retrofitted to recover the natural hydrology (Rossmiller, 2014). In an urban environment, the typical land uses include building roofs, driveways, sidewalks, parking lots, streets, and lawns and open spaces (e.g., public parks). The first five of these listed land uses usually have an impervious surface with near-zero infiltration capacity. The increased runoff from building roofs can be treated using two types of LIDs, namely green roof and rain garden, whereas the increased runoff from driveways, sidewalks, parking lots, and streets may be treated using another two types of LIDs, namely porous pavement and permeable

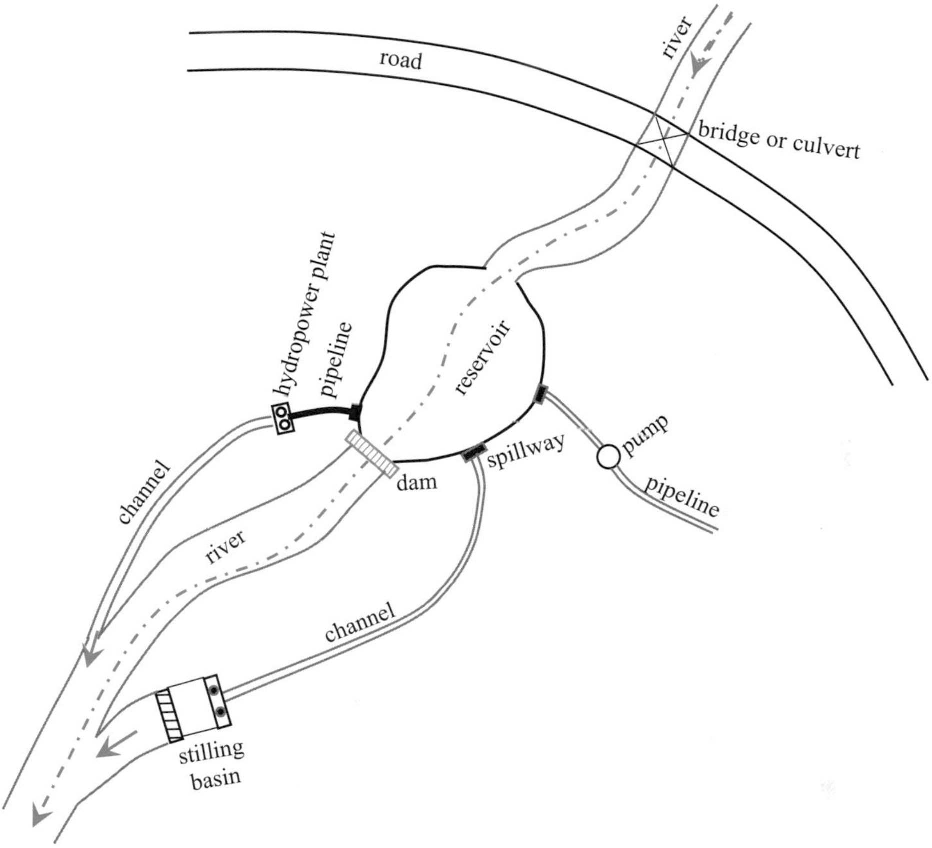

Figure 1.1 **A typical water engineering system showing the most common components.**

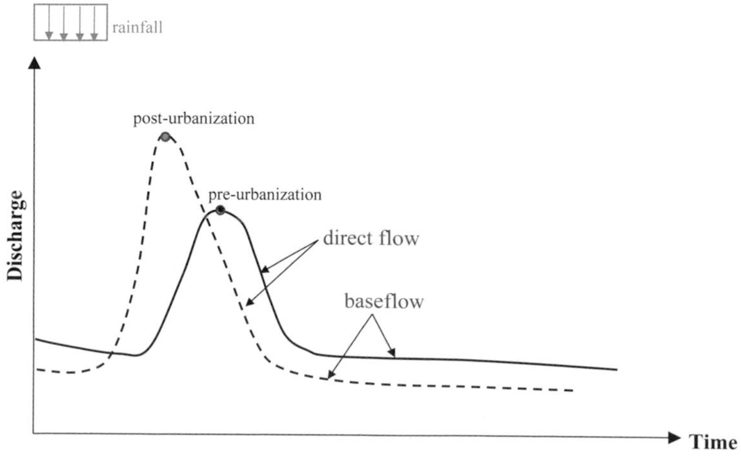

Figure 1.2 **Urbanization impacts on natural hydrology.** The flow hydrograph at a given location is affected by urbanization.

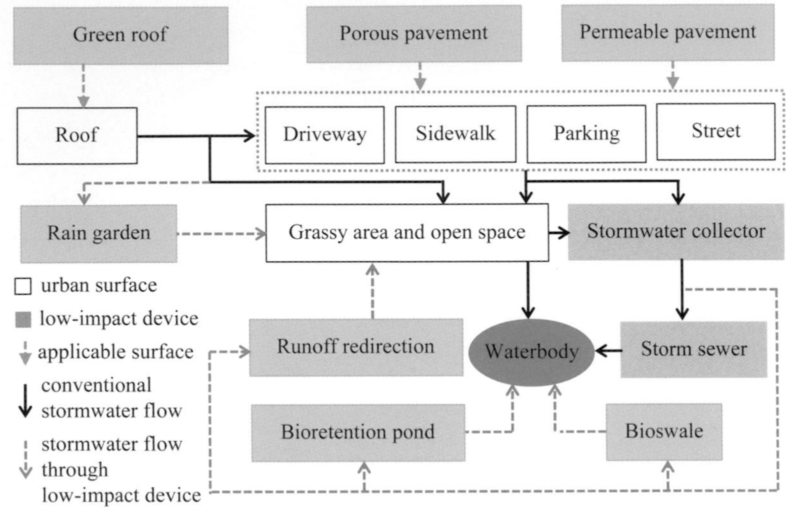

Figure 1.3 **Various features in a typical urban environment.** Land uses are indicated with hollow boxes and low-impact devices are shown with light green shading.

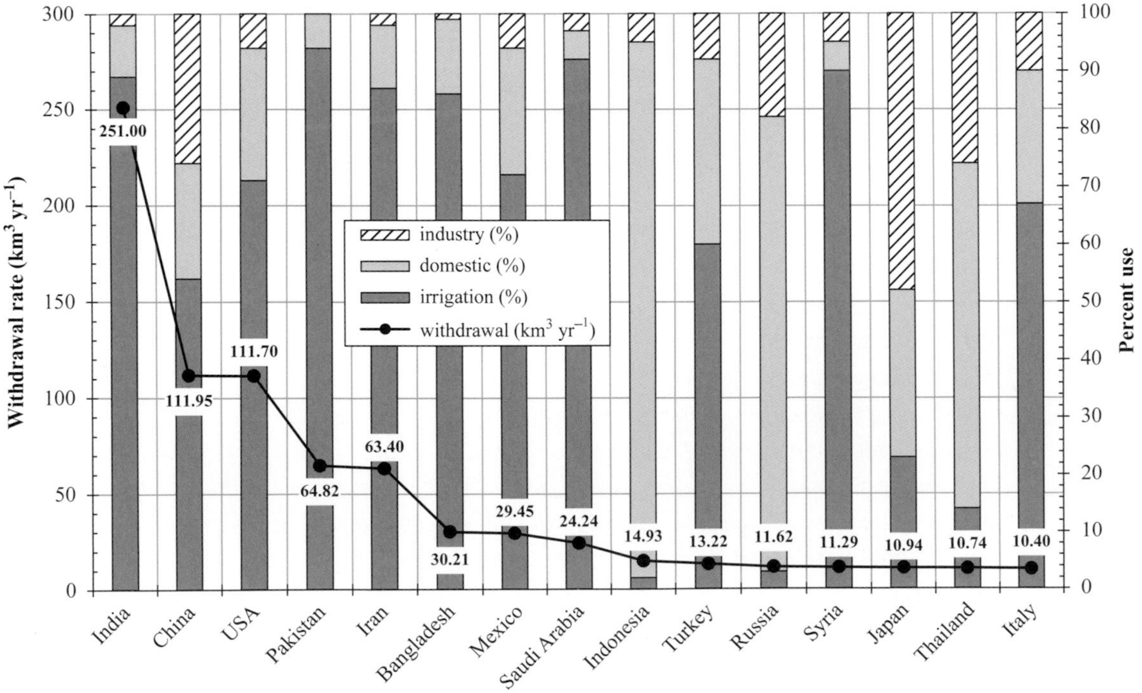

Figure 1.4 **Groundwater exploitation.** Withdrawal rate and uses in 15 countries, which represent nearly 80% of the groundwater withdrawal volume of the world.

pavement. The total runoff from a community of interest can be: (1) directed into three types of LIDs, namely bioswale, bioretention pond, and wet pond; (2) redirected to a pervious area for infiltration, evapotranspiration, and detention; and (3) discharged into a storm sewer system. The outflow from the LIDs may be partially redirected to a pervious area and/or directly released downstream into the storm sewer system and/or a receiving waterway. The runoff, which is redirected to the pervious area and has a reduced volume, will be discharged into the storm sewer, emptying into the receiving waterway. In practice, two or more LIDs, which may either belong to a same type or different types, can be installed in series to treat stormwater runoff to maximize the overall efficiency. In such a case, the outflow from one LID will be the inflow into another downward LID.

Currently, the world withdraws groundwater at an annual rate of 982 km^3 for irrigation, domestic use, and industry, with only 15 countries accounting for about 78.5% of that total, as shown in Figure 1.4 (Todd and Mays, 2005; Margat and van der Gun, 2013). In Indonesia, Thailand, and Russia, groundwater is primarily for domestic use, whereas in Japan, groundwater is mainly used for industry. In another 11 countries, groundwater is mostly used for irrigation. In addition, some countries (e.g., Bahrain and Mongolia) solely rely on groundwater for all water use sectors. Thus, groundwater hydraulics and extraction is an important component of water resources engineering.

1.2 Hydrology and Hydraulics in Water Resources Engineering

For an area of interest, quantifying the amount and spatiotemporal distribution of available water resources is always needed for its sustainable development. In practice, a variety of hydraulic structures, such as reservoirs, ponds, channels, and pipelines, usually need to be constructed to regulate runoff flows as mandated by flood control, drought resistance, and water supply. Also, road crossings, such as bridges and culverts, are imperative for transportation purposes. The design of those structures and crossings relies on the two related science disciplines introduced earlier in this chapter, namely hydrology and hydraulics.

Hydrology studies the occurrence, storage, and movement of the Earth's water. The occurrence concerns precipitation and its partitioning into infiltration, direct runoff, and evapotranspiration, whereas the storage concerns the percent of runoff and infiltrated water that will be detained and/or retained. The detained water will be held back for a limited time (e.g., from minutes to a year), after which it will be released from the holding space. In contrast, the retained water (e.g., film water adsorbed onto soil particles and groundwater in deep confined aquifers) may reside in a space permanently. Some of the detained and retained water can be evaporated back into atmosphere and/or infiltrated into deeper soils, starting a new cycle. The movement of water concerns the physical processes of runoff, infiltration, and evapotranspiration as well as their rates of occurrence. It usually involves quantifying runoff flow paths and velocities, infiltration rates, and evapotranspiration rates as functions of time.

Bras (1999) presented a brief history of hydrology. Precipitation as the original source of stream flow was hypothesized during the first century BC, but such a hypothesis was not supported by quantitative measurements until the seventeenth century. Since then, although the cause-and-effect relationship between precipitation and stream flow was mainly understood in a qualitative way, the understanding ultimately led to the establishment of the water balance concept. However, this fundamental concept did not become quantitative until the twentieth century, when Sherman (1932) enunciated the unit hydrograph method and Horton (1940) put forward the mathematical equations for estimating infiltration, soil moisture, and runoff. This greatly prompted and matured the applications of hydrology in water resources engineering and led to the creation of a new subject field called *engineering hydrology*, which was codified by the textbooks written by Linsley *et al.* (1958, 1982) and Chow (1964) and adopted by various universities to educate engineering hydrologists. In the 1970s, mathematical models were developed by integrating primary hydrologic processes such as infiltration, evapotranspiration, and runoff, making it feasible to predict responses of stream flow to precipitation as influenced by land uses and soil properties. The models had a lumped structure, using one set of parameters to represent a drainage area through its physiographic characteristics, which might have large spatial heterogeneities. Nevertheless, the development of the models was a revolution of the subject of hydrology and its engineering applications.

In the late 1980s and 1990s, recognizing the importance of spatial variability of rainfall, soil properties, land cover and land use, and topography on hydrology, researchers redeveloped the models to have a distributed structure. In this regard, the drainage area of interest is subdivided into a number of hydrologic response units (HRUs), each of which can be assumed to be spatially uniform and homogenous. Since the 2000s, extensive research efforts have been made to predict what the future climate would look like, leading to the development of various General Circulation Models (GCMs) and Regional Climate Models (RCMs) (Laprise, 2008). The GCMs were used to predict the future precipitations and temperatures at a coarse spatial resolution (e.g., 250 to 600 km) at the global scale, while the RCMs were paired with the GCMs to predict the future precipitations and temperatures at a fine spatial resolution (e.g., $\leq$ 50 km) at the regional scale. How to incorporate climate change and its uncertainties into hydrologic engineering design is becoming an important topic that is driven by the practical needs to sustain and improve the resilience of water infrastructure to worsening flooding and drought disasters. Moreover, as an important component of the water cycle, groundwater has been extensively studied since the 1970s. The results have been used widely in practice to direct exploration, utilization, and protection of groundwater resources. The existing models (e.g., MODFLOW) (Harbaugh, 2005) use various algorithms and parameters to represent influences of spatially variable medium properties on groundwater flow and transport.

Hydraulics deals with the conveyance of water through waterways such as pipes and channels. Its practices (e.g., the aforementioned Dujiangyan Irrigation System) began during prehistoric times, but the rationalization of hydraulics as a science was first attempted by Greeks (e.g., Archimedes) in 287 to 212 BC and continued slowly until the time of the Renaissance (i.e., from 1300 to 1600). During those three centuries, hydraulics was primarily acknowledged by observations and empiricisms with minimal uses of mathematics. However, in the seventeenth century, some researchers (e.g., Bernoulli) established the discipline of *hydrodynamics* by describing physical phenomena of flows using the modern mathematics, which produced an array of intimidating equations and methods. This classical hydrodynamics discipline was based on a pure mathematical approach and neglected observations and experimental works. On the other hand, other researchers (e.g., Chezy, Venturi, Darcy, Weisbach, Reynolds, and Froude) devised various apparatuses and experiments to measure flow velocities and resistances and established the discipline of *experimental hydraulics*. In the subsequent two centuries, these two disciplines were far apart and progressed essentially independently, though both aimed to quantitatively describe flow phenomena.

In the twentieth century, researchers (e.g., Prandtl) fused these two superficially interdependent disciplines into one unified science of fluid mechanics, which in turn was applied in almost every branch of engineering. Both the science and its engineering applications were tremendously advanced by the systematic laboratory works of Prandtl, von Kármán, Moody, Colebrook, and Saint-Venant, to name just a few. This textbook presents applications of fluid mechanics in water resources engineering. Since the 1940s, the advent of computers and sensors has led to the development and use of various software packages (e.g., HEC-RAS) to analyze sophisticated hydraulic systems with or without manmade structures. Given the increasing challenges of sustainable development, hydraulics will continue to progress and play an increasingly important role in water management. A detailed overview of the history of hydraulics can be found in Rouse (1983).

PROBLEMS

1.1 By conducting an internet search, identify three prehistoric water resources engineering projects and document their construction years, locations, and major purposes. Are they still being used today?

1.2 Based on an internet search, generate a bar chart showing the mean annual precipitation and evapotranspiration by each state of the USA.

1.3 Based on an internet search, document the mean annual water consumptions by categories (e.g., irrigation, domestic, and industrial) for a selected state in the United States.

1.4 Based on an internet search, document the construction years, locations, materials, heights, storage capacities, major purposes, and accessary structures of the dams for a selected state in the United States.

1.5 Select and sketch a water engineering system as shown in Figure 1.1.

1.6 Based on an internet search, document the impacts of urbanization on water resources for a selected city and the LIDs installed to mitigate the impacts.

1.7 Based on an internet search, document the available groundwater resources and its exploitations for a southwestern state in the United States.

1.8 Based on an internet search, document the impacts of climate change on water resources and possible adaptive measures.

1.9 Describe the roles of hydrology in water resources engineering.

1.10 Describe the roles of hydraulics in water resources engineering.

2 Overview of Hydromechanics

This chapter provides a broad overview of the basics of hydromechanics (also referred to as *fluid mechanics*) that are needed as background for subsequent chapters. The purpose of this chapter is to highlight fundamental concepts and equations rather than to substitute for textbooks of fluid mechanics. The overview covers water properties, hydrostatic pressure and force, conservations of mass, energy and momentum in flowing water, and dimensional analysis and similitude.

2.1 Engineering Properties of Water

Five properties of water are used extensively in the field of water resources and hydraulic engineering. These properties are density, specific weight, dynamic viscosity, kinematic viscosity, and saturation vapor pressure. They are functions of atmospheric pressure and water temperature. Table 2.1 lists the values of the properties at standard atmospheric pressure (Finnemore and Franzini, 2002).

Density is defined as the mass per unit volume of water, whereas specific weight is defined as the weight per unit volume of water. In consistent units, they are related as:

$$\gamma = \rho g \tag{2.1}$$

specific weight [lbf ft^{-3}; N m^{-3}] → γ
gravitational acceleration (= 32.2 ft sec^{-2}; 9.81 m s^{-2}) → g
density [slug ft^{-3}; kg m^{-3}] → ρ

Dynamic viscosity (also known as absolute viscosity) is the shear stress between two layers of water per unit velocity gradient. It can be computed as:

Table 2.1 Properties of water at standard atmospheric pressure.

Temperature, T_w	Density, ρ[1]	Specific weight, γ[2]	Dynamic viscosity, μ[1]	Kinematic viscosity, ν[3]	Saturation vapor pressure, p_v[4]
°C	kg m^{-3}	N m^{-3}	N · s m^{-2}	m^2 s^{-1}	kPa abs
0	1000	9810	1.79×10^{-3}	1.79×10^{-6}	0.611
5	1000	9810	1.51×10^{-3}	1.51×10^{-6}	0.871
10	1000	9810	1.31×10^{-3}	1.31×10^{-6}	1.224
15	999	9800	1.14×10^{-3}	1.14×10^{-6}	1.698
20	998	9790	1.00×10^{-3}	1.00×10^{-6}	2.326
25	997	9781	8.91×10^{-4}	8.94×10^{-7}	3.147
30	996	9771	7.97×10^{-4}	8.00×10^{-7}	4.210
35	994	9751	7.20×10^{-4}	7.24×10^{-7}	5.572
40	992	9732	6.53×10^{-4}	6.58×10^{-7}	7.301
°F	slug ft^{-3}	lbf ft^{-3}	lbf · sec ft^{-2}	ft^2 sec^{-1}	psia
40	1.94	62.4	3.23×10^{-5}	1.66×10^{-5}	0.122
50	1.94	62.4	2.73×10^{-5}	1.41×10^{-5}	0.178
60	1.94	62.4	2.36×10^{-5}	1.22×10^{-5}	0.255
70	1.94	62.4	2.05×10^{-5}	1.06×10^{-5}	0.361
80	1.93	62.1	1.80×10^{-5}	9.33×10^{-6}	0.503
100	1.93	62.1	1.42×10^{-5}	7.36×10^{-6}	0.940
120	1.92	61.8	1.17×10^{-5}	6.09×10^{-6}	1.673
140	1.91	61.5	9.81×10^{-6}	5.14×10^{-6}	2.850

[1] Reproduced from Tables E4.1 and E4.2 of Urroz and Hoeft (2009).
[2] Computed using Eq. (2.1).
[3] Computed using Eq. (2.3).
[4] Computed using Eqs. (2.4) and (2.5).

$$\underset{\text{dynamic viscosity}}{\mu} = \frac{\overset{\text{shear stress between water layers}}{\tau}}{\underset{\text{velocity gradient between water layers}}{\left(\frac{du}{dy}\right)}} \tag{2.2}$$

Kinematic viscosity is the ratio of dynamic viscosity to density of water. It can be computed as:

$$\underset{\text{kinematic viscosity}}{\nu} = \frac{\overset{\text{dynamic viscosity}}{\mu}}{\underset{\text{density}}{\rho}} \tag{2.3}$$

Saturation vapor pressure is the partial pressure of water vapor when the ambient gas mixture is fully saturated (i.e., in equilibrium with solid or liquid water). It is always given as an absolute pressure, which is the summation of the relative (i.e., gauged) pressure and the atmospheric pressure. Saturation vapor pressure can be estimated as (Tetens, 1930):

$$\underbrace{p_\upsilon}_{\text{saturation vapor pressure [kPa abs]}} = 0.61078e^{\frac{17.27T_w}{T_w+237.3}} \quad (T_w: \text{water temperature [°C]}) \tag{2.4}$$

$$\underbrace{p_\upsilon}_{\text{saturation vapor pressure [psia]}} = 0.088586e^{\frac{17.27(T_w-32)}{T_w+395.14}} \quad (T_w: \text{water temperature [°F]}) \tag{2.5}$$

2.2 Hydrostatic Pressure and Force

Water exerts hydrostatic pressures on its confining surface. The integration of the pressures across the surface area produces the corresponding hydrostatic force. This section introduces how to determine hydrostatic pressure and resultant force. In principle, the resultant force on an immersed surface is the integration of the hydrostatic pressures over the surface area.

2.2.1 Pressure

The hydrostatic pressure at a point has the same magnitude in all directions and is always perpendicular to the surface on which the point is located (Figure 2.1). It is computed as:

$$\underbrace{p}_{\text{relative pressure}} = \underbrace{\gamma}_{\text{specific weight of water}} \underbrace{h}_{\text{vertical distance from the free water surface to the point of interest}} \tag{2.6}$$

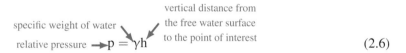

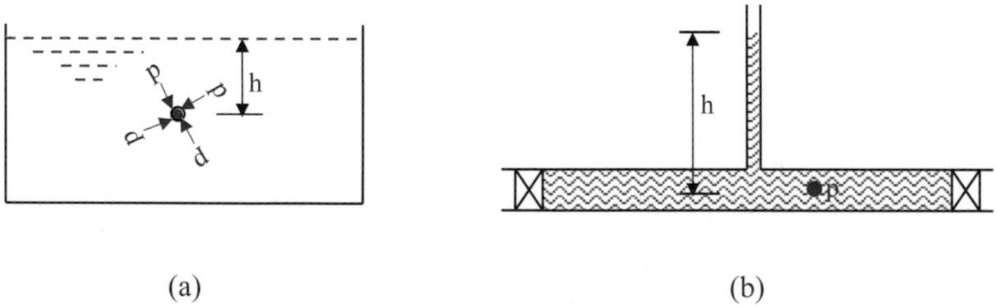

Figure 2.1 **Hydrostatic pressure at a point.** Diagrams show situation for: (a) an open channel; and (b) a closed pipe.

Example 2.1 For the hydraulic structures submerged in water ($\gamma = 62.4\,\text{lbf ft}^{-3}$) shown in Figure 2.2, determine and sketch the hydrostatic pressures on the submerged surfaces.

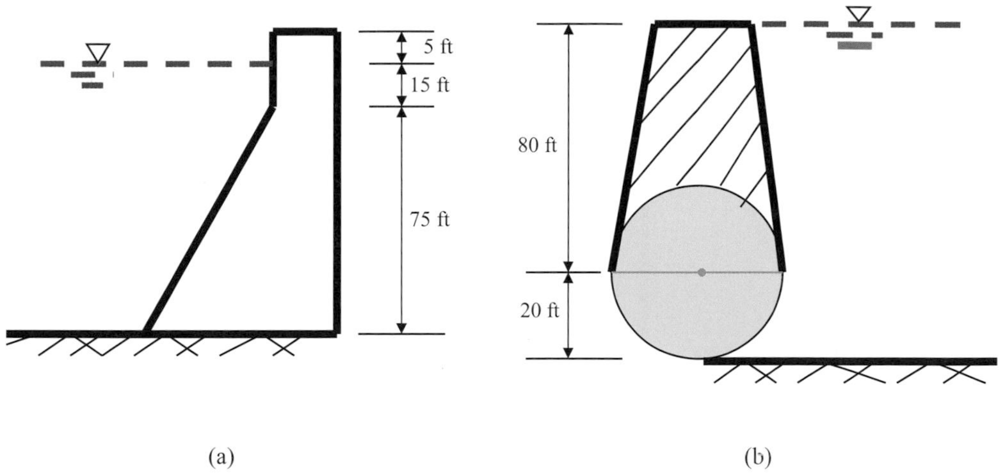

Figure 2.2 **The submerged structures for Example 2.1.** (Not drawn to scale.)

Solution

$\gamma = 62.4\,\text{lbf ft}^{-3}$. Use Eq. (2.6) to compute pressures at different water depths.

For the structure in Figure 2.2a:

At $h = 0\,\text{ft}$, $p = (62.4\,\text{lbf ft}^{-3}) * (0\,\text{ft}) = 0\,\text{lbf ft}^{-2}$.

At $h = 15\,\text{ft}$, $p = (62.4\,\text{lbf ft}^{-3}) * (15\,\text{ft}) = 936\,\text{lbf ft}^{-2}$.

At $h = 15\,\text{ft} + 75\,\text{ft} = 90\,\text{ft}$, $p = (62.4\,\text{lbf ft}^{-3}) * (90\,\text{ft}) = 5616\,\text{lbf ft}^{-2}$.

The pressure distributions between 0 and 15 ft as well as between 15 and 90 ft are linear. The results are shown in Figure 2.3a.

For the structure in Figure 2.2b:

At $h = 0\,\text{ft}$, $p = (62.4\,\text{lbf ft}^{-3}) * (0\,\text{ft}) = 0\,\text{lbf ft}^{-2}$.

At $h = 80\,\text{ft}$, $p = (62.4\,\text{lbf ft}^{-3}) * (80\,\text{ft}) = 4992\,\text{lbf ft}^{-2}$.

At the midpoint of the submerged arch, $h = 80\,\text{ft} + (20\,\text{ft}) * \sin(\pi/4) = 94.14\,\text{ft}$, $p = (62.4\,\text{lbf ft}^{-3}) * (94.14\,\text{ft}) = 5874.3\,\text{lbf ft}^{-2}$.

At $h = 80\,\text{ft} + 20\,\text{ft} = 100\,\text{ft}$, $p = (62.4\,\text{lbf ft}^{-3}) * (100\,\text{ft}) = 6240\,\text{lbf ft}^{-2}$.

The pressure distributions between 0 and 80 ft as well as between 80 and 100 ft are linear. However, the directions of the pressures between 80 and 100 ft vary from point to point and always point to the center of the circle. The results are shown in Figure 2.3b.

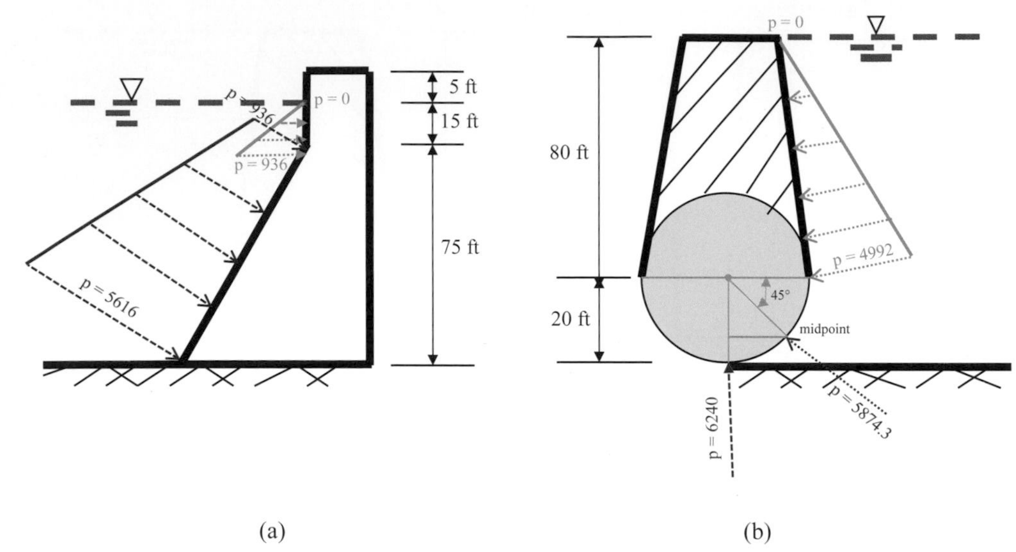

Figure 2.3 **The computed pressures for Example 2.1.**

2.2.2 Force

Force is the integration of pressure across an area. In this regard, the resultant force and its acting location on a submerged plane (Figure 2.4) can be computed as:

$$F = \gamma h_c A \tag{2.7}$$

(resultant force → F; specific weight of water → γ; water depth above centroid of the plane → h_c; surface area of the plane → A)

$$y_p = y_{pc} + \frac{I_c}{y_{pc} A} \tag{2.8}$$

(acting location of resultant force (see Figure 2.4a) → y_p; location of centroid of the plane (see Figure 2.4a) → y_{pc}; momentum of inertia around centroid axis of the plane (see Figure 2.4b) → I_c; surface area of the plane (see Figure 2.4b) → A)

The centroid moment of inertia in Eq. (2.8) depends on the shape of the plane. Table 2.2 presents the formulas for four basic shapes, namely rectangle, triangle, circle, and semicircle. The formulas for more shapes can be found in textbooks of engineering dynamics (e.g., Kasdin and Paley, 2011).

The resultant force acting on a non-planar (e.g., curved) surface can be determined by computing its horizontal and vertical components separately. The horizontal component is computed by applying Eqs. (2.7) and (2.8) to a projected vertical plane of the curved surface, while the vertical component is computed as the weight of water sandwiched by the curved surface and the projected horizontal plane. For the illustration in Figure 2.5, the resultant hydrostatic force can be computed as:

Table 2.2 Centroid moment of inertia for four basic shapes.

Shape	Area, A	Centroid moment of inertia, I_c, around X_c axis
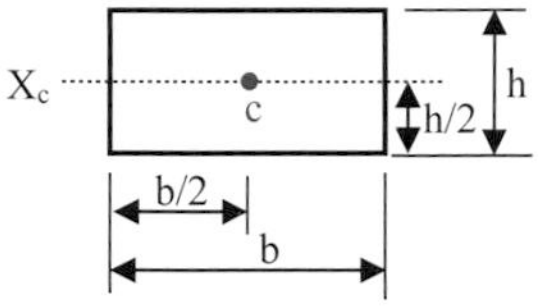	bh	$\frac{bh^3}{12}$
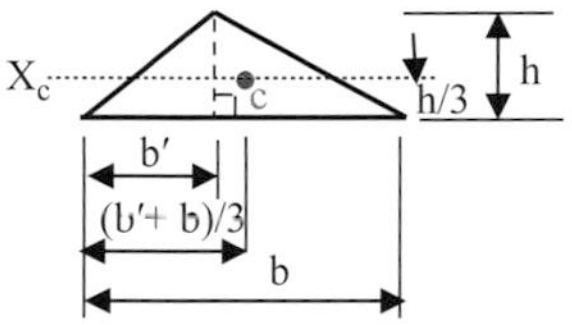	$\frac{bh}{2}$	$\frac{bh^3}{36}$
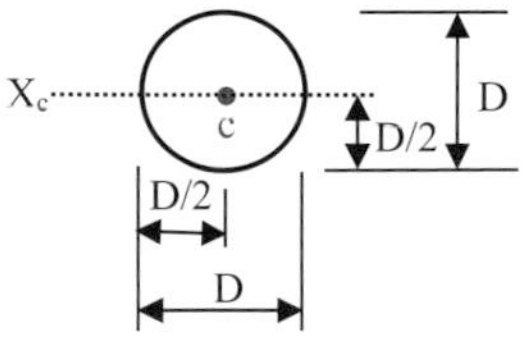	$\frac{\pi D^2}{4}$	$\frac{\pi D^4}{64}$
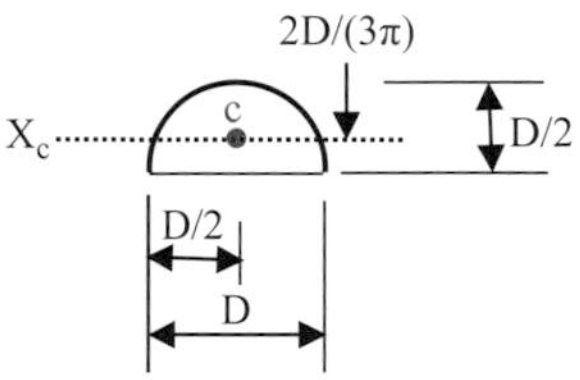	$\frac{\pi D^2}{8}$	$\frac{\pi D^4}{128}$

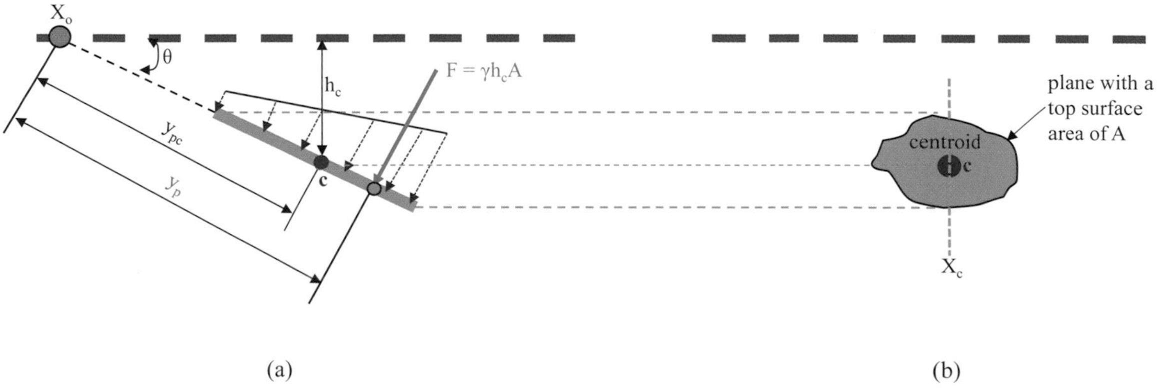

Figure 2.4 Illustration of a submerged plane for computing resultant force. Diagrams show the submerged plane in the: (a) side view; and (b) bird's-eye view. Symbols are as follows: X_o: intersect of the extended plane and water surface; θ: angle between the extended plane and water surface; y_{pc}: distance from the centroid of the plane and X_o; h_c: water depth above the centroid; F: resultant force; y_p: distance from the acting location of F to X_o; A: top surface area of the plane; and X_c: axis to compute moment of inertia in Eq. (2.8).

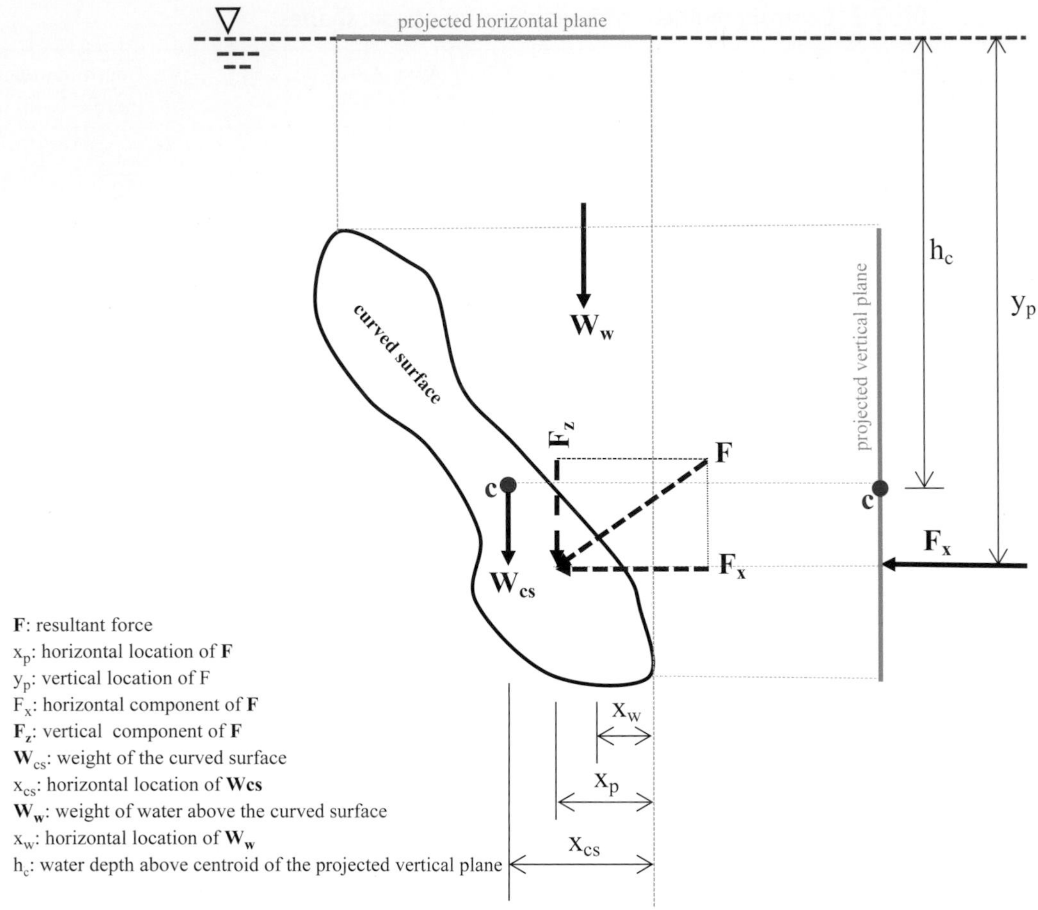

Figure 2.5 Illustration of the computation of resultant force on curved (i.e., non-planar) surface.

\mathbf{F}: resultant force
x_p: horizontal location of \mathbf{F}
y_p: vertical location of F
$\mathbf{F_x}$: horizontal component of \mathbf{F}
$\mathbf{F_z}$: vertical component of \mathbf{F}
$\mathbf{W_{cs}}$: weight of the curved surface
x_{cs}: horizontal location of \mathbf{Wcs}
$\mathbf{W_w}$: weight of water above the curved surface
x_w: horizontal location of $\mathbf{W_w}$
h_c: water depth above centroid of the projected vertical plane

resultant force $\rightarrow F = \sqrt{F_x^2 + F_z^2}$ (2.9)

horizontal component vertical component

specific weight of water

water depth above centroid of
the projected vertical plane

horizontal component of submerged area of the
resultant force $\rightarrow F_x = \gamma h_c A_{pv}$ projected vertical plane (2.10)

water depth above centroid of
the projected vertical plane

momentum of inertia
around centroid axis of the
projected vertical plane

vertical location of
resultant force $\rightarrow y_p = h_c + \dfrac{I_c}{h_c A_{pv}}$ submerged area of the
projected vertical plane (2.11)

$$F_z = W_{cs} + W_w \tag{2.12}$$

where F_z = vertical component of resultant force; W_{cs} = weight of the curved surface; W_w = weight of water sandwiched by the curved surface and the projected horizontal plane

$$x_p = \frac{W_{cs}x_{cs} + W_w x_w}{F_z} \tag{2.13}$$

where x_p = horizontal location of resultant force; W_{cs} = weight of the curved surface; x_{cs} = acting point of W_{cs}; W_w = weight of water sandwiched by the curved surface and the projected horizontal plane; x_w = acting point of W_w; F_z = vertical component of resultant force

Example 2.2 Determine the resultant force acting on the hydraulic structure submerged in water ($\gamma = 62.4\,\text{lbf ft}^{-3}$) shown in Figure 2.2a. The width of the structure is 100 ft and the submerged surfaces are flat. The inclined surface has a slope angle of 60°.

Solution

$\gamma = 62.4\,\text{lbf ft}^{-3}$. Treat the submerged portion of the structure as a curved surface and determine the resultant force using Eqs. (2.9) through (2.13). In reference to Figure 2.6, the computations can be done as follows.

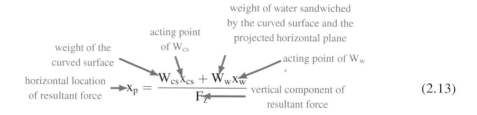

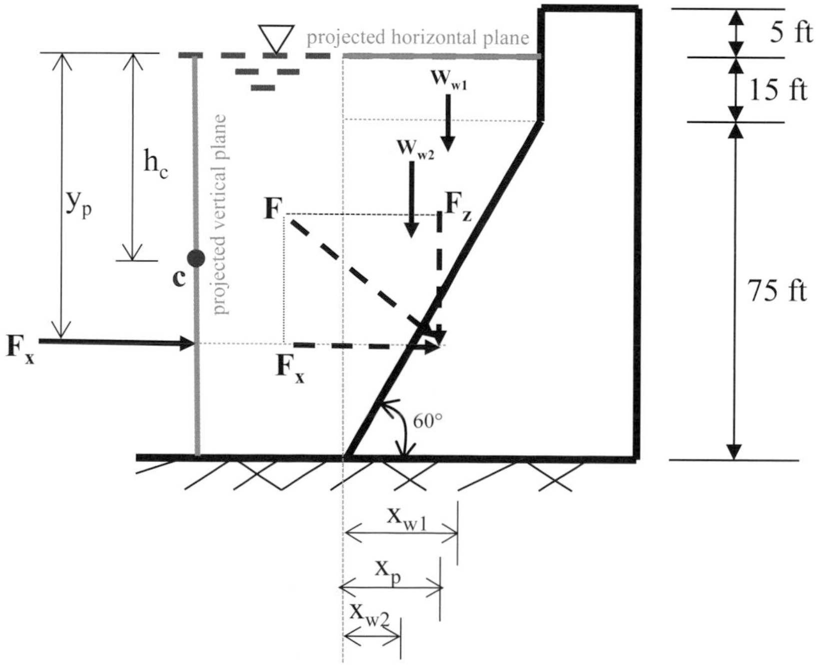

Figure 2.6 **Illustration for solving Example 2.2.**

For the projected vertical plane: rectangle with b = 100 ft and h = 15 ft + 75 ft = 90 ft.

$A_{pv} = (100\,\text{ft}) * (15\,\text{ft} + 75\,\text{ft}) = 9000\,\text{ft}^2$.

From Table 2.2, $I_c = (100\,\text{ft}) * (90\,\text{ft})^3/12 = 6{,}075{,}000\,\text{ft}^4$.

$h_c = (90\,\text{ft})/2 = 45\,\text{ft}$.

Eq. (2.10): $F_x = (62.4\,\text{lbf ft}^{-3}) * (45\,\text{ft}) * (9000\,\text{ft}^2) = 25{,}272{,}000\,\text{lbf}$.

Eq. (2.11): $y_p = 45\,\text{ft} + (6{,}075{,}000\,\text{ft}^4)/[(45\,\text{ft}) * (9000\,\text{ft}^2)] = 60\,\text{ft}$.

For the projected horizontal plane: one rectangle (W_{w1}) and one triangle (W_{w2}).

$W_{w1} = (62.4\,\text{lbf ft}^{-3}) * [(75\,\text{ft})/\tan(60°) * (15\,\text{ft}) * (100\,\text{ft})] = 4{,}052{,}999\,\text{lbf}$.

$x_{w1} = 1/2 * [(75\,\text{ft})/\tan(60°)] = 21.65\,\text{ft}$.

$W_{w2} = (62.4\,\text{lbf ft}^{-3}) * [1/2 * (75\,\text{ft})/\tan(60°) * (75\,\text{ft}) * (100\,\text{ft})] = 10{,}132{,}497\,\text{lbf}$.

$x_{w2} = 1/3 * [(75\,\text{ft})/\tan(60°)] = 14.43\,\text{ft}$.

$F_z = W_{w1} + W_{w2} = 4{,}052{,}998.9\,\text{lbf} + 10{,}132{,}497.2\,\text{lbf} = 14{,}185{,}496\,\text{lbf}$.

Moment theory: Eq. (2.13): $x_p = [(4{,}052{,}999\,\text{lbf}) * (21.65\,\text{ft}) + (10{,}132{,}497\,\text{lbf}) * (14.43\,\text{ft})]/(14{,}185{,}496\,\text{lbf}) = 16.49\,\text{ft}$.

The resultant force:

Eq. (2.9): $F = [(25{,}272{,}000\,\text{lbf})^2 + (14{,}185{,}496\,\text{lbf})^2]^{1/2} = 28{,}981{,}068\,\text{lbf}$.

Acting point: $x_p = 16.49\,\text{ft}$, $y_p = 60\,\text{ft}$.

2.3 Energy in Still and Flowing Water

Total energy in water consists of potential energy, pressure energy, and kinetic energy. It is usually expressed as total energy head, defined as total energy per unit weight of water. In this regard, total energy head is the summation of potential head, pressure head, and velocity head, which are the corresponding energies per unit weight of water. For still (i.e., at-rest) water, the kinematic energy is zero and the total energy is also called hydraulic energy, whereas for flowing water, the summation of pressure energy and kinematic energy is called specific energy. Hereinafter, energy and energy head will be interchangeably used except as otherwise specified. In reference to Figure 2.7, total energy head can be defined as:

$$E = z + \frac{p}{\gamma} + \alpha\frac{V^2}{2g} \qquad (2.14)$$

hydrostatic pressure
energy correction factor
potential head
total energy head
mean velocity
gravitational acceleration
specific weight of water

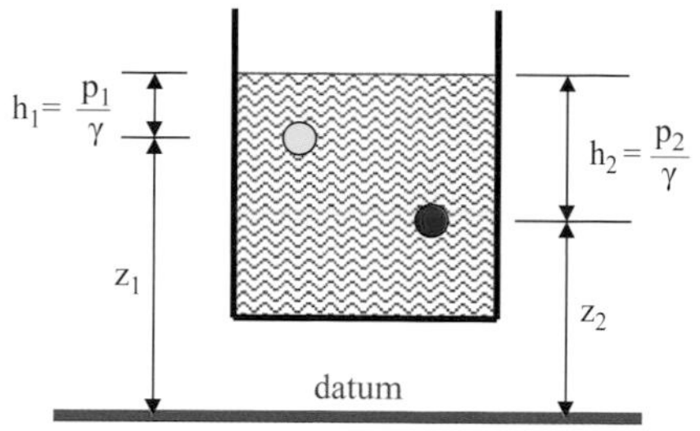

Figure 2.7 **Illustration of potential and pressure energy heads.**

point velocity across A

energy correction factor → $\alpha = \dfrac{\oint_A u^3 dA}{AV^3}$ (2.15)

flow area → ← mean velocity

In Eq. (2.14), $z + \frac{p}{\gamma}$ is defined as hydraulic head, while $\frac{p}{\gamma} + \alpha\frac{V^2}{2g}$ is defined as specific energy head. α reflects the uniformity of velocities across flow area: the more uniform, the smaller the value of α. The minimum value of α is one. It is usually assumed $\alpha = 1.0$ in engineering practice.

Example 2.3 The pressure in a full-flow circular water pipe is 25 psi and the mean velocity is 4.5 ft sec^{-1}. If the 4-ft-diameter pipe is placed at 20 ft above an arbitrary datum, determine the total energy head, hydraulic head, and specific energy head. The specific weight of water is 62.4 lbf ft^{-3}. Assume the energy correction factor $\alpha = 1.0$.

Solution

$\gamma = 62.4\ \text{lbf ft}^{-3}$, $p = 25\ \text{psi} = 25 * 144\ \text{lbf ft}^{-2} = 3600\ \text{lbf ft}^{-2}$, $D = 4\ \text{ft}$, $V = 4.5\ \text{ft sec}^{-1}$, $\alpha = 1.0$, and the pipe is placed at an elevation of 20 ft.

Potential head: it is the elevation of the pipe center $\rightarrow z = 20\ \text{ft} + (4\ \text{ft})/2 = 22\ \text{ft}$.

Pressure head: $p/\gamma = (3600\ \text{lbf ft}^{-2})/(62.4\ \text{lbf ft}^{-3}) = 57.7\ \text{ft}$.

Velocity head: $\alpha V^2/(2g) = 1.0 * (4.5\ \text{ft sec}^{-1})^2/(2 * 32.2\ \text{ft sec}^{-2}) = 0.3\ \text{ft}$.

Total energy head: 22 ft + 57.7 ft + 0.3 ft = 80.0 ft.

Hydraulic head: 22 ft + 57.7 ft = 79.7 ft.

Specific energy head: 57.7 ft + 0.3 ft = 58.0 ft.

Example 2.4 At a location of a diversion channel, the water depth is 6.8 ft and the mean velocity is 2.2 ft sec^{-1}. If the channel bed elevation at this location is 50 ft above an arbitrary datum, determine the total energy head, hydraulic head, and specific energy head. The specific weight of water is 62.4 lbf ft^{-3}. Assume $\alpha = 1.0$.

Solution

$\gamma = 62.4\,\text{lbf ft}^{-3}$, $y = 6.8$ ft, $V = 2.2\,\text{ft sec}^{-1}$, $\alpha = 1.0$, bed elevation is 50 ft.

Potential head: it is the channel bed elevation $\rightarrow z = 50$ ft.

Pressure head: $p/\gamma = y = 6.8$ ft.

Velocity head: $\alpha V^2/(2g) = 1.0 * (2.2\,\text{ft sec}^{-1})^2/(2 * 32.2\,\text{ft sec}^{-2}) = 0.08$ ft.

Total energy head: $50\,\text{ft} + 6.8\,\text{ft} + 0.08\,\text{ft} = 56.88$ ft.

Hydraulic head: $50\,\text{ft} + 6.8\,\text{ft} = 56.8$ ft.

Specific energy head: $6.8\,\text{ft} + 0.08\,\text{ft} = 6.88$ ft.

2.4 Governing Laws of Flowing Water

Flow systems are governed by three laws, namely the continuity equation, energy equation, and momentum equation. This section introduces these three laws and demonstrates how to use them to solve problems of a flow system, which needs to simultaneously apply two or three of the equations.

2.4.1 Continuity Equation

Assuming a constant water density, one can write the continuity equation along the flow path as:

flow area; flow rate

$$\frac{\partial A}{\partial t} + \frac{\partial Q}{\partial x} = q \tag{2.16}$$

net flux per unit length of flow path

time; distance along flow path

For a segment of the flow path between sections ① and ② (e.g., Figure 2.8), if the flow area does not change with time (i.e., $\frac{\partial A}{\partial t} = 0$) and there is no influx or efflux (i.e., $q = 0$), Eq. (2.16) can be integrated to derive a commonly used format of the continuity equation expressed as:

flow rate at section ① ; flow rate at section ②

$$Q_1 = Q_2 \tag{2.17}$$

flow area at section ① ; mean velocity at section ① ; mean velocity at section ② ; flow area at section ②

$$V_1 A_1 = V_2 A_2 \tag{2.18}$$

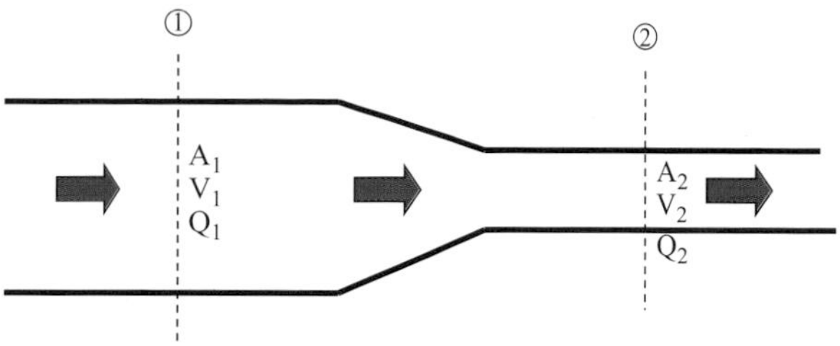

Figure 2.8 **The segment of flow path between two sections ① and ②.** A, V, and Q are flow area, mean velocity, and flow rate, respectively.

Example 2.5 For the river system shown in Figure 2.9, determine the flow rate at section ③.

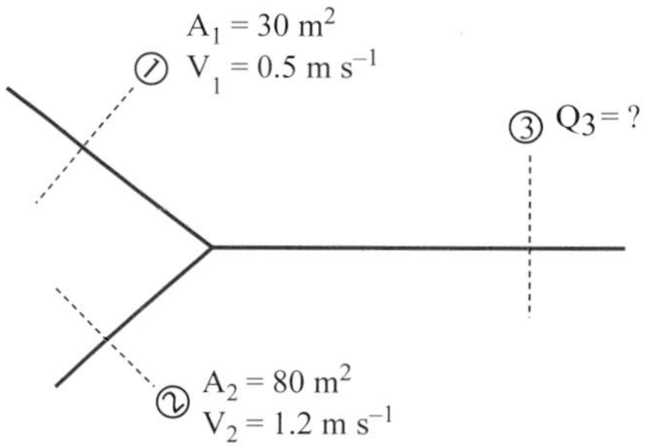

Figure 2.9 **Illustration of the river system for Example 2.5.**

Solution

Based on Eqs. (2.17) and (2.18):

$$Q_3 = Q_1 + Q_2 = V_1A_1 + V_2A_2 = (0.5\,\mathrm{m\,s^{-1}}) * (30\,\mathrm{m^2}) + (1.2\,\mathrm{m\,s^{-1}}) * (80\,\mathrm{m^2}) = 111\,\mathrm{m^3\,s^{-1}}.$$

2.4.2 Energy Equation

For gradually varied flow, one can write the energy equation along the flow path as:

$$\frac{dE}{dx} = -S_f \tag{2.19}$$

(E: total energy head; S_f: energy line gradient; x: distance along flow path)

For a segment of flow path between sections ① and ② (e.g., Figure 2.10), assuming a linear decrease of total energy, one can integrate Eq. (2.19) to get a commonly used format of the energy equation expressed as:

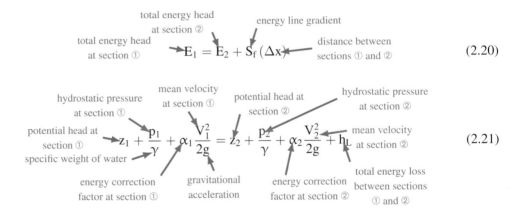

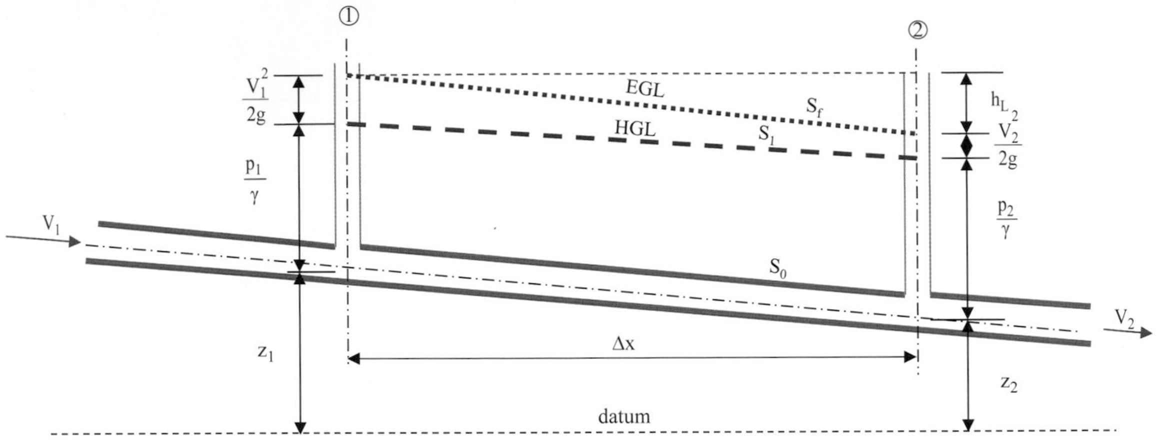

Figure 2.10 **Illustration of channel bed, hydraulic grade line (HGL), and energy grade line (EGL).** Section shown is between ① and ② along the flow path. z, p, and V are elevation, pressure, and mean velocity, respectively, and S_0, S_1, and S_f signify the slopes of the corresponding lines.

$$E_1 = E_2 + S_f(\Delta x) \tag{2.20}$$

total energy head at section ①; total energy head at section ②; energy line gradient; distance between sections ① and ②

$$z_1 + \frac{p_1}{\gamma} + \alpha_1\frac{V_1^2}{2g} = z_2 + \frac{p_2}{\gamma} + \alpha_2\frac{V_2^2}{2g} + h_L \tag{2.21}$$

potential head at section ①; hydrostatic pressure at section ①; specific weight of water; mean velocity at section ①; energy correction factor at section ①; gravitational acceleration; potential head at section ②; hydrostatic pressure at section ②; energy correction factor at section ②; mean velocity at section ②; total energy loss between sections ① and ②

When the energy equation is applied to branch flow paths, energy rather than energy head should be used. This is demonstrated in Example 2.7. Herein, assuming a constant density of water, energy can be represented as the product of energy head and flow rate.

Example 2.6 A short flatbed expansion channel conveys water from a river into a water treatment plant. If the flow areas at the entrance and exit of the channel are the same, but the water depth at the entrance is 4 ft while the water depth at the exit is 2 ft, determine the total energy loss of the expansion channel.

Solution

$y_1 = 4$ ft, $y_2 = 2$ ft, assume $\alpha_1 = \alpha_2$.

Eq. (2.1): $\frac{p_1}{\gamma} = y_1 = 4$ ft, $\frac{p_2}{\gamma} = y_2 = 2$ ft.

Eq. (2.17): $Q_1 = Q_2$.

Same flow areas $\rightarrow A_1 = A_2 \rightarrow V_1 = V_2 \rightarrow \alpha_1 \frac{V_1^2}{2g} = \alpha_2 \frac{V_2^2}{2g}$.

Flatbed $\rightarrow z_1 = z_2$.

Eq. (2.21) between the entrance and exit: canceling the identical terms and rearranging, one can compute the total energy loss: $h_L = 4\,\text{ft} - 2\,\text{ft} = 2\,\text{ft}$.

Example 2.7 For the river system shown in Figure 2.9, section ③ is rectangular with a bottom width of 15 m. If the water depths at section ① , ②, and ③ are 6, 8, and 5 m, respectively, determine the total energy loss. Assume $\alpha_1 = \alpha_2 = \alpha_3 = 1.0$. The channel bed elevations are $z_1 = 110\,\text{m}$, $z_2 = 130\,\text{m}$, and $z_3 = 100\,\text{m}$.

Solution

$A_1 = 30\,\text{m}^2$, $V_1 = 0.5\,\text{m}\,\text{s}^{-1}$, $y_1 = 6\,\text{m}$, $z_1 = 110\,\text{m}$; $A_2 = 80\,\text{m}^2$, $V_2 = 1.2\,\text{m}\,\text{s}^{-1}$, $y_2 = 8\,\text{m}$, $z_2 = 130\,\text{m}$; $b_3 = 15\,\text{m}$, $y_3 = 5\,\text{m}$, $z_3 = 100\,\text{m}$.

$Q_1 = (0.5\,\text{m}\,\text{s}^{-1}) * (30\,\text{m}^2) = 15\,\text{m}^3\,\text{s}^{-1}$.

$Q_2 = (1.2\,\text{m}\,\text{s}^{-1}) * (80\,\text{m}^2) = 96\,\text{m}^3\,\text{s}^{-1}$.

Eqs. (2.17) and (2.18): $Q_3 = 15\,\text{m}^3\,\text{s}^{-1} + 96\,\text{m}^3\,\text{s}^{-1} = 111\,\text{m}^3\,\text{s}^{-1}$.

$A_3 = (15\,\text{m}) * (5\,\text{m}) = 75\,\text{m}^2$.

$V_3 = (111\,\text{m}^3\,\text{s}^{-1})/(75\,\text{m}^2) = 1.48\,\text{m}\,\text{s}^{-1}$.

Energy equation between sections ①, ②, and ③ $\rightarrow Q_1E_1 + Q_2E_2 = Q_3E_3 + E_L$

$(15\,\text{m}^3\,\text{s}^{-1}) * [110\,\text{m} + 6\,\text{m} + (1.0) * (0.5\,\text{m}\,\text{s}^{-1})^2/(2 * 9.81\,\text{m}\,\text{s}^{-2})] +$

$(96\,\text{m}^3\,\text{s}^{-1}) * [130\,\text{m} + 8\,\text{m} + (1.0) * (1.2\,\text{m}\,\text{s}^{-1})^2/(2 * 9.81\,\text{m}\,\text{s}^{-2})] =$

$(111\,\text{m}^3\,\text{s}^{-1}) * [100\,\text{m} + 5\,\text{m} + (1.0) * (1.48\,\text{m}\,\text{s}^{-1})^2/(2 * 9.81\,\text{m}\,\text{s}^{-2})] + E_L$

$\rightarrow 1740.19\,\text{m}^4\,\text{s}^{-1} + 13{,}255.05\,\text{m}^4\,\text{s}^{-1} = 11{,}667.39\,\text{m}^4\,\text{s}^{-1} + E_L$

$\rightarrow E_L = 3327.85\,\text{m}^4\,\text{s}^{-1} \rightarrow h_L = (3327.85\,\text{m}^4\,\text{s}^{-1})/(111\,\text{m}^3\,\text{s}^{-1}) = 29.98\,\text{m}$.

Total energy loss includes friction loss and minor losses. Friction loss is caused by shear stress between flowing water and the wet surface of the waterway, whereas minor losses are due to entrance, exit, contraction, expansion, bending, and fittings (e.g., valve and gate). That is, h_L is computed as:

$$\underbrace{h_L}_{\text{total energy loss}} = \underbrace{h_f}_{\text{friction loss}} + \underbrace{h_{minor}}_{\text{minor loss}} \tag{2.22}$$

Each minor loss is computed as the product of a loss coefficient and velocity head as presented in Table 2.3.

For gradually varied flow (e.g., flow in channels), friction loss can be computed using the Darcy–Weisbach equation, Manning formula, or Hazen–Williams formula, whereas for rapidly varied flow (e.g., hydraulic jump and flow through spillway), friction loss needs to be determined

Table 2.3 Computation of common minor losses.[1]

Description and formula	Sketch	Minor loss coefficient		
Entrance	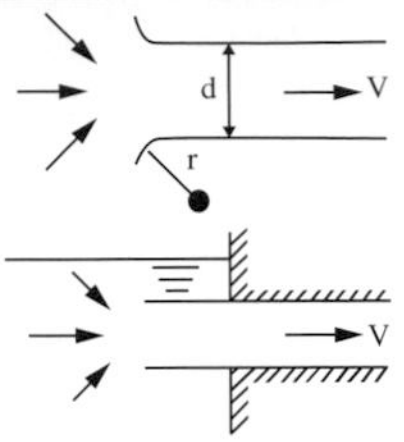	r/d	k_e	
$h_e = k_e \frac{V^2}{2g}$		0.0	0.50	
		0.1	0.12	
		>0.2	0.03	
			1.00	
Contraction	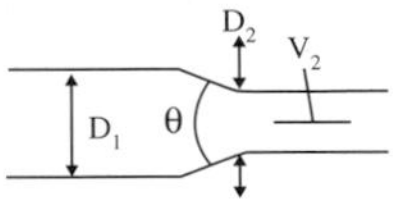	D_2/D_1	k_c	
$h_c = k_e \frac{V_2^2}{2g}$			θ = 60°	θ = 180°
		0.0	0.08	0.50
		0.20	0.08	0.49
		0.40	0.07	0.42
		0.60	0.06	0.32
		0.80	0.05	0.18
		0.90	0.04	0.10
Expansion	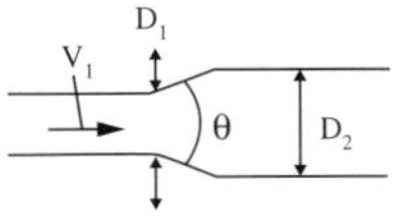	D_1/D_2	k_x	
$h_x = k_x \frac{V_1^2}{2g}$			θ = 10°	θ = 180°
		0.0		1.00
		0.20	0.13	0.92
		0.40	0.11	0.72
		0.60	0.06	0.42
		0.80	0.03	0.16
90° miter bend	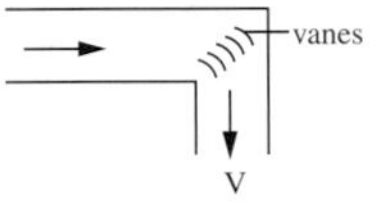	Vanes?	k_b	
$h_b = k_b \frac{V^2}{2g}$		No	1.1	
		Yes	0.2	
Smooth bend	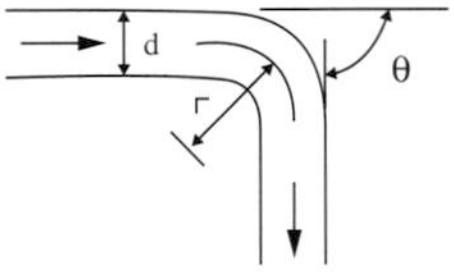	r/d	k_b	
			θ = 45°	θ = 90°
		1	0.10	0.35
		2	0.09	0.19
		4	0.10	0.16
		6	0.12	0.21
Fittings		Type	k_{fit}	
$h_{fit} = k_{fit} \frac{V^2}{2g}$		Global valve – wide open	10.0	
		Angle valve – wide open	5.0	
		Gate valve – wide open	0.2	
		Gate valve – half open	5.6	
		Return bend	2.2	
		Tee	1.8	
		90° elbow	0.9	
		45° elbow	0.4	
Exit	Submerged	$k_{ex} = 1.0$		
$h_{ex} = k_{ex} \frac{V^2}{2g}$	Unsubmerged	$k_{ex} = 0.0$		

[1] Reproduced from Table 5.3 of Roberson *et al.* (1998).

by jointly solving momentum equation (discussed in the next section) and energy equation. The difference between these two types of flows is that the energy grade line (EGL) of gradually varied flow is linear but the EGL of rapidly varied flow is non-linear (i.e., curved). As a theoretical equation, the Darcy–Weisbach equation is applicable for both pressurized and open-channel flow flowing in any geometric shapes of channels. In contrast, as empirical formulas, the Manning formula is only applicable for open-channel flow flowing in any geometric shapes of channels; the Hazen–Williams formula, however, is only applicable for pressurized flow flowing in any geometric shapes of closed conduits. Open-channel flow has a free surface subject to an absolute pressure equal to atmospheric pressure, whereas pressurized flow, on the other hand, is confined in a closed conduit subject to an absolute pressure larger than atmospheric pressure.

The Darcy–Weisbach equation can be written as:

$$h_f = f \frac{L}{4R} \frac{V^2}{2g} \tag{2.23}$$

friction loss (h_f); Darcy friction factor (see Figure 2.11) (f); conveyance length (L); hydraulic radius (see Table 2.4) (R); mean velocity (V); gravitational acceleration (g)

Table 2.4 Hydraulic radius for four cross-sectional shapes.

Cross section	Top width, B	Flow area, A	Wetted perimeter, P_{wet}	Hydraulic radius, $R = A/P_{wet}$	Hydraulic depth, $d = A/B$
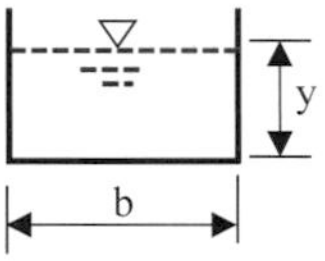	b	by	$b + 2y$	$\frac{by}{b+2y}$	y
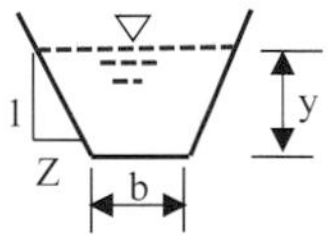	$b + 2Zy$	$(b + Zy)y$	$b + 2y\sqrt{1+Z^2}$	$\frac{(b+Zy)y}{b+2y\sqrt{1+Z^2}}$	$\frac{(b+Zy)y}{b+2Zy}$
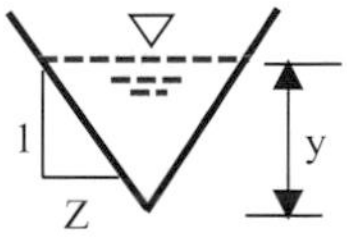	$2Zy$	Zy^2	$2y\sqrt{1+Z^2}$	$\frac{Zy}{2\sqrt{1+Z^2}}$	$\frac{y}{2}$
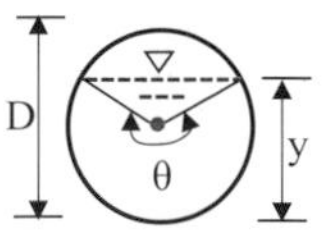	$D\sin\left(\frac{\theta}{2}\right)$	$\frac{D^2}{8}(\theta - \sin\theta)$	$\frac{\theta D}{2}$	$\frac{D}{4}\left(1 - \frac{\sin\theta}{\theta}\right)$	$\frac{D(\theta - \sin\theta)}{8\sin\left(\frac{\theta}{2}\right)}$

$\theta = 2\cos^{-1}\left(\frac{D-2y}{D}\right)$ in radians

hydraulic radius (see Table 2.4); flow area; wetted perimeter

$$R = \frac{A}{P_{wet}} \tag{2.24}$$

In Eq. (2.23), the Darcy friction factor, f, can be determined from the Moody diagram shown in Figure 2.11. It is a function of conduit surface roughness, k_s, and Reynolds number, R_e. k_s depends on the materials and can be determined from existing literature or vendors. In this figure, a small table for k_s is superimposed for convenience. R_e measures the relative magnitude of inertia force to viscous force, and is computed as:

Reynolds number; mean velocity; hydraulic radius; kinematic viscosity

$$R_e = \frac{V(4R)}{\nu} \tag{2.25}$$

The Manning formula can be written as:

mean velocity [ft sec^{-1} for BG; m s^{-1} for SI]; units conversion factor (= 1.486 for BG; 1.0 for SI); hydraulic radius [ft for BG; m for SI]; energy line gradient; Manning's roughness

$$V = \frac{K}{n} R^{\frac{2}{3}} S_f^{\frac{1}{2}} \tag{2.26}$$

Rearranging Eq. (2.26) and considering that S_f is the ratio of friction loss to conveyance length, one can derive the equation expressed as:

friction loss [ft for BG; m for SI]; conveyance length [ft for BG; m for SI]; Manning's roughness; mean velocity [ft sec^{-1} for BG; m s^{-1} for SI]; units conversion factor (= 1.486 for BG; 1.0 for SI); hydraulic radius [ft for BG; m for SI]

$$h_f = L\left(\frac{nV}{KR^{\frac{2}{3}}}\right)^2 \tag{2.27}$$

Manning's roughness (usually called Manning's n) depends on the materials and covers of waterways. However, the determination of Manning's n is more or less subjective. Table 2.5 presents the typical values of Manning's n that are commonly used in practice. Alternatively, equalizing Eqs. (2.23) and (2.27) and rearranging, one can derive the relationship between n and f expressed as:

Manning's roughness; units conversion factor (= 1.486 for BG; 1.0 for SI); hydraulic radius [ft for BG; m for SI]; Darcy friction factor; gravitational acceleration (= 32.2 ft sec^{-2} for BG; 9.81 m s^{-2} for SI)

$$n = \frac{K}{2\sqrt{2g}} R^{\frac{1}{6}} \sqrt{f} \tag{2.28}$$

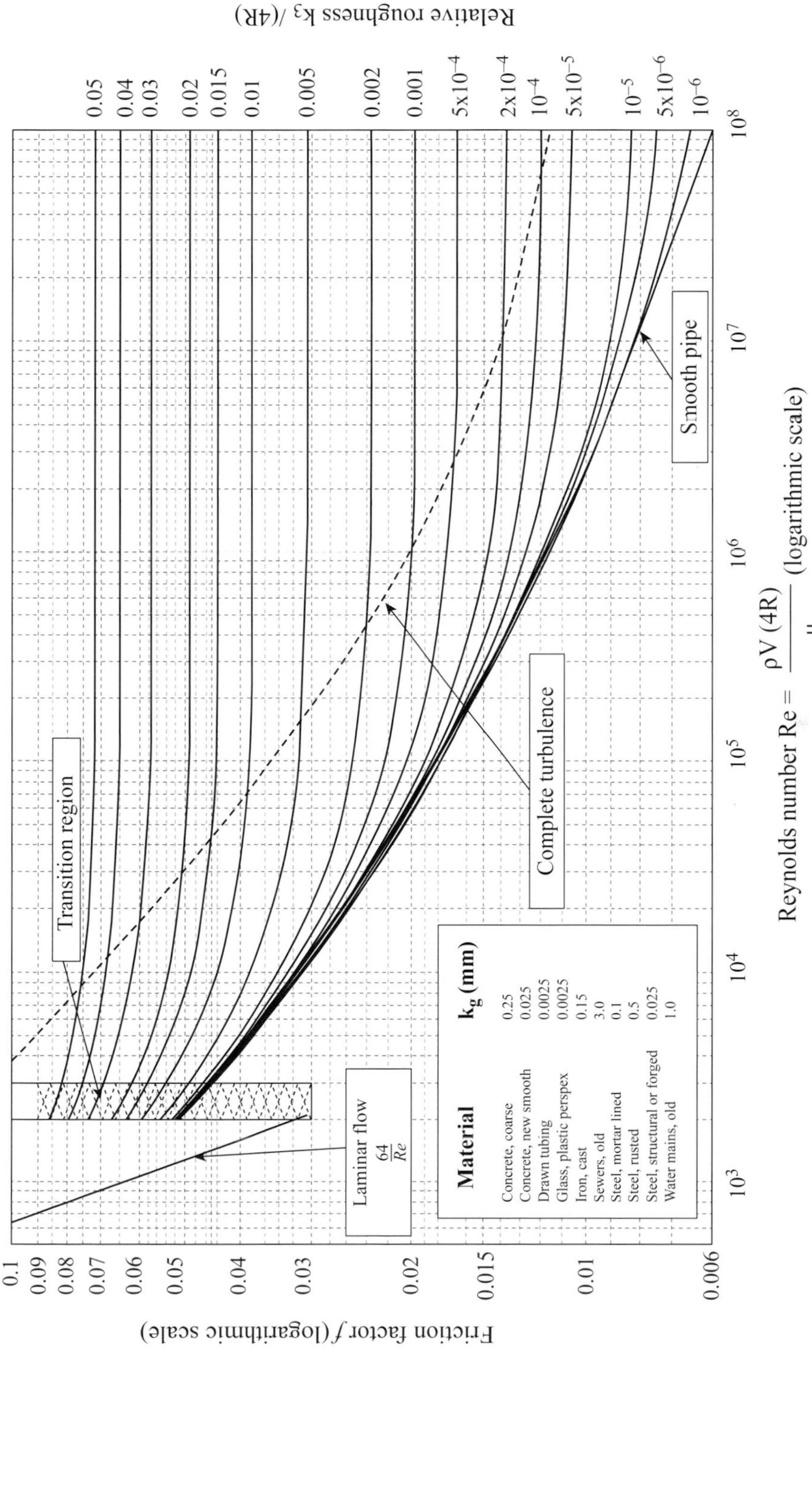

Figure 2.11 Moody diagram. (*Source:* reproduced Figure 1 in Moody, 1944, permitted by Beth Darchi)

Table 2.5 Typical values of Manning's n.[1]

Material and cover	Manning's n	Material and cover	Manning's n
Natural channel		Earth artificial channel	
Clean	0.025 ∼ 0.045	Some vegetation	0.020 ∼ 0.035
Vegetated	0.045 ∼ 0.160	Dense vegetation	0.035 ∼ 0.120
Floodplain		Partially full conduit	
No or scattered brush	0.030 ∼ 0.070	Steel and cast iron	0.010 ∼ 0.016
Dense brush	0.070 ∼ 0.200	Concrete	0.010 ∼ 0.015
Lined artificial channel	0.012 ∼ 0.020	Corrugated metal	0.020 ∼ 0.030

[1] Compiled from various sources (e.g., Urroz and Hoeft, 2009; USDA-NRCS, 2004, 2010).

The Hazen–Williams (i.e., H–W) formula can be written as:

$$V = K_{hw} C_{hw} R^{0.63} S_f^{0.54} \tag{2.29}$$

where: V = mean velocity [ft sec^{-1} for BG; m sec^{-1} for SI]; K_{hw} = units conversion factor (= 1.318 for BG; 0.85 for SI); C_{hw} = H–W coefficient; R = hydraulic radius [ft for BG; m for SI]; S_f = energy line gradient

Rearranging Eq. (2.29) and considering that S_f is the ratio of friction loss to conveyance length, one can derive the equation expressed as:

$$h_f = L\left(\frac{V}{K_{hw} C_{hw} R^{0.63}}\right)^{1.85} \tag{2.30}$$

where: h_f = friction loss [ft for BG; m for SI]; L = conveyance length [ft for BG; m for SI]; V = mean velocity [ft sec^{-1} for BG; m sec^{-1} for SI]; K_{hw} = units conversion factor (= 1.318 for BG; 0.85 for SI); C_{hw} = H–W coefficient; R = hydraulic radius [ft for BG; m for SI]

The H–W coefficient, C_{hw}, reflects the smoothness of conduit. That is, a larger value of C_{hw} is associated with a smaller friction loss, and vice versa. In this regard, C_{hw} has a physical meaning opposite to that of Manning's n and the Darcy friction factor, f. Table 2.6 summarizes the typical

Table 2.6 Typical values of the Hazen–Williams coefficient.[1]

Conduit material	H–W coefficient, C_{hw}	Conduit material	H–W coefficient, C_{hw}
Asbestos cement	140	Galvanized iron	120
Cast iron		Concrete	120 ∼ 140
New	130	Steel	110 ∼ 150
< 20 years old	100 ∼ 130	Tin	130
> 20 years old	64 ∼ 90	Vitrified clay	110 ∼ 140

[1] *Source:* Urroz and Hoeft (2009).

values of C_{hw}. Alternatively, equalizing Eqs. (2.23) and (2.30) and rearranging, one can yield the relationship between C_{hw} and f expressed as:

$$C_{hw} = \frac{1}{K_{hw}(VR)^{0.08}}\left(\frac{8g}{f}\right)^{0.54} \tag{2.31}$$

where: H–W coefficient → C_{hw}; gravitational acceleration (= 32.2 ft sec^{-2} for BG; = 9.81 m s^{-1} for SI) → g; Darcy friction factor → f; units conversion factor (= 1.318 for BG; 0.85 for SI) → K_{hw}; mean velocity [ft sec^{-1} for BG; m sec^{-1} for SI] → V; hydraulic radius [ft for BG; m for SI] → R.

Example 2.8 A 4-ft-diameter 6000-ft-long concrete pipe (roughness height of 0.05 ft) carries 6 cfs water 60°F.

(1) If the average water depth in the pipe is 3.8 ft, determine the friction loss using the Darcy–Weisbach equation and Manning formula.

(2) If the flow in the pipe is pressurized, determine the friction loss using the Darcy–Weisbach equation and Hazen–Williams formula.

Solution

$D = 4\,\text{ft},\ L = 6000\,\text{ft},\ k_s = 0.05\,\text{ft},\ Q = 6\,\text{cfs}$

60°F water → From Table 2.1, $\nu = 1.22 \times 10^{-5}\,\text{ft}^2\,\text{sec}^{-1}$.

(1) Partially full, y = 3.8 ft.

From Table 2.4: $\theta = 2\cos^{-1}\left[\frac{(4\,\text{ft}) - 2*(3.8\,\text{ft})}{4\,\text{ft}}\right] = 5.3811\,\text{rad}$

$$A = \frac{(4\,\text{ft})^2}{8} * [(5.3811\,\text{rad}) - \sin(5.3811\,\text{rad})] = 12.33\,\text{ft}^2$$

$$R = \frac{4\,\text{ft}}{4} * \left[1 - \frac{\sin(5.3811\,\text{rad})}{(5.3811\,\text{rad})}\right] = 1.15\,\text{ft}.$$

$$V = \frac{6\,\text{cfs}}{12.33\,\text{ft}^2} = 0.49\,\text{ft sec}^{-1}$$

Eq. (2.25): $R_e = \frac{\left(0.49\,\text{ft sec}^{-1}\right) * (4 * 1.15\,\text{ft})}{1.22 \times 10^{-5}\,\text{ft}^2\,\text{sec}^{-1}} = 1.85 \times 10^5.$

$$\frac{k_s}{4R} = \frac{0.05\,\text{ft}}{4 * 1.15\,\text{ft}} = 0.01$$

From Figure 2.11: f = 0.0095.

Darcy–Weisbach equation (Eq. (2.23)):

$$h_f = 0.0095 * \frac{6000\,\text{ft}}{4 * 1.15\,\text{ft}} * \frac{\left(0.49\,\text{ft sec}^{-1}\right)^2}{2 * 32.2\,\text{ft sec}^{-2}} = 0.046\,\text{ft}$$

Concrete pipe → From Table 2.5, n = 0.012.

Manning formula (Eq. (2.27)):

$$h_f = (6000\,\text{ft}) * \left(\frac{0.012 * 0.49\,\text{ft sec}^{-1}}{1.486 * (1.15\,\text{ft})^{\frac{2}{3}}} \right)^2 = 0.078\,\text{ft}.$$

(2) Pressurized

$$A = \frac{\pi}{4} * (4\,\text{ft})^2 = 12.57\,\text{ft}^2$$

$$P_{wet} = \pi * (4\,\text{ft}) = 12.57\,\text{ft}$$

$$R = \frac{12.57\,\text{ft}^2}{12.57\,\text{ft}} = 1.0\,\text{ft}$$

$$V = \frac{6\,\text{cfs}}{12.57\,\text{ft}^2} = 0.48\,\text{ft sec}^{-1}$$

$$\text{Eq. (2.25): } R_e = \frac{\left(0.48\,\text{ft sec}^{-1}\right) * (4 * 1.0\,\text{ft})}{1.22 \times 10^{-5}\,\text{ft}^2\,\text{sec}^{-1}} = 1.57 \times 10^5.$$

$$\frac{k_s}{4R} = \frac{0.05\,\text{ft}}{4 * 1.0\,\text{ft}} = 0.0125.$$

From Figure 2.11: $f = 0.01$.

Darcy–Weisbach equation (Eq. (2.23)):

$$h_f = 0.01 * \frac{6000\,\text{ft}}{4 * 1.0\,\text{ft}} * \frac{\left(0.48\,\text{ft sec}^{-1}\right)^2}{2 * 32.2\,\text{ft sec}^{-2}} = 0.054\,\text{ft}$$

Concrete pipe → From Table 2.6, $C_{hw} = 130$.

Hazen–Williams formula (Eq. (2.30)):

$$h_f = (6000\,\text{ft}) * \left(\frac{0.48\,\text{ft sec}^{-1}}{1.318 * 130 * (1.0\,\text{ft})^{0.63}} \right)^{1.85} = 0.11\,\text{ft}.$$

In summary: Roughness height, Manning's n, and the H–W coefficient are very sensitive to the computed friction loss. Unfortunately, these parameters are more or less subjective. In practice, a sensitivity analysis can be conducted by varying the parameters in corresponding reasonable ranges. Nevertheless, good engineering judgement is always crucial.

2.4.3 Momentum Equation

The momentum equation is used to determine the dynamic force or torque of flow water on either stationary (e.g., sluice gate) or rotating (e.g., turbine) objects. It is also used to analyze rapidly varied flow. The momentum equation has two forms: linear and angular. The linear momentum equation is used to determine force, while the angular momentum equation is used to determine torque. Torque is defined as the product of force and its arm length (Kasdin and Paley, 2011). Both forms of the momentum equation are usually applied to control volume (hereinafter designated CV for description purposes).

The linear momentum equation can be written as:

$$\sum \vec{F} = \Delta\left(\rho Q \vec{V}\right) \quad (2.32)$$

operator for computing resultant force ($\sum$); force acting on CV ($\vec{F}$); operator for computing change (Δ); density of water (ρ); flow rate (Q); mean velocity ($\vec{V}$)

Given that both forces and velocities in Eq. (2.32) are vectors, this equation is usually applied along two orthogonal directions such as along the x and y axes. In this regard, all forces are decomposed into the components along the x and y axes, while the velocities are decomposed into the corresponding components as well. The resultant force in the x-axis direction is equal to the change rate of the momentums in the same direction, while the resultant force in the y-axis direction is equal to the change rate of the momentums in the same direction. The resultant force on the object of interest is computed as:

$$F = \sqrt{F_x^2 + F_y^2} \quad (2.33)$$

resultant force on the object (F); resultant force in x-axis direction (F_x); resultant force in y-axis direction (F_y)

The angular momentum equation can be written as:

$$T_{orque} = \Delta\left(\rho Q r^2 \omega\right) \quad (2.34)$$

torque (T_{orque}); operator for computing change (Δ); density of water (ρ); flow rate (Q); arm length (r); angular speed [radian per unit time] (ω)

Example 2.9 A rectangular channel with a bottom width of 1.5 m carries $3.2\,m^3\,s^{-1}$ water ($\rho = 1000\,kg\,m^{-3}$, $\gamma = 9810\,N\,m^{-3}$). The channel bed is flat. A sluice gate with an opening height of 0.5 m creates a flow condition shown in Figure 2.12. The friction on the channel bottom is negligible. Determine the:

(1) Thrust on the gate.
(2) Energy loss between sections ② and ③, where a hydraulic jump occurs.

Solution

$\rho = 1000\,kg\,m^{-3}$, $\gamma = 9810\,N\,m^{-3}$, $Q = 3.2\,m^3\,s^{-1}$, $b = 1.5\,m$.

$y_1 = 2.0\,m$, $y_2 = 0.5\,m$, $z_1 = z_2 = z_3 = 0.0\,m$.

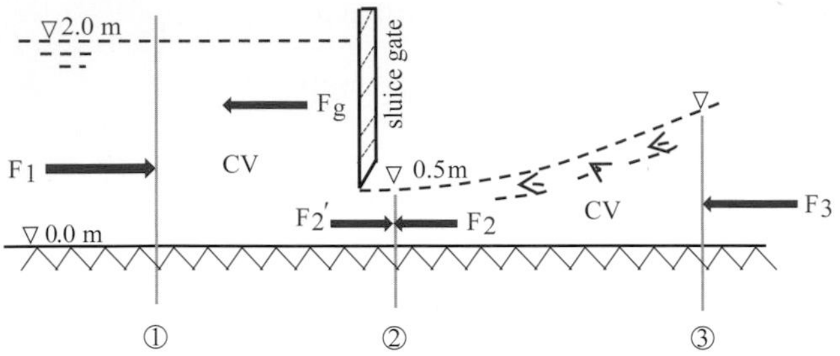

Figure 2.12 **Longitudinal cross section of the gate-controlled channel for Example 2.9.** Red arrows signify forces.

(1) The forces on the CV between sections ① and ② include F_1 (hydrostatic force at ①), F_2 (hydrostatic force at ②), and F_g (reaction force from the sluice gate).

$A_1 = (1.5\,m) * (2.0\,m) = 3.0\,m^2$.

$V_1 = (3.2\,m^3\,s^{-1})/(3.0\,m^2) = 1.07\,m\,s^{-1}$.

$A_2 = (1.5\,m) * (0.5\,m) = 0.75\,m^2$.

$V_2 = (3.2\,m^3\,s^{-1})/(0.75\,m^2) = 4.27\,m\,s^{-1}$.

Eq. (2.7): $F_1 = (9810\,N\,m^{-3}) * (2.0\,m/2) * (3.0\,m^2) = 29{,}430\,N$

$$F_2 = (9810\,N\,m^{-3}) * (0.5m/2) * (0.75\,m^2) = 1839.38\,N.$$

Eq. (2.32): $F_1 - F_2 - F_g = \rho Q(V_2 - V_1)$

$(29{,}430\,N) - (1839.38\,N) - F_g = (1000\,kg\,m^{-3}) * (3.2\,m^3\,s^{-1}) * [(4.27\,m\,s^{-1}) - (1.07\,m\,s^{-1})]$

Solving the equation, one can obtain $F_g = 17{,}350.62\,N$.

The thrust on the gate 17,350.62 N, to the right.

(2) A hydraulic jump is a rapidly varied flow and needs to be analyzed by jointly using the momentum and energy equations.

The forces on the CV between sections ② and ③ include F_2' (hydrostatic force at ②) and F_3 (hydrostatic force at ③).

$A_3 = (1.5\,m) * (y_3) = 1.5y_3\,m^2$.

$V_3 = (3.2\,m^3\,s^{-1})/(1.5y_3\,m^2) = 2.13/y_3\,m\,s^{-1}$.

Eq. (2.7): $F_3 = (9810\,N\,m^{-3}) * (y_3/2\,m) * (1.5y_3\,m^2) = 7357.5(y_3)^2\,N$

$$F_2' = F_2 = 1839.38\,N.$$

Eq. (2.32): $F_2' - F_3 = \rho Q(V_3 - V_2)$

$(1839.38\,N) - [7357.5(y_3)^2\,N] = (1000\,kg\,m^{-3}) * (3.2\,m^3\,s^{-1}) * [(2.13/y_3\,m\,s^{-1}) - (4.27\,m\,s^{-1})]$

Solving the equation by trial-and-error or Excel Solver, yields $y_3 = 1.14\,m$.

Eq. (2.21): $(0.0\,\text{m}) + (0.5\,\text{m}) + [(4.27\,\text{m}\,\text{s}^{-1})^2/(2 * 9.81\,\text{m}\,\text{s}^{-2})] =$
$(0.0\,\text{m}) + (1.14\,\text{m}) + [(2.13/1.14\,\text{m}\,\text{s}^{-1})^2/(2 * 9.81\,\text{m}\,\text{s}^{-2})] + h_L$
Solving the equation, yields $h_L = 0.11$ m.

Example 2.10 A sprinkler has four identical 0.5-in-diameter nozzles (Figure 2.13). The arm length of the nozzles is 0.5 ft. If the water supply rate is 0.01 cfs, determine the rotation speed and torque required to make the sprinkler stationary.

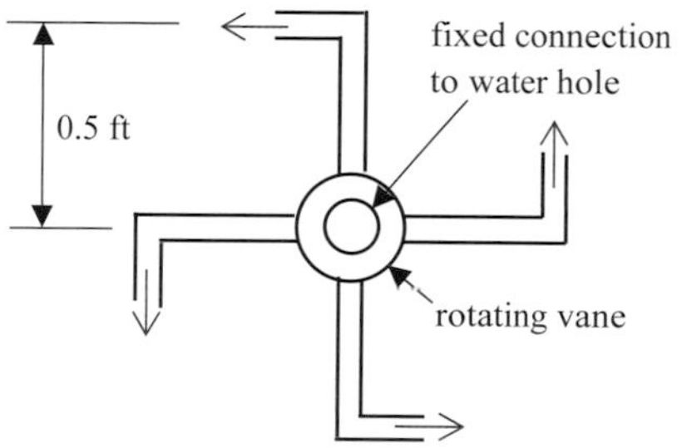

Figure 2.13 **Illustration of the sprinkler for Example 2.10.**

Solution

$\rho = 1.94\,\text{slug ft}^{-3}$, $Q = 0.01$ cfs, $r = 0.5$ ft, $D = 0.5$ in $= 0.0417$ ft.

Flow area of the nozzle: $A = (\pi/4)D^2 = (\pi/4) * (0.0417\,\text{ft})^2 = 0.001366\,\text{ft}^2$.

Velocity of nozzle flow: $V = Q/A = (0.01\,\text{cfs})/(0.001366\,\text{ft}^2) = 7.32\,\text{ft sec}^{-1}$.

Rotation speed: $\omega = V/r = (7.32\,\text{ft sec}^{-1})/(0.5\,\text{ft}) = 14.64\,\text{radian sec}^{-1} = 139.8$ rpm.

Fixed inner cylinder for water supply → angular momentum is zero.

Each nozzle's angular momentum: $(1.94\,\text{slug ft}^{-3}) * [(0.01\,\text{cfs})/4)] * (0.5\,\text{ft})^2 * (14.64\,\text{radian sec}^{-1}) = 0.01803\,\text{lbf} \cdot \text{ft}$.

Eq. (2.34): $T_{orque} = 4 * (0.01803\,\text{lbf} \cdot \text{ft}) - (0\,\text{lbf} \cdot \text{ft}) = 0.07212\,\text{lbf} \cdot \text{ft}$.

Thus, $0.07212\,\text{lbf} \cdot \text{ft}$ torque is required to make the sprinkler stationary.

2.5 Flow Regime

This section introduces flow classifications in terms of time, location, relative magnitude of inertia over friction force, and relative magnitude of inertia over gravity force.

2.5.1 Steady Versus Unsteady Flow

At a given location of waterway, for steady flow, all flow conditions (e.g., velocity and depth) do not vary with time; for unsteady flow, however, at least one flow condition varies with time.

The difference between these two flow regimes is illustrated in Figure 2.14. At a given time, flow conditions may vary from location to location regardless of the flow regimes. After several non-rainy days, the flow in a stream is approximately steady, whereas in a rainy day, the flow in the stream will be unsteady.

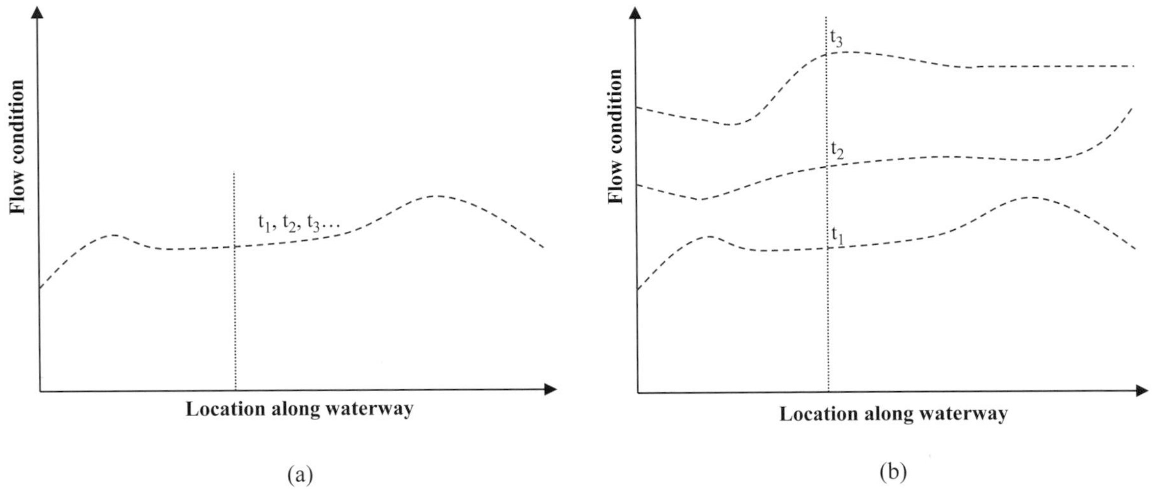

Figure 2.14 **Illustration of flow conditions for: (a) steady flow; and (b) unsteady flow.** Flow condition includes velocity and depth. t_1, t_2, and t_3 signify times.

2.5.2 Uniform Versus Nonuniform Flow

At a given time, for uniform flow, all flow conditions (e.g., velocity and depth) do not vary from location to location along a waterway; for nonuniform flow, however, at least one flow condition varies with location. The difference between these two flow regimes is illustrated in Figure 2.15. At a given location, flow conditions may vary with time regardless of the flow regimes. In reality, uniform flow rarely exists and nonuniform flow is most common.

2.5.3 Laminar Versus Turbulent Flow

In flowing water, the trace made by a single water molecule over a period of time is called the path line, while the mean movement path of a number of water molecules is called the streamline. For laminar flow, all path lines are parallel to the streamline (Figure 2.16a) and the point velocity is the same as the mean velocity. In contrast, for turbulent flow, the path line of a water molecule intersects the path line of another water molecule, and the streamline represents the overall movement path of all water molecules (Figure 2.16b). That is, the point velocity fluctuates around the mean velocity. In practice, the Reynolds number, R_e (Eq. (2.25)) can be computed and used to classify the flow regimes. As shown in the Moody diagram (Figure 2.11), if $R_e < 2000$, the flow is classified as laminar flow, whereas if $R_e > 3000$, the flow is classified as turbulent flow. When R_e is between 2000 and 3000, the flow is classified as transition flow.

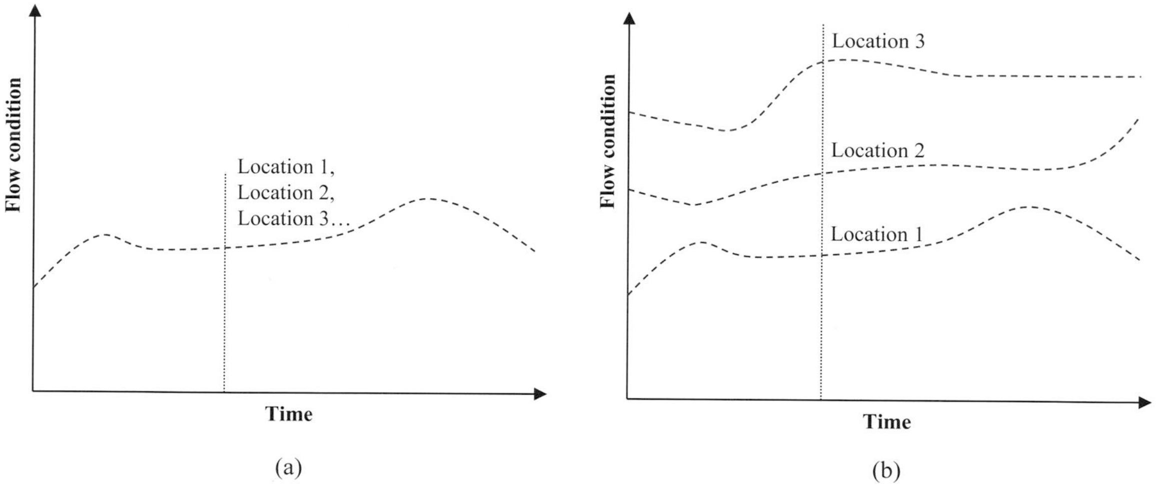

Figure 2.15 **Illustration of: (a) uniform flow, and (b) nonuniform flow.** Flow condition includes velocity and depth. Location 1, Location 2, and Location 3 signify locations along a waterway.

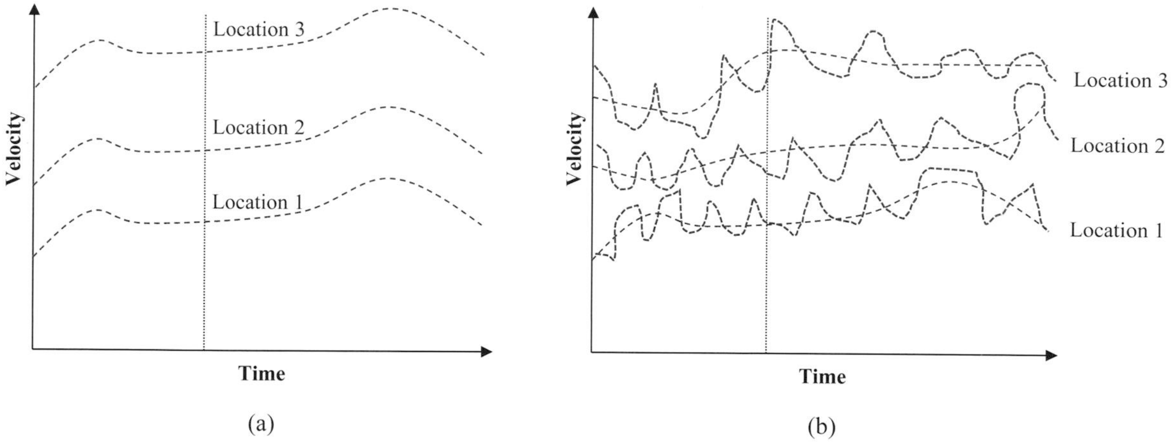

Figure 2.16 **Illustration of: (a) laminar flow, and (b) turbulent flow.** The blue dash line signifies the mean velocity, while the purple dash line signifies the point velocity.

2.5.4 Subcritical Versus Supercritical Flow

This classification is for open-channel flow only. It is based on specific energy at a given location. Assuming an energy correction factor $\alpha = 1.0$, one can simplify Eq. (2.14) to define the specific energy for open-channel flow as:

$$\underbrace{E}_{\text{specific energy}} = \underbrace{y}_{\text{water depth}} + \frac{V^2}{2g} = y + \frac{Q^2}{2gA^2} \qquad (2.35)$$

(V: mean velocity; g: gravitational acceleration; Q: flow rate; A: flow area)

If Q is constant, the relationship of E and y follows the specific energy curve illustrated in Figure 2.17. A given value of E corresponds to two water depths (termed alternate depths). The depth at which E is minimal is defined as the critical depth (usually symbolized as y_c). Flows with a water depth less than y_c are classified as supercritical flow, whereas flows with a water depth greater than y_c are classified as subcritical flow. Flows with a water depth equal to y_c are classified as critical flow. y_c depends on flow rate and cross-sectional geometry, but is independent of channel roughness and bed slope.

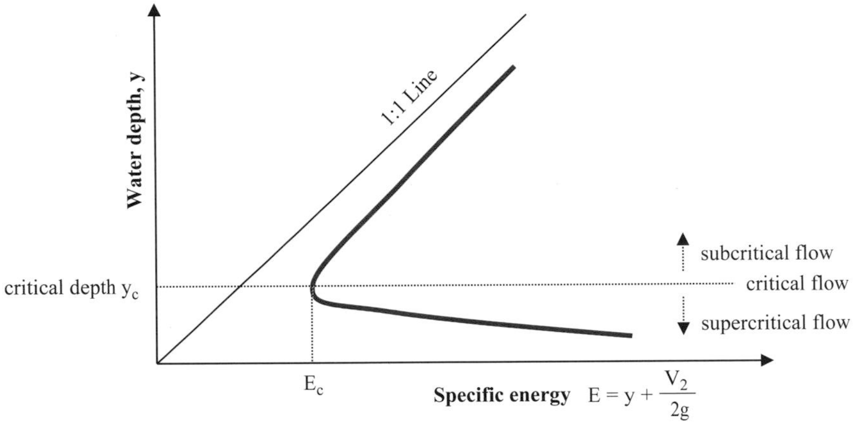

Figure 2.17 **The specific energy curve for a constant flow rate.**

Differentiating Eq. (2.35) with respect to water depth, y, and rearranging, one can derive the equation expressed as:

$$\frac{dE}{dy} = 1 - \frac{Q^2}{gA^3}\frac{dA}{dy} \tag{2.36}$$

(specific energy: E; flow rate: Q; water depth: y; gravitational acceleration: g; flow area: A)

In terms of Figure 2.18, $\frac{dA}{dy} = B$. Substituting this relationship into Eq. (2.36), letting $\frac{dE}{dy} = 0$, and rearranging, one can derive the formula expressed as:

$$\frac{Q^2}{g} = \left(\frac{A^3}{B}\right)_{y_c} \tag{2.37}$$

(flow rate: Q; gravitational acceleration: g; flow area: A; top width: B; critical depth: y_c)

For rectangular channel with a bottom width of b, at critical depth, $A_c = by_c$ and $B_c = b$. Substituting these two relationships into Eq. (2.37) and rearranging, yields:

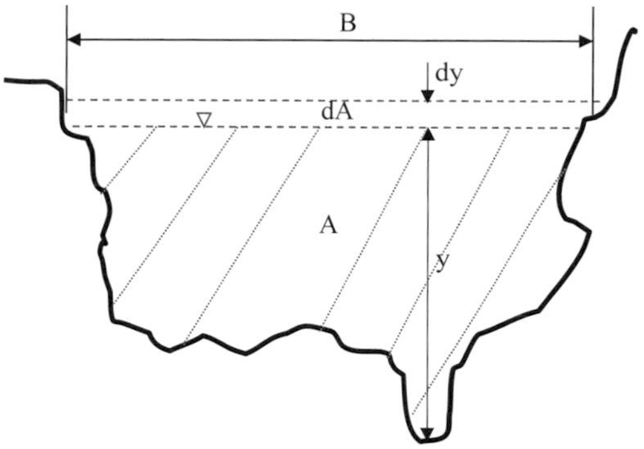

Figure 2.18 Channel cross section illustrating the relationship among changes of water depth, flow area, and top width.

$$y_c = \left[\frac{1}{g}\left(\frac{Q}{b}\right)^2\right]^{\frac{1}{3}} = \left(\frac{q^2}{g}\right)^{\frac{1}{3}} \tag{2.38}$$

(critical depth in rectangular channel: y_c; flow rate: Q; unit-width flow rate: q; gravitational acceleration: g; bottom width: b)

Substituting Eq. (2.38) into Eq. (2.35) and rearranging, yields:

$$E_c = \frac{3}{2}y_c \tag{2.39}$$

(minimum specific energy in rectangular channel: E_c; critical depth in rectangular channel: y_c)

Example 2.11 At a location of a rectangular channel, the water depth is 0.5 m while the mean velocity is $6\,\mathrm{m\,s^{-1}}$. Determine:

(1) The specific energy
(2) The alternate depth
(3) The critical depth
(4) The minimum specific energy
(5) If the flow is critical, subcritical, or critical.

Solution

$y = 0.5\,\mathrm{m}$, $V = 6\,\mathrm{m\,s^{-1}} \rightarrow q = Vy = (6\,\mathrm{m\,s^{-1}}) * (0.5\,\mathrm{m}) = 3\,\mathrm{m^2\,s^{-1}}$

(1) Eq. (2.35): $E = (0.5\,\mathrm{m}) + (6\,\mathrm{m\,s^{-1}})^2/(2 * 9.81\,\mathrm{m\,s^{-2}}) = 2.33\,\mathrm{m}$.
(2) The alternate depth satisfies: $E = y_1 + [(3\,\mathrm{m^2\,s^{-1}})/y_1]^2/(2 * 9.81\,\mathrm{m\,s^{-2}}) = 2.33\,\mathrm{m}$ $\rightarrow y_1 + 0.4587/(y_1)^2 = 2.33 \rightarrow$ Solving the equation by trial-and-error or Excel Solver, one can obtain $y_1 = 2.24\,\mathrm{m}$ (neglect the other two roots).
(3) Rectangular channel $\rightarrow$ Eq. (2.38): $y_c = \left(\frac{(3\,\mathrm{m^2\,s^{-1}})^2}{9.81\,\mathrm{m\,s^{-2}}}\right)^{\frac{1}{3}} = 0.97\,\mathrm{m}$.
(4) Specific energy is minimal at $y_c \rightarrow$ Eq. (2.35): $E_c = (0.97\,\mathrm{m}) + [(3\,\mathrm{m^2\,s^{-1}})/(0.97\,\mathrm{m})]^2/(2 * 9.81\,\mathrm{m\,s^{-2}}) = 1.46\,\mathrm{m}$.

Or Eq. (2.39): $y_c = 3/2 * (0.97\,\text{m}) = 1.46\,\text{m}$.

(5) $y < y_c \rightarrow$ supercritical flow.

If E is constant, Eq. (2.35) can be rewritten as:

$$Q = A\sqrt{2g(E - y)} \tag{2.40}$$

(flow rate: Q; flow area: A; gravitational acceleration: g; specific energy: E; water depth: y)

Differentiating Eq. (2.40) in relation to water depth, y, rearranging, and considering $\frac{dA}{dy} = B$ (Figure 2.18), one can derive:

$$\frac{dQ}{dy} = \frac{-2gBy + g(2BE - A)}{\sqrt{2g(E - y)}} \tag{2.41}$$

(flow rate: Q; water depth: y; gravitational acceleration: g; top width: B; specific energy: E; flow area: A)

Letting $\dfrac{dQ}{dy} = 0$ and rearranging, one can derive the formula expressed as:

$$y_c = E - \left(\frac{A}{2B}\right)_{y_c} \tag{2.42}$$

(critical depth at constant specific energy: y_c; specific energy: E; flow area: A; top width: B; critical depth: y_c)

Substituting Eq. (2.42) into Eq. (2.40), one can obtain the maximum flow rate expressed as:

$$Q_{max} = \left(A\sqrt{\frac{gA}{B}}\right)_{y_c} \tag{2.43}$$

(maximum flow rate at constant specific energy: Q_{max}; flow area: A; gravitational acceleration: g; top width: B; critical depth: y_c)

For a rectangular channel with a bottom width of b, at critical depth, $A_c = by_c$ and $B_c = b$. Substituting these two relationships into Eqs. (2.42) and (2.43) and rearranging, one can have:

$$y_c = \frac{2}{3}E \tag{2.44}$$

(critical depth for rectangular channel with constant specific energy: y_c; specific energy: E) maximum flow rate for rectangular channel with constant specific energy → top width (b), gravitational acceleration (g), specific energy (E)

$$Q_{max} = \frac{2}{3} bE \sqrt{\frac{2}{3} gE} \tag{2.45}$$

Example 2.12 For a 1-ft-wide rectangular channel controlled by a sluice gate, if the constant specific energy is 12 ft, compute y_c and Q_{max}. Also, develop and plot $Q/Q_{max} \sim y/y_c$.

	A	B	C	D	E
1	b =	1	ft		
2	E =	12	ft		
3	g =	32.2	ft sec^{-2}		
4					
5	Eq. (2.42):	y_c =	8	ft	
6	Eq. (2.43):	Q_{max} =	128.4	cfs	
7					
8	y (ft)	A (ft^2)	Eq. (2.38):Q (cfs)	y/y_c	Q/Q_{max}
9	0.0	0.0	0	0.00	0.00
10	0.5	0.5	13.6	0.06	0.11
11	1.0	1.0	26.6	0.13	0.21
12	1.5	1.5	39	0.19	0.30
13	2.0	2.0	50.8	0.25	0.40
14	2.5	2.5	61.8	0.31	0.48
15	3.0	3.0	72.2	0.38	0.56
16	3.5	3.5	81.9	0.44	0.64
17	4.0	4.0	90.8	0.50	0.71
18	4.5	4.5	98.9	0.56	0.77
19	5.0	5.0	106.2	0.63	0.83
20	5.5	5.5	112.5	0.69	0.88
21	6.0	6.0	117.9	0.75	0.92
22	6.5	6.5	122.3	0.81	0.95
23	7.0	7.0	125.6	0.88	0.98
24	7.5	7.5	127.7	0.94	0.99
25	8.0	8.0	128.4	1.00	1.00
26	8.5	8.5	127.6	1.06	0.99
27	9.0	9.0	125.1	1.13	0.97
28	9.5	9.5	120.5	1.19	0.94
29	10.0	10.0	113.5	1.25	0.88
30	10.5	10.5	103.2	1.31	0.80
31	11.0	11.0	88.3	1.38	0.69
32	11.5	11.5	65.3	1.44	0.51
33	12.0	12.0	0	1.50	0.00

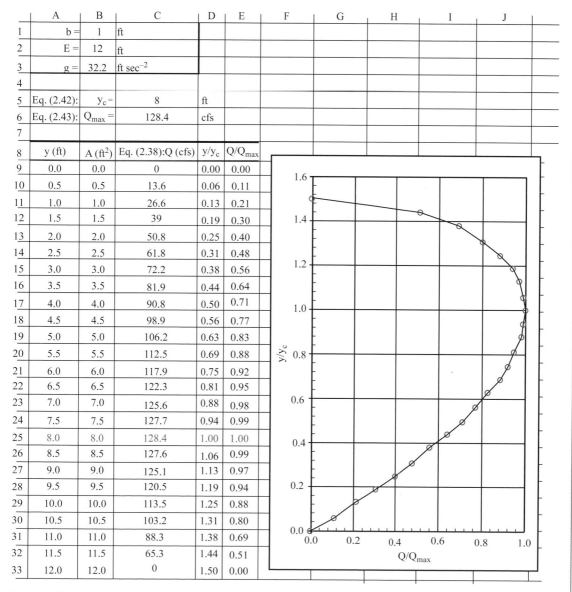

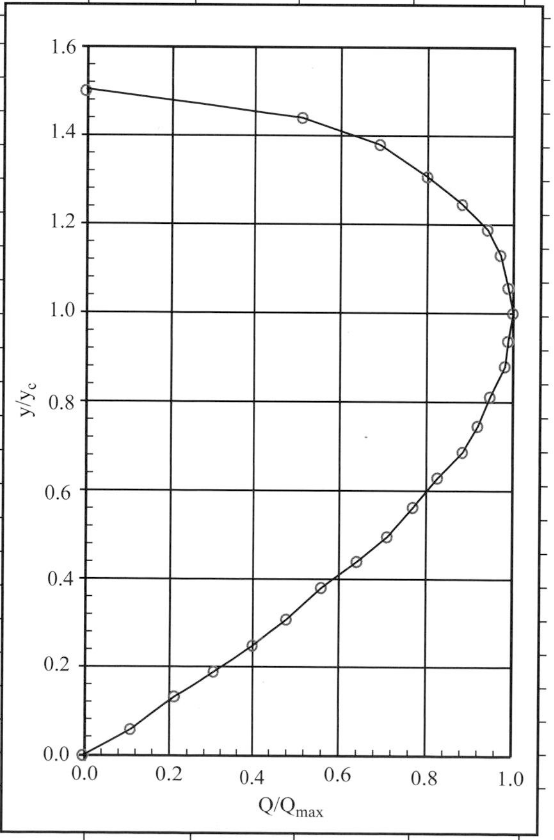

Figure 2.19 **Spreadsheet and plot for Q/Q_{max} versus y/y_c computed in Example 2.12.**

Solution

$b = 1$ ft, $E = 12$ ft. The channel is rectangular.

Eq. (2.44): $y_c = 2/3 * (12\,\text{ft}) = 8\,\text{ft}$.

Eq. (2.45): $Q_{max} = \frac{2}{3} * (1\,\text{ft}) * (12\,\text{ft}) * \sqrt{\frac{2}{3} * (32.2\,\text{ft sec}^{-2}) * (12\,\text{ft})} = 128.4\,\text{cfs}$.

For selected values of y between 0 and 12 ft, use Eq. (2.40) to compute Q. Afterward, compute the ratios of Q/Q_{max} and y/y_c and plot them in an Excel spreadsheet (Figure 2.19). For instance, for $y = 1.0\,\text{ft}$, flow area $A = (1\,\text{ft}) * (1.0\,\text{ft}) = 1.0\,\text{ft}^2$; using Eq. (2.40), one has $Q = \left(1.0\,\text{ft}^2\right) * \sqrt{2 * (32.2\,\text{ft sec}^{-2}) * [(12\,\text{ft}) - (1.0\,\text{ft})]} = 26.6\,\text{cfs} \rightarrow Q/Q_{max} = (26.6\,\text{cfs})/(128.4\,\text{cfs}) = 0.21$, $y/y_c = (1.0\,\text{ft})/(8\,\text{ft}) = 0.13$.

The Froude number measures the relative magnitude of inertia force to gravitational force. It is computed as:

$$F_r = \frac{V}{\sqrt{gd}} \tag{2.46}$$

(Froude number: F_r; mean velocity: V; gravitational acceleration: g; hydraulic depth (ratio of flow area to top width; see Table 2.4): d)

For either constant Q or E, substituting Eq. (2.37) or Eq. (2.42) into Eq. (2.46) and rearranging, one can easily show that $F_r = 1$ at critical depth, y_c. In fact, flows with $F_r < 1$ are classified as subcritical, whereas flows with $F_r > 1$ are classified as supercritical. Flows with $F_r = 1$ are classified as critical.

Example 2.13 Show that the Froude number for critical flows is equal to one.

Solution

For constant Q:

$$\text{Eq. (2.37)} \rightarrow \frac{Q^2}{(A^2)_{y_c}} = g\left(\frac{A}{B}\right)_{y_c} \rightarrow (V^2)_{y_c} = g(d)_{y_c} \rightarrow (F_r)_{y_c} = \frac{(V)_{y_c}}{\sqrt{g(d)_{y_c}}} = 1.$$

For constant E:

$$\text{Eq. (2.42)} \rightarrow y_c = \left[y_c + \frac{(V^2)_{y_c}}{2g}\right] - \left(\frac{A}{2B}\right)_{y_c} \rightarrow \frac{(V^2)_{y_c}}{2g} = \frac{(d)_{y_c}}{2} \rightarrow (F_r)_{y_c} = \frac{(V)_{y_c}}{\sqrt{g(d)_{y_c}}} = 1.$$

Example 2.14 Water flows at $6\,\text{m s}^{-1}$ in a triangular channel (side slope $Z = 1$). If the water depth is 0.5 m, compute the Froude number and classify the flow regime.

Solution

$Z = 1$, $y = 0.5\,\text{m}$, $V = 6\,\text{m s}^{-1}$.

Triangular channel → Table 2.4: hydraulic depth $d = 1/2 * (0.5\,\text{m}) = 0.25\,\text{m}$.

Eq. (2.46): $F_r = \frac{6\,m\,s^{-1}}{\sqrt{(9.81\,m\,s^{-2}) * (0.25\,m)}} = 3.83.$

$F_r > 1 \rightarrow$ the flow is supercritical.

2.6 Dimensional Analysis and Similitude

This section introduces the basics of dimensional and similitude analyses, which make it feasible to develop empirical functional relationships in terms of laboratory experiments.

2.6.1 Dimensional Analysis

The main purpose of dimensional analysis is to reduce the number of independent variables by transforming all variables into dimensionless ratios. Herein, independent variables should be "independent" from each other, that is, one independent variable cannot be derived from any other independent variables. For example, velocity, distance, and time are not "independent" independent variables, whereas velocity, density, and viscosity are "independent" independent variables. There are three basic dimensions, namely mass (M), length (L), and time (T), from which all other dimensions are derived.

Dimensional analysis is based on Buckingham's π theorem (Buckingham, 1914; Finnemore and Franzini, 2002), which is stated as: "If a physical process involves a functional relationship among n variables, which can be expressed in terms of m basic dimensions, it can be reduced to a relation between (n–m) dimensionless variables (or π terms), by choosing m repeating variables, each of which is combined in turn with the remaining variables to form the π terms as products of the variables taken to the appropriate powers. The m repeating variables must contain among them all basic dimensions found in all the variables but cannot themselves form a π term."

Example 2.15 It is hypothesized that the friction factor (f) for a fluid system is a function of its hydraulic radius (R), roughness height (k_S), fluid velocity (V), fluid density (ρ), and fluid dynamic viscosity (μ). State the dimensionless function.

Solution

Five independent variables: R, k_S, V, ρ, and μ (they are "independent").

One dependent variable: f.

Totally six variables (i.e., n = 6).

Three basic dimensions (i.e., m = 3): M, L, and T.

Buckingham's π theorem $\rightarrow$ 3 (n − m = 6 − 3) π terms can be formed.

Choose three independent variables as repeating variables, which contain all three basic dimensions. Herein, R, ρ, and μ are chosen. Of course, different choices may be made as long as all three basic dimensions are contained.

The functional relationship can be written as: $f = F_{n1}(R, k_s, V, \rho, \mu)$.

The dimensional analysis can be done by following the four steps listed in Table 2.7. It results in three dimensionless variables, namely f, $\frac{k_s}{R}$, and $\frac{\rho VR}{\mu}$. The latter two terms represent the relative roughness height and Reynolds number, respectively. Thus, the dimensional relationship is consistent with the Moody diagram shown in Figure 2.11.

Table 2.7 Dimensional analysis for Example 2.15.

Step	Purpose	Operation	Functional relationship in terms of	
			Variables	Dimensions
1	Establish	Express in equation	$f = F_{n1}(R, k_s, V, \rho, \mu)$	$[-] = F_{n1}([L], [L], [L\,T^{-1}], [M\,L^{-3}], [M\,L^{-1}\,T^{-1}])$
2	Cancel [L]	Divide by R or its power	$f = F_{n2}\left(\frac{k_s}{R}, \frac{V}{R}, \rho R^3, \mu R\right)$	$[-] = F_{n2}([-], [T^{-1}], [M], [M\,T^{-1}])$
3	Cancel [M]	Divide by ρR^3 or its power	$f = F_{n3}\left(\frac{k_s}{R}, \frac{V}{R}, \frac{\mu}{\rho R^2}\right)$	$[-] = F_{n3}([-], [T^{-1}], [T^{-1}])$
4	Cancel [T]	Divide by $\frac{\mu}{\rho R^2}$ or its power	$f = F_{n4}\left(\frac{k_s}{R}, \frac{\rho VR}{\mu}\right)$	$[-] = F_{n4}([-], [-])$

2.6.2 Similitude

Similitude is the theory and art of predicting prototype performance from model observations (Crowe *et al.*, 2005). The basic idea is that the forces acting on a model and those acting on the prototype must be in the same ratio (Buckingham, 1915). In practice, if gravitational force is most important, the Froude number needs to be preserved (i.e., the Froude number for the model must be same as the Froude number for the prototype), whereas if viscous (i.e., shear) force is most important, the Reynolds number needs to be preserved (i.e., the Reynolds number for the model must be same as the Reynolds number for the prototype). For a given prototype, either the Froude number or the Reynolds number may be preserved, but it is not feasible to preserve both.

Example 2.16 A 1:100-scale model of a spillway is used to predict the flow condition of the prototype. The model is tested in a laboratory flume.

(1) If the design discharge of the spillway is 10,000 cfs, what flow rate should be used in the model test?

(2) If a flow velocity of 0.5 ft sec^{-1} in the flume is used, what is the corresponding flow velocity through the spillway?

Solution

$L_m/L_p = 0.01$. For spillway design, gravitational force is most important and thus the Froude number should be preserved. That is, $\frac{V_m}{\sqrt{gL_m}} = \frac{V_p}{\sqrt{gL_p}}$.

(1) $Q_p = 10{,}000$ cfs. Rearranging $\rightarrow V_m = V_p\sqrt{\frac{L_m}{L_p}} \rightarrow$

$$Q_m = V_m L_m^2 = V_p L_m^2\sqrt{\frac{L_m}{L_p}} = \left(V_p L_p^2\right)\frac{L_m^2}{L_p^2}\sqrt{\frac{L_m}{L_p}} = Q_p\left(\frac{L_m}{L_p}\right)^{\frac{5}{2}} = (10,000\,\text{cfs}) * 0.01^{\frac{5}{2}} = 0.1\,\text{cfs}.$$

(2) $V_m = 0.5\,\text{ft sec}^{-1}$. Rearranging $\rightarrow V_p = V_m\sqrt{\frac{L_p}{L_m}} = \left(0.5\,\text{ft sec}^{-1}\right) * \sqrt{100} = 5\,\text{ft sec}^{-1}$.

Example 2.17 A 1:20-scale laboratory model of a diversion channel is used to predict the flow condition in the prototype. If the channel is designed to convey $50\,\text{m}^3\,\text{s}^{-1}$, what flow rate should be used in the model test?

Solution

$L_m/L_p = 0.05$, $Q_p = 50\,\text{m}^3\,\text{s}^{-1}$. For diversion channel, viscous force is most important and thus the Reynolds number should be preserved. That is $\frac{V_m L_m}{\nu_m} = \frac{V_p L_p}{\nu_p} \rightarrow$ assuming $\nu_m = \nu_p \rightarrow V_m = V_p\frac{L_p}{L_m} \rightarrow$

$$Q_m = V_m L_m^2 = \left(V_p\frac{L_p}{L_m}\right)L_m^2 = \left(V_p L_p^2\right)\frac{L_m^2}{L_p^2}\frac{L_p}{L_m} = Q_p\left(\frac{L_m}{L_p}\right) \rightarrow Q_m = \left(50\,\text{m}^3\,\text{s}^{-1}\right) * 0.05 = 2.5\,\text{m}^3\,\text{s}^{-1}.$$

PROBLEMS

2.1 If the density of water is $997\,\text{kg m}^{-3}$, what is the specific weight in SI and BG units?

2.2 Based on Table 2.1, what is the percent change of the density of water if its temperature is increased from 15 to 35°C? Does the density of water increase or decrease as its temperature increases?

2.3 If the density and dynamic viscosity of water are $1.94\,\text{slug ft}^{-3}$ and $2.36 \times 10^{-5}\,\text{lbf}\cdot\text{sec ft}^{-2}$, respectively, what is the kinematic viscosity?

2.4 Based on Table 2.1, what is the percent change of the kinematic viscosity of water if its temperature is increased from 40 to 60°F? Does the kinematic viscosity of water increase or decrease as its temperature increases?

2.5 What is the saturation vapor pressure of water at 8°C? Is this relative or absolute pressure?

2.6 What is the saturation vapor pressure of water at 72°F? What is the corresponding water height of the pressure?

2.7 A particle is still at 1.5 ft below the water surface. If the water temperature is 60°F, what is the magnitude and direction of the hydrostatic pressure on the top of the particle? What is the hydrostatic pressure at the bottom of the particle?

2.8 Compute and depict the hydrostatic pressures on the planes submerged in water at 20°C. The planes are drawn in red.

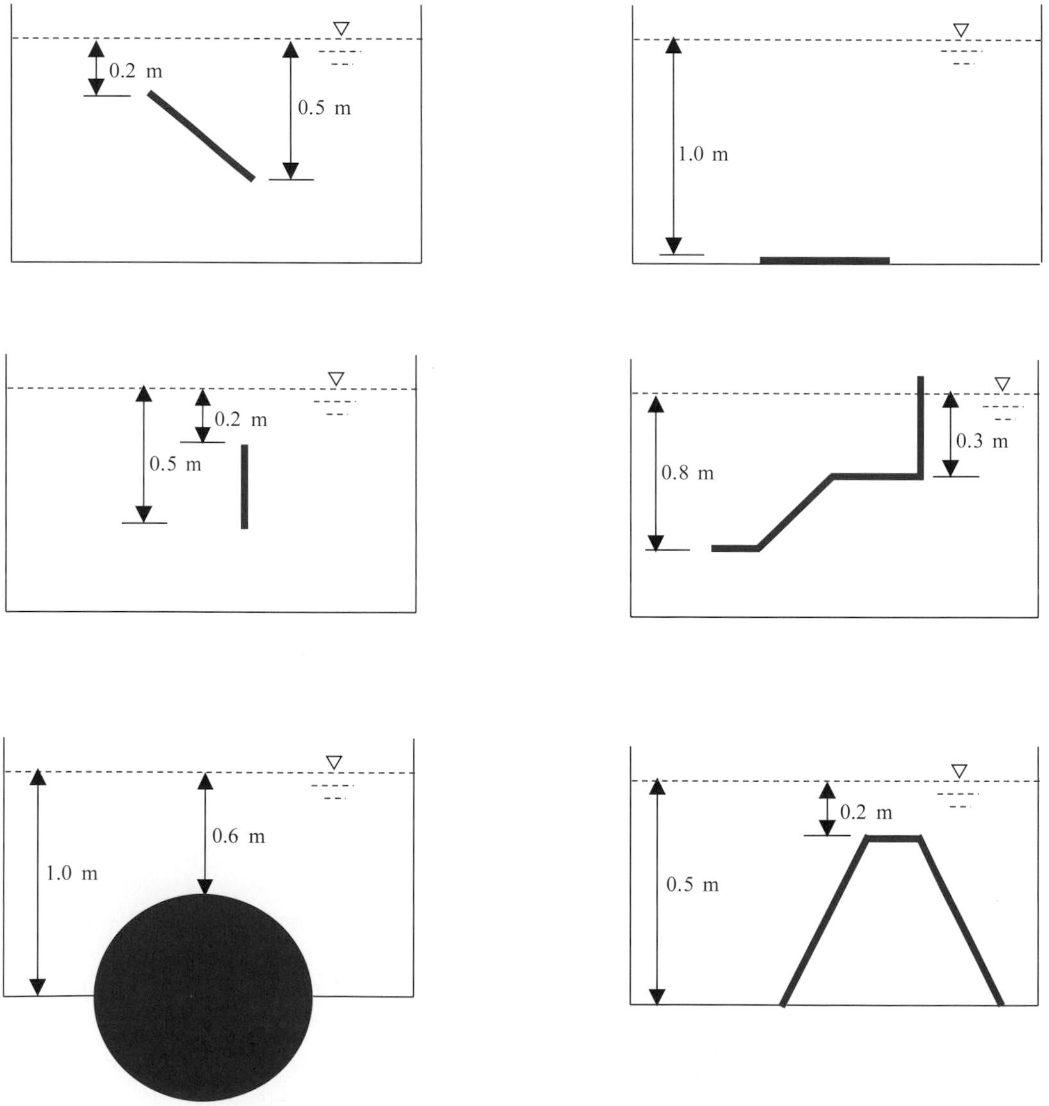

2.9 Compute and depict the resultant forces per unit-width for the following structures submerged in water at 60°F.

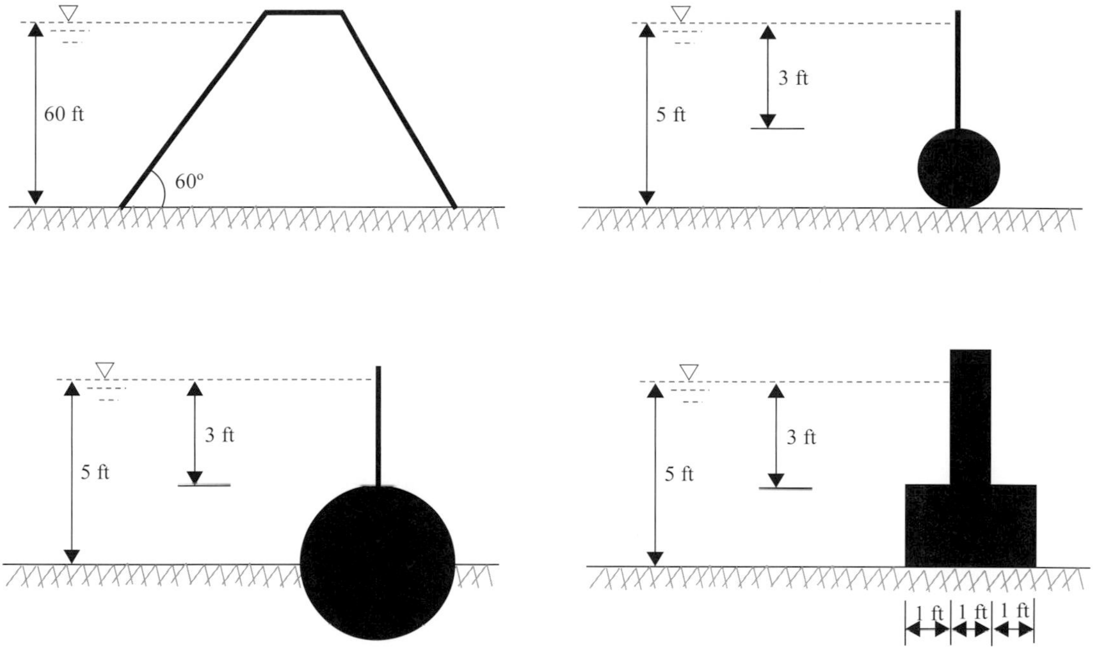

2.10 For the cross sections shown, compute the top widths, cross-sectional areas, flow areas, wetted perimeters, hydraulic radius, water depths, and hydraulic depths.

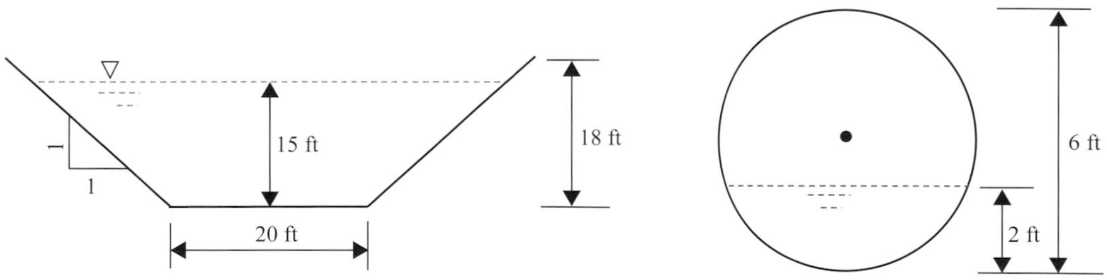

2.11 Given the following water pipe system, determine the mean velocity in pipe ③.

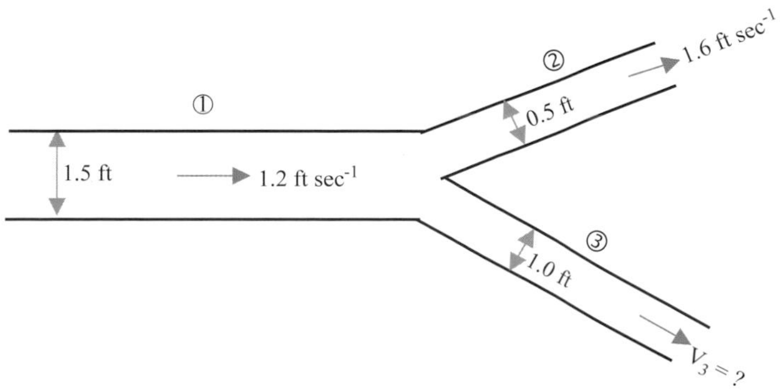

2.12 For the following steady-state river system, if the water losses are negligible, determine the discharges at stations A, B, and C.

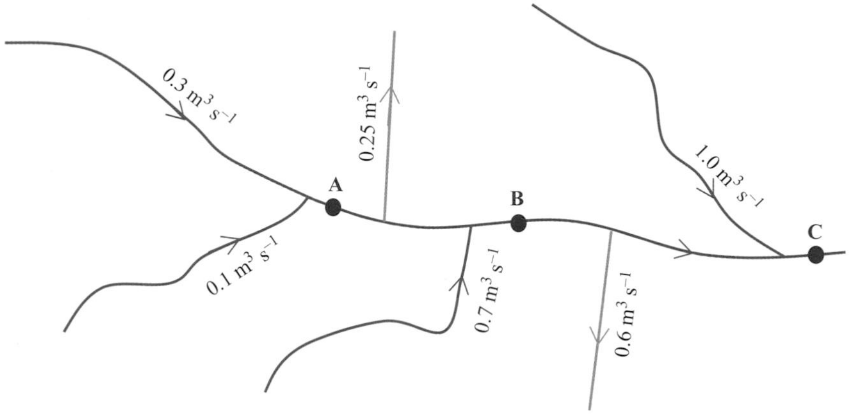

2.13 A concrete-lined rectangular channel (2 ft wide and 100 ft long) carries 200 cfs water 60°F. If the average water depth in the channel is 5 ft, determine the friction loss using the:
(1) Darcy–Weisbach equation
(2) Manning formula.

2.14 A 100-m-long concrete-lined triangular channel (side slope 2.5) carries $50\,m^3\,s^{-1}$ water 20°C. If the average water depth in the channel is 2 m, determine the friction loss using the:
(1) Darcy–Weisbach equation
(2) Manning formula.

2.15 A 100-ft-long 2-ft-diameter cast iron (new) pipe is designed to carry 200 cfs water 60°F. If the pipe flow is pressurized, determine the friction loss using the:
(1) Darcy–Weisbach equation
(2) Hazen–Williams formula.

2.16 A 500-m-long 1.5-m-diameter steel pipe carries $10\,m^3\,s^{-1}$ water 20°C. If the pipe flow is pressurized, determine the friction loss using the:
(1) Darcy–Weisbach equation
(2) Hazen–Williams formula.

2.17 A 1000-m-long 1.5-m-diameter steel pipe (sharp-edge entrance) is used to convey 20°C water from Reservoir A to Reservoir B. Determine the flow rate in the pipe.

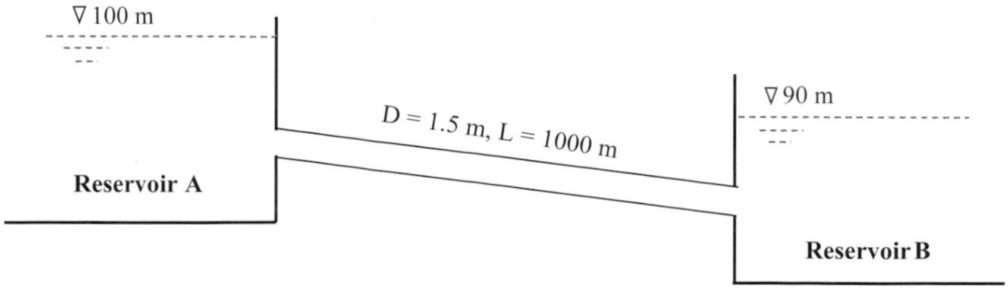

2.18 A 1000-ft-long 2-ft-diameter concrete pipe (sharp-edge entrance) is used to convey 60°F water from Reservoir A to B. At the end of the pipe, a diffuser with an exit diameter of 3.5 ft is installed to reduce discharge loss. If the friction loss in the diffuser is negligible, determine the flow rate in the pipe. Consider the expansion loss of the diffuser as the difference of the pre- and post-expansion velocity heads.

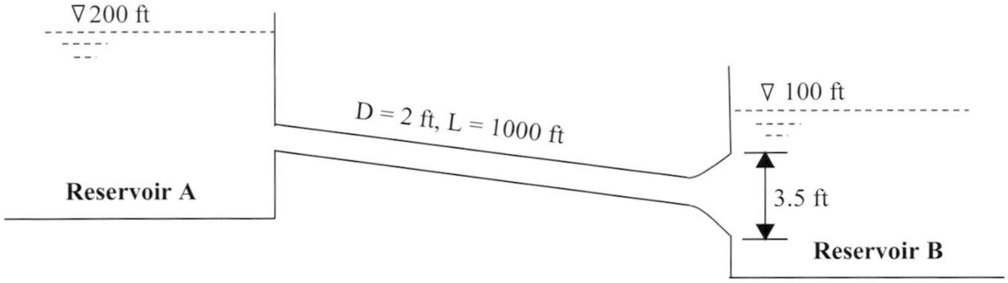

2.19 A 1-m-diameter culvert ($C_{hw} = 100$) conveys water along a 2.5-m-wide rectangular channel. If the entrance and exit loss coefficients are 0.3 and 1.0, respectively, determine the:

(1) Flow rate in the culvert

(2) Pressure at section ②.

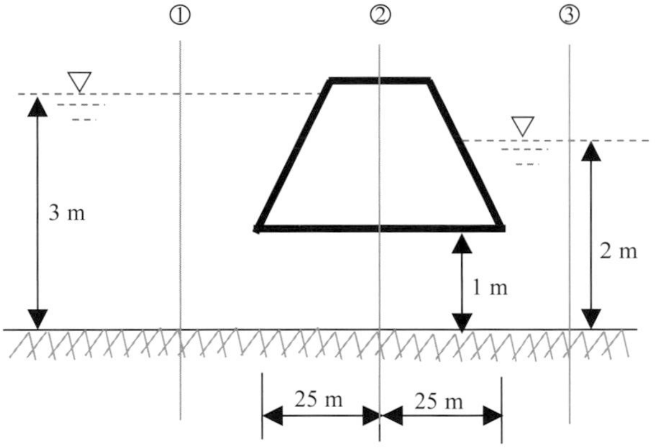

2.20 For the following water system, determine the total energy loss in the channels surrounded by sections ①, ②, and ③.

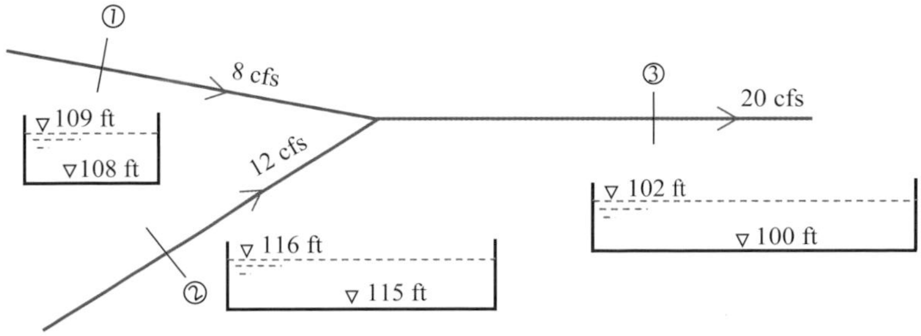

2.21 A 5-ft-wide concrete rectangular channel carries 200 cfs water. At a location, a concrete rise is constructed to control the flow to have an upstream water depth of 10 ft. If the critical water depth is achieved at the rise, determine the thrust on the rise. Neglect any frictions.

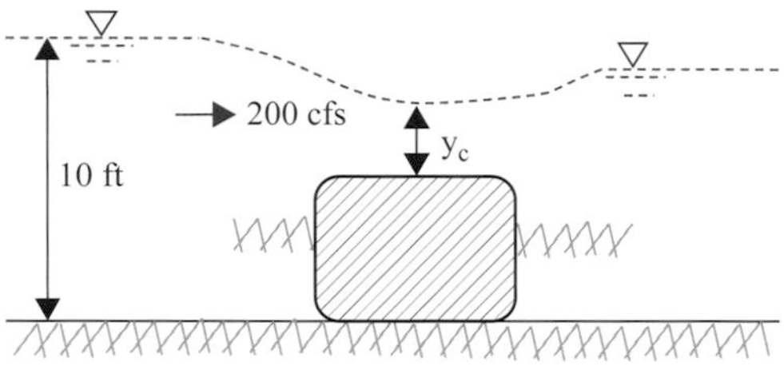

2.22 A 2-m-wide concrete rectangular channel carries 5 $m^3\ s^{-1}$ water with a depth of 5 m. If a dip causes an increased depth of 6.5 m, determine the thrust on the dip. Neglect any frictions.

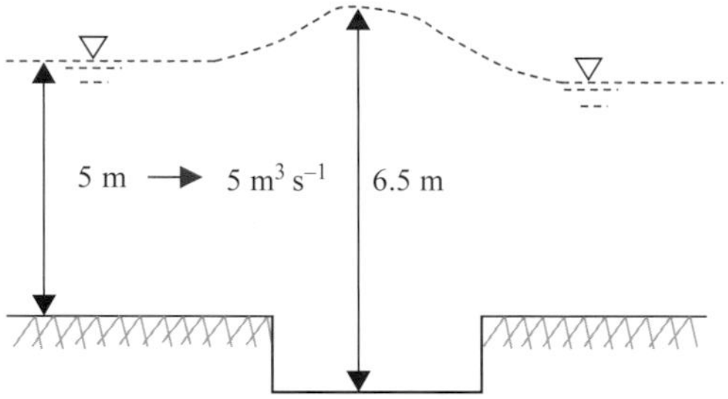

2.23 For the following sprinkler, determine the rotation speed and torque required to make it stationary.

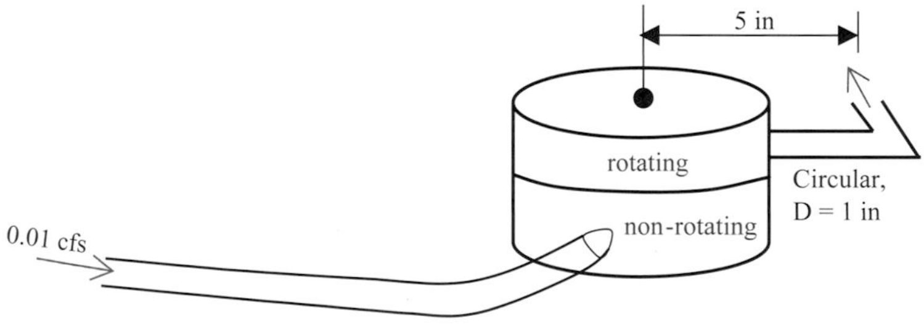

2.24 List three examples of both steady and unsteady flow.

2.25 A 2-ft-diameter pipe carries 10 cfs water 60°F. Is the flow laminar or turbulent? Why?

2.26 A trapezoidal channel (bottom width 5 m and side slope 2.5) conveys 20 $m^3\ s^{-1}$ water at 20°C. If the water depth in the channel is 8 m, determine if the flow is laminar or turbulent.

2.27 A 2-ft-wide rectangular channel carries a uniform flow of 2.5 cfs. If the water depth at a location is 2.5 ft, determine the water depth, velocity, and specific energy at 100 ft downstream of the location.

2.28 A 3-m-wide rectangular channel (bed slope 0.005) carries 6 $m^3\ s^{-1}$ nonuniform flow. If the water depths at two locations 200 m apart are 1.5 and 2.0 m, respectively, determine the energy loss and sketch the channel bed, hydraulic grade line (HGL), and energy grade line (EGL).

2.29 A triangular channel (side slope 1.5) carries 50 cfs.
(1) Develop and plot the specific energy curve.
(2) Compute the critical water depth and minimum specific energy.
(3) Compute the alternate depth paired with 4.5 ft.
(4) Compute the Froude number for a water depth of 2.5 ft.
(5) If the water depth is 1.0 ft, is the flow critical, subcritical, or supercritical? Why?

2.30 A 2-m-wide rectangular channel carries 25 $m^3 s^{-1}$ water. Determine the critical depth and minimum specific energy. If the channel bed slope is increased but the flow rate kept the same, do you think that the critical depth should be changed as well? Why?

2.31 The flow in a 1-ft-wide flume is controlled by a sluice gate to have a constant specific energy of 1.2 ft.
(1) Determine the critical depth (y_c) and maximum flow rate (Q_{max}).
(2) Plot $\frac{Q}{Q_{max}} \sim \frac{y}{y_c}$ and compare it with Figure 2.17.

2.32 The flow in a triangular channel (side slope 1.5) is controlled to have a constant specific energy of 2.5 m.
(1) Determine the critical depth (y_c) and maximum flow rate (Q_{max}).
(2) Plot $\frac{Q}{Q_{max}} \sim \frac{y}{y_c}$ and compare it with Figure 2.17.

2.33 Practical experiences indicate that friction loss (h_f) is a function of flow velocity (V), conveyance length (L), roughness height (k_s), flow area (A), wetted perimeter (P_{wet}), and gravitational acceleration (g). State the dimensionless function.

2.34 A researcher hypothesizes that the fluid drag force (F_d) on a moving object is a function of fluid dynamic viscosity (μ) and specific weight (γ) as well as object frontal area (A_f) and moving speed (V). State the dimensionless function.

2.35 A 1:50-scale model of a canoe is used to predict the draft depth (i.e., the vertical distance between waterline and canoe bottom) of full-size prototype. The test needs to preserve the Froude number, which is computed as $\frac{Q/b}{\sqrt{g y_d A}}$, where Q is the flow rate in the waterway, b is the width of the canoe, y_d is the draft depth of the canoe, and A is the cross-sectional area of the canoe. The canoe will be used in a waterway of 1500 cfs. If the model has a draft depth of 1.5 in in a flume flow of 0.2 cfs, predict the draft depth of the canoe in the waterway.

2.36 A 1:100-scale laboratory model of an irrigation canal is used to predict the flow condition of the prototype. If the model is tested using a flow rate of 0.1 $m^3 s^{-1}$, what flow rate in the canal can be predicted based on the model test?

3 Watershed and Hydrologic Processes

This chapter discusses hydrologic processes and analysis methods related to the design and management of water resources engineering structures. In this regard, the concept of a watershed is introduced and its physical characteristics are described, the qualitative and quantitative aspects of water balance are examined, and the notion of water resources protection and utilization is discussed. Furthermore, the five major hydrologic processes that control runoff volume and flow rate, in particular peak discharge, are explored.

3.1 Watershed and Water Balance

This section introduces the concept of a watershed, describes its most important characteristics, and investigates the components of water balance. It also presents the basic analysis methods of water supply and demand.

3.1.1 Watershed Concepts and Characteristics

A watershed (Figure 3.1) is the land area from which water tends to flow downward due to gravity along a drainage network to a single common point, called the outlet. The drainage network can consist of natural streams, manmade channels, drainage ditches, culverts, and storm sewers. Its boundary is defined by a ridge line (which usually is not linear per se, but rather curvilinear), on which any given point has a topographic elevation higher than the two adjacent points along the straight line passing through the three points and perpendicular to the ridge line (as shown by the I–I′ elevation profile in Figure 3.1). Thus, water inside the watershed cannot flow across its boundary due only to gravity; likewise, water outside the boundary cannot flow into the watershed only due to gravity.

Conventionally, a watershed boundary is delineated manually using a topographic map, which shows elevation contours (e.g., Figure 3.2). Each contour represents the ground locations with the same elevation. A delineated watershed boundary can be influenced both by map precision

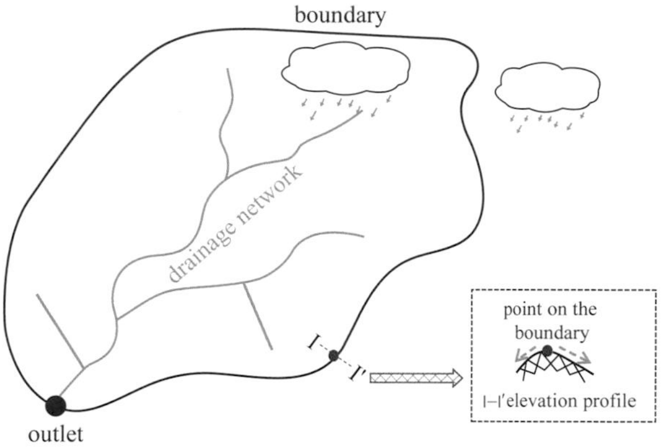

Figure 3.1 **Map of a conceptual watershed and its drainage network.** Section I–I′ on the map is shown in the inset diagram.

and accuracy. Map precision is dependent on the contour interval (i.e., the elevation difference between two adjacent contours): the smaller the interval, the greater the precision. For instance, a map with a 0.5 ft contour interval is more precise than one with a 1.0 ft interval. However, a more precise map does not mean that it is more accurate. Map accuracy is influenced by a number of factors, such as data measurement technique and processing method.

The delineation of contours is usually implemented by two steps. The first step is to identify the contours representing ridges and the contours representing valleys. The V-shaped contours that point toward lower elevations form a ridge, whereas the V-shaped contours that point toward higher elevations form a valley, as shown in Figure 3.2b. The second step is to start from the watershed outlet, where one is interested in knowing runoff volume and streamflow, and draw stepwise straight lines along the topographic gradients between the contours, as illustrated in Figure 3.2c. For any two adjacent contours, the gradient is the steepest slope along the straight line connecting the two geographically closest points on the contours. For instance, the gradient line between the 50 and 60 ft contours in Figure 3.2c is along AB because AB is shortest. For a large watershed, manual delineation can be very tedious and time-consuming as well as subject to delineators' experiences.

Example 3.1 Figure 3.2a shows a topographic map with a contour interval of 10 ft. Delineate the boundary of the watershed with its outlet at W.

Solution

The delineation can be implemented by following three steps. First, identify the contours representing ridges and the contours representing valleys. The V-shaped contours that point toward lower elevations form a ridge, whereas the V-shaped contours that point toward higher elevations form a valley. The identified ridges and valleys are labeled in Figure 3.2b. Second, starting from outlet W draw stepwise straight gradient lines along the two ridges adjacent to the stream that drains the watershed, as shown in Figure 3.2b. Note that any line should always be perpendicular to its two intersecting contours. Finally, make necessary adjustments to ensure that any artificial high-elevation obstructions (e.g., railroads and highways) or localized high spots (i.e., closed contours) are considered as parts of the watershed boundary. The final delineated watershed boundary is shown in Figure 3.2c.

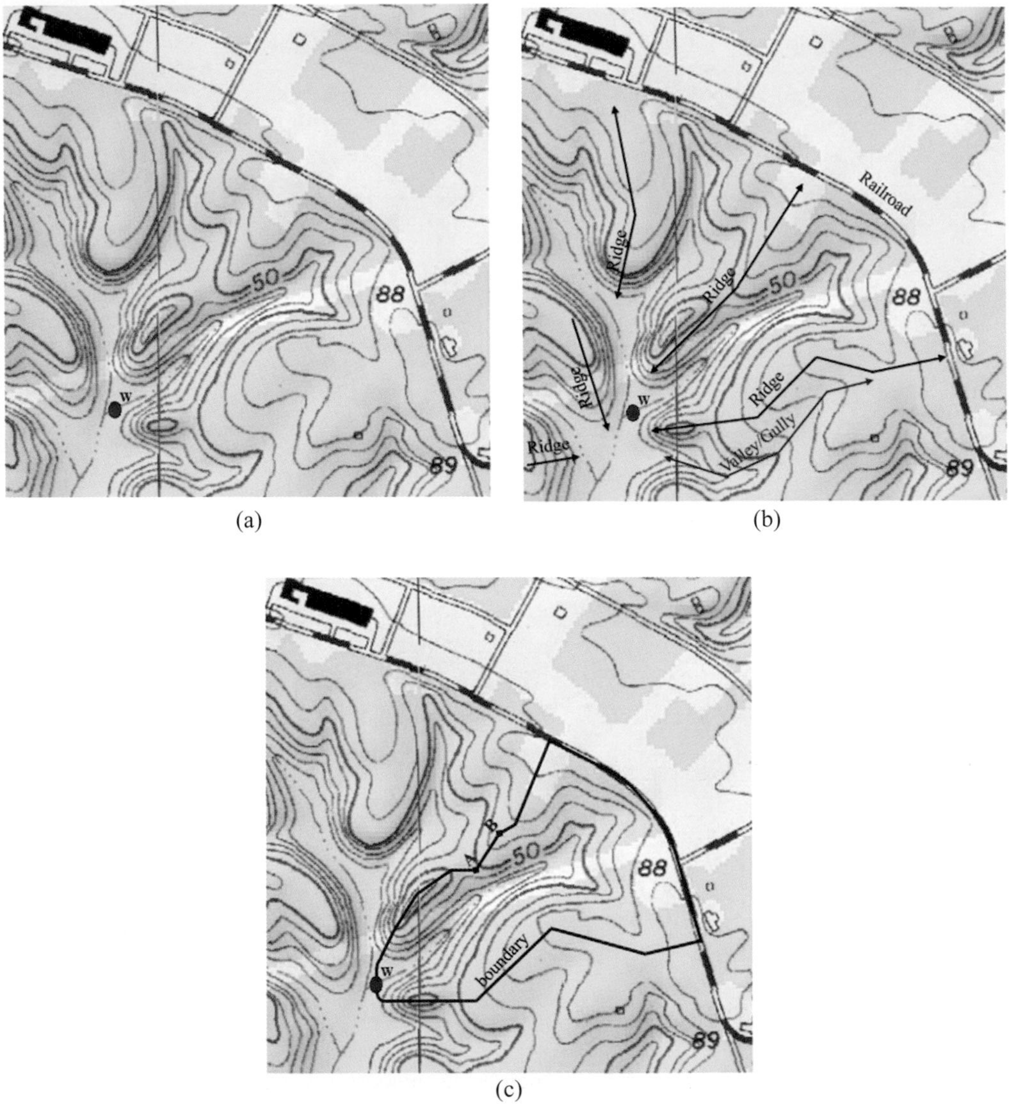

Figure 3.2 **A topographic map showing the process of delineating a watershed.** The contour interval is 10 ft. The solution to Example 3.1 is also shown. (a) Watershed to be delineated has an outlet point W; (b) identification of several ridges and valleys; and (c) delineation of watershed boundary.

In recent decades, high-resolution and high-accuracy digital elevation models (DEMs) have been developed. The DEMs have a raster data structure that represents the ground surface using a number of square grids with a specific dimension or resolution. The DEMs developed using

LiDAR (Light Detection and Ranging) data can have a resolution of centimeters. Currently, LiDAR DEMs with variable (1 to 30 m) spatial resolutions are available for the continental United States, Alaska, Hawaii, Puerto Rico, Guam, and the United States Virgin Islands. The DEMs can be used in a Geographic Information System (GIS) (e.g., ArcMAP by ESRI) (ESRI, 2011) to automatically delineate a watershed boundary by implementing an eight-direction (i.e., D8) algorithm with an assumption that water in a given grid flows to one of eight adjacent grids with the steepest slope.

The details of the algorithm are available in numerous publications (e.g., O'Callaghan and Mark, 1984; Garbrecht and Martz, 1997; Seibert and McGlynn, 2007). To gain practice in the use of this algorithm, an Excel macro subroutine, named *FlowDir_DEM ( )*, can be downloaded from the textbook website https://fs.wp.odu.edu/x4wang/textbook. *FlowDir_DEM ( )* can be used to generate the gradient and "flow-to" matrices using a DEM stored in a spreadsheet by implementing the D8 algorithm developed by Garbrecht and Martz (1997). Figure 3.3 illustrates a 10-m DEM and the delineated watershed boundary using the output matrices of this subroutine.

Example 3.2 Delineate the watershed boundary based on the 10-m DEM given in Figure 3.3a using the D8 algorithm. The number of rows is $M = 4$ and the number of columns is $N = 5$. Elevation values of the DEM are in meters.

Solution

The grid size of the DEM is $a = 10\,\text{m}$. The delineation can be implemented by following four steps.

For each grid, starting from the upper-left grid, calculate slopes from this grid toward its adjacent grids. Let Elv_{ij} be the elevation of the grid in row i and column j ($i = 1, 2, \ldots, M$; $j = 1, 2, \ldots, N$), the slopes between this grid and its adjacent grids are computed as $S_{ij\to mn} = \frac{Elv_{ij} - Elv_{mn}}{L_{ij\to mn}}$, where $m = i - 1, i$, or $i + 1$; $n = j - 1, j$, or $j + 1$; $mn \neq ij$, $m \geq 1$, $n \geq 1$; Elv_{mn} is the elevation of the grid in row m and column n; and $L_{ij\to mn}$ is the geographical distance between grid ij and mn. If $m = i$ or $n = j$, $L_{ij\to mn} = a = 10\,\text{m}$, otherwise, $L_{ij\to mn} = \sqrt{2}a = 10\sqrt{2}\,\text{m}$. A positive value for $S_{ij\to mn}$ indicates that grid ij is higher than grid mn, whereas a negative value for $S_{ij\to mn}$ indicates that grid ij is lower than grid mn. For instance, the slopes from grid 22 toward its adjacent grids 11, 12, 13, 21, 23, 31, 32, and 33 are calculated as: $S_{22\to 11} = \frac{90-80}{10\sqrt{2}} = 0.71$; $S_{22\to 12} = \frac{90-120}{10} = -3.00$; $S_{22\to 13} = \frac{90-130}{10\sqrt{2}} = -2.83$; $S_{22\to 21} = \frac{90-140}{10} = -5.00$; $S_{22\to 23} = \frac{90-85}{10} = 0.50$; $S_{22\to 31} = \frac{90-160}{10\sqrt{2}} = -4.95$; $S_{22\to 32} = \frac{90-100}{10} = -1.00$; $S_{22\to 33} = \frac{90-75}{10\sqrt{2}} = 1.06$

Second, for each grid, calculate its gradient, which is the steepest positive slope from this grid toward one of its adjacent grids. When all slopes are negative, this grid is a "sink," meaning that water from the adjacent grids flow into this grid. Figure 3.3b shows the gradient matrix.

Third, identify the "sink" grid that drains most grids and treat it as the watershed outlet. For instance, grid 22 has a gradient of $\max\{0.71, -3.00, -2.83, -5.00, 0.50, -4.95, -1.00, 1.06\} = 1.06$ and is drained to grid 33. Figure 3.3c shows the "flow-to" matrix.

Finally, draw the watershed boundary by including all grids drained to the outlet. The grids that are not included in the boundary are either drained to another watershed or function as localized storage areas (e.g., isolated ponds or wetlands). The delineated watershed boundary is shown in Figure 3.3c.

Row i =	Column j = 1	2	3	4	5
1	80	120	130	100	85
2	140	90	85	95	100
3	160	100	75	75	120
4	120	150	70	120	130

(a)

Row i =	Column j = 1	2	3	4	5
1	sink	4.00	4.50	1.50	sink
2	6.00	1.06	1.00	2.00	1.77
3	6.00	2.50	0.50	0.35	4.50
4	2.12	8.00	sink	5.00	3.89

(b)

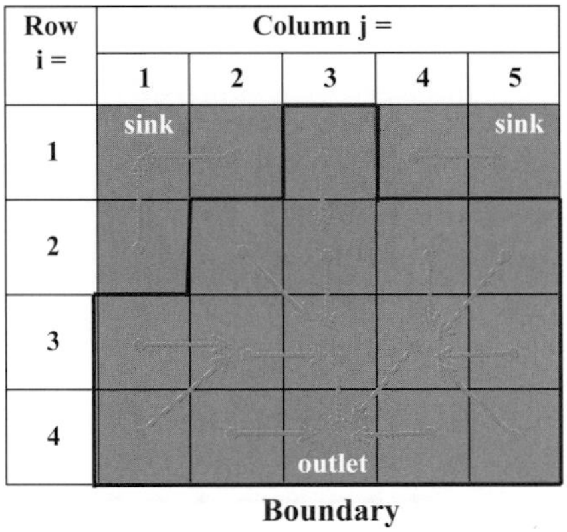

(c)

Figure 3.3 **Solution to Example 3.2 using a 10-m digital elevation model (DEM).** The grids shown represent the: (a) 10-m DEM; (b) gradient matrix; and (c) "flow-to" matrix and watershed boundary. The elevation values are in meters above mean sea level.

Watershed geometric characteristics include drainage area (A), watershed length or longest flow path (L_w), watershed width (B_w), watershed slope (S_w), longest slope length (L_s), and drainage density (D_{dn}). As illustrated in Figure 3.4a, A is the total land area drained to the watershed outlet W_o; L_w is the geographical distance along the main channel from W_o to W_b, the intersect between the "extrapolated" main channel (i.e., the longest flow path) and the boundary; B_w is the width along the transect line perpendicular to the main channel that gives a maximum distance between the two intersects of the transect line and the watershed boundary. S_w is the ratio of the difference of ground elevations at W_b and W_o to L_w, whereas D_{dn} is the ratio of the total length of all streams, rivers, channels, canals, and ditches to A. L_s is the distance along the main channel from the watershed outlet to the point at which the slope decrease percentage is larger than threshold value α_s (e.g., 50%). For instance, in Figure 3.4b, if $\frac{S_{0,2}-S_{0,1}}{S_{0,2}} > \alpha_s$, $L_s = L_1$; otherwise, $L_s = L_1 + L_2$. Given that $S_{0,3}$ is negative because this channel segment has an adverse slope, L_s cannot be longer than $L_1 + L_2$. For a given watershed, these geometric characteristics can be determined by

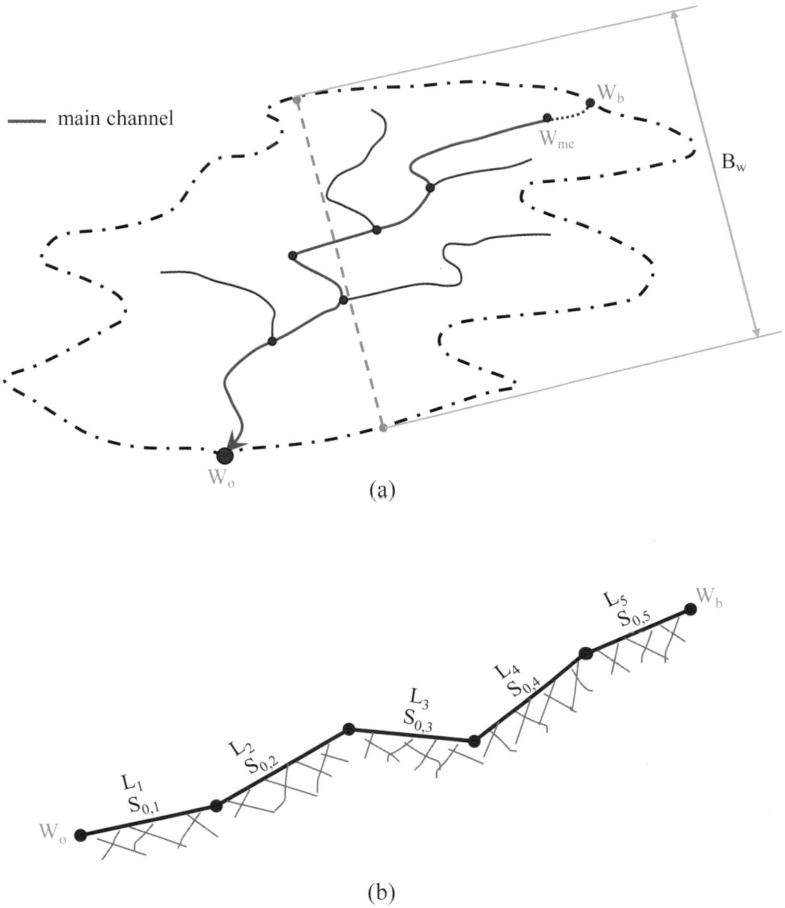

Figure 3.4 **Watershed geometric characteristics.** (a) Sample watershed boundary and drainage network; and (b) an assumed main channel profile. W_o is the watershed outlet, W_{mc} the end of the main channel, and W_b the intersect between the "extrapolated" main channel (i.e., the flow path) and the boundary. The green dashed line signifies the transect defining the watershed width B_w. $S_{01}, S_{02}, \ldots, S_{05}$ are the bed slopes of the main channel segments.

analyzing either a topographic map or DEM. For situations in the absence of other information, when A is given, L_w can be empirically computed as (USDA-NRCS, 2010):

$$\underbrace{L_w}_{\text{watershed length [ft]}} = 209\underbrace{A}_{\text{drainage area [ac]}}^{0.6} \tag{3.1}$$

Moreover, land use and land cover (LULC) and soils are important watershed characteristics as well. LULC can greatly influence water consumption and movement as well as soil erosion and sedimentation. For a given watershed, LULC can vary from time to time, depending on climate conditions (e.g., precipitation) and human activities (e.g., farming and urbanization). In contrast, soils are relatively invariant at the watershed scale though some soil properties (e.g., permeability) may be changed at a local scale due to various disturbances (e.g., compaction). In terms of permeability, in 2004, the US Department of Agriculture (USDA) Natural Resources Conservation Service (NRCS) published the criteria for assignment of hydrologic soil groups (HSGs). In terms of the criteria, soils are classified into HSG A, B, C, and D (Table 3.1), which in this order have an increasing permeability and thus an increasing runoff potential.

Table 3.1 Definition of hydrologic soil groups (HSGs).[1]

HSG	Runoff potential	Definition
A	low	High rate of water transmission; well to excessively drained sands or gravels; high infiltration rate, saturated hydraulic conductivity of the least transmissive layer $K_{sat} > 5.67$ in hr^{-1}
B	moderately low	Moderate rate of water transmission; moderately well to well drained soils with moderately fine to moderately coarse textures; moderate infiltration rate, $K_{sat} = 1.42 \sim 5.67$ in hr^{-1}
C	moderately high	Slow rate of water transmission; soils with a layer that impedes downward movement of water or with moderately fine to fine texture; slow infiltration rate, $K_{sat} = 0.06 \sim 1.42$ in hr^{-1}
D	high	Very slow rate of water transmission; clay soils, soils with a permanent high water table, soils with a claypan or clay layer at or near the surface, and shallow soils over nearly impervious material; very slow infiltration rate, $K_{sat} \leq 0.06$ in hr^{-1}

[1] *Source:* USDA-NRCS (2004).

3.1.2 Estimation of Time Concentration

Two important terms relating to rainfall and water flow are *hyetograph* and *hydrograph*. Whereas a hyetograph is a graphical representation of rainfall intensity over time, a hydrograph is a graphical representation of flow rate over time. When a rainfall hyetograph and its resulting flow hydrograph are plotted together (Figure 3.5), one can define the important times for watershed hydrology. At the beginning of a storm event, the rainfall intensity is smaller than the soil infiltration capacity and thus the rainfall may be lost to canopy interception, depression storage, and/or infiltration. The summation of these losses is termed as initial abstraction. As the storm continues, the soil infiltration capacity tends to decrease and will eventually become equal to the rainfall intensity. After the point of equivalency is reached, a portion of the rainfall (termed as rainfall excess or excess rainfall) will be converted into direct runoff, which in turn will form a flow hydrograph as illustrated in Figure 3.5. Herein, the rainfall excess is equal to the direct runoff depth, computed as the ratio of the total runoff volume to the drainage area. The total runoff volume is computed as the integration of discharge (excluding baseflow) with respect to time from A (i.e., beginning of direct runoff) to B (i.e., end of direct runoff).

The time between A and B is defined as the time base of the direct runoff hydrograph, t_b. The time between A and the end of rainfall is defined as the duration of runoff-producing rain, t_s, and

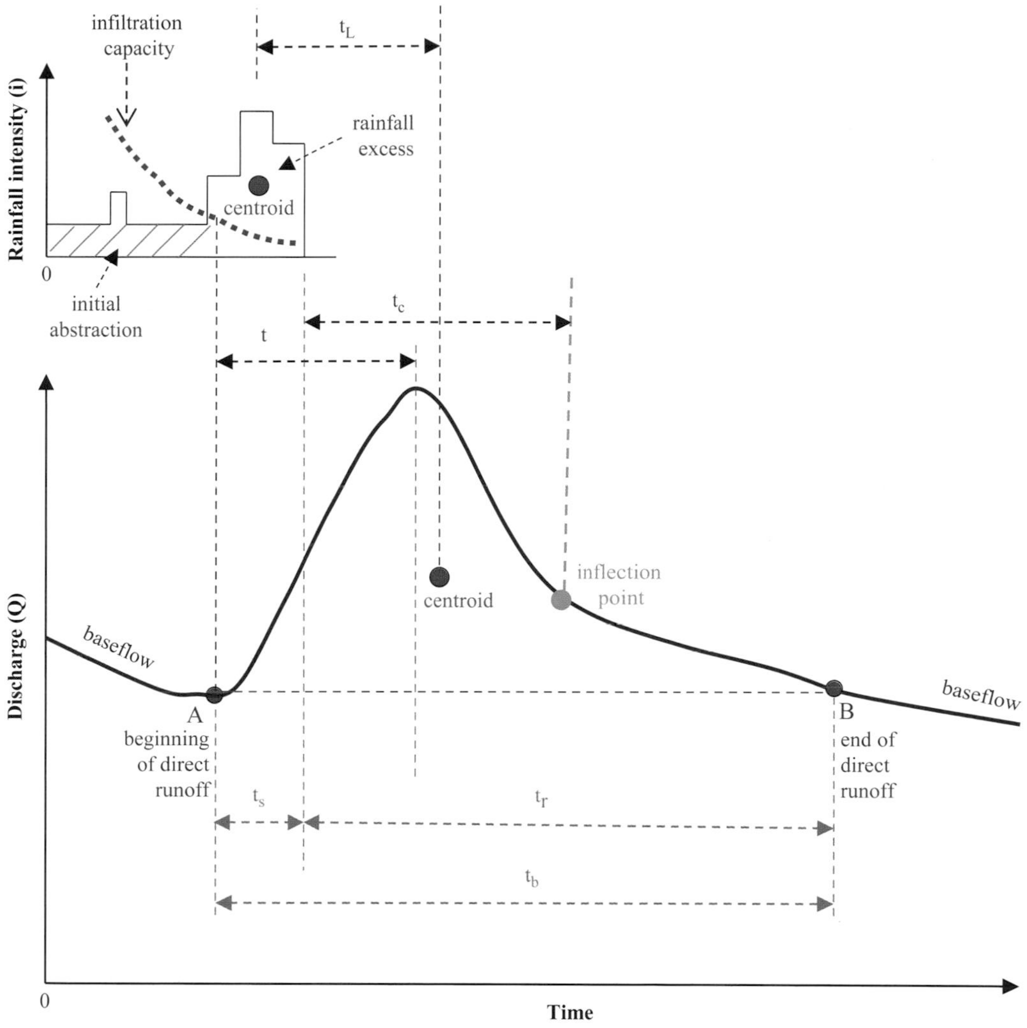

Figure 3.5 **Relationship in time of a rainfall hyetograph (top diagram) and flow hydrograph (bottom diagram).** The linear, stepwise, black curve represents rainfall, whereas the solid blue curve shows discharge. The vertical axes are at time zero. t_s: duration of runoff-producing rain; t_r: release time of rainfall excess; t_b: time base of the direct runoff hydrograph; t_p: time to peak; t_c: time of concentration; and t_L: lag time.

the time between the end of rainfall and B is defined as the release time of rainfall excess, t_r. These three times are subject to $t_b = t_s + t_r$. In addition, the time between A and peak discharge is defined as the time to peak, t_p, whereas the time between the end of rainfall and the inflection point of the recession limb curve of the flow hydrograph is defined as the time of concentration, t_c. At the inflection point, the recession limb curve changes from being concave to convex. In theory, t_c is the travel time of a raindrop from the remotest location to the watershed outlet (e.g., from W_b to W_o in Figure 3.4). That is, t_c is the time it takes for runoff from the entire watershed to concentrate at the outlet. Furthermore, the time between the centroid (i.e., mass center) of the rainfall excess hyetograph and the centroid of the direct runoff hydrograph is defined as the lag time, t_L. Usually, $t_L \approx 0.6t_c$. In practice, t_c can be estimated using the USDA Soil Conservation Service (SCS) Lag equation, the Kirpich formula, the US Federal Aviation Administration (FAA) formula, the Kerby formula, or the SCS Technical Release (TR) 55 approach. Application conditions for these equations and formulas are summarized in Table 3.2, with the algebraic expressions of Eqs (3.2) through (3.11).

Table 3.2 Application conditions of the estimation methods for time of concentration, t_c.

Method	Equation	Application condition	Reference
SCS Lag equation	3.2	Agricultural watersheds	USDA-NRCS (2010)
Kirpich formula	3.3	Urban areas and agricultural watersheds up to 200 ac	Chow *et al.* (1988)
FAA formula	3.4	Urban areas	Singh (1992); Corbitt (1999)
Kerby formula	3.5	Main channel less than 1200 ft, watershed slope less than 1%, and drainage area less than 10 ac	Chin (2000)
SCS TR55 approach	3.6 to 3.11	Urban and agricultural watersheds with a sheet flow path length less than 300 ft	USDA-SCS (1990); USDA-NRCS (2010)

The SCS Lag equation (USDA-SCS, 1990) can be expressed as:

$$t_c = \frac{1.67L_w^{0.8}\left(\frac{1000}{CN_{II}} - 9\right)^{0.7}}{1900S_w^{0.5}} \tag{3.2}$$

time of concentration [hr]; watershed length [ft]; SCS curve number for antecedent moisture condition (AMC) II (see Section 3.2.4); watershed slope [%]

The Kirpich formula (Kirpich, 1940) can be expressed as:

$$t_c = 0.00013\, r_{ki} L_w^{0.77} S_w^{-0.385} \tag{3.3}$$

time of concentration [hr]; Kirpich adjustment factor (Table 3.3); watershed length [ft]; watershed slope [ft ft^{-1}]

The FAA formula (Singh, 1992; Corbitt, 1999) can be expressed as:

$$t_c = 0.00646\,(1.1 - C)\, L_w^{0.5} S_w^{-\frac{1}{3}} \tag{3.4}$$

time of concentration [hr]; runoff coefficient [–] (see Section 4.2.2); watershed length [ft]; watershed slope [ft ft^{-1}]

The Kerby formula (Kerby, 1959) can be expressed as:

$$t_c = 0.01378 \left(\frac{L_w r_{ke}}{S_w^{0.5}} \right)^{0.467} \tag{3.5}$$

where t_c = time of concentration [hr]; L_w = watershed length [ft]; r_{ke} = Kerby coefficient [-] (Table 3.3); S_w = watershed slope [ft ft^{-1}]

Table 3.3 Factors or coefficients for the Kirpich (Eq. (3.3)) and Kerby (Eq. (3.5)) formulas.

Land cover	Kirpich factor, $r_{ki}^{[1]}$	Land cover	Kerby coefficient, $r_{ke}^{[2]}$
Natural overland and grass channels	2.0	Conifer timberland, dense grass	0.80
Bare overland and roadside ditches	1.0	Deciduous timberland	0.60
Concrete/asphalt surfaces	0.4	Moderate-density grass	0.40
Concrete channels	0.2	Low-density grass, bare sod	0.30
		Smooth bare packed soil, free of stones	0.10
		Smooth pavements	0.02

[1] Derived from multiple sources (e.g., Kirpich, 1940; Chow *et al.*, 1988).
[2] Derived from multiple sources (e.g., Kerby, 1959; Chin, 2000).

Routing along the longest flow path of a watershed, overland runoff commonly passes through three types of flows, including sheet flow, shallow concentrated flow, and channel flow. Sheet flow (depth generally $\leq$ 1.0 ft) is flow over land surfaces and occurs in the headwater of streams. After a flow path of up to 300 ft, sheet flow may become shallow concentrated flow. Sheet flow does not have a visible flowing direction, whereas shallow concentrated flow has a noticeably dominant flowing direction. Ultimately, shallow concentrated flow will flow downstream into streams and then out of the watershed outlet as channel flow. The TR55 method (USDA-NRCS, 1986; USDA-SCS, 1990) computes the time of concentration as a summation of the travel times of these three types of flows and can be expressed as:

$$t_c = t_{sf} + t_{scf} + t_{cf} \tag{3.6}$$

where t_c = time of concentration [hr]; t_{sf} = travel time of sheet flow [hr]; t_{scf} = travel time of shallow concentrated flow [hr]; t_{cf} = travel time of channel flow [hr]

$$t_{sf} = \frac{0.007 (n_{sf} L_{sf})^{0.8}}{P_2^{0.5} S_{sf}^{0.4}} \tag{3.7}$$

where t_{sf} = travel time of sheet flow [hr]; n_{sf} = Manning's n for sheet flow path [-] (Table 3.4); L_{sf} = length of sheet flow path [ft]; P_2 = 2-yr 24-hr precipitation for the watershed [in] (see Section 4.1.2); S_{sf} = ground slope of sheet flow path [ft ft^{-1}]

Table 3.4 Manning's roughness coefficient for sheet flow, n_{sf}.[1]

Land surface	n_{sf}
Smooth surface (concrete, asphalt, gravel, or bare soil)	0.011
Fallow (no residue)	0.05
Cultivated soils:	
Residue cover $\leq 20\%$	0.06
Residue cover $> 20\%$	0.17
Grass:	
Short grass prairie	0.15
Dense grasses	0.24
Bermuda grass	0.41
Range (natural)	0.13
Woods:	
Light underbrush	0.40
Dense underbrush	0.80

[1] *Source:* USDA-NRCS (2010).

$$t_{scf} = \frac{L_{scf}}{3600 V_{scf}} \tag{3.8}$$

travel time of shallow concentrated flow [hr] (t_{scf}); length of shallow concentrated flow path [ft] (L_{scf}); average velocity of shallow concentrated flow [ft sec^{-1}] (V_{scf})

$$t_{cf} = \frac{L_{cf}}{3600\ V_{cf}} \tag{3.9}$$

travel time of channel flow [hr] (t_{cf}); length of channel flow path [ft] (L_{cf}); average velocity of channel flow [ft sec^{-1}] (V_{cf})

$$V_{scf} = \begin{cases} 20.328\ S_{scf}^{0.5} & \text{(paved surface)} \\ 16.1345\ S_{scf}^{0.5} & \text{(unpaved surface)} \\ \dfrac{1.12}{n_{scf}} y_{scf}^{\frac{2}{3}}\ S_{scf}^{0.5} & \text{(paved curb/gutter)} \end{cases} \tag{3.10}$$

average velocity of shallow concentrated flow [ft sec^{-1}] (V_{scf}); bed slope of shallow concentrated flow path [ft ft^{-1}] (S_{scf}); Manning's n of curb/gutter (n_{scf}); gutter flow depth [ft] (y_{scf})

$$V_{cf} = \frac{1.49}{n_{cf}} R_{cf}^{\frac{2}{3}} S_{cf}^{0.5} \tag{3.11}$$

average velocity of channel flow [ft sec^{-1}] (V_{cf}); hydraulic radius of bank full open channel or culvert flowing full [ft] (R_{cf}); channel bed slope [ft ft^{-1}] (S_{cf}); Manning's n for channel flow path [-] (Table 3.5) (n_{cf})

Table 3.5 Manning's roughness coefficient for channel flow, n_{cf}.[1]

Material	n_{cf}
Lucite	0.008 ~ 0.010
Glass	0.009 ~ 0.013
Neat cement surface	0.010 ~ 0.013
Wood-stave pipe	0.010 ~ 0.013
Plank flumes, planed	0.010 ~ 0.014
Vitrified sewer pipe	0.010 ~ 0.017
Concrete, precast	0.011 ~ 0.013
Metal flumes, smooth	0.011 ~ 0.015
Cement mortar surfaces	0.011 ~ 0.015
Plank flumes, unplanned	0.011 ~ 0.015
Common-clay drainage tile	0.011 ~ 0.017
Concrete, monolithic	0.012 ~ 0.016
Brick with cement mortar	0.012 ~ 0.017
Cast iron, new	0.013 ~ 0.017
Riveted steel	0.017 ~ 0.020
Cement rubble surfaces	0.017 ~ 0.030
Canals and ditches, smooth earth	0.017 ~ 0.025
Corrugated metal pipe	0.021 ~ 0.030
Metal flumes, corrugated	0.022 ~ 0.030
Canals:	
Dredged in earth, smooth	0.025 ~ 0.033
In rock cuts, smooth	0.025 ~ 0.035
Rough beds and weeds on sides	0.025 ~ 0.040
Rock cuts, jagged and irregular	0.035 ~ 0.045
Natural streams:	
Smoothest	0.025 ~ 0.033
Roughest	0.045 ~ 0.060
Very weedy	0.075 ~ 0.150

[1] *Source:* USDA-NRCS (2010).

Example 3.3 A 100 ac agricultural watershed has all three types of flows from the most remote location to the outlet. The watershed has a composite curve number of $CN_{II} = 72$ and a runoff coefficient of $C = 0.65$. For the sheet flow, $L_{sf} = 100$ ft, $S_{sf} = 0.01$ ft ft^{-1}, $n_{sf} = 0.15$, whereas for the shallow concentrated flow, $L_{scf} = 1000$ ft, $S_{scf} = 0.01$ ft ft^{-1}. The channel flow is conveyed in an earth trapezoidal channel with a bottom width of $b = 5$ ft, side slope of $Z = 2$, bank-full depth of $y_n = 2.5$ ft, $L_{cf} = 2500$ ft, $S_{cf} = 0.005$ ft ft^{-1}, $n_{cf} = 0.03$. Estimate t_c using the five methods described above. Assume $r_{ki} = 2.0$ and $r_{ke} = 0.40$. The 2-yr 24-hr rainfall is $P_2 = 3.5$ in.

Solution

$$L_w = L_{sf} + L_{scf} + L_{ch} = 100\ \text{ft} + 1000\ \text{ft} + 2500\ \text{ft} = 3600\ \text{ft}$$

$$S_w = \frac{L_{sf}S_{sf} + L_{scf}S_{scf} + L_{ch}S_{ch}}{L_{sf} + L_{scf} + L_{ch}}$$
$$= \frac{(100\text{ ft}) \times (0.01) + (1000\text{ ft}) \times (0.01) + (2500\text{ ft}) \times (0.005)}{100\text{ ft} + 1000\text{ ft} + 2500\text{ ft}} = 0.0065\text{ ft ft}^{-1} = 0.65\%.$$

SCS Lag equation (Eq. (3.2)):

$$t_c = \frac{1.67L_w^{0.8}\left(\frac{1000}{CN_{II}} - 9\right)^{0.7}}{1900S_w^{0.5}} = \frac{1.67 \times (3600\text{ ft})^{0.8} \times \left(\frac{1000}{72} - 9\right)^{0.7}}{1900 \times (0.65)^{0.5}} = 2.32\text{ hr.}$$

Kirpich formula (Eq. (3.3)):

$$t_c = 0.00013\ r_{ki}L_w^{0.77}S_w^{-0.385} = 0.00013 \times 2.0 \times (3600\text{ ft})^{0.77} \times (0.0065)^{-0.385} = 0.99\text{ hr.}$$

FAA formula (Eq. (3.4)):

$$t_c = 0.00646\,(1.1 - C)\,L_w^{0.5}S_w^{-\frac{1}{3}} = 0.00646 \times (1.1 - 0.65) \times (3600\text{ ft})^{0.5} \times (0.0065)^{-\frac{1}{3}} = 0.93\text{ hr.}$$

Kerby formula (Eq. (3.5)):

$$t_c = 0.01378\left(\frac{L_w r_{ke}}{S_w^{0.5}}\right)^{0.467} = 0.01378 \times \left(\frac{(3600\text{ ft}) \times 0.40}{(0.0065)^{0.5}}\right)^{0.467} = 1.33\text{ hr.}$$

TR 55 approach (Eqs. (3.6) to (3.11)):

$$t_{sf} = \frac{0.007\,(n_{sf}L_{sf})^{0.8}}{P_2^{0.5}S_{sf}^{0.4}} = \frac{0.007 \times (0.15 \times (100\text{ ft}))^{0.8}}{(3.5\text{ in})^{0.5} \times (0.01)^{0.4}} = 0.21\text{ hr.}$$

The agricultural watershed has unpaved surfaces. Thus:

$$V_{scf} = 16.1345\,S_{scf}^{0.5} = 16.1345 \times (0.0065)^{0.5} = 1.30\text{ ft sec}^{-1}$$
$$t_{scf} = \frac{L_{scf}}{3600\,V_{scf}} = \frac{1000\text{ ft}}{3600 \times \left(1.30\text{ ft sec}^{-1}\right)} = 0.21\text{ hr.}$$

For the bank-full trapezoidal channel:

$$A_{cf} = (b + Zy_n)\,y_n = (5\text{ ft} + 2 \times 2.5\text{ ft}) \times (2.5\text{ ft}) = 25\text{ ft}^2$$
$$P_{wet,cf} = b + 2\sqrt{1 + Z^2}y_n = 5\text{ ft} + 2 \times \sqrt{1 + 2^2} \times (2.5\text{ ft}) = 16.18\text{ ft}$$
$$R_{cf} = \frac{A_{cf}}{P_{wet,cf}} = \frac{25\text{ ft}^2}{16.18\text{ ft}} = 1.55\text{ ft}$$
$$V_{cf} = \frac{1.49}{n_{cf}}R_{cf}^{\frac{2}{3}}S_{cf}^{0.5} = \frac{1.49}{0.03} \times (1.55\text{ ft})^{\frac{2}{3}} \times (0.005)^{0.5} = 4.70\text{ ft sec}^{-1}$$
$$t_{cf} = \frac{L_{cf}}{3600V_{cf}} = \frac{2500\text{ ft}}{3600 \times \left(4.70\text{ ft sec}^{-1}\right)} = 0.15\text{ hr.}$$

The time of concentration is:

$$t_c = t_{sf} + t_{scf} + t_{cf} = 0.21 \text{ hr} + 0.21 \text{ hr} + 0.15 \text{ hr} = 0.57 \text{ hr}.$$

3.1.3 Analysis of Water Balance

For a water resources system of interest, such as a body of water, a watershed, a region, or the globe, the difference of the total input of water into the system from the total output of the water out of the system must be equal to the change of water storage within the system. Such a functional relationship, termed as water balance, is always true regardless of the time length (e.g., from minutes to years). Mathematically, the governing equation of water balance can be written as:

$$\frac{\overset{\text{water storage at time t}}{dS}}{dt} = \overset{\text{total influx}}{I} - \underset{\text{total exflux}}{O} \tag{3.12}$$

In Eq. (3.12), consistent units should be used. Storage S can be provided by depressions, ponds or lakes, channels, soil voids, and shallow aquifers. Influx I includes precipitation and runon from outside the system, whereas exflux O includes evaporation, transpiration, artificial withdraw for consumptive uses (e.g., drinking and irrigation), surface outflow, and baseflow from the shallow aquifer (Figure 3.6).

During a multiyear period (e.g., > 10 yrs), water storage can increase for one year but decrease for another year. Averaged across the time period, the overall change of water storage tends to be zero. It can be assumed that the influx is equal to precipitation and runon, while the exflux is equal to surface outflow and evapotranspiration (defined as the summation of evaporation and transpiration). Herein, evaporation is the vaporization loss of open waters and water from bare soils to the ambient atmosphere, whereas transpiration involves the uptake of soil water by plant roots and transfer through stems and leaves into the ambient atmosphere. Thus, using consistent units, for a long analysis period, Eq. (3.12) may be simplified as:

$$\underset{\text{precipitation}}{P} = \overset{\text{runoff}}{R} + \overset{\text{evapotranspiration}}{ET} \tag{3.13}$$

For the purpose of analysis, a water resources system may need to be partitioned into three interrelated subsystems: (1) surface water; (2) soil water (i.e., soil moisture) in the unsaturated or vadose zone; and (3) groundwater from a shallow aquifer (Figure 3.6). Equation (3.12) is applicable to each of these three subsystems, but variables S, I, and O in the equation need to be interpreted differently for each. For the surface water subsystem, S is the total change of water volume in depressions, ponds, lakes, and channels; I is the summation of precipitation, runon, and baseflow; and O is the summation of evaporation from open waters, surface outflow, and infiltration. For the soil water subsystem, S is the change of water volume in soil pores and cracks; I is the infiltration;

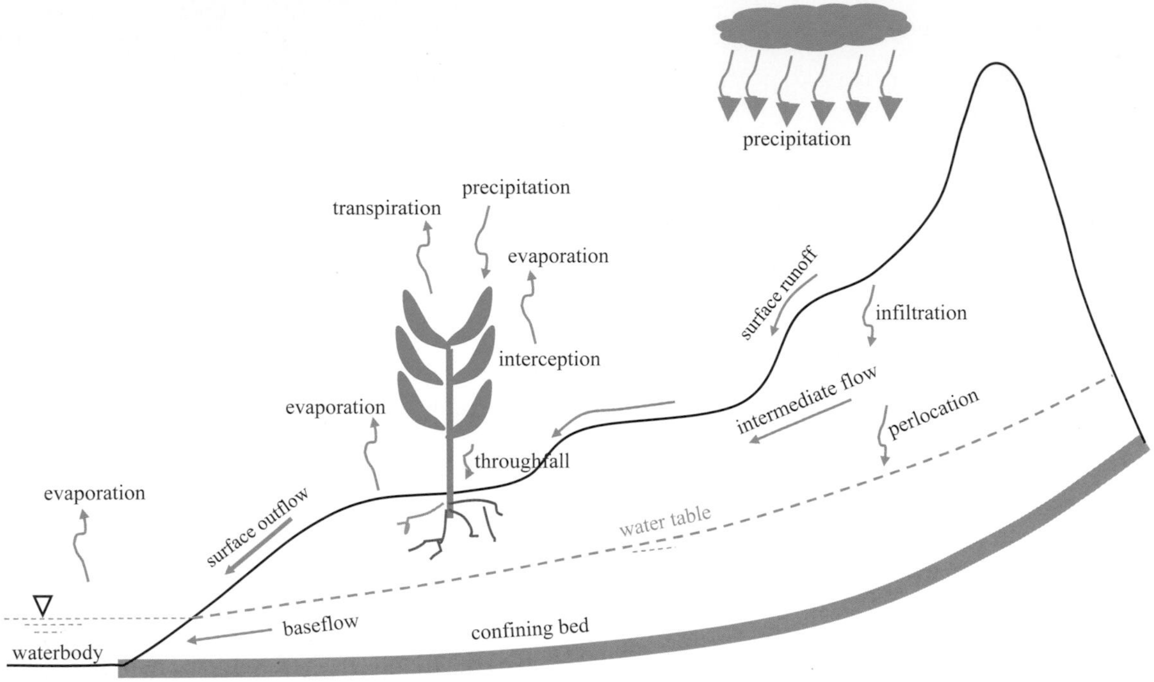

Figure 3.6 **The surface water and groundwater subsystems in a water resources system.**

and O is the summation of evaporation from bare soils, transpiration, and percolation (i.e., seepage into shallow aquifers). For the groundwater subsystem, S is the change of water volume in the shallow aquifers; I is the percolation; and O is the baseflow.

Example 3.4 A small pond was instrumented to measure the water balance variables. The pond has bed and shore materials with a near-zero permeability. The water table is far below the pond bed. During the course of a year, the pond received 36 in precipitation and 60 in inflow, and its storage was increased by 3 in. The outflow was measured to be 55 in. Determine the evaporation for that year.

Solution

Impermeable bed and shore materials $\rightarrow$ no water exchange between the pond and the vadose zone and a shallow aquifer $\rightarrow$ the pond is the system to be analyzed.

The input of the system includes precipitation (P) and inflow (Q_{in}), whereas the output includes outflow (Q_{out}) and evaporation (E). For the analysis time interval of $\Delta t = 1$ yr, the known variables are $\Delta S = 3$ in, $P = 36$ in, $Q_{in} = 60$ in, and $Q_{out} = 55$ in, and the unknown variable is E. The finite-difference format of Eq. (3.12) can be used for the system as follows: $\Delta S = P + Q_{in} - Q_{out} - E$. Rearranging, one can compute the evaporation as: $E = P + Q_{in} - Q_{out} - \Delta S = 36\text{ in} + 60\text{ in} - 55\text{ in} - 3\text{ in} = 38\text{ in}$.

3.1.4 Analysis of Water Supply and Demand

Water resources management can be implemented at spatial scales ranging from a household to a community to an entire country. Regardless of the scale, the water availability is always a function of relative supply and demand (Averyt *et al.*, 2013). The water supply stress index (WaSSI) is the ratio of available water to water demand. On the one hand, surface water and ground water within a spatial domain are the sources for water supply. For a time interval of interest (e.g., annual or multiyear), the water supply can be analyzed using Eq. (3.12) as the summation of water storage (S), surface outflow, and baseflow. On the other hand, water demand can be measured using various approaches, such as total water use, drinking water consumption, non-consumptive use, withdrawal water use, instream use, and water footprint (Wikipedia, 2017). Among these approaches, water footprint has become popular and has been widely adopted, it is herein introduced as the surrogate of water demand.

Hoekstra *et al.* (2011) developed a manual with conceptualization and formulization for assessing water footprint. The manual defines water footprint as the total volume of freshwater used to produce goods and services either consumed by the individual or community or produced by the business. Water use is measured as water volume consumed and/or polluted per unit of time. A water footprint can be calculated for any well-defined group of consumers (e.g., an individual, family, village, city, province, state, or nation) or producers (e.g., a public organization, private enterprise, or economic sector). The water footprint can be classified into blue water footprint, green water footprint, and grey water footprint, depending on the water source and water quality. The blue water footprint is the volume of freshwater from surface water and groundwater that is used to produce goods and services for the individual or community; that is, the freshwater may be lost to evapotranspiration, incorporated into products, or transferred either to green or grey water. The green water footprint is the volume of soil water (i.e., soil moisture) that is used for production or incorporated into products. The grey water footprint is the volume of polluted water associated with the production of goods and services for the individual or community. A water footprint is assessed based on the International Organization for Standardization (ISO) standard for Life Cycle Analysis (LCA) (ISO, 2014), which is a systematic, phased approach to assessing the environmental aspects and potential impacts that are associated with products, processes, or services (Wikipedia, 2017). Life cycle refers to the major activities during the life-span of a product from its manufacture, use, and maintenance, to its final disposal, including the raw material acquisition required to manufacture the product (SAIC, 2006; Stephan *et al.*, 2009). For example, the water footprint for a cup of coffee is about 140 L, the total amount of water that is used to grow, produce, package, and ship the coffee beans as well as make the coffee (WFN, 2014).

The Water Footprint Network, a global platform for collaboration between companies, organizations, and individuals to solve the world's water use, maintains the world's most comprehensive water footprint database, called *WaterStat* (waterfootprint.org/en/resources/water-footprint-statistics). *WaterStat* includes six categories of statistics summarized in Table 3.6. The database provides the blue water footprints, green water footprints, and grey water footprints of common products for countries across the world. For the USA, the water footprints are provided state by state. Tables 3.7 and 3.8 summarize the selected water footprints for both the USA and the world as whole.

Table 3.6 The WaterStat database of the Water Footprint Network.[1]

Item	Description
Product water footprint statistics	Statistics on green, blue, and grey water footprints of crops, derived crop products, biofuels, livestock products, and industrial products. All data are available at national and sub-national level.
Monthly blue water footprint statistics	Statistics on total monthly blue water footprints of production (m^3 per month) at a 30 × 30 arc minute grid scale. The data are averages over the period 1996–2005. The total blue water footprint of production is the sum of blue water footprint for agriculture, industry, and domestic sectors per grid cell.
National water footprint statistics	Statistics on green, blue, and grey water footprints of national production and consumption.
International virtual water flow statistics[2]	Statistics on international virtual water trade flows and on water savings related to international trade.
Water scarcity statistics	Unique datasets showing blue water scarcity in the world on a monthly basis, both at 30 × 30 arc minute resolution and per river-basin scale.
Water pollution level statistics	Statistics on water pollution level per river basin.

[1] Compiled from the information available at http://waterfootprint.org/en/resources/water-footprint-statistics/#CP1.
[2] Virtual water: hidden flow of water in food or other commodities traded from one place to another.

Table 3.7 The national water footprints, in $Mm^3\ yr^{-1}$, for the USA and the world.[1]

Production	USA			World		
	Blue water	Green water	Grey water	Blue water	Green water	Grey water
Crop	95,905	611,971	118,160	899,129	5,771,250	733,179
				4261 (0.21)	**27,352 (0.32)**	**3475 (0.18)**
Grazing	0	120,996	0	0	912,807	0
				0	**4326 (0.30)**	**0**
Animal	3361	0	0	45,923	0	0
				218 (0.26)	**0**	**0**
Industrial	11,030	0	59,937	37,534	0	362,735
				178 (0.19)	**0**	**1719 (0.22)**
Domestic	6544	0	25,557	42,023	0	282,005
				0	**1227 (0.25)**	**199 (0.26)**
Total	116,840	732,967	203,654	1,024,609	6,684,057	1,377,919

[1] The bold number is the global average, whereas the bold number in parentheses is the coefficient variation, which is the ratio of the standard deviation to the global average. The non-bold number is the gross for the USA or the world. Blue water refers to surface water and groundwater, green water to soil water or soil moisture, and grey water to polluted water. Compiled from multiple sources (e.g., Hoekstra and Mekonnen, 2012).

Table 3.8 The product water footprints, averaged in m^3 ton^{-1}, for the USA and the world.[1]

Product	USA			World		
	Blue water	Green water	Grey water	Blue water	Green water	Grey water
Wheat	153	1582	183	347	1292	210
Paddy	760	497	185	341	1146	187
Maize	126	470	180	81	947	194
Sorghum	96	885	106	103	2857	87
Potato	57	74	266	33	191	63
Soybean	175	2685	16	123	3574	65
Tomato	37	62	24	63	108	43

[1] Blue water refers to surface water and groundwater, green water to soil water or soil moisture, and grey water to polluted water. Compiled from multiple sources (e.g., Mekonnen and Hoekstra, 2011).

Example 3.5 A farmer in the USA plans to produce 10 ton wheat, 20 ton maize, 50 ton potato, and 30 ton soybean. On average, the annual available water for the farmer is $100,000 \text{ m}^3$. Estimate the water demand and WaSSI (water supply stress index).

Solution

For a given product, the water demand is computed as the multiplication of the product amount and the corresponding water footprint, which in turn can be determined from Table 3.8.

Blue water demand:

$$(10 \text{ ton}) * \left(153 \text{ m}^3 \text{ ton}^{-1}\right) + (20 \text{ ton}) * \left(126 \text{ m}^3 \text{ ton}^{-1}\right) + (50 \text{ ton}) * \left(57 \text{ m}^3 \text{ ton}^{-1}\right) + (30 \text{ ton}) * \left(175 \text{ m}^3 \text{ ton}^{-1}\right) = 12{,}150 \text{ m}^3.$$

Green water demand:

$$(10 \text{ ton}) * \left(1,582 \text{ m}^3 \text{ ton}^{-1}\right) + (20 \text{ ton}) * \left(470 \text{ m}^3 \text{ ton}^{-1}\right) + (50 \text{ ton}) * \left(74 \text{ m}^3 \text{ ton}^{-1}\right) + (30 \text{ ton}) * \left(2{,}685 \text{ m}^3 \text{ ton}^{-1}\right) = 109{,}470 \text{ m}^3.$$

Grey water demand:

$$(10 \text{ ton}) * \left(183 \text{ m}^3 \text{ ton}^{-1}\right) + (20 \text{ ton}) * \left(180 \text{ m}^3 \text{ ton}^{-1}\right) + (50 \text{ ton}) * \left(266 \text{ m}^3 \text{ ton}^{-1}\right) + (30 \text{ ton}) * \left(16 \text{ m}^3 \text{ ton}^{-1}\right) = 19{,}210 \text{ m}^3.$$

Total water demand:

$$12{,}150 \text{ m}^3 + 109{,}470 \text{ m}^3 + 19{,}210 \text{ m}^3 = 140{,}830 \text{ m}^3.$$

Water supply stress index:

$$\text{WaSSI} = \frac{100{,}000 \text{ m}^3}{140{,}830 \text{ m}^3} = 0.71,$$ indicating that the farmer has a 29% water shortage.

3.1.5 Sustainable Development of Water Resources

Water resources are vital for both socioeconomic and eco-environmental reasons. Their utilization should be sustainable by following a process that meets the needs of the present without compromising the ability to meet those of the future (UN, 1987; Flint, 2004). That is, the process must consider the contemporary and long-term needs of humans as well as the ecological community. Sustainable water resources development should focus on three important aspects of human civilization: economic, socio-political, and eco-environmental (Figure 3.7). However, because of the uncertainties regarding future water availability and socioeconomic situations, various scenarios need to be formulated and analyzed to guide the sustainable use of water resources. The scenarios may need to take into account climate change because of its extensive impacts on available water (Abbot *et al.*, 2015). A scenario is a possible solution to the problem of interest and a combination of decision variables (e.g., amount of water allocated to a crop). Thus, the total number of scenarios that can be formulated is equal to the number of possible combinations of the decision variables.

The development of a sustainable scenario requires an extensive coordination with stakeholders, water resources managers, regulators, politicians, and policy makers. This is crucial for successfully implementing the four iterative steps to develop the best sustainable scenario. First, one or more objectives should be established. For instance, a farmer likely wants to maximize the revenue from planting alternative crops, whereas a water resources manager probably wants to minimize the overall project cost. The objective should be expressed as a function of decision variables. Such a function is termed an objective function. Second, decision variables need to be defined and their possible combinations are enumerated to formulate the scenarios. Third, the "conflicts" among the scenario elements are described as constraints, which in turn are expressed as functions of the decision variables. The scenarios that satisfy the constraints form the feasible region, as illustrated by the shaded area in Figure 3.7. Fourth and finally, the optimal scenario is determined by optimizing the objective function through adjusting the decision variables subject to the constraints. These adjustments can be undertaken using Excel Solver. The optimal scenario is one of the corners of the feasible region. In principle, all scenarios within the feasible region can

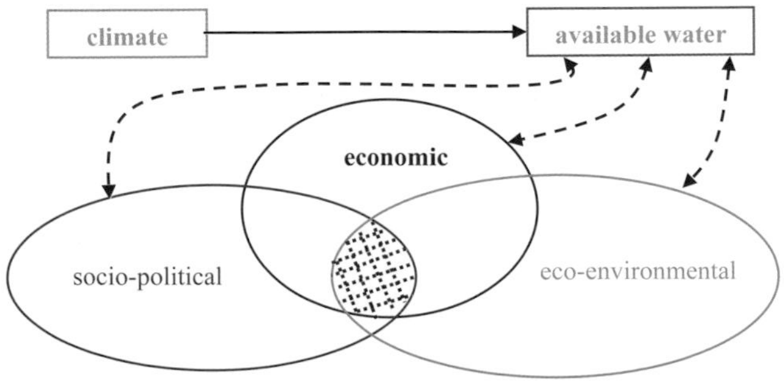

Figure 3.7 **Conceptual framework for sustainable water resources development.** The shaded area is the feasible region for the three important aspects of human civilization: economic, socio-political, and eco-environmental.

be understood as being sustainable; the optimal scenario, however, is the most favorable because it optimizes the primary objective.

Example 3.6 A prediction forecasts that the population in a city will increase by 15% over the next decade. Formulate possible scenarios to meet the future water demand.

Solution

Given the limited information, one can formulate various scenarios as presented in Table 3.9. The constraints are also speculated.

Table 3.9 The sample scenarios and corresponding constraints.

Scenario	Description	Constraints
I	Exploit groundwater	Vulnerability; resilience; subsidence; cost
II	Transfer water from outside	Source; distance; quality; cost
III	Reuse treated wastewater	Treatment plant; acceptability; cost
IV	Take conservation measures	Technology; behavior; education
V	Increase irrigation efficiency	Investment; training; farming culture
VI	Implement best management practice (BMPs)	Selection and placement of BMPs; cost
VII	Optimize industry structure	High-water-demand industries; development

Example 3.7 A farmer, who has 200 ha of agricultural land, considers planting crop A and crop B. The net benefits are \$35 and \$20 per kg for A and B, respectively. In a normal hydrologic year, the yields are 375 and 500 kg ha^{-1} for A and B, respectively. Given the uncertainties of weather and market, each crop will occupy 30 to 70% of the land. Determine the planting areas of crops A and B to maximize the total revenue.

Solution

The decision variables are x_A and x_B, the planting areas of A and B, respectively.

Objective: maximize the total revenue z

$$\max\{z\} = 35*(375*x_A) + 20*(500*x_B) = 13,125x_A + 10,000x_B.$$

Subject to:

$$x_A \geq 30\%*200 = 60 \quad \text{(crop A takes at least 30\% of the land)}$$
$$x_A \leq 70\%*200 = 140 \quad \text{(crop A takes at most 70\% of the land)}$$
$$x_B \geq 30\%*200 = 60 \quad \text{(crop B takes at least 30\% of the land)}$$
$$x_B \leq 70\%*200 = 140 \quad \text{(crop B takes at most 70\% of the land)}$$
$$x_A + x_B = 200 \quad \text{(total planting area equal to the land).}$$

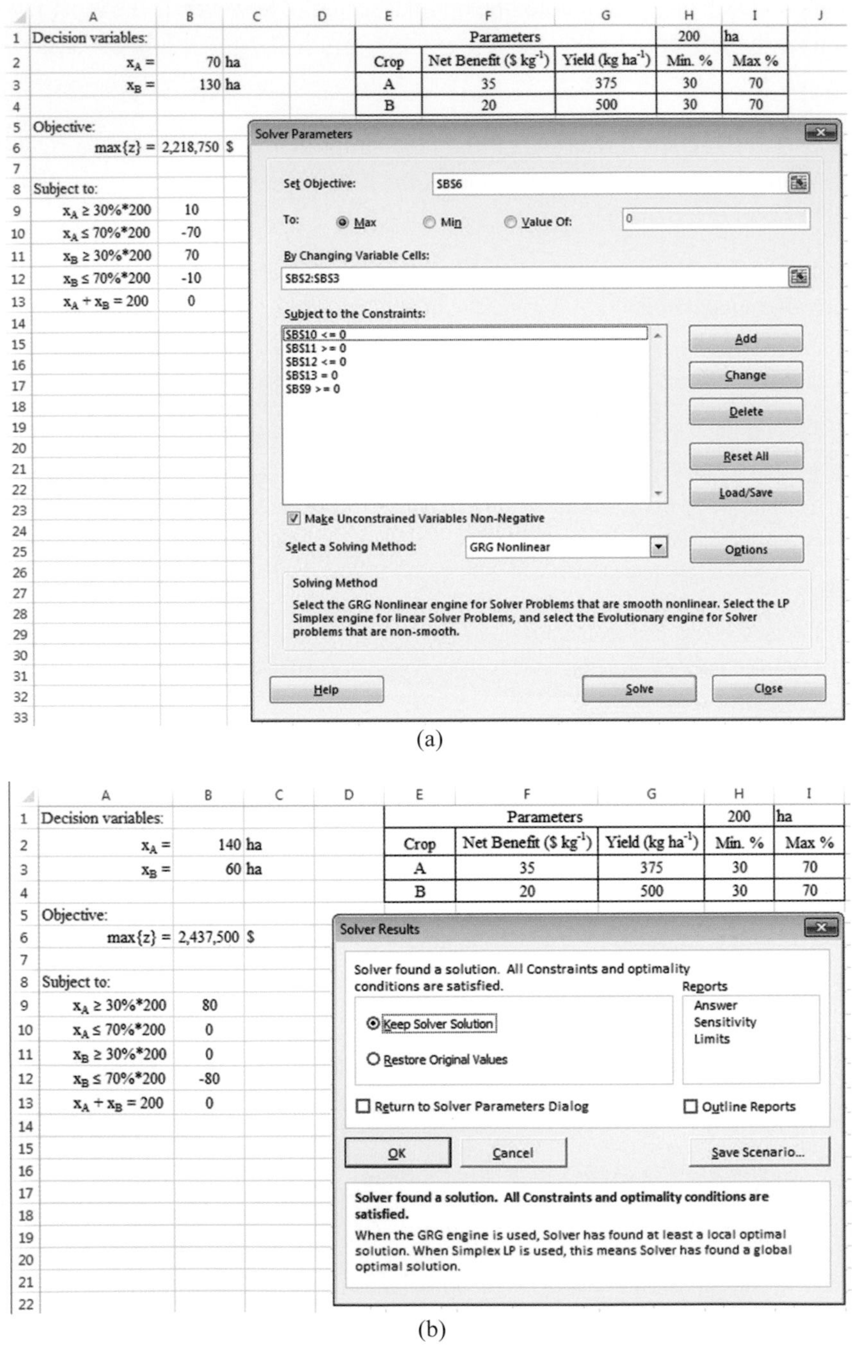

Figure 3.8 **Screenshots of Excel Solver for Example 3.7.** The spreadsheet and dialog box for: (a) setup; and (b) solution.

In Excel Solver, specify the objective function cell in the Set Objective box, check the Max radio button specify the cells of the decision variables in the By Changing Variable Cells box, and add constraints in the Subject to the Constraints box of (Figure 3.8a). Specify initial values of the decision variables that satisfy the constraints (i.e., in the feasible region), such as $x_A = 70$ ha and $x_B = 130$ ha. Click the Solve button to get the optimal solution: $x_A = 140$ ha, $x_B = 60$ ha, and $z = \$2{,}437{,}500$ (Figure 3.8b).

When the five constraints are plotted in the $x_A \sim x_B$ coordinate system, the feasible region can be defined and is shown as the red line in Figure 3.9. Graphically, the initial values of the decision variables ($x_A = 70$ ha, $x_B = 130$ ha) are one point on the line, and the optimal solution ($x_A = 140$ ha, $x_B = 60$ ha) is the lower end of the line. It can be shown that the objective function has a value at any other points on the line smaller than that at the lower end.

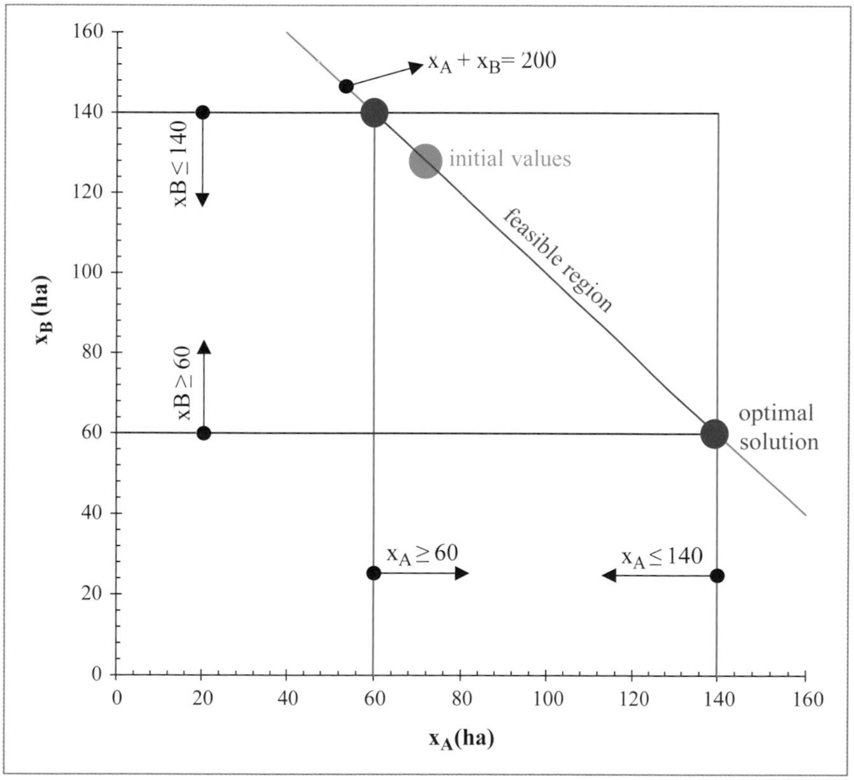

Figure 3.9 **Graphical solution of Example 3.7.**

3.2 Basic Hydrologic Processes

This section introduces the hydrologic cycle and its physical processes and presents the widely used methods for quantifying the processes.

3.2.1 Precipitation

As illustrated in Figure 3.6, precipitation is the primary water source of a hydrologic system. Precipitation occurs when water vapor in air reaches its saturation pressure. It can fall in forms of liquid (i.e., rain), ice crystals (i.e., snow, hail, and sleet), and a mixture thereof, depending on the air temperature at the time (Hendriks, 2010). Precipitation is usually recorded as the equivalent water depth (EWD), determined by melting ice crystals. If only rain occurs, the EWD is the same as the rainfall. Traditionally, precipitation is measured with rain gauges at selected sites. Although various types of gauges (e.g., float and weighting) had been used in the past, the tipping budget gauge (e.g., Figure 3.10) has become popular in recent years (Kifissia, 2017). It funnels the collected rain and melted ice crystals in to a small bucket that tilts and empties each time it fills. The precipitation depth is determined based on the number of bucket tilts. Data recorded at a rain gauge are point values. However, practical applications usually require areal precipitation, which represents the average across a watershed of interest. Three methods, namely simple arithmetic average (Eq. (3.14)), Thiessen polygon area-weighted average (Eq. (3.15)), and isohyet (Eq. (3.16)), can be used to estimate areal precipitation from point values at the selected gauges (Viessman and Lewis, 2002; Wurbs and James, 2002).

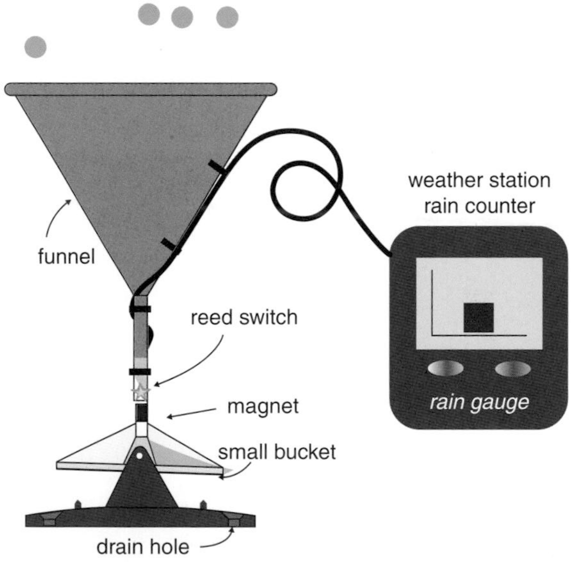

Figure 3.10 **Illustration of a tipping bucket gauge.** (*Source:* Kifissia, 2017, permitted by Fragos Ioannis)

$$\underbrace{P_{areal}}_{\text{areal precipitation}} = \frac{\sum_{i=1}^{N} \overbrace{P_i}^{\text{precipitation at gauge i}}}{\underbrace{N}_{\text{total number of rain gauges within or adjacent to the watershed}}} \tag{3.14}$$

$$P_{areal} = \frac{\sum_{i=1}^{N} (P_i A_i)}{\sum_{i=1}^{N} A_i} \tag{3.15}$$

where: areal precipitation → P_{areal}; total number of rain gauges within or adjacent to the watershed → N; precipitation at gauge i → P_i; area of Thiessen polygon allocated to gauge i → A_i.

$$P_{areal} = \frac{\sum_{j=1}^{M} (\bar{P}_j A_j)}{\sum_{j=1}^{M} A_j} \tag{3.16}$$

where: areal precipitation → P_{areal}; number of isohyetal polygons within the watershed → M; arithmetic average of the isohyets forming isohyetal polygon j → $\bar{P}_j$; area of isohyetal polygon j → A_j.

Thiessen polygons are graphically constructed in three steps. First, connect the adjacent rain gauges to form a triangular network. Second, draw perpendicular bisector lines to the edges of the triangles. Third, draw the polygons from the points of intersection among the bisector lines and between them and the watershed boundary. The summation of the areas of the polygons should be equal to the drainage area of the watershed.

An isohyet is a curve along which the precipitation is constant. The isohyets are graphically constructed by linear interpolation of the point precipitations at the gauges. An isohyetal polygon is formed by the adjacent isohyets and the watershed boundary. The number of polygons depends on the isohyet interval: the smaller the interval, the greater the number of polygons, and vice versa. Again, the summation of the areas of the isohyetal polygons should be equal to the drainage area of the watershed.

Example 3.8 A watershed is monitored with five rain gauges, whose locations are labeled as a, b, c, d, and e in Figure 3.11a. The average annual precipitations at these gauges are also annotated in the figure. Each square grid represents an area of 9 km^2. Estimate the areal average annual precipitation of the watershed using the:

(1) Simple arithmetic average method
(2) Thiessen polygon area-weighted average method
(3) Isohyetal method.

Solution

$N = 5$, $P_a = 45$ mm, $P_b = 57$ mm, $P_c = 69$ mm, $P_d = 66$ mm, $P_e = 78$ mm.

(1) Using Eq. (3.14), one can get:

$$P_{areal} = \frac{P_a + P_b + P_c + P_d + P_e}{N} = \frac{45\text{ mm} + 57\text{ mm} + 69\text{ mm} + 66\text{ mm} + 78\text{ mm}}{5} = 63\text{ mm}.$$

(2) Follow the three steps discussed above to form the Thiessen polygons, as shown in Figure 3.11b. The network is indicated by thin blue solid lines, and the perpendicular bisector

lines are shown in thick green, purple, yellow, and pink solid lines. By counting the number of grids in each Thiessen polygon, one can determine its area. Note that the incomplete grids in a polygon should be visually combined together into one complete grid or more in size. For instance, the polygon that is represented by "gauge a" has 2 complete grids and 13 incomplete grids. Visually, the area represented by the 13 incomplete grids is equivalent to the area of 6.2 complete grids. Thus, the area of the polygon is $(2 + 6.2) * 9\ \text{km}^2 = 73.8\ \text{km}^2$. Similarly, the area of the Thiessen polygons that are represented by gauges b, c, d, and e, respectively, are $(1 + 3.5) * 9\ \text{km}^2 = 40.5\ \text{km}^2$, $3.7 * 9\ \text{km}^2 = 33.3\ \text{km}^2$, $3 * 9\ \text{km}^2 = 27\ \text{km}^2$, and $1.5 * 9\ \text{km}^2 = 13.5\ \text{km}^2$. The total drainage area is $188.1\ \text{km}^2$.
Substituting the values into Eq. (3.15), one can get:

$$P_{\text{areal}} = \frac{(45\ \text{mm}) * (73.8\ \text{km}^2) + (57\ \text{mm}) * (40.5\ \text{km}^2) + (69\ \text{mm}) * (33.3\ \text{km}^2) + (66\ \text{mm}) * (27\ \text{km}^2) + (78\ \text{mm}) * (13.5\ \text{km}^2)}{188.1\ \text{km}^2}$$

$$= 57\ \text{mm}.$$

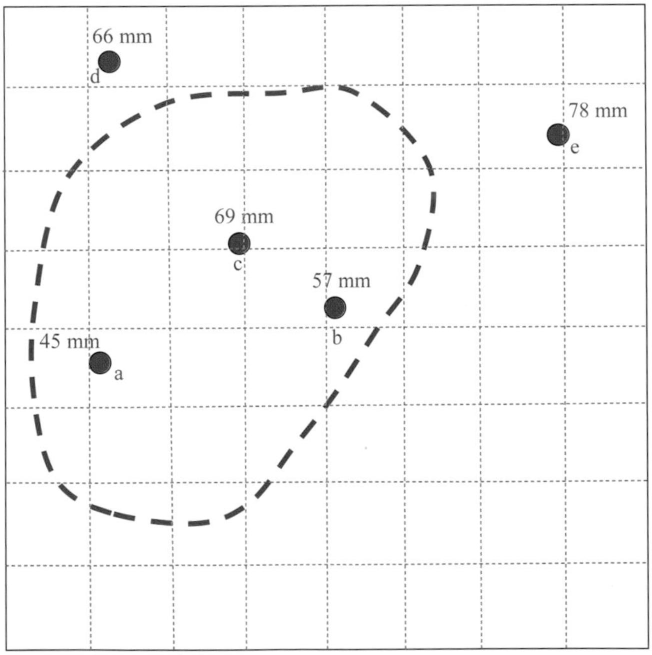

Figure 3.11 Map used for Example 3.8. Each square grid represents a 3×3 km $(9\ \text{km}^2)$ area. The watershed boundary is shown as a dashed red closed loop. (a) Location of rain gauges indicated with prominent blue filled circles, adjacent to which the corresponding average annual precipitations are annotated; (b) construction of Thiessen polygons; and (c) construction of isohyets.

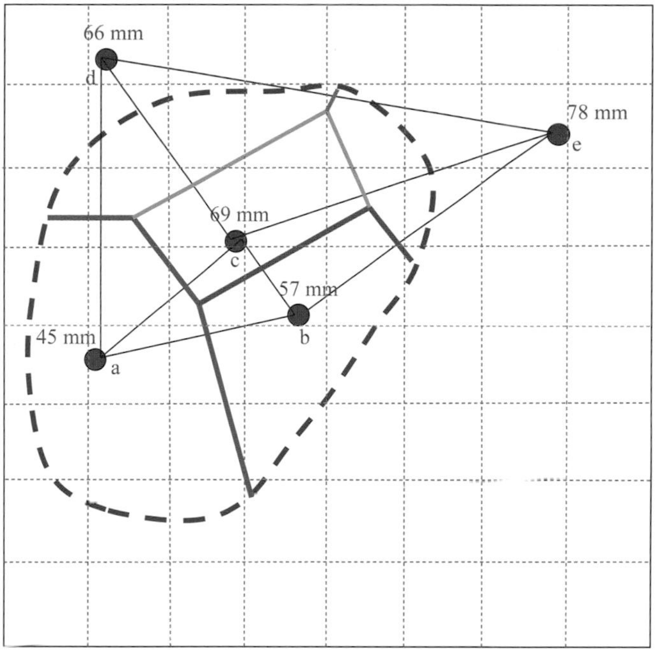

(b)

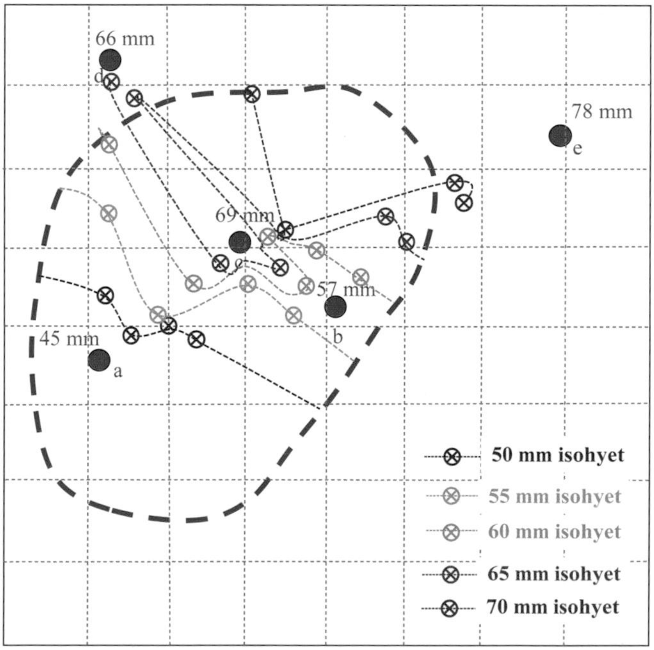

(c)

Figure 3.11 *(Continued)*

(3) The precipitations at the gauges range from 45 to 78 mm. For an isohyet interval of 5 mm, one can construct five isohyets of 50, 55, 60, 65, and 70 mm. The isohyet of 75 mm could be constructed as well. However, because this latter isohyet is outside the watershed and thus has a minimal influence on the computed areal precipitation, it is not used so as to reduce the workload. In this regard, use a ruler to measure the length of the line connecting any two of the five gauges and then determine the point on the line by linear interpolation. Note that the grid size of 3 km can be used as the measurement scale. For example, the length of line ab is measured to be 8.3 km. The point of 50 mm precipitation on line ab is $(8.3\text{ km})/(57\text{ mm} - 45\text{ mm}) * (50\text{ mm} - 45\text{ mm}) = 3.5$ km from gauge a, and the point of 55 mm precipitation on line ab is $(8.3\text{ km})/(57\text{ mm} - 45\text{ mm}) * (55\text{ mm} - 45\text{ mm}) = 6.9$ km from gauge a. There are no points on line ab that have precipitation of 60, 65, or 70 mm. Similarly, the points, which represent the precipitations of the isohyets, on other lines connecting any other two gauges can be determined and are shown in Table 3.10. Based on these points, the corresponding isohyetal curves can be drawn as shown in Figure 3.11c. In addition, by counting the number of grids within each isohyetal polygon as described in (2), the area of the polygon can be determined (Table 3.11). For instance, the number of grids within the polygon below isohyet 50 mm is $2 + 5.2 = 7.2$, so the polygon area is $7.2 * 9\text{ km}^2 = 64.8\text{ km}^2$.

Table 3.10 The isohyetal points on the lines connecting two rain gauges.

Gauge		Length	Precipitation (mm) at		Distance from start gauge (km) to the point of[2]				
Start	End	(km)[1]	Start	End	50 mm	55 mm	60 mm	65 mm	70 mm
a	b	8.3	45	57	3.5	6.9			
	c	6.0		69	1.3	2.5	3.8	5.0	
	d	10.5		66	2.5	5.0	7.5	10.0	
	e	18.0		78	2.7	5.5	8.2	10.9	13.6
b	c	3.0	57	69			0.8	2.0	
	d	11.3		66			3.8	10.0	
	e	11.3		78			1.6	4.3	7.0
c	d	7.5	69	66					
	e	12.0		78					1.3
d	e	16.0	66	78					5.3

[1] Measured from Figure 3.11c with a grid size of 3 km as the scale.

[2] The blank cell indicates that this isohyetal point does not exist between these two gauges. The value is computed by linear interpolation.

Substituting the values into Eq. (3.16), one can get:

$$P_{areal} = \left[\begin{array}{l} (50.0\text{ mm}) * (64.8\text{ km}^2) + (52.5\text{ mm}) * (30.6\text{ km}^2) + (57.5\text{ mm}) * (25.2\text{ km}^2) + \\ (62.5\text{ mm}) * (22.5\text{ km}^2) + (67.5\text{ mm}) * (18.0\text{ km}^2) + (70.0\text{ mm}) * (27.0\text{ km}^2) \end{array} \right] \div (188.1\text{ km}^2) = 57\text{ mm}.$$

Table 3.11 Areas and precipitations of the isohyetal polygons.

Isohyetal precipitation (mm)	Isohyetal polygon		Average precipitation, $\overline{P}_j$ (mm)
	Number of grids[1]	Area (km^2)[2]	
< 50	7.2	64.8	50
50 ~ 55	3.4	30.6	(50 + 55) /2 = 52.5
55 ~ 60	2.8	25.2	(55 + 60)/2 = 57.5
60 ~ 65	2.5	22.5	(60 + 65)/2 = 62.5
65 ~ 70	2.0	18.0	(65 + 70)/2 = 67.5
> 70	3.0	27.0	70
Total	20.9	188.1	57 (by Eq. (3.16))

[1] Counted from Figure 3.11c.

[2] Computed as the product of the number of grids and the grid area of 9 km^2.

3.2.2 Infiltration

Infiltration is the process by which water seeps into soils below the surface of the land (Viessman and Lewis, 2002), as illustrated in Figure 3.6. It can be characterized either by instantaneous infiltration rate or cumulative infiltration amount during a time period. Mathematically, cumulative infiltration is the integration of infiltration rate with respect to time. Infiltration rate is controlled both by the infiltration capacity of soils and the rainfall intensity or available water. If water is always available for infiltration and rainfall intensity is larger than infiltration capacity, infiltration rate is equal to infiltration capacity; otherwise, infiltration rate is equal to rainfall intensity (Figure 3.12). That is, infiltration rate is the minimum of infiltration capacity and rainfall intensity. Direct runoff does not occur unless rainfall intensity is greater than infiltration capacity.

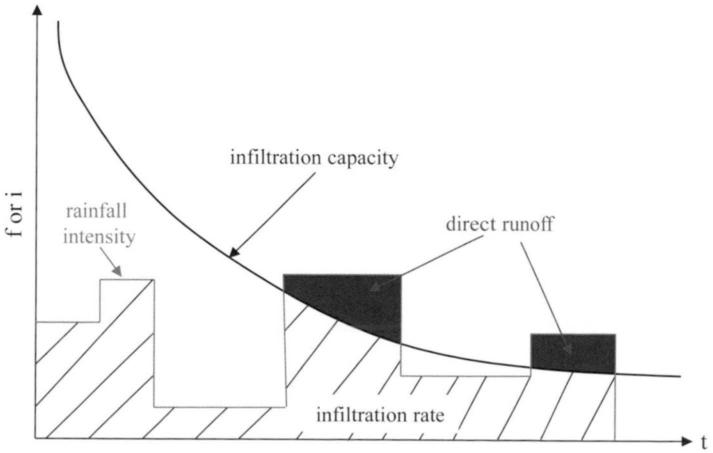

Figure 3.12 **Graphical relationship between infiltration rate, infiltration capacity, and rainfall intensity.** Note that infiltration rate is controlled by infiltration capacity, f, and rainfall intensity, i. Both f and i have units of depth per unit time (e.g., in hr^{-1} or mm hr^{-1}).

Infiltration capacity depends on the physiographic characteristics of the area of interest, such as soil properties, LULC, and depth to water table (i.e., shallow groundwater level). For a given soil, its infiltration capacity tends to decrease with increasing soil moisture. Herein, soil moisture is defined as the ratio of pore water volume to soil volume. Thus, as illustrated in Figure 3.12, infiltration capacity tends to decrease with elapsed time because the already infiltrated water fills pores in the soil, increasing soil moisture. Its asymptote denotes the ultimate infiltration capacity, which is a constant close in value to the saturated hydraulic conductivity of the soil. For practical applications, infiltration capacity is usually estimated using mathematical models. In the literature (e.g., Mishra *et al.*, 1999), existing infiltration models are classified into three groups: physically based, semi-empirical, and empirical. The physically based models are direct derivatives from the mass-conservation and Darcy's laws (Todd and Mays, 2005) with a variety of simplifications and assumptions (Hilpert and Glantz, 2013). The most popular such models were developed by Green and Ampt (1911), Philip (1957, 1969), and Mein and Larson (1973). In contrast, empirical models, which have various forms and complexities (e.g., linear versus nonlinear), are solely based on test or experimental data. Examples of these models include those developed by Kostiakov (1932), Huggins and Monke (1966), Smith (1972), Collis-George (1977), Smith and Parlange (1978), USDA-NRCS (1986), and Parhi *et al.* (2007). Moreover, between the physically based and empirical models are the semi-empirical models, which satisfy continuity and simple hypotheses of water movement in soil. The representatives of such models were developed by Horton (1940), Holtan (1961), Overton (1964), Swartzendruber (1987), Singh and Yu (1990), and Grigorjev and Iritz (1991). Overall, these models perform better for clayey or wet soils than sandy or dry soils (Mishra *et al.*, 2003; Mirzaee *et al.*, 2013; Wang *et al.*, 2017). Nevertheless, the models in one group are not necessary to have a better or poorer performance than those in another group, while in a same group, one model is not necessary to have a better or poorer performance than another (Sihag *et al.*, 2017; Wang *et al.*, 2017). Hereinafter, three models, namely Horton (1940), Philip (1957), and Green and Ampt (1911), are described because they are relatively more widely used than the others. Consistent units should be used in these models.

The Horton model computes infiltration capacity as:

$$\underbrace{f(t)}_{\text{infiltration capacity}} = \underbrace{f_c}_{\text{ultimate infiltration capacity}} + (\underbrace{f_0}_{\text{initial infiltration capacity}} - f_c)e^{-\underbrace{\beta}_{\text{decay rate}}\underbrace{t}_{\text{time}}} \tag{3.17}$$

The Philip model computes infiltration capacity as:

$$\underbrace{f(t)}_{\text{infiltration capacity}} = \frac{\overbrace{S_1}^{\text{soil storativity coefficient}}}{2\sqrt{\underbrace{t}_{\text{time}}}} + \underbrace{f_u}_{\text{ultimate infiltration capacity}} \tag{3.18}$$

soil storativity coefficient $\longrightarrow S_1 = \sqrt{K_{sat}\,(0.95\theta_{sat} - \theta_a)\,(H + \psi_f)}$ (3.19)

soil saturated hydraulic conductivity

initial soil moisture

suction head at wetting front

saturated soil moisture

ponding water depth on ground

soil pore size distribution index bubbling capillary pressure head

suction head at wetting front $\longrightarrow \psi_f = \dfrac{3\lambda + 2}{3\lambda + 1}\left(\dfrac{\psi_b}{2}\right)$ (3.20)

The Green and Ampt (G–A) model computes infiltration capacity as:

suction head at wetting front

saturated soil moisture

ponding water depth on ground

initial soil moisture

saturated hydraulic conductivity

infiltration capacity $\longrightarrow f(t) = \dfrac{K_{sat}}{2}\left[1 + \dfrac{(H + \psi_f)(0.95\theta_{sat} - \theta_a)}{F(t)}\right]$ (3.21)

cumulative infiltration

time

cumulative infiltration

saturated hydraulic conductivity

suction head at wetting front

initial soil moisture

$F(t) = \dfrac{K_{sat}}{2}t + (H + \psi_f)(0.95\theta_{sat} - \theta_a)\ln\left[1 + \dfrac{F(t)}{(H + \psi_f)\,(0.95\theta_{sat} - \theta_a)}\right]$ (3.22)

time

ponding water depth on ground

saturated soil moisture

The parameters of the Horton model, namely f_0, fc, and β, and those of the Philip model, namely S_1 and f_u, can be determined by fitting data on f(t) using a least-squares regression method (Neter *et al.*, 1996). The data can be collected by either laboratory constant and variable head tests (Barnes, 2000) or field infiltrometer experiments (Pitt *et al.*, 1999; ASTM, 2003). The least-squares regression can be implemented in Excel Solver through minimizing the total square error by changing the corresponding parameters. In contrast, S_1 and the parameters of the G–A model can be estimated using site-specific or literature (e.g., Table 3.12) soil water parameters (Assoú-line, 2005). The G–A model assumes that the: (1) soil profile is homogenous; (2) initial soil moisture is uniformly distributed throughout the profile; (3) soils above the wetting front are naturally saturated; and (4) soils below the wetting front have the initial soil moisture. These four assumptions are illustrated in Figure 3.13.

Table 3.12 Mean (standard deviation) of soil water parameters for infiltration capacity.[1]

Soil texture	Wilting point, θ_ψ	Field capacity, θ_{fc}	Saturated soil moisture, θ_{sat}	Bubbling capillary pressure head, ψ_b [m]	Saturated hydraulic conductivity, K_{sat} $\left[m\ d^{-1}\right]$	Dry density, ρ_d $\left[MG\ m^{-3}\right]$	Soil pore size distribution index, $\lambda\,[-]$
Clay	0.302 (0.041)	0.422 (0.033)	0.498 (0.037)	1.25 (1.88)	0.041 (0.034)	1.33 (0.10)	0.09 (0.03)
Clay loam	0.210 (0.027)	0.349 (0.025)	0.473 (0.018)	0.53 (0.42)	0.125 (0.054)	1.40 (0.05)	0.20 (0.08)
Loam	0.128 (0.045)	0.268 (0.048)	0.461 (0.011)	0.28 (0.16)	0.535 (0.374)	1.43 (0.03)	0.31 (0.11)
Loamy sand	0.059 (0.027)	0.123 (0.022)	0.456 (0.015)	0.08 (0.03)	2.237 (0.789)	1.44 (0.04)	1.16 (0.49)
Sand	0.044 (0.017)	0.093 (0.019)	0.467 (0.011)	0.07 (0.01)	3.004 (0.741)	1.41 (0.03)	3.28 (1.20)
Sandy clay	0.252 (0.034)	0.363 (0.037)	0.443 (0.012)	0.37 (0.23)	0.040 (0.033)	1.48 (0.03)	0.24 (0.06)
Sandy clay loam	0.180 (0.032)	0.286 (0.043)	0.438 (0.014)	0.17 (0.11)	0.240 (0.162)	1.49 (0.04)	0.36 (0.14)
Sandy loam	0.083 (0.040)	0.192 (0.038)	0.450 (0.016)	0.13 (0.07)	1.228 (0.704)	1.46 (0.04)	1.41 (0.65)
Silt	0.065 (0.020)	0.314 (0.023)	0.483 (0.007)	0.62 (0.27)	0.486 (0.172)	1.37 (0.02)	0.53 (0.24)
Silty clay	0.274 (0.030)	0.413 (0.017)	0.531 (0.018)	2.00 (2.00)	0.096 (0.021)	1.24 (0.05)	0.19 (0.07)
Silty clay loam	0.208 (0.029)	0.377 (0.019)	0.508 (0.016)	1.00 (0.60)	0.146 (0.038)	1.30 (0.04)	0.51 (0.21)
Silty loam	0.109 (0.051)	0.307 (0.056)	0.484 (0.016)	0.50 (0.30)	0.52 (0.447)	1.37 (0.04)	0.26 (0.08)

[1] The values of ψ_b are from Table 4.4.1 of Charbeneau (2006), whereas the values of λ are estimated using the data and formulas presented in Ghanbarian-Alavijeh *et al.* (2010). The values for the other parameters are calculated using the SPAW model (Saxton *et al.*, 1986; Saxton and Rawls, 2006).

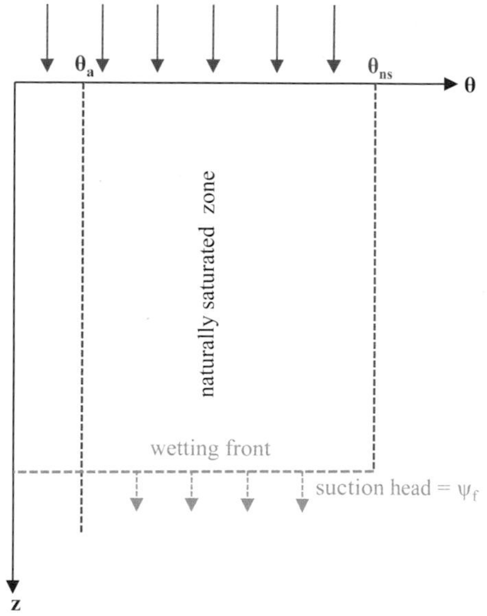

Figure 3.13 **Illustration of the assumptions of the G–A model.** θ: soil moisture; z: soil depth from ground surface; θ_a: initial soil moisture; and θ_{ns}: naturally saturated soil moisture, which is approximately equal to 95% of the saturated soil moisture, θ_{sat}.

Example 3.9 The first two columns ([C1] and [C2]) in Figure 3.14 are the measured data on infiltration capacity from a field infiltrometer experiment. Determine the parameters of the Horton and Philip models for the experiment site.

Solution

For the Horton model, assuming the initial values of the parameters to be $f_0 = 20 \text{ in hr}^{-1}$, $f_c = 12 \text{ in hr}^{-1}$, and $\beta = 2 \text{ hr}^{-1}$, which are in cells I2, I3, and I4, respectively, one can compute the values of infiltration capacity, as shown in the third column ([C3]) of Figure 3.14. The square errors at the measurement times are computed as $\{[C2] - [C3]\}^2$, and their summation is the total square error, as shown in cell D28. For instance, at t = 0.04 hr, the predicted f(t) = $12 \text{ in hr}^{-1} + \left[\left(20 \text{ in hr}^{-1}\right) - \left(12 \text{ in hr}^{-1}\right)\right] e^{-[(2 \text{ hr}-1)*(0.04 \text{ hr})]} = 19.38 \text{ in hr}^{-1}$, and the square error is $\left[\left(32.25 \text{ in hr}^{-1}\right) - \left(19.38 \text{ in hr}^{-1}\right)\right]^2 = 165.64 \text{ in}^2 \text{ hr}^{-2}$. In Solver, specify cell D28 in the Set Objective box, check the Min radio button, specify cells I2:I4 in the By Changing Variable Cells box, and then click the Solve button to get the solution: $f_0 = 31.56 \text{ in hr}^{-1}$, $f_c = 14.87 \text{ in hr}^{-1}$, and $\beta = 2.61 \text{ hr}^{-1}$.

Similarly, for the Philip model, assuming the initial values of the parameters to be $f_u = 12 \text{ in hr}^{-1}$ and $S_1 = 10 \text{ in hr}^{-1/2}$, which are in cells J5 and J6, respectively, one can compute the values of infiltration capacity, as shown in the fifth column ([C4]) of Figure 3.14. The square errors at the measurement times are computed as $\{[C2] - [C4]\}^2$, and their summation is the total square error, as

Measured Data		Horton Model		Philip Model	
t (hr)	f(t) (in hr^{-1})	f(t) (in hr^{-1})	Square Error	f(t) (in hr^{-1})	Square Error
[C1]	[C2]	[C3]	$\{[C2] - [C3]\}^2$	[C4]	$\{[C2] - [C4]\}^2$
0.04	32.25	29.90	5.52	34.27	4.10
0.13	25.50	26.75	1.57	24.43	1.15
0.21	22.50	24.51	4.04	21.81	0.47
0.29	21.75	22.69	0.89	20.37	1.90
0.38	22.50	21.05	2.09	19.33	10.02
0.46	14.25	19.89	31.78	18.68	19.63
0.54	25.50	18.94	43.01	18.18	53.61
0.63	15.00	18.09	9.54	17.73	7.46
0.71	17.25	17.48	0.05	17.41	0.03
0.79	18.00	16.99	1.02	17.14	0.75
0.88	19.50	16.55	8.72	16.87	6.90
0.96	15.00	16.23	1.52	16.67	2.80
1.04	17.25	15.98	1.62	16.50	0.57
1.13	16.50	15.74	0.57	16.32	0.03
1.21	21.75	15.58	38.06	16.18	31.02
1.29	12.75	15.45	7.28	16.05	10.91
1.38	11.25	15.33	16.63	15.92	21.85
1.46	20.25	15.24	25.08	15.82	19.63
1.54	15.75	15.17	0.33	15.72	0.00
1.63	12.00	15.11	9.67	15.62	13.13
1.71	18.00	15.07	8.61	15.54	6.04
1.79	14.25	15.03	0.61	15.47	1.48
1.88	12.75	15.00	5.05	15.38	6.94
1.96	11.25	14.97	13.87	15.32	16.55
Total Square Error			**237.13**		**236.97**

Parameter	Horton Model	Philip Model
f_0 (in hr^{-1})	31.56	
f_c (in hr^{-1})	14.87	
β (hr^{-1})	2.61	
f_u (in hr^{-1})		12.16
S_1 (in $hr^{-1/2}$)		8.85

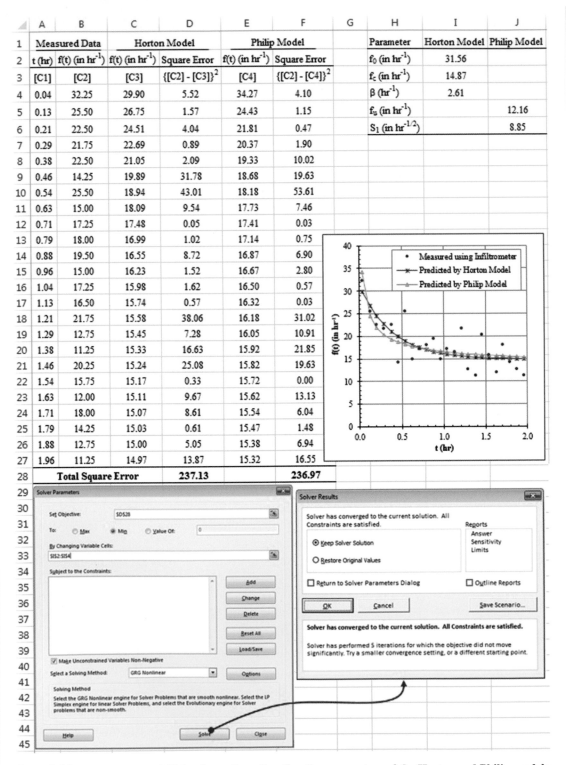

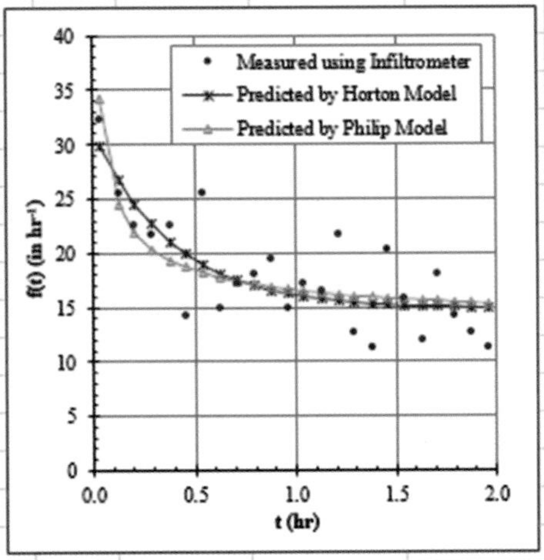

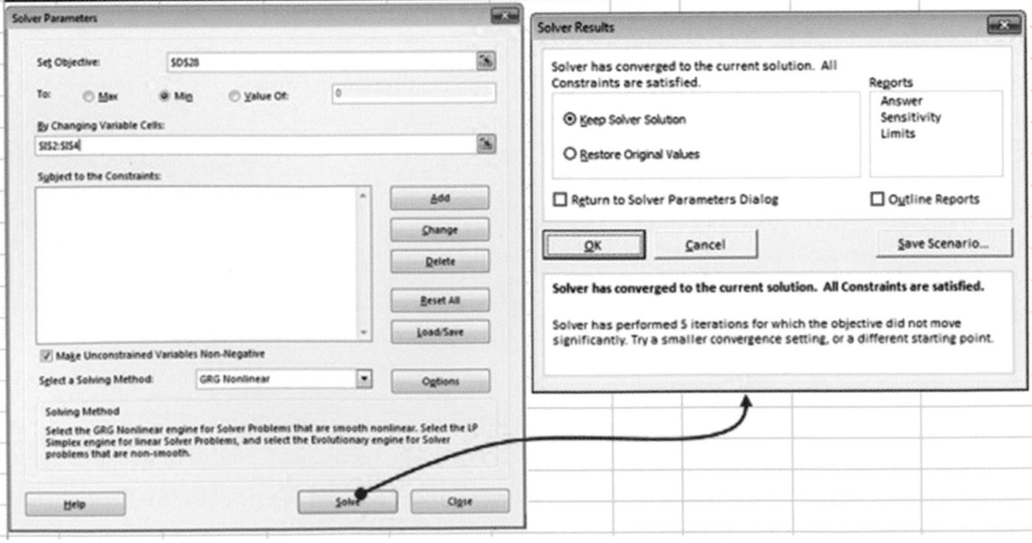

Figure 3.14 **Spreadsheet and dialog boxes for estimating the parameters of the Horton and Philip models in Example 3.9.**

shown in cell F28. For instance, at $t = 0.04$ hr, the predicted $f(t) = (10 \text{ in hr}^{-1/2})/[2*(004 \text{ hr})^{0.5}] + 12 \text{ in hr}^{-1} = 37.00 \text{ in hr}^{-1}$, and the square error is $\left[\left(32.25 \text{ in hr}^{-1}\right) - \left(37.00 \text{ in hr}^{-1}\right)\right]^2 = 22.56 \text{ in}^2 \text{ hr}^{-2}$. In Solver, specify cell F28 in the Set Objective box, check the Min radio button, specify cells J5:J6 in the By Changing Variable Cells box, and then click the Solve button to get the solution: $f_u = 12.16 \text{ in hr}^{-1}$ and $S_l = 8.85 \text{ in hr}^{-1/2}$.

Example 3.10 A soil ($\theta_{sat} = 0.38$, $K_{sat} = 0.029 \text{ m d}^{-1}$, $\psi_b = 0.37$ m, and $\lambda = 0.23$) has an initial soil moisture of 0.12. If the ponding water depth on the ground is negligible, develop the infiltration capacity curve using the G–A model.

Solution

$\theta_a = 0.12$, $H \approx 0$, $\theta_{sat} = 0.38$, $K_{sat} = 0.029 \text{ m d}^{-1}$, $\psi_b = 0.37$ m, and $\lambda = 0.23$.

Using Eq. (3.20): $\psi_f = \dfrac{3*0.23 + 2}{3*0.23 + 1}\left(\dfrac{0.37 \text{ m}}{2}\right) = 0.29$ m.

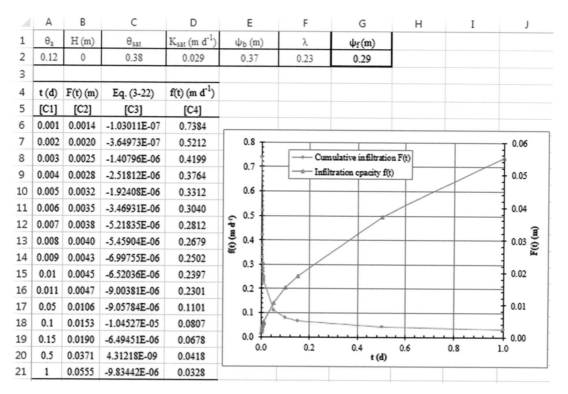

	A	B	C	D	E	F	G
1	θ_a	H (m)	θ_{sat}	K_{sat} (m d^{-1})	ψ_b (m)	λ	ψ_f (m)
2	0.12	0	0.38	0.029	0.37	0.23	0.29
3							
4	t (d)	F(t) (m)	Eq. (3-22)	f(t) (m d^{-1})			
5	[C1]	[C2]	[C3]	[C4]			
6	0.001	0.0014	-1.03011E-07	0.7384			
7	0.002	0.0020	-3.64973E-07	0.5212			
8	0.003	0.0025	-1.40796E-06	0.4199			
9	0.004	0.0028	-2.51812E-06	0.3764			
10	0.005	0.0032	-1.92408E-06	0.3312			
11	0.006	0.0035	-3.46931E-06	0.3040			
12	0.007	0.0038	-5.21835E-06	0.2812			
13	0.008	0.0040	-5.45904E-06	0.2679			
14	0.009	0.0043	-6.99755E-06	0.2502			
15	0.01	0.0045	-6.52036E-06	0.2397			
16	0.011	0.0047	-9.00381E-06	0.2301			
17	0.05	0.0106	-9.05784E-06	0.1101			
18	0.1	0.0153	-1.04527E-05	0.0807			
19	0.15	0.0190	-6.49451E-06	0.0678			
20	0.5	0.0371	4.31218E-09	0.0418			
21	1	0.0555	-9.83442E-06	0.0328			

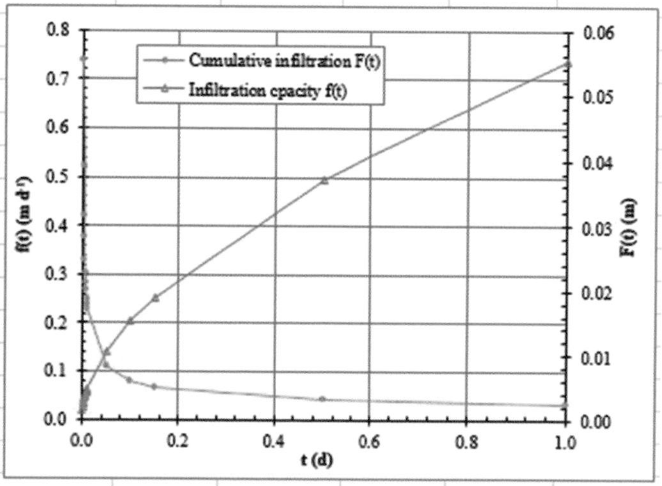

Figure 3.15 **Spreadsheet and dialog boxes for developing the infiltration capacity curve using the G–A model in Example 3.10.**

Let $t = 0.001$ d, using Eq. (3.22), one has:

$$F(t) = \frac{0.029 \text{ m d}^{-1}}{2} * (0.001 \text{ d}) + (0 + 0.29 \text{ m}) * (0.95*0.38 - 0.12) * \ln \left[1 + \frac{F(t)}{(0 + 0.29 \text{ m}) * (0.95*0.28 - 0.12)}\right].$$

Simplifying, produces: $F(t) = 1.45 \times 10^{-5} + 0.06989 \ln \left[1 + \frac{F(t)}{0.06989}\right].$

Using a trial-and-error method or Excel Solver, yields: $F(t) = 0.0014$ m.

Using Eq. (3.21), one has:

$$f(t) = \frac{0.029 \text{ m d}^{-1}}{2} * \left[1 + \frac{(0 + 0.29 \text{ m}) * (0.95*0.38 - 0.12)}{0.0014 \text{ m}}\right] = 0.7384 \text{ m d}^{-1}.$$

For other times, similar calculations can be done.

Figure 3.15 shows the calculations for the selected times, and the cumulative and instantaneous infiltration capacity curves. Suction capillary pressure head Ψ_f is computed in cell G2 using Eq. (3.20). For a given time t, the third column ([C3]) is the objective function formed by moving the right side terms of Eq. (3.22) to the left side. In Solver, specify the corresponding cell in the third column ([C3]) in the Set Objective box, check the Value of: 0 radio button, specify the corresponding cell in second column ([C2]) in the By Changing Variable Cells box and then click the Solve button to get the solution for F(t). The fourth column ([C4]) computes the corresponding infiltration capacity using Eq. (3.21).

As discussed above, the actual infiltration rate is equal to rainfall intensity as long as rainfall intensity is smaller than soil infiltration capacity. The length of raining time needed for infiltration capacity to be equal to rainfall intensity is termed as ponding time, t_{pd}, as illustrated in Figure 3.16a. By ponding time, the amount of infiltration is equal to the amount of rainfall (i.e., the product of rainfall intensity and t_{pd}. The corresponding potential infiltration time, $t_{pd,c}$, is defined as the time period throughout which the integration of infiltration capacity curve is the same as this amount of infiltration. Both t_{pd} and $t_{pd,c}$ can be determined using infiltration models. Regardless of the models, the actual cumulative infiltration and total direct runoff can be computed as:

$$F_{ac}(t_{rain}) = \begin{cases} i t_{pd} + \int_{t_{pd,c}}^{t_{pd,c} + t_{rain} - t_{pd}} f(t) dt & \text{if } t_{pd} < t_{rain} \\ i t_{rain} & \text{if } t_{pd} \geq t_{rain} \end{cases} \tag{3.23}$$

(Labels: actual cumulative infiltration → F_{ac}; rainfall duration → t_{rain}; rainfall intensity → i; ponding time → t_{pd}; infiltration capacity → $f(t)$; potential infiltration time → $t_{pd,c}$.)

$$R(t_{rain}) = i t_{rain} - F_{ac}(t_{rain}) \tag{3.24}$$

(Labels: total direct runoff → R; rainfall duration → t_{rain}; rainfall intensity → i; actual cumulative infiltration → $F_{ac}(t_{rain})$.)

For the Horton model, if rainfall intensity is smaller than ultimate infiltration capacity (i.e., $i < f_c$), ponding will never occur, and ponding time is infinite. Otherwise, in terms of Eq. (3.17) and the conditions for ponding to incept, ponding time and potential infiltration time will satisfy the relationships expressed as:

ultimate infiltration capacity
rainfall intensity
initial infiltration capacity
potential infiltration time

$$\begin{cases} i = f_c + (f_0 - f_c)\, e^{-\beta t_{pd,c}} \\ i t_{pd} = \int_0^{t_{pd,c}} \left[f_c + (f_0 - f_c)\, e^{-\beta t} \right] dt = f_c t_{pd,c} + \dfrac{f_0 - f_c}{\beta} \left(1 - e^{-\beta t_{pd,c}} \right) \end{cases} \tag{3.25}$$

ponding time
decay rate

Solving Eq. (3.25) and rearranging, produces:

initial infiltration capacity
ultimate infiltration capacity
potential infiltration time
ponding time

$$\begin{cases} t_{pd,c} = \dfrac{1}{\beta} \ln \left(\dfrac{f_0 - f_c}{i - f_c} \right) \\ t_{pd} = \dfrac{\beta f_c t_{pd,c} + (f_0 - f_c)\left(1 - e^{-\beta t_{pd,c}}\right)}{\beta i} \end{cases} \tag{3.26}$$

decay rate
rainfall intensity

Substituting Eq. (3.17) into Eq. (3.23), integrating, and rearranging, yields:

rainfall intensity
ultimate infiltration capacity
rainfall duration

$$F_{ac}(t_{rain}) = \begin{cases} (i - f_c)\, t_{pd} + f_c t_{rain} + \dfrac{f_0 - f_c}{\beta} \left[e^{-\beta t_{pd,c}} - e^{-\beta \left(t_{pd,c} + t_{rain} - t_{pd} \right)} \right] & \text{if } t_{pd} < t_{rain} \\ i t_{rain} & \text{if } t_{pd} \geq t_{rain} \end{cases} \tag{3.27}$$

actual cumulative infiltration
ponding time
decay rate
potential infiltration time

For the Philip model, if rainfall intensity is smaller than the ultimate infiltration capacity (i.e., $i < f_u$), ponding will never occur, and ponding time is infinite. Otherwise, in terms of Eq. (3.18) and the conditions for ponding to incept, ponding time and potential infiltration time will satisfy the relationships expressed as:

soil storativity coefficient
ultimate infiltration capacity
rainfall intensity

$$\begin{cases} i = \dfrac{S_1}{2\sqrt{t_{pd,c}}} + f_u \\ i t_{pd} = \int_0^{t_{pd,c}} \left(\dfrac{S_1}{2\sqrt{t}} + f_u \right) dt = S_1 \sqrt{t_{pd,c}} + f_u t_{pd,c} \end{cases} \tag{3.28}$$

ponding time
potential infiltration time

Solving Eq. (3.28) and rearranging, produces:

$$\begin{cases} t_{pd,c} = \left[\dfrac{S_1}{2\left(i - f_u\right)}\right]^2 \\ t_{pd} = \dfrac{S_1\sqrt{t_{pd,c}} + f_u t_{pd,c}}{i} \end{cases} \tag{3.29}$$

soil storativity coefficient
potential infiltration time
ultimate infiltration capacity
ponding time
rainfall intensity

Substituting Eq. (3.18) into Eq. (3.23), integrating, and rearranging, yields:

$$F_{ac}\left(t_{rain}\right) = \begin{cases} \left(i - f_u\right) t_{pd} + f_u t_{rain} + S_1\left(\sqrt{t_{pd,c} + t_{rain} - t_{pd}} - \sqrt{t_{pd,c}}\right) & \text{if } t_{pd} < t_{rain} \\ i t_{rain} & \text{if } t_{pd} \geq t_{rain} \end{cases} \tag{3.30}$$

rainfall intensity
rainfall duration
ponding time
soil storativity coefficient
potential infiltration time
actual cumulative infiltration
ultimate infiltration capacity

For the G–A model, if rainfall intensity is smaller than effective hydraulic conductivity (defined as half of saturated hydraulic conductivity) (i.e., $i < K_{sat}/2$), ponding will never occur, and ponding time is infinite. Otherwise, in terms of Eqs. (3.21) and (3.22) and the conditions for ponding to incept, ponding time and potential infiltration time will satisfy the relationships expressed as:

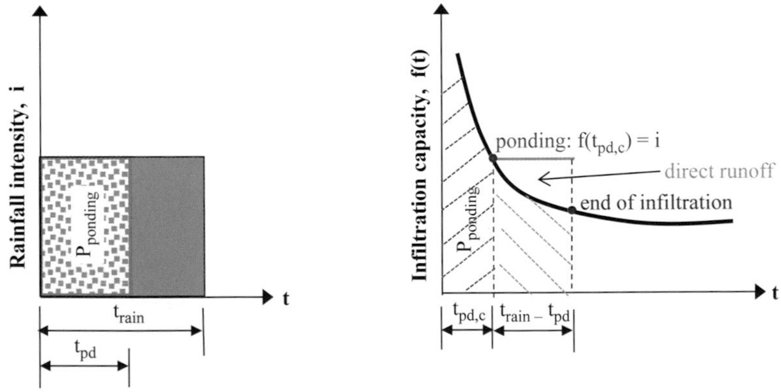

(a) (b)

Figure 3.16 **Actual infiltration and ponding time for a uniform rainfall event.** Diagrams show the: (a) hyetograph of the rainfall event; and (b) infiltration capacity curve. t_{rain}: duration of the rainfall event; t_{pd}: ponding time; $P_{ponding} = i t_{pd}$: rainfall needed for ponding to occur; $t_{pd,c}$: potential infiltration time, defined as the time period throughout which the integration of infiltration capacity curve is the same as $P_{ponding}$. The total actual infiltration from the rainfall event is the summation of the two areas shaded in part (b).

saturated hydraulic conductivity
rainfall intensity
ponding water depth on ground
initial soil moisture

$$\begin{cases} i = \dfrac{K_{sat}}{2}\left[1 + \dfrac{(H + \psi_f)(0.95\theta_{sat} - \theta_a)}{it_{pd}}\right] \\ it_{pd} = \dfrac{K_{sat}}{2} t_{pd,c} + (H + \psi_f)(0.95\theta_{sat} - \theta_a) \ln\left[1 + \dfrac{it_{pd}}{(H + \psi_f)(0.95\theta_{sat} - \theta_a)}\right] \end{cases} \tag{3.31}$$

ponding time
potential infiltration time
suction head at wetting front
saturated soil moisture

Solving Eq. (3.31) and rearranging, produces:

saturated hydraulic conductivity
ponding time
ponding water depth on ground
initial soil moisture

$$\begin{cases} t_{pd} = \dfrac{K_{sat}(H + \psi_f)(0.95\theta_{sat} - \theta_a)}{i(2i - K_{sat})} \\ t_{pd,c} = \dfrac{2}{K_{sat}}\left\{it_{pd} - (H + \psi_f)(0.95\theta_{sat} - \theta_a) \ln\left[1 + \dfrac{it_{pd}}{(H + \psi_f)(0.95\theta_{sat} - \theta_a)}\right]\right\} \end{cases} \tag{3.32}$$

rainfall intensity
suction head at wetting front
saturated soil moisture
potential infiltration time

Substituting Eq. (3.21) into Eq. (3.23) and considering Eq. (3.22), yields:

actual cumulative infiltration
ponding time
potential infiltration time

$$F_{ac}(t_{rain}) = \begin{cases} it_{pd} + \left[F\left(t_{pd,c} + t_{rain} - t_{pd}\right) - F\left(t_{pd,c}\right)\right] & \text{if } t_{pd} < t_{rain} \\ it_{rain} & \text{if } t_{pd} \geq t_{rain} \end{cases} \tag{3.33}$$

rainfall duration
cumulative infiltration capacity as computed by Eq. (3.22)
rainfall intensity

Example 3.11 A site has the following soil water parameters: $\theta_{sat} = 0.43$, $K_{sat} = 2.8$ cm hr^{-1}, $\psi_b = 35$ cm, and $\lambda = 0.56$. It receives a 2-hr storm with an average rainfall intensity of 3.5 cm hr^{-1}. If the initial soil moisture is 0.15 and the ponding water depth on the ground is very small, estimate the actual cumulative infiltration and total direct runoff using the G–A model.

Solution

$$\theta_{sat} = 0.43, K_{sat} = 2.8 \text{ cm hr}^{-1}, \psi_b = 35 \text{ cm}, \lambda = 0.56$$
$$t_{rain} = 2 \text{ hr}, i = 3.5 \text{ cm hr}^{-1}, \theta_a = 0.15, H \approx 0$$

Using Eq. (3.20), yields:

$$\psi_f = \frac{3*0.56 + 2}{3*0.56 + 1}\left(\frac{35 \text{ cm}}{2}\right) = 24.03 \text{ cm.}$$

Using Eq. (3.32), produces:

$$t_{pd} = \frac{(2.8\text{ cm hr}^{-1}) * (0 + 24.03\text{ cm}) * (0.95*0.43 - 0.15)}{(3.5\text{ cm hr}^{-1}) * [2*(3.5\text{ cm hr}^{-1}) - (2.8\text{ cm hr}^{-1})]} = 1.18\text{ hr}$$

$$t_{pd,c} = \frac{2}{2.8\text{ cm hr}^{-1}} * \left\{ \begin{array}{l} (3.5\text{ cm hr}^{-1}) * (1.18\text{ hr}) - (0 + 24.03\text{ cm}) * (0.95*0.43 - 0.15) * \\ \ln\left[1 + \dfrac{(3.5\text{ cm hr}^{-1}) * (1.18\text{ hr})}{(0 + 24.03\text{ cm}) * (0.95*0.43 - 0.15)}\right] \end{array} \right\} = 0.688\text{ hr}$$

$$t_{pd,c} + t_{rain} - t_{pd} = 0.688\text{ hr} + 2\text{ hr} - 1.18\text{ hr} = 1.508\text{ hr}.$$

Given $t_{pd} < t_{rain}$, using Eq. (3.22), one can get:

$$F(t_{pd,c} + t_{rain} - t_{pd}) = \frac{2.8\text{ cm hr}^{-1}}{2} * (1.508\text{ hr}) + (0 + 24.03\text{ cm})(0.95*0.43 - 0.15) * \ln\left[1 + \frac{F(t_{pd,c} + t_{rain} - t_{pd})}{(0 + 24.03\text{ cm})(0.95*0.43 - 0.15)}\right].$$

Simplifying, one has: $F(t_{pd,c} + t_{rain} - t_{pd}) = 2.1112 + 6.2118 * \ln\left[1 + \frac{F(t_{pd,c} + t_{rain} - t_{pd})}{6.2118}\right].$

By a trial-and-error method or Excel Solver, one has: $F(t_{pd,c} + t_{rain} - t_{pd}) = 6.616$ cm

$$F(t_{pd,c}) = \frac{2.8\text{ cm hr}^{-1}}{2} * (0.688\text{ hr}) + (0 + 24.03\text{ cm})(0.95*0.43 - 0.15) * \ln\left[1 + \frac{F(t_{pd,c})}{(0 + 24.03\text{ cm})(0.95*0.43 - 0.15)}\right].$$

Simplifying, one has: $F(t_{pd,c}) = 0.9632 + 6.2118 * \ln\left[1 + \frac{F(t_{pd,c})}{6.2118}\right].$

By a trial-and-error method or Excel Solver, one has: $F(t_{pd,c}) = 4.129$ cm.

Using the first part of Eq. (3.33), one can get the actual cumulative infiltration:

$$F_{ac}(t_{rain}) = (3.5\text{ cm hr}^{-1}) * (1.18\text{ hr}) + [(6.616\text{ cm}) - (4.129\text{ cm})] = 6.617\text{ cm}.$$

Eq. (3.24) → the direct runoff: $R(t_{rain}) = (3.5\text{ cm hr}^{-1}) * (2\text{ hr}) - (6.617\text{ cm}) = 0.383$ cm.

3.2.3 Evapotranspiration

In most engineering applications, evaporation and transpiration are combined into evapotranspiration (ET). For a given location of interest, ET is affected by the factors of climate, environment, and plant morphology as well as available water (Viessman and Lewis, 2002). The climate factors include net solar radiation, air temperature, air relative humidity, and wind speed, and the environment factors include soil texture, soil structure, and soil chemistry (e.g., salinity). The plant factors include plant type, growth stage, height, density, root depth and structure, leaf area, and tissue features. The first two groups of factors affect both evaporation and transpiration, whereas the plant factors affect transpiration. The amount of evaporation that would occur when water is always available to meet the atmospheric demand is defined as potential ET (PET) (Brutsaert, 2005). That is, PET depends on the climate factors only and is independent of the environment and plant factors. In addition, the evapotranspiration that would occur if a soil-plant surface is saturated and there is no limitation on the availability of water is defined as wet-environment evapotranspiration (ET_W) (Wang *et al.*, 2006). Obviously, ET_W is affected by the climate and plant factors, but is independent of the environment factors. Further, the evapotranspiration that depends on the three groups of factors is defined as actual evapotranspiration (AET), which is the case most of the time. These three types of evapotranspiration follow a complementary relationship (Bouchet, 1963; Morton, 1983a, 1986; Hobbins *et al.*, 2001; Wang *et al.*, 2006), illustrated in Figure 3.17 and defined as:

actual evapotranspiration → AET; wet-environment evapotranspiration → ET_W; potential evapotranspiration → PET

$$\mathrm{PET} + \mathrm{AET} \approx 2\mathrm{ET_W} \tag{3.34}$$

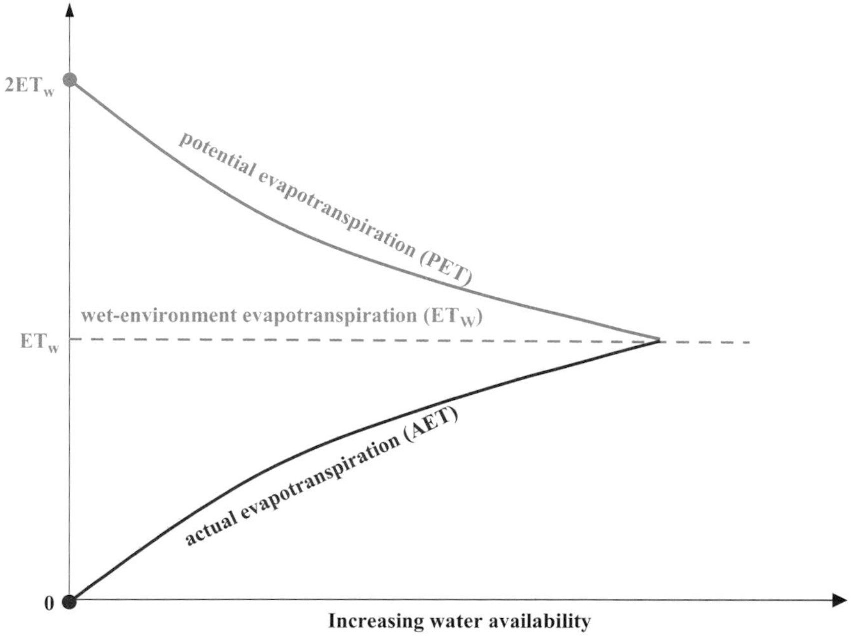

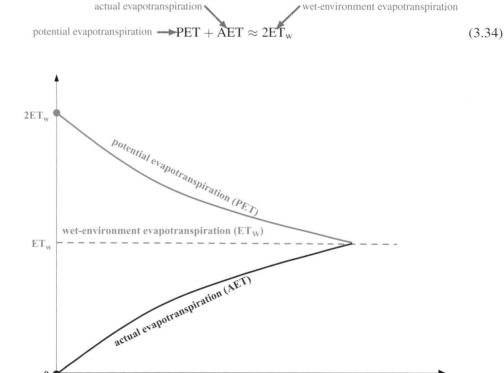

Figure 3.17 **Illustration of the complementary law of evapotranspiration.**

AET can be directly measured using two categories of methods, namely water balance and water vapor transfer (Shuttleworth, 2008). The basic idea of the water balance methods is to measure the inputs and outputs (except for AET) of a selected hydrologic system (e.g., evaporation pan, small basin, or undisturbed soil column), and then solve Eq. (3.12) to estimate AET. In contrast, the basic idea of the water vapor transfer methods is to measure air temperature, humidity and wind speed gradients and/or sap flow in plant trunk, branches, or roots using heat as a tracer, which in turn are used to determine latent heat and its equivalent AET (USGS, 2001). These methods have different strengths and weaknesses and application conditions as summarized by Rana and Katerji (2000) and Shuttleworth (2008). Nevertheless, direct measurement of AET can be prohibitively costly and time consuming. Alternatively, AET can be estimated using Eq. (3.34) as the difference of $2ET_w$ and PET.

The Penman–Monteith equation (Allen *et al.*, 1998) computes ET_w as:

wet-environment evapotranspiration [mm s^{-1}]; slope of vapor pressure curve [kPa °C^{-1}]; net radiation at vegetated surface [MW m^{-2}]; soil heat flux [MW m^{-2}]; air density [kg m^{-3}]; air specific heat [= 10^{-3} MJ kg^{-1} °C^{-1}]; saturation vapor pressure [kPa]; actual vapor pressure [kPa]; aerodynamic resistance to vapor transport [s m^{-1}]; soil-canopy resistance to vapor transport [s m^{-1}]; latent heat of vaporization [= 2.257 MJ kg^{-1}]; psychrometric constant [kPa °C^{-1}]

$$ET_w = \frac{\Delta(R_n - G) + \rho_a c_a\left(\frac{e_s - e_a}{r_a}\right)}{\lambda\left[\Delta + \gamma\left(1 + \frac{r_s}{r_a}\right)\right]} \tag{3.35}$$

In Eq. (3.35), r_s depends on the soil-water properties and vegetation characteristics. It reflects the combined effects of environmental and plant factors on ET. In contrast, r_a depends on atmospheric dynamics and reflects the effects of climate factors on ET. For a geographic location, both r_s and r_a vary from time to time. r_s needs to be estimated using observed site-specific data (e.g., Amer and Hatfield, 2004), whereas r_a and other parameters (Tetens, 1930; Murray, 1967; Brutsaert, 2005; Wang *et al.*, 2014) can be estimated using:

slope of vapor pressure curve [kPa °C^{-1}]; mean air temperature [°C]

$$\Delta = \frac{4098\left(0.6108e^{\frac{17.27T_a}{T_a+237.3}}\right)}{(T_a + 237.3)^2} \tag{3.36}$$

mean air temperature [°C]; minimum air temperature [°C]; maximum air temperature [°C]

$$T_a = \frac{T_{min} + T_{max}}{2} \tag{3.37}$$

soil heat flux [MW m^{-2}]; soil heat capacity [MJ m^{-3} °C^{-1}]; effective soil depth [m]; air temperature at end of time interval [°C]; air temperature at start of time interval [°C]; end of time interval [s]; start of time interval

$$G = c_s z_s \frac{T_{a,i} - T_{a,i-1}}{t_i - t_{i-1}} \tag{3.38}$$

air density [kg m^{-3}]; atmospheric pressure [kPa]; dry air gas constant [= 287.058 J kg^{-1} K^{-1}]; mean daily air temperature [°C]

$$\rho_a = \frac{1000p_a}{R_a(T_a + 273.15)} \tag{3.39}$$

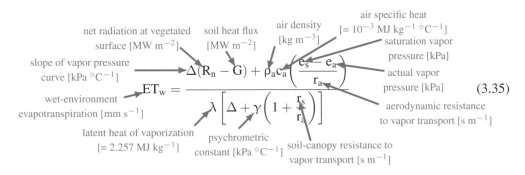

$$e_s = 0.611\left(10^{\frac{7.5T_a}{T_a+237.3}}\right) \tag{3.40}$$

saturation vapor pressure [kPa]
mean daily air temperature [°C]

$$e_a = e_s\left(\frac{RH}{100}\right) \tag{3.41}$$

actual vapor pressure [kPa]
saturation vapor pressure [kPa]
air relative humidity [%]

$$\gamma = \frac{p_a}{1405.72} \tag{3.42}$$

psychrometric constant [kPa °C^{-1}]
atmospheric pressure [kPa]

$$r_a = \frac{\ln\left(\frac{z_m - d_0}{z_{0m}}\right)\ln\left(\frac{z_h - d_0}{z_{0h}}\right)}{\kappa^2 u_{zm}} \tag{3.43}$$

aerodynamic resistance to vapor transport [s m^{-1}]
above-ground height of wind measurement [m]
zero plane displacement height [m]
above-ground height of humidity measurement [m]
roughness length governing momentum transfer [m]
roughness length governing heat and vapor transfer [m]
von Karman constant [= 0.41]
wind speed at above-ground height z_m [m s^{-1}]

$$d_0 = \begin{cases} 0 & \text{bare surface} \\ 0.67h_{plant} & \text{vegetated surface} \end{cases} \tag{3.44}$$

zero plane displacement height [m]
plant height [m]

$$z_{0m} = \begin{cases} h_{soil} & \text{bare surface} \\ 0.123h_{plant} & \text{vegetated surface} \end{cases} \tag{3.45}$$

roughness length governing momentum transfer [m]
soil surface roughness height [m]
plant height [m]

$$z_{0h} = \begin{cases} h_{soil} & \text{bare surface} \\ 0.0123h_{plant} & \text{vegetated surface} \end{cases} \tag{3.46}$$

roughness length governing heat and vapor transfer [m]
soil surface roughness height [m]
plant height [m]

Setting canopy resistance $r_s = 0$, Eq. (3.35) can be adopted to compute PET as:

$$PET = \frac{\Delta(R_n - G) + \rho_a c_a \left(\frac{e_s - e_a}{r_a}\right)}{\lambda[\Delta + \gamma]} \tag{3.47}$$

potential evapotranspiration [mm s^{-1}]
slope of vapor pressure curve [kPa °C^{-1}]
net radiation at vegetated surface [MW m^{-2}]
soil heat flux [MW m^{-2}]
air density [kg m^{-3}]
air specific heat [= 10^{-3} MJ kg^{-1} °C^{-1}]
saturation vapor pressure [kPa]
actual vapor pressure [kPa]
aerodynamic resistance to vapor transport [s m^{-1}]
latent heat of vaporization [= 2.257 MJ kg^{-1}]
psychrometric constant [kPa °C^{-1}]

Once ET_w and PET are computed using Eqs. (3.35) and (3.47), respectively, Eq. (3.34) can be rearranged to compute AET. In addition, other models have also been presented in existing literature to estimate ET_w and PET (McMahon *et al.*, 2013). Among them, the models proposed by Morton (1983a,b) have been used extensively in Australia, where historical wind data were unavailable until recent times, to generated time series of estimated historical PET (McVicar *et al.*, 2008). Further, PET can be estimated as the product of Class A pan evaporation and a pan coefficient (< 1.0) (Farnsworth and Thompson, 1982; Farnsworth *et al.*, 1982), and AET is estimated as the product of PET and a crop coefficient (Allen *et al.*, 1998). A Class A pan is a cylinder with a diameter of 47.5 in and a depth of 10 in (HyQuest, 2016). It measures open-water evaporation as the water depth drop at a time interval of interest (e.g., daily or monthly). The details of using Class A pan data to estimate ET are referenced to relevant publications (e.g., Farnsworth and Thompson, 1982; Farnsworth *et al.*, 1982; Allen *et al.*, 1998).

Example 3.12 On a day of interest, a potato field has the following climate factors: $R_n = 0.00065$ MW m^{-2}, $T_a = 28$°C, RH $= 60\%$, $p_a = 100$ kPa, $u_{zm} = 10$ m s^{-1}. Both u_{zm} and RH are measured at 2 m above the ground surface. In terms of the environment and plant factors, it is determined that the soil-canopy resistance to vapor transport is $r_s = 25$ s m^{-1}. The soil heat flux is negligible. The average height of the potato on that day is 0.5 m. Estimate the actual evapotranspiration rate.

Solution

$$R_n = 0.00065 \text{ MW m}^{-2}, T_a = 28°\text{C}, \text{RH} = 60\%, p_a = 100 \text{ kPa}, u_{zm} = 10 \text{ m s}^{-1},$$
$$z_m = z_h = 2 \text{ m}, G = 0, r_s = 25 \text{ s m}^{-1}$$

Using Eq. (3.36), yields:

$$\Delta = \frac{4098\left(0.6108 e^{\frac{17.27 * 28°C}{28°C + 237.3}}\right)}{(28°C + 237.3)^2} = 0.2201 \text{ kPa °C}^{-1}.$$

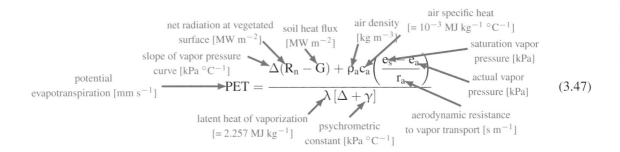

Using Eq. (3.39), yields:

$$\rho_a = \frac{1000*(100\,\text{kPa})}{\left(287.058\,\text{J}\,\text{kg}^{-1}\,\text{K}^{-1}\right)*(28°\text{C}+273.15)} = 1.1568\,\text{kg m}^{-3}.$$

Using Eq. (3.40), yields:

$$e_s = 0.611\left(10^{\frac{7.5*28°\text{C}}{28°\text{C}+237.3}}\right) = 3.7809\,\text{kPa}.$$

Using Eq. (3.41), yields:

$$e_a = (3.7809\,\text{kPa})*\left(\frac{60}{100}\right) = 2.2685\ \text{kPa}.$$

Using Eq. (3.42), yields:

$$\gamma = \frac{100\,\text{kPa}}{1405.72} = 0.0711\,\text{kPa}\,°\text{C}^{-1}.$$

Using Eq. (3.44), yields:

$$d_0 = 0.67*(0.5\,\text{m}) = 0.335\,\text{m}.$$

Using Eq. (3.45), yields:

$$z_{0m} = 0.123*(0.5\,\text{m}) = 0.0615\,\text{m}.$$

Using Eq. (3.46), yields:

$$z_{0h} = 0.0123*(0.5\,\text{m}) = 0.0062\,\text{m}.$$

Using Eq. (3.43), yields:

$$r_a = \frac{\ln\left(\frac{2\,\text{m}-0.335\,\text{m}}{0.0615,\text{m}}\right)\ln\left(\frac{2\,\text{m}-0.335\,\text{m}}{0.0062\,\text{m}}\right)}{0.41^2*(10\,\text{m s}^{-1})} = 10.9749\,\text{s m}^{-1}.$$

Using Eq. (3.35), produces:

$$ET_w = \left[\begin{array}{l}\left(0.2201\ \text{kPa}\ °\text{C}^{-1}\right)*\left(0.00065\ \text{MW m}^{-2}-0\right)+\left(1.1568\ \text{kg m}^{-3}\right)\\ *\left(10^{-3}\ \text{MJ kg}^{-1}\ °\text{C}^{-1}\right)*\left(\frac{3.7809\ \text{kPa}-2.2685\ \text{kPa}}{10.9749\ \text{s m}^{-1}}\right)\end{array}\right]$$

$$\div\left[\left(2.257\ \mathrm{MJ\ kg^{-1}}\right)*\left(0.2201\ \mathrm{kPa\ ^\circ C^{-1}}+\left(0.0711\ \mathrm{kPa\ ^\circ C^{-1}}\right)*\left(1+\frac{25\ \mathrm{s\ m^{-1}}}{10.9749\ \mathrm{s\ m^{-1}}}\right)\right)\right]$$

$$= 0.000296\ \mathrm{mm\ s^{-1}} = 25.57\ \mathrm{mm\ d^{-1}}.$$

Using Eq. (3.47), produces:

$$\mathrm{PET}=\left[\begin{array}{l}\left(0.2201\ \mathrm{kPa^\circ C^{-1}}\right)*\left(0.00065\ \mathrm{MW\ m^{-2}}-0\right)+\left(1.1568\ \mathrm{kg\ m^{-3}}\right)\\ *\left(10^{-3}\ \mathrm{MJ\ kg^{-1}\ ^\circ C^{-1}}\right)*\left(\dfrac{3.7809\ \mathrm{kPa}-2.2685\ \mathrm{kPa}}{10.9749\ \mathrm{s\ m^{-1}}}\right)\end{array}\right]$$

$$\div\left[\left(2.257\ \mathrm{MJ\ kg^{-1}}\right)*\left(0.2201\ \mathrm{kPa^\circ C^{-1}}+0.0711\ \mathrm{kPa^\circ C^{-1}}\right)\right]$$

$$= 0.000460\ \mathrm{mm\ s^{-1}} = 39.74\ \mathrm{mm\ d^{-1}}.$$

Finally, using Eq. (3.34), produces:

$$\mathrm{AET} = 2*\left(0.000296\ \mathrm{mm\ s^{-1}}\right)-\left(0.000460\ \mathrm{mm\ s^{-1}}\right) = 0.000132\ \mathrm{mm\ s^{-1}} = 11.40\ \mathrm{mm\ d^{-1}}$$

3.2.4 Surface Runoff

As illustrated in Figures 3.6 and 3.12, surface runoff occurs once rainfall intensity exceeds soil infiltration capacity. It is sometimes termed as direct runoff or excess rainfall. Determination of direct runoff from rainfall is fundamental to a range of hydrologic applications (Viessman and Lewis, 2002; Wang and Melesse, 2005), such as engineering design (Debo and Reese, 2003), flood forecasting (Nash and Sutcliffe, 1970; Hapuarachchi *et al.*, 2011), and assessing the effects of watershed best-management practices (Arnold *et al.*, 1998, 2011; Wang and Melesse, 2006; Wang *et al.*, 2008; Wang *et al.*, 2010a; Nalbantis *et al.*, 2011). In practice, direct runoff is usually determined using rainfall–runoff models, such as the SCS-CN method (USDA-NRCS, 2004) and its modified versions (Wang *et al.*, 2008, 2012). Either BG or SI units can be used if consistent.

The SCS-CN method computes direct runoff as:

$$Q=\begin{cases}\dfrac{(P-I_a)^2}{(P-I_a)+S} & P>I_a\\ 0 & P\le I_a\end{cases} \tag{3.48}$$

direct runoff → Q; rainfall → P; initial abstraction → I_a; potential maximum retention after runoff begins → S

$$S=\begin{cases}25.4\left(\dfrac{1000}{CN}-10\right) & [\mathrm{mm}]\\ \dfrac{1000}{CN}-10 & [\mathrm{in}]\end{cases} \tag{3.49}$$

potential maximum retention after runoff begins → S; curve number [-] → CN

$$\underset{\text{[mm or in]}}{\text{initial abstraction}} \rightarrow I_a = 0.2S \leftarrow \underset{\text{after runoff begins [mm or in]}}{\text{potential maximum retention}} \qquad (3.50)$$

The initial abstraction, I_a, includes canopy storage or interception, infiltration during early parts of the storm, and surface depression storage. In Eq. (3.48), for a rainfall event P, the total maximum possible loss is given by the summation of I_a and S. The value of CN depends on LULC, HSGs (Table 3.1), and antecedent moisture conditions (AMCs) or antecedent runoff conditions (ARCs). The SCS defines three ARCs: ARC I is for dry conditions that correspond to wilting point, whereas ARC III is for wet conditions that correspond to field capacity. ARC II is for average conditions that correspond to a soil moisture value between the wilting point and field capacity. Hjelmfelt (1991) stated that ARC II should be interpreted as the soil moisture condition associated with the median curve number CN_{II}, whereas ARC I and ARC III are defined as the soil moisture conditions associated with the curve numbers of 10% and 90% exceedance frequencies, respectively. The curve numbers for ARC I (CN_I) and for ARC III (CN_{III}) can be estimated based on CN_{II} (Gray *et al.*, 1982; Silveira *et al.*, 2000; USDA-NRCS, 2004). The potential maximum retention for ARC I, S_I, can be computed by substituting CN_I for CN in Eq. (3.49), whereas the potential maximum retention for ARC III, S_{III}, can be computed by substituting CN_{III} for CN. The potential maximum retention for ARC II, S_{II}, can be computed by substituting CN_{II} for CN in Eq. (3.49). Tables 3.13 and 3.14 give CN_{II} for urban and rural areas, respectively. Values for CN_I and CN_{II} can be computed using Eqs. (3.51) and (3.52).

The curve number for ARC I can be computed as (USDA-NRCS, 2004):

$$\underset{\text{runoff condition I [-]}}{\text{curve number for antecedent}} \rightarrow CN_I = \overset{\text{curve number for antecedent runoff condition II [-]}}{CN_{II}} - \frac{20(100 - CN_{II})}{100 - CN_{II} + e^{[2.533 - 0.0636(100 - CN_{II})]}} \qquad (3.51)$$

The curve number for ARC III can be computed as (USDA-NRCS, 2004):

$$\underset{\text{runoff condition III [-]}}{\text{curve number for antecedent}} \rightarrow CN_{III} = \overset{\text{curve number for antecedent runoff condition II [-]}}{CN_{II}} e^{[0.00673(100 - CN_{II})]} \qquad (3.52)$$

Table 3.13 Typical values of CN_{II} for urban areas.[1]

Cover description		CN_{II} for HSG			
Cover type and hydrologic condition	Imperviousness (%)	A	B	C	D
Fully developed (vegetation established)					
Open space (lawns, parks, golf courses, cemeteries, etc.):					
Poor condition (grass cover < 50%)		68	79	86	89
Fair condition (grass cover 50 to 75%)		49	69	79	84
Good condition (grass cover > 75%)		39	61	74	80
Impervious areas:					
Paved parking lots, roofs, drive ways, etc. (excl. ROW)		98	98	98	98
Streets and roads					
Paved: curbs and storm sewers (excl. ROW)		98	98	98	98
Paved: open ditches (incl. ROW)		83	89	92	93
Gravel (incl. ROW)		76	85	89	91
Dirt (incl. ROW)		72	82	87	89
Western desert urban areas:					
Natural desert landscaping (pervious areas only)		63	77	85	88
Artificial desert landscaping (impervious weed barrier, desert shrub with 1- to 2-inch sand or gravel mulch and basin borders)		96	96	96	96
Urban districts:					
Commercial and business	85	89	92	94	95
Industrial	72	81	88	91	93
Residential districts by average lot size:					
1/8 ac or less (town houses)	65	77	85	90	92
1/4 ac	38	61	75	83	87
1/3 ac	30	57	72	81	86
1/2 ac	25	54	70	80	85
1 ac	20	51	68	79	84
2 ac	12	46	65	77	82
Newly graded areas (pervious areas only, no vegetation)		77	86	91	94

[1] *Source:* USDA-SCS (1989);CN_{II}: curve number for antecedent runoff condition (ARC) II (USDA-NRCS, 2004); HSG: hydrologic soil group; ROW: right-of-way; excl.: excluding; incl.: including.

Example 3.13 A 0.5 ac residential lot has C soil. The lot is in ARC II when a 30-min storm event with a rainfall of 0.8 in occurs. Determine the direct runoff.

Solution

P = 0.8 in. From Table 3.13, $CN_{II} = 80$.

BG units, Eq. (3.49): $S = \frac{1000}{80} - 10 \text{ in} = 2.5 \text{ in}.$

Eq. (3.50): $I_a = 0.2 * (2.5 \text{ in}) = 0.5 \text{ in}.$

$P > I_a$, Eq. (3.48): $Q = \frac{(0.8 \text{ in} - 0.5 \text{ in})^2}{(0.8 \text{ in} - 0.5 \text{ in}) + 2.5 \text{ in}} = 0.03 \text{ in}.$

Table 3.14 Typical values of CN_{II} for rural areas.[1]

Cover type and hydrologic condition	CN_{II} for HSG			
	A	B	C	D
Pasture, grassland, or range-continuous forage for grazing				
Poor (< 50% ground cover or heavily grazed with no mulch)	68	79	86	89
Fair (50 to 75% ground cover and not heavily grazed)	49	69	79	84
Good (> 75% ground cover and lightly or occasionally grazed)	39	61	74	80
Meadow-continuous grass, protected from grazing and generally mowed for hay	30	58	71	78
Brush-brush weed-grass mixture with brush the major element				
Poor (< 50% ground cover)	48	67	77	83
Fair (50 to 75% ground cover)	35	56	70	77
Good (> 75% ground cover)	30	48	65	73
Woods-grass combination (orchard or tree farm) (50% woods and 50% grass or pasture)				
Poor (< 50% ground cover)	57	73	82	86
Fair (50 to 75% ground cover)	43	65	76	82
Good (> 75% ground cover)	32	58	72	79
Woods				
Poor (forest litter, small trees, and brush are destroyed by heavily grazing or regular burning)	45	66	77	83
Fair (woods are grazed but not burned, and some forest litter covers the soil)	36	60	73	79
Good (woods are protected from grazing, and litter and brush adequately cover the soil)	30	55	70	77
Farmsteads-buildings, lanes, driveways, and surrounding lots	59	74	82	86

[1] *Source:* USDA-SCS (1989); CN_{II}: curve number for antecedent runoff condition (ARC) II (USDA-NRCS, 2004); HSG: hydrologic soil group; ROW: right-of-way; excl.: excluding; incl.: including.

Example 3.14 A 110 km^2 rural watershed has three HSGs and two land covers as shown in Figure 3.18. The areas for the unique combinations of the soils and land covers are also shown. The watershed is in ARC I when a 30-min storm occurs, with rainfall of 35, 60, and 20 mm in the first, second, and third 10 min periods, respectively. Determine the direct runoff from the rainfall in each 10 min period as well as the total direct runoff from the storm.

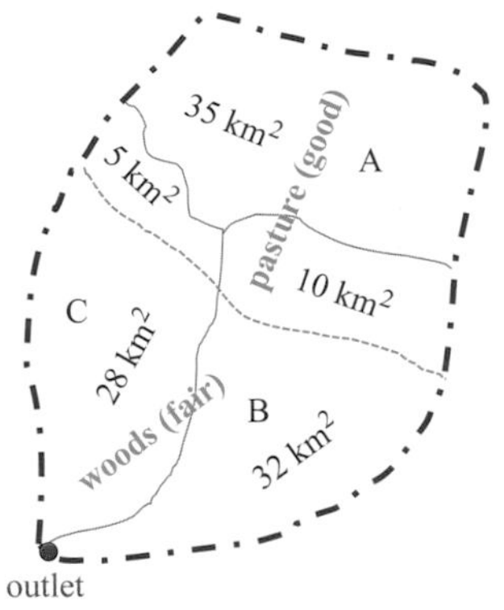

Figure 3.18 **Illustration of the watershed for Example 3.14.**

Solution

From Table 3.14, one can find the values of CN_{II} for the unique combinations of the HSGs and land covers, as given in Table 3.15. Because the watershed is in ARC I, one needs to use Eq. (3.51) to compute CN_I for each combination. For example, for the combination of pasture (good) and HSG A, $CN_{II} = 39$ from Table 3.15, and $CN_I = 39 - \frac{20*(100-39)}{100-39+e^{[2.533-0.0636*(100-39)]}} = 19$ (rounded to a whole number!). Similarly, the values of CN_I for the other combinations can be determined and are shown in Table 3.15. Moreover, for each combination, the product of its area and CN_I is computed. The summation of the products of the CN_I and areas for all five combinations is divided by the watershed area of 110 km^2 to get the "composite" curve number for the watershed, designated $CN_{I,c}$ for description purposes. The value of $CN_{I,c}$ is used to compute direct runoff. Herein, $CN_{I,c} = (665 + 410 + 270 + 1280 + 1484) \div 110 = 37$ (rounded to a whole number!).

SI units, Eq. (3.49): $S = 25.4*\left(\frac{1000}{37} - 10\right)$ mm = 432.5 mm.

Eq. (3.50): $I_a = 0.2*432.5$ mm = 86.5 mm.

Table 3.15 Determination of the composite curve number for Example 3.14.

Land cover	HSG	Area (km^2)	CN_{II}	CN_I	Area $*CN_I$(km^2)
Pasture (good)	A	35	39	19	665
Pasture (good)	B	10	61	41	410
Pasture (good)	C	5	74	54	270
Woods (fair)	B	32	60	40	1280
Wood (fair)	C	28	73	53	1484

For the first 10 min: P = 35 mm.

Because P $< I_a$, no direct runoff is generated.

For the second 10 min: P = 60 mm.

Because the ARC for the second 10 min is not known, the computation must be done from the beginning of the storm. The runoff of the second 10 min is calculated as the difference of the runoff for the first 20 min and the runoff for the first 10 min.

The rainfall in the first 20 min is 35 mm + 60 mm = 95 mm $> I_a$, thus:

$$\text{Eq. (3.48): } Q = \frac{(95\text{ mm} - 86.5\text{ mm})^2}{(95\text{ mm} - 86.5\text{ mm}) + 432.5\text{ mm}} = 0.16\text{ mm}.$$

The runoff in the second 10 min is 0.16 mm − 0 mm = 0.16 mm.

For the third 10 min: P = 20 mm.

Again, because the ARC for the third 10 min is not known, the computation must be done from the beginning of the storm. The runoff of the third 10 min is calculated as the difference of the runoff for the 30 min and the runoff for the first 20 min.

The rainfall in the 30 min is 35 mm + 60 mm + 20 mm = 115 mm $> I_a$, thus:

$$\text{Eq. (3.48):} Q = \frac{(115\text{ mm} - 86.5\text{ mm})^2}{(115\text{ mm} - 86.5\text{ mm}) + 432.5\text{ mm}} = 1.76\text{ mm}.$$

The runoff in the third 10 min is 1.76 mm − 0.16 mm = 1.60 mm.

Summary: At the beginning of the storm event, the watershed is dry, as indicated by the large initial abstraction of 86.5 mm. The rainfall in the first 10 min is totally lost to the initial abstraction, generating no direct runoff but replenishing soil moisture. As a result, the AMC for the second 10 min becomes wetter though it cannot be determined because of insufficient information. The rainfall in the second 10 min generates some direct runoff. The rainfall losses in the first 20 min makes the AMC for the third 10 min further wetter. This is reason why the runoff in the third 10 min is much larger than the runoff in the second 10 min, while the rainfall in the third 10 min is much less than the rainfall in the second 10 min.

The SCS-CN method has two unrealistic assumptions. The first assumption is that soil moistures with a value between the wilting point and field capacity are assumed to be associated with a single value for CN_{II} or S_{II} (Hjelmfelt *et al.*, 1982; Cronshey, 1983; Hjelmfelt, 1987, 1991; van Mullem, 1992). That is, both CN_{II} and S_{II} are assumed not to vary with soil moisture, although daily CN does change with soil moisture. This can result in unreasonable sudden jumps in estimated runoff (Michel *et al.*, 2005; Ali *et al.*, 2010; Sahu *et al.*, 2010). The second assumption of initial abstraction (Eq. (3.50)) has no scientific basis, is very ambiguous, and has raised inconclusive arguments (Silveira *et al.*, 2000; Baltas *et al.*, 2007; Sahu *et al.*, 2007; Shi *et al.*, 2009; Wang *et al.* 2010a). To overcome these two limitations, Wang *et al.* (2008, 2012) proposed two improved models, namely MoCN and MoRE. Overall, the MoRE model performs better than the MoCN model, whereas both models perform better than the SCS-CN method (Wang *et al.*, 2012).

The MoCN model computes direct runoff using Eq. (3.48), with S and I_a computed as:

$$S = S_I \left[1 - \left(\frac{\theta - \theta_\psi}{\theta_{sat} - \theta_\psi} \right)^b \right] \tag{3.53}$$

(S: potential maximum retention after runoff begins; S_I: potential maximum retention for ARC I; θ: soil moisture; θ_ψ: wilting point; θ_{sat}: saturated soil moisture; b: exponent constant)

$$b = \frac{\ln\left(1 - \frac{S_{III}}{S_I}\right)}{\ln\left(\frac{\theta_{fc}}{\theta_{sat}}\right)} \tag{3.54}$$

(b: exponent constant; S_{III}: potential maximum retention for ARC III; S_I: potential maximum retention for ARC I; θ_{fc}: field capacity; θ_{sat}: saturated soil moisture)

$$I_a = \lambda_1 \cdot S \cdot \left(\frac{P}{P+S}\right)^{\alpha} \tag{3.55}$$

(I_a: initial abstraction; λ_1: calibration constant (0.09 ~ 11.36); S: potential maximum retention after runoff begins; P: rainfall; α: calibration constant (0 ~ 2.82))

The MoRE model computes direct runoff as:

$$Q = \begin{cases} P\left(10^{\left[-0.143\left[\log_{10}\left(\frac{S}{P}\right)\right]^2 - 0.598\log_{10}\left(\frac{S}{P}\right) - 0.296\right]}\right) & \text{when } \frac{S}{P} \geq 1.0 \\ P\left(10^{\left[-0.296\exp^{\left[1.837\cdot\log_{10}\left(\frac{S}{P}\right)\right]}\right]}\right) & \text{when } 0 < \frac{S}{P} < 1.0 \\ P & \text{when } \frac{S}{P} = 0 \end{cases} \tag{3.56}$$

(Q: direct runoff; P: rainfall; S: potential maximum retention after runoff begins (Eqs. 3.53 and 3.54))

In Eq. (3.53), θ varies between θ_ψ and θ_{fc}. If measured values are not available, θ can be estimated as (Wang *et al.*, 2012):

$$\theta = \begin{cases} \theta_\psi & \text{for } P_5 < 12.5 \text{ mm (dormant season) or } P_5 < 35.5 \text{ mm (growing season)} \\ \theta_\psi + \frac{\theta_{fc} - \theta_\psi}{15.5} \cdot (P_5 - 12.5) & \text{for } 12.5 \text{ mm} \leq P_5 < 28.0 \text{ mm (dormant season)} \\ \theta_\psi + \frac{\theta_{fc} - \theta_\psi}{18.0} \cdot (P_5 - 35.5) & \text{for } 35.5 \text{ mm} \leq P_5 < 53.5 \text{ mm (growing season)} \\ \theta_{fc} & \text{for } P_5 \geq 28.0 \text{ mm (dormant season) or } P_5 \geq 53.5 \text{ mm (growing season)} \end{cases} \tag{3.57}$$

(θ: soil moisture; θ_ψ: wilting point; P_5: antecedent 5-d rainfall (mm))

Example 3.15 The soil in a watershed has the properties: $\theta_\psi = 0.08$, $\theta_{fc} = 0.35$, and $\theta_{sat} = 0.46$. The watershed has a composite ARC II curve number of $CN_{II} = 78$. A storm with a rainfall of 35 mm occurs during the growing season. During the five days before the storm begins, the total rainfall is 45 mm. Estimate the direct runoff from this storm using the:

(1) MoCN model
(2) MoRE model
(3) SCS-CN method.

Solution

$\theta_\psi = 0.08$, $\theta_{fc} = 0.35$, $\theta_{sat} = 0.46$, $CN_{II} = 78$, $P = 35$ mm, $P_5 = 45$ mm

Eq. (3.51): $CN_I = 78 - \dfrac{20*(100-78)}{100-78+e^{[2.533-0.0636*(100-78)]}} = 60$ (rounded to a whole number!).

Eq. (3.52): $CN_{III} = 78*e^{[0.00673*(100-78)]} = 90$ (rounded to a whole number!).

SI units, Eq. (3.49): $S_I = 25.4*\left(\dfrac{1000}{60} - 10\right)\text{ mm} = 169.3\text{ mm}$

$$S_{III} = 25.4*\left(\frac{1000}{90} - 10\right)\text{ mm} = 28.2\text{ mm}.$$

Eq. (3.54): $b = \dfrac{\ln\left(1 - \dfrac{28.2\text{ mm}}{169.3\text{ mm}}\right)}{\ln\left(\dfrac{0.35}{0.46}\right)} = 0.667.$

$P_5 = 45$ mm in growing season, Eq. (3.57): $\theta = 0.08 + \dfrac{0.35-0.08}{18.0}*(45-35.5) = 0.22.$

Eq. (3.53): $S = (169.3\text{ mm})*\left[1 - \left(\dfrac{0.22-0.08}{0.46-0.08}\right)^{0.667}\right] = 82.3\text{ mm}.$

(1) Assume $\lambda_1 = 0.5$ and $\alpha = 2.5$

Eq. (3.55): $I_a = 0.5*(82.3\text{ mm})*\left(\dfrac{35\text{ mm}}{35\text{ mm} + 82.3\text{ mm}}\right)^{2.5} = 2.0\text{ mm}.$

$P > I_a$, Eq. (3.48): $Q = \dfrac{(35\text{ mm} - 2.0\text{ mm})^2}{(35\text{ mm} - 2.0\text{ mm}) + 82.3\text{ mm}} = 9.4\text{ mm}.$

(2) $\dfrac{S}{P} = \dfrac{82.3\text{ mm}}{35\text{ mm}} = 2.35 > 1.0$, using Eq. (3.56), one can get:

$$Q = (35\text{ mm})*\left(10^{\left[-0.143[\log_{10}(2.35)]^2 - 0.598\log_{10}(2.35) - 0.296\right]}\right) = 10.2\text{ mm}.$$

(3) Based on the definition of ARC, because $\theta = 0.22$ is between $\theta_\psi = 0.08$ and $\theta_{fc} = 0.35$, the watershed is judged to have ARC II.

SI units, Eq. (3.49): $S_{II} = 25.4*\left(\dfrac{1000}{78} - 10\right)\text{ mm} = 71.6\text{ mm}.$

Eq. (3.50): $I_a = 0.2*71.6\ \text{mm} = 14.3\ \text{mm}.$

Eq. (3.48): $Q = \dfrac{(35\ \text{mm} - 14.3\ \text{mm})^2}{(35\ \text{mm} - 14.3\ \text{mm}) + 71.6\ \text{mm}} = 4.6\ \text{mm}.$

Summary: The MoCN model is sensitive to λ_1 and α because they closely influence the computed value of I_a. In practice, for a watershed of interest, these two parameters need to be estimated using observed data on rainfall and runoff by model calibration. In contrast, the MoRE model does not require I_a be determined and thus it is less subjective than the MoCN model and the SCS-CN method. As mentioned above, the SCS-CN method gives a same value of direct runoff for all values of θ between θ_ψ and θ_{fc}. That is, for this example, the runoff for $\theta = 0.12$ will be predicted to be the same as the runoff for $\theta = 0.32$, which is not true in reality. Thus, the SCS-CN method may be good for long-term average conditions, although it tends to overestimate the runoff when θ approaches θ_ψ but underestimate the runoff when θ approaches θ_{fc}.

3.2.5 Streamflow

Streamflow is the flow of water in streams, rivers, and artificial channels (USGS, 2016). It consists of surface runoff from overland, intermediate flow from shallow soils, and baseflow from shallow and/or deep aquifers (Figure 3.6). Streamflow is usually studied using a hydrograph, a chart showing flow rate (or discharge) versus time at a river station (e.g., Figure 3.5). In practice, the flow hydrograph at an upstream station needs to be routed through hydraulic structures (e.g., channel, reservoir, and pond) and/or natural channels to derive the corresponding flow hydrograph at a downstream station. Such computations are termed as flow or flood routing, and involve three interrelated physical processes, namely storing, translation, and attenuation. As a river is followed downstream, some water is either permanently or temporarily stored in channels, reservoirs, and ponds, and the corresponding flow hydrographs obtained at progressively downstream locations change shape. From upstream to downstream, the hydrograph shape is broadened and flattened into a new "fatter" hydrograph shape with a reduced peak but a longer time base (Figure 3.19). The processes responsible for these changes in shape are governed by Saint–Venant equations (Roberson *et al.*, 1998; Bedient *et al.*, 2012), which can be numerically solved using either a finite-difference or finite-element approach (Strelkoff, 1970; Akbari and Firoozi, 2010). Depending on how the Saint–Venant equations are simplified, various routing methods (Table 3.16) have been proposed and applied (Miller, 1984). In the following context, three selected methods, namely Modified Puls, Kinematic Wave, and Muskingum–Cunge, are introduced from an application perspective.

The Modified Puls method can be expressed as:

$$\left(\frac{\overset{\text{storage volume at time t}}{S_t}}{\underset{\text{computational time step}}{\Delta t}} + \frac{\overset{\text{outflow at time t}}{Q_{O,t}}}{2}\right) = \frac{\overset{\text{inflow at time t-1}}{Q_{I,t-1}} + \overset{\text{inflow at time t}}{Q_{I,t}}}{2} + \left(\frac{\overset{\text{storage volume at time t-1}}{S_{t-1}}}{\Delta t} - \frac{\overset{\text{outflow at time t-1}}{Q_{O,t-1}}}{2}\right) \tag{3.58}$$

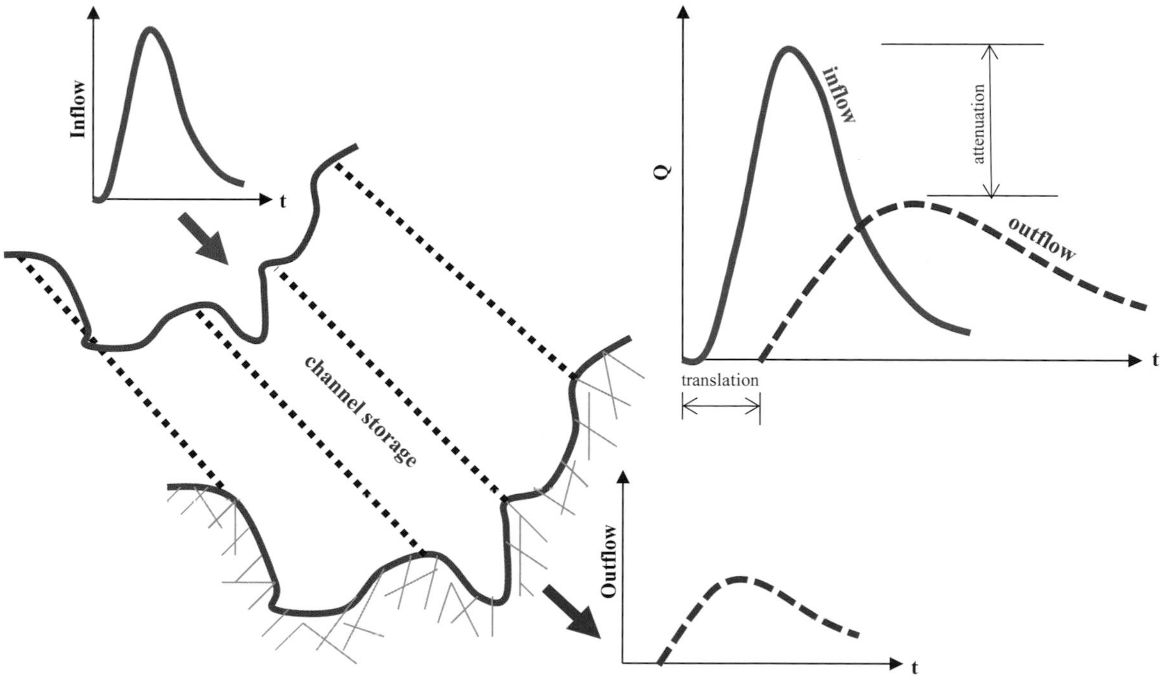

Figure 3.19 **Illustration of flow routing through a channel.**

Table 3.16 **Summary of commonly used flow routing methods.**

Category	Method name	Most applicable for	Physical process
Hydrologic routing	Lag	Small pond/reservoir	Translation
	Modified Puls	Pond/reservoir	Storing, translation, and attenuation
	Muskingum	Channel	Storing, translation, and attenuation
	Straddle Stagger (average-lag)	Channel	Translation
Hydraulic routing	Kinematic Wave	Overland and Channel	Storing, translation, and attenuation
	Muskingum–Cunge	Channel	Storing, translation, and attenuation
	Dynamic Wave	Channel	Storing, translation, and attenuation

Example 3.16 The storage–outflow relation of a small detention pond and the inflow hydrograph from a storm are given in Table 3.17. If the beginning storage of the pond is 2500 m^3, determine the outflow hydrograph using the Modified Puls method.

Solution

Use a computational time step $\Delta t = 1$ hr $= 3600$ s, which is same as the tabulation time interval of the inflow hydrograph to ease the computations.

Table 3.17 Storage vs. outflow, and the inflow hydrograph.

S (m^3)	Q_o ($m^3 s^{-1}$)	t (hr)	Q_I (m^3 s^{-1})
2500	0	0	0
5000	0.5	1	5
10,000	3.2	2	20
15,000	8.5	3	18
20,000	12.6	4	10
65,000	18.8	5	5
		6	0

First, develop $[S/\Delta t + Q_o/2]$ vs. S and Q_o from Table 3.17, leading to the Characteristic Curve of the Pond shown in the left panel of Figure 3.20. As a sample calculation, for $S = 2500$ m^3, $Q_o = 0$ m^3 s^{-1}, $[S/\Delta t + Q_o/2] = [(2500\ \text{m}^3)/(3600\ \text{s}) + (0\ \text{m}^3\ \text{s}^{-1})/2] = 0.6944\ \text{m}^3\ \text{s}^{-1}$. A similar calculation can be done for each of the other five pairs of S and Q_o.

Second, do the flow routing computations using Eq. (3.58), which leads to the Flow Routing using the Modified Puls method and the outflow hydrograph shown in the right panel of Figure 3.20.

Δt =	1	hr	3600	s
S_0 =	2500	m^3		

Characteristic Curve of the Pond

S (m^3)	Q_o (m^3 s^{-1})	$[S/\Delta t + Q_o/2]$ (m^3 s^{-1})
2500	0	0.6944
5000	0.5	1.6389
10000	3.2	4.3778
15000	8.5	8.4167
20000	12.6	11.8556
65000	18.8	27.4556

Flow Routing using the Modified Puls Method

t (hr)	Q_I (m^3 s^{-1})	$[Q_{I,t-1} + Q_{I,t}]/2$ (m^3 s^{-1})	$[S_{t-1}/\Delta t - Q_{o,t-1}/2]$ (m^3 s^{-1})	$[S_t/\Delta t + Q_{o,t}/2]$ (m^3 s^{-1})	$Q_{o,t}$ (m^3 s^{-1})	S_t (m^3)
0	0		0.6944		0.0	2500.0000
1	5	2.5	1.1777	3.1944	2.0	7839.6437
2	20	12.5	0.3656	13.6777	13.3	25256.0577
3	18	19	3.7732	19.3656	15.6	41663.4615
4	10	14	2.7972	17.7732	15.0	37070.0000
5	5	7.5	-0.4238	10.2972	10.7	17734.1592
6	0	2.5	1.1606	2.0762	0.9	5798.3132

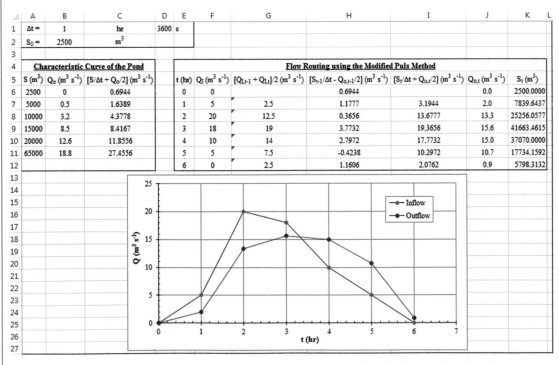

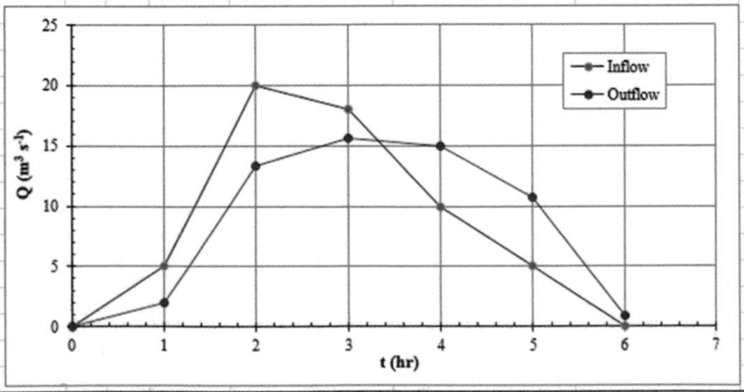

Figure 3.20 **Flow routing computations and associated plots using the Modified Puls method for Example 3.16.**

For t = 0 hr: Given $S_0 = 2500\,m^3 \rightarrow$ linear interpolation of the characteristic curve: $Q_{o,0} = 0\,m^3\,s^{-1} \rightarrow S_0/\Delta t - Q_{o,0}/2 = (2500\,m^3)/(3600\,s) + (0\,m^3\,s^{-1})/2 = 0.6944\,m^3\,s^{-1}$.

For t = 1 hr: $[Q_{I,0} + Q_{I,1}]/2 = [(0\,m^3\,s^{-1}) + (5\,m^3\,s^{-1})]/2 = 2.5\,m^3 s^{-1} \rightarrow$ Eq. (3.58): $S_1/\Delta t + Q_{o,1}/2 = (2.5\,m^3\,s^{-1}) + (0.6944\,m^3\,s^{-1}) = 3.1944\,m^3\,s^{-1} \rightarrow$ linear interpolation of the characteristic curve: $Q_{o,1} = (0.5\,m^3\,s^{-1}) + [(3.2\,m^3\,s^{-1}) - (0.5\,m^3\,s^{-1})]/[(4.3778\,m^3\,s^{-1}) - (1.6389\,m^3\,s^{-1})] * [(3.1944\,m^3\,s^{-1}) - (1.6389\,m^3\,s^{-1})] = 2.0\,m^3\,s^{-1} \rightarrow$ linear interpolation of the characteristic curve: $S_1 = (5000\,m^3) + [(10000\,m^3) - (5000\,m^3)]/[(4.3778\,m^3\,s^{-1}) - (1.6389\,m^3 s^{-1})] * [(3.1944\,m^3\,s^{-1}) - (1.6389\,m^3\,s^{-1})] = 7839.6437\,m^3 \rightarrow S_1/\Delta t - Q_{o,1}/2 = (7839.6437\,m^3)/(3600\,s) - (2.0\,m^3\,s^{-1})/2 = 1.1777\,m^3\,s^{-1}$.

For t = 2 hr: $[Q_{I,1} + Q_{I,2}]/2 = [(5\,m^3\,s^{-1}) + (20\,m^3\,s^{-1})]/2 = 12.5\,m^3\,s^{-1} \rightarrow$ Eq. (3.58): $S_2/\Delta t + Q_{o,2}/2 = (12.5\,m^3\,s^{-1}) + (1.1777\,m^3\,s^{-1}) = 13.6777\,m^3\,s^{-1} \rightarrow$ linear interpolation of the characteristic curve: $Q_{o,2} = (12.6\,m^3\,s^{-1}) + [(18.8\,m^3\,s^{-1}) - (12.6\,m^3\,s^{-1})]/[(27.4556\,m^3\,s^{-1}) - (11.8556\,m^3\,s^{-1})] * [(13.6777\,m^3\,s^{-1}) - (11.8556\,m^3\,s^{-1})] = 13.3\,m^3\,s^{-1} \rightarrow$ linear interpolation of the characteristic curve: $S_2 = (20000\,m^3) + [(65000\,m^3) - (20000\,m^3)]/[(27.4556\,m^3\,s^{-1}) - (11.8556\,m^3\,s^{-1})] * [(13.6777\,m^3\,s^{-1}) - (11.8556\,m^3\,s^{-1})] = 25256.0577\,m^3 \rightarrow S_2/\Delta t - Q_{o,2}/2 = (25256.0577\,m^3)/(3600\,s) - (13.3\,m^3\,s^{-1})/2 = 0.3656\,m^3\,s^{-1}$.

For t = 3, 4, 5, and 6 hr: similar computations can be done to get $Q_{o,t}$ and S_t.

The Kinematic Wave method assumes uniform flow with a Froude number, $F_r < 2$. It assumes that the flow rate is a power function of flow area, which in turn is a function of time. Usually, when consistent units are used, the power function can be expressed as:

$$Q = \alpha A^m \quad (3.59)$$

flow rate → Q; coefficient related to geometry, roughness, and slope of flow path → α; flow area → A; exponent related to geometry, roughness, and slope of flow path → m

Substituting Eq. (3.59) into the continuity equation (Eq. (2.16)) and rearranging, leads to:

$$\frac{\partial A}{\partial t} + \alpha m A^{m-1} \frac{\partial A}{\partial x} = q \quad (3.60)$$

flow area → A; time → t; coefficient related to geometry, roughness, and slope of flow path → α; exponent related to geometry, roughness, and slope of flow path → m; travel distance → x; net flux per unit length of flow path → q

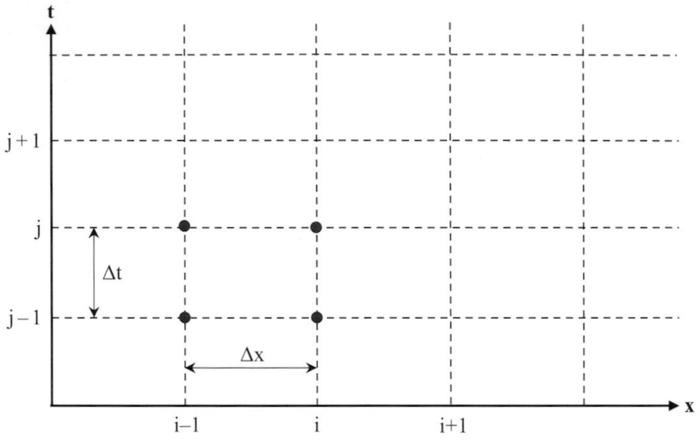

Figure 3.21 **The discretization scheme of flow path (x) and time domain (t).** The information at the blue dots are used to determine the information at the red dot.

For the discretization scheme shown in Figure 3.21, the flow path is subdivided into smaller segments with a distance interval of Δx, and the computational time step is Δt. Let i be the node number of the flow path and j be the node number of the time domain. Writing Eq. (3.60) into a finite-difference format and rearranging, one produces:

flow area at distance i and time j; flow area at distance i and time j−1; net flux per unit length of flow path at distance i and time j−1; net flux per unit length of flow path at distance i and time j; computational time step; coefficient related to geometry, roughness, and slope of flow path; exponent related to geometry, roughness, and slope of flow path; distance interval; flow area at distance i−1 and time j−1

$$A_i^j = A_i^{j-1} + \left(\frac{q_i^{j-1} + q_i^j}{2}\right)(\Delta t) - \alpha m\left(\frac{\Delta t}{\Delta x}\right)\left(\frac{A_i^{j-1} + A_{i-1}^{j-1}}{2}\right)^{m-1}\left(A_i^{j-1} - A_{i-1}^{j-1}\right) \tag{3.61}$$

Substituting Eq. (3.61) into Eq. (3.59), the expression for the flow rate is derived as:

flow rate at distance i and time j; coefficient related to geometry, roughness, and slope of flow path; exponent related to geometry, roughness, and slope of flow path; flow area at distance i and time j

$$Q_i^j = \alpha\left(A_i^j\right)^m \tag{3.62}$$

The numerical stability condition of Eqs. (3.61) and (3.62) is expressed as:

coefficient related to geometry,
roughness, and slope of flow path

distance interval

computational $\dfrac{\Delta x}{\Delta t} < \alpha m \left(A_i^{j-1} \right)^{m-1}$ flow area at distance
time step i and time j

$\qquad\qquad\qquad\qquad\qquad\qquad\qquad\qquad\qquad$ (3.63)

exponent related to geometry,
roughness, and slope of flow path

In Eqs. (3.59) to (3.63), α and m can be determined in Microsoft Excel by fitting a power trendline to pairs of measured or computed flow rate versus flow area. These equations can be applied both for channel and overland flow routing. For an overland flow path, it can be treated as a wide-shallow channel. The overland flow routing can be implemented by simply replacing flow area A by water depth y as well as interpreting Q as flow rate per unit width, and q as net flux per unit area, of the overland.

Example 3.17 The measured values of flow area and flow rate for a 1200-ft-long channel with a uniform cross section are shown in the left panel of Figure 3.22. A storm event generates an inflow hydrograph shown in the right panel (E8:F24) of the figure. Determine the flow hydrograph at the lower end of the channel using the Kinematic Wave method with $\Delta t = 10$ min and $\Delta x = 400$ ft. Assume the beginning flow area is 10 ft^2 and there is no lateral inflow along the entire channel.

Solution
Channel node i = 0, 1, 2, and 3; time node j = 0, 1, 2, ..., 16

$$\Delta t = 10 \min = 600 \sec, \Delta x = 400 \text{ ft}, \Delta x/\Delta t = 0.6667 \text{ ft sec}^{-1}, A_i^0 = 10 \text{ ft}^2.$$

First, determine α and m by plotting the measured flow rate (Q) versus flow area (A) in an Excel spreadsheet and then adding a power trendline. This results in $\alpha = 0.000312$ and m = 2.58, as shown in Figure 3.22.

Second, do the flow routing sub-reach by sub-reach, leading to the results shown in the right panel of Figure 3.22. The sample computations are as follows.

For the upper end of the channel (i = 0): Given the inflow hydrograph $Q_0^j \rightarrow$ linear interpolation of the measured A ~ Q curve to get the corresponding flow areas A_0^j. The values for Q_0^j are in cells F8:F24, and the values for A_0^j are in cells G8:G24. For instance, at time t = 20 min (i.e., j = 2), $Q_0^2 = 20$ cfs$\rightarrow A_0^2 = \left(10 \text{ ft}^2\right) + \left[\left(100 \text{ ft}^2\right) - \left(10 \text{ ft}^2\right)\right] / [(78 \text{ cfs}) - (0.1 \text{ cfs})] * [(20 \text{ cfs}) - (0.1 \text{ cfs})] = 33.0 \text{ ft}^2$. Similar interpolations can be done for other times.

α =	3.12E-04	Δt =	10	min	600	sec
m =	2.58	Δx =	400	ft		
		Δx/Δt =	0.6667	ft sec⁻¹		
		L =	1200	ft		

Measured Values of A (ft²)	Q (cfs)
10	0.1
100	78
200	225
300	600
380	1500
450	2000

		i = 0		i = 1			i = 2			i = 3		
j	t (min)	Q_0^j (cfs)	A_0^j (ft2)	q_1^j (ft² sec⁻¹)	A_1^j (ft²)	Q_1^j (cfs)	q_2^j (ft² sec⁻¹)	A_2^j (ft²)	Q_2^j (cfs)	q_3^j (ft² sec⁻¹)	A_3^j (ft²)	Q_3^j (cfs)
0	0	0	10.0	0.00	10.0	0.1	0	10.0	0.1	0	10.0	0.1
1	10	10	21.4	0.03	19.0	0.6	0	10.0	0.1	0	10.0	0.1
2	20	20	33.0	0.05	43.3	5.2	0	10.7	0.1	0	10.0	0.1
3	30	15	27.2	0.04	66.4	15.7	0	17.9	0.5	0	10.0	0.1
4	40	10	21.4	0.03	66.8	15.9	0	39.5	4.1	0	10.6	0.1
5	50	5	15.7	0.01	57.1	10.6	0	57.1	10.6	0	16.3	0.4
6	60	3	13.4	0.01	48.5	7.0	0	57.1	10.6	0	30.9	2.2
7	70	2	12.2	0.01	44.9	5.7	0	51.6	8.2	0	43.4	5.2
8	80	1	11.0	0.00	40.0	4.2	0	47.9	6.8	0	47.8	6.7
9	90	0	10.0	0.00	34.2	2.8	0	44.1	5.5	0	47.9	6.8
10	100	0	10.0	0.00	30.3	2.1	0	40.2	4.3	0	46.0	6.1
11	110	0	10.0	0.00	27.5	1.6	0	36.9	3.4	0	43.3	5.2
12	120	0	10.0	0.00	25.3	1.3	0	34.2	2.8	0	40.7	4.4
13	130	0	10.0	0.00	23.6	1.1	0	31.9	2.4	0	38.3	3.8
14	140	0	10.0	0.00	22.2	0.9	0	30.0	2	0	36.2	3.3
15	150	0	10.0	0.00	21.0	0.8	0	28.4	1.8	0	34.3	2.9
16	160	0	10.0	0.00	20.0	0.7	0	27.0	1.5	0	32.7	2.5

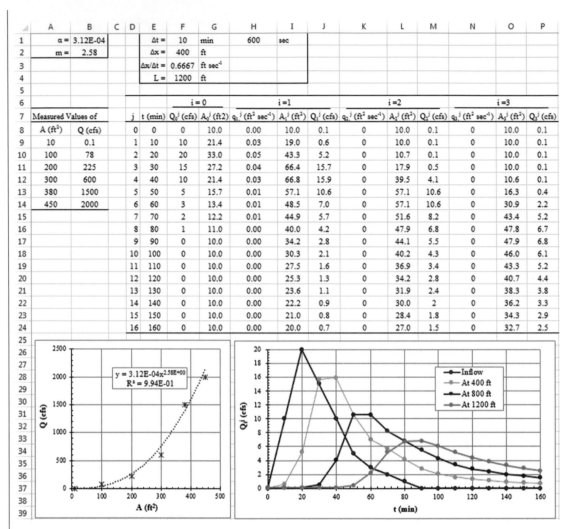

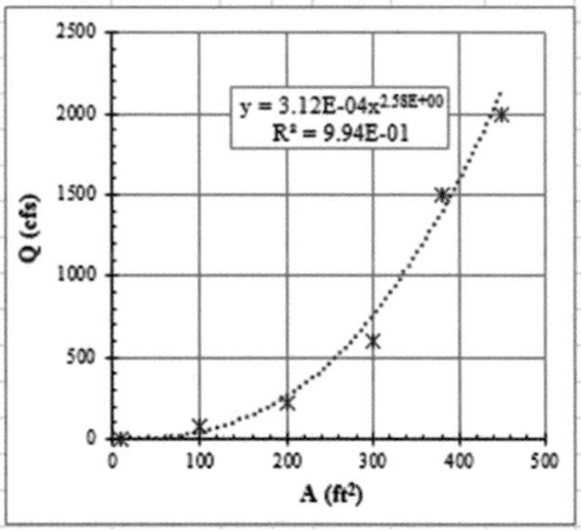

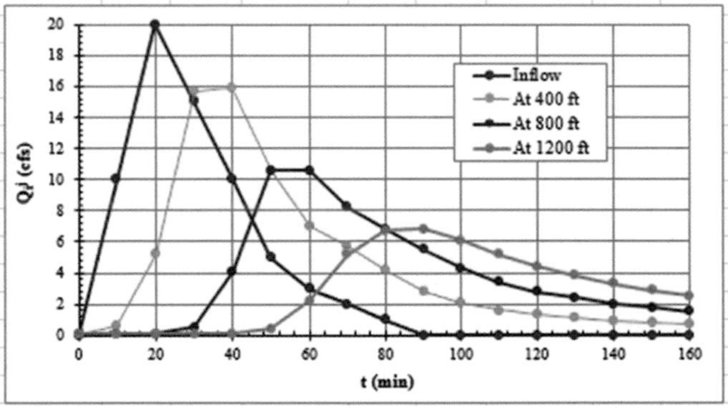

Figure 3.22 **Spreadsheet and associated plots for channel routing using the Kinematic Wave method for Example 3.17.**

At 400 ft (i = 1): The flow rate per unit length of the channel $q_1^j = Q_0^j/\Delta x \rightarrow$ compute A_1^j using Eq. (3.61) → compute Q_1^j using Eq. (3.62) → check stability Eq. (3.63). The values for q_1^j are in cells H8:H24, for A_1^j in cells I8:I24, and for Q_1^j in cells J8:J24. For instance, at time t = 20 min (i.e., j = 2), q_1^2 = (20 cfs) / (400 ft) = 0.05 ft² sec⁻¹.

$$A_1^2 = 19.0\,\text{ft}^2 + \left(\frac{0.03\,\text{cfs}+0.05\text{cfs}}{2}\right)*600\,\text{sec} - 0.000312*2.58*\left(\frac{1}{0.6667\,\text{ft sec}^{-1}}\right)\left(\frac{19.0\,\text{ft}^2+21.4\,\text{ft}^2}{2}\right)^{2.58-1} * \left(19.0\,\text{ft}^2 - 21.4\,\text{ft}^2\right) = 43.3\,\text{ft}^2$$

$Q_1^2 = 0.000312*\left(43.3\ \text{ft}^2\right)^{2.58} = 5.2\,\text{cfs}$. Similar computations can be done for other times.

At 800 ft ($i = 2$): Zero lateral inflows $\rightarrow q_2^j = 0$. The remaining computations are same as those at 400 ft ($i = 1$). The values for q_2^j are in cells K8:K24, for A_2^j in cells L8:L24, and for Q_2^j in cells M8:M24.

At 1200 ft ($i = 3$): Zero lateral inflows $\rightarrow q_3^j = 0$. The remaining computations are same as those at 400 ft ($i = 1$). The values for q_3^j are in cells N8:N24, for A_3^j in cells O8:O24, and for Q_3^j in cells P8:P24.

Example 3.18 A concrete-lined parking lot has a downgradient length of 300 ft. The measured values of runoff depth and unit-width flow rate for the parking lot are shown in the left panel of Figure 3.23. A 30-min storm event has rainfall intensities shown in the right panel (E8:F24) of the figure. Determine the flow hydrograph at the lower end of the parking lot using the Kinematic Wave method with $\Delta t = 5$ min and $\Delta x = 100$ ft. Assume that the parking lot is dry at the beginning of the storm and there is no lateral inflow into the parking lot.

Solution

Parking lot node $i = 0, 1, 2,$ and 3; time node $j = 0, 1, 2, \ldots, 16$.

$$\Delta t = 5\,\text{min} = 300\,\text{sec},\ \Delta x = 100\,\text{ft},\ \Delta x/\Delta t = 0.3333\,\text{ft sec}^{-1}$$

There is no ponding at the upper end of the parking lot $\rightarrow y_0^j = 0$.

The parking lot is dry at the beginning $\rightarrow y_i^0 = 0\,\text{ft}, Q_i^0 = 0\,\text{ft}^2\,\text{sec}^{-1}$.

The parking lot is concrete-lined, so the infiltration is zero. At time j, the net flux per unit area of the parking lot is equal to the rainfall intensity i_r^j.

First, determine α and m by plotting the measured unit-width flow rate (q) versus runoff depth (y) in an Excel spreadsheet and then adding a power trendline. This results in $\alpha = 0.267$ and $m = 1.40$, as shown in Figure 3.23.

Second, do the flow routing segment by segment, leading to the results shown in the right panel of Figure 3.23. The sample computations are as follows.

For the upper end of the parking lot ($i = 0$): Given runoff depth $y_0^j = 0$ ft $\rightarrow$ unit-width flow rate $Q_0^j = 0$ cfs.

At 100 ft ($i = 1$): The unit-area net flux $q_1^j = i_r^j \rightarrow$ compute y_1^j using Eq. (3.61) $\rightarrow$ compute Q_1^j using Eq. (3.62) $\rightarrow$ check stability Eq. (3.63). The values for q_1^j are in cells H8:H24, for y_1^j in cells I8:I24, and for Q_1^j in cells J8:J24. For instance, at time $t = 10$ min (i.e., $j = 2$), $q_1^2 = 2\,\text{in hr}^{-1} = 4.63 \times 10^{-5}\,\text{ft sec}^{-1}$,

	A	B	C	D	E	F	G	H	I	J	K	L	M	N	O	P
1	α =	2.67E-01			Δt =	5	min	300	sec							
2	m =	1.40			Δx =	100	ft									
3					Δx/Δt =	0.3333	ft sec^{-1}									
4					L =	300	ft									
5																
6							i = 0		i = 1			i = 2			i = 3	
7	Measured Values of			j	t (min)	i_r^j (in hr^{-1})	y_0^j (ft)	q_1^j (ft sec^{-1})	y_1^j (ft)	Q_1^j (ft^2 sec^{-1})	q_2^j (ft sec^{-1})	y_2^j (ft)	Q_2^j (ft^2 sec^{-1})	q_3^j (ft sec^{-1})	y_3^j (ft)	Q_3^j (ft^2 sec^{-1})
8	y (ft)	Q (ft^2 sec^{-1})		0	0	0	0	0.00E+00	0	0	0.00E+00	0	0	0.00E+00	0	0
9	0.1	0.01		1	5	1	0	2.31E-05	3.47E-03	9.60E-05	2.31E-05	3.47E-03	9.60E-05	2.31E-05	3.47E-03	9.60E-05
10	0.2	0.03		2	10	2	0	4.63E-05	1.36E-02	6.49E-04	4.63E-05	1.39E-02	6.69E-04	4.63E-05	1.39E-02	6.69E-04
11	0.3	0.05		3	15	3	0	6.94E-05	2.89E-02	1.87E-03	6.94E-05	3.12E-02	2.08E-03	6.94E-05	3.12E-02	2.08E-03
12	0.4	0.08		4	20	3	0	6.94E-05	4.37E-02	3.34E-03	6.94E-05	5.14E-02	4.18E-03	6.94E-05	5.20E-02	4.26E-03
13	0.5	0.1		5	25	1	0	2.31E-05	4.70E-02	3.69E-03	2.31E-05	6.27E-02	5.53E-03	2.31E-05	6.57E-02	5.90E-03
14	0.6	0.12		6	30	0.5	0	1.16E-05	4.04E-02	2.99E-03	1.16E-05	6.24E-02	5.49E-03	1.16E-05	6.98E-02	6.42E-03
15				7	35	0	0	0	3.27E-02	2.22E-03	0	5.66E-02	4.79E-03	0	6.87E-02	6.29E-03
16				8	40	0	0	0	2.56E-02	1.58E-03	0	4.89E-02	3.90E-03	0	6.42E-02	5.72E-03
17				9	45	0	0	0	2.06E-02	1.16E-03	0	4.19E-02	3.14E-03	0	5.88E-02	5.05E-03
18				10	50	0	0	0	1.69E-02	8.80E-04	0	3.59E-02	2.53E-03	0	5.30E-02	4.37E-03
19				11	55	0	0	0	1.41E-02	6.83E-04	0	3.09E-02	2.05E-03	0	4.75E-02	3.75E-03
20				12	60	0	0	0	1.19E-02	5.40E-04	0	2.68E-02	1.68E-03	0	4.24E-02	3.20E-03
21				13	65	0	0	0	1.02E-02	4.34E-04	0	2.33E-02	1.39E-03	0	3.78E-02	2.73E-03
22				14	70	0	0	0	8.80E-03	3.54E-04	0	2.05E-02	1.15E-03	0	3.38E-02	2.33E-03
23				15	75	0	0	0	7.67E-03	2.92E-04	0	1.80E-02	9.67E-04	0	3.03E-02	2.00E-03
24				16	80	0	0	0	6.74E-03	2.44E-04	0	1.60E-02	8.18E-04	0	2.72E-02	1.72E-03

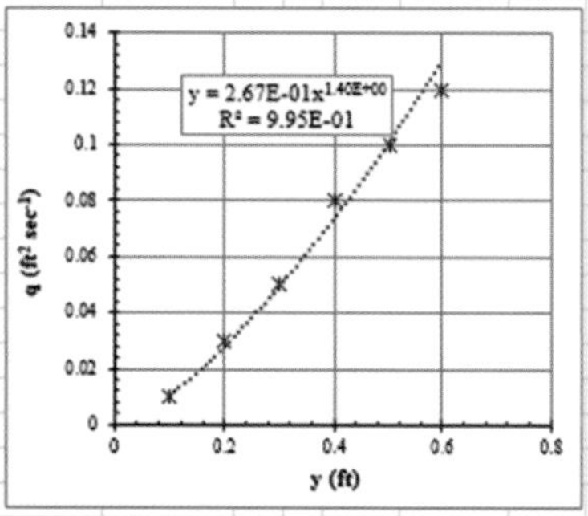

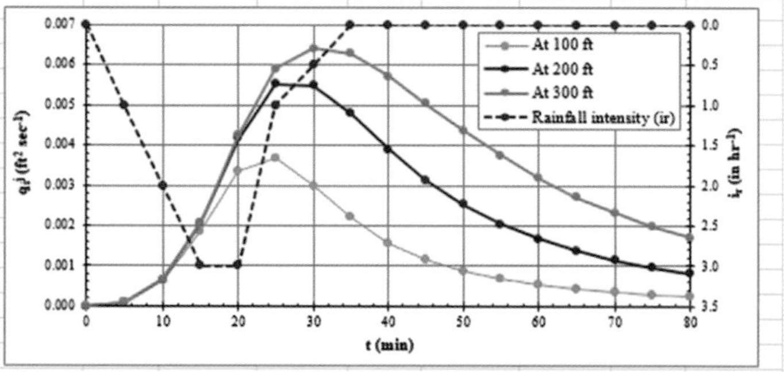

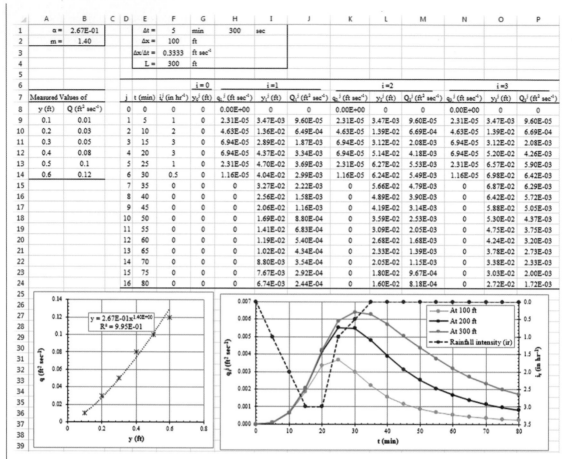

Figure 3.23 **Spreadsheet and associated plots for overland routing using the Kinematic Wave method for Example 3.18.**

$$y_1^2 = 3.47 \times 10^{-3}\,\text{ft} + \left(\frac{2.31 \times 10^{-5}\,\text{ft sec}^{-1} + 4.63 \times 10^{-5}\,\text{ft sec}^{-1}}{2}\right) * 300\,\text{sec} -$$

$$0.267 * 1.40 * \left(\frac{1}{0.3333\,\text{ft sec}^{-1}}\right)\left(\frac{3.47 \times 10^{-3}\,\text{ft} + 0\,\text{ft}}{2}\right)^{1.40-1} * \left(3.47 \times 10^{-3}\,\text{ft} - 0\,\text{ft}\right) = 0.014\,\text{ft}$$

$Q_1^2 = 0.267 * (0.014\,\text{ft})^{1.40} = 0.000649\,\text{ft}^2\ \text{sec}^{-1}$. Similar computations can be done for other times.

At 200 ft ($i = 2$): The unit-area net flux $q_2^j = i_r^j$. The remaining computations are the same as those at 100 ft ($i = 1$). The values for q_2^j are in cells K8:K24, for y_2^j in cells L8:L24, and for Q_2^j in cells M8:M24.

At 300 ft ($i = 3$): The unit-area net flux $q_3^j = i_r^j$. The remaining computations are the same as those at 100 ft ($i = 1$). The values for q_3^j are in cells N8:N24, for y_3^j in cells O8:O24, and for Q_3^j in cells P8:P24.

The Muskingum–Cunge method also uses the discretization scheme shown in Figure 3.21. It computes the unknown flow rate at distance i and time j (the red dots in the figure) as the weighted average of the known flow rates at (i – 1, j – 1), (i – 1, j), and (i, j – 1) (the blue dots in the figure) plus a constant. The weights will not change from time to time and from location to location if the top width and hydraulic depth (i.e., the ratio of flow area to top width) do not vary with water depth (i.e., for steady uniform rectangular channel flow); otherwise, the weights need to be recalculated for each location at each time. The method can be expressed as:

$$Q_i^j = C_0 + C_1 Q_{i-1}^{j-1} + C_2 Q_{i-1}^{j} + C_3 Q_i^{j-1} \tag{3.64}$$

Q_i^j: flow rate at distance i and time j; C_0: constant related to lateral inflow; C_1: first weight; Q_{i-1}^{j-1}: flow rate at distance i – 1 and time j – 1; C_2: second weight; Q_{i-1}^{j}: flow rate at distance i – 1 and time j; C_3: third weight; Q_i^{j-1}: flow rate at distance i and time j – 1

$$C_0 = \frac{2q(\Delta x)(\Delta t)}{\Delta t + 2K_t(1 - K_x)} \tag{3.65}$$

C_0: constant related to lateral inflow; q: average lateral inflow per unit length of channel; Δx: distance interval; Δt: computational time step; K_t: travel-time parameter; K_x: diffusion parameter

$$C_1 = \frac{\Delta t + 2K_t K_x}{\Delta t + 2K_t(1 - K_x)} \tag{3.66}$$

C_1: first weight; Δt: computational time step; K_t: travel-time parameter; K_x: diffusion parameter

$$C_2 = \frac{\Delta t - 2K_t K_x}{\Delta t + 2K_t(1 - K_x)} \tag{3.67}$$

C_2: second weight; Δt: computational time step; K_t: travel-time parameter; K_x: diffusion parameter

$$C_3 = \frac{-\Delta t + 2K_t(1 - K_x)}{\Delta t + 2K_t(1 - K_x)} \tag{3.68}$$

C_3: third weight; Δt: computational time step; K_t: travel-time parameter; K_x: diffusion parameter

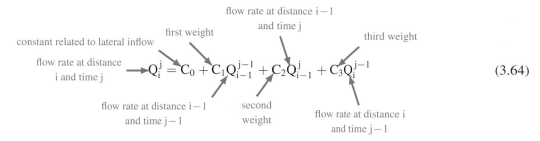

$$K_t = \frac{\Delta x}{c} \tag{3.69}$$

(travel-time parameter K_t; distance interval Δx; wave celerity c)

$$K_x = \frac{1}{2} - \frac{Q_{0,p}}{2BS_0(\Delta x)\,c} \tag{3.70}$$

(diffusion parameter K_x; peak of inflow hydrograph $Q_{0,p}$; channel top width B; channel bed slope S_0; distance interval Δx; wave celerity c)

$$c = \sqrt{g\left(\frac{A}{B}\right)} \tag{3.71}$$

(wave celerity c; gravitational acceleration g; flow area A; channel top width B)

The stability condition of the Muskingum–Cunge method is expressed as:

$$\frac{\Delta x}{\Delta t} \le c \tag{3.72}$$

(distance interval Δx; computational time step Δt; wave celerity c)

Example 3.19 A 900-ft-long diversion channel ($S_0 = 0.0005$) has an almost constant hydraulic depth (i.e., ratio of flow area to top width or A/B) of 3.5 ft. A flow hydrograph at the inlet of the channel is tabulated in cells C9:C25 in Figure 3.24. The average channel top width is $B = 50$ ft. Determine the flow hydrograph at the lower end of the channel using the Muskingum–Cunge method with $\Delta t = 10$ min and $\Delta x = 300$ ft. Assume that the channel is dry at the beginning and receives an average lateral inflow of 0.005 $ft^2 sec^{-1}$.

Solution

Channel node $i = 0, 1, 2$, and 3; time node $j = 0, 1, 2, \ldots, 16$.

$$\Delta t = 10\,\text{min} = 600\,\text{sec},\ \Delta x = 300\,\text{ft},\ \Delta x/\Delta t = 0.5\,\text{ft sec}^{-1},\ q = 0.005\,\text{ft}^2\text{sec}^{-1},\ S_0 = 0.0005.$$

The channel is dry at the beginning $\rightarrow Q_i^{\,0} = 0\,\text{ft}^3\text{sec}^{-1}$.

Cells C9:C25 in Figure 3.24 $\rightarrow Q_{0,p} = 20\,\text{cfs}$.

Constant $B = 50\,\text{ft}$ and $A/B = 3.5 \rightarrow$ constant weights.

Eq. (3.71): $c = \sqrt{\left(32.2\,\text{ft sec}^{-2}\right) * (3.5\,\text{ft})} = 10.616\,\text{ft sec}^{-1}$.

Eq. (3.72): $\Delta x/\Delta t < c \rightarrow$ stable.

	A	B	C	D	E	F	G	H	I	J	K	L	M	N
1	Δt =	10	min	600	sec	B =	50	ft	c =	10.616	ft sec^{-1}	q =	0.005	ft^2 sec^{-1}
2	Δx =	300	ft			A/B =	3.5		$Q_{0,p}$ =	20	cfs	C_0 =	2.833	cfs
3	Δx/Δt =	0.5	ft sec^{-1}			S_0 =	0.0005		K_x =	0.374		C_1 =	0.978	
4	L =	900	ft			g =	32.2	ft sec^{-2}	K_t =	28.259	sec	C_2 =	0.911	
5												C_3 =	-0.889	
6														

	A	B	C	D	E	F
7			i = 0	i =1	i =2	i =3
8	j	t (min)	Q_0^j (cfs)	Q_1^j (cfs)	Q_2^j (cfs)	Q_3^j (cfs)
9	0	0	0	0	0	0
10	1	10	10	11.9	13.7	15.3
11	2	20	20	20.2	20.7	21.5
12	3	30	15	18.1	20.6	22.8
13	4	40	10	10.5	11.8	13.5
14	5	50	5	7.8	9.8	11.3
15	6	60	3	3.5	5.0	6.9
16	7	70	2	4.5	5.9	6.9
17	8	80	1	1.7	3.5	5.6
18	9	90	0.5	2.7	3.9	4.8
19	10	100	0.5	1.4	3.3	5.4
20	11	110	0.5	2.6	3.6	4.5
21	12	120	0.5	1.5	3.5	5.5
22	13	130	0.5	2.5	3.4	4.5
23	14	140	0.5	1.6	3.7	5.5
24	15	150	0.5	2.4	3.3	4.5
25	16	160	0.5	1.7	3.8	5.5

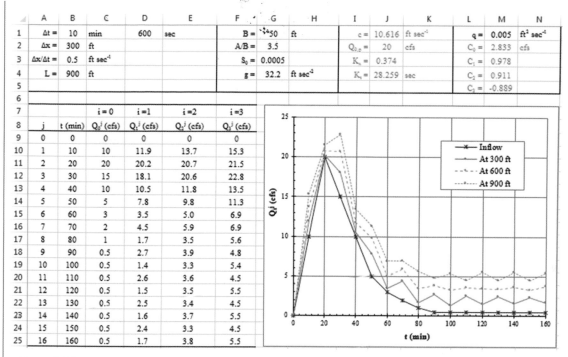

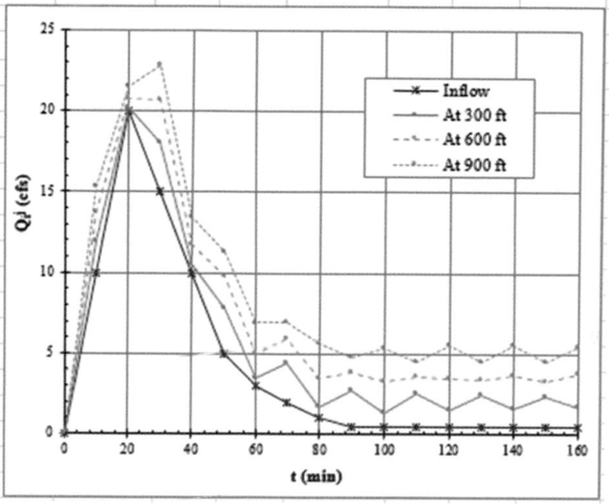

Figure 3.24 **Spreadsheet and associated plots for channel routing using the Muskingum–Cunge method for Example 3.19.**

$$\text{Eq. (3.70): } K_x = \frac{1}{2} - \frac{20\,\text{cfs}}{2*(50\,\text{ft})*0.0005*(300\,\text{ft})*\left(10.616\,\text{ft sec}^{-1}\right)} = 0.374.$$

$$\text{Eq. (3.69): } K_t = \frac{300\,\text{ft}}{10.616\,\text{ft sec}}{}^{-1} = 28.259\,\text{sec}.$$

$$\text{Eq. (3.65): } C_0 = \frac{2*\left(0.005\,\text{ft}^2\,\text{sec}^{-1}\right)*(300\,\text{ft})*(600\,\text{sec})}{(600\,\text{sec}) + 2*(28.259\,\text{sec})*(1-0.374)} = 2.833\,\text{cfs}.$$

$$\text{Eq. (3.66): } C_1 = \frac{(600\,\text{sec}) + 2*(28.259\,\text{sec})*0.374}{(600\,\text{sec}) + 2*(28.259\,\text{sec})*(1-0.374)} = 0.978.$$

$$\text{Eq. (3.67): } C_2 = \frac{(600\,\text{sec}) - 2*(28.259\,\text{sec})*0.374}{(600\,\text{sec}) + 2*(28.259\,\text{sec})*(1-0.374)} = 0.911.$$

$$\text{Eq. (3.68): } C_3 = \frac{(-600\,\text{sec}) + 2*(28.259\,\text{sec})*(1-0.374)}{(600\,\text{sec}) + 2*(28.259\,\text{sec})*(1-0.374)} = -0.889.$$

At 300 ft (i.e., $i = 1$): compute Q_1^j using Eq. (3.64). For instance, for $t = 10$ min (i.e., $j = 1$), $Q_1^{\,1} = 2.833\text{ cfs} + 0.978*(0\text{ cfs}) + 0.911*(10\text{ cfs}) + (-0.889)*(0\text{ cfs}) = 11.9\text{ cfs}$. Similar computations can be done for other times as shown in Figure 3.24.

At 600 ft (i.e., $i = 2$): compute Q_2^j using Eq. (3.64). For instance, for $t = 20$ min (i.e., $j = 2$), $Q_2^{\,2} = 2.833\text{ cfs} + 0.978*(11.9\text{ cfs}) + 0.911*(20.2\text{ cfs}) + (-0.889)*(13.7\text{ cfs}) = 20.7\text{ cfs}$. Similar computations can be done for other times as shown in Figure 3.24.

At 900 ft (i.e., $i = 3$): compute Q_3^j using Eq. (3.64). For instance, for $t = 20$ min (i.e., $j = 2$), $Q_3^{\,2} = 2.833\text{ cfs} + 0.978*(11.9\text{ cfs}) + 0.911*(20.2\text{ cfs}) + (-0.889)*(13.7\text{ cfs}) = 20.7\text{ cfs}$. Similar computations can be done for other times as shown in Figure 3.24.

PROBLEMS

3.1 For the following USGS quadrangle map, identify and highlight the ridges and valleys or gullies.

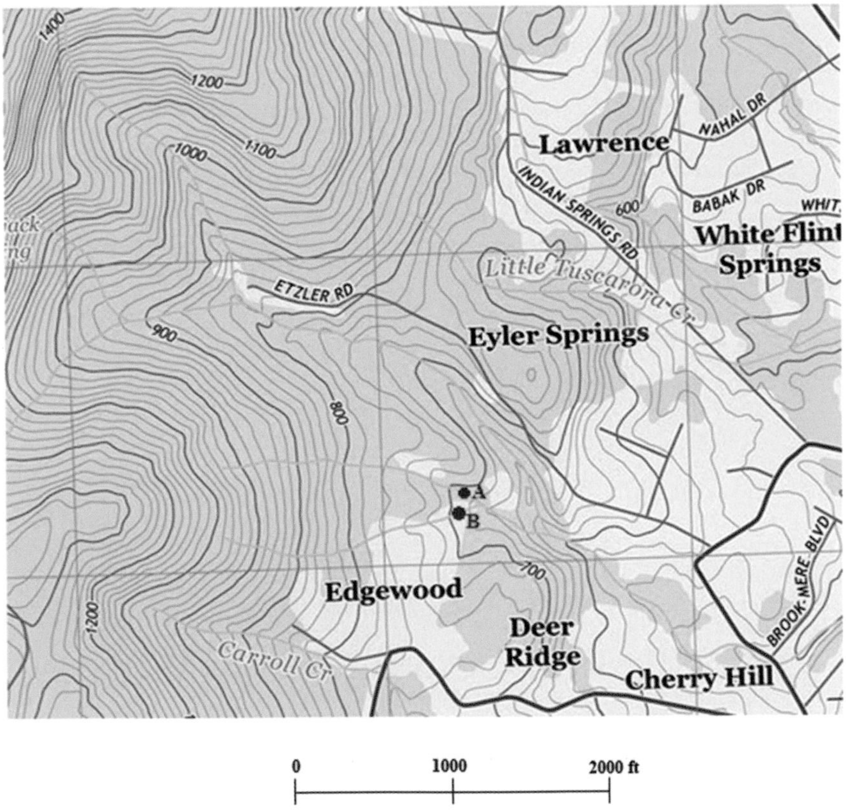

Source: Quadrangle map courtesy of the US Geological Survey.

3.2 For the USGS quadrangle map of Problem 3.1, delineate the watersheds with their outlets at A and B, respectively. Also, determine the geometric characteristics of each of the two watersheds, including drainage area, longest flow path length, width, slope, slope length, and drainage density.

3.3 The following topographic map is superimposed by 30-m grids to develop a digital elevation model (DEM). The elevations on the map have units of meters.

(1) Develop the DEM.

(2) Determine the flow directions using the D8 algorithm manually.

(3) Use subroutine *FlowDir_DEM( )* to check your answer in (2).

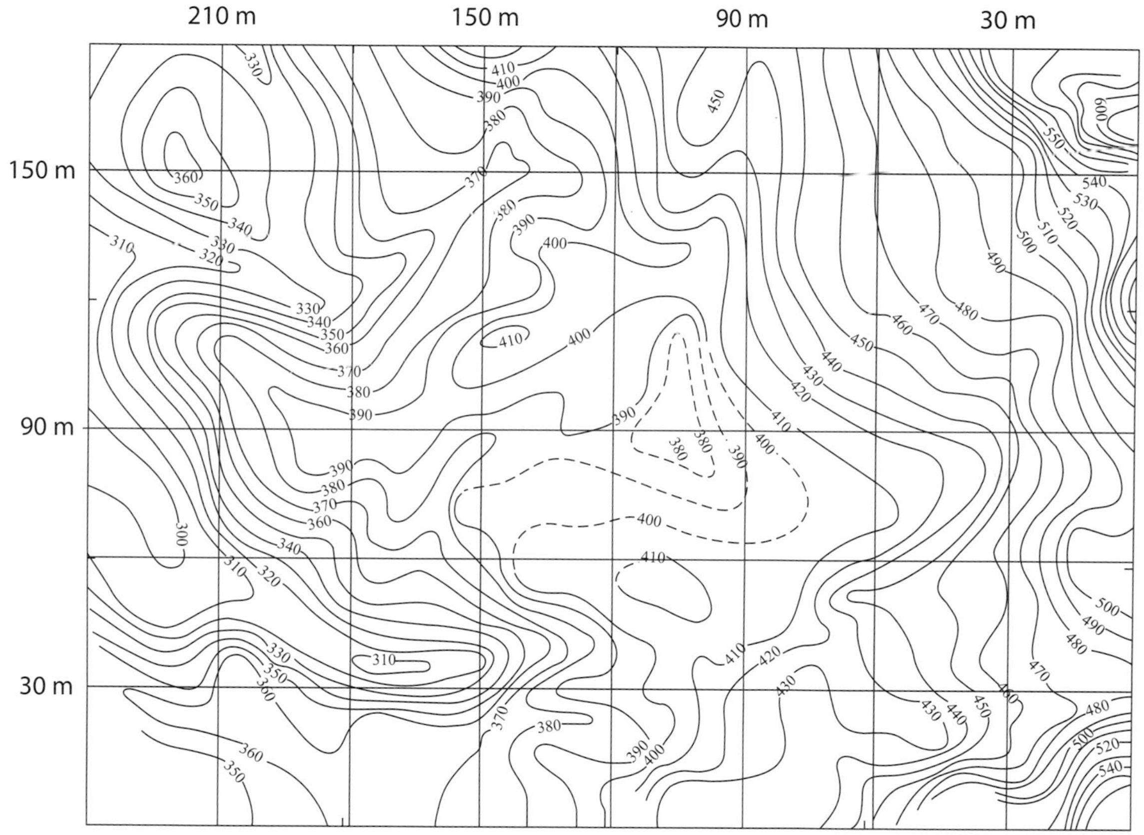

Source: Quadrangle map courtesy of the US Geological Survey.

3.4 A large agricultural watershed has a longest flow path length of 2300 ft, a slope of 0.2%, a composite curve number of 75, and a runoff coefficient of 0.16. Estimate the time of concentration using the SCS Lag equation and the FAA formula, respectively.

3.5 A concrete-paved parking lot (surface slope 0.5%) has a sheet flow path length of 150 ft, followed by a 50-ft-long gutter (bed slope 0.2% and depth 0.5 ft). The 2-year 24-hr rainfall is 3.5 in. Estimate the time of concentration using the Kirpich formula, the Kerby formula, and the TR-55 method, respectively.

3.6 Use the TR-55 method to estimate the time of concentration for following watershed if the 2-year 24-hr rainfall is 2.5 in.

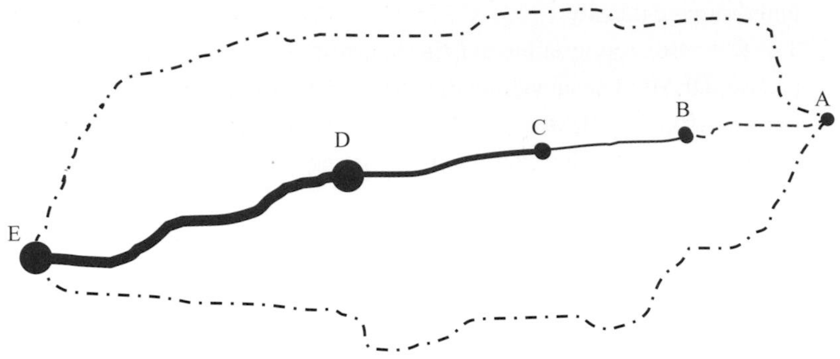

Flow path	Description	Slope	Length (ft)
A → B	Overland flow, dense grasses	0.02	350
B → C	Shallow concentrated flow, unpaved surfaces	0.05	500
C → D	Channel flow, rectangular channel, bottom width 2 ft, depth 1.5 ft, Manning's n = 0.04	0.03	1000
D → E	Channel flow, trapezoidal channel, bottom width 4 ft, depth 3 ft, side slope 2:1, Manning's n = 0.035	0.005	1500

3.7 For a long time period, if a watershed receives 300 mm precipitation and has a 200 mm evapotranspiration annually, determine the annual average runoff.

3.8 During a five-day non-rainy period, a detention pond has an average inflow of 1.5 cfs and outflow of 0.8 cfs. The seepage rate at the pond bottom is 2.8 in hr^{-1}, while the evaporation rate from the pond water surface is 0.2 in hr^{-1}. If the pond has a rectangular shape with a water surface area of $10,000\,ft^2$, determine the change of water level. Is it increased or decreased? Why?

3.9 In a year, a cone-shaped lake (side slope 2.5) has an average inflow of 0.1 $m^3\ s^{-1}$ and outflow of 0.03 $m^3\ s^{-1}$. The average evaporation rate from the lake surface is 3.5 mm hr^{-1}, while the seepage loss is negligible. If the lake has an initial depth of 3.0 m at the beginning of the year, determine the water depth at the end of the year. Neglect the water added into the lake directly from precipitation.

3.10 A farmer in the USA plans to produce 20 ton wheat, 35 ton maize, and 50 ton tomato. The annual average available water for the farmer is $50,000\,m^3$. Estimate the water demand and water supply stress index (WaSSI).

3.11 For the farmer in Problem 3.10, how much more water is demanded than the global average?

3.12 A company has $300,000\,m^3$ water to be supplied to five potential users A through E. As an agreement, the supply to a user must be at least 2000 m^3, while the total supply to A, B, and C must account for more than 75% of the water. If the net revenues for the company from the users (in the order of A to E) are \$2.2, \$3.2, \$1.8, \$4.8, and \$3.9 per m^3 water, respectively, determine the supplies to the users to maximize the overall revenue.

3.13 A farmer, who has 350 ac agricultural land, considers planting crop A and crop B. The net benefits are \$12 and \$18 per lb for A and B, respectively. In a normal hydrologic year, the yields are 300 and 400 lb ac^{-1} for A and B, respectively. Given the uncertainties of weather and market, each crop will occupy 40 to 65% of the land. Also, in order to maintain the long-term productivity, at least 10% of the land needs to be idle. Determine the planting areas of crops A and B to maximize the total revenue and graphically show the feasible region and optimal solution.

3.14 For the following watershed, estimate the areal precipitation using the:

(1) Simple arithmetic average method
(2) Thiessen polygon area-weighted average method
(3) Isohyetal method.

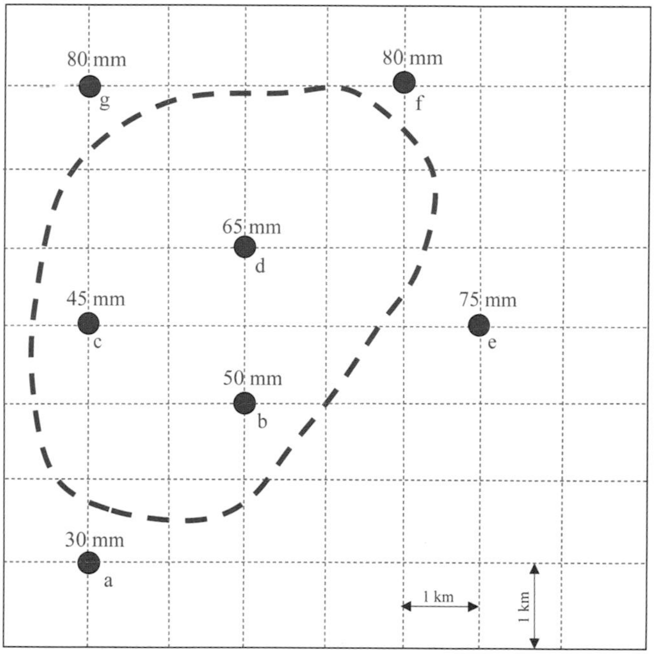

3.15 For a site with loam soils, the results from a field infiltrometer test are shown in the following table.

(1) Determine the parameters of the Horton and Philip models for the site
(2) Estimate the actual cumulative infiltration and total direct runoff resulting from a 1-hr storm with an average rainfall intensity of 45 mm hr^{-1} using these two models

Time (min)	0	2	5	10	20	30	40	60
Infiltration rate (mm hr^{-1})	240	140	120	102	66	42	27	27

3.16 For the site in Problem 3.15, if the initial soil moisture is 0.15, develop the infiltration capacity curve using the G–A model, using the means of the soil water parameters listed in Table 3.12. Also, plot the developed curve versus the measured values on the same graph to visually assess the performance of the G–A model.

3.17 A site has following soil water parameters: $\theta_{sat} = 0.38$, $K_{sat} = 0.1\,\mathrm{m\,d^{-1}}$, $\psi_b = 0.5\,\mathrm{m}$, and $\lambda = 0.28$. The site receives a 1-hr storm with an average rainfall intensity of 8.5 mm $\mathrm{hr^{-1}}$. If the initial soil moisture is 0.2 and the ponding water depth on the ground is negligible, estimate the actual cumulative infiltration and total direct runoff using the:

(1) G–A model

(2) Philip model.

3.18 On a given day, a steppe grassland has the following climate factors: $R_n = 0.00018\,\mathrm{MW\,m^{-2}}$, $T_{min} = 8.0°\mathrm{C}$, $T_{max} = 18.9°\mathrm{C}$, RH = 63%, $p_a = 89\,\mathrm{kPa}$, $u_{zm} = 6.8\,\mathrm{m\,s^{-1}}$. Both u_{zm} and RH are measured at 2 m above the ground surface. In terms of the environment and plant factors, it is determined that the soil-canopy resistance to vapor transport is $r_s = 15\,\mathrm{s\,m^{-1}}$. The soil heat flux is negligible. The average height of the steppe grass on that day is 0.15 m. Estimate the:

(1) Wet-environment evapotranspiration (ET_w)

(2) Potential evapotranspiration (PET)

(3) Actual evapotranspiration (AET).

3.19 On a given day, an agricultural land (maize) has the following climate factors: $R_n = 0.0004\,\mathrm{MW\,m^{-2}}$, $T_{min} = 12.3°\mathrm{C}$, $T_{max} = 35.7°\mathrm{C}$, RH = 45%, $p_a = 89\,\mathrm{kPa}$, $u_{zm} = 2.3\,\mathrm{m\,s^{-1}}$. Both u_{zm} and RH are measured at 3 m above the ground surface. In terms of the environment and plant factors, it is determined that the soil-canopy resistance to vapor transport is $r_s = 30\,\mathrm{s\,m^{-1}}$. The soil heat flux is negligible. The average height of the maize on that day is 1.75 m. Estimate the:

(1) Wet-environment evapotranspiration (ET_w)

(2) Potential evapotranspiration (PET)

(3) Actual evapotranspiration (AET).

3.20 If a drainage basin with a composite curve number of 75 receives 2.5 in rainfall, determine the direct runoff using the SCS-CN method.

3.21 If the curve number of a drainage area for ARC II is 65, what are the curve numbers for ARC I and III?

3.22 A drainage basin has a composite curve number of 78 when it receives a 30-min storm with rainfall of 20, 15, and 5 mm in the first, second, and third 10 min interval, respectively. Determine the direct runoff from the rainfall in each 10 min interval as well as the total direct runoff from the storm using the SCS-CN method.

3.23 A 50 $\mathrm{mi^2}$ watershed has three HSGs and seven land covers as shown in the following figure. The watershed receives a 2-hr storm with rainfall of 2.3 and 0.5 in in the first and second hour, respectively. Determine the:

(1) Composite curve number of the watershed for ARC II

(2) Direct runoffs for each hourly rainfall and the whole storm if the watershed is in ARC II at the inception of the storm

(3) Direct runoffs for each hourly rainfall and the whole storm if the watershed is in ARC I at the inception of the storm

(4) Direct runoffs for each hourly rainfall and the whole storm if the watershed is in ARC III at the inception of the storm.

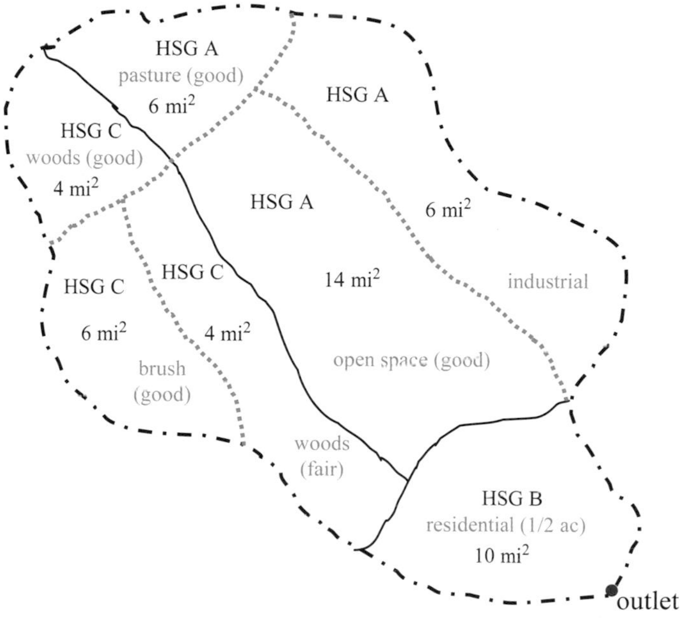

3.24 The soil in a watershed has the following properties: $\theta_{\psi} = 0.02$, $\theta_{fc} = 0.25$, and $\theta_{sat} = 0.38$. The watershed has a composite ARC II curve number of $CN_{II} = 80$. A storm with a rainfall of 30 mm occurs in the dormant season. During the five days before the storm begins, the total rainfall is 25 mm. Estimate the direct runoff from this storm using the MoRE and SCS-CN models.

3.25 For the watershed and storm in Problem 3.24, evaluate how parameters λ_1 and α of the MoCN model affect its estimated direct runoff from this storm. Assuming that the estimation of the MoRE model is accurate, what are the best values for λ_1 and α?

3.26 A detention pond has a storage–discharge relationship of $Q_o = 0.00025S$, where Q_o (in cfs) is the discharge, and S (in ft^3) is the storage. For a computational time step of $\Delta t = 1$ hr, use the Modified Puls method to route the inflow hydrograph shown in the following table through the detention pond. Assume an initial storage of 50 ft^3.

t (hr)	0	1	2	3	4	5	6	7
Q_I (cfs)	0	120	290	500	350	200	90	0

3.27 A reservoir has a storage–discharge relationship of $Q_o = 6.38\ln(S) - 41.795$, where Q_o (in m^3 s^{-1}) is the discharge, and S (in m^3) is the storage. The reservoir has a dead storage of 770 m^3, below which its discharge is zero. For a computational time step of $\Delta t = 1$ hr, use the Modified Puls method to route the inflow hydrograph shown in the following table through the reservoir. Assume the dead storage is at the beginning.

t (hr)	0	1	2	3	4	5	6	7
Q_I ($m^3\ s^{-1}$)	0	10	22	18	12	8	4	0

3.28 A 600-m-long channel with a uniform cross section has a flow area versus discharge relationship (i.e., Q ~ A) as shown in the following table. Route the inflow hydrograph shown in the following table through the channel using the Kinematic Wave method with $\Delta t = 20$ min and $\Delta x = 150$ m. Assume the channel is dry at the beginning and there is no lateral inflow along the entire channel.

A (m^2)	0	460	1050	1760	2600	3560	4650	5860	7200
Q ($m^3\ s^{-1}$)	0.0	7.5	25.3	53.0	91.4	141.3	203.5	279.1	368.8

t (min)	0	20	40	60	80	100	120	140	160
Q_I ($m^3\ s^{-1}$)	0	20	45	38	26	18	10	6	0

3.29 Redo Problem 3.28 if the lateral inflows along the second and last 150-m segments are as shown in the following table.

t (min)	0	20	40	60	80	100	120	140	160
Lateral inflow along the second 150-m sediment ($m^2\ s^{-1}$)	0	0.01	0.01	0.01	0.02	0.02	0.01	0.01	0
Lateral inflow along the last segment ($m^2\ s^{-1}$)	0	0.02	0.03	0.04	0.03	0.02	0.01	0.01	0

3.30 An impervious drainage area has a downgradient length of 600 ft. It has a runoff depth versus unit-width flow rate relationship (i.e., q ~ y) as shown in the following table. A 30-min storm has the instantaneous rainfall intensities (i_r^j) also given in the following table. Determine the flow hydrograph at the lower end of the drainage area using the Kinematic Wave method with $\Delta t = 5$ min and $\Delta x = 150$ ft. Assume that the drainage area is dry at the beginning of the storm and there is no lateral inflow into this drainage area.

y (in)	0.0	1.0	2.0	3.0	4.0	5.0	6.0	7.0	8.0
Q ($ft^2\ sec^{-1}$)	0.00	0.06	0.20	0.39	0.63	0.91	1.23	1.59	1.99

t (min)	0	5	10	15	20	25	30
i_r^j ($in\ hr^{-1}$)	0	1.0	3.0	2.5	1.0	1.5	0.0

3.31 A 1000-ft-long canal has a bed slope of 0.001, an average top width of 30 ft, and a hydraulic depth of 4.5 ft. Determine the flow hydrograph at the lower end of the canal using the Muskingum–Cunge method with $\Delta t = 20$ min and $\Delta x = 200$ ft. Assume that the canal is dry at the beginning and receives an average lateral inflow of 0.002 $ft^2\ sec^{-1}$.

t (min)	0	20	40	60	80	100	120	140	160
Q_0^j (cfs)	0	700	1575	1330	910	630	350	210	0

3.32 A 5000-m-long trapezoidal channel (Manning's $n = 0.025$) has a bottom width of 10 m, a side slope of 2.5, and a bed slope of 0.002. Route the inflow hydrograph shown in the following table through the channel using the Muskingum–Cunge method with $\Delta t = 30$ min and $\Delta x = 1000$ m. Assume zero lateral inflow and uniform flow in the channel. The channel is dry at the beginning.

t (min)	0	30	60	90	120	150	180	210	240	270	300	330	360	390
$Q_I\ (m^3\ s^{-1})$	0	1	3	9	26	15	14	8	5	4	3	2	1	0

4 Synthetic Storm and Design Discharge

Water resources engineering structures (e.g., storm sewers and channels), which will be introduced in the subsequent chapters, need to be designed to accommodate for minimally required values of runoff volume and/or flow rate. In practice, these values can either be estimated empirically or derived through statistical analysis of observed data on rainfall and streamflow. They are associated with either a real or a synthetic storm. Such a storm, which is called a design storm, generates a corresponding design peak discharge and design flow hydrograph. Unlike a real storm, a synthetic storm is a hypothetical rainfall event that preserves some statistics (e.g., mean) of the rainfall data, while it might not actually occur in the past and will probably not occur in the future as well. This chapter presents commonly used methods for formulating a design storm, estimating the design peak discharge, and creating a design flow hydrograph.

4.1 Design Storm Formulation

Design storm refers to a rainfall event that is used to determine the minimally required values of runoff volume and/or flow rate to size a water resources engineering structure of interest. This section introduces the methods for formulating a design storm, which can be an historical, synthetic, or probable maximum event.

4.1.1 Historical Event

If long-term subdaily (e.g., 15-min or 1-hr) rainfall data are available, an historical storm event may be selected as a design storm by assuming that such a storm would occur again. Herein, a storm event is defined as a continuous raining process that is proceeded and followed by non-rainy periods of at least one time interval of the rainfall data. In this regard, when 15-min rainfall data are used, storm events are separated by non-rainy periods of 15 minutes or longer; when 1-hr rainfall data are used, storm events are separated by non-rainy periods of one hour or longer. In practice, the length of non-rainy periods may be empirically specified in terms of design requirements. Also, the interval rainfall during non-rainy periods can be specified to be either equal to zero or less than a small number (e.g., the lower measurable limit of a rain gauge where the data are used). For the rainfall data shown in Figure 4.1, the storm events separated by non-rainy intervals with zero rainfall are different from those by non-rainy intervals with 0.1 in or less rainfall (Table 4.1). The design storm can be a storm event with the longest period, a maximum cumulative rainfall, or a maximum intensity, depending on whether runoff volume or runoff peak is the major concern of hydrologic design. For the storm events listed in Table 4.1, if runoff volume is the major concern, the storm event from 0 : 15 ~ 2 : 45 or 1 : 15 ~ 2 : 45, which have a largest cumulative rainfall, may be selected as the design storm, whereas, if runoff peak is the major concern, the storm event from 5 : 15 ~ 5 : 45 or 5 : 30 ~ 5 : 45, which have a maximum intensity, should be selected as the design storm. If rainy period is the major concern, storm event 3 or 2 can be selected as the design storm.

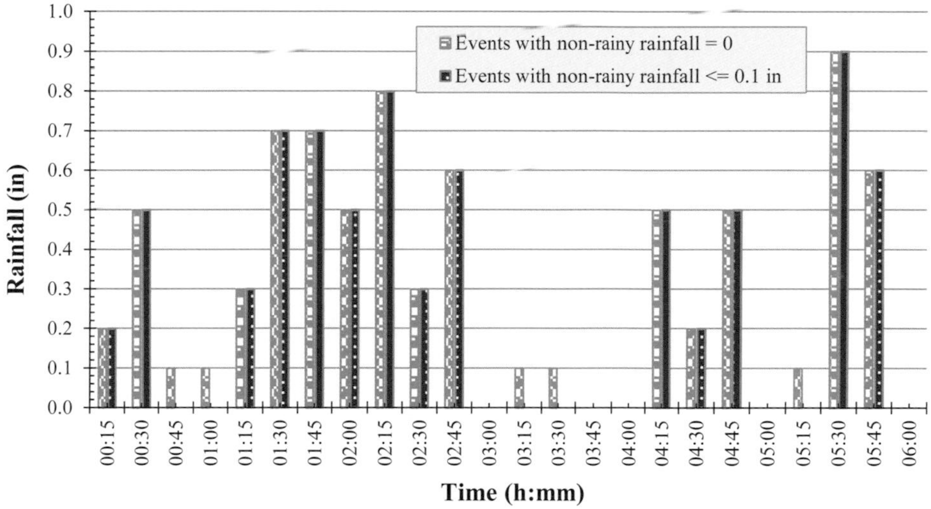

Figure 4.1 **Hyetograph.** Sample for defining storm events from 15-min rainfall data.

Table 4.1 Characteristics of the storm events shown in Figure 4.1.

No. of storm event	Non-rainy rainfall = 0			Non-rainy rainfall ≤ 0.1 in		
	Period	Cumulative rainfall (in)	Maximum intensity (in hr^{-1})	Period	Cumulative rainfall (in)	Maximum intensity (in hr^{-1})
1	0 : 15 ~ 2 : 45	4.8	3.2	0 : 15 ~ 0 : 30	0.7	2.0
2	3.15 ~ 3 : 30	0.2	0.4	1 : 15 ~ 2 : 45	3.9	3.2
3	4 : 15 ~ 4 : 45	1.2	2.0	4 : 15 ~ 4 : 45	1.2	2.0
4	5 : 15 ~ 5 : 45	1.6	3.6	5 : 30 ~ 5 : 45	1.5	3.6

The advantages of using an historical event as the design storm is that it did occur in the past and its selection is straightforward. However, there are three major disadvantages. First, site-specific long-term subdaily rainfall data are needed, which may not be available for most areas of interest. Second, the design event is likely not to occur again in the future, making it hard to judge whether the design is sufficient for future conditions. Third, the characteristics of storm events, namely period, cumulative rainfall, and maximum intensity, can often be different from one area to another, making it infeasible to compare the designs in different areas as well as to standardize the selection criteria of the design storm.

4.1.2 Intensity–Duration–Frequency Curves

An intensity–duration–frequency (IDF) curve (e.g., Figure 4.2) is a statistically based functional relationship between rainfall intensity, duration, and exceedance probability (i.e., frequency). Herein, for a given point on the curve, rainfall intensity is the ratio of total rainfall within the duration to duration. Frequency is defined as the percent chance (i.e., probability) for a rainfall intensity to be equaled or exceeded at any given time. Its reciprocal is defined as recurrence or return period. For instance, in Figure 4.2, the 2-year 10-min rainfall intensity is 36 mm hr^{-1}. The total rainfall within 10 min is 6 mm$[= (36 \text{ mm hr}^{-1}) * (10\,\text{min}) * (1 \text{ hr}/(60\,\text{min}))]$. The return period of the rainfall intensity is 2-year, and the corresponding frequency is $50\%(= 1/2 * 100\%)$. It is important to note that 2-year does not mean that the 36 mm hr^{-1} intensity will occur once every two years; in fact, it can occur at any given time.

The IDF curves are site-specific and can be presented in graphical or tabular formats or with use of the "Steel Formula." These three formats provide equivalent information with a compatible accuracy. Figure 4.2 displays typical examples of IDF curves presented in a graphical format. When tabulated for selected durations, IDF curves can be presented in tabular format. The Steel Formula, a best fit to an IDF curve, can be expressed as:

$$i = \frac{a}{(t + b)^c} \tag{4.1}$$

rainfall intensity [in hr^{-1} for BG; mm hr^{-1} for SI] → i
site-specific frequency-related coefficients → a, b, c
duration [min] → t

In practice, IDF curves can be obtained from federal, state, and local government agencies. For instance, Figure 4.3 shows the IDF curves for the cities of Norfolk and Richmond in Virginia, USA, obtained from the Virginia Department of Conservation and Recreation (VDCR, 1999). In addition, the IDF curves for a location of interest in the United States can be obtained from the National Weather Service (NWS) Precipitation Frequency Data Server (PFDS) website (https://hdsc.nws.noaa.gov/hdsc/pfds/pfds_map_cont.html). After selecting the location by latitude and longitude, address, or by viewing an appropriate map, one can extract the IDF curves in both tabular and graphical formats (e.g., Figure 4.4). One can switch between the two formats by clicking the PF tabular and PF graphical tabs below the map. Further, when long-term subdaily rainfall data are available, IDF curves can be developed by assuming a statistical distribution of the rainfall extremes presented by the data. In this regard, three distributions, namely Fréchet (1927),

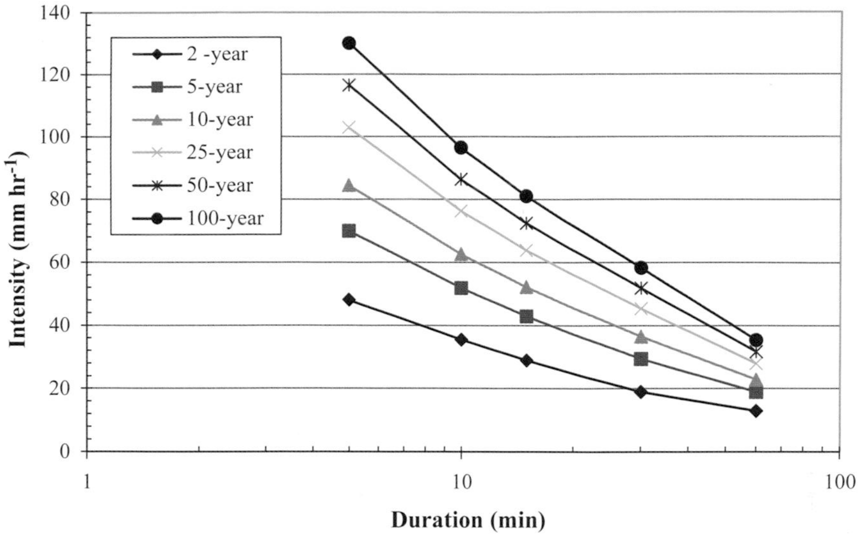

Figure 4.2 **Examples of intensity–duration–frequency curves.**

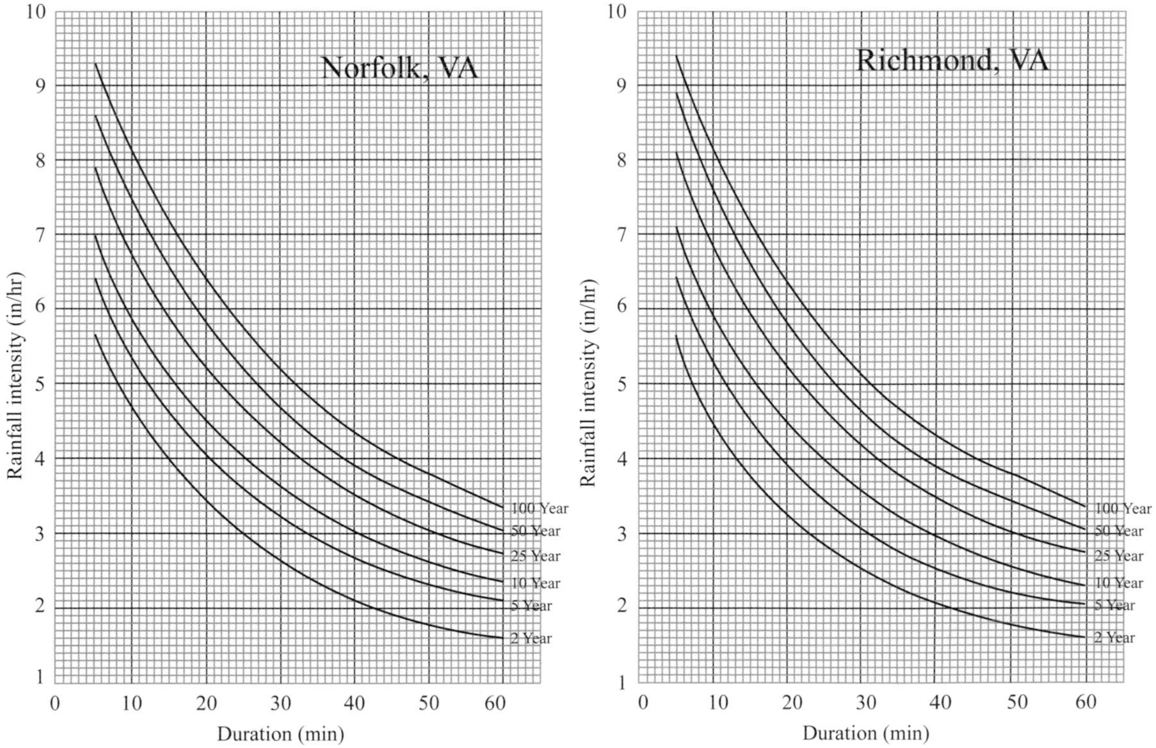

Figure 4.3 **Intensity–duration–frequency (IDF) curves for the cities of Norfolk and Richmond, Virginia.** (*Source:* VDCR, 1999)

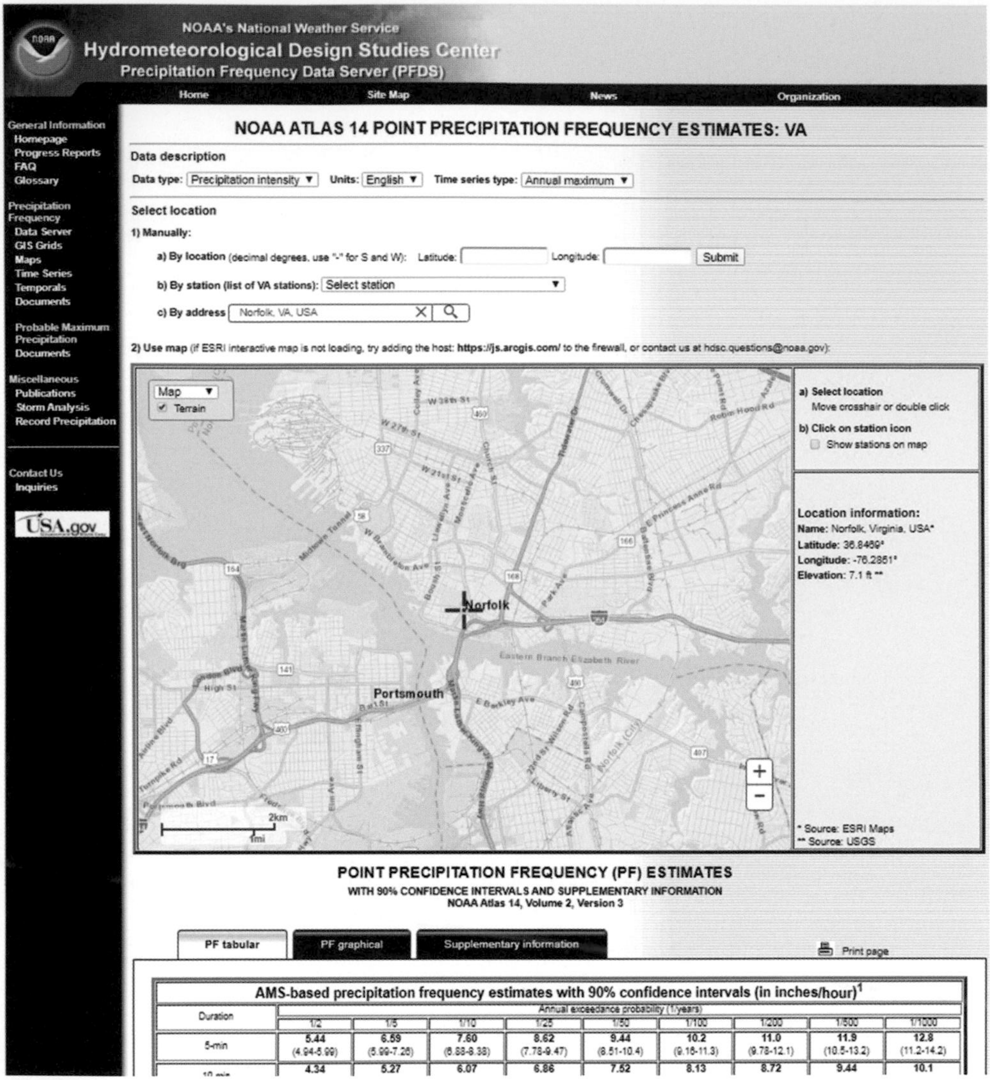

Figure 4.4 **Screenshot from NOAA website showing the process of extracting point precipitation frequency estimates.** The URL of the website is https://hdsc.nws.noaa.gov/hdsc/pfds/pfds_map_cont.html.

Gumbel (1958), and Weibull (1951) have been commonly used. The probability theories of these distributions are discussed by Longin (2016).

For the Gumbel distribution, the chance for a rainfall amount or intensity (x_p) to be equaled or exceeded is computed as:

rainfall for the frequency of interest

exponent (see Eq. (4.3))

frequency of rainfall x_p →

$$\Pr\left(x \geq x_p\right) = 1 - e^{-e^{-b}} \tag{4.2}$$

$$b = \frac{1}{0.7797s}\left(x_p - \bar{x} + 0.45s\right) \tag{4.3}$$

exponent (see Eq. (4.2)) → b
standard deviation of observed rainfall extreme values for a duration of interest → s
rainfall for the frequency of interest → x_p
mean of observed rainfall extreme values for a duration of interest → $\bar{x}$

For a given frequency, the point of a duration on an IDF curve of interest can be determined by implementing three steps. First, compute $\bar{x}$ and s using the observed rainfall extreme values for the duration. Second, compute exponent b using Eq. (4.2). Third, compute x_p using Eq. (4.3). These three steps should be repeated to determine the points for other durations. The triads of x_p, duration, and frequency can be tabulated or plotted to represent the IDF curve. Also, Eq. (4.1) can be fitted to the triads using a least-square method to determine a, b, and c to represent the IDF curve using the Steel Formula. In this regard, Microsoft Excel Solver is a handy tool for executing the least-square method.

For the Fréchet distribution, the chance for a rainfall amount or intensity (x_p) to be equaled or exceeded is computed as:

$$\Pr\left(x \geq x_p\right) = 1 - e^{-\left(\frac{x_p - m}{S_s}\right)^{-\alpha}} \tag{4.4}$$

frequency of rainfall x_p → $\Pr(x \geq x_p)$
rainfall for the frequency of interest → x_p
location parameter → m
scale parameter → S_s
shape parameter (> 2) → α

$$\bar{x} = m + S_s\left[\Gamma\left(1 - \frac{1}{\alpha}\right)\right] \tag{4.5}$$

mean of observed rainfall extreme values for a duration of interest → $\bar{x}$
location parameter → m
scale parameter → S_s
gamma function (Table 4.2; Excel: GAMMA( )) → Γ
shape parameter (> 2) → α

$$x_m = m + \frac{S_s}{\sqrt[\alpha]{\ln(2)}} \tag{4.6}$$

median of observed rainfall extreme values for a duration of interest → x_m
location parameter → m
scale parameter → S_s
shape parameter (> 2) → α

$$s = S_s\sqrt{\Gamma\left(1 - \frac{2}{\alpha}\right) - \left[\Gamma\left(1 - \frac{1}{\alpha}\right)\right]^2} \tag{4.7}$$

standard deviation of observed rainfall extreme values for a duration of interest → s
scale parameter → S_s
gamma function (Table 4.2; Excel: GAMMA( )) → Γ
shape parameter → α

As with the case of the Gumbel distribution above, for a given frequency, the point of a duration on an IDF curve of interest can be determined by implementing three steps. First, compute $\bar{x}$, x_m, and s using the observed rainfall extreme values for the duration. Second, determine parameters m, S_s, and α by simultaneously solving Eqs. (4.5) to (4.7) as shown in the following. Third, compute

Table 4.2 Gamma functions for selected values of α in Eqs. (4.5) and (4.7).

α	$\Gamma\left(1-\frac{1}{\alpha}\right)$	$\Gamma\left(1-\frac{2}{\alpha}\right)$	α	$\Gamma\left(1-\frac{1}{\alpha}\right)$	$\Gamma\left(1-\frac{2}{\alpha}\right)$	α	$\Gamma\left(1-\frac{1}{\alpha}\right)$	$\Gamma\left(1-\frac{2}{\alpha}\right)$
2.01	1.76	200.43	2.16	1.65	12.99	2.31	1.57	6.99
2.02	1.76	100.43	2.17	1.65	12.26	3.31	1.30	2.24
2.03	1.75	67.10	2.18	1.64	11.61	4.31	1.20	1.66
2.04	1.74	50.44	2.19	1.63	11.03	5.31	1.15	1.44
2.05	1.73	40.45	2.20	1.63	10.51	6.31	1.12	1.33
2.06	1.72	33.78	2.21	1.62	10.03	7.31	1.10	1.26
2.07	1.72	29.03	2.22	1.62	9.60	8.31	1.09	1.21
2.08	1.71	25.46	2.23	1.61	9.21	9.31	1.07	1.18
2.09	1.70	22.69	2.24	1.61	8.85	10.31	1.07	1.16
2.10	1.69	20.47	2.25	1.60	8.52	11.31	1.06	1.14
2.11	1.69	18.65	2.26	1.60	8.22	12.31	1.05	1.12
2.12	1.68	17.14	2.27	1.59	7.94	13.31	1.05	1.11
2.13	1.67	15.86	2.28	1.58	7.68	14.31	1.05	1.10
2.14	1.67	14.77	2.29	1.58	7.43	15.31	1.04	1.09
2.15	1.66	13.82	2.30	1.57	7.20	16.31	1.04	1.09

x_p using Eq. (4.4). These three steps should be repeated to determine the points for other durations. The triads of x_p, duration, and frequency can be tabulated or plotted to represent the IDF curve. Also, Eq. (4.1) can be fitted to the triads using a least-square method to determine a, b, and c to represent the IDF curve using the Steel Formula. Again, Microsoft Excel Solver is a handy tool for executing the least-square method.

Subtracting Eq. (4.6) from Eq. (4.5) and rearranging, yields:

$$S_s = \frac{\bar{x} - x_m}{\Gamma\left(1-\frac{1}{\alpha}\right) - \frac{1}{\sqrt[\alpha]{\ln 2}}} \tag{4.8}$$

S_s: scale parameter; $\bar{x}$: mean of observed rainfall extreme values for a duration of interest; x_m: median of observed rainfall extreme values for a duration of interest; Γ: gamma function (Table 4.2; Excel: GAMMA( )); α: shape parameter

Substituting Eq. (4.8) into Eq. (4.7), produces:

$$s = \frac{\bar{x} - x_m}{\Gamma\left(1-\frac{1}{\alpha}\right) - \frac{1}{\sqrt[\alpha]{\ln 2}}}\sqrt{\Gamma\left(1-\frac{2}{\alpha}\right) - \left[\Gamma\left(1-\frac{1}{\alpha}\right)\right]^2} \tag{4.9}$$

s: standard deviation of observed rainfall extreme values for a duration of interest; $\bar{x}$: mean of observed rainfall extreme values for a duration of interest; x_m: median of observed rainfall extreme values for a duration of interest; Γ: gamma function (Table 4.2; Excel: GAMMA( )); α: shape parameter

Table 4.3 Gamma functions for selected values of k in Eqs. (4.11) and (4.12).

k	$\Gamma\left(1+\frac{1}{k}\right)$	$\Gamma\left(1+\frac{2}{k}\right)$	k	$\Gamma\left(1+\frac{1}{k}\right)$	$\Gamma\left(1+\frac{2}{k}\right)$	k	$\Gamma\left(1+\frac{1}{k}\right)$	$\Gamma\left(1+\frac{2}{k}\right)$
0.50	2.00	24.00	0.65	1.37	6.61	0.80	1.13	3.32
0.51	1.93	21.34	0.66	1.34	6.23	1.80	0.89	1.05
0.52	1.87	19.09	0.67	1.32	5.89	2.80	0.89	0.91
0.53	1.81	17.16	0.68	1.30	5.58	3.80	0.90	0.89
0.54	1.75	15.51	0.69	1.28	5.29	4.80	0.92	0.89
0.55	1.70	14.09	0.70	1.27	5.03	5.80	0.93	0.89
0.56	1.66	12.85	0.71	1.25	4.79	6.80	0.93	0.90
0.57	1.61	11.77	0.72	1.23	4.57	7.80	0.94	0.91
0.58	1.57	10.83	0.73	1.22	4.37	8.80	0.95	0.91
0.59	1.54	10.00	0.74	1.20	4.18	9.80	0.95	0.92
0.60	1.50	9.26	0.75	1.19	4.01	10.80	0.95	0.92
0.61	1.47	8.61	0.76	1.18	3.85	11.80	0.96	0.93
0.62	1.44	8.02	0.77	1.17	3.71	12.80	0.96	0.93
0.63	1.42	7.50	0.78	1.15	3.57	13.80	0.96	0.93
0.64	1.39	7.04	0.79	1.14	3.44	14.80	0.97	0.94

To simultaneously solve Eqs. (4.5) to (4.7), use a trial-and-error method or Excel Solver to solve Eq. (4.9) to obtain α; substitute α into Eq. (4.8) to get S_s; and then substitute α and S_s into either Eq. (4.5) or (4.6) to obtain m.

For the Weibull distribution, the chance for a rainfall amount or intensity (x_p) to be equaled or exceeded is computed as:

rainfall for the frequency of interest; frequency of rainfall x_p; shape parameter (> 0); scale parameter (> 0)

$$\Pr\left(x \geq x_p\right) = e^{-\left(\frac{x_p}{\lambda}\right)^k} \tag{4.10}$$

gamma function (Table 4.3; Excel: GAMMA( )); mean of observed rainfall extreme values for a duration of interest; scale parameter (> 0); shape parameter (> 0)

$$\bar{x} = \lambda\left[\Gamma\left(1+\frac{1}{k}\right)\right] \tag{4.11}$$

gamma function (Table 4.3; Excel: GAMMA( )); standard deviation of observed rainfall extreme values for a duration of interest; scale parameter (> 0); shape parameter (> 0)

$$s = \lambda\sqrt{\Gamma\left(1+\frac{2}{k}\right) - \left[\Gamma\left(1+\frac{1}{k}\right)\right]^2} \tag{4.12}$$

As discussed above for the Gumbel and Fréchet distributions, for the Weibull distribution for a given frequency, the point of a duration on an IDF curve of interest can be determined by implementing three steps. First, compute $\bar{x}$ and s using the observed rainfall extreme values for the duration. Second, determine parameters k and λ by simultaneously solving Eqs. (4.11) and (4.12) as shown in the following. Third, compute x_p using Eq. (4.10). These three steps should be repeated to determine the points for other durations. The triads of x_p, duration, and frequency can be tabulated or plotted to represent the IDF curve. Also, Eq. (4.1) can be fitted to the triads using a least-square method to determine a, b, and c to represent the IDF curve using the Steel Formula. As described above, Microsoft Excel Solver is a handy tool for executing the least-square method.

Dividing Eq. (4.12) by Eq. (4.11) and rearranging, yields:

gamma function (Table 4.3; Excel: GAMMA( ))
standard deviation of observed rainfall extreme values for a duration of interest
shape parameter (> 0)
mean of observed rainfall extreme values for a duration of interest

$$\frac{s}{\bar{x}} = \sqrt{\frac{\Gamma\left(1+\frac{2}{k}\right)}{\left[\Gamma\left(1+\frac{1}{k}\right)\right]^2} - 1} \tag{4.13}$$

To simultaneously solve Eqs. (4.11) and (4.12), use a trial-and-error method or Excel Solver to solve Eq. (4.13) to obtain k; and then substitute k into either Eq. (4.11) or (4.12) to determine λ.

Microsoft Excel VBA functions for these three distributions are available on the book website (https://fs.wp.odu.edu/x4wang/textbook). For a given duration, the functions compute rainfall for a frequency of interest. For the Gumbel distribution, the function is *IDF_Gumbel (xbar As Double, s As Double, Pr As Double)*, where xbar is the mean, s is the standard deviation, and Pr is the frequency of interest. For the Fréchet distribution, the function is *IDF_Frechet(xbar As Double, xm As Double, s As Double, Pr As Double)*, where xm is the median, and the other inputs are the same as those of *IDF_Gumbel( )*. For the Weibull distribution, the function is *IDF_Weibull*(xbar As Double, s As Double, Pr As Double), where the inputs are exactly the same as those of *IDF_Gumbel( )*.

Example 4.1 A weather station has 5-min rainfall data from 1951 to 2000. The data are used to generate 10-, 15-, 30-, and 60-min rainfall data. The annual maximums for these five durations are determined, formulating five datasets, each of which has 50 extreme values for the 50 record years. The datasets are used to compute the means, medians, and standard deviations, presented in Table 4.4. For the selected return periods of 2-, 5-, 10-, 25-, 50-, and 100-year, develop the IDF curves by assuming that the extreme values follow a:

(1) Gumbel distribution
(2) Fréchet distribution
(3) Weibull distribution.

Table 4.4 Statistics of the annual maximum durational rainfalls for Example 4.1.

Statistics	Duration (mm)				
	5	10	15	30	60
Mean, $\bar{x}$ (mm)	4.35	6.44	7.92	10.51	13.51
Median, x_m (mm)	4.00	5.80	7.55	9.60	12.50
Standard deviation, s (mm)	2.07	3.08	3.95	5.96	6.78

Solution

(1) For a return period, compute exponent b using Eq. (4.2) and then compute values of x_p for the five durations. The intensity of a given duration is the ratio of the responding value of x_p to the duration, which in turn is multiplied by 60 to have units of mm hr^{-1}. The computations can be done by hand or in Excel using the *IDF_Gumbel( )* function. The results are presented in Table 4.5 and Figure 4.5.

Table 4.5 Computed rainfall (in mm) assuming a Gumbel distribution.

Duration (min)	Return period (year)					
	2	5	10	25	50	100
	Frequency					
	50%	20%	10%	4%	2%	1%
5	4.01	5.84	7.05	8.58	9.72	10.84
10	5.93	8.66	10.46	12.74	14.42	16.10
15	7.27	10.76	13.07	15.99	18.16	20.31
30	9.53	14.80	18.29	22.69	25.96	29.20
60	12.40	18.39	22.36	27.37	31.09	34.78
b[1]	0.37	1.50	2.25	3.20	3.90	4.60

[1] Exponent computed using Eq. (4.2).

For instance, for 2-year return period, the corresponding frequency is $Pr = 1/2 * 100\% = 50\%$. Eq. (4.2): $b = -\ln[-\ln(1 - Pr)] = -\ln[-\ln(1 - 50\%)] = 0.37$. For the duration of 5 min, Eq. (4.3): $x_p = 0.7797 * (2.07\text{ mm}) * 0.37 + (4.35\text{ mm}) - 0.45 * (2.07\text{ mm}) = 4.01$ mm. The corresponding intensity is $(4.01\text{ mm})/(5\text{ min}) * (60\text{ min}/1\text{ hr}) = 48.12\text{ mm hr}^{-1}$. For the duration of 10 min, Eq. (4.3): $x_p = 0.7797 * (3.08\text{ mm}) * 0.37 + (6.44\text{ mm}) - 0.45 * (3.08\text{ mm}) = 5.93$ mm. The corresponding intensity is $(5.93\text{ mm})/(10\text{ min}) * (60\text{ min}/1\text{ hr}) = 35.61\text{ mm hr}^{-1}$. For the other durations, the values of x_p can be computed similarly. Moreover, the values of x_p for the other return periods and the durations can be computed by following the same procedures.

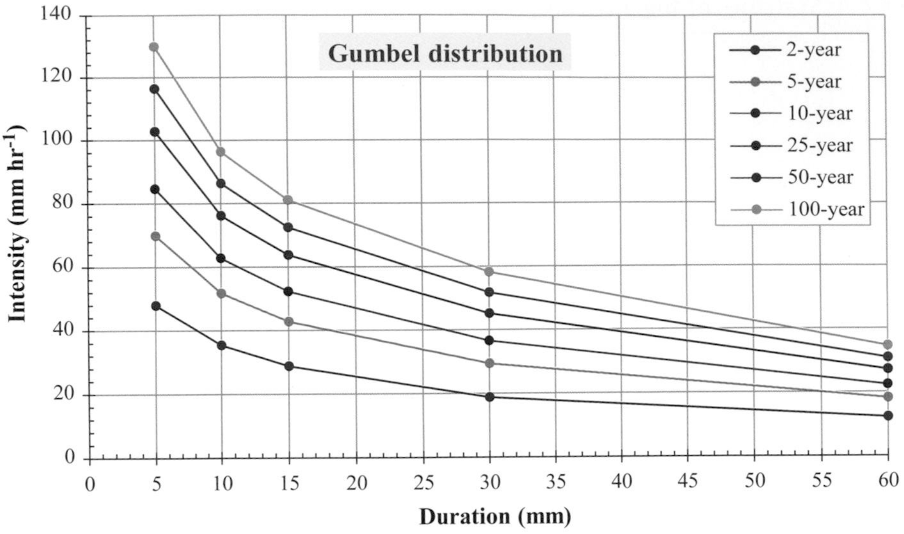

Figure 4.5 **Results for part (1) of Example 4.1.** The IDF curves are created by assuming a Gumbel distribution.

(2) For a duration, compute α using Eq. (4.9) and then S_s using Eq. (4.8), and then substitute α and S_s into either Eq. (4.5) or (4.6) to compute m. Afterward, for each return period or frequency, compute x_p using Eq. (4.4) as well as the intensity as the ratio of x_p to the duration, which in turn is multiplied by 60 to have units of mm hr^{-1}. The computations can be done by hand or in Excel using the function of *IDF_Frechet ( )*. The results are presented in Table 4.6 and Figure 4.6.

Table 4.6 Computed rainfall (mm) assuming a Fréchet distribution.

Duration (min)	Return period (year)						Parameter[1]		
	2	5	10	25	50	100	α	S_s (mm)	m (mm)
	Frequency								
	50%	20%	10%	4%	2%	1%			
5	4.00	4.63	5.27	6.44	7.70	9.41	2.23	0.81	3.05
10	5.80	7.06	8.27	10.46	12.72	15.72	2.44	1.83	3.68
15	7.55	8.16	8.80	10.02	11.36	13.24	2.06	0.70	6.72
30	9.60	11.20	12.83	15.86	19.15	23.64	2.18	1.98	7.25
60	12.50	14.27	16.07	19.43	23.08	28.08	2.17	2.17	9.93

[1] α: shape parameter; S_s: scale parameter; m: location parameter. See Eq. (4.4).

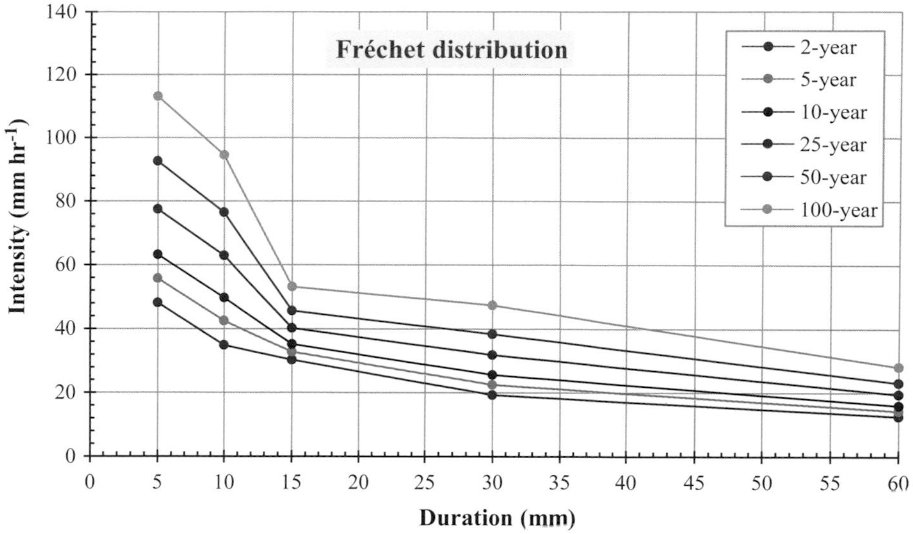

Figure 4.6 **Results of part (2) of Example 4.1.** The IDF curves are created by assuming a Fréchet distribution.

For instance, for the duration of 5 min, using Excel Solver to solve Eq. (4.9): $2.07\text{ mm} = \dfrac{(4.35\text{ mm}) - (4.00\text{ mm})}{\Gamma\left(1 - \frac{1}{\alpha}\right) - \frac{1}{\sqrt[\alpha]{\ln 2}}}\sqrt{\Gamma\left(1 - \frac{2}{\alpha}\right) - \left[\Gamma\left(1 - \frac{1}{\alpha}\right)\right]^2}$, yields $\alpha = 2.23$; Eq. (4.8) and Table 4.2: $S_s = \dfrac{(4.35\text{ mm}) - (4.00\text{ mm})}{\Gamma\left(1 - \frac{1}{2.23}\right) - \frac{1}{\sqrt[2.23]{\ln 2}}} = \dfrac{0.35\text{ mm}}{1.61 - 1.18} = 0.81\text{ mm}$; Eq. (4.6): $m = (4.00\text{ mm}) - \dfrac{0.81\text{ mm}}{\sqrt[2.23]{\ln(2)}} = 3.05\text{ mm}$. For 2-year return period, the frequency is $Pr = 1/2 * 100\% = 50\%$. Eq. (4.4): $x_p = 3.05\text{ mm} + (0.81\text{ mm}) * e^{-\frac{\ln[-\ln(1-50\%)]}{2.23}} = 4.00\text{ mm}$; for the other return periods, compute values of x_p using Eq. (4.4). Moreover, the values of x_p for the other durations and the return periods can be computed by following the same procedures.

(3) For a duration, compute k using Eq. (4.13) and then substitute k into either Eq. (4.11) or (4.12) to compute λ. Afterward, for each return period or frequency, compute x_p using Eq. (4.10) as well as the intensity as the ratio of x_p to the duration, which in turn is multiplied by 60 to have units of mm hr^{-1}. The computations can be done by hand or in Excel using the function of *IDF_Weibull ( )*. The results are presented in Table 4.7 and Figure 4.7.

Table 4.7 Computed rainfall (mm) assuming a Weibull distribution.

Duration (min)	Return period (year) 2	5	10	25	50	100	Parameter[1] k	λ (mm)
	Frequency 50%	20%	10%	4%	2%	1%		
5	4.16	6.09	7.15	8.32	9.08	9.78	2.22	4.91
10	6.16	9.02	10.61	12.35	13.49	14.53	2.21	7.27
15	7.51	11.21	13.29	15.58	17.09	18.47	2.11	8.94
30	9.68	15.35	18.67	22.44	24.97	27.30	1.83	11.83
60	12.80	19.15	22.73	26.68	29.28	31.66	2.09	15.25

[1] k: shape parameter; λ: scale parameter. See Eq. (4.10).

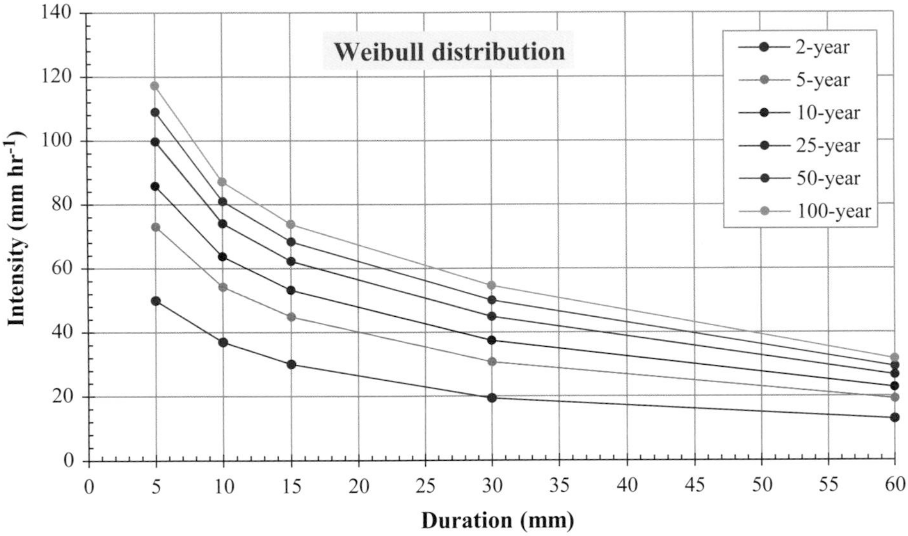

Figure 4.7 Results of part (3) of Example 4.1. The IDF curves are created by assuming a Weibull distribution.

For instance, for the duration of 5 min, using Excel Solver to solve Eq. (4.13): $\frac{2.07\ \text{mm}}{4.35\ \text{mm}} = \sqrt{\frac{\Gamma\left(1+\frac{2}{k}\right)}{\left[\Gamma\left(1+\frac{1}{k}\right)\right]^2} - 1}$, yields $k = 2.22$; Eq. (4.11) and Table 4.3: $\lambda = \frac{4.35\ \text{mm}}{\Gamma\left(1+\frac{1}{2.22}\right)} = \frac{4.35\ \text{mm}}{0.885} = 4.91$ mm. For 2-year return period, the frequency is $Pr = 1/2 * 100\% = 50\%$. Eq. (4.10): $x_p = (4.91\ \text{mm}) * e^{\frac{\ln[-\ln(50\%)]}{2.22}} = 4.16$ mm; for the other return periods, compute values of x_p using Eq. (4.10). Moreover, the values of x_p for the other durations and the return periods can be computed by following the same procedures.

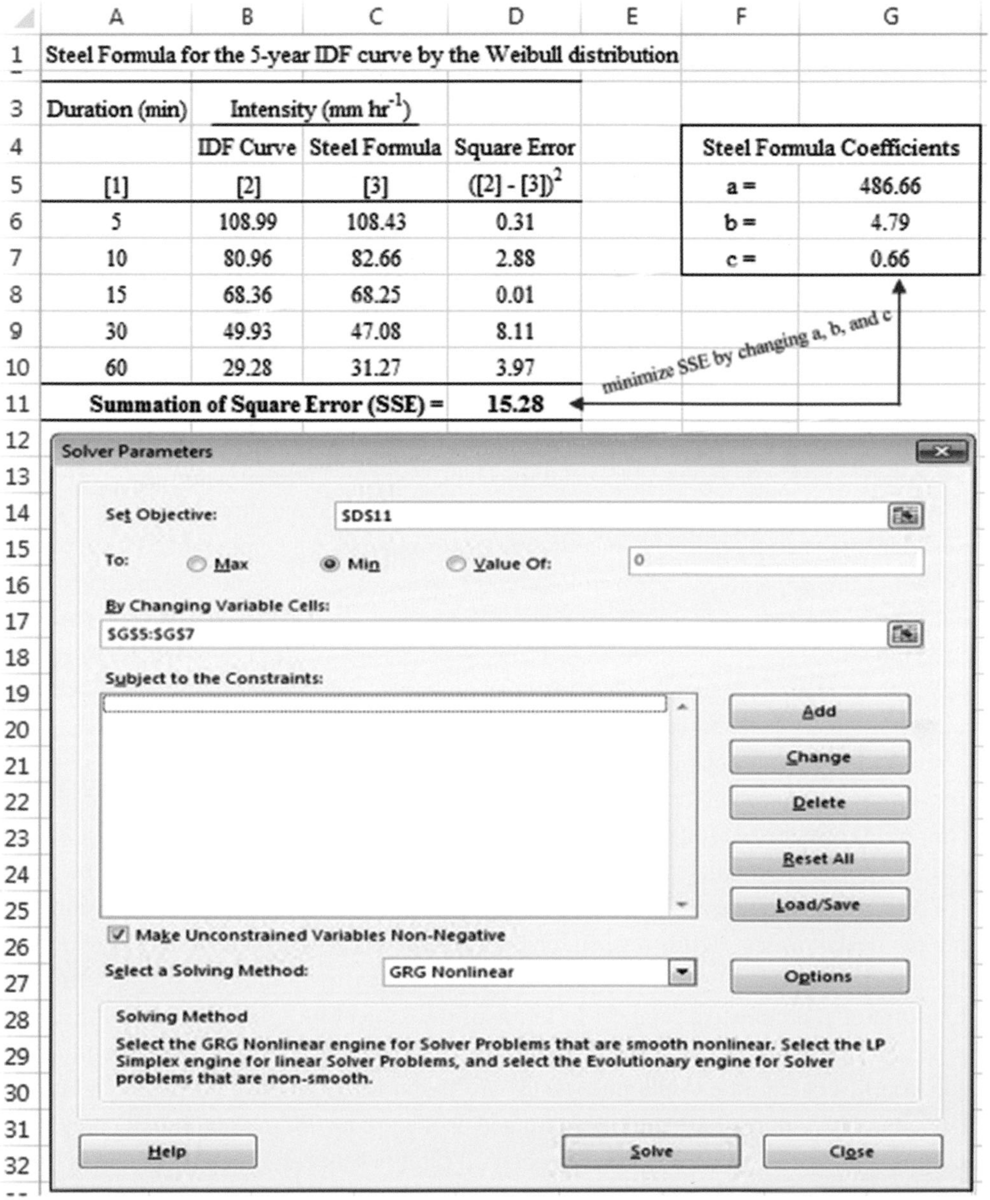

Steel Formula for the 5-year IDF curve by the Weibull distribution

Duration (min)	Intensity (mm hr^{-1})		
	IDF Curve	Steel Formula	Square Error
[1]	[2]	[3]	([2] - [3])^2
5	108.99	108.43	0.31
10	80.96	82.66	2.88
15	68.36	68.25	0.01
30	49.93	47.08	8.11
60	29.28	31.27	3.97
Summation of Square Error (SSE) =			15.28

Steel Formula Coefficients	
a =	486.66
b =	4.79
c =	0.66

Figure 4.8 **Portion of screen and spreadsheet showing the conversion of an IDF curve into a Steel Formula.** Partial solution to problems in Example 4.1.

As an example, Figure 4.8 shows the conversion of the 50-year IDF curve created by the Weibull distribution into the corresponding Steel Formula using Excel Solver. In the spreadsheet, column B

([2]) is the durational intensities from Figure 4.7, and column C ([3]) is the durational intensities computed using the Steel Formula (Eq. (4.2)). Column D is square error, and cell D11 stores the summation of square error (i.e., = SUM (D6:D10)). For arbitrary initial values of the Steel Formula coefficients, such as a = 100, b = 10, and c = 2, use Solver to minimize the summation of square error (SSE) by changing a, b, and c. Check the Make Unconstrained Variables Non-Negative box to ensure that the coefficients are positive. The solution is $a = 486.66\ \text{mm hr}^{-1}$, $b = 4.79$ min, and $c = 0.66$. The Steel Formula is $i = \frac{486.66}{(t + 4.79)^{0.66}}$, where i is the rainfall intensity in mm hr^{-1}; and t is the duration in min.

4.1.3 Synthetic Storm

A synthetic storm is a hypothetical rainfall event that preserves some statistics (e.g., mean) of the rainfall data. Because it is a conceptual construct, no synthetic storm might occur in history and may occur in the future. However, it captures the rainfall features of a particular locality (Prodanovic and Simonovic, 2004) and thus has been conventionally used for hydrologic design. A synthetic storm for a designated frequency and a specified duration, which are usually selected based on site-specific design standards, can be formulated at a computational time interval Δt by following the procedures shown in Figure 4.9. The design standards depend on the importance and acceptable risk of the project of interest, and are regulated by relevant guidelines of local government.

When a balanced triangular distribution is preferred, for a given cumulative time t, the average rainfall intensity is determined using the site-specific Steel Formula or IDF curve. The intensity is multiplied by t to obtain the corresponding cumulative rainfall. The interval rainfall between time t and $t - \Delta t$ is computed as the difference of the corresponding values of cumulative rainfall. The values of interval rainfall are rearranged to have a bell-shaped distribution (e.g., Figure 4.10), which in turn is multiplied by an areal adjustment factor (Figure 4.11) to derive the design storm.

For a storm duration of 24 hr, the NRCS 24-hr rainfall distribution can be assumed to formulate the design storm. In this regard, the durational average rainfall intensity is determined using the site-specific Steel Formula or IDF curve. The intensity is multiplied by 24 hr to determine the durational cumulative rainfall. In terms of the project location, the rainfall distribution type is identified from the NRCS map (Figure 4.12). For the distribution type, the rainfall distribution coefficients at time t with interval Δt are found from the NRCS standard table (Table 4.8). The cumulative rainfall at time t is computed as the product of the durational cumulative rainfall and the corresponding coefficient. The interval rainfall between time t and $t - \Delta t$ is computed as the difference of the corresponding values of cumulative rainfall. The interval rainfall is multiplied by an areal adjustment factor (Figure 4.11) to obtain the design storm.

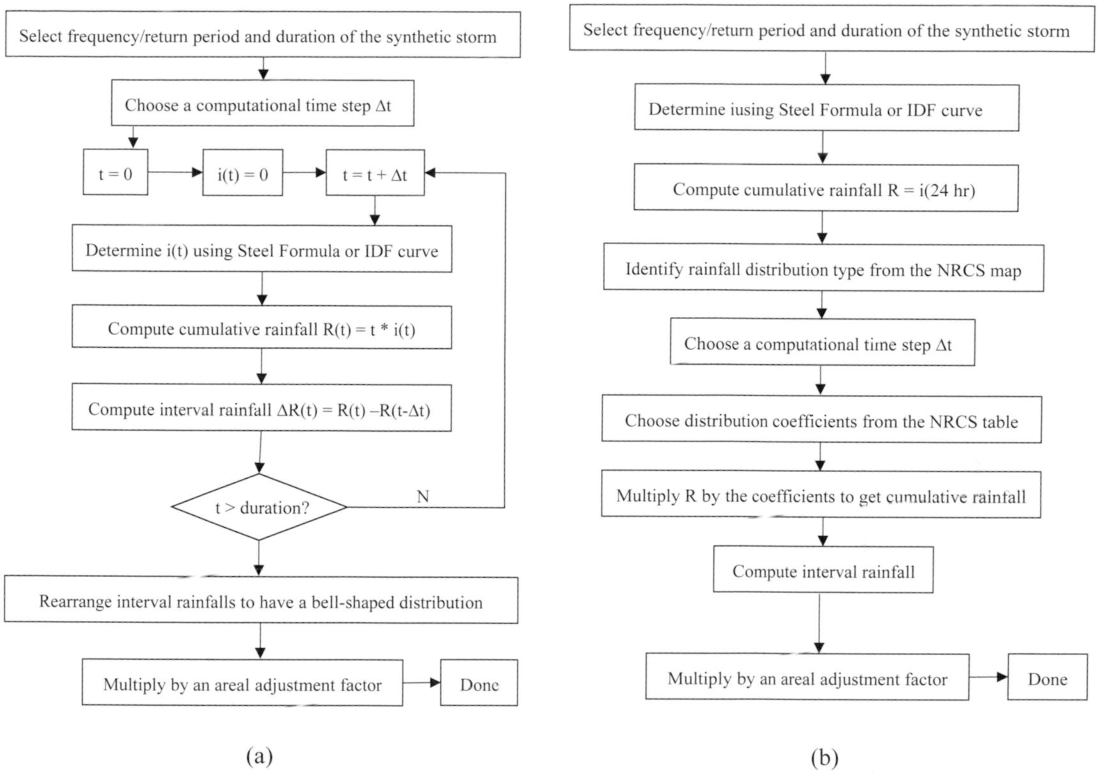

Figure 4.9 **Synthetic storm formulation.** Flowchart for: (a) balanced triangular distribution, and (b) Natural Resources Conservation Service (NRCS) 24-hr rainfall distribution. i(t) is the average rainfall intensity spanning time period t.

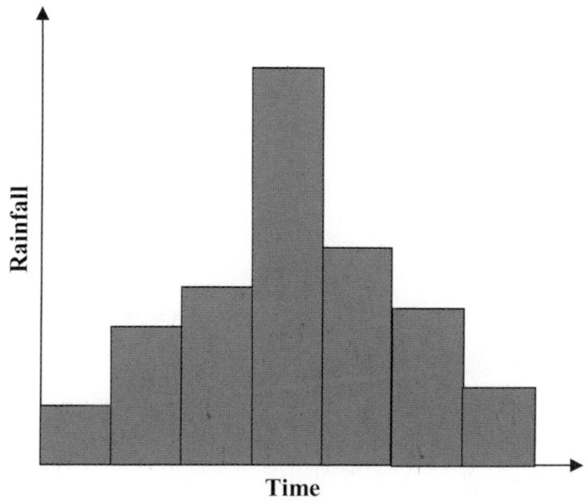

Figure 4.10 **Balanced triangular distribution of rainfall during a hypothetical storm event.** The histogram is an example of the storm hyetograph.

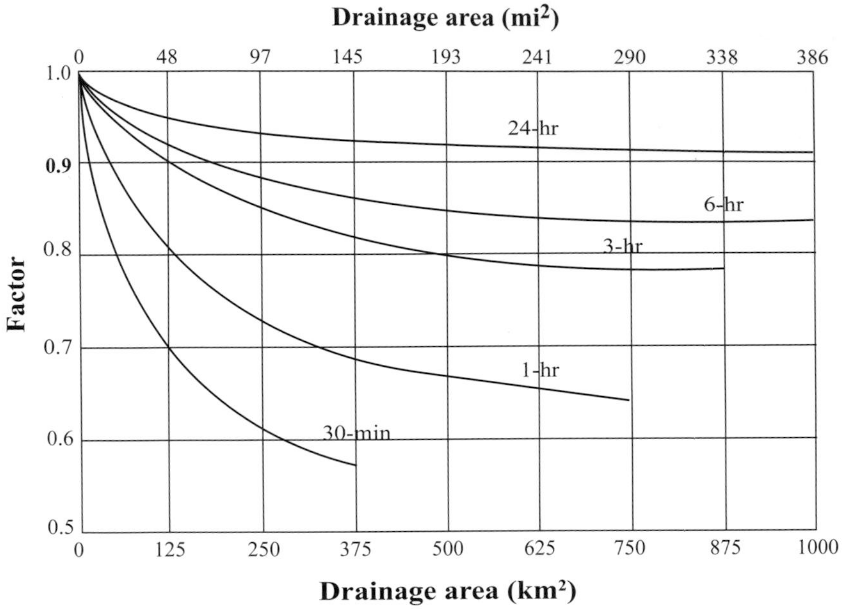

Figure 4.11 **Nomogram for determining areal adjustment factors in the formation of design storms.** (*Sources:* Miller *et al.*, 1973; Myers and Zehr, 1980)

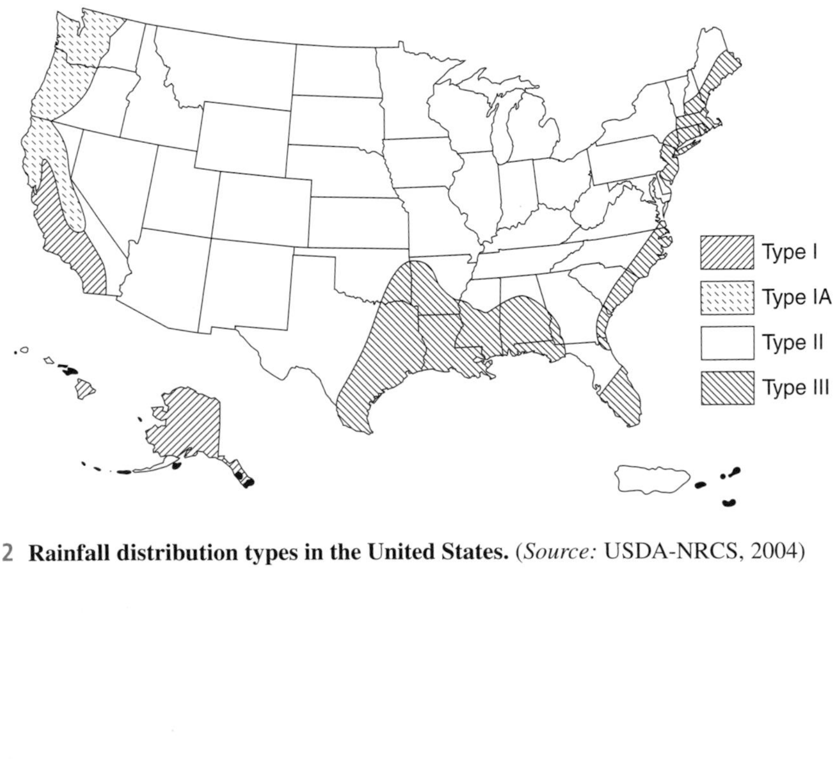

Figure 4.12 **Rainfall distribution types in the United States.** (*Source:* USDA-NRCS, 2004)

Table 4.8 NRCS 24-hr cumulative rainfall distribution coefficients.[1]

Time (hr)	I	IA	II	III	Time (hr)	I	IA	II	III
0.5	0.008	0.010	0.0053	0.0050	12.5	0.706	0.683	0.7351	0.7020
1.0	0.017	0.020	0.0108	0.0100	13.0	0.728	0.701	0.7724	0.7500
1.5	0.026	0.035	0.0164	0.0150	13.5	0.748	0.719	0.7989	0.7835
2.0	0.035	0.050	0.0223	0.0200	14.0	0.766	0.736	0.8197	0.8110
2.5	0.045	0.067	0.0284	0.0252	14.5	0.783	0.753	0.8380	0.8341
3.0	0.055	0.082	0.0347	0.0308	15.0	0.799	0.769	0.8538	0.8542
3.5	0.065	0.098	0.0414	0.0367	15.5	0.815	0.785	0.8676	0.8716
4.0	0.076	0.116	0.0483	0.0430	16.0	0.830	0.800	0.8801	0.8860
4.5	0.087	0.135	0.0555	0.0497	16.5	0.844	0.815	0.8914	0.8984
5.0	0.099	0.156	0.0632	0.0568	17.0	0.857	0.830	0.9019	0.9095
5.5	0.112	0.180	0.0712	0.0642	17.5	0.870	0.844	0.9115	0.9194
6.0	0.126	0.206	0.0797	0.0720	18.0	0.882	0.858	0.9206	0.9280
6.5	0.140	0.237	0.0887	0.0806	18.5	0.893	0.871	0.9291	0.9358
7.0	0.156	0.268	0.0984	0.0905	19.0	0.905	0.844	0.9371	0.9432
7.5	0.174	0.310	0.1089	0.1016	19.5	0.916	0.896	0.9446	0.9503
8.0	0.194	0.425	0.1203	0.1140	20.0	0.926	0.908	0.9519	0.9570
8.5	0.219	0.480	0.1328	0.1284	20.5	0.936	0.920	0.9588	0.9634
9.0	0.254	0.520	0.1467	0.1458	21.0	0.946	0.932	0.9653	0.9694
9.5	0.303	0.550	0.1625	0.1659	21.5	0.956	0.944	0.9717	0.9752
10.0	0.515	0.577	0.1808	0.1890	22.0	0.965	0.956	0.9777	0.9808
10.5	0.583	0.601	0.2042	0.2165	22.5	0.974	0.967	0.9836	0.9860
11.0	0.624	0.624	0.2351	0.2500	23.0	0.983	0.978	0.9892	0.9909
11.5	0.655	0.645	0.2833	0.2980	23.5	0.992	0.989	0.9947	0.9956
12.0	0.682	0.664	0.6632	0.5000	24.0	1.000	1.000	1.0000	1.0000

[1] *Source:* USDA-NRCS (2004).

Example 4.2 The IDF curves shown in Figure 4.7 are applicable for a 500 km^2 watershed located in Dallas, TX. Use a computational time interval of 1.0 hr to develop a 50-year 24-hr synthetic storm by assuming:

(1) Balanced triangular distribution
(2) NRCS 24-hr rainfall distribution.

Solution

From Example 4.1, the Steel Formula for the 50-year IDF curve is $i = \dfrac{486.66}{(t + 4.79)^{0.66}}$, where i is rainfall intensity in mm hr^{-1}; and t is time in min. From Figure 4.11, the areal adjustment factor for a drainage area of 500 km^2 and a duration of 24 hr is 0.92. The computations can be done in an Excel spreadsheet.

(1) The computations and formulated storm are shown in Figure 4.13. The values for $t = 0$ are zeros. As a sample, let us look at the computations for $t = 1$ hr $= 60$ min: rainfall intensity $i = 486.66/(60\text{ min} + 4.79)^{0.66} = 31.02$ mm hr^{-1}; cumulative rainfall $=$ (31.02 mm hr^{-1})$*$

	A	B	C	D	E	F	G	H
1	**Balanced triangular distribution**					Areal adjustment factor =		0.92
2								
3	Time		Intensity	Cumulative	Interval	Bell-Shaped Distribution		Areal-Adjusted
4	t		i	Rainfall	Rainfall	Rainfall	Rank	Rainfall
5	(hr)	(min)	(mm hr-1)	(mm)	(mm)	(mm)		(mm)
6	0	0	0.00	0.00	0.00	0.00	25	0
7	1	60	31.02	31.02	31.02	1.47	23	1.35
8	2	120	20.13	40.26	9.24	1.50	21	1.38
9	3	180	15.53	46.59	6.33	1.65	19	1.52
10	4	240	12.90	51.60	5.01	1.77	17	1.63
11	5	300	11.16	55.80	4.20	1.92	15	1.77
12	6	360	9.91	59.46	3.66	2.14	13	1.97
13	7	420	8.97	62.79	3.33	2.37	11	2.18
14	8	480	8.22	65.76	2.97	2.73	9	2.51
15	9	540	7.61	68.49	2.73	3.33	7	3.06
16	10	600	7.10	71.00	2.51	4.20	5	3.86
17	11	660	6.67	73.37	2.37	6.33	3	5.82
18	12	720	6.30	75.60	2.23	31.02	1	28.54
19	13	780	5.98	77.74	2.14	9.24	2	8.5
20	14	840	5.70	79.80	2.06	5.01	4	4.61
21	15	900	5.44	81.60	1.80	3.66	6	3.37
22	16	960	5.22	83.52	1.92	2.97	8	2.73
23	17	1020	5.01	85.17	1.65	2.51	10	2.31
24	18	1080	4.83	86.94	1.77	2.23	12	2.05
25	19	1140	4.66	88.54	1.60	2.06	14	1.9
26	20	1200	4.51	90.20	1.66	1.80	16	1.66
27	21	1260	4.36	91.56	1.36	1.66	18	1.53
28	22	1320	4.23	93.06	1.50	1.60	20	1.47
29	23	1380	4.11	94.53	1.47	1.47	22	1.35
30	24	1440	4.00	96.00	1.47	1.36	24	1.25

Figure 4.13 **Computations for the storm using a balanced triangular distribution.** Solution for part (1) of Example 4.2.

(1 hr) = 31.02 mm; interval rainfall = (31.02 mm) − (0 mm) = 31.02 mm. Similarly, the values of rainfall intensity, cumulative rainfall, and interval rainfall for t = 2, 3, . . . , 24 hr can be computed, as shown in columns C, D, and E, respectively.

The values in column E are re-ordered so that the largest value is positioned at t = 12 hr, the second largest value at t = 13 hr, the third largest value at t = 11 hr, the fourth largest value at t = 14 hr, the fifth largest value at t = 10 hr, and so on, leading to the bell-shaped distribution indicated by the values in column F.

Column F is multiplied by the areal adjustment factor of 0.92 to determine the values of areal-adjusted rainfall, which is the 50-year 24-hr synthetic storm, assuming a balanced triangular distribution.

(2) The computations and formulated storm are shown in Figure 4.14. The values for $t = 0$ are zeros. The storm duration is $t = 24$ hr $= 1440$ min; so the durational rainfall intensity is $i = 486.66/(1440\,\text{min} + 4.79)^{0.66} = 4.00\ \text{mm hr}^{-1}$ and the durational rainfall is $R = (4.00\ \text{mm hr}^{-1}) * (24\ \text{hr}) = 96$ mm. Dallas, in east-central TX, has Type II rainfall distribution and rainfall distribution coefficients, provided in Table 4.8 and listed in column N.

	M	N	O	P	Q	R
1	NRCS 24-hr rainfall distribution				Areal adjustment factor =	0.92
2	24-hr intensity =	4.00	mm hr^{-1}	24-hr rainfall =	96.00	mm
3						
4	Time t	Coefficient from	Cumulative Rainfall	Interval Rainfall	Areal-Adjusted Rainfall	
5	(hr)	Table 3-25	(mm)	(mm)	(mm)	
6	0	0.0000	0.00	0.00	0.00	
7	1	0.0108	1.04	1.04	0.96	
8	2	0.0223	2.14	1.10	1.01	
9	3	0.0347	3.33	1.19	1.09	
10	4	0.0483	4.64	1.31	1.21	
11	5	0.0632	6.07	1.43	1.32	
12	6	0.0797	7.65	1.58	1.45	
13	7	0.0984	9.45	1.80	1.66	
14	8	0.1203	11.55	2.10	1.93	
15	9	0.1467	14.08	2.53	2.33	
16	10	0.1808	17.36	3.28	3.02	
17	11	0.2351	22.57	5.21	4.79	
18	12	0.6632	63.67	41.10	37.81	
19	13	0.7724	74.15	10.48	9.64	
20	14	0.8197	78.69	4.54	4.18	
21	15	0.8538	81.96	3.27	3.01	
22	16	0.8801	84.49	2.53	2.33	
23	17	0.9019	86.58	2.09	1.92	
24	18	0.9206	88.38	1.80	1.66	
25	19	0.9371	89.96	1.58	1.45	
26	20	0.9519	91.38	1.42	1.31	
27	21	0.9653	92.67	1.29	1.19	
28	22	0.9777	93.86	1.19	1.09	
29	23	0.9892	94.96	1.10	1.01	
30	24	1.0000	96.00	1.04	0.96	

Figure 4.14 **Computations for the storm using the NRCS 24-hr rainfall distribution.** Solution for part (2) of Example 4.2.

Column N is multiplied by R = 96 mm to get the values of cumulative rainfall (column O), which in turn are used to compute the values of interval rainfall (column P). As a sample, let us look at the computations for t = 1 hr: the distribution coefficient is 0.0108; the cumulative rainfall is $(0.0108) * (96\ \text{mm}) = 1.04\ \text{mm}$; the interval rainfall is $(1.04\ \text{mm}) - (0\ \text{mm}) = 1.04\ \text{mm}$; and the areal-adjusted rainfall is $(1.04\ \text{mm}) * (0.92) = 0.96\ \text{mm}$. Similarly, the values of cumulative rainfall, interval rainfall, and areal-adjusted rainfall for $t = 2, 3, \ldots, 24$ hr can be computed, as shown in columns O, P, and Q, respectively. Column Q is the 50-year 24-hr synthetic storm derived by assuming NRCS 24-hr rainfall distribution.

In summary: Figure 4.15 compares the synthetic storm of a balanced triangular distribution with that of a NRCS 24-hr rainfall distribution. The two storms are comparable except for the peaks. For this example, the NRCS distribution yields a larger peak than the balanced triangular distribution. Albeit, the durational rainfall for the two distributions is the same.

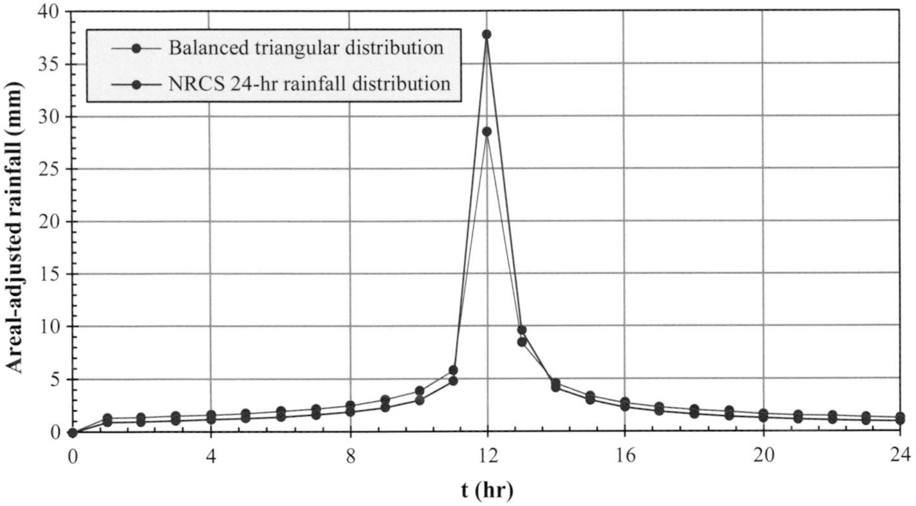

Figure 4.15 **The synthetic storms formulated in Example 4.2.**

4.1.4 Probable Maximum Storm

Probable maximum precipitation (PMP) is the theoretical maximum precipitation for a given duration under modern meteorological conditions. Such a precipitation is likely to happen over a design watershed, or a storm area of a given size, at a certain time of year (WMO, 2009). The PMP is primarily considered to be the precipitation resulting from a storm induced by the optimal dynamic factor (usually the precipitation efficiency) and the maximum moisture factor simultaneously. There are six methods of PMP estimation currently used, namely the: (1) local method (local storm maximization or local model); (2) transposition method (storm transposition or transposition model); (3) combination method (temporal and spatial maximization of storm combination model); (4) inferential method (theoretical model or ratiocination model); (5) generalized method (generalized estimation); and (6) statistical method (statistical estimation). Once PMP is determined, the probable maximum storm (PMS) can be formulated using either

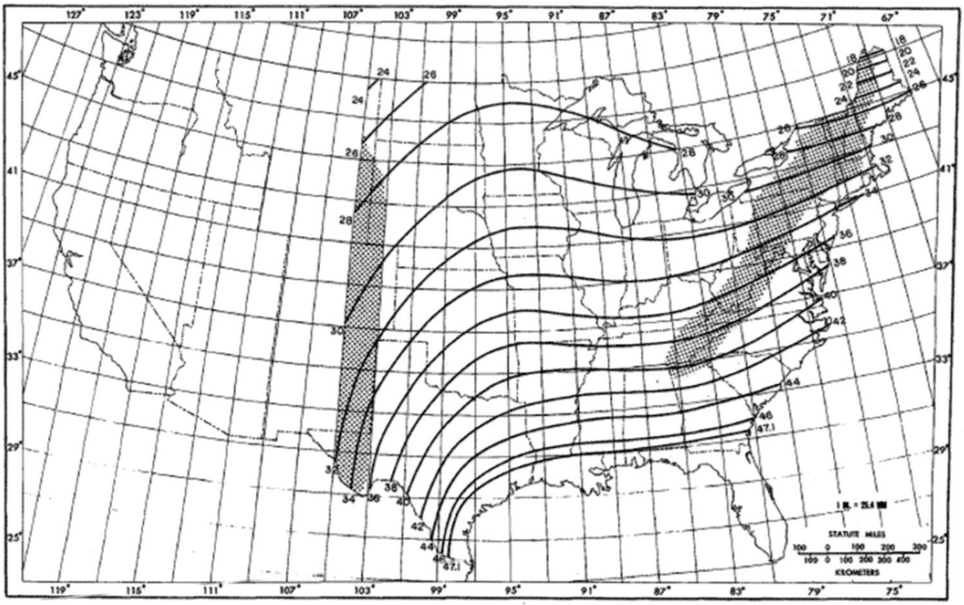

Figure 4.16 **Probable maximum precipitation contours shown on a map of the United States for a 24-hr duration and a drainage area of 10 mi^2.** The contours have units of in. (*Source:* NOAA, 1978)

the balanced triangular distribution or NRCS 24-hr rainfall distribution method, as discussed in Section 4.1.3.

NOAA (1978) developed PMP maps for the contiguous United States for five durations (6, 12, 24, 48, and 72 hr) and six area sizes (10, 200, 1000, 5000, 10,000, and 20,000 mi^2). Given that the 24-hr duration is most commonly used in practice, the six maps for this duration are reproduced and shown in Figures 4.16 to 4.21.

4.2 Estimating the Design Peak Discharge

Peak discharge is an important parameter to size hydraulic structures such as channels and culverts. This section presents the commonly used estimation methods of design peak discharge.

4.2.1 Maximum Historical Value

If the magnitude of the historical maximum peak discharge can be determined, it may be used as the design discharge (Wang *et al.*, 2010b). There are two methods to determine the maximum peak discharge. The first method simply computes the maximum value of recorded annual peak discharges, and the second method uses flood marks on trees, rocks, bridges, and buildings (e.g., Figure 4.22) to estimate the maximum peak values (Acreman, 1989). A flood mark is any line indicative of the maximum water height during an event. Such a design discharge is not associated with a return period or frequency.

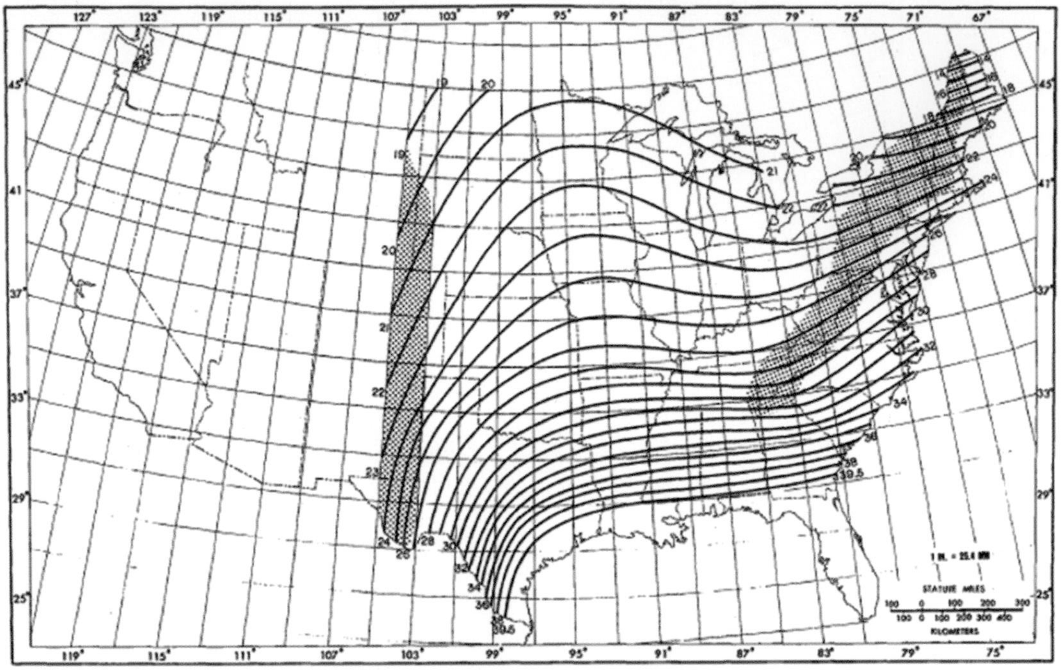

Figure 4.17 **Probable maximum precipitation contours shown on a map of the United States for a 24-hr duration and a drainage area of 200 mi^2.** The contours have units of in. (*Source:* NOAA, 1978)

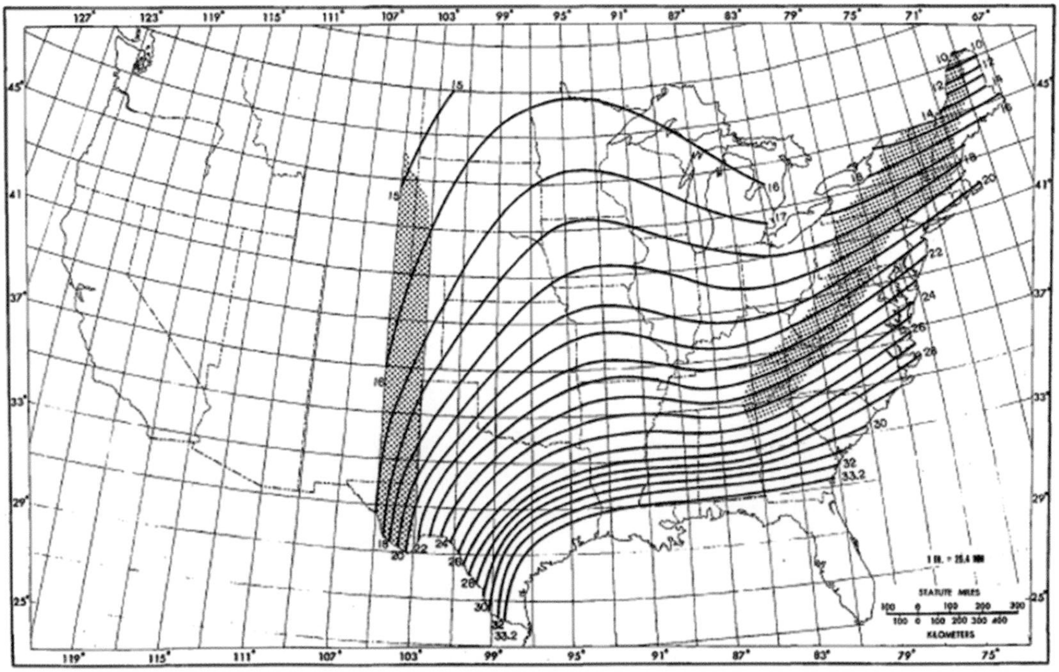

Figure 4.18 **Probable maximum precipitation contours shown on a map of the United States for a 24-hr duration and a drainage area of 1000 mi^2.** The contours have units of in. (*Source:* NOAA, 1978)

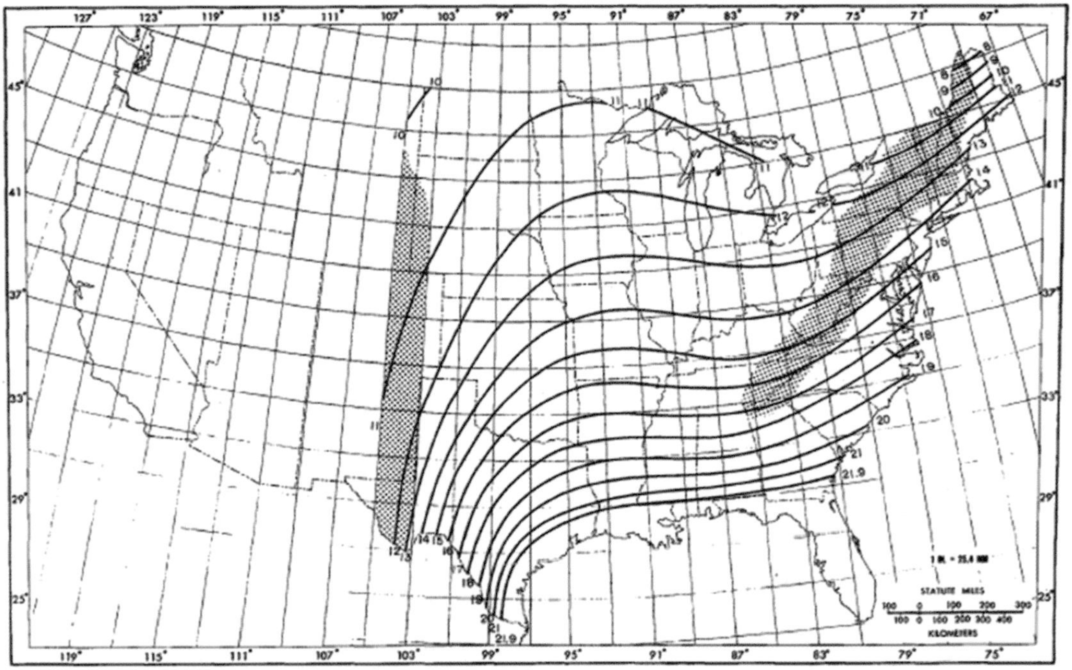

Figure 4.19 **Probable maximum precipitation contours shown on a map of the United States for a 24-hr duration and a drainage area of 5000 mi^2.** The contours have units of in. (*Source:* NOAA, 1978)

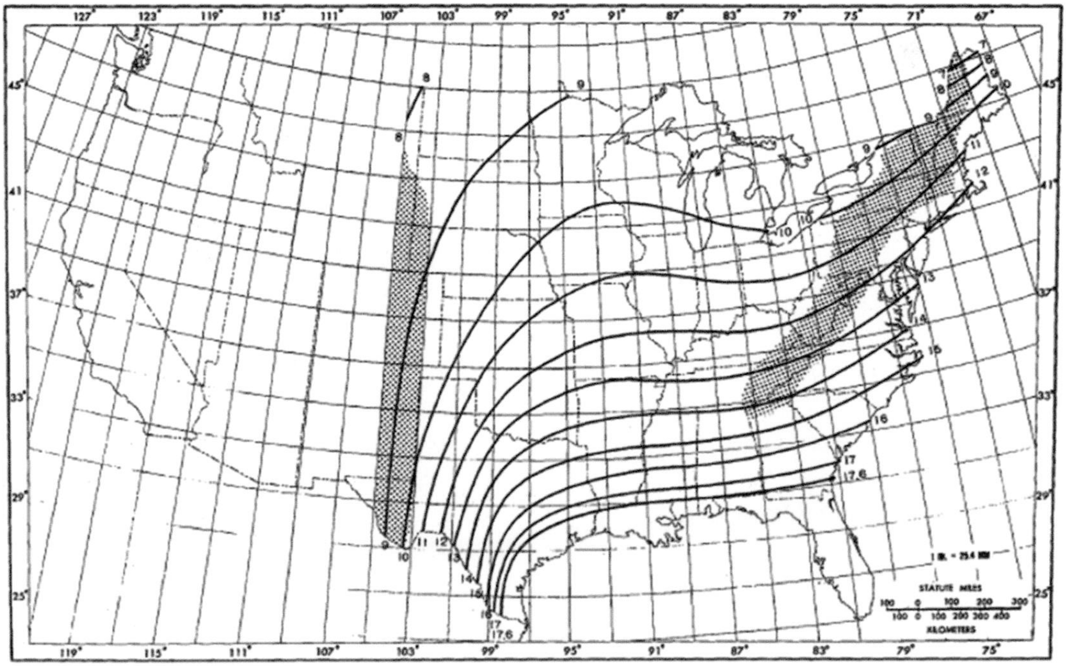

Figure 4.20 **Probable maximum precipitation contours shown on a map of the United States for a 24-hr duration and a drainage area of 10,000 mi^2.** The contours have units of in. (*Source:* NOAA, 1978)

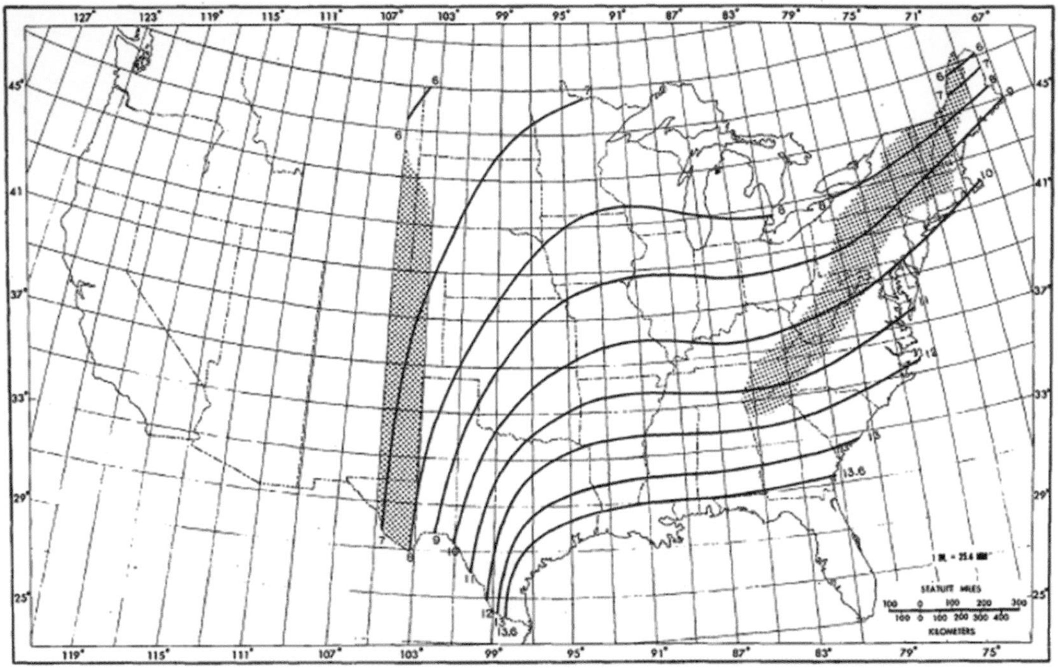

Figure 4.21 **Probable maximum precipitation contours shown on a map of the United States for a 24-hr duration and a drainage area of 20,000 mi^2.** The contours have units of in. (*Source:* NOAA, 1978)

Figure 4.22 **Sample flood marks.** (*Source:* kumikomini / E+ / Getty Images (left); Martin Siepmann / Getty Images (right))

4.2.2 Modeled Value

Two approaches have been widely used to estimate the design peak discharge. These are the Rational Equation and the Soil Conservation Service (SCS) Graphical Peak Discharge Method (Mulvaney, 1851; Turazza, 1880; Kuichling, 1889; Chow, 1964; USDA-NRCS, 1986, Wang *et al.*, 2012). Whereas the Rational Equation uses the rainfall intensity of a design storm to compute the peak discharge, the SCS Graphical Peak Discharge Method uses the durational rainfall of a design storm to compute the peak discharge. Herein, it is assumed that the frequency of the design storm is the same as the frequency of the computed peak discharge though this assumption may be invalid in the real world (Wang *et al.*, 2010b). The Rational Equation is usually used for drainage areas of 20 ac or smaller, while the SCS Graphical Peak Discharge Method can be applied for larger watersheds (Wang *et al.*, 2012).

The Rational Equation can be expressed as:

$$Q_p = KCiA \tag{4.14}$$

- Q_p: peak discharge [cfs for BG; $m^3\ s^{-1}$ for SI]
- C: runoff coefficient (Table 4.9)
- K: units conversion factor (= 1.0 for BG; 0.002778 for SI)
- i: rainfall intensity is relative to time of concentration t_c [in hr^{-1} for BG; mm hr^{-1} for SI]
- A: drainage area [ac for BG; ha for SI]

In Eq. (4.14), the runoff coefficient C depends on four factors, namely land use, hydrologic soil group (HSG) (Table 3.1), overland slope, and storm magnitude. For a given combination of the first three factors, the larger C value should be used for storms with a return period of 25-year or larger, while the smaller C value should be used for storms with a smaller return period. For instance, for the combination of pasture, HSG B, and 2.5% overland slope, $C = 0.28$ for storms of smaller than 25-year, while $C = 0.34$ for storms of 25-year or larger. In addition, the rainfall intensity i should be determined relative to the time of concentration, t_c, of the watershed of interest, when the entire drainage area starts to contribute runoff. Thus, if an IDF curve or a Steel Formula is used to determine i, the duration should be t_c. However, for a real storm with a duration of D and an average intensity of i, if $D \geq t_c$, i and A should be used to compute Q_p; otherwise, i is used but A must be proportionally reduced by a ratio (e.g., D/t_c). For instance, if a 48-ac watershed has a $t_c = 30$ min, the duration of 30 min should be used to determine i from an IDF curve or a Steel Formula. For a 1-hr real storm with a rainfall intensity of 1.5 in hr^{-1}, $i = 1.5$ in hr^{-1} and $A = 48$ ac should be used in Eq. (4.14). In contrast, for a 20-min real storm with the same rainfall intensity, $i = 1.5$in hr^{-1} and $A = [(20\,\text{min})/(30\,\text{min})] * (48\ \text{ac}) = 32$ ac need to be used.

Example 4.3 The land uses of a 10 ac urban catchment (HSG C and 3% overland slope) consist of 2 ac commercial land, 1 ac street, 4 ac open space, and 3 ac parking lot. The IDF curves shown in Figure 4.5 are applicable to the catchment. If the time of concentration of the catchment is 30 min, estimate the 50-year peak discharge using the Rational Equation.

Solution

$A = (10\ \text{ac}) * [(1\ \text{ha})/(2.471\ \text{ac})] = 4.05$ ha

Table 4.9 Runoff coefficient for the Rational Equation.[1]

Land use		HSG A			HSG B			HSG C			HSG D			Return period of storm
		< 2%	42 ∼ 6%	> 6%	< 2%	2 ∼ 6%	> 6%	< 2%	2 ∼ 6%	> 6%	< 2%	2 ∼ 6%	> 6%	
Rural														
	cultivated	0.08	0.13	0.16	0.11	0.15	0.21	0.14	0.19	0.26	0.18	0.23	0.31	< 25-year
		0.14	0.18	0.22	0.16	0.21	0.28	0.20	0.25	0.34	0.24	0.29	0.41	≥ 25-year
	pasture	0.12	0.20	0.30	0.18	0.28	0.37	0.24	0.34	0.44	0.30	0.40	0.50	< 25-year
		0.15	0.25	0.37	0.23	0.34	0.45	0.30	0.42	0.52	0.37	0.50	0.62	≥ 25-year
	meadow	0.10	0.16	0.25	0.14	0.22	0.30	0.20	0.28	0.36	0.24	0.30	0.40	< 25-year
		0.14	0.22	0.30	0.20	0.28	0.37	0.26	0.35	0.44	0.30	0.40	0.50	≥ 25-year
	forest	0.05	0.08	0.11	0.08	0.11	0.14	0.10	0.13	0.16	0.12	0.16	0.20	< 25-year
		0.08	0.11	0.14	0.10	0.14	0.18	0.12	0.16	0.20	0.15	0.20	0.25	≥ 25-year
Urban														
	residential													
	1/8 ac	0.25	0.28	0.31	0.27	0.30	0.35	0.30	0.33	0.38	0.33	0.36	0.42	< 25-year
		0.33	0.37	0.40	0.35	0.39	0.44	0.38	0.42	0.49	0.41	0.45	0.54	≥ 25-year
	1/4 ac	0.22	0.26	0.29	0.24	0.29	0.33	0.27	0.31	0.36	0.30	0.34	0.40	< 25-year
		0.30	0.34	0.37	0.33	0.37	0.42	0.36	0.40	0.47	0.38	0.42	0.52	≥ 25-year
	1/3 ac	0.19	0.23	0.26	0.22	0.26	0.30	0.25	0.29	0.34	0.28	0.32	0.39	< 25-year
		0.28	0.32	0.35	0.30	0.35	0.39	0.33	0.38	0.45	0.36	0.40	0.50	≥ 25-year
	1/2 ac	0.16	0.20	0.24	0.19	0.23	0.28	0.22	0.27	0.32	0.26	0.30	0.37	< 25-year
		0.25	0.29	0.32	0.28	0.32	0.36	0.31	0.35	0.42	0.34	0.38	0.48	≥ 25-year
	1 ac	0.14	0.19	0.22	0.17	0.21	0.26	0.20	0.25	0.31	0.24	0.29	0.35	< 25-year
		0.22	0.26	0.29	0.24	0.28	0.34	0.28	0.32	0.40	0.31	0.35	0.46	≥ 25-year

Table 4.9 *(Continued)*

Land use	HSG A			HSG B			HSG C			HSG D			Return period of storm
	< 2%	42 ~ 6%	> 6%	< 2%	2 ~ 6%	> 6%	< 2%	2 ~ 6%	> 6%	< 2%	2 ~ 6%	> 6%	
industrial	0.67	0.68	0.68	0.68	0.68	0.69	0.68	0.69	0.69	0.69	0.69	0.70	< 25-year
	0.85	0.85	0.86	0.85	0.86	0.86	0.86	0.86	0.87	0.86	0.86	0.88	≥ 25-year
commercial	0.71	0.71	0.72	0.71	0.72	0.72	0.72	0.72	0.72	0.72	0.72	0.72	< 25-year
	0.88	0.88	0.89	0.89	0.89	0.89	0.89	0.89	0.90	0.89	0.89	0.90	≥ 25-year
street	0.70	0.71	0.72	0.71	0.72	0.74	0.72	0.73	0.76	0.73	0.75	0.78	< 25-year
	0.76	0.77	0.79	0.80	0.82	0.84	0.84	0.85	0.89	0.89	0.91	0.95	≥ 25-year
open space	0.05	0.10	0.14	0.08	0.13	0.19	0.12	0.17	0.24	0.15	0.21	0.28	< 25-year
	0.11	0.16	0.20	0.14	0.19	0.26	0.18	0.23	0.32	0.22	0.27	0.39	≥ 25-year
parking	0.85	0.86	0.87	0.85	0.86	0.87	0.85	0.86	0.87	0.85	0.86	0.87	< 25-year
	0.95	0.96	0.97	0.95	0.96	0.97	0.95	0.96	0.97	0.95	0.96	0.97	≥ 25-year

[1] Reproduced from USDA-NRCS (1986) and VDCR (1999).

Table 4.10 Adjustment factor, F_p, for pond and swamp areas across watershed.[1]

Percentage of pond and/or swamp areas	F_p
0	1.00
0.2	0.97
1.0	0.87
3.0	0.75
5.0	0.72

[1] *Source:* USDA-NRCS (1986).

HSG C, 3% slope, and 50-year storm → Table 4.9: runoff coefficients for commercial, street, open space, and parking are 0.89, 0.85, 0.23, and 0.96, respectively → composite runoff coefficient for the catchment $C = [(2\text{ ac}) * 0.89 + (1\text{ ac}) * 0.85 + (4\text{ ac}) * 0.23 + (3\text{ ac}) * 0.96]/[(2\text{ ac}) + (1\text{ ac}) + (4\text{ ac}) + (3\text{ ac})] = 0.643$.

$t_c = 30$ min, 50-year storm → Figure 4.5: $i = 46\text{ mm hr}^{-1}$ → Eq. (4.14): $Q_p = 0.002778 * 0.643 * (46\text{ mm hr}^{-1}) * (4.05\text{ ha}) = 0.33\text{ m}^3\text{ s}^{-1}$.

Example 4.4 If the catchment in Example 4.3 receives a 10-min storm on one day and a 1-hr storm on another day, both storms have a rainfall intensity of 2.3 in hr^{-1}, estimate the peak discharges using the Rational Equation.

Solution

$C = 0.643$, $t_c = 30$ min. Use Eq. (4.14) for these calculations.

10-min storm $< t_c \rightarrow i = 2.3\text{ in hr}^{-1}$, $A = [(10\text{ min})/(30\text{ min})] * (10\text{ ac}) = 10/3\text{ ac} \rightarrow Q_p = 0.643 * (2.3\text{ in hr}^{-1}) * (10/3\text{ ac}) = 4.93$ cfs.

1-hr storm $> t_c \rightarrow i = 2.3\text{ in hr}^{-1}$, $A = 10\text{ ac} \rightarrow Q_p = 0.643 * (2.3\text{ in hr}^{-1}) * (10\text{ ac}) = 14.79$ cfs.

The SCS Graphical Peak Discharge Method computes peak discharge as:

$$Q_p = q_u Q A F_p \tag{4.15}$$

peak discharge [cfs] → Q_p
unit peak from Figure 4.23 [csm in^{-1}] → q_u
excess rainfall from Eq. (3.48) [in] → Q
drainage area [mi^2] → A
adjustment factor for pond/swamp areas from Table 4.10 [-] → F_p

In Eq. (4.15), q_u is the peak discharge per unit drainage area per unit excess rainfall and has units of csm in^{-1}, which stands for cfs $\text{mi}^{-2}\text{ in}^{-1}$. F_p takes into account the storage and attenuation effects of ponds, swamps, marshes, and wetlands on runoff peak. This method can be implemented as the following seven steps: (1) compute initial abstraction I_a using Eq. (3.50) or (3.55) and the ratio of I_a to rainfall P (i.e., I_a/P); (2) compute excess rainfall Q using Eq. (3.48); (3) estimate time of concentration t_c using Eqs. (3.2) to (3.11); (4) identify the rainfall distribution type from Figure 4.12; (5) determine unit peak q_u from Figure 4.23 by linear interpolation or extrapolation; (6) determine F_p from Table 4.10 by linear interpolation or extrapolation; and (7) compute Q_p using Eq. (4.15).

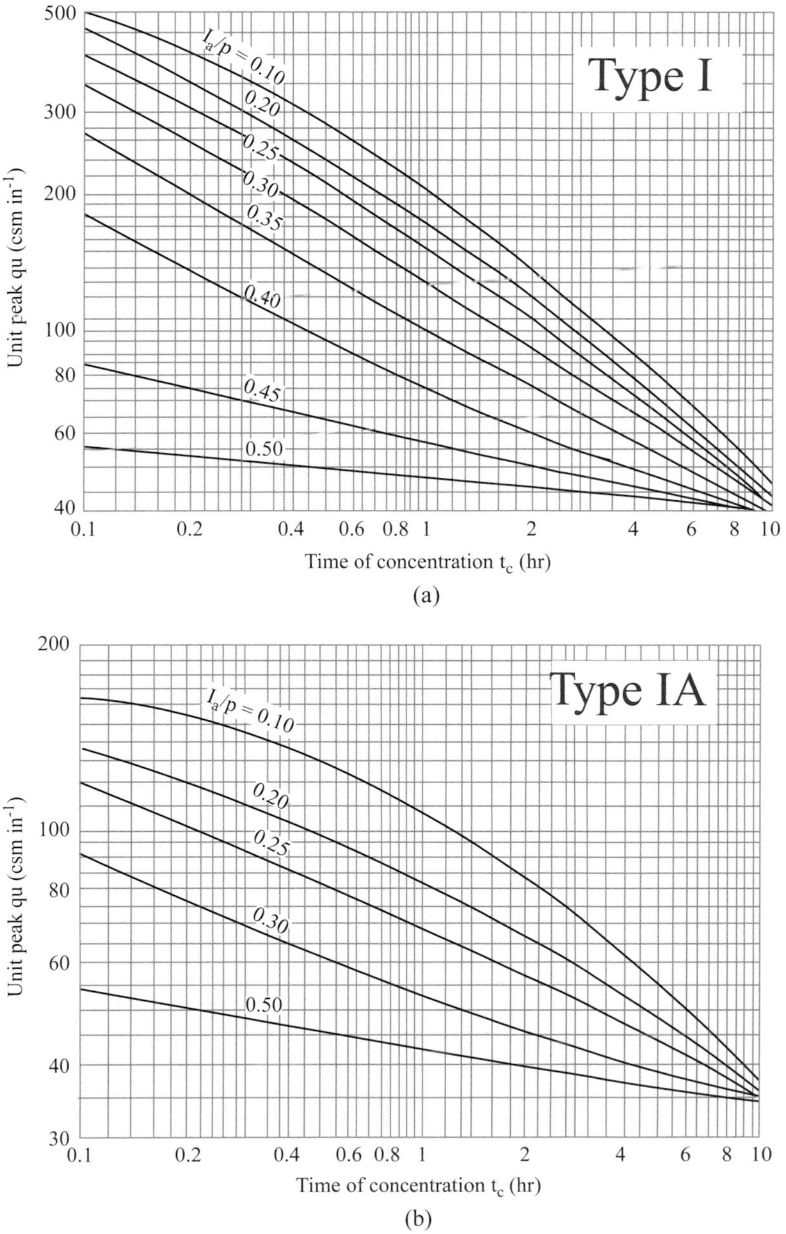

Figure 4.23 **Determination of unit peak discharge, q_u.** Nomograms for rainfall distribution Type: (a) I; (b) IA; (c) II; and (d) III. (*Source:* USDA-NRCS, 1986)

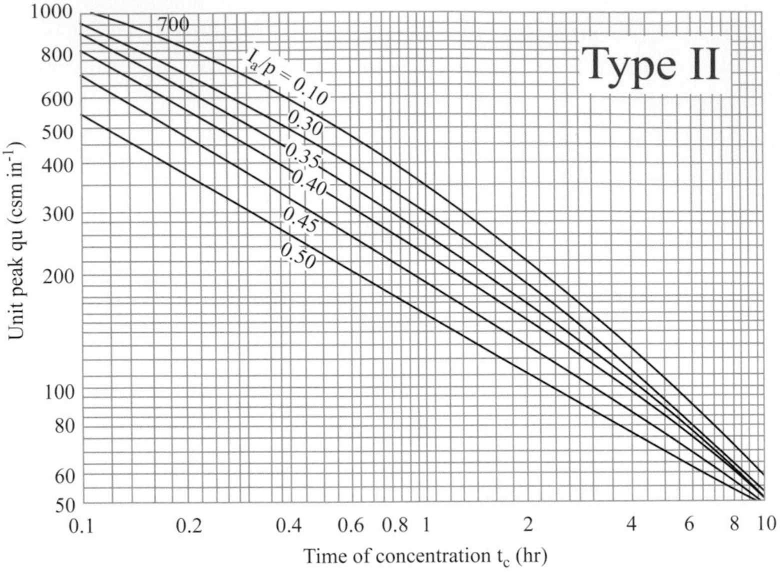

(c)

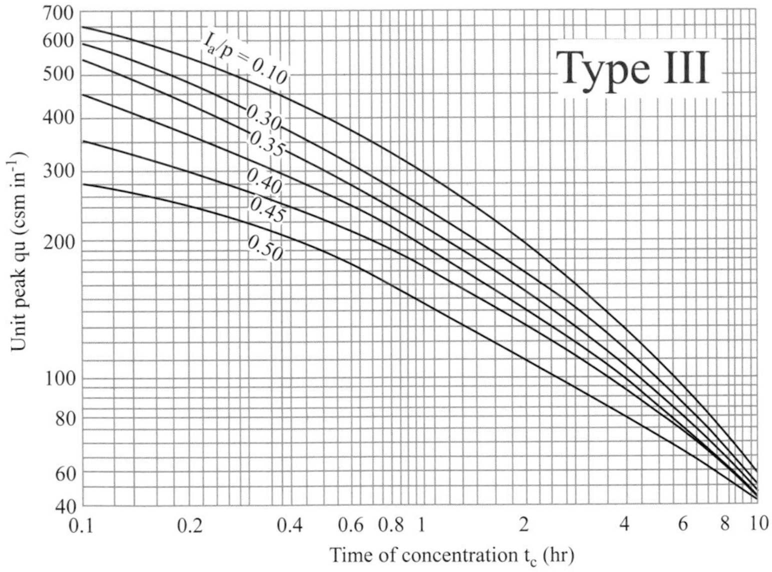

(d)

Figure 4.23 *(Continued)*

Example 4.5 A 130 mi^2 watershed, located in coastal Virginia, has a composite curve number of 83, a time of concentration of $t_c = 1.5$ hr, and 5% pond and swamp areas. If the 50-year 1-hr rainfall is 3 in, determine the peak discharge using the SCS Graphical Peak Discharge Method.

Solution

$A = 130\ mi^2$, $CN = 83$, $t_c = 1.5$ hr, $P = 3$ in. Follow the seven steps.

Step 1: Eq. (3.49) $S = (1000/83 - 10)in = 2.05\ in \rightarrow$ Eq. (3.50) $I_a = 0.2 * (2.05\ in) = 0.42\ in \rightarrow I_a/P = (0.42\ in)/(3\ in) = 0.14$.

Step 2: Eq. (3.48) $Q = [(3\ in) - (0.42\ in)]^2 / [(3\ in) - (0.42\ in) + (2.1\ in)] = 1.42$ in.

Step 3: Given $t_c = 1.5$ hr.

Step 4: Figure 4.12 $\rightarrow$ Type III rainfall distribution for coastal Virginia.

Step 5: Figure 4.23d $\rightarrow q_u = 240\ csm\ in^{-1}$ for $I_a/P = 0.30$ and $q_u = 300\ csm\ in^{-1}$ for $I_a/P = 0.10 \rightarrow$ linear interpolation, for $I_a/P = 0.14$, $q_u = (240\ csm\ in^{-1}) + [(300\ csm\ in^{-1}) - (240\ csm\ in^{-1})]/(0.10 - 0.30) * (0.14 - 0.30) = 288\ csm\ in^{-1}$.

Step 6: Table 4.10 $\rightarrow F_p = 0.72$ for 5% pond and swamp areas.

Step 7: Eq. (4.15) $\rightarrow Q_p = (288\ csm\ in^{-1}) * (1.42\ in) * (130\ mi^2) * (0.72) = 38,279$ cfs.

4.2.3 Flood Frequency Analysis

Flood frequency is the chance (i.e., probability) for a flood magnitude to be equaled or exceeded in any given year. Its reciprocal is defined as the return period, which has units of year. Algebraically, the two are related as:

$$\text{return period} \longrightarrow T = \frac{1}{Pr} \longleftarrow \text{flood frequency} \tag{4.16}$$

In engineering applications, the commonly used return periods (flood frequencies) include 2-year (50%), 5-year (20%), 10-year (10%), 50-year (2%), 100-year (1%), and 500-year (0.2%), though other values are also used. If data on annual peak discharge are available for a number of (usually 10 or more) years, flood magnitudes for several selected frequencies can be determined by fitting a statistical distribution to the data. The plot showing flood magnitudes versus frequencies is called a flood frequency curve. The magnitude for a frequency of interest is the design peak discharge. Various flood frequency analysis methods have been reported in literature (Wang *et al.*, 2010b). The method documented in Bulletin 17B (USGS, 1982) is presented because it has been most widely used in practice. This method assumes that the 10 base logarithms of peak discharges follow a log-Pearson III distribution. It is implemented by following six steps:

Step 1: Sort the peak discharges $Q_1, Q_2, \ldots, Q_n$, for n years of record in descending order (i.e., from the largest to the smallest), forming $Q'_1 < Q'_2 < \ldots < Q'_n$.

Step 2: Assign consecutive ranks of $j = 1, 2, \ldots, n$ to $Q'_1 < Q'_2 < \ldots < Q'_n$.

Step 3: Compute the empirical frequency of $Q'_j (j = 1, 2, \ldots, n))$ as:

$$\Pr\left(Q \geq Q'_j\right) = \frac{j}{n+1}(100\%) \tag{4.17}$$

empirical frequency of discharge Q'_j → $\Pr(Q \geq Q'_j)$; rank of discharge Q'_j → j; years of record → n

Step 4: Fit a log-Pearson III distribution to the pairs of $(\Pr(Q \geq Q'_j), Q'_j)$ as follows:

$$x_j = \log_{10}\left(Q'_j\right) \tag{4.18}$$

base-10 logarithm of discharge Q'_j → x_j; discharge → Q'_j

$$\bar{x} = \frac{\sum_{j=1}^{n} x_j}{n} \tag{4.19}$$

mean → $\bar{x}$; base-10 logarithm of discharge Q'_j → x_j; years of record → n

$$S_x = \sqrt{\frac{\sum_{j=1}^{n}\left(x_j - \bar{x}\right)^2}{n-1}} \tag{4.20}$$

standard deviation → S_x; base-10 logarithm of discharge Q'_j → x_j; mean → $\bar{x}$; years of record → n

$$y = \bar{x} + KS_x \tag{4.21}$$

base-10 logarithm of fitted discharge → y; mean → $\bar{x}$; frequency-related factor → K; standard deviation → S_x

$$K = \frac{2}{a}\left\{\left[\left(z - \frac{a}{6}\right)\frac{a}{6} + 1\right]^3 - 1\right\} \tag{4.22}$$

frequency-related factor → K; skew coefficient → a; standard normal deviate for the frequency (Table 4.11) → z

Table 4.11 The standard normal deviate for flood frequency.[1]

Pr (%)	99	95	90	80	50	20	10	4	2	1	0.5
T (year)	1.0101	1.0526	1.1111	1.25	2	5	10	25	50	100	200
z	−2.326	−1.645	−1.282	−0.842	0	0.842	1.282	1.751	2.054	2.326	2.576

[1] Pr: frequency; T: return period; and z: standard normal deviate.

In Eq. (4.22), a can be either the station or weighted skew coefficient. The station skew coefficient can be computed as:

$$\underbrace{G}_{\text{station skew coefficient}} = \frac{\overbrace{n}^{\text{years of record}}}{(n-1)(n-2)} \sum_{j=1}^{n} \left(\frac{\overbrace{x_j}^{\text{base-10 logarithm of discharge } Q'_j} - \overbrace{\bar{x}}^{\text{mean}}}{\underbrace{S_x}_{\text{standard deviation}}} \right)^3 \tag{4.23}$$

The weighted skew coefficient considers the regional distribution and is computed as:

$$\underbrace{G_w}_{\text{weighted skew coefficient}} = \frac{0.302\overbrace{G}^{\text{station skew coefficient (Eq. (3.95))}} + \overbrace{C_R}^{\text{weight of regional distribution (Eq. (3.97))}}\overbrace{G_R}^{\text{regional skew coefficient (Figure 4.24)}}}{0.302 + C_R} \tag{4.24}$$

$$\underbrace{C_R}_{\text{weight of regional distribution}} = 10^{\left[\underbrace{A}_{\text{station constant (Eq. (3.98))}} - \underbrace{B}_{\text{station constant (Eq. (3.99))}} \log_{10}\left(\frac{\overbrace{n}^{\text{years of record}}}{10}\right)\right]} \tag{4.25}$$

$$\underbrace{A}_{\text{station constant}} = \begin{cases} -0.33 + 0.08\,|\overbrace{G}^{\text{station skew coefficient (Eq. (3.95))}}| & \text{if } |G| \le 0.9 \\ -0.52 + 0.30\,|G| & \text{if } |G| > 0.9 \end{cases} \tag{4.26}$$

$$\underbrace{B}_{\text{station constant}} = \begin{cases} 0.94 - 0.26\,|\overbrace{G}^{\text{station skew coefficient (Eq. (3.95))}}| & \text{if } |G| \le 1.5 \\ 0.55 & \text{if } |G| > 1.5 \end{cases} \tag{4.27}$$

The Excel functions for Eqs. (4.18), (4.19), (4.20), and (4.23) are *log10( ), average( ), stdev( ),* and *skew( )*, to compute base-10 logarithm (x_j), mean ($\bar{x}$), standard deviation (S_x), and station skew coefficient (G), respectively. The standard normal deviate z in Eq. (4.22) can be computed using function *norm.inv( )* in Excel.

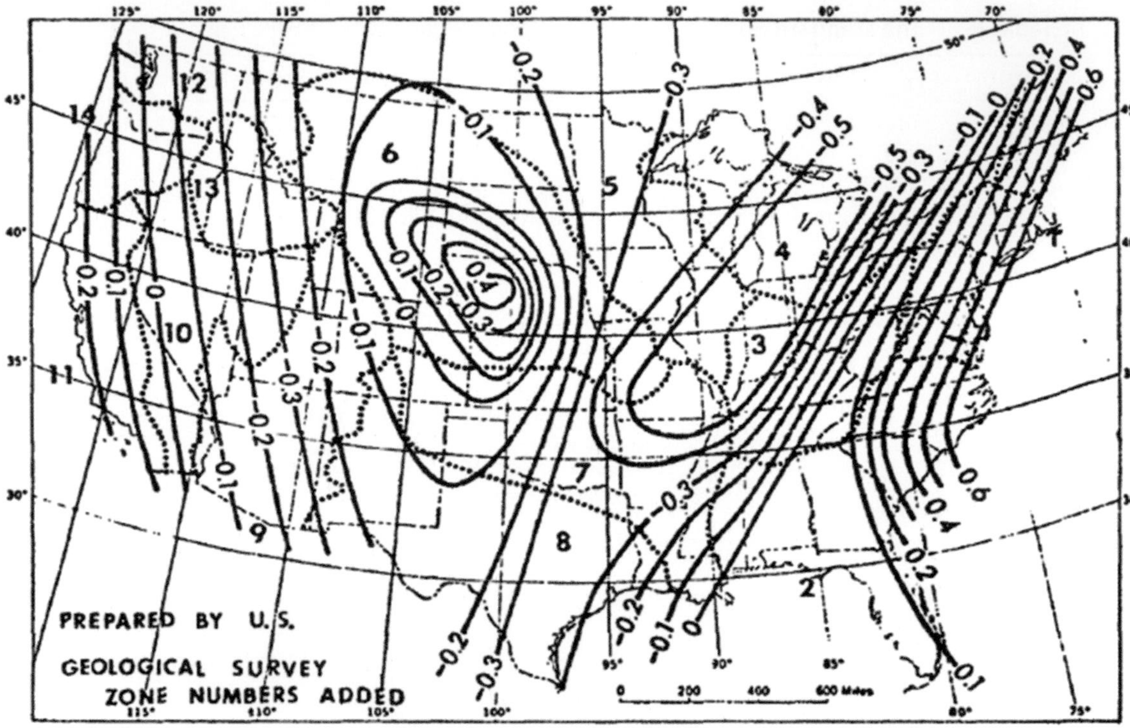

Figure 4.24 **Contours of regional skew coefficients of annual maximum streamflow logarithms shown on a map of the United States.** (*Source:* USGS, 1982)

Step 5: Determine the 95% confidence limits as:

lower limit → y_{ll}; base-10 logarithm of fitted discharge → y; lower-limit coefficient (Table 4.12) → k_{ll}; standard deviation → S_x

$$y_{ll} = y - k_{ll}S_x \tag{4.28}$$

upper limit → y_{ul}; base-10 logarithm of fitted discharge → y; upper-limit coefficient (Table 4.12) → k_{ul}; standard deviation → S_x

$$y_{ul} = y + k_{ul}S_x \tag{4.29}$$

Step 6: Antilog transform to obtain frequency flow 10^y and its 95% confidence upper limit $10^{y_{ul}}$ and lower limit $10^{y_{ll}}$, and then plot these values versus frequency to create the flood frequency curve.

Table 4.12 The coefficients for 95% confidence limits of the flood frequency curve.[1]

Years of record, n	Upper-limit coefficient, k_{ul}, for frequency (%) of						
	99.9	**99**	**90**	**50**	**10**	**1**	**0.1**
5	1.22	1.00	0.76	0.95	2.12	3.41	4.41
10	0.94	0.76	0.57	0.58	1.07	1.65	2.11
15	0.80	0.65	0.48	0.46	0.79	1.19	1.52
20	0.71	0.58	0.42	0.39	0.64	0.97	1.23
30	0.60	0.49	0.35	0.31	0.50	0.74	0.93
40	0.53	0.43	0.31	0.27	0.42	0.61	0.77
50	0.49	0.39	0.28	0.24	0.36	0.54	0.67
70	0.42	0.34	0.24	0.20	0.30	0.44	0.55
100	0.37	0.29	0.21	0.17	0.25	0.36	0.45
	0.1	**1**	**10**	**50**	**90**	**99**	**99.9**
	Lower-limit coefficient, k_{ll}, for frequency of						

[1] *Source:* USGS (1982).

Example 4.6

The data on annual peak discharges at a streamflow gauge are shown in columns A and B in Figure 4.25. Complete a flood frequency analysis.

Solution

n = 20. Use the Excel spreadsheet shown in Figure 4.25 to undertake the analysis.

The sorted data in descending order are in column C, and the consecutive ranks are in column D. The empirical frequency for a given rank is computed using Eq. (4.17) and stored in column E. For instance, $\Pr(Q \geq 4718 \text{ cfs}) = 1/(20+1) * 100\% = 4.76\%$, while $\Pr(Q \geq 4489 \text{ cfs}) = 2/(20+1) * 100\% = 9.52\%$.

Compute base-10 logarithms of Q'_j using Eq. (4.18); the results are shown in column F. Compute the mean $\bar{y}$ = *average(F3:F22)*, standard deviation S_y = *stdev(F3:F22)*, and station skew coefficient G = *skew(F3:F22)*. The computed values are shown in cells F24, F25, and F26, respectively. Let a = G to use the station skew coefficient in the analysis.

Use Eq. (4.22) to compute the frequency-related factor K for the empirical frequency of column E, with z value computed using the *norm.inv( )* function. For instance, the formula in cell G3 is "= 2/$F 26*(((NORM.INV(1-E3/100,0,1)–F$26/6) * (F$26/6) + 1) ∧ 3 – 1)." The results are shown in column G.

Use Eq. (4.21) to compute the base-10 logarithm of the fitted discharge; the results are shown in column H. For instance, cell H3 = $3.19 + 1.41 * 0.41 = 3.77$.

Determine coefficients for 95% confidence limits from Table 4.12 using a linear interpolation method. For instance, cell I3 = $0.58 + (0.42 - 0.58)/(10 - 1) * (4.76 - 1) = 0.51$, and cell K3 = $0.64 + (0.97 - 0.64)/(1 - 10) * (4.76 - 10) = 0.83$. The results for k_{ll} and k_{ul} are shown in columns I and K, respectively.

	A	B	C	D	E	F	G	H	I	J	K	L	M	N	O
1	Measured Peaks		Sorted	Rank	Empirical	Log10(Q_j')	Factor	Fitted	Lower Limit		Upper Limit			Antilog	
2	Year	Q_i (cfs)	Q_j' (cfs)	j	Pr(Q≥Q_j') (%)	x_j	K	y	k_{ll}	y_{ll}	k_{ul}	y_{ul}	Q (cfs)	Q_{ll} (cfs)	Q_{ul} (cfs)
3	1997	2209	4718	1	4.76	3.67	1.41	3.77	0.51	3.56	0.83	4.11	5888	3631	12882
4	1998	1767	4489	2	9.52	3.65	1.19	3.68	0.43	3.50	0.66	3.95	4786	3162	8913
5	1999	3175	3843	3	14.29	3.58	1.02	3.61	0.42	3.44	0.61	3.86	4074	2754	7244
6	2000	3742	3819	4	19.05	3.58	0.88	3.55	0.41	3.38	0.58	3.79	3548	2399	6166
7	2001	1117	3742	5	23.81	3.57	0.75	3.50	0.41	3.33	0.55	3.73	3162	2138	5370
8	2002	185	3175	6	28.57	3.50	0.63	3.45	0.41	3.28	0.52	3.66	2818	1905	4571
9	2003	876	3144	7	33.33	3.50	0.52	3.40	0.40	3.24	0.49	3.60	2512	1738	3981
10	2004	1058	2974	8	38.1	3.47	0.41	3.36	0.40	3.20	0.46	3.55	2291	1585	3548
11	2005	2974	2555	9	42.86	3.41	0.30	3.31	0.40	3.15	0.43	3.49	2042	1413	3090
12	2006	3144	2209	10	47.62	3.34	0.19	3.27	0.39	3.11	0.40	3.43	1862	1288	2692
13	2007	294	1767	11	52.38	3.25	0.07	3.22	0.40	3.06	0.39	3.38	1660	1148	2399
14	2008	4489	1558	12	57.14	3.19	-0.05	3.17	0.43	2.99	0.40	3.33	1479	977	2138
15	2009	3819	1117	13	61.9	3.05	-0.18	3.12	0.46	2.93	0.40	3.28	1318	851	1905
16	2010	781	1058	14	66.67	3.02	-0.31	3.06	0.49	2.86	0.40	3.22	1148	724	1660
17	2011	620	876	15	71.43	2.94	-0.46	3.00	0.52	2.79	0.41	3.17	1000	617	1479
18	2012	4718	781	16	76.19	2.89	-0.63	2.93	0.55	2.70	0.41	3.10	851	501	1259
19	2013	1558	620	17	80.95	2.79	-0.82	2.85	0.58	2.61	0.41	3.02	708	407	1047
20	2014	3843	468	18	85.71	2.67	-1.06	2.76	0.61	2.51	0.42	2.93	575	324	851
21	2015	468	294	19	90.48	2.47	-1.37	2.63	0.66	2.36	0.43	2.81	427	229	646
22	2016	2555	185	20	95.24	2.27	-1.87	2.42	0.83	2.08	0.51	2.63	263	120	427
23					n =	20									
24					xbar =	3.19									
25					S_x =	0.41									
26					a = G =	-0.78									

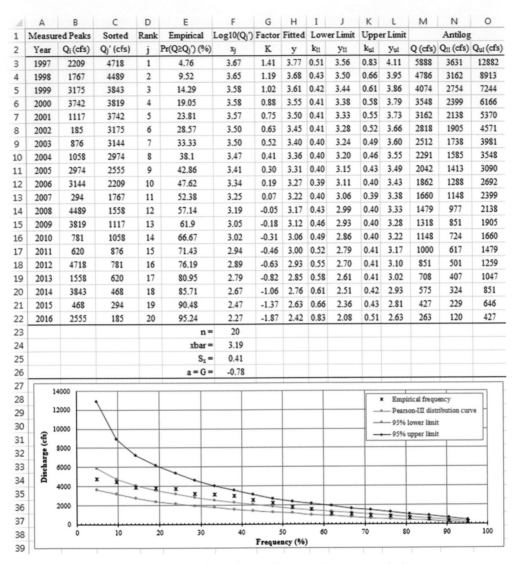

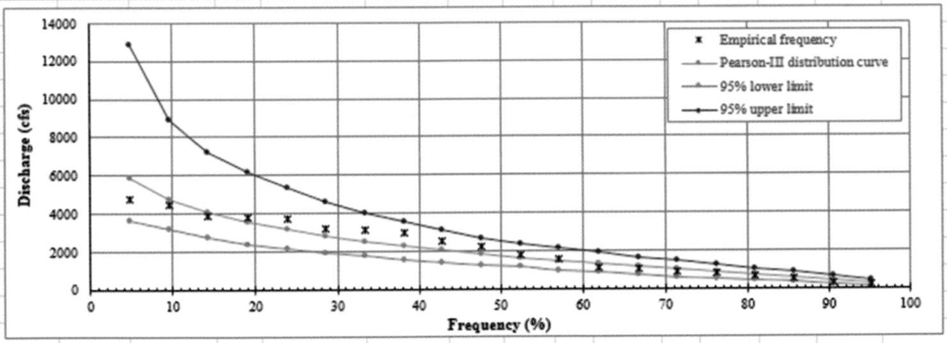

Figure 4.25 **Data and plot for Example 4.6.** Excel spreadsheet for frequency analysis.

Use Eq. (4.28) to compute lower limit and Eq. (4.29) upper limit. The results are shown in columns J and L, respectively. For instance, cell J3 $= 3.77 - 0.51 * 0.41 = 3.56$, and cell L3 $= 3.77 + 0.83 * 0.41 = 4.12$.

Antilog transform columns H, J, and L to obtain the fitted discharge in column M, its upper limit in column N, and its lower limit in column O. For instance, cell M3 $= 10^{3.77} = 5888$ cfs, cell N3 $= 10^{3.56} = 3631$ cfs, and cell O3 $= 10^{4.11} = 12,882$ cfs.

Plot columns C, M, N, and O versus column E to generate the frequency curves. Reformatting may be needed to make the plot more readable and professional.

4.3 Formulating a Design Flow Hydrograph

To design a detention or retention structure (e.g., reservoirs and ponds), in addition to the design peak discharge, a design flow hydrograph is also needed to determine the required storage capacity. This section presents commonly used methods to formulate the design flow hydrograph.

4.3.1 Historical Flow Hydrograph

A flow hydrograph is needed for determining the required storage capacity of water resources structures (e.g., detention or retention ponds and reservoirs). If the observed subdaily flow data are available for multiple storm events spanning a number of years, an historical flow hydrograph may be selected as the design flow hydrograph. The selection needs to consider three aspects, namely time base (t_b), time to peak (t_p), and magnitude of peak (Q_p) (Figure 3.5). For structures with a larger storage capacity, a hydrograph with larger values of t_b and t_p is preferred, whereas for structures with a smaller storage capacity, a hydrograph with a larger value of Q_p is more appropriate. For a given value of Q_p, a hydrograph with a smaller value of t_p/t_b should be used to design structures with a smaller storage capacity, a hydrograph with a larger value of t_p/t_b, however, should be used to design structures with a larger storage capacity. Herein, the basic assumption of using an historical flow hydrograph as the design flow hydrograph is that it can provide the required protection of structures for future hydrologic conditions (Wang *et al.*, 2010b).

4.3.2 Unit Flow Hydrograph

A unit hydrograph (UH) is the flow hydrograph generated by a unit excess rainfall. That is, the direct runoff depth represented by UH is exactly 1 in for BG or 1 mm for SI. Note that direct runoff is excess rainfall rather than total rainfall. The time base of a UH is equal to the duration of its producing unit excess rainfall, the UH, however, can be tabulated in any time interval. For instance, a UH that results from a 1-hr unit excess rainfall has a time base of 1 hr, but can be tabulated in a time interval of 30 min, 1 hr, or 2 hr, and so on.

For watersheds with observed data on rainfall and streamflow, a storm event and its resulting flow hydrograph may be analyzed to derive a UH. First, subtract the baseflow from the observed streamflow hydrograph to generate a direct flow hydrograph (e.g., the portion above line AB in Figure 3.5). Second, determine the amount and duration of the excess rainfall by subtracting any losses from the observed rainfall. Third, divide the amount of excess rainfall by the direct flow hydrograph ordinates to get the UH with a time base equal to the duration of the excess rainfall.

Example 4.7 A 4-hr storm generates the flow hydrograph shown in columns A and B in Figure 4.26. The rainfall intensities of the storm are shown in columns G and H in the figure. If the constant rainfall loss rate is 10 mm hr^{-1} (column I in the figure) and the constant baseflow rate is 3 m^3 s^{-1} (column C in the figure), derive a UH. What is the time base of the UH?

Time (hr)	Streamflow ($m^3 s^{-1}$)	Baseflow ($m^3 s^{-1}$)	Direct Flow ($m^3 s^{-1}$)	2-hr UH ($m^3 s^{-1} mm^{-1}$)	Time (hr)	Rainfall ($mm\ hr^{-1}$)	Loss ($mm\ hr^{-1}$)	Excess Rainfall ($mm\ hr^{-1}$)
0	3	3	0	0	0	10	10	0
1	3	3	0	0	1	10	10	0
2	8	3	5	0.33	1	15	10	5
3	17	3	14	0.93	2	15	10	5
4	23	3	20	1.33	2	20	10	10
5	28	3	25	1.67	3	20	10	10
6	25	3	22	1.47	3	5	10	0
7	19	3	16	1.07	4	5	10	0
8	12	3	9	0.6	Amount of excess rainfall (mm)			15
9	9	3	6	0.4				
10	3	3	0	0				
11	3	3	0	0				

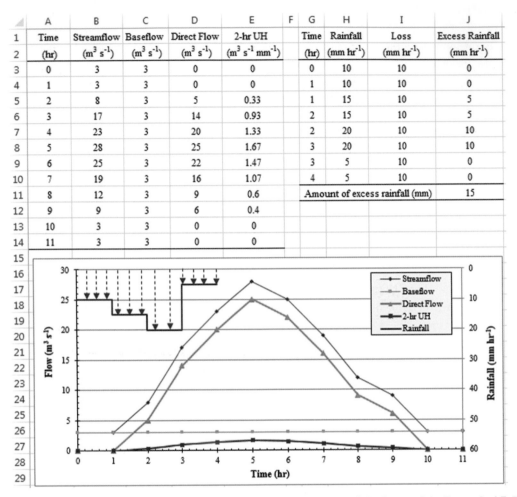

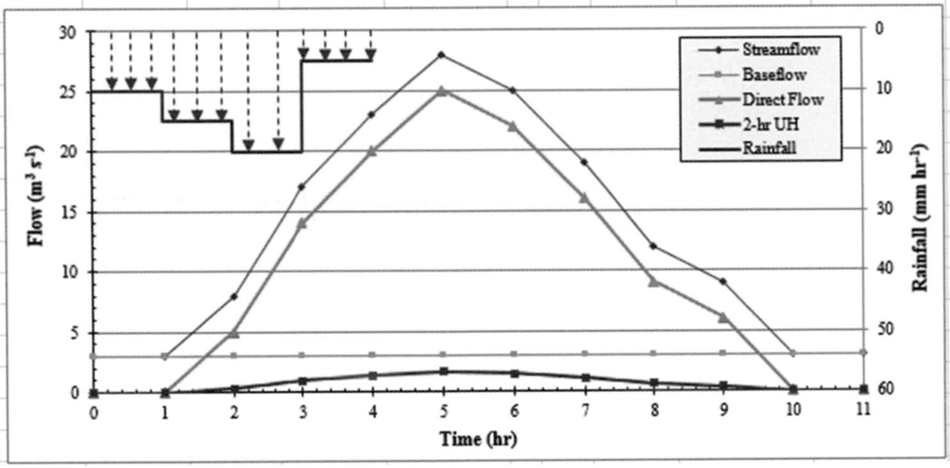

Figure 4.26 **Generation of a flow hydrograph and the derivation of a unit hydrograph in Example 4.7.** The Excel spreadsheet and associated plot are shown.

Solution

First, subtract column C from column B to determine the ordinates of the direct flow hydrograph as shown in column D. Second, subtract column I from column H to obtain the values of excess rainfall as shown in column J. The excess rainfall has a total amount of 15 mm and a duration of 2 hr. Third, divide column D (direct flow hydrograph ordinates) by 15 mm to get the UH ordinates as shown in column E. Because the excess rainfall has a duration of 2 hr, the UH has a time base of 2 hr.

For watersheds without sufficient rainfall and flow data, synthetic UHs, such as SCS UH (USDA-NRCS, 1986, 2010) and Snyder UH (Snyder, 1938), can be used.

Table 4.13 The SCS UH dimensionless ordinates.[1]

t/t_p	q/q_p	t/t_p	q/q_p	t/t_p	q/q_p
0	0	1.0	1.00	2.4	0.18
0.1	0.015	1.1	0.98	2.6	0.13
0.2	0.075	1.2	0.92	2.8	0.098
0.3	0.16	1.3	0.84	3.0	0.075
0.4	0.28	1.4	0.75	3.5	0.036
0.5	0.43	1.5	0.66	4.0	0.018
0.6	0.60	1.6	0.56	4.5	0.009
0.7	0.77	1.8	0.42	5.0	0.004
0.8	0.89	2.0	0.32		
0.9	0.97	2.2	0.24		

[1] *Source:* USDA-NRCS (2010). t_p: time to peak; q_p: peak of UH; t and q: pairs of time and flow on UH.

A SCS UH with a time base of D can be derived by following five steps:

Step 1: Estimate the time of concentration t_c using the SCS Lag equation (Eq. (3.2)).
Step 2: Compute the lag time as $t_L = 0.6t_c$.
Step 3: Compute the time to peak as $t_p = D/2 + t_L$.
Step 4: Compute the UH peak as $q_p = \frac{484A}{t_p}$, where q_p is the peak with units of cfs in^{-1}; A is the drainage area (mi^2), and t_p is the time to peak with units of hr.
Step 5: Multiply t_p and q_p by the SCS dimensionless ordinates (Table 4.13) to derive the UH.

A Snyder UH is derived by fitting a smooth curve through seven points (Figure 4.27) and then adjusting the curve to represent exactly 1 in direct runoff. A Snyder UH with a time base of D can be derived by following five steps:

Step 1: Estimate the time lag as:

$$t_L = C_t \left(L L_{ca}\right)^{0.3} \tag{4.30}$$

lag time [hr] → t_L; basin coefficient (0.4 to 8.0) → C_t; watershed length [mi] → L; length from outlet to centroid of watershed [mi] → L_{ca}

Step 2: Compute the time base as:

$$t_b = \begin{cases} 3 + \dfrac{t_L}{8} & \text{for large watershed} \\ (3 \sim 5)\, t_L & \text{for small watershed} \end{cases} \tag{4.31}$$

time base [hr] → t_b; lag time [hr] → t_L

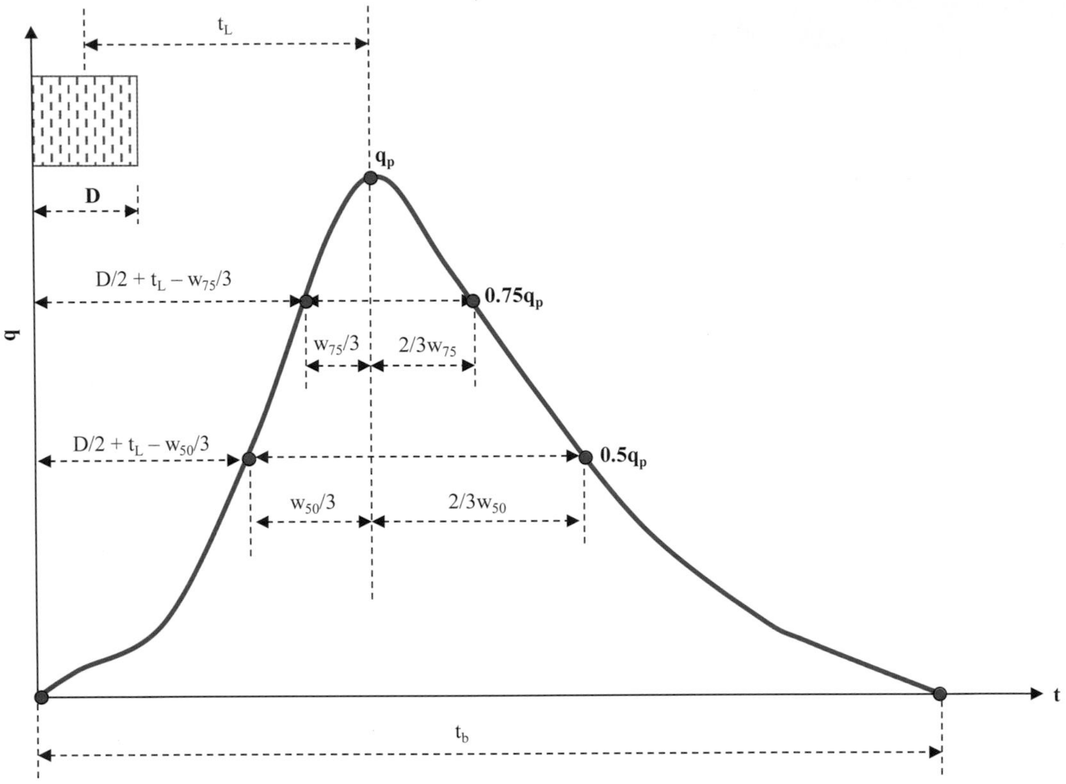

Figure 4.27 **Snyder unit hydrograph derivation.** The seven points (in red) are fitted with time base D. t_L: lag time; q_p: peak of the UH; w_{50}: time width at 50%q_p; w_{75}: time width at 75%q_p; t and q: time and ordinate pairs on the UH.

Step 3: Compute the UH peak as:

$$\underbrace{q_p}_{\text{UH peak [cfs in}^{-1}]} = \frac{14,080\overbrace{C_p}^{\text{storage coefficient (0.4 to 0.8)}}\underbrace{A}_{\text{drainage area [mi}^2]}}{5.5\underbrace{D}_{\text{UH time base [hr]}} + 21\underbrace{t_L}_{\text{lag time [hr]}}} \tag{4.32}$$

Step 4: Compute the time widths at 50 and 75% of the UH peak as:

$$\underbrace{w_{50}}_{\text{time width at 50\%}q_p\text{ [hr]}} = 770\left(\frac{\overbrace{A}^{\text{drainage area [mi}^2]}}{\underbrace{q_p}_{\text{UH peak [cfs in}^{-1}]}}\right)^{1.1} \tag{4.33}$$

drainage area [mi²]

time width at 75%q_p [hr] → $$w_{75} = 440\left(\frac{A}{q_p}\right)^{1.1} \quad (4.34)$$

UH peak [cfs in^{-1}]

Step 5: Fit a smooth curve through the seven red points as shown in Figure 4.27.

Example 4.8 A 60 mi^2 drainage area has the following characteristics: longest flow path length of 10 mi, flow path length from outlet to centroid of 3.2 mi, overland slope of 0.05, curve number of 82, basin coefficient of 1.35, and storage coefficient of 0.6. Derive a 1-hr:

(1) SCS UH
(2) Snyder UH.

Solution

$A = 60$ mi^2, $L = L_w = 10$ mi, $L_{ca} = 3.2$ mi, $S_w = 0.05$, $CN_{II} = 82$, $C_t = 1.35$, $C_p = 0.6$, $D = 1$ hr

(1) Eq. (3.2): $t_c = 1.67 * [(10\text{ mi}) * (5280\text{ ft mi}^{-1})]^{0.8} * (1000/82 - 9)^{0.7}/(1900 * 5^{0.5}) = 5.32$ hr.
Lag time: $t_L = 0.6 * (5.32\text{ hr}) = 3.19$ hr.
Time to peak: $t_p = (1\text{ hr})/2 + 3.19\text{ hr} = 3.69$ hr.
UH peak: $q_p = 484 * (60\text{ mi}^2)/(3.69\text{ hr}) = 7870\text{ cfs in}^{-1}$.
Multiply the ordinates in Table 4.13 by t_p and q_p to obtain the UH, which is done in Excel as shown in Figure 4.28. The 1-hr UH is given in columns D and E in the figure.

(2) Eq. (4.30): $t_L = 1.35 * [(10\text{ mi}) * (3.2\text{ mi})]^{0.3} = 3.81$ hr.
Small watershed → Eq. (4.31): $t_b = 4 * (3.81\text{ hr}) = 15.24$ hr.
Eq. (4.32): $q_p = 14,080 * 0.6 * (60\text{ mi}^2)/[5.5 * (1\text{ hr}) + 21 * (3.81\text{ hr})] = 5928\text{ cfs in}^{-1}$.
Eq. (4.33): $w_{50} = 770 * \left[(60\text{ mi}^2)/(5928\text{ cfs in}^{-1})\right]^{1.1} = 4.92$ hr.
Eq. (4.34): $w_{75} = 440 * [(60\text{ mi}^2)/(5928\text{ cfs in}^{-1})]^{1.1} = 2.81$ hr.
The seven points (in columns I and J in Figure 4.28) are determined as:

$$\begin{aligned}
\text{1st}: t &= 0\text{ hr}, q = 0\text{ cfs in}^{-1} \\
\text{2nd}: t &= (1\text{ hr}/2) + (3.81\text{ hr}) - (4.92\text{ hr}/3) = 2.67\text{ hr} \\
q &= 50\% * (5928\text{ cfs in}^{-1}) = 2964\text{ cfs in}^{-1} \\
\text{3rd}: t &= (1\text{ hr}/2) + (3.81\text{ hr}) - (2.81\text{ hr}/3) = 3.37\text{ hr} \\
q &= 75\% * (5928\text{ cfs in}^{-1}) = 4446\text{ cfs in}^{-1} \\
\text{4th}: t &= (1\text{ hr}/2) + (3.81\text{ hr}) = 4.31\text{ hr} \\
q &= q_p = 5928\text{ cfs in}^{-1} \\
\text{5th}: t &= (1\text{ hr}/2) + (3.81\text{ hr}) + 2/3 * (2.81\text{ hr}) = 6.18\text{ hr} \\
q &= 75\% * (5928\text{ cfs in}^{-1}) = 4446\text{ cfs in}^{-1}
\end{aligned}$$

$$6\text{th}: \mathrm{t} = (1\ \mathrm{hr}/2) + (3.81\ \mathrm{hr}) + 2/3 * (4.92\ \mathrm{hr}) = 7.59\ \mathrm{hr}$$

$$\mathrm{q} = 50\% * (5928\ \mathrm{cfs\ in}^{-1}) = 2964\ \mathrm{cfs\ in}^{-1}$$

$$7\text{th}: \mathrm{t} = \mathrm{t_b} = 15.24\ \mathrm{hr}, \mathrm{q} = 0\ \mathrm{cfs\ in}^{-1}.$$

	A	B	C	D	E	I	J
1	t_p =	3.69	hr				
2	q_p =	7870	cfs in^{-1}				
3	Table 3-30			SCS UH		Snyder UH	
4	t/t_p	q/q_p		t (hr)	q (cfs in^{-1})	t (hr)	q (cfs in^{-1})
5	0	0		0.00	0.00	0	0
6	0.1	0.015		0.37	118.05	2.67	2964
7	0.2	0.075		0.74	590.25	3.37	4446
8	0.3	0.16		1.11	1259.20	4.32	5928
9	0.4	0.28		1.48	2203.60	6.18	4446
10	0.5	0.43		1.85	3384.10	7.59	2964
11	0.6	0.6		2.21	4722.00	15.24	0
12	0.7	0.77		2.58	6059.90		
13	0.8	0.89		2.95	7004.30		
14	0.9	0.97		3.32	7633.90		
15	1	1		3.69	7870.00		
16	1.1	0.98		4.06	7712.60		
17	1.2	0.92		4.43	7240.40		
18	1.3	0.84		4.80	6610.80		
19	1.4	0.75		5.17	5902.50		
20	1.5	0.66		5.54	5194.20		
21	1.6	0.56		5.90	4407.20		
22	1.8	0.42		6.64	3305.40		
23	2	0.32		7.38	2518.40		
24	2.2	0.24		8.12	1888.80		
25	2.4	0.18		8.86	1416.60		
26	2.6	0.13		9.59	1023.10		
27	2.8	0.098		10.33	771.26		
28	3	0.075		11.07	590.25		
29	3.5	0.036		12.92	283.32		
30	4	0.018		14.76	141.66		
31	4.5	0.009		16.61	70.83		
32	5	0.004		18.45	31.48		

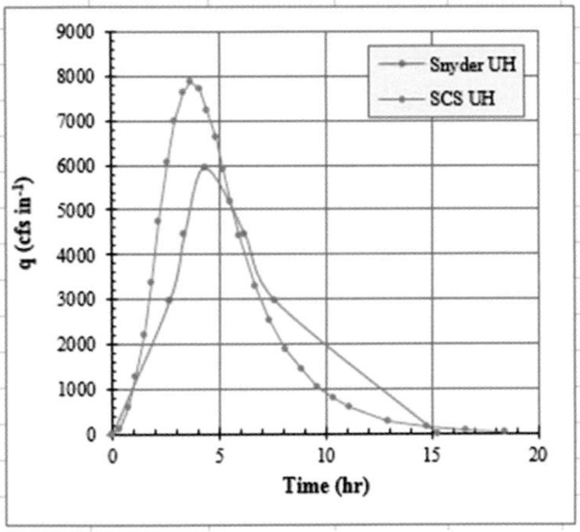

Figure 4.28 **Spreadsheet and derived unit hydrographs in Example 4.8.**

The flow hydrograph from a storm can be formulated by multiplying the excess rainfall of the storm and a UH. Note that the time base of the UH must be the same as the duration of the excess rainfall. In practice, a UH with time base D_1 can be converted into a UH with time base D_2 using the S-curve method as illustrated in Figure 4.29. The S-curve is the flow hydrograph generated by an infinite number of D_1-durarion excess rainfalls, all of which have an intensity of $1/D_1$. Afterward, the S-curve is shifted to the right by D_2 and the difference between the S-curve and the shifted S-curve is multiplied by D_1/D_2 to derive the UH with time base D_2.

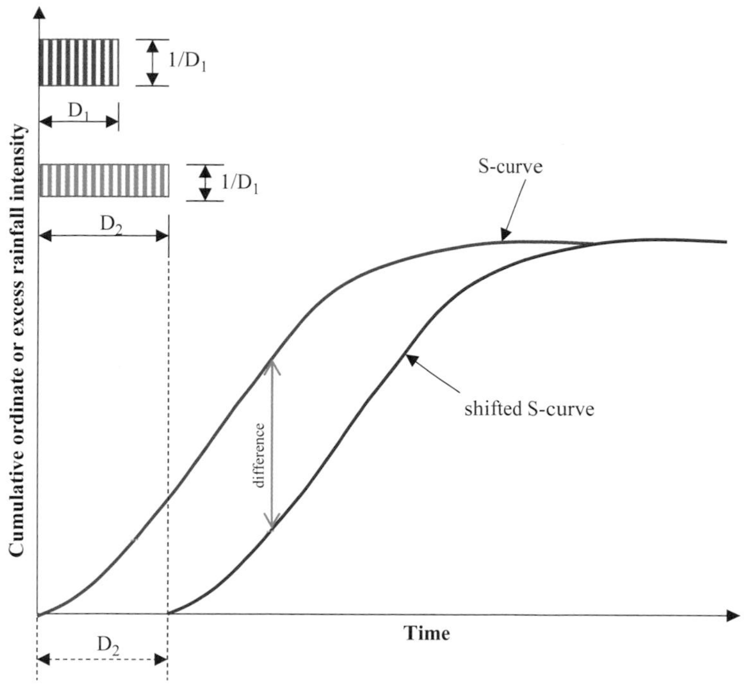

Figure 4.29 **The S-curve method for conversion of a unit hydrograph.** Diagram shows the conversion of a UH with time base D_1 to a UH with time base D_2.

Example 4.9 The 2-hr UH for a watershed is given in columns A and B in Figure 4.30. The watershed receives a 4-hr storm, which generates an excess rainfall of 1.5 in for the first two hours and 3.2 in for the second two hours.

(1) Determine the flow hydrograph from the storm
(2) Develop a 4-hr UH based on the 2-hr UH.

Solution

(1) For each of the 2-hr periods, the flow hydrograph is derived by multiplying the excess rainfall by the UH. Note that the flow hydrograph for the second 2-hr period should be lagged by 2 hr. The results for the first 2-hr period (Flow1) and second 2-hr period (Flow2) are in columns E and F in Figure 4.30, respectively. For instance, for t = 1 hr, Flow1 = $(1.5 \text{ in}) * (40 \text{ cfs in}^{-1}) = 60$ cfs, while Flow2 = 0 because the second 2-hr rain has not started yet. For t = 2 hr, Flow1 = $(1.5 \text{ in}) * (80 \text{ cfs in}^{-1}) = 120$ cfs, and Flow2 = $(3.2 \text{ in}) * (0 \text{ cfs in}^{-1}) = 0$ cfs.

The flow hydrograph from the storm, Q, is the summation of Flow1 and Flow2. The results are in column G. For instance, for t = 3 hr, Q = 240 cfs + 128 cfs = 368 cfs. The UH, Flow1, Flow2, and Q are plotted in Figure 4.30.

Time Period (hr)	0 ~ 2	2 ~ 4
Excess rainfall (in)	1.5	3.2

2-hr UH			Flow Hydrograph from the 4-hr Storm			
t	q		t	Flow1	Flow2	Q
(hr)	(cfs in^{-1})		(hr)	(cfs)	(cfs)	(cfs)
0	0		0	0		0
1	40		1	60		60
2	80		2	120	0	120
3	160		3	240	128	368
4	320		4	480	256	736
5	280		5	420	512	932
6	240		6	360	1024	1384
7	200		7	300	896	1196
8	160		8	240	768	1008
9	120		9	180	640	820
10	80		10	120	512	632
11	40		11	60	384	444
12	0		12	0	256	256
13	0		13	0	128	128
14	0		14	0	0	0

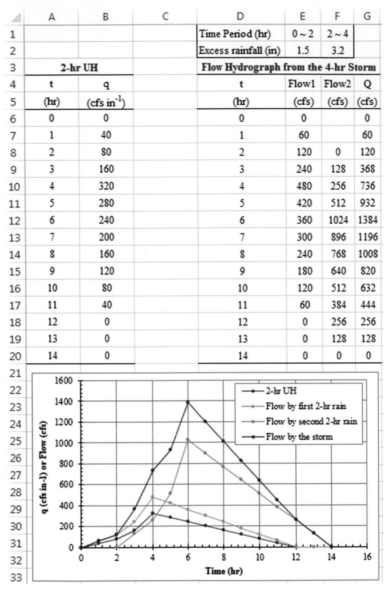

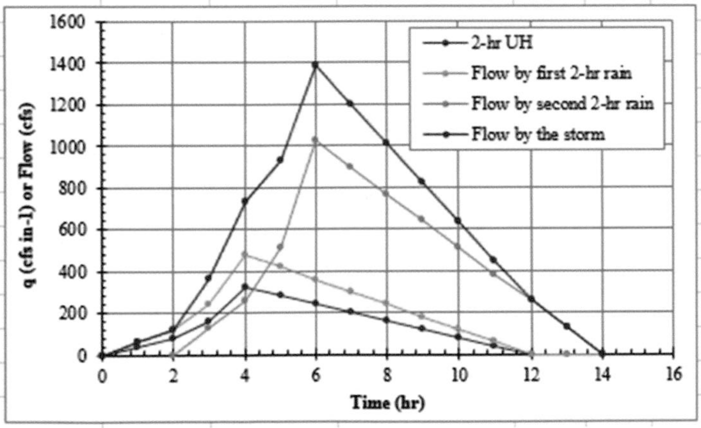

Figure 4.30 **Unit hydrograph ordinates for part (1) of Example 4.9.** Excel spreadsheet to compute the flow hydrograph.

	J	K	L	M	N	O	P	Q	R	S	T
1								D_1 =	2	hr	
2								D_2 =	4	hr	
3											
4	t	2-hr UH	Lag 2 hr	Lag 4 hr	Lag 6 hr	Lag 8 hr	Lag 10 hr	Lag 12 hr	S-Curve	Shifted S-curve	4-hr UH
5	(hr)	(cfs in^{-1})								(by 3 hr)	(cfs in^{-1})
6	0	0							0		0
7	1	40							40		20
8	2	80	0						80		40
9	3	160	40						200		100
10	4	320	80	0					400	0	200
11	5	280	160	40					480	40	220
12	6	240	320	80	0				640	80	280
13	7	200	280	160	40				680	200	240
14	8	160	240	320	80	0			800	400	200
15	9	120	200	280	160	40			800	480	160
16	10	80	160	240	320	80	0		880	640	120
17	11	40	120	200	280	160	40		840	680	80
18	12	0	80	160	240	320	80	0	880	800	40
19	13	0	40	120	200	280	160	40	840	800	20
20	14	0	0	80	160	240	320	80	880	880	0

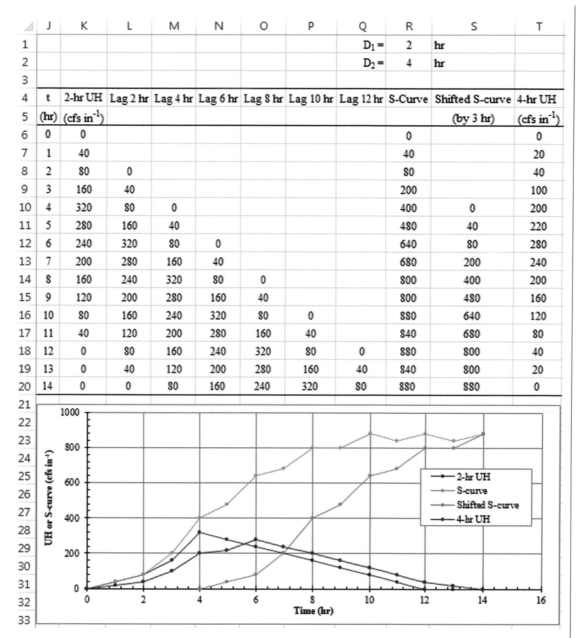

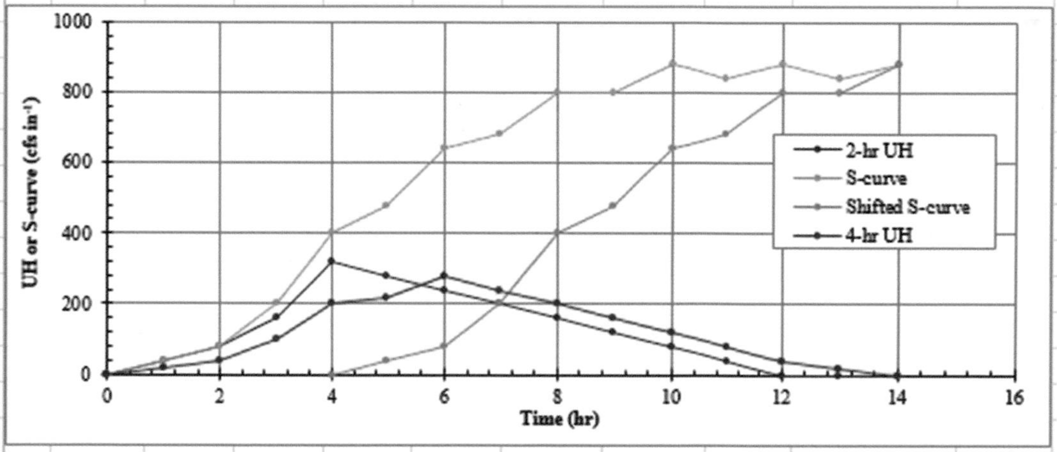

Figure 4.31 **Unit hydrograph ordinates for part (2) of Example 4.9.** Excel spreadsheet to derive the 4-hr UH.

(2) As illustrated in columns K to P of Figure 4.31, the 2-hr UH is lagged by 2, 4, 6, 8, 10, and 12 hr. The S-curve is the summation of the 2-hr UH and the lagged ordinates, resulting in column R. To derive the 4-hr UH, the S-curve is shifted by 4 hr to get the shifted S-curve, and then

the difference between these two curves is multiplied by 2/4 to get column T. For instance, for $t = 5$ hr, the ordinate of the 4-hr UH is: $2/4 * (480 - 40)\text{cfs in}^{-1} = 220 \text{ cfs in}^{-1}$. The 2-hr UH, S-curve, shifted S-curve, and 4-hr UH are plotted in Figure 4.31.

4.3.3 Modeled Flow Hydrograph

The Modified Rational method assumes a trapezoidal flow hydrograph as illustrated in Figure 4.32. For a storm with a duration longer than the time of concentration (i.e., $D > t_c$), the peak discharge Q_p is assumed to occur at time t_c and can be estimated using Eq. (4.14). Q_p will be maintained until the end of rainfall, after which the flow will recess and becomes zero at time $D+t_c$. However, if $D \leq t_c$, the hydrograph will have a triangular shape, with the maximum discharge occurring at time D and then linearly decreasing to zero at time $D + t_c$. As with the Rational Equation, the Modified Rational method is better applicable for small (< 20 ac) than large watersheds.

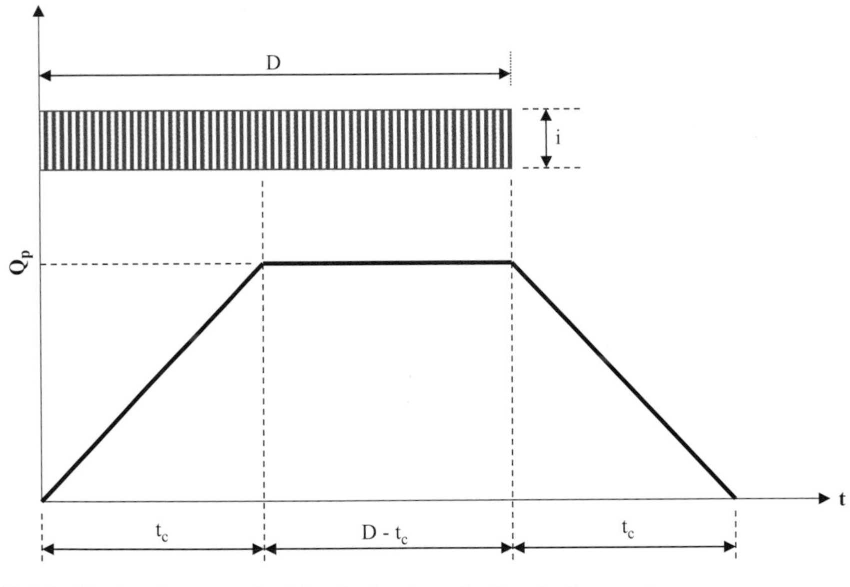

Figure 4.32 **Modified rational method for derivation of a flow hydrograph.**

Example 4.10 A 10 ac watershed has a time of concentration $t_c = 20$ min and a composite runoff coefficient $C = 0.6$. Use the Modified Rational equation to develop a flow hydrograph from a 1-hr storm with an average intensity of 1.2 in hr^{-1}.

Solution

$A = 10$ ac, $C = 0.6$, $i = 1.2 \text{ in hr}^{-1}$, $D = 1 \text{ hr} = 60 \text{ min}$, $t_c = 20 \text{ min}$

The storm duration $> t_c \rightarrow$ Eq. (4.14): $Q_p = 0.6 * (1.2 \text{ in hr}^{-1}) * (10 \text{ ac}) = 7.2$ cfs.

The trapezoidal flow hydrograph is shown in Figure 4.33.

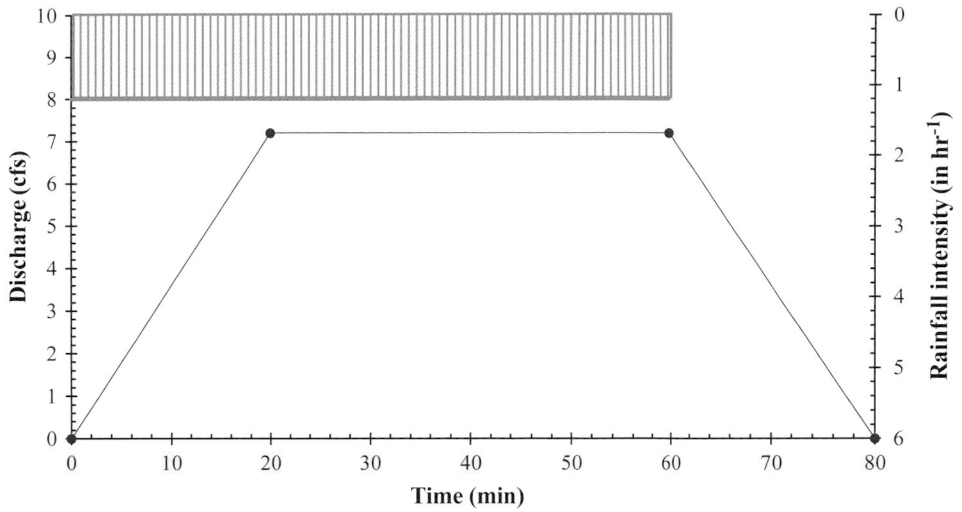

Figure 4.33 **The flow hydrograph for Example 4.10.**

For large watersheds, the approach of SCS TR-55 ordinates can be used to develop flow hydrograph by following five steps:

Step 1: Compute the initial abstraction I_a using Eqs. (3.49) and (3.50) and excess rainfall using Eq. (3.48) $\rightarrow I_a/P$.

Step 2: Estimate the time of concentration t_c using Eq. (3.2) and channel travel time t_{cf} using Eqs. (3.9) and (3.11).

Step 3: Identify the rainfall distribution type from Figure 4.12.

Step 4: In terms of the rainfall distribution type, t_{cf}, t_c, and I_a/P, look up the SCS tables to select the hydrograph ordinates. Figure 4.34 is the table for Type II distribution, $t_c = 0.5$ hr, and $I_{a/P} = 0.1, 0.3$, and 0.5. More tables are available in USDA-NRCS (1986). The ordinates have units of csm in^{-1}, which stands for cfs mi^{-2} in^{-1}.

Step 5: Develop the flow hydrograph by multiplying the excess rainfall, drainage area, and SCS ordinates.

Example 4.11 A 0.3 mi^2 watershed has a Type II rainfall distribution, time of concentration $t_c = 0.5$ hr, channel travel time $t_{cf} = 1.5$ hr, and a composite curve number CN = 65. If the 2-year 24-hr rainfall is 3.5 in, develop a flow hydrograph using the approach of SCS TR-55 ordinates.

Solution

$A = 0.3$ mi^2, $t_c = 0.5$ hr, $t_{cf} = 1.5$ hr, CN = 75, P = 3.5 in, Type II distribution

Eq. (3.49): $S = (1000/65 - 10)\text{in} = 5.38$ in.

Eq. (3.50): $I_a = 0.2 * (5.38 \text{ in}) = 1.08 \text{ in} \rightarrow I_{a/P} = (1.08 \text{ in})/(3.5 \text{ in}) = 0.30$.

Exhibit 5-II: Tabular hydrograph unit discharges (csm/in) for type II rainfall distribution—continued

TRVL TIME (hr) / HYDROGRAPH TIME (HOURS)

TRVL TIME (hr)	11.0	11.3	11.6	11.9	12.0	12.1	12.2	12.3	12.4	12.5	12.6	12.7	12.8	13.0	13.2	13.4	13.6	13.8	14.0	14.3	14.6	15.0	15.5	16.0	16.5	17.0	17.5	18.0	19.0	20.0	22.0	26.0
IA/P = 0.10												*** TC = 0.5 HR ***																IA/P = 0.10				
0.0	17	23	32	57	94	170	308	467	529	507	402	297	226	140	96	74	61	53	47	41	36	32	29	26	23	21	20	19	16	14	12	0
.10	16	22	30	51	80	140	252	395	484	499	434	343	265	162	108	80	65	55	49	42	36	33	29	26	23	21	20	19	16	14	12	0
.20	14	19	25	38	47	69	116	207	332	434	477	449	378	238	149	101	77	62	53	45	39	34	30	27	24	22	20	19	17	14	12	0
.30	13	18	24	35	43	60	97	170	278	382	446	448	401	270	171	114	83	66	56	46	40	34	31	27	24	22	20	19	17	15	12	0
.40	12	15	21	29	33	40	53	83	141	233	332	408	434	361	243	157	107	79	64	51	43	36	32	28	25	22	21	20	17	15	12	0
.50	11	15	20	28	31	37	48	71	118	194	286	367	412	378	271	178	119	86	68	53	44	37	32	29	25	23	21	20	17	15	12	0
.75	9	11	14	19	21	24	27	31	37	49	74	118	182	319	374	328	244	169	117	76	56	43	35	31	28	25	22	21	18	16	12	1
1.0	7	9	12	16	17	19	21	24	27	32	40	55	83	188	309	359	322	245	172	102	68	49	38	32	29	26	23	21	19	16	12	1
1.5	5	7	8	11	12	13	14	15	17	19	21	23	27	43	89	175	269	322	309	225	140	77	49	38	32	29	25	23	20	17	13	5
2.0	3	4	6	7	8	8	9	10	10	11	12	14	15	18	23	35	65	123	202	297	280	181	88	52	39	33	29	26	21	19	14	10
2.5	2	3	4	5	5	6	6	7	7	8	9	9	10	12	15	18	24	36	66	150	244	278	171	87	52	39	33	29	23	20	15	11
3.0	1	1	2	3	3	4	4	4	5	5	6	6	7	8	9	11	13	16	20	37	86	198	263	182	96	56	40	33	26	21	16	11
IA/P = 0.30												*** TC = 0.5 HR ***																IA/P = 0.30				
0.0	0	0	0	1	9	53	157	314	433	439	379	299	237	159	118	95	81	71	65	56	50	46	42	38	34	31	30	28	25	22	19	0
.10	0	0	0	0	1	6	37	117	248	372	416	391	330	218	150	113	92	79	70	60	53	47	43	39	35	32	30	29	26	22	19	0
.20	0	0	0	0	1	4	26	87	194	313	382	388	349	244	167	122	97	82	72	62	54	48	43	39	35	32	30	29	26	22	19	0
.30	0	0	0	0	0	0	3	19	64	151	259	341	372	316	223	156	117	94	80	67	58	50	45	41	36	33	31	29	26	23	19	0
.40	0	0	0	0	0	0	2	13	47	116	211	298	354	328	245	172	127	100	83	69	59	51	45	41	37	33	31	29	26	23	19	0
.50	0	0	0	0	0	0	0	1	9	34	89	170	255	341	303	225	161	120	96	76	64	54	47	42	38	34	31	30	27	24	19	0
.75	0	0	0	0	0	0	0	1	4	14	41	89	152	270	305	268	207	155	118	87	70	57	48	44	39	35	32	30	27	24	19	0
1.0	0	0	0	0	0	0	0	0	0	0	2	7	22	98	212	295	285	237	181	120	88	67	53	46	42	38	34	31	28	25	19	2
1.5	0	0	0	0	0	0	0	0	0	0	0	0	0	5	30	95	183	249	265	217	152	96	66	53	46	41	37	34	30	26	20	8
2.0	0	0	0	0	0	0	0	0	0	0	0	0	0	0	0	3	18	59	125	221	245	182	105	69	54	47	42	38	32	28	22	16
2.5	0	0	0	0	0	0	0	0	0	0	0	0	0	0	0	0	1	5	21	84	174	230	172	103	69	54	46	42	34	30	23	18
3.0	0	0	0	0	0	0	0	0	0	0	0	0	0	0	0	0	0	0	1	13	56	157	217	163	101	68	53	46	37	31	25	18
IA/P = 0.50												*** TC = 0.5 HR ***																IA/P = 0.50				
0.0	0	0	0	0	0	2	26	89	170	217	229	200	179	144	119	104	93	85	78	70	64	59	55	51	46	43	41	40	36	32	28	0
.10	0	0	0	0	0	0	1	18	65	135	190	216	205	170	137	115	101	91	83	74	67	61	56	52	47	44	42	40	36	32	28	0
.20	0	0	0	0	0	0	1	12	47	106	162	198	203	178	145	121	105	94	85	76	68	61	57	52	48	44	42	40	37	32	28	0
.30	0	0	0	0	0	0	0	1	8	34	82	135	177	194	168	139	117	102	92	80	71	63	58	54	49	45	43	41	37	33	28	0
.40	0	0	0	0	0	0	0	0	6	25	63	111	155	189	174	146	122	106	94	82	73	64	58	54	50	45	43	41	37	33	28	0
.50	0	0	0	0	0	0	0	0	4	18	48	90	133	184	177	152	128	110	97	84	74	65	59	55	50	45	43	41	38	33	28	0
.75	0	0	0	0	0	0	0	0	1	7	22	47	80	142	169	164	144	124	108	91	79	68	61	56	51	47	44	42	38	34	28	0
1.0	0	0	0	0	0	0	0	0	0	0	1	3	11	51	112	155	166	154	134	109	91	76	65	59	54	49	45	43	39	35	28	2
1.5	0	0	0	0	0	0	0	0	0	0	0	0	0	2	16	50	97	136	154	145	121	95	75	64	58	54	49	45	41	37	29	10
2.0	0	0	0	0	0	0	0	0	0	0	0	0	0	0	0	4	18	47	86	134	146	125	94	75	64	58	53	49	42	39	31	21
2.5	0	0	0	0	0	0	0	0	0	0	0	0	0	0	0	0	0	3	11	44	95	140	127	97	77	65	58	54	45	41	33	26
3.3	0	0	0	0	0	0	0	0	0	0	0	0	0	0	0	0	0	0	1	7	29	86	135	122	95	76	65	58	49	43	35	27

RAINFALL TYPE = II *** TC = 0.5 HR *** SHEET 5 OF 10

(210-VI-TR-55, Second Ed., June 1986) 5-33

Figure 4.34 **The SCS TR-55 approach to developing a flow hydrograph.** Tabulated hydrograph ordinates. (*Source:* USDA-NRCS, 1986)

Eq. (3.48): $Q = [(3.5\text{ in}) - (1.08\text{ in})]^2/[(3.5\text{ in}) - (1.08\text{ in}) + (5.38\text{ in})] = 0.75$ in.

For $I_a/P = 0.30$, $t_c = 0.5$ hr, and $t_{cf} = 1.5$ hr, find the SCS ordinates from Figure 4.34 and develop the flow hydrograph by multiplying $Q = 0.75$ in, $A = 0.3$ mi^2, and the ordinates in column B, as shown in Figure 4.35. For instance, for $t = 13.2$ hr, the ordinate is 30 csm in^{-1}, and thus the hydrograph flow is $(0.75\text{ in}) * (0.3\text{ mi}^2) * (30\text{ csm in}^{-1}) = 6.8$ cfs.

	A	B	C
1	A =	0.3	mi^2
2	I_a/P =	0.30	
3	Q =	0.75	in
4			
5	t	Ordinate	Flow
6	(hr)	(csm in^{-1})	(cfs)
7	12.8	0	0.0
8	13	5	1.1
9	13.2	30	6.8
10	13.4	95	21.4
11	13.6	183	41.2
12	13.8	249	56.0
13	14	265	59.6
14	14.3	217	48.8
15	14.6	152	34.2
16	15	96	21.6
17	15.5	66	14.9
18	16	53	11.9
19	16.5	46	10.4
20	17	41	9.2
21	17.5	37	8.3
22	18	34	7.7
23	19	30	6.8
24	20	26	5.9
25	22	20	4.5
26	26	8	1.8

	E	F
5	t	Rainfall Intensity
6	(hr)	(in hr^{-1})
7	0	3.5
8	1	3.5

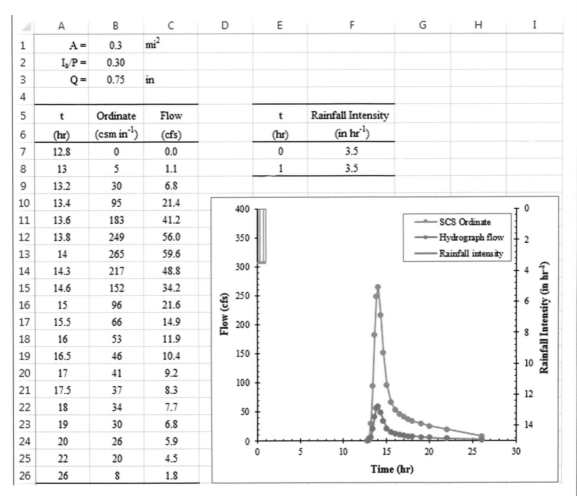

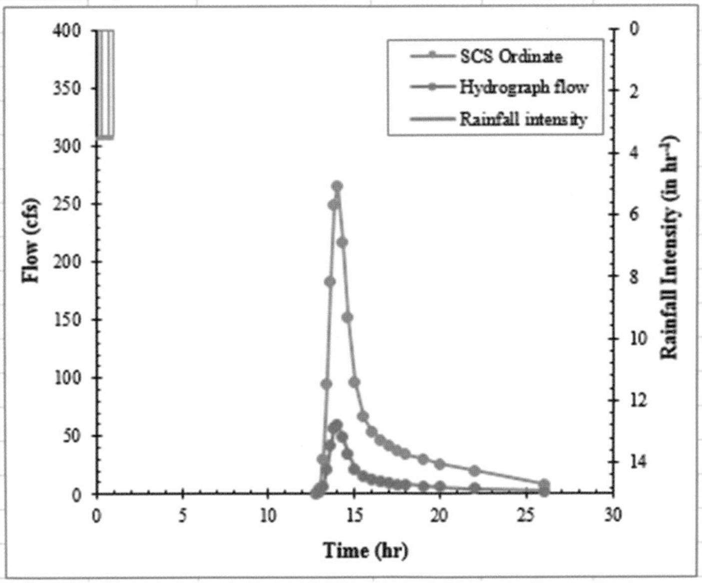

Figure 4.35 **Development of the flow hydrograph of Example 4.11.**

4.3.4 Hydrologic Simulation Models

Various computer programs and simulation models have been developed to facilitate the hydrologic analyses and computations discussed above. They can be classified in terms of modeling objective, temporal domain, spatial domain, and structure (Figure 4.36). For rainfall–runoff modeling, HEC-HMS, TR-55, HSPF, SWAT, and SWMM are widely used, whereas for frequency analysis, HEC-FFA and PEAKFQ are commonly used. The typical models for flow routing are

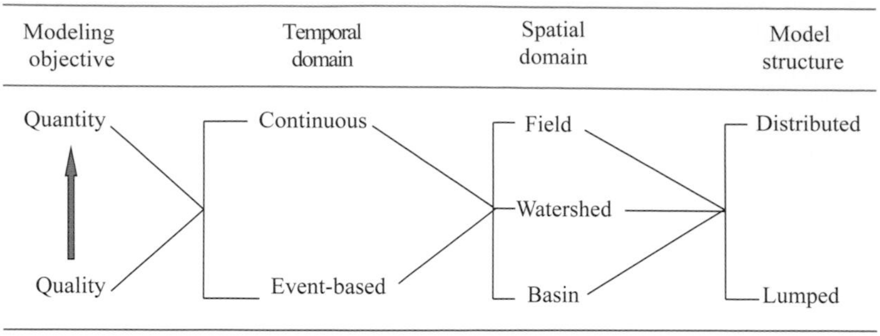

Figure 4.36 **Classification of hydrologic models.**

HEC-RAS and HY8. In addition, SWAT, SWMM, and HSPF are also used for modeling the transport of contaminants by runoff. The details of these programs and models are beyond the scope of this book. The reader who is interested in a specific model is referred to the relevant documents available on the corresponding websites, which can be found online using the model name as the keyword.

PROBLEMS

4.1 The annual series of maximum rainfall depths (in mm) for durations of 5, 10, 15, 30, and 45 minutes obtained from a recording rain gauge are tabulated in the following table. For the selected return periods of 2-, 5-, 10-, 25-, 50-, and 100-year, develop the IDF curves by assuming that the extreme rainfall follows a:

(1) Gumbel distribution
(2) Fréchet distribution
(3) Weibull distribution.

Recorded rainfall depths (mm).

Year	5 min	10 min	15 min	30 min	45 min
1982	2.6	3.5	4.3	5.2	8.4
1983	5.1	7.7	8.5	14.7	28.2
1984	2.9	5.1	6.6	7.4	10.2
1985	3.2	5.1	5.4	16.1	23.9
1986	2.2	3.2	3.0	10.1	17.1
1987	9.6	10.9	12.0	23.4	29.3
1988	3.8	6.4	6.8	8.3	10.0
1989	8.0	9.0	12.1	22.4	33.1
1990	3.5	5.4	6.8	7.2	10.4
1991	5.1	7.7	7.7	24.8	35.0
1992	2.9	5.4	8.1	13.2	18.5

Year	5 min	10 min	15 min	30 min	45 min
1993	4.2	6.7	6.8	20.1	20.5
1994	8.0	9.0	12.0	16.2	21.6
1995	2.2	4.5	5.0	8.4	13.4
1996	4.2	6.1	9.2	10.0	13.4
1997	3.5	4.2	5.1	11.2	13.6
1998	3.2	4.5	6.9	15.2	21.9
1999	4.8	7.0	7.1	15.8	24.2
2000	4.5	7.0	8.4	17.5	20.2
2001	3.8	5.8	6.2	12.9	17.5
2002	2.9	4.5	5.9	6.3	8.3
2003	4.5	6.7	9.0	15.4	17.6
2004	3.2	4.5	7.6	7.9	10.7
2005	7.0	13.1	18.0	25.0	39.3
2006	2.6	4.2	6.6	8.1	12.4
2007	2.9	4.8	6.7	10.6	18.0
2008	3.2	5.4	8.3	11.4	22.0
2009	5.4	7.7	10.4	16.9	28.2
2010	2.6	3.8	4.7	10.0	13.6
2011	4.8	5.8	7.7	8.3	13.7
2012	5.8	7.7	8.3	18.3	24.5
2013	4.2	8.6	9.5	25.5	35.1
2014	5.8	7.7	9.0	16.3	27.1
2015	4.5	5.8	6.0	17.0	32.4
2016	2.6	3.5	4.1	10.9	11.2
2017	1.6	3.2	3.2	8.7	9.6

4.2 Convert the following IDF curves of Problem 4.1 into Steel Formulas:
(1) The 2- and 50-year IDF curves by the Gumbel distribution
(2) The 5- and 10-year IDF curves by the Fréchet distribution
(3) The 25- and 100-year IDF curves by the Weibull distribution.

4.3 The 50-year IDF curve for a 125 km^2 catchment located in the Midwest is defined as a Steel Formula of $i = \frac{485}{(t + 4.5)^{0.6}}$, where i is the average rainfall intensity in mm hr^{-1} and t is the duration in min. Use a computational time step of 1 hr to develop a 50-year 24-hr synthetic design storm for the catchment using the following methods:
(1) Balanced triangular distribution
(2) 24-hr NRCS rainfall distribution.

4.4 The 100-year IDF curve for a 50 km^2 drainage area in Iowa is defined as a Steel Formula of $i = \frac{500}{(t + 4.8)^{0.5}}$, where i is the average rainfall intensity in mm hr^{-1} and t is the duration in min. For a computational time step of 1 hr, develop a 100-year 6-hr synthetic design storm using the balanced triangular distribution methods. Can the NRCS 24-hr rainfall distribution method be used? Why?

4.5 What are the values of 24-hr PMP for 10 and 200 mi^2 drainage areas in Richmond, VA? If a Steel Formula of $i = \frac{150}{(t + 3.5)^c}$ is applicable, where i is the average rainfall intensity in in hr^{-1}, t is the duration in min, and c is the PMP-specific shape factor, for a computational time step of 1 hr, develop the PMSs for these two drainage areas using the following methods:

(1) Balanced triangular distribution
(2) 24-hr NRCS rainfall distribution.

4.6 By conducting an internet search, document an engineering project that is designed using the historical maximum peak discharge.

4.7 Explain the variables of the Rational Equation (Eq. (3.86)). In selecting a T-year design storm intensity, why are the rainfall duration and time of concentration equated? Note the situations for which the duration is both less than and greater than time of concentration.

4.8 The time of concentration of a 2 ac catchment is 20 min. If the composite runoff coefficient of the catchment is 0.6, use the Rational Equation to determine the peak discharge from the:

(1) 10-min storm with a rainfall intensity of 3 in hr^{-1}
(2) 40-min storm with a rainfall intensity of 1 in hr^{-1}
(3) 20-min storm with a rainfall intensity of 2 in hr^{-1}.

4.9 Consider a plan to build single-family dwellings on a 17 ac site (HSG B) that is currently pastureland. The site is to be broken into 120 lots of 0.125 ac with a 3000 ft^2 home on each lot. The remaining 2 ac will be devoted to streets and roadways. Assume a time of concentration of 10 min and use a 50-year 24-hr design storm with a rainfall intensity of 3.5 in hr^{-1}. If the site has an overland slope of less than 2%, determine:

(1) The peak discharge before development
(2) The peak discharge after development
(3) The design discharge for sizing a storm sewer.

4.10 A 5.2 ac urban drainage area (HSG C) in Norfolk, VA consists of 30% open space and 70% parking. It has an overland slope of 0.5% and a time of concentration of 20 min. Determine the peak discharge from a 50-year storm.

4.11 The parking lots shown in map view below are located in Richmond, VA. If the storm sewer pipeline is designed using a 50-year storm, determine the design peak discharges for pipe ab, bc, and cd, respectively, using the Rational Equation. The average velocity, V, of sheet flow in a parking lot can be estimated as $V = \frac{\gamma y^2 S}{3\mu}$, where γ is the specific weight of runoff water, y is the sheet flow depth, S is the slope of the parking lot, and μ is the dynamic viscosity of runoff water. Assume the runoff water temperature is 60°F and the flow travel times in the pipes are minimal.

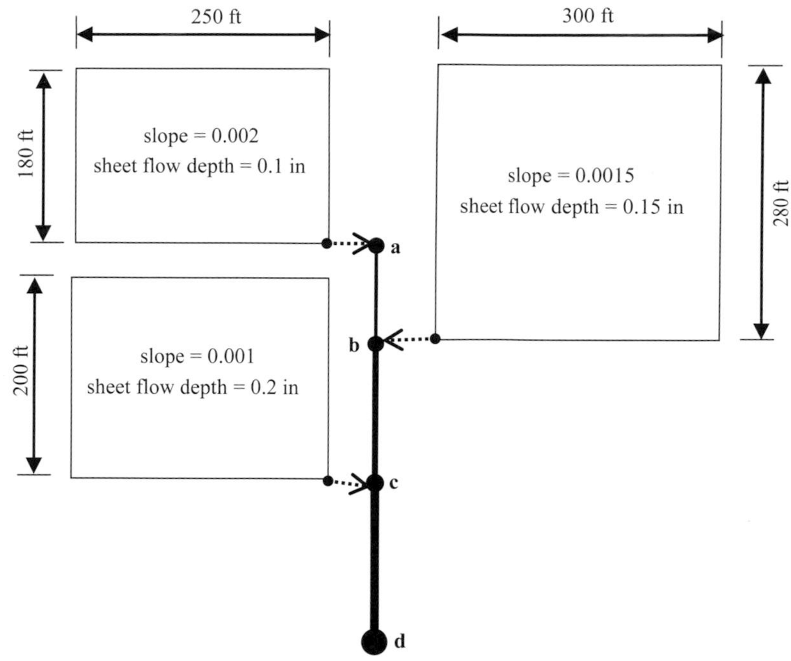

4.12 A 60 ac urban watershed in Norfolk, VA has a 15-min time of concentration and a composite curve number of 85. It has 0.5% pond and swamp areas. If the 50-year, 24-hr rainfall is 2.5 in, determine the peak discharge using the SCS Graphical Peak Discharge Method.

4.13 A 10 mi^2 watershed, located in North Dakota, has a composite curve number of 68. It has a 2-hr time of concentration and 3% pond and swamp areas. If the watershed receives 4.5 in rainfall, determine the peak discharge using the SCS Graphical Peak Discharge Method.

4.14 The annual peak discharges at USGS 01671100 Little River near Doswell, VA are listed in the following table.

Year	Peak (cfs)	Year	Peak (cfs)
1961	4430	1988	1010
1962	1840	1989	2610
1963	850	1990	850
1964	730	1991	520
1965	790	1992	2990
1966	515	1993	3910
1967	1020	1994	1050
1968	12,000	1995	2210
1969	880	1996	959
1970	1780	1997	3380
1971	8300	1998	466
1972	2290	1999	1040
1973	2610	2000	161
1974	4890	2001	2670
1975	2480	2002	1870
1976	1090	2003	1340

1977	4120	2004	873
1978	4120	2005	1880
1979	2000	2006	1010
1980	164	2007	528
1981	1490	2008	1380
1982	1360	2009	1900
1983	3470	2010	2560
1984	3240	2011	559
1985	2240	2012	3680
1986	2340	2013	1130
1987	1110	2014	1100

Complete a flood frequency analysis using the:
(1) Station skew coefficient
(2) Weighted skew coefficient.

4.15 By conducting an internet search, document an engineering project that is designed using the historical flow hydrograph.

4.16 At the outlet of a watershed, the flow hydrograph from a 4-hr storm is recorded in the following table. If the baseflow is 100 m^3 s^{-1} and the average infiltration loss rate is 5 mm hr^{-1}, derive a UH for the watershed and determine its time base.

Time (hr)	0	1	2	3	4	5	6	7	8	9	10	11
Discharge (m^3 s^{-1})	100	180	450	800	1200	900	650	350	300	250	150	100

Duration (hr)	0 ~ 1	1 ~ 2	2 ~ 3	3 ~ 4
Rainfall intensity (mm hr^{-1})	5	20	30	10

4.17 For the flow and rainfall data in Problem 4.16, if the baseflow is 100 m^3 s^{-1} and the average infiltration loss rate is 20 mm hr^{-1}, derive a UH for the watershed and determine its time base.

4.18 Convert the 30-min UH shown in the following table into the corresponding 1- and 2-hr UHs using the S-curve method. Plot the UHs, S-curve, and shifted S-curve on the same graph.

Time (hr)	0	0.5	1.0	1.5	2.0	2.5	3.0	3.5	4.0
UH ordinate (cfs in^{-1})	0	1	3	5	8	6	3	1	0

4.19 Convert the 4-hr UH shown in the following table into a 2-hr UH using the S-curve method. Plot the UHs, S-curve, and shifted S-curve on the same graph.

Time (hr)	0	2	4	6	8	10	12
UH ordinate (cfs in^{-1})	0	88	353	565	671	600	388
Time (hr)	14	16	18	20	22	24	
UH ordinate (cfs in^{-1})	247	106	71	21	7	0	

4.20 The 1-hr UH for a watershed is given in the following table. Determine the flow hydrograph if it receives a:

(1) 1-hr storm that generates 8 mm excess rainfall
(2) 2-hr storm that generates 10 mm excess rainfall
(3) 3-hr storm that generates 5, 15, and 3 mm excess rainfall in the first, second, and third hour, respectively.

Time (hr)	0	1	2	3	4	5	6	7	8	9	10	11
Discharge ($m^3 s^{-1} mm^{-1}$)	0	8	35	70	110	80	55	25	20	15	5	0

4.21 A 6 mi^2 watershed has a longest flow path length of 5.8 mi, slope of 0.002, basin coefficient of 2.0, storage coefficient of 0.6, and a composite curve number of 65. The length along the main stream from the outlet to the projected point of the watershed centroid is 3.2 mi. Derive a 1-hr:

(1) SCS UH
(2) Snyder UH.

4.22 If thc watershed in Problem 4.21 receives a 2-hr storm with an excess rainfall of 2.5 in, determine the flow hydrograph using a:

(1) SCS UH
(2) Snyder UH.

4.23 The time of concentration of a 5 ac catchment is 30 min. If the composite runoff coefficient of the catchment is 0.6, use the Modified Rational Equation to develop the flow hydrograph from a 40-min storm with a rainfall intensity of 1.2 in hr^{-1}.

4.24 A 10 ac drainage basin with a time of concentration of 1 hr receives a 30-min storm with a rainfall intensity of 2.8 in hr^{-1}. If the composite runoff coefficient is 0.8, develop and sketch a flow hydrograph using the Modified Rational method.

4.25 A 5 mi^2 watershed located in Richmond, VA has a time of concentration of 30 min and a channel travel time of 12 min. Its composite curve number is 69. Develop and plot a 10-year flow hydrograph at the watershed outlet using the SCS TR-55 ordinates.

4.26 A 6 km^2 watershed has HSG B and Type II rainfall distribution. Its land cover consists of 40% good pasture, 30% woods–grass combination with a fair condition, 20% poor woods, and 10% residential districts with 1-ac dwelling lots. The watershed has a longest flow path length of 1520 m and an average slope of 0.36. The 1368-m-long main stream has an approximately trapezoidal channel (Manning's n = 0.03) with a bottom width of 10 m, side slope of 2.5, bed slope of 0.008, and a bank-full depth of 2 m. Use the SCS TR-55 ordinates to develop a flow hydrograph from a 57.4-mm storm. Assume an ARC II condition.

4.27 Develop an example of hydrologic analysis using the WinTR-55 computer program (www.nrcs.usda.gov/wps/portal/nrcs/detailfull/national/home/?cid=stelprdb1042901).

4.28 Classify HEC-HMS, HSPF, SWAT, SWMM, and HEC-RAS in accordance with Figure 4.36.

5 Hydraulic Machinery

Pumps and turbines, two commonly used hydraulic machines, are important components in water resources engineering systems. Pumps are used to transfer water, whereas turbines are used to harvest water energy to generate electricity. This chapter discusses the characteristics and selection of these two hydraulic machines from the perspective of hydraulic engineering. The purpose of this chapter is to present the basics for water resources and hydraulic engineers to understand the performance of these machines and how they can be selected to best match a hydraulic engineering system. How these machines are designed is a subject for other disciplines, such as mechanical engineering, and is thus beyond the scope of this book. In addition to the differences and similarities between pumps and turbines, this chapter also discusses the selection considerations and characteristics of these two machines.

5.1 Overview of Pumps and Turbines

Pumps are devices that lift water by converting mechanical energy into water energy. A pump consumes electronic power to generate rotatory motion, through which the power is transferred into a water system so that the water can be transported from one location to another between which a negative energy gradient exists. Pumps are widely used in water resources and environmental engineering, such as in water supply, flood control, irrigation, wastewater treatment, and groundwater exploration. For instance, as illustrated in Figure 5.1, a pump is needed to transfer the water in Reservoir A through a pipeline upward to Reservoir B. The head against by the pump is $\Delta H = z_B - z_A$, while the head that needs to be added by the pump should be at least $h_p = \Delta H + h_L$, where h_L is the total energy loss, including friction and minor losses (Eq. (2.22)). h_p is also called the total head against which the pump operates. Given that the velocity heads in the two reservoirs are relatively minimal and thus can be assumed to be zero, the energy equation between the two reservoir surfaces (i.e., sections ① and ②) can be written as:

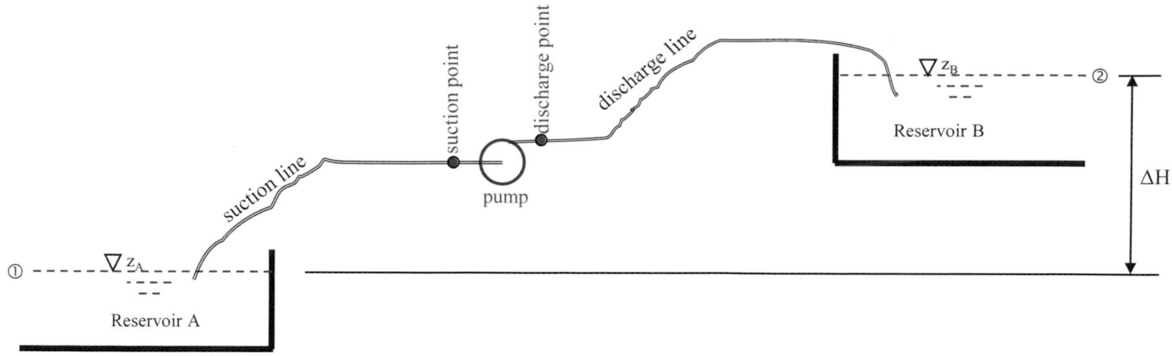

Figure 5.1 **Two reservoirs connected by a pipeline that includes a pump.** Water is pumped from Reservoir A to B, whose water surfaces have a difference in elevation of ΔH.

$$h_p = \Delta H + h_L = z_B - z_A + h_L \tag{5.1}$$

(total head against by pump: h_p; head against by pump: ΔH; total energy loss (Eq. (2.22)): h_L; water surface elevation of Reservoir B: z_B; water surface elevation of Reservoir A: z_A)

In contract to pumps, turbines are power-generation devices that work by harvesting potential and/or pressure energy in a water system. The water energy drives a turbine to rotate, which in turn drives an electronic generator to synchronically rotate and generate power.

Example 5.1 For the water system shown in Figure 5.1, if $z_A = 50\,\text{m}$, $z_B = 80\,\text{m}$, the suction line is 200 m long, the discharge line is 1500 m long, both the suction and discharge lines have a diameter of 0.5 m and a Darcy friction factor of 0.012, the entrance and exit loss coefficients are 0.12 and 1.0, respectively, the pump is located at an elevation of 58 m, and the water transfer rate is 0.2 $m^3\,s^{-1}$, determine the:

(1) Head against by the pump
(2) Total head against (i.e., required minimum head to be added) by the pump
(3) Pressure head at the suction point
(4) Pressure head at the discharge point.

Solution

$z_A = 50\,\text{m}$, $z_B = 80\,\text{m}$, $L_s = 200\,\text{m}$, $L_d = 1500\,\text{m}$, $D = 0.5\,\text{m}$, $f = 0.012$, $k_e = 0.12$, $k_d = 1.0$, $z_{pump} = 58\,\text{m}$, $Q = 0.2\,\text{m}^3\,\text{s}^{-1}$

(1) $\Delta H = z_B - z_A = (80\,\text{m}) - (50\,\text{m}) = 30\,\text{m}$.
(2) Cross-sectional area of the pipeline: $A = \pi/4 * (0.5\,\text{m})^2 = 0.1963\,\text{m}^2$.
Mean flow velocity in the pipeline: $V = (0.2\,\text{m}^3\,\text{s}^{-1})/(0.1963\,\text{m}^2) = 1.0188\,\text{m}\,\text{s}^{-1}$.

Darcy–Weisbach equation (Eq. (2.23)) → friction loss: $h_f = 0.012*[(200\,m+1500\,m)/(0.5\,m)]*[(1.0188\,m\,s^{-1})^2/(2*9.81\,m\,s^{-2})] = 2.16\,m$.
Entrance loss (Table 2.3): $h_e = (0.12)*(1.0188\,m\,s^{-1})^2/(2*9.81\,m\,s^{-2}) = 0.0063\,m$.
Discharge loss (Table 2.3): $h_d = (1.0)*(1.0188\,m\,s^{-1})^2/(2*9.81\,m\,s^{-2}) = 0.053\,m$.
Total loss (Eq. (2.22)): $h_L = (2.16\,m) + (0.0063\,m) + (0.053\,m) = 2.2193\,m$.
Total head against: $h_p = (30\,m) + (2.2193\,m) = 32.2193\,m \approx 32.22\,m$.

(3) Eq. (2.23) → friction loss in the suction line $h_{f,s} = 0.012*[(200\,m)/(0.5\,m)]*(1.0188\,m\,s^{-1})^2/(2*9.81\,m\,s^{-2}) = 0.25\,m$.
Entrance loss (Table 2.3): $h_e = (0.12)*(1.0188\,m\,s^{-1})^2/(2*9.81\,m\,s^{-2}) = 0.0063\,m$.
Energy equation (Eq. (2.21)) between section ① and the suction point: $(50\,m)+(0\,m)+(0\,m) = (58\,m) + p_s/\gamma + (1.0188\,m\,s^{-1})^2/(2*9.81\,m\,s^{-2}) + (0.25\,m + 0.0063\,m) \rightarrow p_s/\gamma = -8.31\,m$.

(4) Eq. (2.23) → friction loss in the discharge line $h_{f,d} = 0.012*[(1500\,m)/(0.5\,m)]*(1.0188\,m\,s^{-1})^2/(2*9.81\,m\,s^{-2}) = 1.91\,m$.
Discharge loss (Table 2.3): $h_d = (1.0)*(1.0188\,m\,s^{-1})^2/(2*9.81\,m\,s^{-2}) = 0.053\,m$.
Energy equation (Eq. (2.21)) between the discharge point and section ②: $(58\,m) + p_d/\gamma + (1.0188\,m\,s^{-1})^2/(2*9.81\,m\,s^{-2}) = (80\,m) + (0\,m) + (0\,m) + (1.91\,m + 0.053\,m) \rightarrow p_d/\gamma = 23.91\,m$.
In summary: the head added by the pump is equal to the difference between the pressure head at the discharge point and that at the suction point, as verified by $h_p = (23.91\,m) - (-8.31\,m) = 32.22\,m$.

5.2 Pumps

Various types and models of pumps are available in the market. They can have distinctly different operation performances. This section discusses pump characteristics, classifications, and selections in accordance with a hydraulic or water resources engineering system.

5.2.1 Design Considerations

From the hydraulic engineering point of view, the eight parameters listed in Table 5.1 are fundamental for pump design. For a given pump, the total head against, h_p, and discharge, Q, follow an inverse functional relationship. That is, when a larger head is added, a smaller discharge can be delivered, and vice versa. Such a relationship is described by pump performance curves, which are discussed in detail in Section 5.2.3. In addition, the pump size, D, is a design parameter because it dictates, and is dictated by, the location and space where the pump can be installed. For instance, in urban areas with a limited space available, pumps with a smaller D are preferred, whereas in rural areas with a large open space, pumps with a larger D may be used. Also, the rotation speed, n, can be restricted in residential areas because a larger speed can increase noise to an unacceptable level. Changing D and/or n will change the h_p versus Q relationship, as discussed in Section 5.2.4, increasing or decreasing the head to be added and the discharge to be delivered by the same pump.

Table 5.1 Pump design parameters.

No.	Parameter	Symbol	Description
1	Total head against	h_p	Required minimum head to be added
2	Discharge	Q	Delivery capacity
3	Size	D	Impeller diameter
4	Rotation speed	n	Revolution per unit time
5	Water kinematic viscosity	ν	Affected by constituents and temperature of water
6	Brake power	P_{brake}	Power required to drive pump
7	Efficiency	η	Ratio of output to brake power
8	Power transferred	$P_{transferred}$	Power actually transferred to water system

Further, the kinematic viscosity of the water to be transferred dictates which types of pumps can be used (see Section 5.2.2) and affects the performance curve of a given pump. Overall, both the head to be added and the discharge to be delivered by a pump tend to decrease as the water viscosity increases.

The last three parameters listed in Table 5.1 determine the required power to drive a pump and how much of the supplied power can be transferred to the water system. The required power is called brake power, P_{brake}. The power that can be supplied depends on the existing local power network grid, making it inappropriate to use pumps with a power demand (i.e., P_{brake}) beyond the network capacity. In addition, the efficiency of a pump indicates the percentage of P_{brake} that is available after the pump to be transferred to the water system. The power after the pump is called output power, P_{out}. For a pump, a lower efficiency means that the pump is subject to a larger energy loss due to internal friction, wasting energy and driving up the cost of operating the pump. Usually, pump efficiency is a parabolic function of discharge: increasing and then decreasing with increase of discharge. The paired values of h_p and Q that respond to the maximum efficiency is termed the best efficiency point (BEP). Further, P_{out} can be subject to major (i.e., friction) and minor losses (see Eq. (2.22)) in the water system (e.g., the pipeline in Figure 5.1). As a result, the power actually transferred to the system, $P_{transferred}$, is less than P_{out}. Various relationships among the parameters are provided in Eqs. (5.2) through (5.7).

The output power can be computed as:

discharge; specific weight of water; output power; total head against (Eq. (5.1))

$$P_{out} = \gamma Q h_p \quad (5.2)$$

The efficiency is defined as:

output power; efficiency; brake power

$$\eta = \frac{P_{out}}{P_{brake}} (100\%) \quad (5.3)$$

The power transferred to water is computed as:

$$\underbrace{P_{transferred}}_{\text{power transferred to water}} = \underbrace{\gamma}_{\text{specific weight of water}} \underbrace{Q}_{\text{discharge}} \underbrace{(\Delta H)}_{\text{head against}} \tag{5.4}$$

The power loss due to internal friction in the pump is computed as:

$$\underbrace{\Delta P_{pump}}_{\text{power loss from pump itself}} = \underbrace{P_{brake}}_{\text{brake power}} - \underbrace{P_{out}}_{\text{output power}} = (1 - \underbrace{\eta}_{\text{efficiency}})\, P_{brake} \tag{5.5}$$

The power loss due to the water system is computed as:

$$\underbrace{\Delta P_{ws}}_{\text{power loss from water system}} = \underbrace{P_{out}}_{\text{output power}} - \underbrace{P_{transferred}}_{\text{power transferred to water}} = \underbrace{\eta}_{\text{efficiency}} \underbrace{P_{brake}}_{\text{brake power}} - \underbrace{\gamma}_{\text{specific weight of water}} \underbrace{Q}_{\text{discharge}} \underbrace{(\Delta H)}_{\text{head against}} \tag{5.6}$$

Finally, combining Eqs. (5.5) and (5.6), yields the following equation for gross power loss:

$$\underbrace{\Delta P}_{\text{gross power loss}} = \underbrace{\Delta P_{pump}}_{\text{power loss from pump itself}} + \underbrace{\Delta P_{ws}}_{\text{power loss from water system}} = \underbrace{P_{brake}}_{\text{brake power}} - \underbrace{\gamma}_{\text{specific weight of water}} \underbrace{Q}_{\text{discharge}} \underbrace{(\Delta H)}_{\text{head against}} \tag{5.7}$$

Example 5.2 If the pump in Example 5.1 has an efficiency of 0.75, determine the:

(1) Output power
(2) Brake power
(3) Power transferred to water
(4) Power loss due to the pump itself
(5) Power loss due to the pipeline
(6) Gross power loss.

Solution

Given $\eta = 0.75$

Example 5.1: $\Delta H = 30\,m$, $h_p = 32.22\,m$, $h_L = 2.22\,m$, $Q = 0.2\,m^3\,s^{-1}$

(1) Eq. (5.2): $P_{out} = (9.81\,kN\,m^{-3}) * (0.2\,m^3\,s^{-1}) * (32.22\,m) = 63.22\,kW$.

(2) Eq. (5.3): $P_{brake} = (63.22\,kW)/0.75 = 84.29\,kW$.

(3) Eq. (5.4): $P_{transferred} = (9.81\,kN\,m^{-3}) * (0.2\,m^3\,s^{-1}) * (30\,m) = 58.86\,kW$.

(4) Eq. (5.5): $\Delta P_{pump} = (84.29\,kW) - (63.22\,kW) = 21.07\,kW$.

(5) Eq. (5.6): $\Delta P_{ws} = (63.22\,kW) - (58.86\,kW) = 4.36\,kW$.

(6) Eq. (5.7): $\Delta P = (21.07\,kW) + (4.36\,kW) = 25.43\,kW$.

Example 5.3 As shown in Figure 5.2, a pump ($\eta = 0.82$) is used to pump groundwater into a water supply pond at 3.5 $ft^3\,sec^{-1}$. If the bending and discharge loss coefficients are 0.1 and 1.0, respectively, determine the:

(1) Output power
(2) Brake power
(3) Power transferred to water
(4) Power loss duc to the pump itself
(5) Power loss due to the pipeline
(6) Gross power loss.

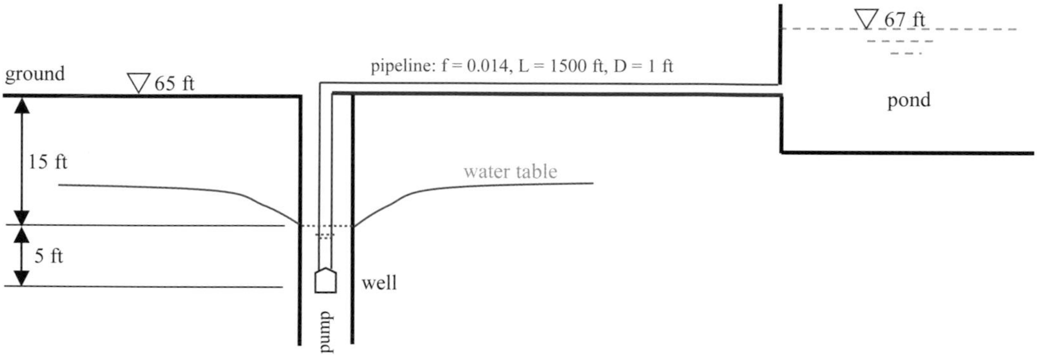

Figure 5.2 **Schematic of the well, pump, pipeline, and pond for Example 5.3.**

Solution

$\eta = 0.82$, $Q = 3.5\,ft^3\,sec^{-1}$, $L = 1500\,ft$, $D = 1\,ft$, $f = 0.014$, $k_b = 0.1$, $k_d = 1.0$

Cross-sectional area of the pipeline: $A = \pi/4 * (1\,ft)^2 = 0.7854\,ft^2$.

Mean flow velocity in the pipeline: $V = (3.5\,ft^3\,sec^{-1})/(0.7854\,ft^2) = 4.4563\,ft\,sec^{-1}$.

Darcy–Weisbach equation (Eq. (2.23)) → friction loss: $h_f = 0.014 * (1500\,ft)/(1\,ft)] * (4.4563\,ft\,sec^{-1})^2/(2 * 32.2\,ft\,sec^{-2})] = 6.48\,ft$.

Bending loss (Table 2.3): $h_b = (0.1) * (4.4563\,ft\,sec^{-1})^2/(2 * 32.2\,ft\,sec^{-2}) = 0.031\,ft$.

Discharge loss (Table 2.3): $h_d = (1.0) * (4.4563\,ft\,sec^{-1})^2/(2 * 32.2\,ft\,sec^{-2}) = 0.31\,ft$.

Total loss (Eq. (2.22)): $h_L = (6.48\,\text{ft}) + (0.031\,\text{ft}) + (0.31\,\text{ft}) = 6.82\,\text{ft}$.

Energy equation (Eq. (2.21)) between the pump and the pond water surface:

$$(65\,\text{ft} - 15\,\text{ft} - 5\,\text{ft}) + (5\,\text{ft}) + (0\,\text{ft}) + h_p = (67\,\text{ft}) + (0\,\text{ft}) + (0\,\text{ft}) + (6.82\,\text{ft})$$

$$h_p = 23.82\,\text{ft}.$$

Head against (Eq. (5.1)): $\Delta H = (23.82\,\text{ft}) - (6.82\,\text{ft}) = 17.00\,\text{ft}$.

(1) Eq. (5.2): $P_{out} = (62.4\,\text{lbf ft}^{-3}) * (3.5\,\text{ft}^3\,\text{sec}^{-1}) * (23.82\,\text{ft}) = 5202.29\,\text{lbf-ft sec}^{-1}$.

(2) Eq. (5.3): $P_{brake} = (5202.29\,\text{lbf-ft sec}^{-1})/0.82 = 6344.26\,\text{lbf-ft sec}^{-1}$.

(3) Eq. (5.4): $P_{transferred} = (62.4\,\text{lbf ft}^{-3}) * (3.5\,\text{ft}^3\,\text{sec}^{-1}) * (17.00\,\text{ft}) = 3712.80\,\text{lbf-ft sec}^{-1}$.

(4) Eq. (5.5): $\Delta P_{pump} = (6344.26\,\text{lbf-ft sec}^{-1}) - (5202.29\,\text{lbf-ft sec}^{-1}) = 1141.97\,\text{lbf-ft sec}^{-1}$.

(5) Eq. (5.6): $\Delta P_{ws} = (5202.29\,\text{lbf-ft sec}^{-1}) - (3712.80\,\text{lbf-ft sec}^{-1}) = 1489.49\,\text{lbf-ft sec}^{-1}$.

(6) Eq. (5.7): $\Delta P = (1141.97\,\text{lbf-ft sec}^{-1}) + (1489.49\,\text{lbf-ft sec}^{-1}) = 2631.46\,\text{lbf-ft sec}^{-1}$.

5.2.2 Characteristics and Classifications

In terms of how water is transferred, water pumps can be classified into two categories: centrifugal or displacement (Figure 5.3). A centrifugal pump uses a rotating impeller to move water into the pump and pressurize the discharge flow (Fristam, 2014). Centrifugal pumps can be further subdivided into three types, namely radial-, axial-, and mixed-flow pumps. A radial-flow pump (Figure 5.4a) sucks water into its suction eye and then moves the water into its volute (a small gradually widening channel between the outer tips of the impeller and the casing), where the water is pressurized and discharged out of the pump along the direction perpendicular to the pump axis. Radial-flow pumps can add a high-pressure head, and tend to deliver a low flow rate. In contrast, an axial-flow pump (Figure 5.4b) sucks water into its impeller along the radial direction and then moves the water to its center, where the water is pressurized and discharged out of the pump along its axial direction. Axial-flow pumps may not be able to add a high-pressure head, but they can deliver a large flow rate. A mixed-flow pump (Figure 5.4c), which shares the features of radial- and axial-flow pumps, sucks water into its eye along the axial direction as well as into its impeller

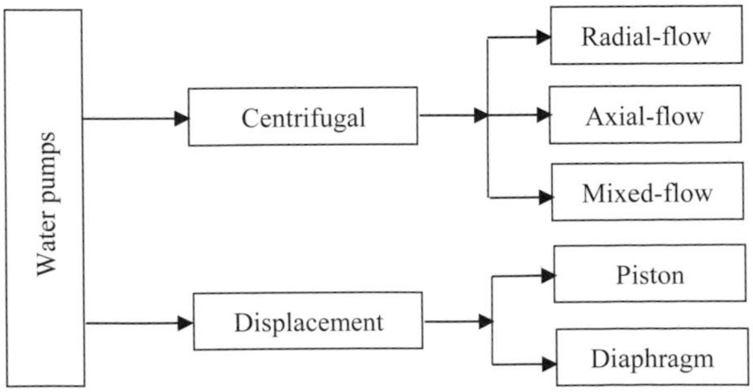

Figure 5.3 **The classification of water pumps.**

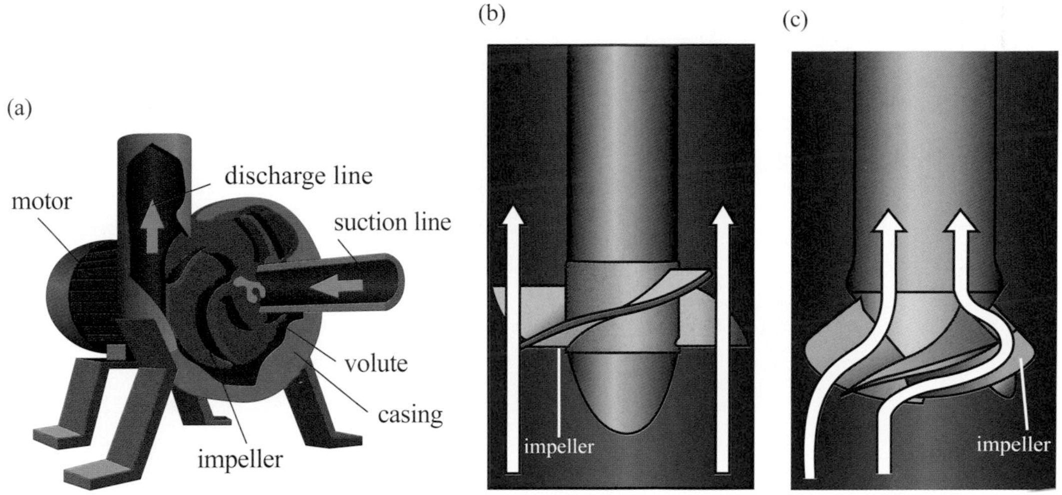

Figure 5.4 **Sketches of centrifugal pump types.** (a) Radial-flow pump; (b) axial-flow pump; and (c) mixed-flow pump. (*Source:* screenshots from www.youtube.com/watch?v=guATpg2drs4. Reproduced with permission by Rashid Majeed.)

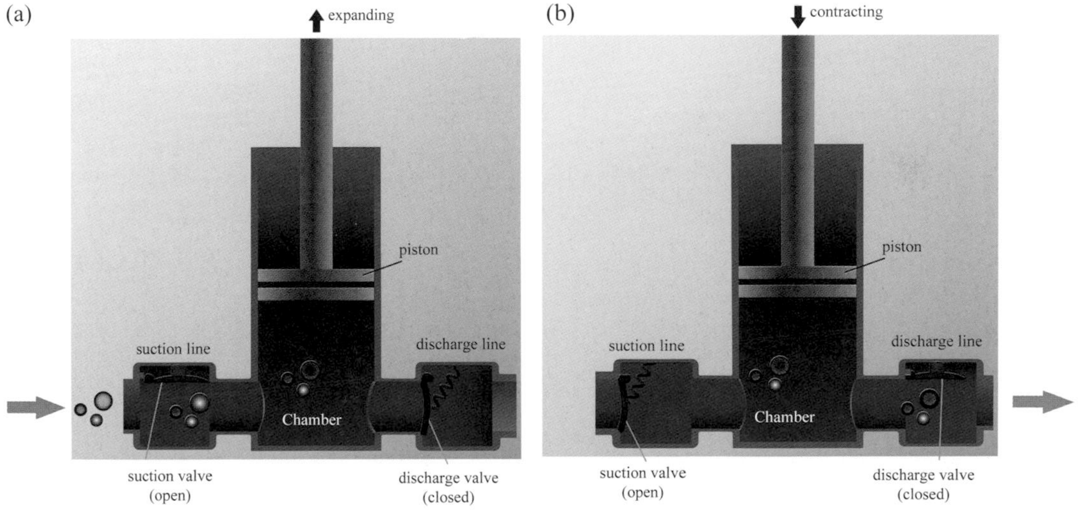

Figure 5.5 **Sketches of a piston pump.** Diagrams show pump chamber: (a) expanding; and (b) contracting. (*Source:* screenshots from www.youtube.com/watch?v=4OJTN0M1DBk. Reproduced with permission by Adam Balogh.)

along the radial direction. Mixed-flow pumps can add a high-pressure head and at the same time deliver a large flow rate.

Displacement pumps deliver a fixed amount of water at a specific time interval through an expanding and contracting chamber regulated by a series of check valves. Depending on the expansion/contraction mechanism, displacement pumps can be further subdivided into two types,

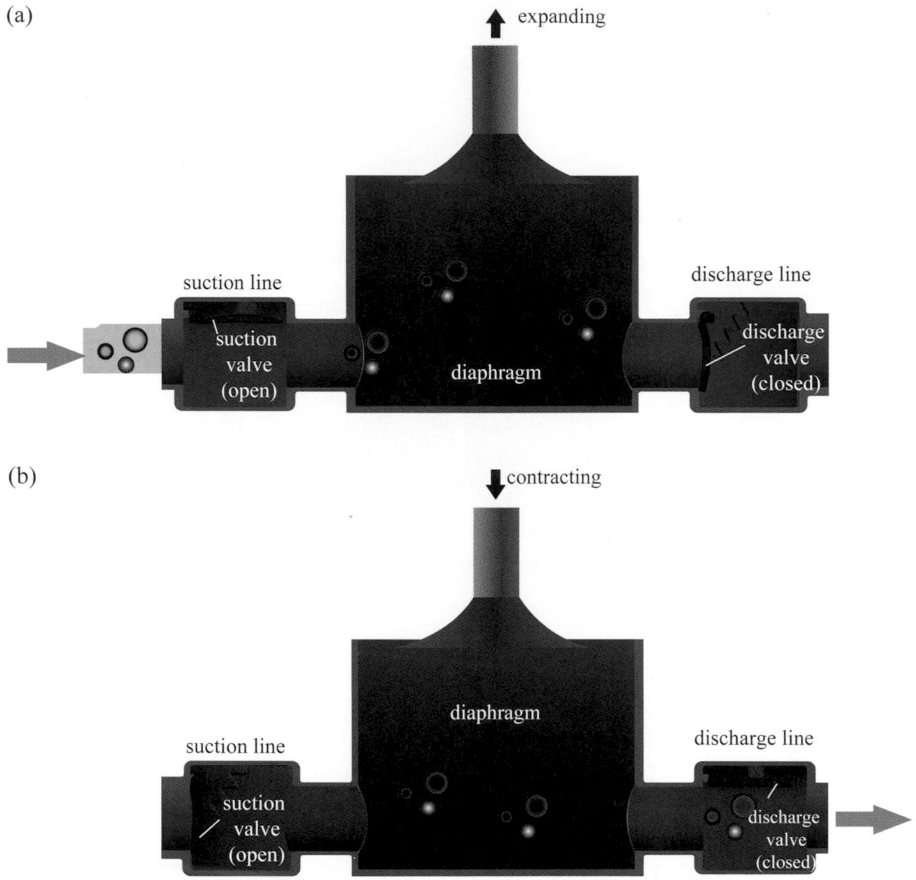

Figure 5.6 **Sketches of a diaphragm pump.** Diagrams show pump diaphragm: (a) expanding; and (b) contracting. (*Source:* screenshots from www.youtube.com/watch?v=4OJTN0M1DBk. Reproduced with permission by Adam Balogh.)

namely piston and diaphragm. For a piston pump, when its piston is pulled to expand the chamber (Figure 5.5a), the suction check valve is opened while the discharge check valve is closed, sucking water into the chamber. Once the chamber is filled, the piston is pushed to contract the chamber (Figure 5.5b), the suction check valve is closed while the discharge check valve is opened, moving the water out of (and thus emptying) the chamber. Similarly, for a diaphragm pump, when the flexible diaphragm is expanded (Figure 5.6a), the suction check valve is opened while the discharge check valve is closed, sucking water into the diaphragm. Once the diaphragm is filled, the diaphragm is contracted (Figure 5.6b), the suction check valve is closed while the discharge check valve is opened, squeezing the water out of (and thus emptying) the diaphragm.

Table 5.2 compares the characteristics of the two categories of pumps. Overall, centrifugal pumps are preferred when a large continuous discharge is of interest, whereas displacement pumps may be appropriate when a large pressure rise per stage is required. In addition, both displacement and centrifugal (open or semi-open casing) pumps are good for waste water or water with a high viscosity. Centrifugal pumps with a closed casing can be used to pump clear water only. Further,

Table 5.2 Characteristics of centrifugal versus displacement pumps.

Characteristic	Centrifugal	Displacement
Discharge	high	low
Discharge stream	continuous	pulsing
Pressure rise per stage	low	high
Constant quantity over operating range	pressure	discharge
Appropriateness for viscous water	yes (open and semi-open casing) no (closed casing)	yes
Self-priming[1]	yes (closed casing)	no

[1] Automatically adding water into the pump casing to evacuate any trapped air while to create a water seal inside the casing.

to avoid possible damages from cavitation (see Section 5.2.6), centrifugal pumps with a closed casing are usually self-priming (Gongol and Gongol, 2008), automatically adding water into the casing to evacuate trapped air while creasing a water seal inside the casing. Self-priming pumps cannot operate without water in the casing.

5.2.3 Performance Curves

For a given pump, its dimensional performance curves (e.g., Figure 5.7) show the functional relationships among head capacity (ΔH_p), brake power (P_{brake}), efficiency (η), and discharge capacity (Q_p). Such performance curves are pump-specific; that is, they are different from pump to pump. As shown in Figure 5.7, ΔH_p decreases with increase of Q_p, whereas P_{brake} increases with increase of Q_p. Similarly, η increases with increase of Q_p, reaches a peak at a value of Q_p, and then decreases with further increase of Q_p. At the peak efficiency, the point defined by the corresponding paired values of ΔH_p and Q_p is called best efficiency point (BEP). At the BEP of a pump, the power loss due to the internal frictions of the pump is minimized and thus a maximum percentage

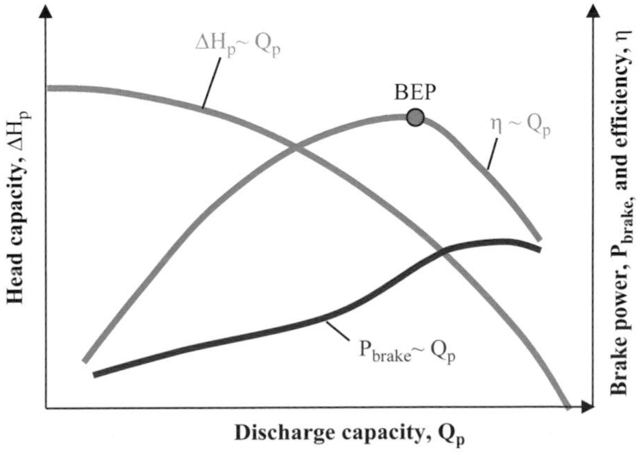

Figure 5.7 Dimensional performance curves for a hypothetical pump. BEP: best efficiency point.

of the brake power can be potentially transferred to water. In addition, for a given pump, if its impeller size and/or rotation speed are changed, its performance curves will be changed as well (see Section 5.2.4).

Example 5.4 The performance curves for a centrifugal pump with various sizes of impellers are shown in Figure 5.8. For the 8.5-in-diameter impeller,

(1) Determine the head added when the pump delivers 42 gpm
(2) Determine the discharge delivered when the pump adds a head of 110 ft
(3) Develop and plot the efficiency versus discharge capacity ($\eta \sim Q_p$) curve
(4) Determine the peak efficiency and BEP.

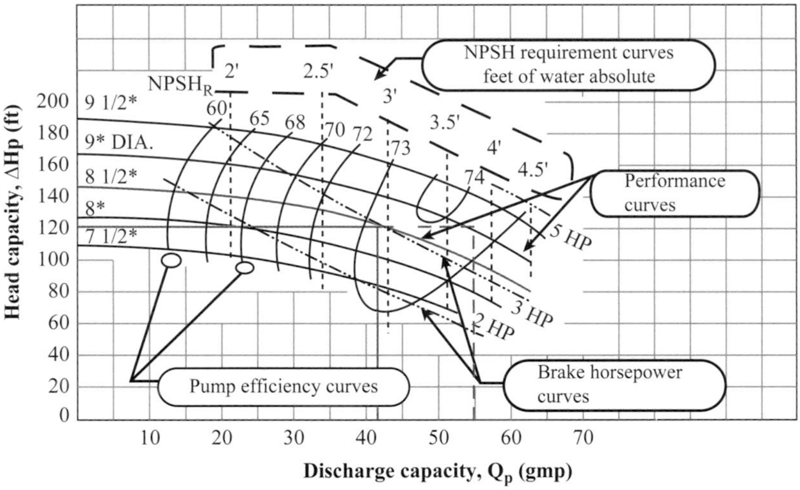

Figure 5.8 **The pump performance curves used in Example 5.4 and 5.5.** (*Source:* Satterfield, 2013)

Solution

The answers should be based on the curve highlighted in pink.

(1) Draw a vertical line at $Q_p = 42$ gpm to intersect with the $\Delta H_p \sim Q_p$ curve for the 8.5-in-diameter impeller, and then draw a horizontal line to intersect with the left y-axis, giving the head added $\Delta H_p = 120$ ft.

(2) Draw a horizontal line starting from the left y-axis at $\Delta H_p = 110$ ft to intersect with the $\Delta H_p \sim Q_p$ curve for the 8.5-in-diameter impeller, and then draw a vertical line to intersect with the x-axis, giving the discharge delivered $Q_p = 50$ gpm.

(3) From the intersections between the efficiency contours and the $\Delta H_p \sim Q_p$ curve for the 8.5-in-diameter impeller, draw vertical lines to intersect with the x-axis, giving the discharges, as illustrated in Figure 5.9. The results are shown in Table 5.3 and plotted in Figure 5.10.

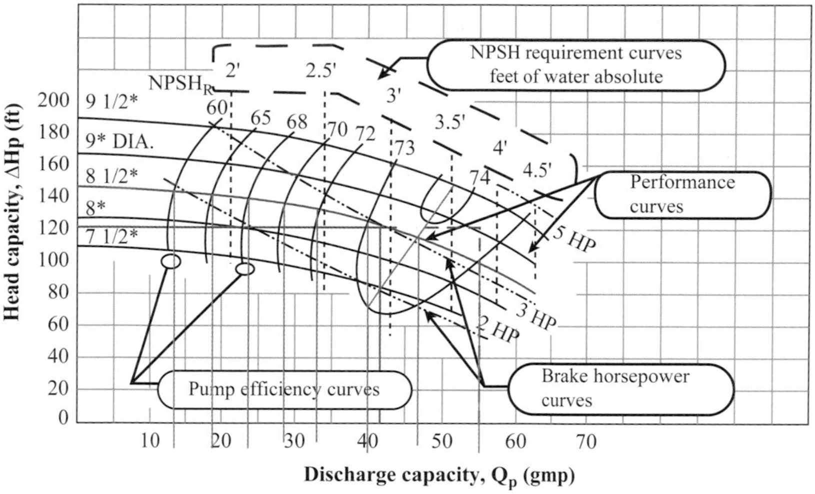

Figure 5.9 **Developing the efficiency versus discharge capacity curve for part (3) of Example 5.4.** (*Source:* Satterfield, 2013)

Table 5.3 Pump efficiency versus discharge capacity for Example 5.4.

Efficiency, η (%)	60	65	68	70	72	73	73.3	73
Discharge, Q_p (gpm)	13.0	18.5	24.0	28.0	33.0	40.0	47.0	55.0

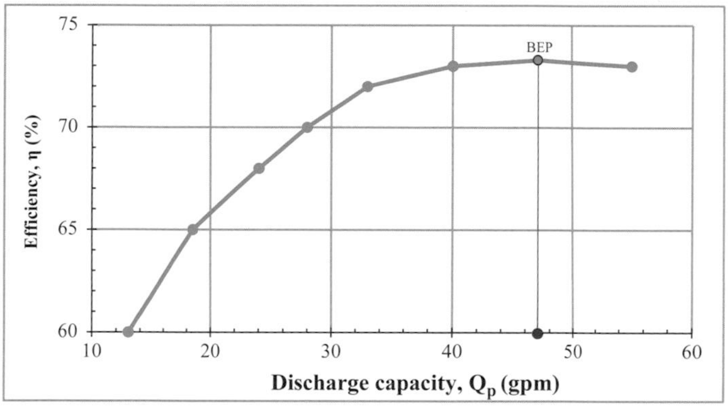

Figure 5.10 **Pump efficiency versus discharge capacity for part (3) of Example 5.4.** BEP: best efficiency point.

(4) From Table 5.3 and Figure 5.10, the peak efficiency is 73.3%.
At the BEP → Figure 5.10: $Q_p = 47$ gpm → Figure 5.8: $\Delta H_p = 115$ ft.

Example 5.5 For the performance curves shown in Figure 5.8, determine the relationship between:

(1) Head capacity and impeller size for a discharge capacity of 42 gpm
(2) Discharge capacity and impeller size for a head capacity of 120 ft.

Solution

(1) Draw a vertical line at $Q_p = 42$ gpm to intersect with the $\Delta H_p \sim Q_p$ curves for the impellers with different sizes, draw horizontal lines from the intersections to intersect with the left y-axis, which yield the head capacities. The results are summarized in Table 5.4. Plotting and regressing $\Delta H_p/82$ on D/7.5 (Figure 5.11a), one derives $\Delta H_p/82 = 1.03(D/7.5)^{2.72}$.

Table 5.4 Head capacity versus impeller size for Example 5.5.

Impeller diameter, D (in)	7.5	8.0	8.5	9.0	9.5
Head capacity, ΔH_p (ft)	82	100	120	140	160
D/7.5	–	1.07	1.13	1.20	1.27
$\Delta H_p/82$	–	1.22	1.46	1.71	1.96

(2) Draw a horizontal line at $\Delta H_p = 120$ ft to intersect with the $\Delta H_p \sim Q_p$ curves for the impellers with different sizes, draw vertical lines from the intersections to intersect with the x-axis, which yield the discharge capacities. Note that the horizontal line does not intersect with the performance curve for the 7.5-in impeller, meaning that the pump with this impeller size cannot

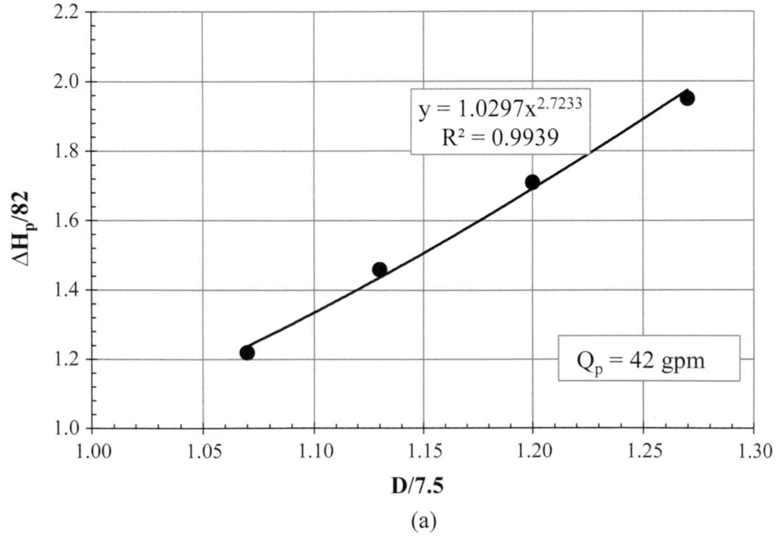

Figure 5.11 Relationship between impeller size and standardized: (a) head capacity; and (b) discharge capacity for Example 5.5.

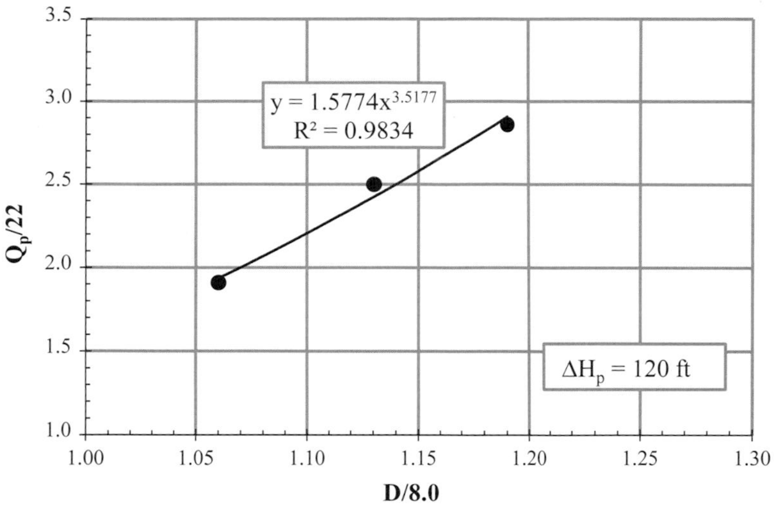

(b)

Figure 5.11 *(Continued)*

add 120 ft head and will automatically shut itself off. The results are summarized in Table 5.5. Plotting and regressing $Q_p/22$ on D/8.0 (Figure 5.11b), one derives $Q_p/22 = 1.58(D/8.0)^{3.52}$.

Table 5.5 Discharge capacity versus impeller diameter for Example 5.5.

Impeller diameter, D (in)	7.5	8.0	8.5	9.0	9.5
Discharge capacity, Q_p (gpm)	–	22	42	55	63
D/8	–	1.00	1.06	1.13	1.19
$Q_p/22$	–	1.00	1.91	2.50	2.86

Besides dimensional performance curves, dimensionless performance curves (Roberson *et al.*, 1998), which show the functional relationships among three dimensionless coefficients (namely head coefficient, discharge coefficient, and power coefficient) and efficiency, are also used in practice. These three coefficients can be derived (see Example 5.6) using dimensional analysis (Kuritza *et al.*, 2017), discussed in Section 2.6. Formulas for the three coefficients are provided in Eqs. (5.8) through (5.10).

The head coefficient is computed as:

$$C_H = \frac{g\left(\Delta H_p\right)}{n^2 D^2} \tag{5.8}$$

gravitational acceleration → g; head capacity → ΔH_p; head coefficient → C_H; rotation speed in revolution per unit time → n; impeller diameter → D

The discharge coefficient is computed as:

$$C_Q = \frac{Q_p}{nD^3} \tag{5.9}$$

discharge coefficient (C_Q); discharge capacity (Q_p); rotation speed in revolution per unit time (n); impeller diameter (D)

The power coefficient is computed as:

$$C_P = \frac{P_{brake}}{\rho n^3 D^5} \tag{5.10}$$

power coefficient (C_P); brake power (P_{brake}); density of water (ρ); rotation speed in revolution per unit time (n); impeller diameter (D)

Example 5.6 Perform dimensional analysis to derive the:

(1) Head coefficient, C_H (Eq. (5.8))
(2) Discharge coefficient, C_Q (Eq. (5.9))
(3) Power coefficient, C_P (Eq. (5.10)).

Solution

(1) Based on the principles of physics, the maximum pressure attainable by a pump is a function, f_1, of four independent variables expressed as:

$$\rho g\left(\Delta H_p\right) = f_1\left(\rho, n, D, Q_p\right) \tag{5.11}$$

density of water (ρ); gravitational acceleration (g); head capacity (ΔH_p); rotation speed in revolution per unit time (n); impeller diameter (D); discharge capacity (Q_p)

In Eq. (5.11), there are five variables, four of which are independent variables (ρ, n, D, and Q_p) that are independent from each other. The five variables have three basic dimensions, namely time [T], mass [M], and length [L]. Thus, one can derive two dimensionless terms. Let ρ, n, and D be the repeating variables because they have all three basic dimensions. The dimensions of the variables are as follows:

$\Delta H_p : [L]$
$\rho : [M L^{-3}]$
$g : [L T^{-2}]$
$n : [T^{-1}]$
$D : [L]$
$Q_p : [L^3 T^{-1}]$.

First, cancel [M] by dividing by ρ. Second, cancel [L] by dividing by D^2 or D^3. Finally, cancel [T] by dividing by n^3 or n. These steps result in a new function, f_2, expressed as:

gravitational acceleration; head capacity; discharge capacity

$$\frac{g\,(\Delta H_p)}{n^2 D^2} = f_2\left(\frac{Q_p}{nD^3}\right) \tag{5.12}$$

rotation speed in revolution per unit time; impeller diameter

The dimensionless variable on the left side of Eq. (5.12) is defined as the head coefficient, C_H.

(2) The dimensionless variable on the right side of Eq. (5.12) is defined as the discharge coefficient, C_Q.

(3) Based on the principles of physics, the brake power to drive a pump is a function, f_3, of four independent variables, expressed as:

density of water; rotation speed in revolution per unit time

brake power; discharge capacity

$$P_{brake} = f_3\,(\rho, n, D, Q_p) \tag{5.13}$$

impeller diameter

In Eq. (5.13), there are five variables, four of which are independent variables (ρ, n, D, and Q_p) that are independent from each other. The five variables have three basic dimensions, namely time [T], mass [M], and length [L]. Thus, one can derive two dimensionless terms. Let ρ, n, and D be the repeating variables because they have all three basic dimensions. The dimensions of the variables are as follows:

P_{brake} : $[M\,L^2\,T^{-3}]$
ρ : $[M\,L^{-3}]$
n : $[T^{-1}]$
D : $[L]$
Q_p : $[L^3\,T^{-1}]$.

First, cancel [M] by dividing by ρ. Second, cancel [L] by dividing by D^2 or D^3. Finally, cancel [T] by dividing by n^3 or n. These steps result in a new function, f_4, expressed as:

brake power; discharge capacity

$$\frac{P_{brake}}{\rho n^3 D^5} = f_4\left(\frac{Q_p}{nD^3}\right) \tag{5.14}$$

density of water; impeller diameter

rotation speed in revolution per unit time

The dimensionless variable on the left side of Eq. (5.14) is defined as power coefficient C_P.

Dimensionless performance curves are pump type-specific (Timár, 2005). That is, each pump type (e.g., radial-flow pumps) shares a set of dimensionless performance curves regardless of rotation speeds and impeller sizes. Such performance curves can be developed from laboratory experiments. Figure 5.12a shows the dimensionless performance curves for radial-flow pumps, while Figure 5.12b shows the dimensionless performance curves for axial-flow pumps.

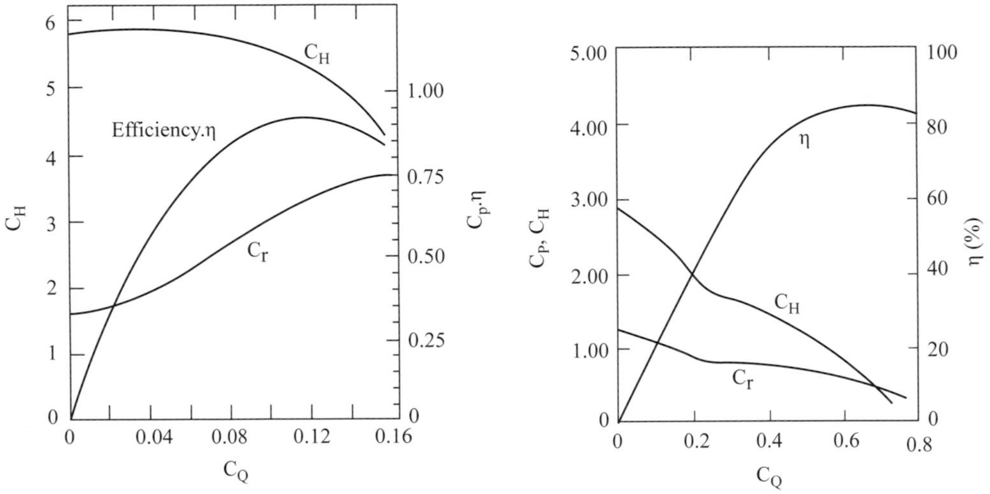

Figure 5.12 **Dimensionless performance curves.** Nomogram for: (a) radial-flow pumps; and (b) axial-flow pumps. (*Source:* Figures 8.9 and 8.5 in Roberson *et al.*, 1998)

Example 5.7 If a radial-flow pump has an impeller diameter of 30 cm and a rotation speed of 600 rpm, develop and plot its dimensional performance curves.

Solution

$D = 30\,\text{cm} = 0.3\,\text{m}$, $n = 600\,\text{rpm} = 10\,\text{rps}$, radial-flow pump

For arbitrarily selected values of C_Q ranging from 0.0 to 0.15, use Eq. (5.9) to calculate the corresponding values of Q_p and read Figure 5.12a to determine the corresponding values of C_H, C_P, and η. The values of C_H are used in Eq. (5.8) to calculate the corresponding values of ΔH_p, while the values of C_P are used in Eq. (5.10) to calculate the corresponding values of P_{brake}. The calculations are done in the Excel spreadsheet shown in Figure 5.13. The values from Figure 5.12a are in columns A to D, and the calculated values are in columns E to G. The dimensional performance curves are also shown in Figure 5.23.

For instance, for $C_Q = 0.02$, one can find $C_H = 5.82$, $C_P = 0.35$, and $\eta = 0.32$ from Figure 5.12a. Eq. (5.9) $\rightarrow Q_p = 0.02 * (10\,\text{rps}) * (0.3\,\text{m})^3 = 0.0054\,\text{m}^3\,\text{s}^{-1}$; Eq. (5.8) $\rightarrow \Delta H_p = 5.82 * (10\,\text{rps})^2 * (0.3\,\text{m})^2/(9.81\,\text{m}\,\text{s}^{-2}) = 5.34\,\text{m}$; and Eq. (5.10) $\rightarrow P_{brake} = 0.35 * (1000\,\text{kg}\,\text{m}^{-3}) * (10\,\text{rps})^3 * (0.3\,\text{m})^5 = 850.5\,\text{W} = 0.8505\,\text{kW}$. Similar calculations can be completed for other values of C_Q listed in column A.

	A	B	C	D	E	F	G
1	D =	30	cm =	0.3	m		
2	n =	600	rpm =	10	rps		
3	g =	9.81	m s^{-1}				
4	ρ =	1000	kg m^{-3}				
5							
6	C_Q	C_H	C_P	η	Q_p (m^3 s^{-1})	ΔH_p (m)	P_{brake} (kW)
7					(Eq. 5.9)	(Eq. 5.8)	(Eq. 5.10)
8	0.00	5.80	0.34	0.00	0.0000	5.32	0.8262
9	0.02	5.82	0.35	0.32	0.0054	5.34	0.8505
10	0.04	5.83	0.39	0.58	0.0108	5.35	0.9477
11	0.06	5.81	0.47	0.74	0.0162	5.33	1.1421
12	0.08	5.72	0.55	0.83	0.0216	5.25	1.3365
13	0.10	5.59	0.62	0.90	0.0270	5.13	1.5066
14	0.12	5.29	0.68	0.93	0.0324	4.85	1.6524
15	0.14	4.80	0.74	0.90	0.0378	4.40	1.7982
16	0.15	4.50	0.75	0.88	0.0405	4.13	1.8225

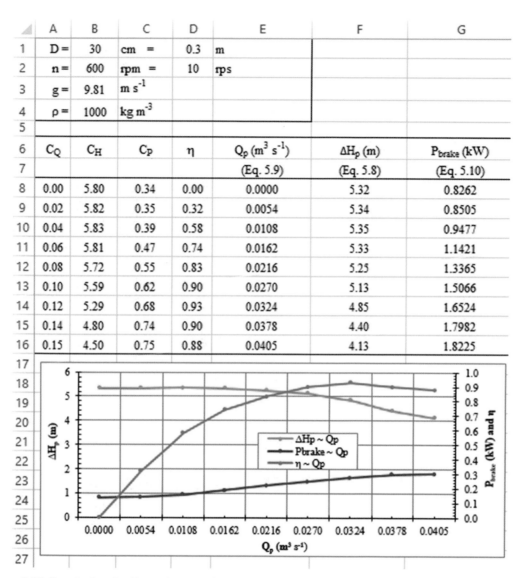

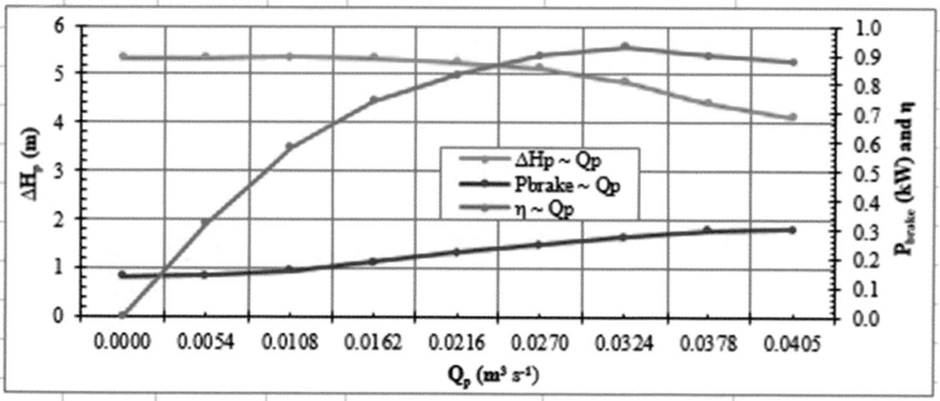

Figure 5.13 **Developing the dimensional performance curves in Example 5.7.**

Example 5.8 If a 1-ft-diameter axial-flow pump is discharging 5 cfs at a rotation speed of 840 rpm, determine the:

(1) Head capacity
(2) Brake power
(3) Efficiency
(4) Output power and power transferred to the water.

Solution

$D = 1\,ft$, $n = 840\,rpm = 14\,rps$, $Q_p = 5\,cfs$, $g = 32.2\,ft\,sec^{-2}$, $\rho = 1.94\,slug\,ft^{-3}$

Eq. (5.9): $C_Q = (5\,cfs)/[(14\,rps) * (1\,ft)^3] = 0.36 \rightarrow$ Figure 5.12b (axial-flow pump): $C_H = 1.60$, $C_P = 0.85$, $\eta = 0.62$.

(1) Eq. (5.8): $\Delta H_p = 1.60 * (14\,rps)^2 * (1\,ft)^2/(32.2\,ft\,sec^{-2}) = 9.74\,ft$.
(2) Eq. (5.10): $P_{brake} = 0.85 * (1.94\,slug\,ft^{-3}) * (14\,rps)^3 * (1\,ft)^5 = (4524.86\,lbf\text{-}ft\,sec^{-1}) * [(1\,hp)/(550\,lbf\text{-}ft\,sec^{-1})] = 8.23\,hp$.
(3) $C_Q = 0.36 \rightarrow$ Figure 5.12b: $\eta = 0.62$.
(4) Eq. (5.3): $P_{out} = 0.62 * (8.23\,hp) = 5.10\,hp$.
Eq. (5.4): $P_{transferred} = [(1.94\,slug\,ft^{-3}) * (32.2\,ft\,sec^{-2})] * (5\,cfs) * (9.74\,ft) = (3042.19\,lbf\text{-}ft\,sec^{-1}) * [(1\,hp)/(550\,lbf\text{-}ft\,sec^{-1})] = 5.53\,hp$.

For a pump used to pump water with various kinematic viscosities, the corresponding performance curves should be used. As illustrated in Figure 5.14, the head capacity and discharge capacity will be smaller for a larger viscosity than those for a lower viscosity, whereas the brake power will be greater for a larger viscosity (Roberson *et al.*, 1998).

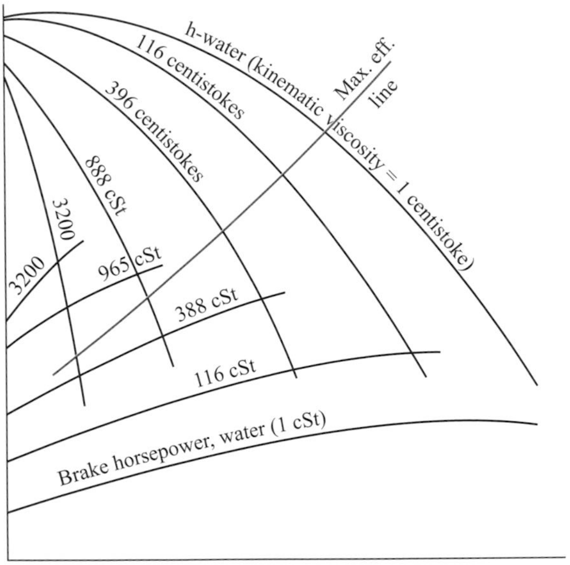

Figure 5.14 **Pump performance curves for selected kinematic viscosities.** (*Source:* reproduced from Figure 15.14 in Finnemore and Franzini, 2002)

5.2.4 Affinity Laws of Pumps

For a given pump, if its impeller size and/or rotation speed are changed, its dimensional performance curves will be changed as well. The new dimensional performance curves can be derived from the original dimensional performance curves using affinity laws. Given that the dimensionless performance curves are independent of impeller size and rotation speed (i.e., pump-type specific), the affinity laws can be formulated in terms of Eqs. (5.8) to (5.10). In this regard, if the impeller size is changed from D_1 to D_2 and the rotation speed from n_1 to n_2, one will have the relationships expressed as:

head coefficient; gravitational acceleration; head capacity; rotation speed in revolution per unit time; impeller diameter

$$C_H = \frac{g\left(\Delta H_{p,1}\right)}{n_1^2 D_1^2} = \frac{g\left(\Delta H_{p,2}\right)}{n_2^2 D_2^2} \Rightarrow \frac{\Delta H_{p,2}}{\Delta H_{p,1}} = \frac{n_2^2 D_2^2}{n_1^2 D_1^2} \tag{5.15}$$

discharge coefficient; discharge capacity; rotation speed in revolution per unit time; impeller diameter

$$C_Q = \frac{Q_{p,1}}{n_1 D_1^3} = \frac{Q_{p,2}}{n_2 D_2^3} \Rightarrow \frac{Q_{p,2}}{Q_{p,1}} = \frac{n_2 D_2^3}{n_1 D_1^3} \tag{5.16}$$

power coefficient; brake power; density of water; rotation speed in revolution per unit time; impeller diameter

$$C_P = \frac{P_{brake,1}}{\rho n_1^3 D_1^5} = \frac{P_{brake,2}}{\rho n_2^3 D_2^5} \Rightarrow \frac{P_{brake,2}}{P_{brake,1}} = \frac{n_2^3 D_2^5}{n_1^3 D_1^5} \tag{5.17}$$

Example 5.9 For the pump in Example 5.7, develop the dimensional performance curves if the:

(1) Impeller size is reduced to 20 cm
(2) Impeller size is increased to 40 cm
(3) Rotation speed is decreased to 480 rpm
(4) Rotation speed is increased to 720 rpm
(5) Impeller size is increased to 40 cm and the rotation speed is increased to 720 rpm.

Solution

$D_1 = 30\,\text{cm} = 0.3\,\text{m}$, $n_1 = 600\,\text{rpm} = 10\,\text{rps}$

The development of the performance curves can be done using Eqs. (5.15) to (5.17) in the Excel spreadsheet as shown in Figures 5.15 to 5.19. In these figures, columns A to D are the original performance curves, whereas columns D to G are the new performance curves.

(1) $D_2 = 20\,\text{cm} = 0.2\,\text{m}$, $n_2 = n_1 = 10\,\text{rps}$. The results are shown in Figure 5.15.

For instance, for the paired values on the original performance curves of $Q_{p,1} = 0.0054\,\text{m}^3\,\text{s}^{-1}$, $\Delta H_{p,1} = 5.34\,\text{m}$, and $P_{brake,1} = 0.8505\,\text{kW}$, the corresponding paired values on the new performance curves are computed as follows:

Eq. (5.16): $Q_{p,2} = (0.0054\,\text{m}^3\,\text{s}^{-1}) * [(10\,\text{rps}) * (0.2\,\text{m})^3]/[(10\,\text{rps}) * (0.3\,\text{m})^3] = 0.0016\,\text{m}^3\,\text{s}^{-1}$

	A	B	C	D	E	F	G
1	$D_1 =$	30	cm =	0.3	m		
2	$n_1 =$	600	rpm =	10	rps		
3	$D_2 =$	20	cm =	0.2	m		
4	$n_2 =$	600	rpm =	10	rps		
5							
6	$Q_{p,1}$ ($m^3\ s^{-1}$)	$\Delta H_{p,1}$ (m)	$P_{brake,1}$ (kW)	η	$Q_{p,2}$ ($m^3\ s^{-1}$)	$\Delta H_{p,2}$ (m)	$P_{brake,2}$ (kW)
7					(Eq. 5.16)	(Eq. 5.15)	(Eq. 5.17)
8	0.0000	5.32	0.8262	0.00	0.0000	2.36	0.1088
9	0.0054	5.34	0.8505	0.32	0.0016	2.37	0.1120
10	0.0108	5.35	0.9477	0.58	0.0032	2.38	0.1248
11	0.0162	5.33	1.1421	0.74	0.0048	2.37	0.1504
12	0.0216	5.25	1.3365	0.83	0.0064	2.33	0.1760
13	0.0270	5.13	1.5066	0.90	0.0080	2.28	0.1984
14	0.0324	4.85	1.6524	0.93	0.0096	2.16	0.2176
15	0.0378	4.40	1.7982	0.90	0.0112	1.96	0.2368
16	0.0405	4.13	1.8225	0.88	0.0120	1.84	0.2400

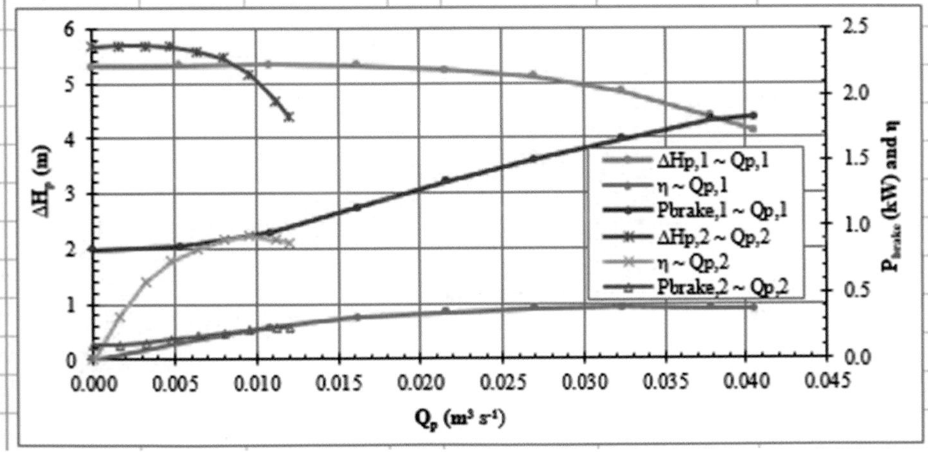

Figure 5.15 **Original and new performance curves computed in part (1) of Example 5.9.**

Eq. (5.15): $\Delta H_{p,2} = (5.34\,\text{m}) * [(10\,\text{rps})^2 * (0.2\,\text{m})^2]/[(10\,\text{rps})^2 * (0.3\,\text{m})^2] = 2.37\,\text{m}$.

Eq. (5.17): $P_{brake,2} = (0.8505\,\text{kW}) * [(10\,\text{rps})^3 * (0.2\,\text{m})^5]/[(10\,\text{rps})^3 * (0.3\,\text{m})^5]$
$= 0.112\,\text{kW}$.

(2) $D_2 = 40\,\text{cm} = 0.4\,\text{m}$, $n_2 = n_1 = 10\,\text{rps}$. The results are shown in Figure 5.16.
For instance, for the paired values on the original performance curves of $Q_{p,1} = 0.0054\,\text{m}^3\,\text{s}^{-1}$, $\Delta H_{p,1} = 5.34\,\text{m}$, and $P_{brake,1} = 0.8505\,\text{kW}$, the corresponding paired values on the new performance curves are computed as follows:

	A	B	C	D	E	F	G
1	D_1 =	30	cm =	0.3	m		
2	n_1 =	600	rpm =	10	rps		
3	D_2 =	40	cm =	0.4	m		
4	n_2 =	600	rpm =	10	rps		
5							
6	$Q_{p,1}$ ($m^3\ s^{-1}$)	$\Delta H_{p,1}$ (m)	$P_{brake,1}$ (KW)	η	$Q_{p,2}$ ($m^3\ s^{-1}$)	$\Delta H_{p,2}$ (m)	$P_{brake,2}$ (kW)
7					(Eq. 5.16)	(Eq. 5.15)	(Eq. 5.17)
8	0.0000	5.32	0.8262	0.00	0.0000	9.46	3.4816
9	0.0054	5.34	0.8505	0.32	0.0128	9.49	3.5840
10	0.0108	5.35	0.9477	0.58	0.0256	9.51	3.9936
11	0.0162	5.33	1.1421	0.74	0.0384	9.48	4.8128
12	0.0216	5.25	1.3365	0.83	0.0512	9.33	5.6320
13	0.0270	5.13	1.5066	0.90	0.0640	9.12	6.3488
14	0.0324	4.85	1.6524	0.93	0.0768	8.62	6.9632
15	0.0378	4.40	1.7982	0.90	0.0896	7.82	7.5776
16	0.0405	4.13	1.8225	0.88	0.0960	7.34	7.6800

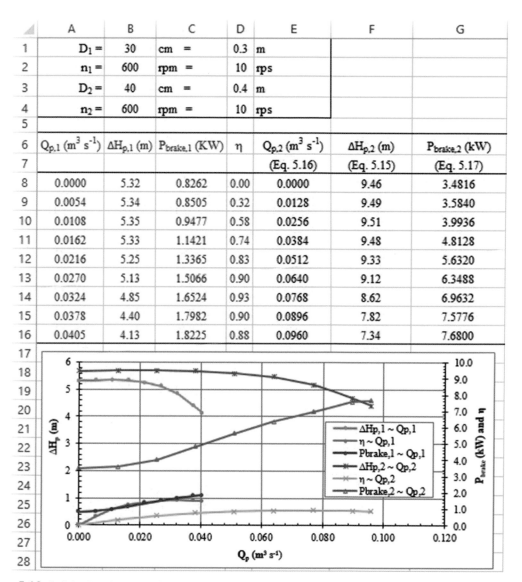

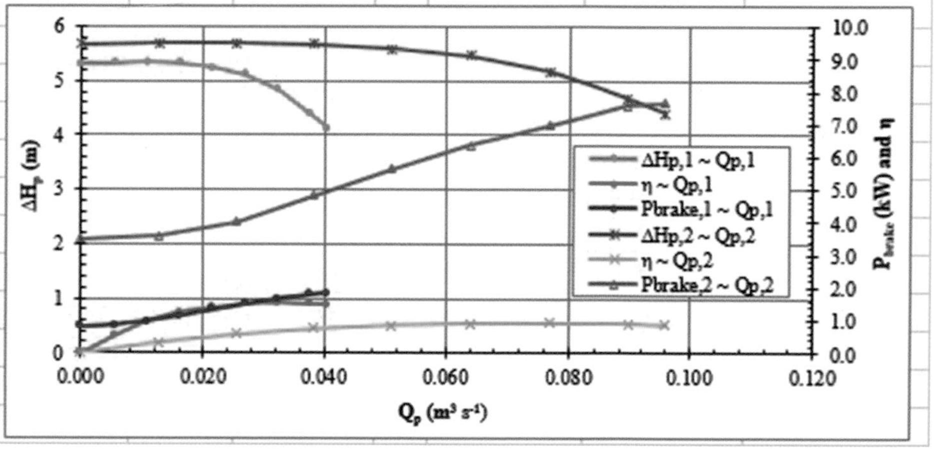

Figure 5.16 **Original and new performance curves computed in part (2) of Example 5.9.**

Eq. (5.16): $Q_{p,2} = (0.0054\,m^3\,s^{-1}) * [(10\,rps) * (0.4\,m)^3]/[(10\,rps) * (0.3\,m)^3] = 0.0128\,m^3\,s^{-1}$

Eq. (5.15): $\Delta H_{p,2} = (5.34\,m) * [(10\,rps)^2 * (0.4\,m)^2]/[(10\,rps)^2 * (0.3\,m)^2] = 9.49\,m$

Eq. (5.17): $P_{brake,2} = (0.8505\,kW) * [(10\,rps)^3 * (0.4\,m)^5]/[(10\,rps)^3 * (0.3\,m)^5] = 3.584\,kW$.

(3) $D_2 = 30\,cm = 0.3\,m$, $n_2 = 480\,rpm = 8\,rps$. The results are shown in Figure 5.17.
For instance, for the paired values on the original performance curves of $Q_{p,1} = 0.0054\,m^3\,s^{-1}$, $\Delta H_{p,1} = 5.34\,m$, and $P_{brake,1} = 0.8505\,kW$, the corresponding paired values on the new performance curves are computed as follows:

	A	B	C	D	E	F	G
1	D_1 =	30	cm =	0.3	m		
2	n_1 =	600	rpm =	10	rps		
3	D_2 =	30	cm =	0.3	m		
4	n_2 =	480	rpm =	8	rps		
5							
6	$Q_{p,1}$ ($m^3\ s^{-1}$)	$\Delta H_{p,1}$ (m)	$P_{brake,1}$ (KW)	η	$Q_{p,2}$ ($m^3\ s^{-1}$)	$\Delta H_{p,2}$ (m)	$P_{brake,2}$ (kW)
7					(Eq. 5.16)	(Eq. 5.15)	(Eq. 5.17)
8	0.0000	5.32	0.8262	0.00	0.0000	3.40	0.4230
9	0.0054	5.34	0.8505	0.32	0.0043	3.42	0.4355
10	0.0108	5.35	0.9477	0.58	0.0086	3.42	0.4852
11	0.0162	5.33	1.1421	0.74	0.0130	3.41	0.5848
12	0.0216	5.25	1.3365	0.83	0.0173	3.36	0.6843
13	0.0270	5.13	1.5066	0.90	0.0216	3.28	0.7714
14	0.0324	4.85	1.6524	0.93	0.0259	3.10	0.8460
15	0.0378	4.40	1.7982	0.90	0.0302	2.82	0.9207
16	0.0405	4.13	1.8225	0.88	0.0324	2.64	0.9331

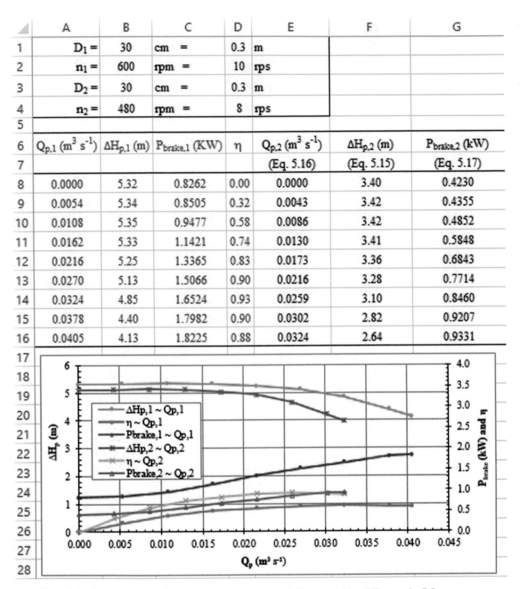

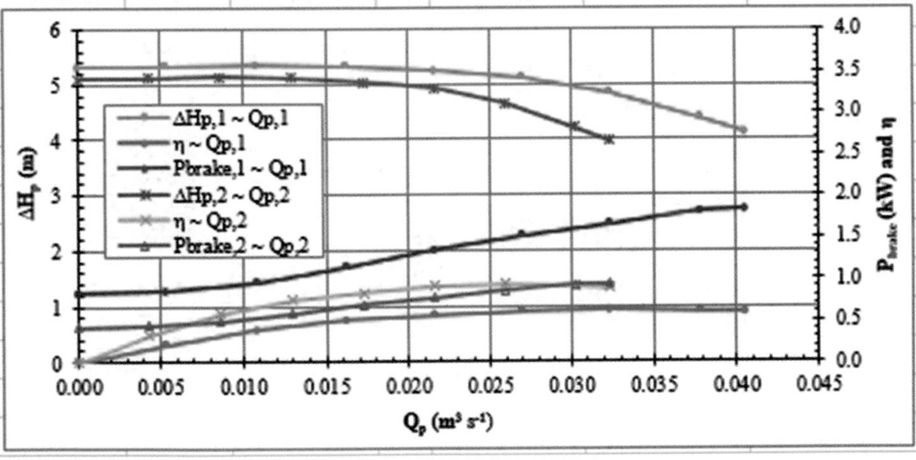

Figure 5.17 **Original and new performance curves computed in part (3) of Example 5.9.**

Eq. (5.16): $Q_{p,2} = (0.0054\,m^3\,s^{-1}) * [(8\,rps) * (0.3\,m)^3]/[(10\,rps) * (0.3\,m)^3] = 0.00432\,m^3\,s^{-1}$

Eq. (5.15): $\Delta H_{p,2} = (5.34\,m) * [(8\,rps)^2 * (0.3\,m)^2]/[(10\,rps)^2 * (0.3\,m)^2] = 3.42\,m$

Eq. (5.17): $P_{brake,2} = (0.8505\,kW) * [(8\,rps)^3 * (0.3\,m)^5]/[(10\,rps)^3 * (0.3\,m)^5] = 0.4355\,kW$.

(4) $D_2 = 30\,cm = 0.3\,m$, $n_2 = 720\,rpm = 12\,rps$. The results are shown in Figure 5.18.

	A	B	C	D	E	F	G
1	$D_1 =$	30	cm =	0.3 m			
2	$n_1 =$	600	rpm =	10 rps			
3	$D_2 =$	30	cm =	0.3 m			
4	$n_2 =$	720	rpm =	12 rps			
5							
6	$Q_{p,1}$ ($m^3\ s^{-1}$)	$\Delta H_{p,1}$ (m)	$P_{brake,1}$ (KW)	η	$Q_{p,2}$ ($m^3\ s^{-1}$)	$\Delta H_{p,2}$ (m)	$P_{brake,2}$ (kW)
7					(Eq. 5.16)	(Eq. 5.15)	(Eq. 5.17)
8	0.0000	5.32	0.8262	0.00	0.0000	7.66	1.4277
9	0.0054	5.34	0.8505	0.32	0.0065	7.69	1.4697
10	0.0108	5.35	0.9477	0.58	0.0130	7.70	1.6376
11	0.0162	5.33	1.1421	0.74	0.0194	7.68	1.9735
12	0.0216	5.25	1.3365	0.83	0.0259	7.56	2.3095
13	0.0270	5.13	1.5066	0.90	0.0324	7.39	2.6034
14	0.0324	4.85	1.6524	0.93	0.0389	6.98	2.8553
15	0.0378	4.40	1.7982	0.90	0.0454	6.34	3.1073
16	0.0405	4.13	1.8225	0.88	0.0486	5.95	3.1493

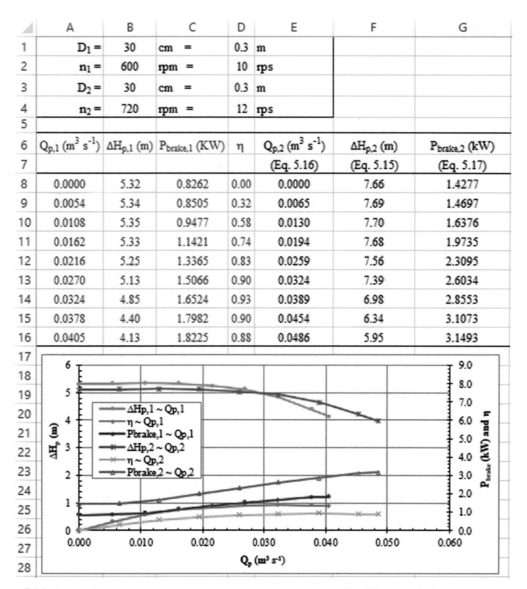

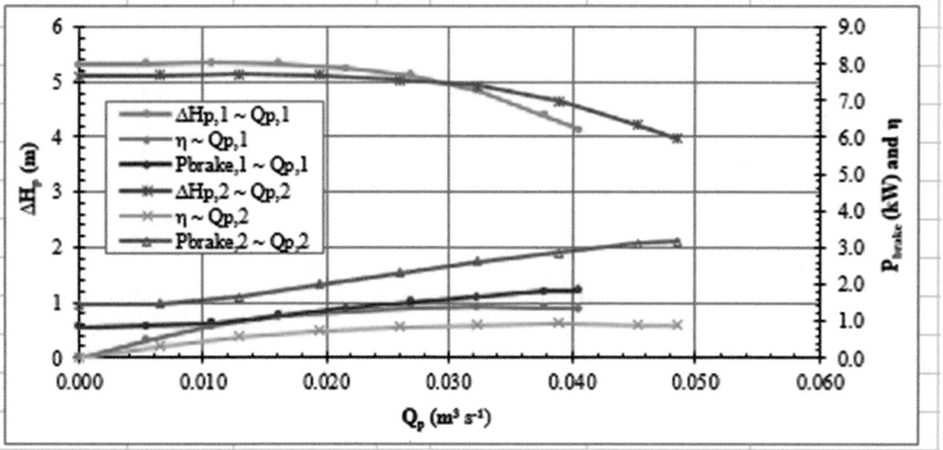

Figure 5.18 **Original and new performance curves computed in part (4) of Example 5.9.**

For instance, for the paired values on the original performance curves of $Q_{p,1} = 0.0054\ m^3\ s^{-1}$, $\Delta H_{p,1} = 5.34$ m, and $P_{brake,1} = 0.8505$ kW, the corresponding paired values on the new performance curves are computed as follows:

Eq. (5.16): $Q_{p,2} = (0.0054\ m^3\ s^{-1}) * [(12\ rps) * (0.3\ m)^3]/[(10\ rps) * (0.3\ m)^3] = 0.0065\ m^3\ s^{-1}$

Eq. (5.15): $\Delta H_{p,2} = (5.34\ m) * [(12\ rps)^2 * (0.3\ m)^2]/[(10\ rps)^2 * (0.3\ m)^2] = 7.69\ m$

Eq. (5.17): $P_{brake,2} = (0.8505\ kW) * [(12\ rps)^3 * (0.3\ m)^5]/[(10\ rps)^3 * (0.3\ m)^5] = 1.4697\ kW.$

	A	B	C	D	E	F	G
1	D_1 =	30	cm =	0.3	m		
2	n_1 =	600	rpm =	10	rps		
3	D_2 =	40	cm =	0.4	m		
4	n_2 =	720	rpm =	12	rps		
5							
6	$Q_{p,1}$ (m^3 s^{-1})	$\Delta H_{p,1}$ (m)	$P_{brake,1}$ (KW)	η	$Q_{p,2}$ (m^3 s^{-1})	$\Delta H_{p,2}$ (m)	$P_{brake,2}$ (kW)
7					(Eq. 5.16)	(Eq. 5.15)	(Eq. 5.17)
8	0.0000	5.32	0.8262	0.00	0.0000	13.62	6.0162
9	0.0054	5.34	0.8505	0.32	0.0154	13.67	6.1932
10	0.0108	5.35	0.9477	0.58	0.0307	13.70	6.9009
11	0.0162	5.33	1.1421	0.74	0.0461	13.64	8.3165
12	0.0216	5.25	1.3365	0.83	0.0614	13.44	9.7321
13	0.0270	5.13	1.5066	0.90	0.0768	13.13	10.9707
14	0.0324	4.85	1.6524	0.93	0.0922	12.42	12.0324
15	0.0378	4.40	1.7982	0.90	0.1075	11.26	13.0941
16	0.0405	4.13	1.8225	0.88	0.1152	10.57	13.2710

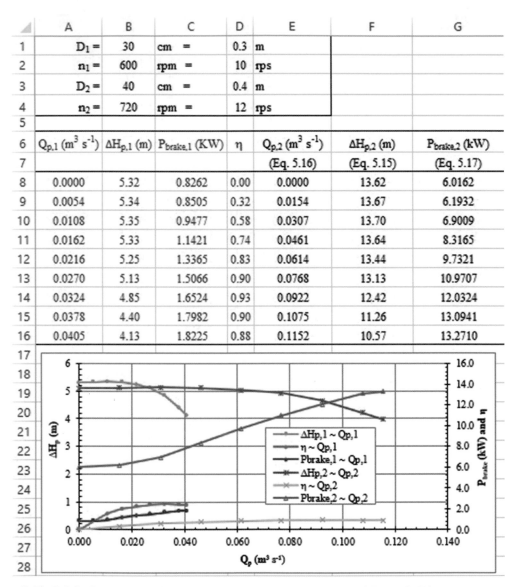

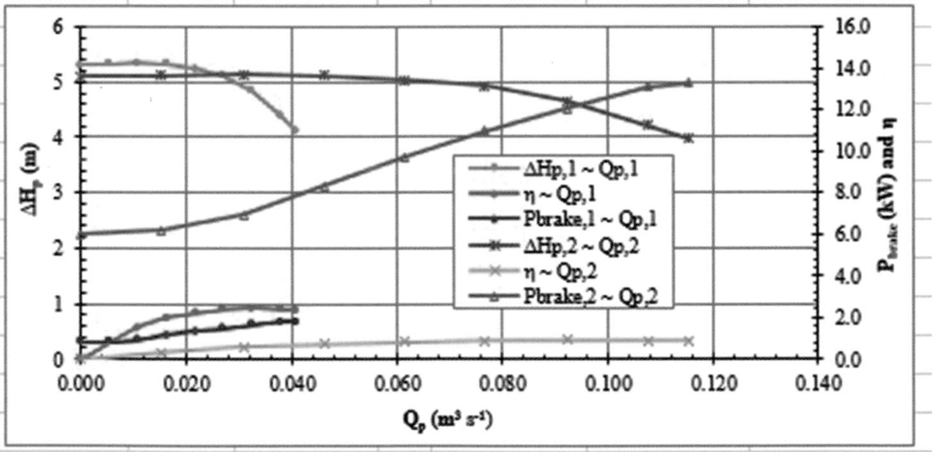

Figure 5.19 **Original and new performance curves computed in part (5) of Example 5.9.**

(5) $D_2 = 40\,\text{cm} = 0.4\,\text{m}$, $n_2 = 720\,\text{rpm} = 12\,\text{rps}$. The results are shown in Figure 5.19. For instance, for the paired values on the original performance curves of $Q_{p,1} = 0.0054\,\text{m}^3\,\text{s}^{-1}$, $\Delta H_{p,1} = 5.34\,\text{m}$, and $P_{brake,1} = 0.8505\,\text{kW}$, the corresponding paired values on the new performance curves are computed as follows:

Eq. (5.16): $Q_{p,2} = (0.0054\,\text{m}^3\,\text{s}^{-1}) * [(12\,\text{rps}) * (0.4\,\text{m})^3]/[(10\,\text{rps}) * (0.3\,\text{m})^3] = 0.0154\,\text{m}^3\,\text{s}^{-1}$

Eq. (5.15): $\Delta H_{p,2} = (5.34\,\text{m}) * [(12\,\text{rps})^2 * (0.4\,\text{m})^2]/[(10\,\text{rps})^2 * (0.3\,\text{m})^2] = 13.67\,\text{m}$

Eq. (5.17): $P_{brake,2} = (0.8505\,kW) * [(12\,rps)^3 * (0.4\,m)^5]/[(10\,rps)^3 * (0.3\,m)^5] = 6.1932\,kW.$

Example 5.10 A pump operating at 1500 rpm delivers 350 gpm against a total head of 180 ft. If the total head against is increased to 230 ft by simply increasing the rotation speed, determine the:

(1) New rotation speed
(2) New discharge
(3) Percent increase of brake power.

Solution

$D_2 = D_1$, $n_1 = 1500\,rpm$, $Q_{p,1} = 350\,gpm$, $\Delta H_{p,1} = 180\,ft$, $\Delta H_{p,2} = 230\,ft$

(1) Eq. (5.15): $n_2 = (1500\,rpm) * [(230\,ft)/(180\,ft)]^{1/2} = 1696\,rpm.$

(2) Eq. (5.16): $Q_{p,2} = (350\,gpm) * [(1696\,rpm)/(1500\,rpm)] = 396\,gpm.$

(3) Eq. (5.17): $P_{brake,2} = P_{brake,1} * [(1696\,rpm)^3/(1500\,rpm)^3] = 1.45P_{brake,1}.$
Percent increase of brake power $= (1.45P_{brake,1} - P_{brake,1})/P_{brake,1} * 100\% = 45\%.$

5.2.5 Selection of Pumps

Combining Eqs. (5.8) and (5.9) to eliminate impeller size D, one can derive the formula for dimensionless specific speed, which is expressed as:

rotation speed in revolution per unit time
discharge capacity
discharge coefficient
head capacity
dimensionless specific speed
head coefficient
gravitational acceleration

$$n_s = \frac{(C_Q)^{\frac{1}{2}}}{(C_H)^{\frac{3}{4}}} = \frac{n\,(Q_p)^{\frac{1}{2}}}{g^{\frac{3}{4}}\,(\Delta H_p)^{\frac{3}{4}}} \tag{5.18}$$

In practice, specific speed is not dimensionless and calculated differently in BG and SI units. Nevertheless, it is always computed using Q_p and ΔH_p at BEP. The formulas for the two units systems are given as:

rotation speed [rpm]
discharge capacity at BEP [gpm]
specific speed in BG units
head capacity at BEP [ft]

$$n_{s,BG} = \frac{n_{rev}\,(Q_{p,BEP})^{\frac{1}{2}}}{(\Delta H_{p,BEP})^{\frac{3}{4}}} \tag{5.19}$$

$$n_{s,SI} = \frac{n_{rad}\left(Q_{p,BEP}\right)^{\frac{1}{2}}}{g^{\frac{3}{4}}\left(\Delta H_{p,BEP}\right)^{\frac{3}{4}}} \tag{5.20}$$

rotation speed [rad s^{-1}]; discharge capacity at BEP [m^3 s^{-1}]; specific speed in SI units; gravitational acceleration (9.81 m s^{-2}); head capacity at BEP [m]

Example 5.11 Show that specific speeds in BG and SI units satisfy $n_{s,BG} = 2734 n_{s,SI}$.

Solution

$1\ \text{rad s}^{-1} = [(1\ \text{rev})/(2\pi\ \text{rad})] * [(60\ \text{s})/(1\ \text{min})] = 9.5493\ \text{rpm}$

$1\ \text{m}^3\ \text{s}^{-1} = [(1\ \text{gal})/(0.0037854\ \text{m}^3)] * [(60\ \text{s})/(1\ \text{min})] = 15,850.3725\ \text{gpm}$

$1\ \text{m} = (1\ \text{ft})/(0.3048\ \text{m}) = 3.2808\ \text{ft}.$

Let $Q_{p,BEP}$ have a units of gpm and $\Delta H_{p,BEP}$ have a units of ft. To calculate $n_{s,SI}$, one can plug $n_{rev}/9.5493$, $Q_{p,BEP}/15,850.3725$, $\Delta H_{p,BEP}/3.2808$, and $g = 9.81\ \text{m sec}^{-2}$ into Eq. (5.20) and then rearrange it as:

$$n_{s,SI} = \frac{\frac{n_{rev}}{9.5493}\left(\frac{Q_{p,BEP}}{15,850.3725}\right)^{\frac{1}{2}}}{(9.81)^{\frac{3}{4}}\left(\frac{\Delta H_{p,BEP}}{3.2808}\right)^{\frac{3}{4}}} = 3.65798 \times 10^{-4}\frac{n_{rev}\left(Q_{p,BEP}\right)^{\frac{1}{2}}}{\left(\Delta H_{p,BEP}\right)^{\frac{3}{4}}} = 3.65798 \times 10^{-4} n_{s,BG}.$$

Dividing the constant by the two sides, one can easily show that $n_{s,BG} = 2734 n_{s,SI}$.

Example 5.12 For the pump in Figure 5.13, compute its specific speeds in BG and SI units.

Solution

From Figure 5.13, $n_{rev} = 600\ \text{rpm}$, $n_{rad} = 62.83\ \text{rad s}^{-1}$

At BEP: $Q_{p,BEP} = 0.0324\ \text{m}^3\ \text{s}^{-1} = 513.55\ \text{gpm}$, $\Delta H_{p,BEP} = 4.8\ \text{m} = 15.75\ \text{ft}$.

Eq. (5.19): $n_{s,BG} = [(600\ \text{rpm}) * (513.55\ \text{gpm})^{1/2}]/(15.75\ \text{ft})^{3/4} = 1720.$

Eq. (5.20): $n_{s,SI} = [(62.83\ \text{rad s}^{-1}) * (0.0324\ \text{m}^3\ \text{s}^{-1})^{1/2}]/[(9.81\ \text{m s}^{-2})^{3/4} * (4.8\ \text{m})^{3/4}] = 0.629.$

Once the specific speed of a pump is computed, its type, peak efficiency, and peripheral-velocity factor can be determined from Figure 5.20. The pump size can be computed using the peripheral-velocity factor as:

$$D = \frac{60\phi_e}{\pi n_{rev}}\sqrt{2g\left(\Delta H_{p,BEP}\right)} \tag{5.21}$$

impeller diameter [ft in BG; m in SI]; peripheral-velocity factor (Figure 5.20); head capacity at BEP [ft in BG; m in SI]; constant pi; rotation speed [rpm]; gravitational acceleration (32.2 ft sec^{-2} or 9.81 m s^{-2})

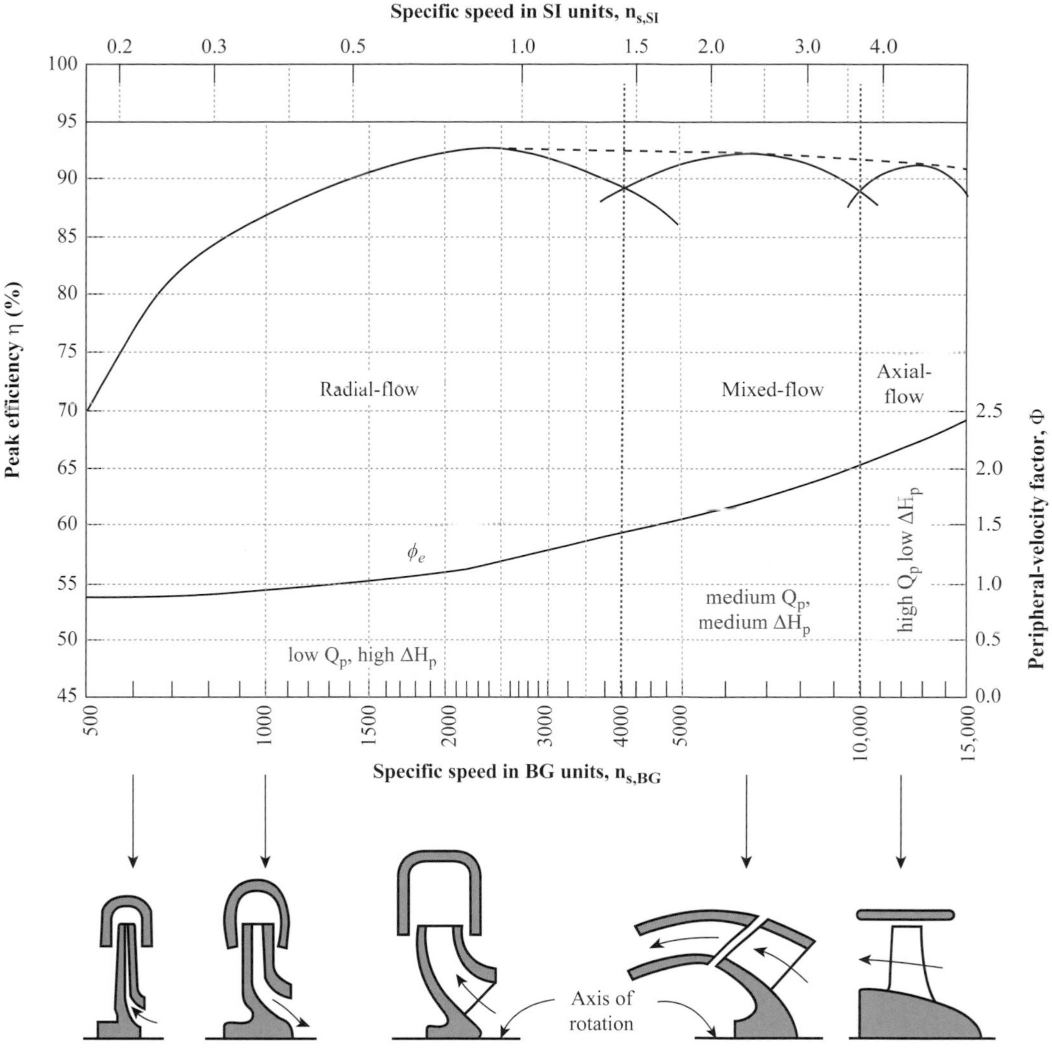

Figure 5.20 **Centrifugal pump selection.** Nomogram showing pump type versus specific speed. (*Source:* Figure 8.15 in Roberson *et al.*, 1998)

Example 5.13 Determine the type, peak efficiency, impeller size of the pump in Example 5.12.

Solution

Example 5.12: $n_{s,BG} = 1720 \rightarrow$ Figure 5.20: radial-flow, $\eta = 0.92$, $\phi_e = 1.05$ $n_{rev} = 600\,\text{rpm}$, $\Delta H_{p,BEP} = 15.75\,\text{ft}$

Eq. (5.21): $D = 60 * 1.05/(\pi * 600\,\text{rpm}) * [2 * (32.2\,\text{ft sec}^{-2}) * (15.75\,\text{ft})]^{1/2} = 1.06\,\text{ft}.$

Example 5.14 A water supply system needs a pump to add a total head of 65 ft and deliver 15 MGD to its users. If the pump has a rotation speed of 1500 rpm and operates at its BEP, determine its type, peak efficiency, and impeller size.

Solution

$n_{rev} = 1500\,\text{rpm},\ \Delta H_{p,BEP} = 65\,\text{ft}$

$Q_{p,BEP} = (15\,\text{MGD}) * [(10^6\,\text{gal})/(1\,\text{MG})] * [(1\,\text{d})/(24\,\text{hr})] * [(1\,\text{hr})/(60\,\text{min})] = 10,416.7\,\text{gpm}.$

Eq. (5.19): $n_{s,BG} = [(1500\,\text{rpm}) * (10,416.7\,\text{gpm})^{1/2}]/(65\,\text{ft})^{3/4} = 6688.$

Figure 5.20: mixed-flow pump, $\eta = 0.92$, $\phi_e = 1.75$.

Eq. (5.21): $D = 60 * 1.75/(\pi * 1500\,\text{rpm}) * [2 * (32.2\,\text{ft sec}^{-2}) * (65\,\text{ft})]^{1/2} = 1.44\,\text{ft}.$

5.2.6 Operation Point and Cavitation

For a given pump, its $\Delta H_p \sim Q_p$ curve defines pairs of head and discharge that can be provided by the pump. When the pump is installed in a hydraulic system, its operation point is the intersection between the $\Delta H_p \sim Q_p$ curve and the system curve that describes how head loss varies with discharge (i.e., $h_p \sim Q$) as defined by Eq. (5.1). As that performance curve is pump-specific and does not change regardless of where the pump is used, the system curve reflects the intrinsic features of a hydraulic system (e.g., pipe layout and size) only and is independent of the specific pump installed in the system. That is, the performance curve defines the capacity of a pump, whereas the system curve defines the demand of a hydraulic system. The operation point of a pump satisfies both the capacity and demand in terms of head and discharge.

Example 5.15 Develop and plot the system curve for Example 5.1.

Solution

$z_A = 50\,\text{m},\ z_B = 80\,\text{m},\ L_s = 200\,\text{m},\ L_d = 1500\,\text{m},\ D = 0.5\,\text{m},\ f = 0.012,\ k_e = 0.12,\ k_d = 1.0$

The cross-sectional area of the pipeline: $A = \pi/4 * (0.5\,\text{m})^2 = 0.1963\,\text{m}^2$.

The mean flow velocity in the pipeline: $V = Q/(0.1963\,\text{m}^2) = 5.0942Q\,\text{m s}^{-1}$.

Darcy–Weisbach equation (Eq. (2.23)) → friction loss: $h_f = 0.012 * [(200\,\text{m} + 1500\,\text{m})/(0.5\,\text{m})] * [(5.0942Q\,\text{m s}^{-1})^2/(2 * 9.81\,\text{m s}^{-2})] = 53.9651Q^2\,\text{m}$.

Entrance loss (Table 2.3): $h_e = (0.12) * (5.0942Q\,\text{m s}^{-1})^2/(2 * 9.81\,\text{m s}^{-2}) = 0.1587Q^2\,\text{m}$.

Discharge loss (Table 2.3): $h_d = (1.0) * (5.0942Q\,\text{m s}^{-1})^2/(2 * 9.81\,\text{m s}^{-2}) = 1.3227Q^2\,\text{m}$.

Total loss (Eq. (2.22)): $h_L = (53.9651Q^2 + 0.1587Q^2 + 1.3227Q^2)\,\text{m} = 55.4465Q^2\,\text{m}$.

	A	B
1	Q ($m^3 s^{-1}$)	h_p (m)
2	0.00	30.00
3	0.05	30.14
4	0.10	30.55
5	0.15	31.25
6	0.20	32.22
7	0.25	33.47
8	0.30	34.99
9	0.35	36.79
10	0.40	38.87
11	0.45	41.23
12	0.50	43.86
13	0.55	46.77
14	0.60	49.96
15	0.65	53.43
16	0.70	57.17
17	0.75	61.19
18	0.80	65.49
19	0.85	70.06
20	0.90	74.91
21	0.95	80.04
22	1.00	85.45

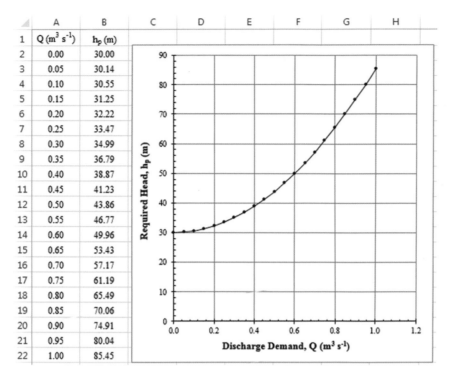

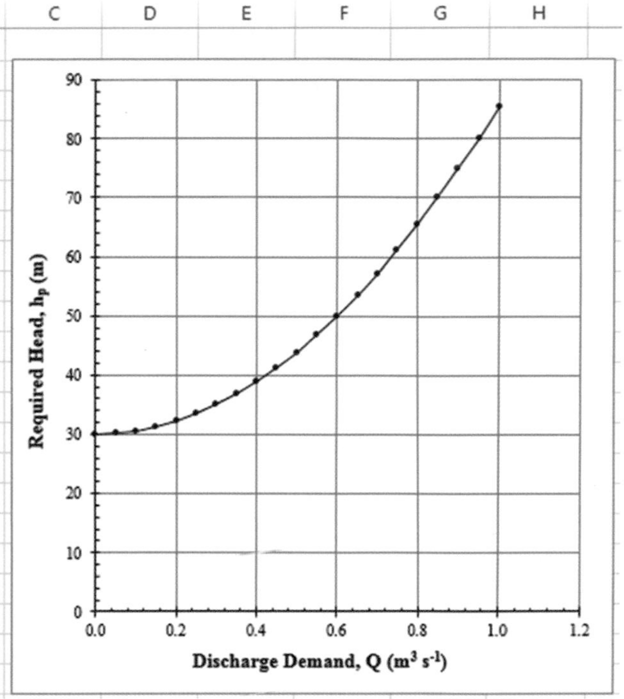

Figure 5.21 **The Excel spreadsheet for developing the system curve in Example 5.15.**

Eq. (5.1): $h_p = (80\,m) - (50\,m) + (55.4465Q^2\,m) = (30 + 55.4465Q^2)\,m$.

For selected values of Q, compute the corresponding values of h_p and then plot $h_p \sim Q$. This is done in the Excel spreadsheet shown in Figure 5.21. For instance, for $Q = 0.00\,m^3\,s^{-1}$, $h_p = (30 + 55.4465 * 0.00^2)\,m = 30.00\,m$; for $Q = 0.05\,m^3\,s^{-1}$, $h_p = (30 + 55.4465 * 0.05^2)\,m = 30.14\,m$; and so on.

Example 5.16 For the hydraulic system of Example 5.15, determine the operation point and efficiency of the pump with performance curves given in columns D to F in Figure 5.22.

Solution

Plot the system curve (Figure 5.21) and the performance curves on the same graph as shown in Figure 5.22. The operation point is the intersection between the system curve and the $\Delta H_p \sim Q_p$ curve. The efficiency can be determine by reading the value of η at the discharge corresponding to the operation point. The operation point is: $Q = Q_p = 0.81\,m^3\,s^{-1}$ and $h_p = \Delta H_p = 66\,m$; the efficiency is $\eta = 0.93$. For this case, because the operation point is the same as the BEP, the pump may be a best choice for the hydraulic system.

	A	B	C	D	E	F
1	System Curve			Pump Performance Curve		
2	Q (m^3 s^{-1})	h_p (m)		Q_p (m^3 s^{-1})	ΔH_p (m)	η
3	0.00	30.00		0.0000	90.00	0.10
4	0.05	30.14		0.0500	89.91	0.20
5	0.10	30.55		0.1500	89.21	0.40
6	0.15	31.25		0.3000	86.85	0.70
7	0.20	32.22		0.4000	84.40	0.83
8	0.25	33.47		0.6000	77.40	0.90
9	0.30	34.99		0.8000	67.60	0.93
10	0.35	36.79		0.9000	61.65	0.92
11	0.40	38.87		1.0000	55.00	0.90
12	0.45	41.23				
13	0.50	43.86				
14	0.55	46.77				
15	0.60	49.96				
16	0.65	53.43				
17	0.70	57.17				
18	0.75	61.19				
19	0.80	65.49				
20	0.85	70.06				
21	0.90	74.91				
22	0.95	80.04				
23	1.00	85.45				

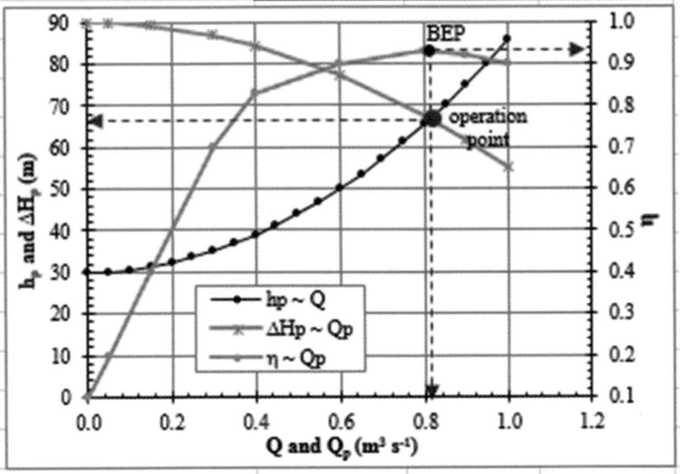

Figure 5.22 **Spreadsheet and determination of the operation point for Example 5.16.**

For a hydraulic system with a pump (e.g., Figure 5.1), if the pressure at the suction point is too low, the liquid water can be converted into vapor to form bubbles (Crowe *et al.*, 2005). Once the bubbles are transported to a location with a higher pressure, they may implode, damaging the hydraulic system and/or the pump. Such a phenomenon is referred to as cavitation and must be avoided in all engineered projects. In practice, the net positive suction head (NPSH) (Eq. (5.22)), which is the difference between the absolute hydrostatic pressure head at the suction point and the saturation vapor pressure head for the water pumped, is computed and compared with a critical value (hereinafter signified by $NPSH_c$). Also, the cavitation number (Eq. (5.23)), the ratio of NPSH to h_p, can be computed and compared with a critical value (hereinafter signified by σ_c). Cavitation is likely to occur if $NPSH < NPSH_c$ and/or $\sigma < \sigma_c$. For a given pump, its values for $NPSH_c$ and σ_c are provided by the manufacturer and can be found on a label pinned on the pump. Moreover, three approaches may be considered to avoid cavitation. First, the pump is positioned at a lower-elevation location to increase the pressure at the suction point. Second, a different pump with $NPSH_c > NPSH$ and $\sigma_c > \sigma$ is used. Third, the hydraulic system is redesigned to reduce the mean flow velocity and friction and minor losses, making h_p smaller. In this regard, a pipeline with a larger size and a smaller roughness height (Figure 2.11) may be a better choice, while an entrance with a smooth transition and fewer bends and fittings (Table 2.3) may be adopted.

$$\text{NPSH} = \frac{p_{s,abs}}{\gamma} - \frac{p_v}{\gamma} = \left(\frac{p_{atm}}{\gamma} + \frac{p_s}{\gamma}\right) - \frac{p_v}{\gamma} \tag{5.22}$$

(NPSH: net positive suction head; $p_{s,abs}$: absolute pressure at suction point; p_v: saturation vapor pressure of water; p_{atm}: atmospheric or barometric pressure; p_s: relative pressure at suction point; γ: specific weight of water)

$$\sigma = \frac{\text{NPSH}}{h_p} \tag{5.23}$$

(σ: cavitation number; NPSH: net positive suction head; h_p: total head against by pump (Eq. (5.1)))

Example 5.17 If the pump ($NPSH_c = 4.5$ m) in Example 5.1 pumps water 15°C, determine:

(1) The NPSH
(2) The σ
(3) Whether cavitation will occur. If so, propose and verify a simple method to avoid cavitation.

Solution

Assume a standard barometric pressure of $p_{atm} = 101.3$ kPa abs

Table 2.1 → for 15°C water: $\gamma = 9.8\ \text{kN m}^{-3}$, $p_v = 1.698$ kPa abs.

(1) From Example 5.1, $p_s/\gamma = -8.31$ m.
Eq. (5.22): $\text{NPSH} = [(101.3\ \text{kPa abs})/(9.8\ \text{kN m}^{-3}) + (-8.31\ \text{m})] - [(1.698\ \text{kPa abs})/(9.8\ \text{kN m}^{-3})] = 1.85\ \text{m}$.

(2) From Example 5.1, $h_p = 32.22$ m.
Eq. (5.23): $\sigma = (1.85\ \text{m})/(32.22\ \text{m}) = 0.057$.

(3) $(\text{NPSH} = 1.85\ \text{m}) < (\text{NPSH}_c = 4.5\ \text{m}) \rightarrow$ cavitation will occur.
Let us determine if cavitation will be avoided by lowering the pump installation position.
At the allowable highest installation position $z_{pump,c}$, $\text{NPSH} = \text{NPSH}_c = 4.5$ m.
Eq. (5.22): $4.5\ \text{m} = [(101.3\ \text{kPa})/(9.8\ \text{kN m}^{-3}) + p_s/\gamma] - [(1.698\ \text{kPa abs})/(9.8\ \text{kN m}^{-3})]$
$= (10.16 + p_s/\gamma)\ \text{m}$
$\rightarrow p_s/\gamma = -5.66\ \text{m}$.
In Figure 5.1, energy equation Eq. (2.21) between ① and the suction point:

$$(50\ \text{m}) + (0\ \text{m}) + (0\ \text{m}) = z_{pump,c} + (-5.66\ \text{m}) + (1.0188\ \text{m s}^{-1})^2/(2 * 9.81\ \text{m s}^{-2}) + (0.25\ \text{m} + 0.0063\ \text{m}) \rightarrow z_{pump,c} = 55.35\ \text{m}.$$

Thus, if the pump is reinstalled at a position 2.65 m lower, cavitation will be avoided.

5.2.7 Pumps in Parallel and Series

If the required discharge of a hydraulic system is too large to be supplied by a single pump, two or more identical or different pumps can be installed in parallel (e.g., Figure 5.23) to meet the requirement. For a given head capacity, the overall discharge capacity will be the summation of the corresponding discharge capacities of the individual pumps. Note that all parallel pumps have the same total head against. Considering this basic fact, a combined performance curve of the pumps can be developed. The operation point is determined as the intersection between the combined performance curve and the system curve.

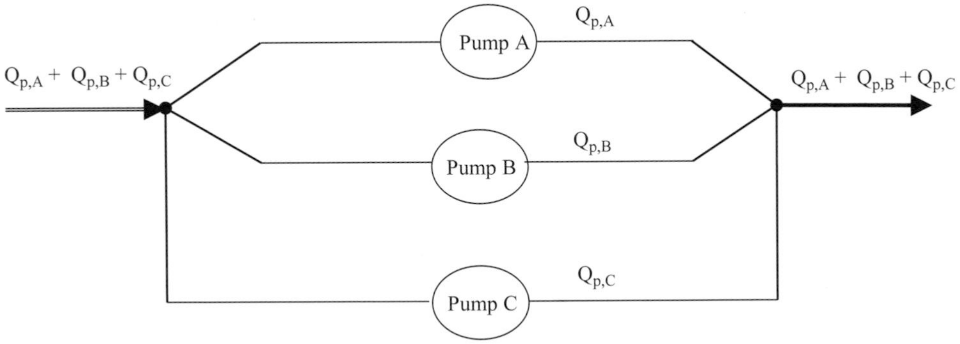

Figure 5.23 **Schematic diagram of three pumps installed in parallel.** $Q_{p,A}$, $Q_{p,B}$, and $Q_{p,C}$ are the discharge capacities of Pumps A, B, and C, respectively, for a given total head against.

Example 5.18 The lower panel of Figure 5.24 shows the performance curves ($\Delta H_p \sim Q_p$) for Pumps A, B, and C. Develop the combined performance curve if the three pumps are installed in parallel.

Solution

For a selected value of ΔH_p between 0 and 53 ft, read the corresponding values of Q_p from the $\Delta H_p \sim Q_p$ curves for the three pumps and then sum the values to obtain the discharge capacity on the combined performance curve at this selected ΔH_p. Do this for several selected values of ΔH_p (as shown in columns M to Q in the upper panel of Figure 5.24) and plot the generated pairs of (Q_p, ΔH_p) to create the combined performance curve shown in the lower panel of the figure. For instance, for $\Delta H_p = 5$ ft, the values of Q_p for Pumps A, B, and C are read (linearly interpolated as needed) to be 137.9, 187.5, and 239.6 gpm, respectively; thus, the corresponding combined discharge capacity is $Q_p = (137.9 + 187.5 + 239.6)$ gpm $= 565.0$ gpm. Note that when a selected value of ΔH_p is larger than the maximum head capacity of a pump, the pump will automatically shut itself off, leading to a zero discharge from this pump.

Combined Performance Curve (Installed in Parallel)				
ΔH_p (ft)	Q_p (gpm)			
	Pump A	Pump B	Pump C	Combined
0	150.0	200.0	250.0	600.0
5	137.9	187.5	239.6	565.0
10	127.1	175.0	229.2	531.3
15	108.3	159.4	218.8	486.5
20	85.0	142.9	208.3	436.2
25	25.0	125.0	197.2	347.2
30	0.0	107.1	183.3	290.4
35	0.0	75.0	167.9	242.9
40	0.0	0.0	150.0	150.0
45	0.0	0.0	125.0	125.0
50	0.0	0.0	75.0	75.0
52	0.0	0.0	25.0	25.0
53	0.0	0.0	0.0	0.0

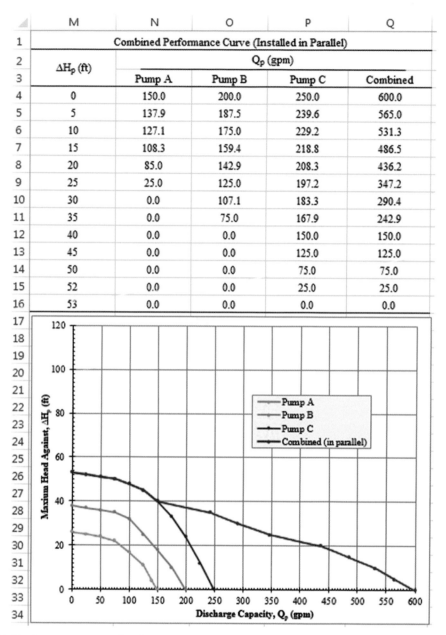

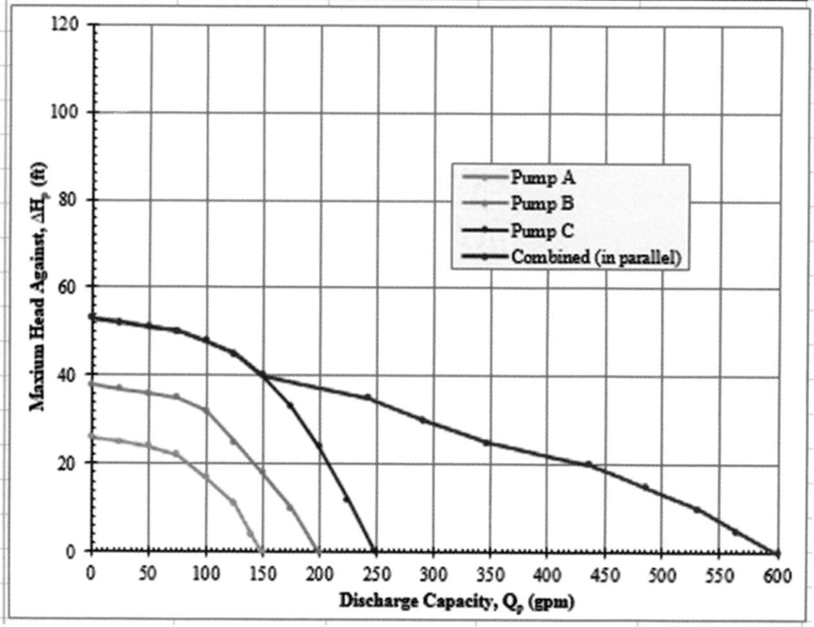

Figure 5.24 **Performance curves for three pumps used in Example 5.18.** Also shown are the supporting spreadsheet data and the combined performance curve for a configuration in which the pumps are installed in parallel.

Example 5.19 The water system shown in Figure 5.1 has a total head against of 20 ft. If the three pumps in Example 5.18 are installed in parallel in the system to transfer water from Reservoir A to B, determine the transfer rate by:

(1) The three pumps combined
(2) Each of the pumps.

Solution

At the operation point, $h_p = \Delta H_p = 20$ ft.

(1) Reading the combined performance curve in Figure 5.24, one can get $Q_p = 435$ gpm.
(2) Reading the performance curves of the individual pumps, one can get $Q_p = 85$ gpm for Pump A; $Q_p = 140$ gpm for Pump B; and $Q_p = 210$ gpm for Pump C.

If the required head increase of a hydraulic system is too large to be provided by a single pump, two or more identical or different pumps can be installed in series (e.g., Figure 5.25) to meet the requirement. For a given discharge, the overall head capacity will be the summation of the corresponding head capacities of the individual pumps. Note that all series pumps have the same discharge. In terms of this basic fact, a combined performance curve of the pumps can be developed. The operation point is determined as the intersection between the combined performance curve and the system curve.

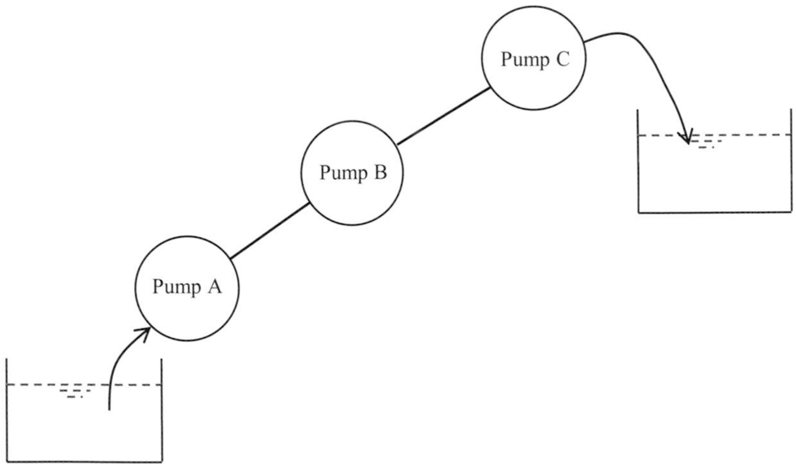

Figure 5.25 **Schematic diagram of three pumps installed in series.** For a given discharge, the overall head capacity is the summation of the head capacities of the individual pumps.

Example 5.20 If the three pumps in Example 5.18 are installed in series, develop the combined performance curve.

Q_p (gpm)	Combined Performance Curve (Installed in Series) ΔH_p (ft) Pump A	Pump B	Pump C	Combined
0	26.0	38.0	53.0	117.0
20	25.2	37.2	52.2	114.6
40	24.4	36.4	51.4	112.2
60	23.2	35.6	50.6	109.4
80	21.0	34.4	49.6	105.0
100	17.0	32.0	48.0	97.0
120	12.2	26.4	45.6	84.2
140	4.0	20.8	42.0	66.8
160	0.0	14.8	37.2	52.0
180	0.0	8.0	31.2	39.2
200	0.0	0.0	24.0	24.0
220	0.0	0.0	14.4	14.4
250	0.0	0.0	0.0	0.0

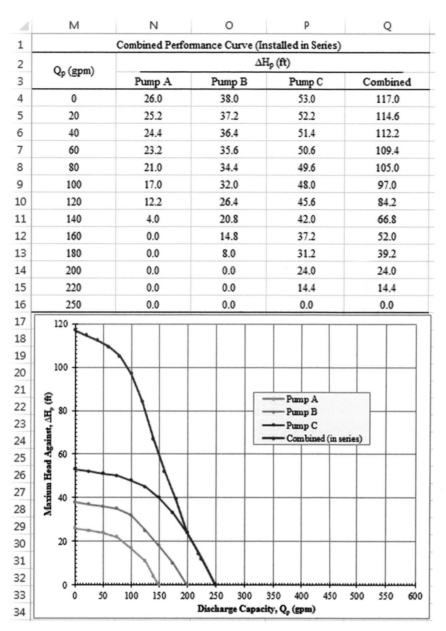

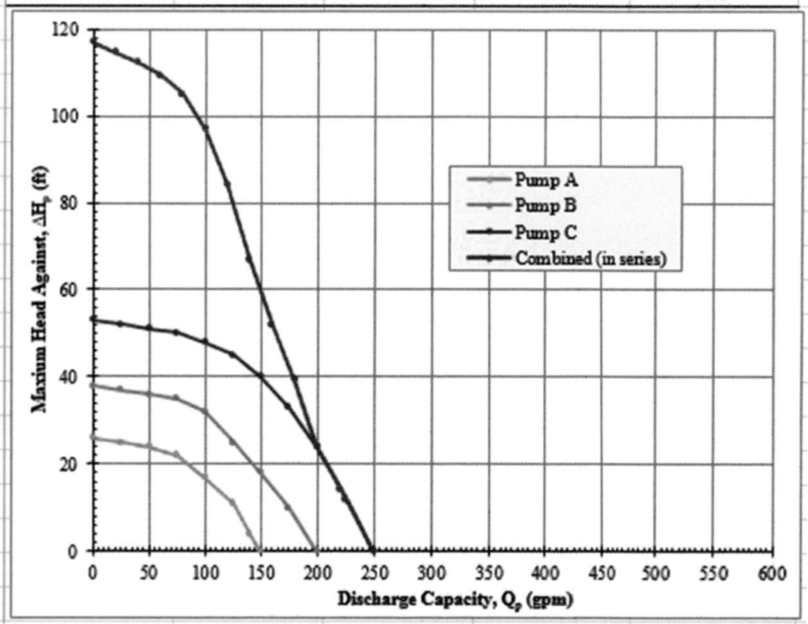

Figure 5.26 **Performance curves for three pumps used in Example 5.20.** Also shown are the supporting spreadsheet data and the combined performance curve for a configuration in which the pumps are installed in series.

Solution

For a selected value of Q_p between 0 and 250 gpm, read the corresponding values of ΔH_p from the $\Delta H_p \sim Q_p$ curves for the three pumps and then sum the values to determine the overall head capacity on the combined performance curve at this selected Q_p. Complete this procedure for several selected values of Q_p (as shown in columns M to Q in the upper panel of Figure 5.26) and plot the generated pairs of (Q_p, ΔH_p) to obtain the combined performance curve shown in the lower panel of Figure 5.26. For instance, for $Q_p = 20$ gpm, the values of ΔH_p for Pumps A, B, and C are read (linearly interpolated as needed) to be 25.2, 37.2, and 52.2 ft, respectively; thus, the corresponding overall head capacity is $\Delta H_p = (25.2 + 37.2 + 52.2)$ ft $= 114.6$ ft. Note that when a selected value of Q_p is larger than the largest discharge capacity of a pump, the pump will automatically shut itself off, leading to a zero head increase by this pump.

Example 5.21 If the three pumps in Example 5.20 are installed in series in the water system shown in Figure 5.1 to transfer water from Reservoir A to B at 100 gpm, determine the total head against by:

(1) The three pumps combined
(2) Each of the pumps.

Solution

At the operation point, $Q = Q_p = 100$ gpm.

(1) Reading the combined performance curve in Figure 5.26, one can get $\Delta H_p = 97$ ft.
(2) Reading the performance curves of the individual pumps, one can get $\Delta H_p = 17$ ft for Pump A; $\Delta H_p = 32$ ft for Pump B; and $\Delta H_p = 48$ ft for Pump C.

5.3 Turbines

Turbines convert potential, pressure, and/or kinetic energies of flowing water into mechanical energy, which in turn drives a power generator to generate electricity (Roberson *et al.*, 1998). This section discusses turbine characteristics, types, and selections in accordance with a hydraulic or water resources engineering system.

5.3.1 Types

Depending on the primary type of energy to be exploited, turbines can be classified into two categories, namely impulse (Pelton) and reaction (Francis and Kaplan) (Table 5.6). Pelton turbines, driven by impulse force, are appropriate for high-head and low-flow-rate conditions. A Pelton turbine mainly harvests potential energy when a source of water at a higher altitude is conveyed through a penstock (i.e., pipeline) downward to a location at a lower altitude, where the

Table 5.6 **Types and characteristics of turbines.**

Characteristic	**Impulse (Pelton) turbine**	**Reaction turbine**	
		Kaplan	**Francis**
Mechanism	Jet flow impinges on vanes (buckets) of the turbine runner	Pressured flow fills the chamber of the turbine runner to induce reaction force by changing flow directions	
Favorable conditions	High head (> 1500 ft) and steady low flow rate	Low head (6 to 100 ft) and high flow rate (>2500 cfs)	Medium head (90 to 1500 ft) and medium flow rate
Propeller	Curved vanes/ buckets	Adjustable blades leading to smooth inflow	Fixed/adjustable blades leading to rotating inflow
Primary energy exploited	Potential	Pressure	
Installation position	At a lowing location with jet flow from nozzle of penstock	On run of river	

water flows out of a nozzle to form high-velocity jet flow striking the turbine buckets (i.e., vanes) (Figure 5.27a). In contrast, Kaplan turbines, driven by reaction force, are good for low-head and high-flow-rate conditions. A Kaplan turbine primarily harvests pressure energy induced by pressurized water flowing through a chamber (Figure 5.27b). Francis turbines, driven both by reaction and impulse forces, work well for medium head and medium flow rate conditions. A Francis turbine harvests potential and pressure energies (Figure 5.27c). Figure 5.28 shows the flow conditions appropriate for these three types of turbines.

5.3.2 Power by a Pelton Turbine

Based on the principles of physics, the power generated by a Pelton turbine is computed as the dot product of impulse force acting on the buckets and tangential speed at the tip of the propeller. The tangential speed is computed as:

diameter of propeller [ft in BG; m in SI]

tangential speed of runner [ft sec^{-1} in BG; m s^{-1} in SI]

$$u = \omega \left(\frac{D}{2} \right) \tag{5.24}$$

angular speed of runner [rad s^{-1}]

The impulse force can be determined by applying the linear momentum equation (Eq. (2.32)) to water striking the bucket (control volume) illustrated in Figure 5.29. The runner is usually designed to have a sufficient number of buckets to enable jet flow to strike at least one bucket at any time, and an optimal budget angle, β_2, is selected to minimize any energy loss of the striking

(a)

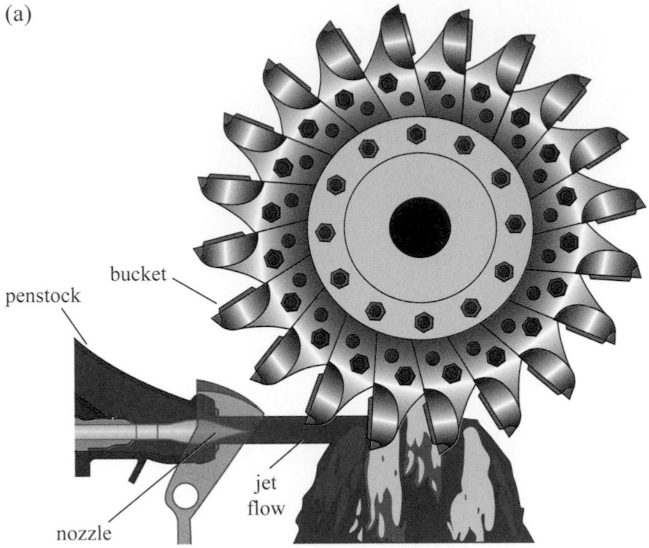

(b)

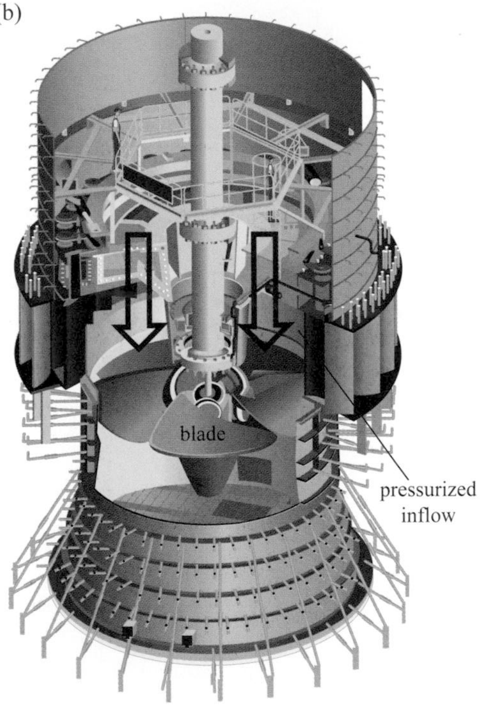

Figure 5.27 **Diagrams of turbine types.** Screenshot for the runner of: (a) impulse or Pelton turbine; (b) Kaplan turbine; and (c) Francis turbine. (*Sources:* images reproduced with permission by: (a) Dr. Seyhan Ersoy at www.mekanizmalar.com; (b) Mr. Marc Pieprzytza at Voith Group, www.voith.com; and (c) Mr. Gopal Mishra at The Constructor, https://theconstructor.org/practical-guide/francis-turbines-components-application/2900.)

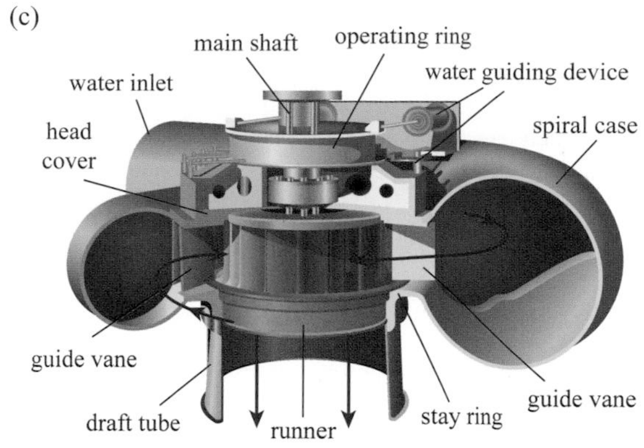

Figure 5.27 *(Continued)*

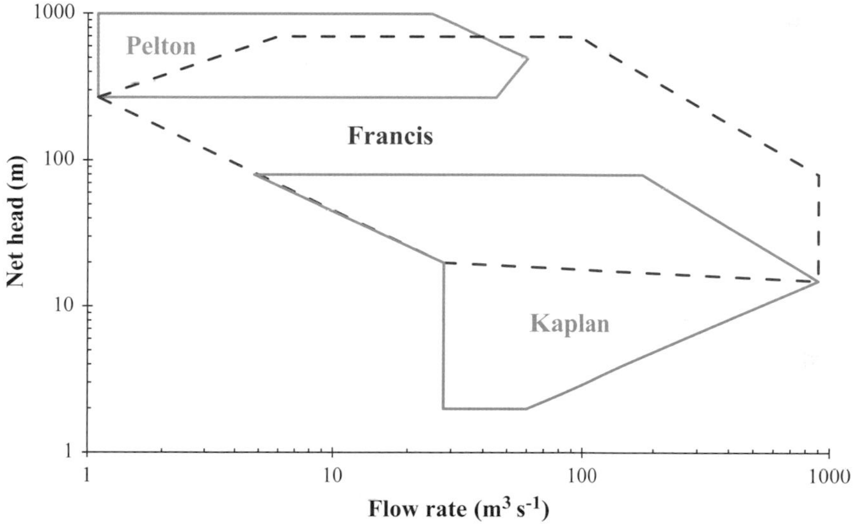

Figure 5.28 **Flow conditions appropriate for three types of turbines.**

flow. That is, the ratio of relative leaving water velocity, υ_2, to relative jet flow velocity, υ_1, is usually minimized to a constant for a given Pelton turbine when it is made by the manufacture. υ_2 can be computed as:

$$\upsilon_2 = \alpha\upsilon_1 = \alpha\left(V_j - u\right) \tag{5.25}$$

where: relative leaving water velocity → υ_2; velocity ratio (< 1.0) → α; relative jet flow velocity → υ_1; absolute jet flow velocity → V_j; tangential speed of runner → u.

Additional parameters can likewise be calculated. The impulse force in the vertical direction does not generate any power because buckets do not move in that direction. In contrast, the impulse force in the horizontal direction generates power and can be determined as:

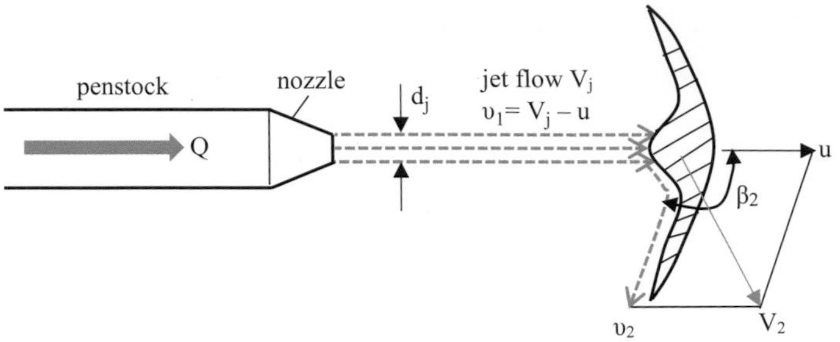

Figure 5.29 **Schematic diagram of a Pelton turbine.** Nozzle jet flow impinging on a bucket. Q: flow rate; d_j: jet flow diameter; V_j: absolute jet flow velocity; υ_1: relative jet flow velocity; u: tangential speed of runner; υ_2: relative leaving water velocity; V_2: absolute leaving water velocity; and β_2: bucket angle.

$$F_x = \rho Q\left[V_j - (u + \upsilon_2 \cos\beta_2)\right] \tag{5.26}$$

(F_x: impulse force in horizontal direction; ρ: density of water; Q: flow rate; V_j: absolute jet flow velocity; u: tangential speed of runner; υ_2: relative leaving water velocity; β_2: bucket angle)

The power harvested by a Pelton turbine is computed as:

$$P_{harvested} = F_x u \tag{5.27}$$

($P_{harvested}$: power harvested; F_x: impulse force in horizontal direction; u: tangential speed of runner)

The jet flow power is computed as:

$$P_j = \gamma Q \frac{V_j^2}{2g} \tag{5.28}$$

(P_j: power in jet flow; γ: specific weight of water; Q: flow rate; V_j: jet flow velocity; g: gravitational acceleration)

The efficiency of a Pelton turbine is computed as:

$$\eta = \frac{P_{harvested}}{P_j} 100\% \tag{5.29}$$

(η: efficiency of Pelton turbine; $P_{harvested}$: power harvested; P_j: power in jet flow)

For a given hydraulic system (i.e., when jet flow rate and velocity are fixed), the condition for a Pelton turbine to reach its maximum efficiency is:

$$u = \frac{V_j}{2} \tag{5.30}$$

(u: tangential speed of runner (Eq. (5.24)); V_j: jet flow velocity)

Substituting Eqs. (5.25), (5.26), and (5.30) into Eq. (5.27) and rearranging, one can calculate the maximum power that can be harvested by the turbine as:

$$P_{harvested,max} = \frac{\gamma Q\,(1 - \alpha \cos \beta_2)}{2} \frac{V_j^2}{2g} \tag{5.31}$$

(maximum power harvested: $P_{harvested,max}$; specific weight of water: γ; flow rate: Q; velocity ratio (Eq. (5.25)): α; bucket angle (Figure 5.29): β_2; jet flow velocity: V_j; gravitational acceleration: g)

Finally, for a given hydraulic system, the net head is equal to the jet flow velocity head, which represents the maximum amount of available water energy that can be harvested by a Pelton turbine. That is, net head can be computed as:

$$\Delta H = \frac{V_j^2}{2g} \tag{5.32}$$

(net head on Pelton turbine: ΔH; jet flow velocity: V_j; gravitational acceleration: g)

Example 5.22 Prove Eqs. (5.30) and (5.31).

Solution

Substituting Eq. (5.25) into (5.26), one can get:

$$\begin{aligned} F_x &= \rho Q \left[V_j - \left(u + \alpha\left(V_j - u\right)\cos\beta_2\right)\right] = \rho Q\left[\left(V_j - u\right) - \alpha\left(V_j - u\right)\cos\beta_2\right] \\ &= \rho Q\,(1 - \alpha\cos\beta_2)\left(V_j - u\right). \end{aligned}$$

Substituting F_x into Eq. (5.27), one has:

$$P_{harvested} = \rho Q\,(1 - \alpha\cos\beta_2)\left(V_j - u\right)u = \rho Q\,(1 - \alpha\cos\beta_2)\left(V_j u - u^2\right).$$

Differentiating $P_{harvested}$ with respect to u, one has:

$$\frac{dP_{harvested}}{du} = \rho Q\,(1 - \alpha\cos\beta_2)\left(V_j - 2u\right).$$

Letting $\frac{dP_{harvested}}{du} = 0$ and dividing $\rho Q\,(1 - \alpha\cos\beta_2)$ by the two sides, one has:
$V_j - 2u = 0 \rightarrow u = V_j/2$, indicating that Eq. (5.30) is true when $P_{harvested}$ is maximized.
Substituting $u = V_j/2$ back into the above equation for $P_{harvested}$, one has:

$$P_{harvested,max} = \rho Q(1 - \alpha\cos\beta_2)[V_j(V_j/2) - (V_j/2)^2] = \rho Q(1 - \alpha\cos\beta_2)[V_j^2/4]$$

$= [(\rho g)Q(1 - \alpha\cos\beta_2)]/2[V_j^2/(2g)] = [\gamma Q(1 - \alpha\cos\beta_2)]/2[V_j^2/(2g)]$, which is Eq. (5.31).

Example 5.23 The jet flow from a reservoir impinges a Pelton turbine (Figure 5.30). The penstock has an entrance loss coefficient of 0.03; energy loss in the nozzle is negligible. The turbine has a velocity ratio of $\alpha = 0.92$ and a bucket angle of $\beta_2 = 165°$. Determine the:

(1) Jet velocity, flow rate, and net head
(2) Power harvested if the turbine runner rotates at 120 rpm
(3) Maximum power harvested and best efficiency.

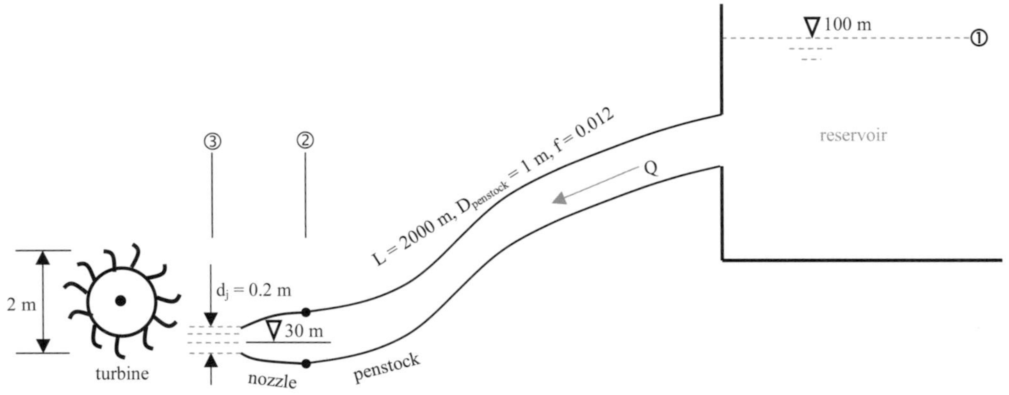

Figure 5.30 **The water system with a Pelton turbine used in Example 5.23.**

Solution

$L = 2000\,\text{m}$, $D_{penstock} = 1\,\text{m}$, $f = 0.012$, $k_e = 0.03$, $d_j = 0.2\,\text{m}$

$D = 2\,\text{m}$, $\beta_2 = 165°$, $\alpha = 0.92$, $\rho = 1000\,\text{kg m}^{-3}$, $\gamma = 9.81\,\text{kN m}^{-3}$

Penstock cross-sectional area: $A_{penstock} = (\pi/4) * (1\,\text{m})^2 = 0.7854\,\text{m}^2$.

Jet flow area: $A_j = (\pi/4) * (0.2\,\text{m})^2 = 0.03142\,\text{m}^2$.

(1) Penstock flow velocity: $V_{penstock} = Q/A_{penstock} = Q/(0.7854\,\text{m}^2) = 1.2732Q\,\text{m s}^{-1}$.
Jet flow velocity: $V_j = Q/A_j = Q/(0.03142\,\text{m}^2) = 31.8269Q\,\text{m s}^{-1}$.
Based on the Darcy–Weisbach equation (Eq. (2.23)), friction loss in the penstock:

$$h_f = (0.012) * (2000\,\text{m})/(1\,\text{m}) * (1.2732Q\,\text{m s}^{-1})^2/[2 * (9.81\,\text{m s}^{-2})] = 1.9829Q^2\,\text{m}.$$

The penstock entrance loss (Table 2.3):

$$h_e = 0.03 * (1.2732Q\,\text{m s}^{-1})^2/[2 * (9.81\,\text{m s}^{-2})] = 0.002479Q^2\,\text{m}.$$

The total head loss (Eq. (2.22)) between section ① and ③:

$$h_L = (1.9829Q^2 + 0.002479Q^2)\,\text{m} = 1.9854Q^2\,\text{m}.$$

The energy equation (Eq. (2.21)) between section ① and ③:

$$100\,\text{m} + 0\,\text{m} + 0\,\text{m} = 30\,\text{m} + 0\,\text{m} + (31.8269Q\,\text{m s}^{-1})^2/[2 * (9.81\,\text{m s}^{-2})] + 1.9854Q^2\,\text{m}$$

$70 = 53.6139Q^2 \rightarrow Q^2 = 70/53.6139\,(m^3\,s^{-1})^2 = 1.3056\,(m^3\,s^{-1})^2 \rightarrow Q = 1.14\,m^3\,s^{-1}$
$\rightarrow V_j = 31.8269 * 1.14\,m\,s^{-1} = 36.28\,m\,s^{-1}$.

Eq. (5.32): $\Delta H = (36.28\,m\,s^{-1})^2/[2 * (9.81\,m\,s^{-2})] = 67.09\,m$.

(2) $\omega = (120\,rpm) * [(2\pi\,rad)/(1\,rev)] * [(1\ min)/(60\,s)] = 12.5664\,rad\,s^{-1}$

$u = (12.5664\,rad\,s^{-1}) * [(2\,m)/2] = 12.57\,m\,s^{-1}$.

Eq. (5.25): $\upsilon_2 = 0.92 * [(36.28\,m\,s^{-1}) - (12.57\,m\,s^{-1})] = 21.81\,m\,s^{-1}$.

Eq. (5.26): $F_x = (1000\,kg\,m^{-3}) * (1.14\,m^3\,s^{-1}) * \{(36.28\,m\,s^{-1}) - [(12.57\,m\,s^{-1}) + (21.81\,m\,s^{-1}) * \cos 165°]\} = 51{,}045.6\,N$.

Eq. (5.27): $P_{harvested} = (51{,}045.6\,N) * (12.57\,m\,s^{-1}) = 641{,}643.2\,W = 641.64\,kW$.

(3) Eq. (5.30): $u = (36.28\,m\,s^{-1})/2 = 18.14\,m\,s^{-1}$.

Eq. (4.31): $P_{harvested,max} = [(9.81\,kN\,m^{-3}) * (1.14\,m^3\,s^{-1}) * (1 - 0.92 * \cos 165°)]/2 * [(36.28\,m\,s^{-1})^2/(2 * 9.81\,m\,s^{-2})] = 708.49\,kW$.

Eq. (5.28): $P_j = (9.81\,kN\,m^{-3}) * (1.14\,m^3\,s^{-1}) * [(36.28\,m\,s^{-1})^2/(2 * 9.81\,m\,s^{-2})] = 750.26\,kW$.

Eq. (5.29): $\eta = (708.49\,kW)/(750.26\,kW) * 100\% = 94.4\%$.

5.3.3 Power by a Francis Turbine

For Francis turbines, stay vanes and guide vanes are used to ensure that water can uniformly flow into its runner from 360 degrees (i.e., all directions; Figure 5.31a). While the runner is rotating, the radial flows flow along surfaces of the blades with no separation and ultimately empty into a draft tube along the axial direction of the runner. Note that the absolute and relative radial flow velocities are the same. The radial flow velocities can be computed as:

$$V_{1r} = \frac{Q}{2\pi r_1 B} \tag{5.33}$$

(radial velocity of entering flow → V_{1r}; flow rate → Q; outer radius of runner → r_1; height of runner → B)

$$V_{2r} = \frac{Q}{2\pi r_2 B} \tag{5.34}$$

(radial velocity of leaving flow → V_{2r}; flow rate → Q; inner radius of runner → r_2; height of runner → B)

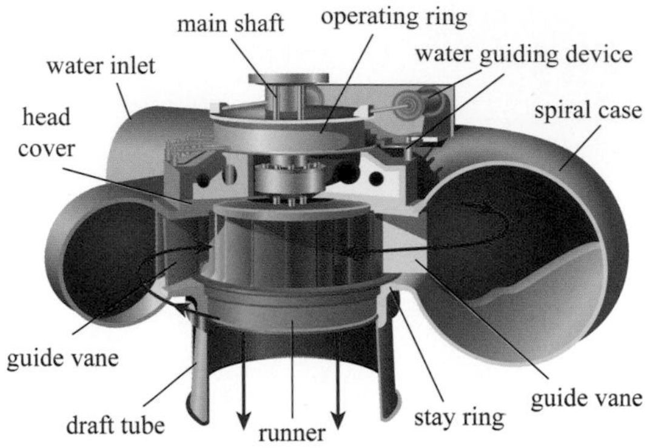

(a)

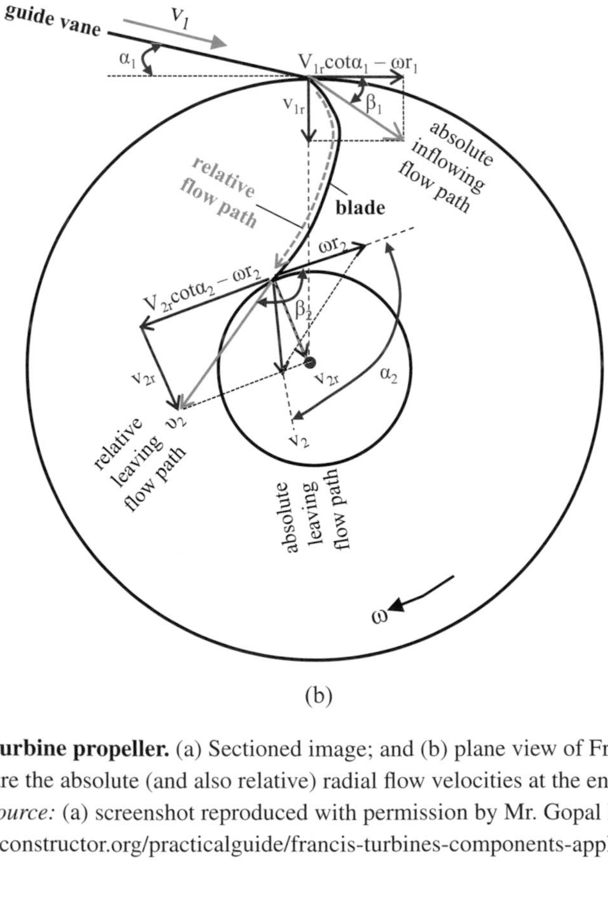

(b)

Figure 5.31 **Francis turbine propeller.** (a) Sectioned image; and (b) plane view of Francis turbine propeller. Note that V_{1r} and V_{2r} are the absolute (and also relative) radial flow velocities at the entering and leaving points, respectively. (*Source:* (a) screenshot reproduced with permission by Mr. Gopal Mishra at The Constructor, https://theconstructor.org/practicalguide/francis-turbines-components-application/2900)

Based on the geometry shown in Figure 5.31b, one can figure out the relationships expressed as:

$$\alpha_1 = \cot^{-1}\left(\frac{\omega r_1}{V_{1r}} + \cot\beta_1\right) \tag{5.35}$$

angular speed of runner in radian per unit time (ω); outer radius of runner (r_1); angle between guide vane and tangential direction at entering point (Figure 5.31b) (α_1); radial velocity of entering flow (V_{1r}); blade angle at entering point (Figure 5.31b) (β_1)

$$\alpha_2 = \cot^{-1}\left(\frac{\omega r_2}{V_{2r}} + \cot\beta_2\right) \tag{5.36}$$

angular speed of runner in radian per unit time (ω); inner radius of runner (r_2); angle between absolute flow path and tangential direction at leaving point (Figure 5.31b) (α_2); radial velocity of leaving flow (V_{2r}); blade angle at leaving point (Figure 5.31b) (β_2)

$$V_1 = \frac{V_{1r}}{\sin\alpha_1} \tag{5.37}$$

absolute flow velocity on guide vane (V_1); radial velocity of entering flow (V_{1r}); angle between guide vane and tangential direction at entering point (Figure 5.31b) (α_1)

$$V_2 = \frac{V_{2r}}{\sin\alpha_2} \tag{5.38}$$

absolute velocity of leaving flow (V_2); radial velocity of leaving flow (V_{2r}); angle between absolute flow path and tangential direction at leaving point (Figure 5.31b) (α_2)

Expressions for torque and power can also be derived as follows. Based on the angular momentum equation Eq. (2.34), the torque acting on the runner of a Francis turbine can be determined as:

$$T_{orque} = \rho Q\left(r_1 V_1 \cos\alpha_1 - r_2 V_2 \cos\alpha_2\right) \tag{5.39}$$

torque on runner (T_{orque}); density of water (ρ); flow rate (Q); outer radius of runner (r_1); absolute flow velocity on guide vane (V_1); angle between guide vane and tangential direction at entering point (Figure 5.31b) (α_1); inner radius of runner (r_2); absolute velocity of leaving flow (V_2); angle between absolute flow path and tangential direction at leaving point (Figure 5.31b) (α_2)

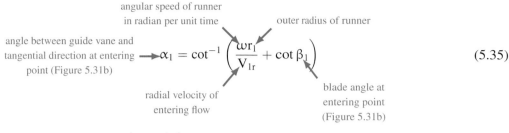

The power harvested by a Francis turbine is computed as:

$$P_{harvested} = T_{orque}\,\omega \tag{5.40}$$

(power harvested → $P_{harvested}$; torque on runner → T_{orque}; angular speed of runner in radian per unit time → ω)

For a given hydraulic system, the net head on a Francis turbine is equal to the difference between the total head in the inflow pipe (i.e., $z_B + \frac{p_B}{\gamma} + \frac{V_B^2}{2g}$) and that in the tailrace (i.e., $\frac{V_{trace}^2}{2g}$), as illustrated in Figure 5.32. The power in the chamber flow is computed as:

$$P_{chamber} = \gamma Q(\Delta H) \tag{5.41}$$

(power in chamber flow → $P_{chamber}$; specific weight of water → γ; flow rate → Q; net head (Figure 5.32) → ΔH)

Finally, the efficiency of a Francis turbine is computed as:

$$\eta = \frac{P_{harvested}}{P_{chamber}} 100\% \tag{5.42}$$

(efficiency of turbine → η; power harvested → $P_{harvested}$; power in chamber flow → $P_{chamber}$)

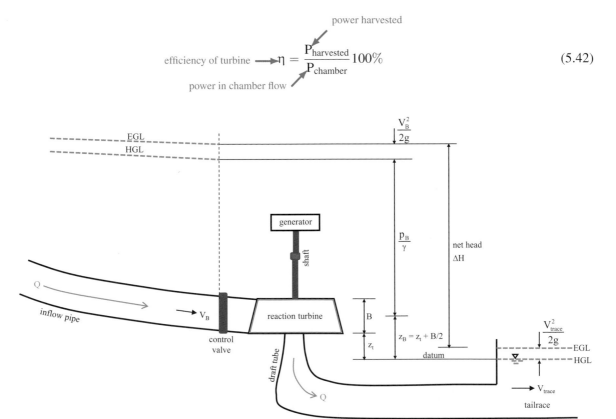

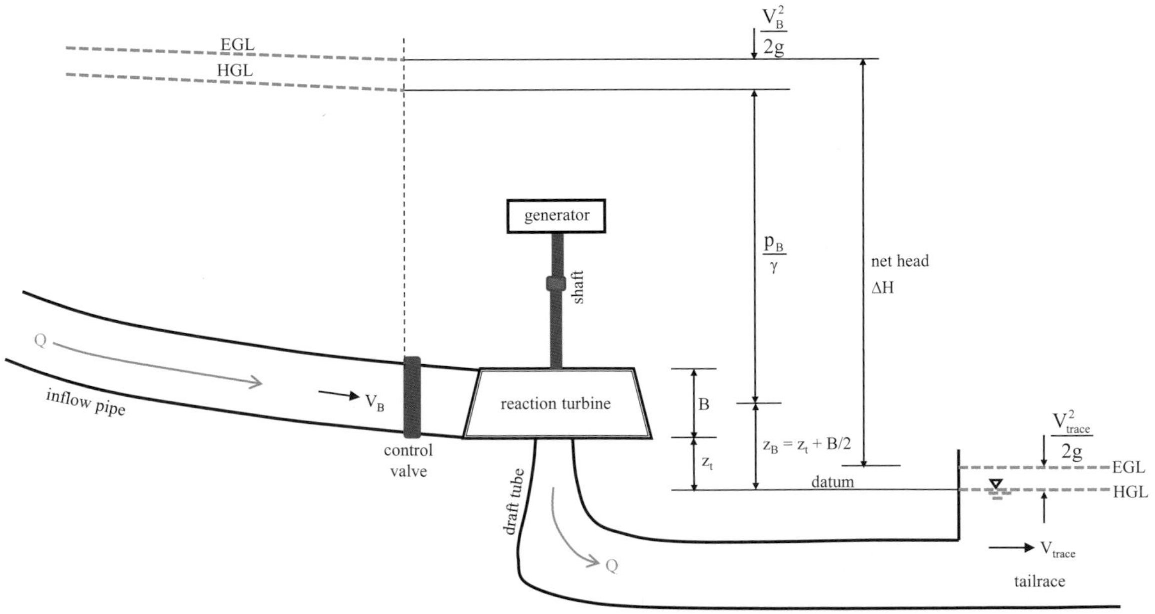

Figure 5.32 **Schematic diagram of Francis turbine installation and net head.**

If the pressure inside a Francis turbine is low, cavitation may occur. In Figure 5.32, assuming $V_{trace} = 0$ and an exit loss coefficient of 1.0, the pressure head inside the turbine can be determined by applying the energy equation (expressed as (Eq. (2.21))) between the horizontal centerline of the turbine and the tailrace water surface. The difference between absolute pressure head and saturation vapor pressure head is divided by net head to define the cavitation number as:

$$\sigma = \frac{\left(\frac{p_{atm}}{\gamma} - z_B\right) - \frac{p_v}{\gamma}}{\Delta H} \qquad (5.43)$$

where σ = cavitation number; p_{atm} = barometric pressure; γ = specific weight of water; z_B = vertical distance from horizontal centerline of turbine to tailrace water surface; p_v = saturation vapor pressure; ΔH = net head.

Cavitation will occur if σ is smaller than a critical value, σ_c, That is, the cavitation condition is $\sigma < \sigma_c$. Figure 5.33 shows a nomogram for determining σ_c.

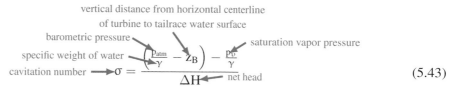

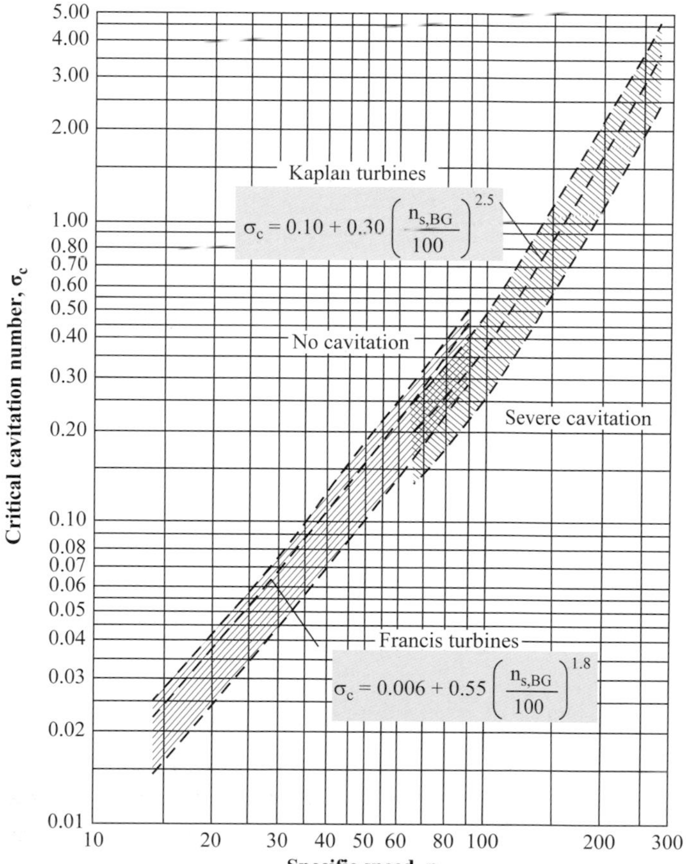

Figure 5.33 **Nomogram for determining critical cavitation number.** Specific speed is computed using Eq. (5.46). (*Source:* Figure 29 in Davis and Sorensen, 1969)

Example 5.24 The characteristics of a Francis turbine are: $B = 1.2\,\text{ft}$, $r_1 = 5.5\,\text{ft}$, $r_2 = 4.5\,\text{ft}$, $\beta_1 = 80°$, $\beta_2 = 155°$, and $\omega = 65\,\text{rpm}$. The turbine is installed in a hydraulic system (water 60°F) with a net head of $\Delta H = 35\,\text{ft}$ and a flow rate of $Q = 200\,\text{cfs}$. The turbine centerline is 5 ft above the tailrace water surface. If nonseparation flow conditions are maintained, calculate the:

(1) Absolute velocities of flows on the guide vanes and at the leaving points
(2) Torque on the runner
(3) Power harvested
(4) Efficiency
(5) Cavitation number and check whether cavitation will occur.

Solution

$\omega = (65\,\text{rpm}) * [(2\pi\,\text{rad})/(1\,\text{rev})] * [(1\,\text{min})/(60\,\text{sec})] = 6.8068\,\text{rad sec}^{-1}$ $\rho = 1.94\,\text{slug ft}^{-3}$, $\gamma = 62.4\,\text{lbf ft}^{-3}$

(1) Eq. (5.33): $V_{1r} = (200\,\text{cfs})/[(2\pi) * (5.5\,\text{ft}) * (1.2\,\text{ft})] = 4.82\,\text{ft sec}^{-1}$.

Eq. (5.34): $V_{2r} = (200\,\text{cfs})/[(2\pi) * (4.5\,\text{ft}) * (1.2\,\text{ft})] = 5.89\,\text{ft sec}^{-1}$.

Eq. (5.35): $\alpha_1 = \cot^{-1}[(6.8068\,\text{rad sec}^{-1}) * (5.5\,\text{ft})/(4.82\,\text{ft sec}^{-1}) + \cot 80°] = 7.18°$.

Eq. (5.36): $\alpha_2 = \cot^{-1}[(6.8068\,\text{rad sec}^{-1}) * (4.5\,\text{ft})/(5.89\,\text{ft sec}^{-1}) + \cot 155°] = 18.12°$.

Eq. (5.37): $V_1 = (4.82\,\text{ft sec}^{-1})/\sin 7.18° = 38.56\,\text{ft sec}^{-1}$.

Eq. (5.38): $V_2 = (5.89\,\text{ft sec}^{-1})/\sin 18.12° = 18.94\,\text{ft sec}^{-1}$.

(2) Eq. (5.39): $T_{orque} = (1.94\,\text{slug ft}^{-3}) * (200\,\text{cfs}) * [(5.5\,\text{ft}) * (38.56\,\text{ft sec}^{-1}) * \cos 7.18° - (4.5\,\text{ft}) * (18.94\,\text{ft sec}^{-1}) * \cos 18.12°] = 50,212.53\,\text{lbf-ft}$.

(3) Eq. (5.40): $P_{harvested} = (50,212.53\,\text{lbf-ft}) * (6.8068\,\text{rad sec}^{-1}) = (341,786.67\,\text{lbf-ft sec}^{-1}) * [(1\,\text{hp})/(550\,\text{lbf-ft sec}^{-1})] = 621.43\,\text{hp}$.

(4) Eq. (5.41): $P_{chamber} = (62.4\,\text{lbf ft}^{-3}) * (200\,\text{cfs}) * (35\,\text{ft}) = (436,800\,\text{lbf-ft sec}^{-1}) * [(1\,\text{hp})/(550\,\text{lbf-ft sec}^{-1})] = 794.18\,\text{hp}$.

Eq. (5.42): $\eta = (621.43\,\text{hp})/(794.18\,\text{hp}) * 100\% = 78.2\%$.

(5) 60°F water → Table 2.1: $p_\upsilon = (0.255\,\text{psia}) * [(144\,\text{in}^2)/(1\,\text{ft}^2)] = 36.72\,\text{lbf ft}^{-2}$ abs
$\gamma = 62.4\,\text{lbf ft}^{-3}$.
Assume $p_{atm} = (14.7\,\text{psia}) * [(144\,\text{in}^2)/(1\,\text{ft}^2)] = 2116.8\,\text{lbf ft}^{-2}$ abs.

$z_B = 5\,\text{ft}$.

Eq. (5.43): $\sigma = [(2116.8\,\text{lbf ft}^{-2})/(62.4\,\text{lbf ft}^{-3}) - (5\,\text{ft}) - (36.72\,\text{lbf ft}^{-2})/(62.4\,\text{lbf ft}^{-3})]/(35\,\text{ft}) = 0.81$.
From Figure 5.33, the maximum of σ_c for Francis turbine is 0.5, which is smaller than $\sigma = 0.81$, thus no cavitation will occur.

Example 5.25 For the hydraulic system and turbine installation scheme in Figure 5.32,

(1) Determine the pressure head inside the turbine
(2) Show Eq. (5.43).

Solution

Assume that the flow velocity in the tailrace is negligible (i.e., $V_{trace} \approx 0$)

(1) Let p_{tb} and V_{tb} be the pressure and velocity inside the turbine, respectively.
Neglect any friction losses $\rightarrow$ the total head loss is equal to the exit loss.
Let V_{tb} be the flow velocity at the draft tube entrance, the total head loss can be computed as:

$$h_L = (1.0) * (V_{tb})^2/(2g) = (V_{tb})^2/(2g).$$

The energy equation (Eq. (2.21)) between the turbine horizontal centerline and the tailrace surface is:

$$z_B + p_{tb}/\gamma + (V_{tb})^2/(2g) = 0 + 0 + 0 + (V_{tb})^2/(2g).$$

Simplifying this equation, one can determine the pressure head inside the turbine: $p_{tb}/\gamma = -z_B$.

(2) The absolute pressure head inside the turbine: $p_{atm}/\gamma + p_{tb}/\gamma = p_{atm}/\gamma - z_B$.
The cavitation number is: $\sigma = [(p_{atm}/\gamma - z_B) - p_\upsilon/\gamma]/\Delta H$, which is Eq. (5.43).

5.3.4 Synchronization with an Electrical Generator

The angular speed of a turbine must be equal to that of the electrical generator. For a given generator, its angular speed depends on the number of magnetic poles that induce a voltage. The number of poles is an even integer between 12 and 96.

In the United States, the angular speed and number of poles are related as:

$$\underbrace{n}_{\text{angular speed of generator [rpm]}} = \frac{7200}{\underbrace{N_{pole}}_{\text{number of poles (even integer of } 12 \sim 96)}} \tag{5.44}$$

In other countries, the angular speed and number of poles are related as:

$$\underbrace{n}_{\text{angular speed of generator [rpm]}} = \frac{6000}{\underbrace{N_{pole}}_{\text{number of poles (even integer of } 12 \sim 96)}} \tag{5.45}$$

5.3.5 Turbine Selection and Cavitation

As with pumps, turbines are selected in terms of specific speed, calculated at maximum efficiency as (Roberson *et al.*, 1998):

angular speed of runner [rpm]

power harvested [hp]

specific speed in BG units $\longrightarrow n_{s,BG} = \dfrac{n\sqrt{P_{harvested}}}{(\Delta H)^{\frac{5}{4}}}$

(5.46)

net head (Eq. (5.32)
or Figure 5.32) [ft]

angular speed of runner [rad s^{-1}]

flow rate [m^3 s^{-1}]

specific speed in SI units $\longrightarrow n_{s,SI} = \dfrac{n\sqrt{Q}}{[g\,(\Delta H)]^{\frac{3}{4}}}$

(5.47)

net head (Eq. (5.32) or Figure 5.32) [m]

gravitational acceleration
$(= 9.81 \text{ m s}^{-2})$

Once a hydraulic system has been designed, the net head, ΔH, and flow rate, Q, will be known. The turbine that mostly fits the hydraulic system can be selected by following the steps shown in Figure 5.34 and explained herein.

Step 1: in terms of ΔH, identify the type of turbine from Figure 5.35.

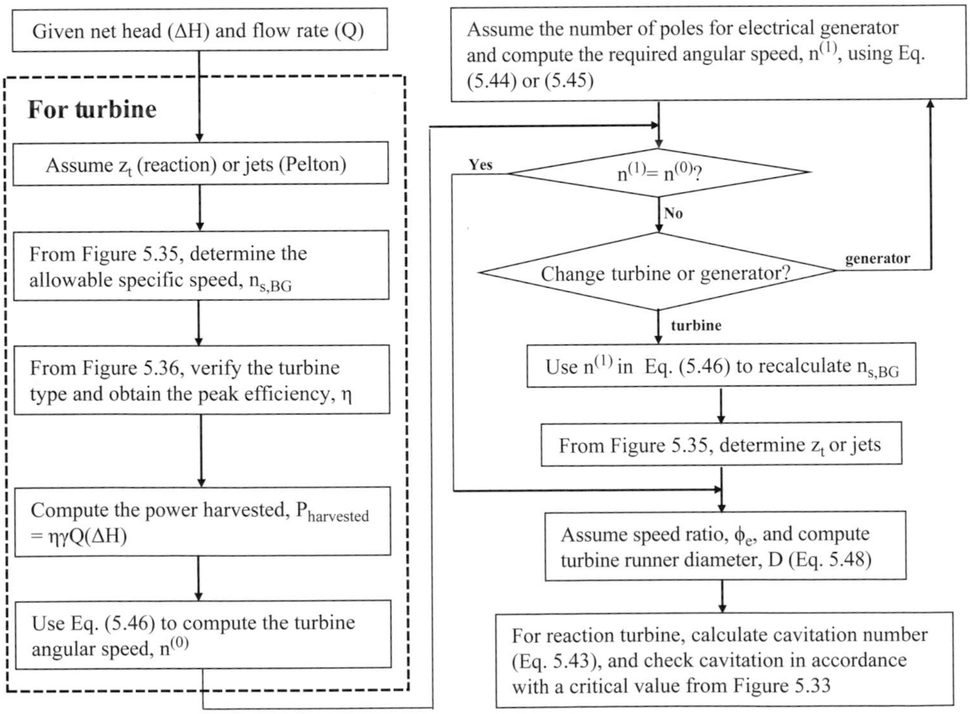

Figure 5.34 Turbine selection flowchart. Selection procedure is presented in boxes and in text.

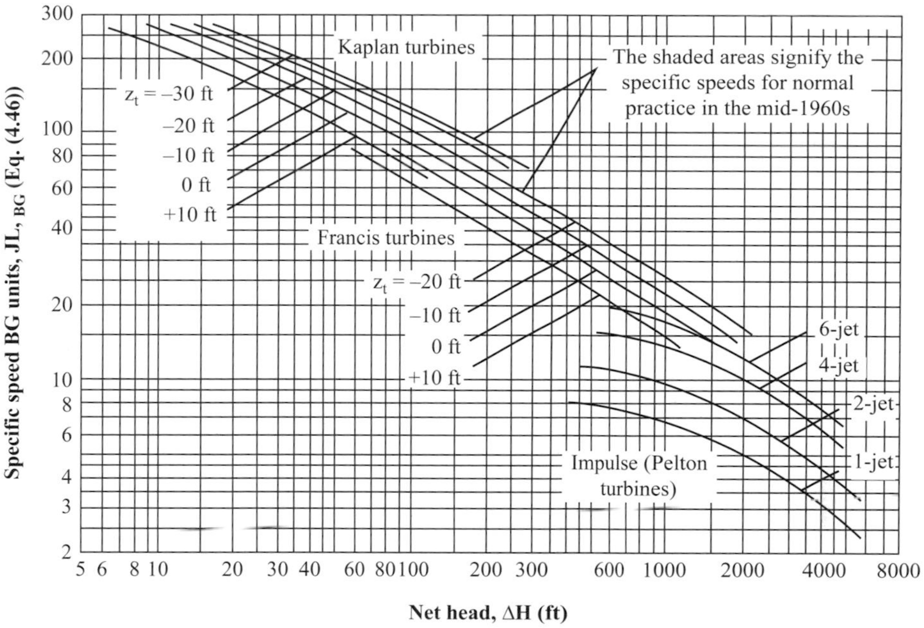

Figure 5.35 **Nomogram of specific speed versus net head for different types of turbines.** (*Source:* Figure 30 in Davis and Sorensen, 1969)

Step 2: based on site-specific conditions, propose the turbine installation scheme and then determine either z_t (vertical distance from turbine bottom to tailrace water surface) if the turbine is Francis (or Kaplan) or the number of jets if the turbine is Pelton.

Step 3: in terms of z_t and ΔH, use the corresponding curve in Figure 5.35 to determine the specific speed, $n_{s,BG}$.

Step 4: verify the turbine type and determine the peak efficiency of the turbine from Figure 5.36.

Step 5: compute the power harvested and the angular speed using Eq. (5.46).

Step 6: check if the turbine angular speed matches the rotation speed of the electrical generator with an assumed number of poles. If not, there will be two choices: either forcing the turbine to rotate at the same speed as the generator or forcing the generator to rotate at the same speed as the turbine. For the first choice, use the generator rotation speed to recalculate $n_{s,BG}$, which in turn is used to re-determine either z_t or the number of jets from Figure 5.35. For the second choice, compute the number of poles using Eq. (5.44) or (5.45) with the turbine rotation speed. If the computed number of poles is an even integer between 12 and 96, it is fine; otherwise, the first choice has to be adopted.

Step 7: select a value of speed ratio from Table 5.7 and determine the propeller size using Eq. (5.48).

Step 8: calculate the cavitation number using Eq. (5.43) and check if cavitation will occur in accordance with a critical value from Figure 5.33. If so, redesign the hydraulic system or lower z_t, and repeat Steps 1 to 8.

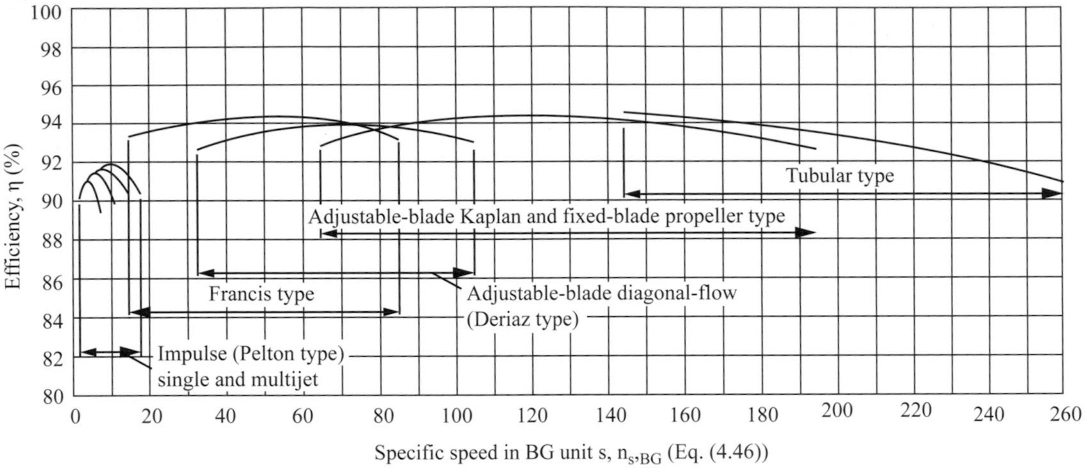

Figure 5.36 **Nomogram of efficiency versus specific speed for different types of turbines.** (*Source:* Figure 25 in Davis and Sorensen, 1969)

Table 5.7 Speed ratio, ϕ_e, for different types of turbines.

Turbine	ϕ_e
Pelton	$0.43 \sim 0.47$
Francis	$0.5 \sim 1.0$
Kaplan	$1.5 \sim 3.0$

The turbine size and speed ratio are functionally related as:

$$D = \frac{\phi_e \sqrt{2g(\Delta H)}}{\pi (n/60)} \tag{5.48}$$

where: turbine propeller diameter (D); speed ratio (Table 5.7) (ϕ_e); gravitational acceleration (g); net head (ΔH); angular speed of runner [rpm] (n)

Example 5.26 A hydropower plant in the United States has a net head of 300 ft and a flow rate of 5000 cfs. If the water temperature is 60°F, select an appropriate turbine with a 2-ft-high propeller. Assume a standard barometric pressure $p_{atm} = 14.7$ psia.

Solution

$\Delta H = 300$ ft, $Q = 5000$ cfs, $B = 2$ ft, $p_{atm} = 14.7$ psia $= 2116.8$ lbf ft^{-2} abs

Table 2.1: 60°F water→$\gamma = 62.4$ lbf ft^{-3}, $p_\upsilon = 0.255$ psia $= 36.72$ lbf ft^{-2} abs.

The selection is done by implementing the eight steps shown in Figure 5.34:

Figure 5.35: $\Delta H = 300\,\text{ft} \rightarrow$ Francis turbine $\rightarrow$ assume $z_t = 10\,\text{ft} \rightarrow n_{s,BG} = 32$.

Figure 5.36: $n_{s,BG} = 32 \rightarrow$ Francis (fixed blades) $\rightarrow \eta = 94\%$.

Eq. (5.41): $P_{harvested} = 94\% * (62.4\,\text{lbf ft}^{-3}) * (5000\,\text{cfs}) * (300\,\text{ft}) = (87,940,000\,\text{lbf-ft sec}^{-1}) * [(1\,\text{hp})/(550\,\text{lbf-ft sec}^{-1})] = 159,970.9\,\text{hp}$.

Eq. (5.46): $n^{(0)} = [32 * (300\,\text{ft})^{5/4}]/(159,970.9\,\text{hp})^{1/2} = 100.0\,\text{rpm}$.

Eq. (5.44): assume number of poles $N_{pole} = 72 \rightarrow n^{(1)} = 7200/72\,\text{rpm} = 100.0\,\text{rpm}$

$$n^{(0)} \approx n^{(1)} \rightarrow n = (100.0 + 100.0)/2\,\text{rpm} = 100.0\,\text{rpm}.$$

Table 5.7: choose $\phi_e = 0.75$ for the turbine.

Eq. (5.48): $D = 0.75 * [2 * (32.2\,\text{ft sec}^{-2}) * (300\,\text{ft})]^{1/2}/[\pi * (100.0\,\text{rpm})/60] = 20.0\,\text{ft}$

$$z_B = z_t + B/2 = 10\,\text{ft} + (2\,\text{ft})/2 = 11\,\text{ft}.$$

Eq. (5.43): $\sigma = \{[(2116.8\,\text{lbf ft}^{-2})/(62.4\,\text{lbf ft}^{-3}) - (11\,\text{ft})] - [(36.72\,\text{lbf ft}^{-2})/(62.4\,\text{lbf ft}^{-3})]\}/(300\,\text{ft}) = 0.074$.

Figure 5.33: $n_{s,BG} = 32 \rightarrow \sigma_c = 0.074$.

$\sigma \approx \sigma_c \rightarrow$ It is just fine. To be conservative, it may be better to lower the turbine to $z_t = 0$.

In summary, the selected turbine is: Francis (fixed blades); $D = 20.0\,\text{ft}$, $\eta = 94\%$, $n_{s,BG} = 32$, $z_t = 10\,\text{ft}$, $z_B = 11\,\text{ft}$, $\sigma = 0.074$.

5.3.6 Affinity Laws of Turbines

For a given turbine, if its propeller size and/or rotation speed are changed, the power harvested by the turbine will be changed accordingly. The turbine efficiency does not change with rotation speed, whereas it does change with propeller size in terms of the functional relationship expressed as (Crowe *et al.*, 2005):

$$\eta_2 = 1 - (1 - \eta_1)\left(\frac{D_1}{D_2}\right)^{0.2} \tag{5.49}$$

(new efficiency η_2; old efficiency η_1; old propeller size D_1; new propeller size D_2)

If the rotation speed remains the same but the propeller size is changed, because the water power of a given hydraulic system is fixed, the relationship between power harvested and efficiency can be expressed as:

$$\frac{P_{harvested,2}}{P_{harvested,1}} = \frac{\eta_2}{\eta_1} \tag{5.50}$$

(new power harvested $P_{harvested,2}$; old power harvested $P_{harvested,1}$; new efficiency η_2; old efficiency η_1)

Given the fact that a turbine installed in a hydraulic system has a constant specific speed regardless of the rotation speed and propeller size of the turbine, Eq. (5.46) can be used to derive how power harvested changes with rotation speed. If the propeller size remains the same, such a functional relationship can be expressed as:

new power harvested; old power harvested; old rotation speed [rpm]; new rotation speed [rpm]

$$\frac{P_{harvested,2}}{P_{harvested,1}} = \frac{n_1^2}{n_2^2} \tag{5.51}$$

If both the propeller size and rotation speed are changed, multiplying Eq. (5.50) and (5.51) and then taking the square root, one can derive the equation expressed as:

new power harvested; old power harvested; old rotation speed; new rotation speed; new efficiency; old efficiency

$$\frac{P_{harvested,2}}{P_{harvested,1}} = \frac{n_1}{n_2}\sqrt{\frac{\eta_2}{\eta_1}} \tag{5.52}$$

Example 5.27 Prove Eq. (5.50).

Solution

Based on Eq. (5.46), at both rotation speed n_1 and n_2, one has:

$$n_{s,BG} = \frac{n_1\sqrt{P_{harvested,1}}}{(\Delta H)^{\frac{5}{4}}} = \frac{n_2\sqrt{P_{harvested,2}}}{(\Delta H)^{\frac{5}{4}}} \rightarrow \frac{P_{harvested,2}}{P_{harvested,1}} = \frac{n_1^2}{n_2^2}, \text{ which is Eq. (5.50).}$$

Example 5.28 A turbine (propeller size $D_1 = 20$ ft and rotation speed $n_1 = 80$ rpm) generates 100,000 hp. If the turbine has an efficiency of $\eta_1 = 0.89$, determine the power generated if the:

(1) Propeller size is decreased to $D_2 = 15$ ft and the rotation speed is unchanged
(2) Propeller size is increased to $D_2 = 25$ ft and the rotation speed is unchanged
(3) Rotation speed is decreased to $n_2 = 75$ rpm and the propeller size is unchanged
(4) Rotation speed is increased to $n_2 = 85$ rpm and the propeller size is unchanged
(5) Propeller size and rotation speed are decreased to $D_2 = 15$ ft and $n_2 = 75$ rpm
(6) Propeller size and rotation speed are increased to $D_2 = 25$ ft and $n_2 = 85$ rpm.

Solution

(1) Eq. (5.49): $\eta_2 = 1 - (1 - 0.89) * [(20\,\text{ft})/(15\,\text{ft})]^{0.2} = 0.883$.
Eq. (5.50): $P_{harvested,2} = (0.883/0.89) * (100,000\,\text{hp}) = 99,214\,\text{hp}$.

(2) Eq. (5.49): $\eta_2 = 1 - (1 - 0.89) * [(20\,\text{ft})/(25\,\text{ft})]^{0.2} = 0.895$.
Eq. (5.50): $P_{\text{harvested},2} = (0.895/0.89) * (100{,}000\,\text{hp}) = 100{,}562\,\text{hp}$.
(3) Eq. (5.51): $P_{\text{harvested},2} = [(80\,\text{rpm})^2/(75\,\text{rpm})^2] * (100{,}000\,\text{hp}) = 113{,}778\,\text{hp}$.
(4) Eq. (5.51): $P_{\text{harvested},2} = [(80\,\text{rpm})^2/(85\,\text{rpm})^2] * (100{,}000\,\text{hp}) = 88{,}581\,\text{hp}$.
(5) Eq. (5.49): $\eta_2 = 1 - (1 - 0.89) * [(20\,\text{ft})/(15\,\text{ft})]^{0.2} = 0.883$.
Eq. (5.52): $P_{\text{harvested},2} = [(80\,\text{rpm})/(75\,\text{rpm})] * (0.883/0.89)^{1/2} * (100{,}000\,\text{hp}) = 106{,}246\,\text{hp}$.
(6) Eq. (5.49): $\eta_2 = 1 - (1 - 0.89) * [(20\,\text{ft})/(25\,\text{ft})]^{0.2} = 0.895$.
Eq. (5.52): $P_{\text{harvested},2} = [(80\,\text{rpm})/(85\,\text{rpm})] * (0.895/0.89)^{1/2} * (100{,}000\,\text{hp}) = 94{,}382\,\text{hp}$.

PROBLEMS

5.1 State the difference between head against and total head against by a pump. How much head must the pump add?

5.2 For the water system shown in Figure 5.1, if $z_A = 30$ ft, $z_B = 100$ ft, both the 500-ft-long suction line and 6000-ft-long discharge line have a diameter of 2 ft and a Darcy friction factor of 0.013, the entrance and exit loss coefficients are 0.12 and 1.0, respectively, the pump is positioned at an elevation of 45 ft, and the water transfer rate is 6.5 cfs, determine the:
(1) Head against by the pump
(2) Total head against by the pump
(3) Pressure head at the suction point
(4) Pressure head at the discharge point
(5) Head against and total head against by the pump when the water transfer rate is increased to 7.0 cfs.

5.3 A pump is used to pump water from a lower to a higher reservoir. If the water temperature is increased or decreased, how will the head added and the discharge delivered by the pump change?

5.4 For the pump in Problem 5.2, determine the:
(1) Output power
(2) Power transferred to water
(3) Power loss due to the pipeline
(4) Power loss due to the pump itself if the pump has an efficiency of 0.85.

5.5 For Example 5.3, if the pumping rate is increased to $5.2\,\text{ft}^3\,\text{sec}^{-1}$, determine the:
(1) Output power
(2) Brake power
(3) Power transferred to water
(4) Power loss due to the pump itself
(5) Power loss due to the pipeline
(6) Gross power loss.

5.6 List the types of pumps that can be used to pump water with substantial sediments.

5.7 What type of pump is usually used to pump groundwater?

5.8 For the pump considered in Figure 5.8 with an impeller diameter of 9 in, determine the:
(1) Head added when the pump delivers 50 gpm
(2) Discharge delivered when the pump raises the head by 150 ft
(3) Peak efficiency
(4) BEP.

5.9 For the pump considered in Figure 5.8 with an impeller diameter of 8 in, develop and plot the:
(1) Efficiency versus discharge capacity curve
(2) Brake power versus discharge capacity curve.

5.10 If the pump considered in Figure 5.8 rotates at 2000 rpm, develop and plot the $C_H \sim C_Q$ and $\eta \sim C_Q$ curves for the 7.5- and 9.5-in-diameter impellers.

5.11 If a radial-flow pump has an impeller diameter of 10 in and a rotation speed of 1200 rpm, develop and plot its dimensional curves of $\Delta H_p \sim Q_p$, $P_{brake} \sim Q_p$, and $\eta \sim Q_p$.

5.12 If an axial-flow pump has an impeller diameter of 10 in and a rotation speed of 1200 rpm, develop and plot its dimensional curves of $\Delta H_p \sim Q_p$, $P_{brake} \sim Q_p$, and $\eta \sim Q_p$.

5.13 If a radial-flow pump with an impeller diameter of 50 cm and a rotation speed of 1500 rpm operates at its peak efficiency, determine the:
(1) Discharge delivered and head added
(2) Power required.

5.14 If an axial-flow pump with an impeller diameter of 50 cm and a rotation speed of 1500 rpm operates at its peak efficiency, determine the:
(1) Discharge delivered and head added
(2) Power required.

5.15 A pump has an impeller diameter of 50 cm and a rotation speed of 1000 rpm. Its current performance curves are: $\Delta H_p = -1389Q_p^2 + 30.13Q_p + 5.25$ and $P_{brake} = 0.8e^{21.8Q_p}$, where ΔH_p is the head capacity in m, Q_p is the discharge capacity in $m^3\,s^{-1}$, and P_{brake} is the brake power in kW. On a same coordinate system, plot the current and new performance curves if the:
(1) Impeller diameter is increased to 90 cm
(2) Impeller diameter is decreased to 30 cm
(3) Rotation speed is increased to 1200 rpm
(4) Rotation speed is decreased to 800 rpm
(5) Impeller diameter is increased to 90 cm and the rotation speed is increased to 1200 rpm
(6) Impeller diameter is increased to 90 cm and the rotation speed is decreased to 800 rpm
(7) Impeller diameter is decreased to 30 cm and the rotation speed is increased to 1200 rpm
(8) Impeller diameter is decreased to 30 cm and the rotation speed is decreased to 800 rpm.

5.16 A pump operating at 1000 rpm delivers a discharge of 250 gpm against a total head of 135 ft. If the discharge is increased to 350 gpm by simply increasing the rotation speed, determine the:
(1) New rotation speed
(2) New total head against
(3) Percent increase of brake power.

5.17 A pump operating at 1000 rpm delivers a discharge of 250 gpm against a total head of 135 ft. If the total head against is increased to 200 ft by simply increasing the rotation speed, determine the:

(1) New rotation speed
(2) New discharge
(3) Percent increase of brake power.

5.18 A 25-cm-diameter pump operating at 1580 rpm delivers a discharge of $30\,\mathrm{L\,s^{-1}}$ against a total head of 5.5 m. As requested by the customer, the total head against needs to be increased to 8.5 m. Determine the:

(1) New rotation speed and discharge as well as percent increase of brake power if the head increase is realized by simply increasing the rotation speed
(2) New impeller diameter and discharge as well as percent increase of brake power if the head increase is realized by simply increasing the impeller size
(3) New discharge and total head against as well as percent increase of brake power if the impeller diameter is increased to 35 cm and the rotation speed is increased to 1700 rpm.

5.19 At a rotation speed of 1800 rpm, the performance curves of a pump are provided in the following table. Determine the:

(1) Dimensionless specific speed
(2) Specific speed in SI units
(3) Specific speed in BG units
(4) Type and peak efficiency of the pump
(5) Impeller diameter of the pump.

$Q_p\,(\mathrm{m^3\,s^{-1}})$	ΔH_p (m)	η
0.0000	47.88	0.00
0.0160	48.06	0.32
0.0320	48.15	0.58
0.0490	47.97	0.74
0.0650	47.25	0.83
0.0810	46.17	0.90
0.0970	43.65	0.93
0.1130	39.60	0.90
0.1220	37.17	0.88

5.20 A pump has a specific speed (in BG units) of 12,000. At its BEP, the head and discharge capacities are 100 ft and 5000 gpm. Determine the following for the pump:

(1) Type and peak efficiency
(2) Impeller diameter
(3) Rotation speed.

5.21 For the water system shown in Figure 5.1, $z_A = 5$ ft, $z_B = 155$ ft, the suction line is 150 ft long, the discharge line is 500 ft long, both the suction and discharge lines have a diameter of 3 in and a Darcy friction factor of 0.0125, the entrance loss is negligible, the exit loss coefficient is 1.0, the pump in Figure 5.8 is used and positioned at a zero elevation. Assume a water temperature of 60°F.

(1) Develop the system curve
(2) Determine the optimal impeller diameter
(3) For the impeller in (2), determine the operation point, efficiency, and brake power
(4) For the impeller in (2), determine the net positive suction head (NPSH) and cavitation number (σ)
(5) For the impeller in (2), determine whether cavitation will occur.

5.22 A pump is used to pump water at 20°C from a reservoir at an elevation of 350 m to another reservoir at a higher elevation of 430 m through a 30-cm-diameter concrete pipe. The pipe is 1000 m long and has a Darcy friction factor of 0.016. If the pump has a performance curve of $\Delta H_p = -280Q_p^2 + 100$, where ΔH_p is the head capacity in m and Q_p is the discharge capacity in $m^3\,s^{-1}$, determine the operation point of the pump. Neglect any minor losses.

5.23 A pump has a performance curve of $\Delta H_p = -125Q_p^2 + 300$, where ΔH_p is the head capacity in ft and Q_p is the discharge capacity in cfs. On a same coordinate system, plot the individual and combined performance curve if two such pumps are installed in:

(1) Parallel
(2) Series.

5.24 The following table shows the performance curves for Pumps A, B, and C. On a same coordinate system, plot the individual and combined performance curve if:

(1) Pumps A and B are installed in parallel
(2) Pumps A and C are installed in parallel
(3) Pumps B and C are installed in parallel
(4) The three pumps are installed in parallel.

Pump A		Pump B		Pump C	
$Q_p\,(m^3s^{-1})$	$\Delta H_p\,(m)$	$Q_p\,(m^3s^{-1})$	$H_p\,(m)$	$Q_p\,(m^3s^{-1})$	$\Delta H_p\,(m)$
0.0000	9.46	0.0000	13.62	0.0000	5.32
0.0128	9.49	0.0154	13.67	0.0054	5.34
0.0256	9.51	0.0307	13.70	0.0108	5.35
0.0384	9.48	0.0461	13.64	0.0162	5.33
0.0512	9.33	0.0614	13.44	0.0216	5.25
0.0640	9.12	0.0768	13.13	0.0270	5.13
0.0768	8.62	0.0922	12.42	0.0324	4.85
0.0896	7.82	0.1075	11.26	0.0378	4.40
0.0960	7.34	0.1152	10.57	0.0405	4.13
0.1000	7.00	0.1200	10.00	0.0500	3.00
0.1100	6.00	0.1300	8.50	0.0600	1.50
0.1200	5.00	0.1400	6.00	0.0660	0.00
0.1300	3.00	0.1500	3.00		
0.1400	0.00	0.1600	0.00		

5.25 For the three pumps in Problem 5.24, on a same coordinate system, plot the individual and combined performance curve if:
(1) Pumps A and B are installed in series
(2) Pumps A and C are installed in series
(3) Pumps B and C are installed in series
(4) The three pumps are installed in series.

5.26 If Pumps A and B in Problem 5.24 are installed in series to deliver a discharge of $0.1\ m^3\ s^{-1}$, determine the total head against by:
(1) Each of the two pumps
(2) The two pumps combined.

5.27 If Pumps A and B in Problem 5.24 are installed in parallel against a total head of 8 m, determine the discharge by:
(1) Each of the two pumps
(2) The two pumps combined.

5.28 For the pumps in Problem 5.24, when Pumps B and C are installed in series to deliver $0.1\ m^3\ s^{-1}$, determine the total head against by each of the two pumps.

5.29 For the pumps in Problem 5.24, when Pumps A and C are installed in parallel against a total head of 12 m, determine the discharge by each of the two pumps.

5.30 List the three commonly used turbines and their application conditions.

5.31 For a Pelton or impulse turbine:
(1) Explain it is important to have sufficient number of buckets
(2) Illustrate the bucket angle, β_2
(3) Plot the ratio of power harvested to the maximum power versus u/V_j, where u is the tangential speed of runner, and V_j is the absolute jet flow velocity.

5.32 A Pelton turbine has a 12-ft-diameter propeller, a bucket angle of 165°, and a velocity ratio of $\alpha = 0.95$. The turbine, impinged by a 10 cfs jet flow with a diameter of 3 in, rotates at 300 rpm. Determine the:
(1) Horizontal impulse force and the torque on the propeller
(2) Power harvested and efficiency
(3) Maximum power harvested.

5.33 A Pelton turbine has a 1.5-m-diameter propeller, a bucket angle of 150°, and a velocity ratio of $\alpha = 0.92$. The turbine, impinged by a $1.2\ m^3\ s^{-1}$ jet flow with a diameter of 20 cm, rotates at 250 rpm. Determine the:
(1) Horizontal impulse force and the torque on the propeller
(2) Power harvested and efficiency
(3) Maximum power harvested.

5.34 The characteristics of a Francis turbine are: $B = 2.5$ ft, $r_1 = 8.2$ ft, $r_2 = 6.8$ ft, $\beta_1 = 75°$, $\beta_2 = 165°$, and $\omega = 120$ rpm. The turbine is installed in a chamber with a net head of $\Delta H = 50$ ft and a flow rate of $Q = 500$ cfs. Its centerline is 3 ft above the tailrace water surface. Nonseparation flow conditions are maintained. If the water temperature is 50°F, determine the:

(1) Power harvested
(2) Power in the chamber flow and efficiency of the pump
(3) Cavitation number and check whether cavitation will occur.

5.35 The characteristics of a Francis turbine are: $B = 1.2\,m$, $r_1 = 3.0\,m$, $r_2 = 2.5\,m$, $\beta_1 = 80°$, $\beta_2 = 155°$, and $\omega = 100\,rpm$. The turbine is installed in a chamber with a net head of $\Delta H = 40\,m$ and a flow rate of $Q = 15\,m^3\,s^{-1}$. Its centerline is 1.5 m above the tailrace water surface. Nonseparation flow conditions are maintained. If the water temperature is 20°C, determine the:
(1) Power harvested
(2) Power in the chamber flow and efficiency of the pump
(3) Cavitation number and check whether cavitation will occur.

5.36 The reaction turbine of a hydropower plant is driven by a torque of 8000 lbf-ft and is connected to generator to generate 280 hp. Determine the number of poles if the plant is located in:
(1) The United States
(2) Another country.

5.37 A hydropower plant in the United States has a net head of 50 ft and a flow rate of 5000 cfs. If the water temperature is 60°F, select an appropriate turbine with a 6-ft-high propeller. Assume a standard atmospheric pressure of 14.7 psia.

5.38 A hydropower plant in the United States has a net head of 2500 ft and a flow rate of 820 cfs. If the water temperature is 60°F, select an appropriate turbine. Assume a standard atmospheric pressure of 14.7 psia.

5.39 A hydropower plant in the United States has a net head of 360 ft and a flow rate of 1000 cfs. If the water temperature is 60°F, select an appropriate turbine with a 3-ft-high propeller. Assume a standard atmospheric pressure of 14.7 psia.

5.40 A turbine (propeller size $D_1 = 5\,m$ and rotation speed $n_1 = 150\,rpm$) generates 80,000 kW. If the turbine has an efficiency of $\eta_1 = 0.85$, determine the power generated if the:
(1) Propeller size is decreased to $D_2 = 4\,m$ and the rotation speed is unchanged
(2) Propeller size is increased to $D_2 = 6\,m$ and the rotation speed is unchanged
(3) Rotation speed is decreased to $n_2 = 100\,rpm$ and the propeller size is unchanged
(4) Rotation speed is increased to $n_2 = 200\,rpm$ and the propeller size is unchanged
(5) Propeller size and rotation speed are decreased to $D_2 = 4\,m$ and $n_2 = 100\,rpm$
(6) Propeller size and rotation speed are increased to $D_2 = 6\,m$ and $n_2 = 200\,rpm$.

6 Open Channel Flow and Channel Design

Open channel flow, which has a free water surface subject to atmosphere pressure, occurs in both natural and manmade channels. Its analysis is imperative to the design and management of water resources engineering projects such as levees and channels. This chapter discusses water surface profile classification and computation, flow measuring, and channel design. First, it discusses the principles of hydraulic and energy grade lines, normal depth, and critical depth. Second, it discusses the water surface profile of gradually varied flow. Third, it discusses the water surface profile and energy loss of rapidly varied flow. Herein, along a homogenous channel segment, which has a fixed cross section, a constant roughness, and an invariant bed slope, gradually varied flow has a linear water surface, whereas rapidly varied flow has a curved water surface. Fourth, the chapter discusses weirs, flumes, and ultrasonic flow meters used for flow measurement. Finally, it discusses the methods of designing non-erosive, erosive, and semi-erosive channels.

6.1 Hydraulic and Energy Grade Lines

At a given location of an open channel, as discussed in Sections 2.3 and 2.4.2, the water surface elevation, which represents the hydraulic energy head, is equal to the summation of the channel bed elevation and the water depth, and the total energy head is equal to the summation of the hydraulic energy head and the velocity head. As illustrated in Figure 6.1, the hydraulic grade line (HGL) defines how the water surface elevation varies along the flow direction in the channel and the energy grade line (EGL) defines how total energy head varies along the flow direction in the channel. The EGL is governed by Eq. (2.19), whereas the HGL is governed by the equation expressed as:

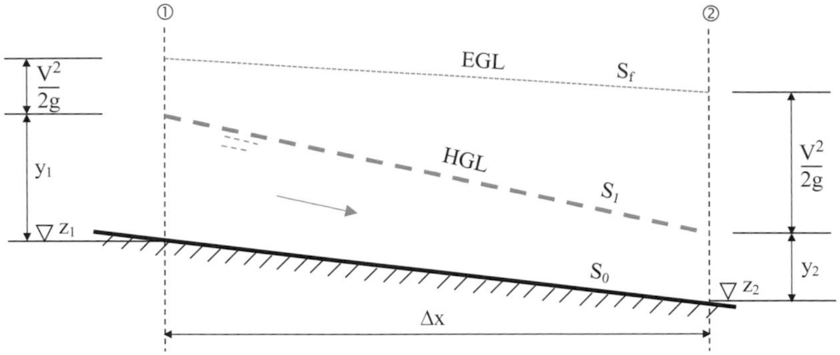

Figure 6.1 Longitudinal cross section of an open channel. Various parameters are shown, including the hydraulic grade line (HGL) and the energy grade line (EGL).

$$\frac{d(z+y)}{dx} = -S_1 \tag{6.1}$$

(channel bed elevation: z; water depth: y; HGL slope: S_1; distance along flow direction: x)

For any segment of an open channel, in terms of the energy equation expressed as Eq. (2.21), the HGL and EGL can be conceptually sketched. It is convenient to first sketch the EGL and then the HGL, both starting from the furthest upstream point of the channel. This sketch does not have to be drawn to a scale, but it must be conceptually correct. For instance, sketching the EGL must illustrate its dips due to minor (e.g., entrance, contraction, expansion, fitting, and exit) losses and its inclinations due to friction losses, whereas sketching HGL must illustrate its ups and downs due to varying velocity heads. Without addition of external energy (e.g., by pump), along the flow direction downstream, it is impossible for the EGL to rise or have a positive slope because energy cannot be created. It is possible, however, for the HGL to rise or have a positive slope if the channel expands, causing the velocity to continually decrease.

Example 6.1 A canal is used to divert water from Lake A to Lake B (Figure 6.2). Sketch the EGL and HGL of the water system.

Solution

First, let us sketch the EGL (thin dashed line in Figure 6.2). In Lake A, the EGL is at the lake water surface if the water velocity approaching the channel is negligible, but is above the lake water surface if the approaching velocity is non-negligible. At the entrance of the gate opening, due to entrance loss, the EGL has a sudden drop. From the gate to section ①, due to friction loss, the EGL is linearly inclined downstream. From sections ① to ②, due to contraction and friction losses, the EGL is curved with an overall downward gradient. From sections ② to ③, due to friction loss, the EGL is linearly

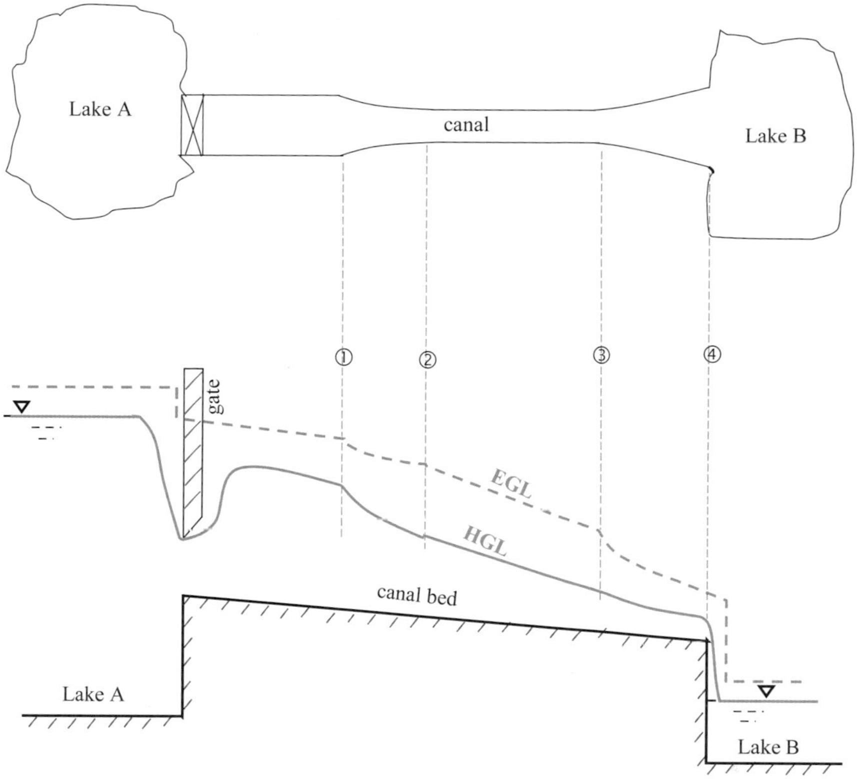

Figure 6.2 **Map and cross section of a canal connecting two lakes as used in Example 6.1.** See text for explanation of how the cross section is constructed.

inclined downstream. From sections ③ to ④, due to expansion and friction losses, the EGL is curved with an overall downward gradient. At section ④, the EGL plunges down. In Lake B, the EGL is at the lake water surface if the leaving velocity is negligible, while it is above the lake water surface if the leaving velocity is non-negligible.

Second, let us sketch the HGL (the thin solid line in Figure 6.2). Note that, at any given location along the channel, the difference between the EGL and HGL is equal to the velocity head. In Lake A, the HGL is at the lake water surface, sharply dropping to the gate crest. After the gate, the HGL rises steeply and then gradually declines. From sections ① to ②, the HGL is curved, with an overall increasing vertical distance away from the EGL because the velocity tends to increase along the downstream direction. From sections ② to ③, because the canal cross section is invariant and thus the velocity is constant, the HGL is parallel to the EGL. From sections ③ to ④, the HGL is curved, with an overall decreasing vertical distance away from the EGL because the velocity tends to decrease along the downstream direction. From section ④ to the water surface of Lake B, the HGL plunges down with a nappe shape (i.e., a curtain of water). In Lake B, the HGL is at the lake water surface.

6.2 Normal Depth and Critical Depth

Normal depth, y_n, is the water depth of uniform flow in an open channel. Given that, for uniform flow, the water depth and velocity do not vary along the channel, the EGL, HGL, and channel bed are parallel to each other and thus they have the same slope. Thus, one can substitute the channel bed slope for the EGL slope in the Manning formula (Eq. (2.26)) to calculate y_n. In this regard, normal depth does not exist in open channels with an adverse (i.e., negative) or zero bed slope, for which the Manning formula is mathematically invalid.

Example 6.2 A 5-ft-wide rectangular channel (n = 0.015) is laid on a slope of 0.0005. If the channel carries 50 cfs, determine the normal depth.

Solution

$b = 5\,\text{ft},\ S_0 = 0.0005,\ n = 0.015,\ Q = 50\,\text{cfs}$

Normal depth, y_n, is computed for uniform flow $\rightarrow S = S_0 = 0.0005$.

Flow area: $A = by_n = 5y_n\,\text{ft}^2$.

Wetted perimeter: $P_{wet} = b + 2y_n = 5 + 2y_n\,\text{ft}$.

Hydraulic radius: $R = A/P_{wet} = (5y_n)/(5 + 2y_n)\,\text{ft}$.

Mean velocity: $V = Q/A = 50/(5y_n)\,\text{ft sec}^{-1} = 10/y_n\,\text{ft sec}^{-1}$.

Manning formula (Eq. (2.26)):

$$10/y_n\,\text{ft sec}^{-1} = 1.486/0.015 * [(5y_n)/(5 + 2y_n)\,\text{ft}]^{2/3} * (0.0005)^{1/2}$$

$$\rightarrow (5 + 2y_n)^{2/3} y_n^{-5/3} = 0.6477 \rightarrow \text{using trial-and-error or Excel Solver : } y_n = 3.51\,\text{ft}.$$

Example 6.3 A triangular channel (n = 0.013) with a side slope of 1.2 is laid on a slope of 0.01. If the channel carries $5\,\text{m}^3\,\text{s}^{-1}$, determine the normal depth.

Solution

$Z = 1.2,\ S_0 = 0.01,\ n = 0.013,\ Q = 5\,\text{m}^3\,\text{s}^{-1}$

Normal depth, y_n, is computed for uniform flow $\rightarrow S = S_0 = 0.01$.

Top width: $B = 2Zy_n = 2 * 1.2y_n\,\text{m} = 2.4y_n\,\text{m}$.

Flow area: $A = \tfrac{1}{2}By_n = \tfrac{1}{2} * (2.4y_n)y_n\,\text{m}^2 = 1.2{y_n}^2\,\text{m}^2$.

Wetted perimeter: $P_{wet} = 2y_n(1 + Z^2)^{1/2} = 2y_n(1 + 1.2^2)^{1/2}\,\text{m} = 3.124y_n\,\text{m}$.

Hydraulic radius: $R = A/P_{wet} = (1.2{y_n}^2)/(3.124y_n)\,\text{m} = 0.3841y_n\,\text{m}$.

Mean velocity: $V = Q/A = 5/(1.2{y_n}^2)\,\text{m s}^{-1} = 4.1667/{y_n}^2\,\text{m s}^{-1}$.

Manning formula (Eq. (2.26)):

$$4.1667/{y_n}^2\,\text{m s}^{-1} = 1/0.013 * [0.3841y_n\,\text{m}]^{2/3} * (0.01)^{1/2}$$

$$\rightarrow y_n^{8/3} = 1.0251 \rightarrow y_n = 1.0251^{3/8} = 1.01\,\text{m}.$$

Example 6.4 A trapezoidal channel ($n = 0.025$) with a bottom width of 15 m and a side slope of 2.5 is laid on a slope of 0.002. If the channel carries $30\,m^3\,s^{-1}$, determine the normal depth.

Solution

$b = 15\,m, Z = 2.5, S_0 = 0.002, n = 0.025, Q = 30\,m^3\,s^{-1}$

Normal depth, y_n, is computed for uniform flow $\rightarrow S = S_0 = 0.002$.

Top width: $B = b + 2Zy_n = 15\,m + 2 * 2.5y_n\,m = 15 + 5.0y_n\,m.$

Flow area: $A = {}^1\!/_2(b + B)y_n = {}^1\!/_2 * (15 + 15 + 5.0y_n)y_n\,m^2 = 15y_n + 2.5{y_n}^2\,m^2.$

Wetted perimeter: $P_{wet} = b + 2y_n(1 + Z^2)^{1/2} = 15\,m + 2y_n(1 + 2.5^2)^{1/2}\,m = 15 + 5.3852y_n\,m.$

Hydraulic radius: $R = A/P_{wet} = (15y_n + 2.5{y_n}^2)/(15 + 5.3852y_n)\,m.$

Mean velocity: $V = Q/A = 30/(15y_n + 2.5{y_n}^2)\,m\,s^{-1}.$

Manning formula (Eq. (2.26)):

$$30/(15y_n + 2.5{y_n}^2)\,m\,s^{-1} = 1/0.025 * [(15y_n + 2.5{y_n}^2)/(15 + 5.3852y_n)\,m]^{2/3} * (0.002)^{1/2}$$

$$\rightarrow (15y_n + 2.5{y_n}^2)^{5/3}(15 + 5.3852y_n)^{-2/3} = 16.7705$$

$\rightarrow$trial-and-error or Excel Solver : $y_n = 1.03\,m.$

Example 6.5 A 48-in-diameter circular channel ($n = 0.012$) is laid on a slope of 0.005. If the channel carries 100 cfs, determine the normal depth.

Solution

$D = 48\,in = 4\,ft, S_0 = 0.005, n = 0.012, Q = 100\,cfs$

Normal depth, y_n, is computed for uniform flow $\rightarrow S = S_0 = 0.005$.

Flow area (Table 2.4): $A = (D^2/8)(\theta - \sin\theta) = [(4\,ft)^2/8](\theta - \sin\theta) = 2(\theta - \sin\theta)\,ft^2.$

Wetted perimeter (Table 2.4): $P_{wet} = \theta(D/2) = \theta[(4\,ft)/2] = 2\theta\,ft.$

Hydraulic radius: $R = A/P_{wet} = [2(\theta - \sin\theta)]/(2\theta)\,ft = (\theta - \sin\theta)/\theta\,ft.$

Mean velocity: $V = Q/A = 100/[2(\theta - \sin\theta)]\,ft\,sec^{-1} = 50/(\theta - \sin\theta)\,ft\,sec^{-1}.$

Manning formula (Eq. (2.26)):

$$50/(\theta - \sin\theta)\,ft\,sec^{-1} = 1.486/0.012 * [(\theta - \sin\theta)/\theta\,ft]^{2/3} * (0.005)^{1/2}$$

$$\rightarrow (\theta - \sin\theta)^{5/3}\theta^{-2/3} = 5.7102 \rightarrow \text{trial-and-error or Excel Solver} : \theta = 4.1788\,rad.$$

Table 2.4: $\theta = 2\cos^{-1}[(D - 2y_n)/D] \rightarrow 4.1788\,rad = 2\cos^{-1}[(4 - 2y_n)/4]$

$$\rightarrow 4 - 2y_n = 4 * \cos(4.1788/2\,rad) = -1.9828 \rightarrow y_n = 2.99\,ft.$$

As discussed in Section 2.5.4, for an open channel, its critical depth, y_c, is the water depth at which the specific energy is minimal. If the total discharge is constant, y_c can be computed using Eq. (2.37) or (2.38), whereas if the specific energy is constant, y_c can be computed using Eq. (2.42) or (2.44). Eqs. (2.37) and (2.42) are applicable for any channel shape, whereas Eqs. (2.38) and (2.44) are only applicable for rectangular channels. Both y_n and y_c are functions of discharge and channel geometry (i.e., cross-sectional shape and dimensions). In addition, y_n also varies with channel roughness and bed slope, but y_c is independent of these channel characteristics.

Example 6.6 For the rectangular channel in Example 6.2, determine its critical depth.

Solution

$b = 5\,\text{ft}$, $Q = 50\,\text{cfs}$

Unit-width flow rate: $q = (50\,\text{cfs})/(5\,\text{ft}) = 10\,\text{ft}^2\,\text{sec}^{-1}$.

Eq. (2.38): $y_c = [(10\,\text{ft}^2\,\text{sec}^{-1})^2/(32.2\,\text{ft}\,\text{sec}^{-2})]^{1/3} = 1.46\,\text{ft}$.

Example 6.7 For the triangular channel in Example 6.3, determine its critical depth.

Solution

$Z = 1.2$, $Q = 5\,\text{m}^3\,\text{s}^{-1}$

Top width: $B_c = 2Zy_c = 2 * 1.2y_c\,\text{m} = 2.4y_c\,\text{m}$.

Flow area: $A_c = \frac{1}{2}B_c y_c = \frac{1}{2} * (2.4y_c)y_c\,\text{m}^2 = 1.2{y_c}^2\,\text{m}^2$.

Eq. (2.37): $(5\,\text{m}^3\,\text{s}^{-1})^2/(9.81\,\text{m}\,\text{s}^{-2}) = (1.2{y_c}^2\,\text{m}^2)^3/(2.4y_c\,\text{m}) \rightarrow {y_c}^5 = 3.5395 \rightarrow y_c = 1.29\,\text{m}$.

Example 6.8 For the trapezoidal channel in Example 6.4, determine its critical depth.

Solution

$b = 15\,\text{m}$, $Z = 2.5$, $Q = 30\,\text{m}^3\,\text{s}^{-1}$

Top width: $B_c = b + 2Zy_c = 15\,\text{m} + 2 * 2.5y_c\,\text{m} = 15 + 5.0y_c\,\text{m}$.

Flow area: $A_c = \frac{1}{2}(b + B_c)y_c = \frac{1}{2} * (15 + 15 + 5.0y_c)y_c\,\text{m}^2 = 15y_c + 2.5{y_c}^2\,\text{m}^2$.

Eq. (2.37): $(30\,\text{m}^3\,\text{s}^{-1})^2/(9.81\,\text{m}\,\text{s}^{-2}) = (15y_c + 2.5{y_c}^2\,\text{m}^2)^3/(15 + 5.0y_c\,\text{m})$

$$\rightarrow \text{trial-and-error or Excel Solver}: y_c = 0.71\,\text{m}.$$

Example 6.9 For the circular channel in Example 6.5, determine its critical depth.

Solution

$D = 48\,\text{in} = 4\,\text{ft}$, $Q = 100\,\text{cfs}$

Top width (Table 2.4): $B_c = D\sin(\theta_c/2) = 4\sin(\theta_c/2)\,\text{ft}$.

Flow area (Table 2.4): $A_c = (D^2/8)(\theta_c - \sin\theta_c) = [(4\,ft)^2/8](\theta_c - \sin\theta_c) = 2(\theta_c - \sin\theta_c)\,ft^2$.

Eq. (2.37): $(100\,cfs)^2/(32.2\,ft\,sec^{-2}) = [2(\theta_c - \sin\theta_c)\,ft^2]^3/(4\sin(\theta_c/2)\,ft)$

$$\rightarrow(\theta_c - \sin\theta_c)^3[\sin(\theta_c/2)]^{-1} = 155.2795$$

$$\rightarrow\text{trial-and-error or Excel Solver}: \theta_c = 4.2232\,rad.$$

Table 2.4: $\theta_c = 2\cos^{-1}[(D - 2y_c)/D]\rightarrow 4.2232\,rad = 2\cos^{-1}[(4 - 2y_c)/4]$

$$\rightarrow 4 - 2y_c = 4 * \cos(4.2232/2\,rad) = -2.0593\rightarrow y_c = 3.03\,ft.$$

Example 6.10 A rectangular channel is controlled by a sluice gate. If the specific energy of the approaching flow is a constant of 7.5 m, determine the critical depth.

Solution

$E = 7.5\,m$

Eq. (2.44): $y_c = 2/3E = 2/3 * (7.5\,m) = 5.0\,m$.

Example 6.11 A trapezoidal channel has a bottom width of 10 ft and a side slope of 1.5. If the channel is controlled to have an approaching flow with a constant specific energy of 25 ft, determine the critical depth.

Solution

$E = 25\,ft$, $b = 10\,ft$, $Z = 1.5$

Top width: $B_c = b + 2Zy_c = 10 + 2 * 1.5y_c\,ft = 10 + 3.0\,y_c\,ft$.

Flow area: $A_c = (b + Zy_c)y_c = (10 + 1.5y_c)y_c\,ft^2$.

Eq. (2.42): $y_c = 25\,ft - [(10 + 1.5y_c)y_c\,ft^2]/[2 * (10 + 3.0y_c\,ft)]$

$$\rightarrow 7.5y_c^2 - 120y_c - 500 = 0\rightarrow\text{quadratic formula}: y_c = 19.43\,ft.$$

6.3 Gradually Varied Flow

Gradually varied flow is one type of nonuniform flows, whose water depth smoothly varies (i.e., has no abrupt change) along flow direction. Mathematically, the first-order derivative of water depth with respect to distance along flow direction exists.

6.3.1 Classification of Water Surface Profiles

A water surface profile defines how water depth varies along a channel. In reality, the bed slope of the channel may be positive, horizontal, or adverse (Figure 6.3). If the bed slope is positive, the

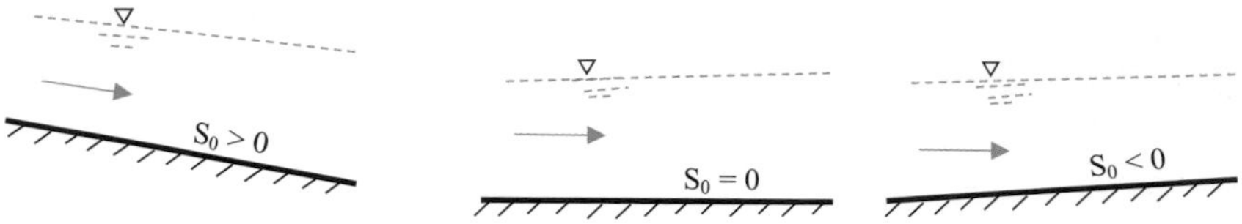

(a) (b) (c)

Figure 6.3 Schematic cross-section diagrams illustrating channel bed slopes. (a) Positive slope; (b) horizontal slope; and (c) adverse slope.

Table 6.1 Classifications of water surface profiles in open channel.[1]

Slope (bed slope) (profile shape)	Zone I ($y > \max\{y_n, y_c\}$)	Zone II ($\min\{y_n, y_c\} \leq y \leq \max\{y_n, y_c\}$)	Zone III ($y \leq \min\{y_n, y_c\}$)
Horizontal ($S_0 = 0$) (bending shape)	N/A	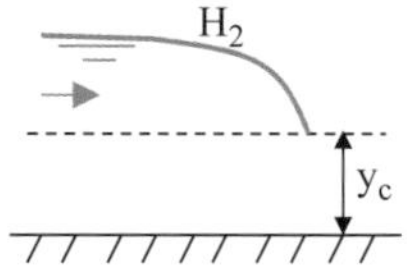	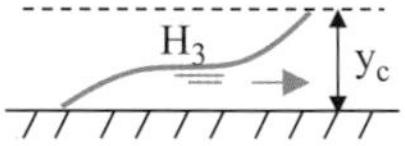
Adverse ($S_0 < 0$) (bending shape)	N/A	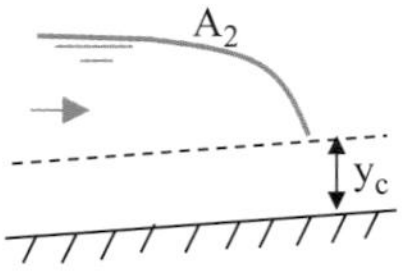	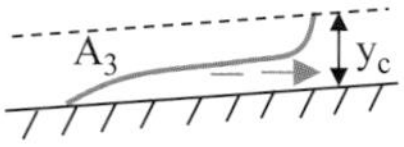
Mild ($S_0 > 0$, $y_n > y_c$) (bending shape)	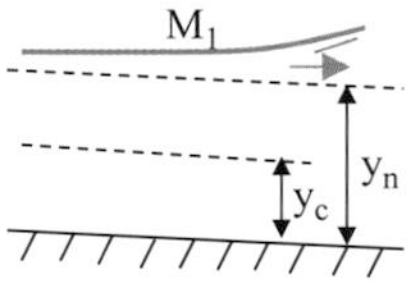	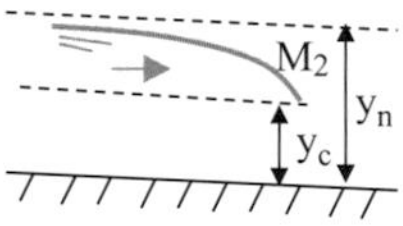	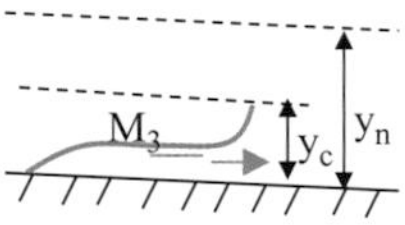
Critical ($S_0 > 0$, $y_n = y_c$) (stretching shape)	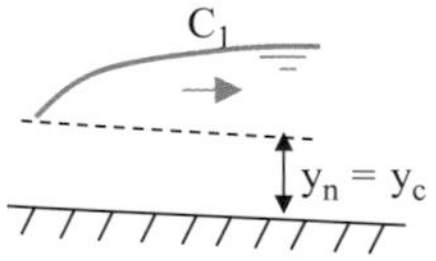	N/A	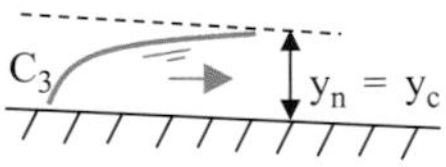
Steep ($S_0 > 0$, $y_n < y_c$) (stretching shape)	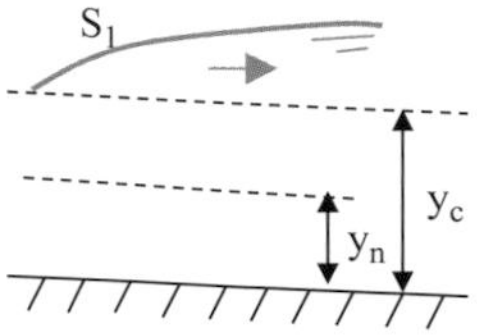	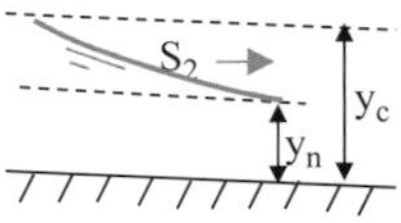	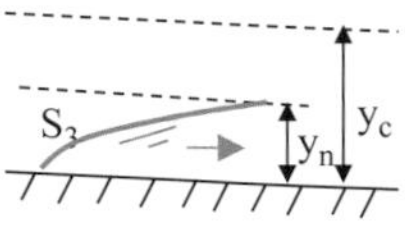

[1] y_n: normal depth; y_c: critical depth; y: actual water depth; S_0: channel bed slope.

bed elevation drops along the flow direction, whereas if the bed slope is adverse, the bed elevation rises. A horizontal slope means its bed elevation is the same throughout the channel. As discussed above, normal depth, y_n, only exists for channels with a positive bed slope, while critical depth, y_c, exists regardless of the channel bed slopes. In terms of the relative magnitude of y_n and y_c, a channel with a positive bed slope can be further classified as having a mild, critical, or steep slope. When $y_n > y_c$, the channel is classified as having a mild slope, whereas when $y_n < y_c$, this same channel is classified as having a steep slope. When $y_n = y_c$, the channel is classified as having a critical slope. Such classifications depend on the discharge in the channel.

In terms of the relative magnitude of actual water depth, y, to y_n and y_c, three zones (hereinafter designated Zones I, II, and III for descriptive purposes) can be defined as follows: $y > \max\{y_n, y_c\}$ in Zone I, $\min\{y_n, y_c\} \leq y \leq \max\{y_n, y_c\}$ in Zone II, and $y \leq \min\{y_n, y_c\}$ in Zone III, where $\max\{\,\}$ and $\min\{\,\}$ are the maximum and minimum functions, respectively. Based on combinations of these slopes and zones, water surface profiles are classified into 12 types (Table 6.1), designated H2, H3, A2, A3, M1, M2, M3, C1, C3, S1, S2, and S3. Herein, H, A, M, C, and S signify horizontal bed slope, adverse bed slope, mild slope, critical slope, and steep slope, respectively, and 1, 2, and 3 signify Zone I, II, and III, respectively. Overall, H, A, and M profiles have a "bending" shape, tending to be plateaued or flattened towards upstream, whereas C and S profiles have a "stretching" shape, tending to be plateaued or flattened towards downstream.

Example 6.12 In Figure 6.4, flood water flows through a spillway, which has an upstream leading mild-slope channel and a downstream stilling basin followed by a horizontal channel and then a steep-slope channel. If the channels, stilling basin and spillway have an identical rectangular cross section, sketch the water surface profiles.

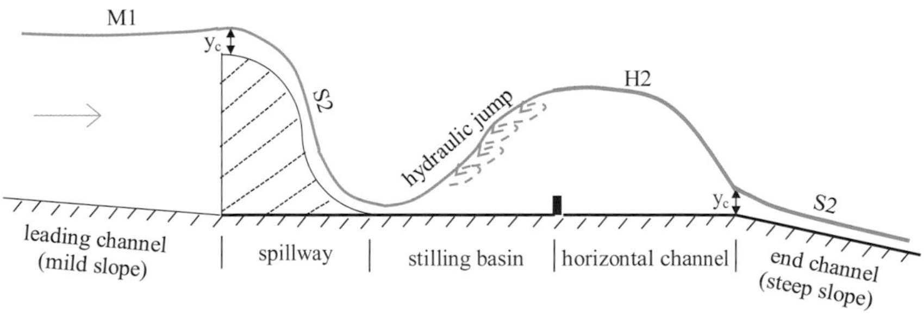

Figure 6.4 **The water surface profiles used in Example 6.12.**

Solution

Identical rectangular cross section → constant critical depth, y_c. The leading channel has a mild slope and thus its normal depth of $y_{n,1} > y_c$, whereas the spillway nappe and the end channel have a steep slope and thus their normal depths are smaller than y_c (i.e., $y_{n,2} < y_c$ and $y_{n,3} < y_c$). A hydraulic jump (see Section 6.4.1) will occur in the stilling basin.

When the spillway starts to release flood water, the actual water depth in the leading channel is larger than y_{n1} and thus the water surface profile should be M1. On the crest of the spillway, the actual water depth is usually equal to y_c but larger than y_{n2}, so the spillway nappe has a S2 water surface profile. After the hydraulic jump, the actual water depth at the entrance of the horizontal channel is larger than y_c, leading to an H2 water surface profile. At the break of the channel slopes, the actual water depth is usually equal to y_c and thus the end channel has a S2 water surface profile.

Example 6.13 Classify the slopes of the channels in:

(1) Example 6.2
(2) Example 6.3
(3) Example 6.4
(4) Example 6.5.

Solution

(1) Example 6.2 $\rightarrow y_n = 3.51$ ft; Example 6.6 $\rightarrow y_c = 1.46$ ft.
$y_n > y_c \rightarrow$ mild slope.

(2) Example 6.3 $\rightarrow y_n = 1.01$ m; Example 6.7 $\rightarrow y_c = 1.29$ m.
$y_n < y_c \rightarrow$ steep slope.

(3) Example 6.4 $\rightarrow y_n = 1.03$ m; Example 6.8 $\rightarrow y_c = 0.71$ m.
$y_n > y_c \rightarrow$ mild slope.

(4) Example 6.5 $\rightarrow y_n = 2.25$ ft; Example 6.9 $\rightarrow y_c = 2.26$ ft.
$y_n \approx y_c \rightarrow$ critical slope.

Example 6.14 A 10-ft-wide rectangular channel ($n = 0.013$) is laid on a slope of 0.001. If the water depth at the channel entrance is 1.2 ft, classify the water surface profile when the channel carries:

(1) 30 cfs
(2) 100 cfs
(3) 50 cfs.

Solution

$b = 10$ ft, $n = 0.013$, $S_0 = 0.001$, $y = 1.2$ ft

Determine the normal depth, $y_n \rightarrow$ uniform flow $\rightarrow S = S_0 = 0.001$.

Flow area: $A = by_n = 10y_n$ ft^2.

Wetted perimeter: $P_{wet} = b + 2y_n = 10 + 2y_n$ ft.

Hydraulic radius: $R = A/P_{wet} = (10y_n)/(10 + 2y_n)$ ft $= (5y_n)/(5 + y_n)$ ft.

(1) Mean velocity: $V = Q/A = 30/(10y_n)\ \text{ft sec}^{-1} = 3/y_n\ \text{ft sec}^{-1}$.
Manning formula (Eq. (2.26)):

$$3/y_n\ \text{ft sec}^{-1} = 1.486/0.013 * [(5y_n)/(5+y_n)\ \text{ft}]^{2/3} * (0.001)^{1/2}$$
$$\rightarrow (5+y_n)^{2/3} y_n^{-5/3} = 3.5232 \rightarrow \text{trial-and-error or Excel Solver}: y_n = 0.96\ \text{ft}.$$

Determine the critical depth, y_c
Unit-width flow rate: $q = (30\ \text{cfs})/(10\ \text{ft}) = 3\ \text{ft}^2\ \text{sec}^{-1}$.
Eq. (2.38): $y_c = [(3\ \text{ft}^2\ \text{sec}^{-1})^2/(32.2\ \text{ft sec}^{-2})]^{1/3} = 0.65\ \text{ft}$.

Classify the water surface profile:
$y_n > y_c \rightarrow$ mild slope; $y > y_n \rightarrow$ M1 profile.

(2) Mean velocity: $V = Q/A = 100/(10y_n)\ \text{ft sec}^{-1} = 10/y_n\ \text{ft sec}^{-1}$.
Manning formula (Eq. (2.26)):

$$10/y_n\ \text{ft sec}^{-1} = 1.486/0.013 * [(5y_n)/(5+y_n)\ \text{ft}]^{2/3} * (0.001)^{1/2}$$
$$\rightarrow (5+y_n)^{2/3} y_n^{-5/3} = 1.0570 \rightarrow \text{trial-and-error or Excel Solver}: y_n = 2.12\ \text{ft}.$$

Determine the critical depth, y_c
Unit-width flow rate: $q = (100\ \text{cfs})/(10\ \text{ft}) = 10\ \text{ft}^2\ \text{sec}^{-1}$.
Eq. (2.38): $y_c = [(10\ \text{ft}^2\ \text{sec}^{-1})^2/(32.2\ \text{ft sec}^{-2})]^{1/3} = 1.46\ \text{ft}$.
Classify the water surface profile:
$y_n > y_c \rightarrow$ mild slope; $y < y_c \rightarrow$ M3 profile.

(3) Mean velocity: $V = Q/A = 50/(10y_n)\ \text{ft sec}^{-1} = 5/y_n\ \text{ft sec}^{-1}$.
Manning formula (Eq. (2.26)):

$$5/y_n\ \text{ft sec}^{-1} = 1.486/0.013 * [(5y_n)/(5+y_n)\ \text{ft}]^{2/3} * (0.001)^{1/2}$$
$$\rightarrow (5+y_n)^{2/3} y_n^{-5/3} = 2.1139 \rightarrow \text{trial-and-error or Excel Solver}: y_n = 1.34\ \text{ft}.$$

Determine the critical depth, y_c
Unit-width flow rate: $q = (50\ \text{cfs})/(10\ \text{ft}) = 5\ \text{ft}^2\ \text{sec}^{-1}$.
Eq. (2.38): $y_c = [(5\ \text{ft}^2\ \text{sec}^{-1})^2/(32.2\ \text{ft sec}^{-2})]^{1/3} = 0.92\ \text{ft}$.
Classify the water surface profile:
$y_n > y_c \rightarrow$ mild slope; $y_c < y < y_n \rightarrow$ M2 profile.

In summary: For the same channel, the water surface profile can be different for different discharges. In fact, this is also true for different entrance water depths. For instance, if the entrance depth were 3.5 ft, the profiles in all three questions would be classified as M1, whereas if the entrance depth were 0.5 ft, the profiles would be classified as M3.

6.3.2 Computation of Water Surface Profiles

For a channel segment (e.g., Figure 6.1) with constant bed and EGL slopes and Manning's n, if the channel has a gradually varied flow, the energy equation can be rearranged (see Example 6.17) to derive the governing equation for a water surface profile. The water surface profile governing equation (WSPGE) can be expressed in two formats, designated hereinafter Format I and II for descriptive purposes.

The Format I WSPGE is described in terms of specific energy as:

specific energy; channel bed slope

$$\frac{dE}{dx} = S_0 - S_f \tag{6.2}$$

distance along flow direction; EGL slope

The Format II WSPGE is described in terms of Froude number as:

water depth; channel bed slope; EGL slope

$$\frac{dy}{dx} = \frac{S_0 - S_f}{1 - F_r^2} \tag{6.3}$$

distance along flow direction; Froude number

For computational purposes, Eq. (6.2) can be written into finite-difference format as:

specific energy at section ①; specific energy at section ②

distance along flow direction

$$\Delta x = \frac{E_1 - E_2}{S_f - S_0} \tag{6.4}$$

EGL slope; channel bed slope

Likewise, Eq. (6.3) can be written into finite-difference format as:

water depth at section ①; water depth at section ②

distance along flow direction; Froude number

$$\Delta x = \frac{(y_1 - y_2)\left(1 - F_r^2\right)}{S_f - S_0} \tag{6.5}$$

EGL slope; channel bed slope

Segments of streams and river reaches have the same physical meaning; thus, they can be interchangeably used in practice. To compute a water surface profile along a river reach, the reach needs to be subdivided into a number of subreaches, each of which should have an approximately constant bed slope and Manning's n (Figure 6.5). The water surface profile along each subreach can be computed by Eq. (6.4) or (6.5) either using a one- or multi-step method. The one-step method treats the entire subreach as the basic computational unit and assumes that the water depth linearly varies within the subreach. It assigns Δx to be equal to the length of the subreach and computes y_1 or y_2, whichever is unknown. The multi-step method, however, further subdivides the subreach

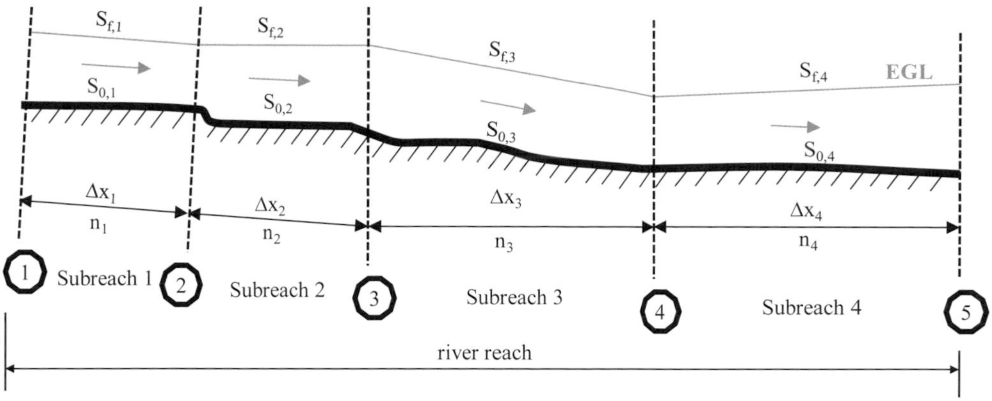

Figure 6.5 **Cross-sectional view of a river reach divided into subreaches.** This type of diagram is useful for water surface profile computations.

into a number of sub-subreaches to capture any nonlinear variations of water depth within the subreach, and applies the one-step method to each sub-subreach. Obviously, the one-step method has lower computational load and accuracy than the multi-step method. In practice, for subreaches with near linearly varied water depths, the one-step method is a good choice, whereas for subreaches with nonlinearly-varied water depths, the multi-step method should be used. Note that a water surface profile needs to be classified before it can be computed.

When Eq. (6.4) or Eq. (6.5) is used for a subreach or sub-subreach, starting from section ①, where the water depth, y_1, is known, one can assume a value for y_2 within the range of the water depths of the appropriate classified profile and then compute the right side of the equation. Such an assumption–computation process needs to be iterated until the computed value for the right side is equal to Δx for an acceptable tolerance limit. This iteration can be automatically done in Excel Solver.

The computations using Eq. (6.4) are implemented by following these five steps:

Step 1: For section ①, compute A_1, $P_{wet,1}$, R_1, V_1, and E_1.
Step 2: For section ②, compute A_2, $P_{wet,2}$, R_2, V_2, and E_2.
Step 3: Compute the average hydraulic radius $\overline{R} = \frac{R_1 + R_2}{2}$, and the average mean velocity $\overline{V} = \frac{V_1 + V_2}{2}$.
Step 4: Use $\overline{R}$ and $\overline{V}$ in the Manning formula (Eq. (2.26)) to compute S_f.
Step 5: Use Eq. (6.4) to compute Δx.

The computations using Eq. (6.5) are implemented by following these six steps:

Step 1: For section ①, compute A_1, B_1, d_1, and V_1.
Step 2: For section ②, compute A_2, B_2, d_2, and V_2.

Step 3: Compute the average hydraulic radius $\overline{R} = \dfrac{R_1 + R_2}{2}$, the average mean velocity $\overline{V} = \dfrac{V_1 + V_2}{2}$, and the average hydraulic depth $\overline{d} = \dfrac{d_1 + d_2}{2}$.

Step 4: Use $\overline{R}$ and $\overline{V}$ in the Manning formula (Eq. (2.26)) to compute S_f.

Step 5: Use $\overline{V}$ and $\overline{d}$ in Eq. (2.46) to compute Froude number F_r.

Step 6: Use Eq. (6.5) to compute Δx.

Example 6.15 A 50-ft-long rectangular channel ($n = 0.015$, $S_0 = 0.005$, $b = 8$ ft) carries 200 cfs. If the water depth at the end of the channel is 4.5 ft, use the one-step method to compute the water surface profile by:

(1) Eq. (6.4)
(2) Eq. (6.5).

Solution

$b = 8$ ft, $n = 0.015$, $S_0 = 0.005$, $L = 50$ ft, $Q = 200$ cfs, $y_1 = 4.5$ ft

Before computing the water surface profile, one needs to classify it.

Determine the normal depth, y_n: $\rightarrow$ uniform flow $\rightarrow S = S_0 = 0.005$.

Flow area: $A = by_n = 8y_n\ \text{ft}^2$.
Wetted perimeter: $P_{wet} = b + 2y_n = 8 + 2y_n$ ft.
Hydraulic radius: $R = A/P_{wet} = 8y_n/(8 + 2y_n)\ \text{ft} = 4y_n/(4 + y_n)$ ft.
Mean velocity: $V = Q/A = (200\ \text{cfs})/(8y_n)\ \text{ft sec}^{-1} = 25/y_n\ \text{ft sec}^{-1}$.
Manning formula (Eq. (2.26)):

$$25/y_n\ \text{ft sec}^{-1} = 1.486/0.015 * [4y_n/(4 + y_n)\ \text{ft}]^{2/3} * 0.005^{1/2}$$

$$\rightarrow (4 + y_n)^{2/3} y_n^{-5/3} = 0.7061 \rightarrow \text{trial-and-error or Excel Solver}: y_n = 2.63\ \text{ft}.$$

Determine the critical depth, y_c:

Unit-width flow rate: $q = (200\ \text{cfs})/(8\ \text{ft}) = 25\ \text{ft}^2\ \text{sec}^{-1}$.
Eq. (2.38): $y_c = [(25\ \text{ft}^2\ \text{sec}^{-1})^2/(32.2\ \text{ft sec}^{-2})]^{1/3} = 2.69$ ft.

Classify the water surface profile:

$y_n < y_c \rightarrow$ steep slope; $y_1 > y_c \rightarrow$ S1 profile $\rightarrow y_c < y_2 < y_1$ (i.e., 2.69 ft $< y_2 <$ 4.5 ft).

At section ①:

$$A_1 = (8\ \text{ft}) * (4.5\ \text{ft}) = 36\ \text{ft}^2$$

$$P_{wet,1} = 8\ \text{ft} + 2 * (4.5\ \text{ft}) = 17\ \text{ft}$$

$$R_1 = (36\ \text{ft}^2)/(17\ \text{ft}) = 2.1176\ \text{ft}$$

$$V_1 = (200\ \text{cfs})/(36\ \text{ft}^2) = 5.5556\ \text{ft sec}^{-1}$$

$$E_1 = 4.5\ \text{ft} + (5.5556\ \text{ft sec}^{-1})^2/(2 * 32.2\ \text{ft sec}^{-2}) = 4.9793\ \text{ft}.$$

At section ②:

$$A_2 = 8y_2\,ft^2$$
$$P_{wet,2} = 8 + 2y_2\,ft$$
$$R_2 = (8y_2\,ft^2)/(8 + 2y_2\,ft) = 4y_2/(4 + y_2)\,ft$$
$$V_2 = (200\,cfs)/(8y_2\,ft^2) = 25/y_2\,ft\,sec^{-1}$$
$$E_2 = y_2 + (25/y_2)^2/(2 * 32.2)\,ft = y_2 + 9.7050/{y_2}^2\,ft.$$

For the channel:

$$\overline{R} = \frac{2.1176 + 4y_2/(4 + y_2)}{2}\,ft = 1.0588 + 2y_2/(4 + y_2)\,ft$$
$$\overline{V} = \frac{5.5556 + 25/y_2}{2}\,ft\,sec^{-1} = 2.7778 + 12.5/y_2\,ft\,sec^{-1}.$$

Use the Manning formula (Eq. (2.26)) to compute the EGL slope:

$$2.7778 + 12.5/y_2 = 1.486/0.015 * [1.0588 + 2y_2/(4 + y_2)]^{2/3}{S_f}^{1/2}$$
$$\rightarrow 0.02804 + 0.1262/y_2 = [1.0588 + 2y_2/(4 + y_2)]^{2/3}{S_f}^{1/2}$$
$$\rightarrow S_f = (0.02804 + 0.1262/y_2)^2 * [1.0588 + 2y_2/(4 + y_2)]^{-4/3}$$

(1) Eq. (6.4): $\Delta x = [4.9793 - (y_2 + 9.7050/{y_2}^2)]/(S_f - 0.005)\,ft = -50\,ft.$
Solve this equation in Excel Solver, as shown in Figure 6.6 $\rightarrow y_2 = 4.26\,ft.$

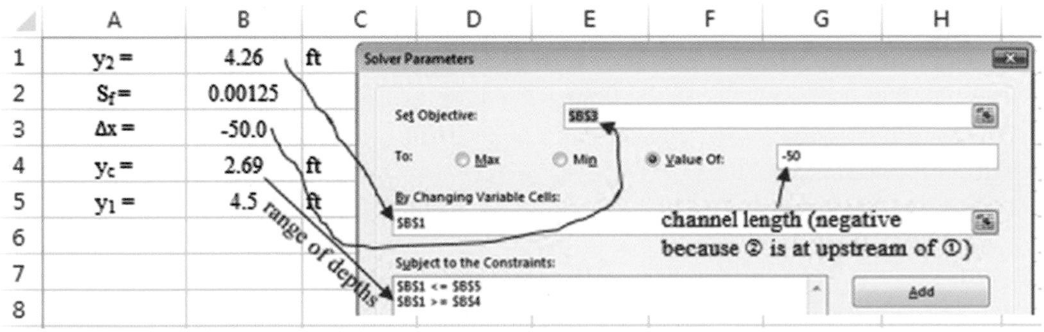

Figure 6.6 **Excerpt of the Excel Solver setup used in part (1) of Example 6.15.**

(2) $F_r = \frac{2.7778 + 12.5/y_2}{\sqrt{32.2 * \frac{4.5 + y_2}{2}}} = \frac{2.7778 + 12.5/y_2}{16.1 * (4.5 + y_2)} \rightarrow F_r^2 = \frac{(2.7778 + 12.5/y_2)^2}{16.1 * (4.5 + y_2)}$

Eq. (6.5): $\Delta x = \frac{(4.5 - y_2)\left[1 - \frac{(2.7778 + 12.5/y_2)^2}{16.1 * (4.5 + y_2)}\right]}{S_f - 0.005}\,ft = -50\,ft.$

Solve this equation in Excel Solver, as shown in Figure 6.7 $\rightarrow y_2 = 4.26\,ft.$

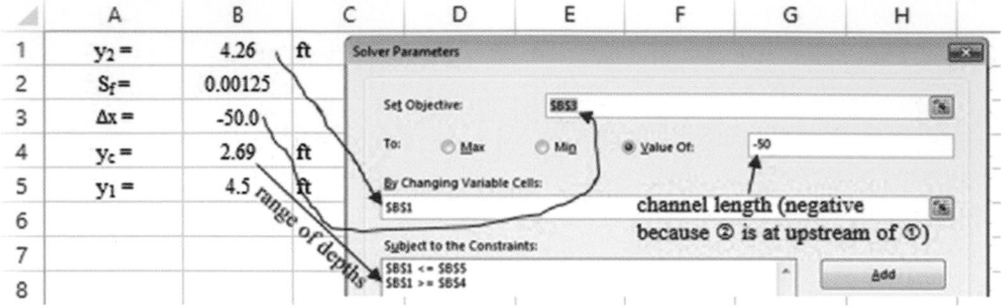

Figure 6.7 **Excerpt of the Excel Solver setup used in part (2) of Example 6.15.**

Example 6.16 For the channel in Example 6.15, use the multi-step method to compute the water surface profile by Eq. (6.4).

Solution

$b = 8\,\text{ft}$, $n = 0.015$, $S_0 = 0.005$, $\Delta x = 50\,\text{ft}$, $Q = 200\,\text{cfs}$

To avoid trial-and-error calculations at each step, for step i, let $y_{1,i} = y_{2,i-1}$, assume a value of $y_{2,i}$ between 2.69 and 4.5 ft, and then use Eq. (6.4) to compute Δx_i. Compute the cumulative distance $\Delta x = \sum_{1}^{i} \Delta x_i$ and check if Δx is equal to −50 ft. If $\Delta x < -50\,\text{ft}$, adjust up $y_{2,i}$ until $\Delta x = -50\,\text{ft}$, which can be done in Excel Solver. Otherwise, let $i = i + 1$ and repeat the above computations. The computations and results are shown in Figure 6.8.

For step i = 1, let $y_{1,1} = 4.5\,\text{ft}$ and assume $y_{2,1} = 4.4\,\text{ft}$.

$$A_1 = (8\,\text{ft}) * (4.5\,\text{ft}) = 36\,\text{ft}^2$$

$$P_{wet,1} = 8\,\text{ft} + 2 * (4.5\,\text{ft}) = 17\,\text{ft}$$

$$R_1 = (36\,\text{ft}^2)/(17\,\text{ft}) = 2.1176\,\text{ft}$$

$$V_1 = (200\,\text{cfs})/(36\,\text{ft}^2) = 5.5556\,\text{ft sec}^{-1}$$

$$E_1 = 4.5\,\text{ft} + (5.5556\,\text{ft sec}^{-1})^2/(2 * 32.2\,\text{ft sec}^{-2}) = 4.9793\,\text{ft}$$

$$A_2 = (8\,\text{ft}) * (4.4\,\text{ft}) = 35.2\,\text{ft}^2$$

$$P_{wet,2} = 8\,\text{ft} + 2 * (4.4\,\text{ft}) = 16.8\,\text{ft}$$

$$R_2 = (35.2\,\text{ft}^2)/(16.8\,\text{ft}) = 2.0952\,\text{ft}$$

$$V_2 = (200\,\text{cfs})/(35.2\,\text{ft}^2) = 5.6818\,\text{ft sec}^{-1}$$

$$E_2 = 4.4\,\text{ft} + (5.6818\,\text{ft sec}^{-1})^2/(2 * 32.2) = 4.9013\,\text{ft}$$

$$\overline{R} = \frac{2.1176 + 2.0952}{2}\,\text{ft} = 2.1064\,\text{ft}$$

$$\overline{V} = \frac{5.5556 + 5.6818}{2}\ \text{ft sec}^{-1} = 5.6187\ \text{ft sec}^{-1}.$$

Use the Manning formula (Eq. (2.26)) to compute the EGL slope:

$$5.6187\ \text{ft sec}^{-1} = 1.486/0.015 * (2.1064\ \text{ft})^{2/3} S_f^{1/2} \rightarrow S_f^{1/2} = 0.03452 \rightarrow S_f = 0.001191$$

Eq. (6.4): $\Delta x_1 = (4.9793 - 4.9013)/(0.001191 - 0.005)\ \text{ft} = -20.5\ \text{ft}$.
$\Delta x = 0 + (-20.5\ \text{ft}) = -20.5\ \text{ft} > -50\ \text{ft} \rightarrow \text{let } i = 1 + 1 = 2$.

For step $i = 2$, let $y_{1,2} = y_{2,1} = 4.4$ ft and assume $y_{2,2} = 4.3$ ft.
Repeating the similar computations with step $i = 1$, one can get $\Delta x_2 = -20.5$ ft.

$$\Delta x = (-20.5\ \text{ft}) + (-20.5\ \text{ft}) = -41.0\ \text{ft} > -50\ \text{ft},\ \text{let } i = 2 + 1 = 3.$$

For step $i = 3$, let $y_{1,3} = y_{2,2} = 4.3$ ft and assume $y_{2,3} = 4.2$ ft.
Repeating the similar computations with step $i = 2$, one can get $\Delta x_3 = -20.4$ ft.
$\Delta x = (-41.0\ \text{ft}) + (-20.4\ \text{ft}) = -61.4\ \text{ft} < -50\ \text{ft} \rightarrow$ adjust $y_{2,3}$ until $\Delta x = -50$ ft.
The adjustment is done using Excel Solver as shown in Figure 6.8 $\rightarrow y_{2,3} = 4.256$ ft.

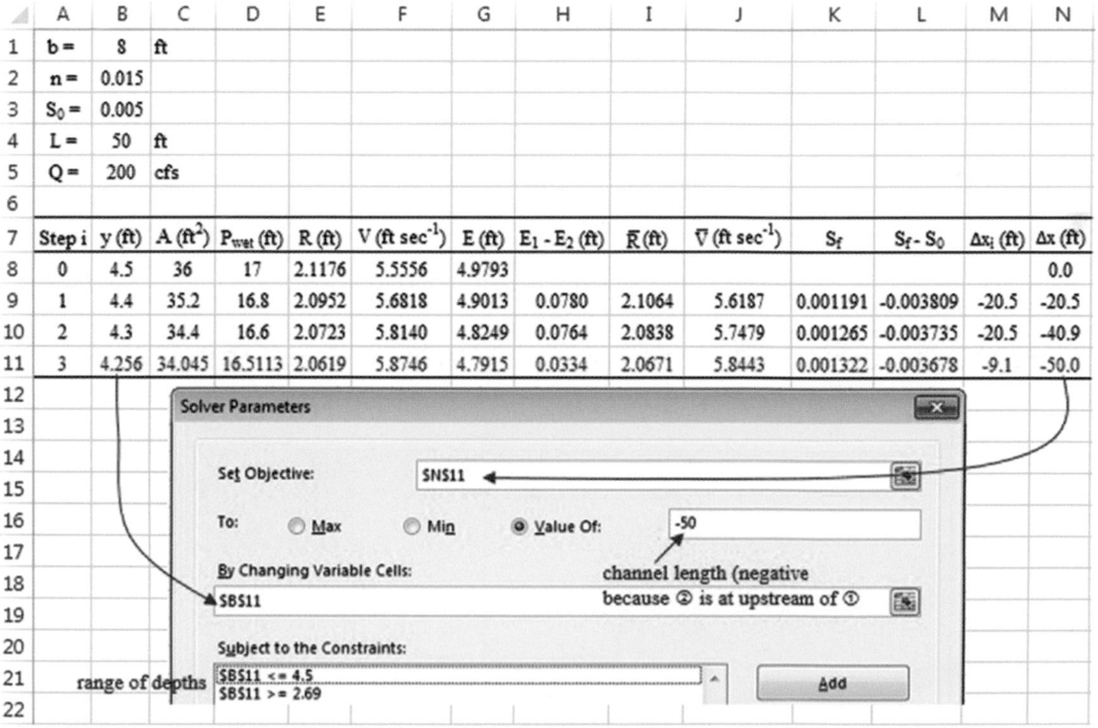

	A	B	C	D	E	F	G	H	I	J	K	L	M	N
1	b =	8	ft											
2	n =	0.015												
3	S_0 =	0.005												
4	L =	50	ft											
5	Q =	200	cfs											
6														
7	Step i	y (ft)	A (ft^2)	P_{wet} (ft)	R (ft)	V (ft sec^{-1})	E (ft)	$E_1 - E_2$ (ft)	$\overline{R}$ (ft)	$\overline{V}$ (ft sec^{-1})	S_f	$S_f - S_0$	Δx_i (ft)	Δx (ft)
8	0	4.5	36	17	2.1176	5.5556	4.9793							0.0
9	1	4.4	35.2	16.8	2.0952	5.6818	4.9013	0.0780	2.1064	5.6187	0.001191	-0.003809	-20.5	-20.5
10	2	4.3	34.4	16.6	2.0723	5.8140	4.8249	0.0764	2.0838	5.7479	0.001265	-0.003735	-20.5	-40.9
11	3	4.256	34.045	16.5113	2.0619	5.8746	4.7915	0.0334	2.0671	5.8443	0.001322	-0.003678	-9.1	-50.0

Figure 6.8 **Excerpt of the Excel spreadsheet for the water surface profile computation using the multi-step method for Example 6.16.**

Example 6.17 Derive:

(1) Eq. (6.2)
(2) Eq. (6.3).

Solution

Referencing Figure 6.1, for channels with a bed slope of $S_0 < 0.2$ (usually true), one has $z_1 - z_2 \approx S_0(\Delta x)$, $h_f \approx S_f(\Delta x)$, $\Delta y = y_2 - y_1$, $\Delta\left(\frac{V^2}{2g}\right) = \frac{V_2^2}{2g} - \frac{{V_1}^2}{2g}$.

The energy equation (Eq. (2.21)) between sections ① and ② can be rewritten as:

$$S_0(\Delta x) = \Delta y + \Delta\left(\frac{V^2}{2g}\right) + S_f(\Delta x).$$

Moving $S_f(\Delta x)$ to the left side and then dividing Δx by the two sides, one can have:

$$S_0 - S_f = \frac{\Delta y}{\Delta x} + \frac{\Delta\left(\frac{V^2}{2g}\right)}{\Delta x}.$$

(1) Combining the two terms on the right side into one term and noting that $E = y + \frac{V^2}{2g}$, one has: $S_0 - S_f = \frac{\Delta E}{\Delta x}$, which is Eq. (6.2) when $\Delta x \to 0$.

(2) $V = Q/A \to \frac{V^2}{2g} = \frac{(Q/A)^2}{2g} = \frac{Q^2}{2g}A^{-2} \to \frac{d\left(\frac{V^2}{2g}\right)}{dx} = \frac{Q^2}{2g}\left(-2A^{-3}\right)\frac{dA}{dx} = -\frac{Q^2}{gA^3}\frac{dA}{dx}$.
From Figure 6.9: hydraulic depth $d = A/B \to A = Bd$; $dA/dx = [(dy)B]/dx = B(dy/dx)$

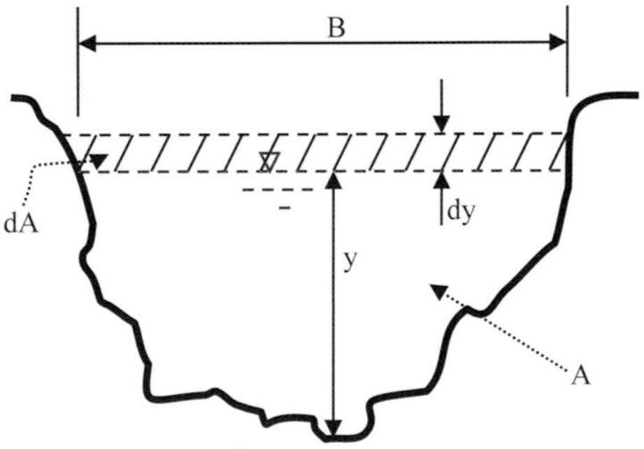

Figure 6.9 **Schematic diagram of a cross section of a generic channel.** Figure shows the geometric relationships for changes of water depth and flow area that are used in part (2) of Example 6.17.

$$\frac{Q^2}{gA^3} = \frac{Q^2}{A^2}\frac{1}{gA} = \frac{V^2}{g\,(Bd))} = \frac{1}{B}\frac{V^2}{gd} = \frac{1}{B}F_r^2 \rightarrow \frac{d\left(\frac{V^2}{2g}\right)}{dx} = -\left(\frac{1}{B}F_r^2\right)\left(B\frac{dy}{dx}\right) = -F_r^2\frac{dy}{dx}.$$

When $\Delta x \rightarrow 0$: $S_0 - S_f = \frac{dy}{dx} + \frac{d\left(\frac{V^2}{2g}\right)}{dx} = \frac{dy}{dx} - F_r^2\frac{dy}{dx} = \left(1 - F_r^2\right)\frac{dy}{dx}.$

Dividing $\left(1 - F_r^2\right)$ by the two sides, one can obtain Eq. (6.3).

6.4 Rapidly Varied Flow

In contrast with gradually varied flow, rapidly varied flow is a type of nonuniform flow, where the water depth varies steeply, or even abruptly, along the flow direction. Mathematically, the first-order derivative of water depth with respect to distance along the flow direction may or may not exist.

6.4.1 Hydraulic Jump

A hydraulic jump is a type of rapidly varied flow, which has currents with an overall forward flow direction and localized backward flow directions (Figure 6.10). It can dissipate a large amount of flow energy. The energy dissipation is mainly caused by collisions of water molecules rather than by friction between flowing water and the surface of the channel conveying it. Thus, the Darcy–Weisbach equation (Eq. 2.23)) and the Manning formula (Eq. (2.26)), which are commonly used to compute friction loss in open channel flow, are not applicable for a hydraulic jump. However, the energy equation (Eq. (2.21)) always holds between the two channel sections within which a hydraulic jump occurs. To determine the energy dissipation caused by the hydraulic jump, one needs to first use a linear momentum equation (Eq. (2.32)) to determine the pre- and post-jump water depths (called conjugate depths) and then solve the energy equation for loss.

For a hydraulic jump to occur between sections ① and ② (Figure 6.10), two conditions must be satisfied. The first condition is that the flow at the pre-jump section (i.e., section ①) must be supercritical, while the flow at the post-jump section (i.e., section ②) must be subcritical. The second condition is that the moment equation holds between these two sections. If either of these two conditions is not satisfied, no hydraulic jump will occur. For channels that satisfy the first condition only, because the moment at section ① is larger than the moment at section ②, ripples moving downward may be formed but no currents (i.e., back flows) can be created, whereas for channels that satisfy the second condition only, the difference of the flow energies at these two sections is not large enough to create currents.

Given that the length of a hydraulic jump is usually short and thus the friction force on the channel surface is negligible, only two hydrostatic forces at sections ① and ② are considered

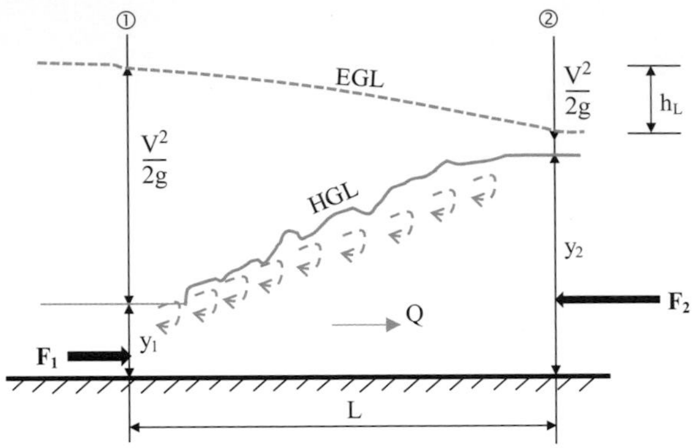

Figure 6.10 Longitudinal diagram of a hydraulic jump. The conjugate water depths are the pre-jump depth, y_1, and the post-jump depth, y_2.

in the moment equation. That is, the conjugate water depths can be determined by solving the following equations:

$$F_1 - F_2 = \rho Q (V_2 - V_1) \tag{6.6}$$

hydrostatic force at section ② ; hydrostatic force at section ① ; flow rate ; mean velocity at section ① ; density of water ; mean velocity at section ②

$$F_1 = \gamma h_{c,1} A_1 \tag{6.7}$$

water depth above centroid of flow area ; flow area at section ① ; hydrostatic force at section ① ; specific weight of water

$$F_2 = \gamma h_{c,2} A_2 \tag{6.8}$$

water depth above centroid of flow area ; flow area at section ② ; hydrostatic force at section ② ; specific weight of water

For rectangular channels with width b, noting that $h_{c,1} = y_1/2$, $h_{c,2} = y_2/2$, $A_1 = by_1$, $A_2 = by_2$, $V_1 = Q/(by_1)$, and $V_2 = Q/(by_2)$, one can rearrange (see Example 6.20) Eqs. (6.6) through (6.8) to derive the following equation:

$$\frac{y_2}{y_1} = \frac{-1 + \sqrt{1 + 8\,(F_{r,1})^2}}{2} \tag{6.9}$$

Froude number at section ① ; post-jump water depth ; pre-jump water depth

The length of hydraulic jump, L, can be computed as (Hager, 1992):

(length of hydraulic jump; y_1: pre-jump water depth; tanh: hyperbolic tangent; $F_{r,1}$: Froude number at section ①)

$$L = 220 y_1 \tanh\left(\frac{F_{r,1} - 1}{22}\right) \tag{6.10}$$

(hyperbolic tangent; χ: variable of $(F_{r,1}-1)/22$)

$$\tanh(\chi) = \frac{e^{\chi} - e^{-\chi}}{e^{\chi} + e^{-\chi}} \tag{6.11}$$

In practice, L can also be estimated as (Roberson *et al.*, 1998):

(length of hydraulic jump; y_2: post-jump water depth)

$$L = (4 \sim 6) y_2 \tag{6.12}$$

Finally, the energy dissipation by hydraulic jump is computed as:

(energy dissipation; z_1: channel bed elevation at section ①; y_1: pre-jump water depth; V_1: mean velocity at section ①; g: gravitational acceleration; z_2: channel bed elevation at section ②; y_2: post-jump water depth; V_2: mean velocity at section ②)

$$h_L = \left(z_1 + y_1 + \frac{V_1^2}{2g}\right) - \left(z_2 + y_2 + \frac{V_2^2}{2g}\right) \tag{6.13}$$

Example 6.18 A 15-ft-wide horizontal rectangular channel carries 450 cfs. If the depth at a section is 2.0 ft, determine the:

(1) Conjugate depth
(2) Length of the hydraulic jump
(3) Energy dissipation by the hydraulic jump.

Solution

$b = 15$ ft, $Q = 450$ cfs, $y_1 = 2.0$ ft, rectangular channel

Unit-width flow rate: $q = (450\,\text{cfs})/(15\,\text{ft}) = 30\,\text{ft}^2\,\text{sec}^{-1}$.

Eq. (2.38): $y_c = [(30\,\text{ft}^2\,\text{sec}^{-1})^2/[2 * 32.2\,\text{ft sec}^{-2}) * (2.0\,\text{ft})]^{1/2} = 2.41$ ft.

$y_1 < y_c \rightarrow$ supercritical flow at section ① $\rightarrow$ the conjugate depth $y_2 > y_c = 2.41$ ft.

(1) $V_1 = q/y_1 = (30\,\text{ft}^2\,\text{sec}^{-1})/(2.0\,\text{ft}) = 15\,\text{ft sec}^{-1}$
Eq. (2.46): $F_{r,1} = (15\,\text{ft sec}^{-1})/[(32.2\,\text{ft sec}^{-2}) * (2.0\,\text{ft})]^{1/2} = 1.87$.
Eq. (6.9): $y_2/y_1 = [-1 + (1 + 8 * 1.87^2)^{1/2}]/2 = 2.19 \rightarrow y_2 = 2.19 * (2.0\,\text{ft}) = 4.38$ ft.

(2) $\chi = (1.87 - 1)/22 = 0.03955$

Eq. (6.11): $\tanh(\chi) = (e^{0.03955} - e^{-0.03955})/(e^{0.03955} + e^{-0.03955}) = 0.03953$.
Eq. (6.10): $L = 220 * (2.0\,ft) * 0.03953 = 17.4\,ft$.

(3) $V_2 = q/y_2 = (30\,ft^2\,sec^{-1})/(4.38\,ft) = 6.85\,ft\,sec^{-1}$
Horizontal channel $\rightarrow z_1 = z_2$.
Eq. (6.13): $h_L = [2.0\,ft + (15\,ft\,sec^{-1})^2/(2 * 32.2\,ft\,sec^{-2})] - [4.38\,ft + (6.85\,ft\,sec^{-1})^2/(2 * 32.2\,ft\,sec^{-2})] = 0.39\,ft$.

Example 6.19 A horizontal triangular channel with a side slope of 1.75 carries 30 cfs. If the depth at a section is 1.2 ft, determine the:

(1) Conjugate depth
(2) Length of the hydraulic jump
(3) Energy dissipation by the hydraulic jump.

Solution

$Z = 1.75$, $Q = 30\,cfs$, $y_1 = 1.2\,ft$, triangular channel.

$$\text{At the critical depth } y_c: \text{ top width } B_c = 2 * (1.75)y_c = 3.5y_c\,ft$$
$$\text{flow area } A_c = 1/2 * (3.5y_c)y_c = 1.75{y_c}^2\,ft^2.$$

Eq. (2.37): $(30\,cfs)^2/(32.2\,ft\,sec^{-2}) = (1.75{y_c}^2\,ft^2)^3/(3.5y_c\,ft) = 1.53125{y_c}^5$

$$\rightarrow {y_c}^5 = 18.2533 \rightarrow y_c = 1.79\,ft.$$

$y_1 < y_c \rightarrow$ supercritical flow at section ① $\rightarrow$ the conjugate depth $y_2 > y_c = 1.79\,ft$.

(1) $A_1 = 1/2 * [2 * (1.75 * 1.2\,ft)] * (1.2\,ft) = 2.52\,ft^2$
$V_1 = (30\,cfs)/(2.52\,ft^2) = 11.91\,ft\,sec^{-1}$
$h_{c,1} = 1/3 * (1.2\,ft) = 0.4\,ft$.
Eq. (6.7): $F_1 = (62.4\,lbf\,ft^{-3}) * (0.4\,ft) * (2.52\,ft^2) = 62.8992\,lbf$.
$A_2 = 1/2 * [2 * (1.75y_2)]y_2 = 1.75{y_2}^2\,ft^2$
$V_2 = (30\,cfs)/(1.75{y_2}^2\,ft^2) = 17.1429/{y_2}^2\,ft\,sec^{-1}$
$h_{c,2} = 1/3y_2 ft$.
Eq. (6.8): $F_2 = (62.4\,lbf\,ft^{-3}) * (1/3y_2 ft) * (1.75{y_2}^2\,ft^2) = 36.4{y_2}^3\,lbf$.
Eq. (6.6): $(62.8992 - 36.4{y_2}^3)\,lbf = (1.94\,slug\,ft^{-3}) * (30\,cfs) * (17.1429/{y_2}^2 - 11.91)\,ft\,sec^{-1} \rightarrow 36.4{y_2}^3 + 997.7168/{y_2}^2 - 756.0612 = 0$

$$\rightarrow \text{trial-and-error or Excel Solver}: y_2 = 2.55\,ft.$$

(2) $B_1 = 2 * (1.75) * (1.2\,ft) = 4.2\,ft$
$d_1 = (2.52\,ft^2)/(4.2\,ft) = 0.6\,ft$
$F_{r,1} = (11.91\,ft\,sec^{-1})/[(32.2\,ft\,sec^{-2}) * (0.6\,ft)]^{1/2} = 2.71$
$\chi = (2.71 - 1)/22 = 0.07773$.

Eq. (6.11): $\tanh(\chi) = (e^{0.07773} - e^{-0.07773})/(e^{0.07773} + e^{-0.07773}) = 0.07757$.
Eq. (6.10): $L = 220 * (1.2\,ft) * 0.07757 = 20.5\,ft$.

(3) $V_2 = 17.1429/2.55^2\,ft\,sec^{-1} = 2.64\,ft\,sec^{-1}$
Horizontal channel $\rightarrow z_1 = z_2$.
Eq. (6.13): $h_L = [1.2\,ft + (11.91\,ft\,sec^{-1})^2/(2 * 32.2\,ft\,sec^{-2})] - [2.55\,ft + (2.64\,ft\,sec^{-1})^2/(2 * 32.2\,ft\,sec^{-2})] = 0.74\,ft$.

Example 6.20 Derive Eq. (6.9).

Solution

Rectangular channel with width b

At section ①: $A_1 = by_1$

$$V_1 = Q/(by_1)$$
$$h_{c,1} = y_1/2$$
$$F_1 = \gamma(y_1/2)(by_1) = (\gamma b y_1^2)/2.$$

At section ②: $A_2 = by_2$

$$V_2 = Q/(by_2)$$
$$h_{c,2} = y_2/2$$
$$F_2 = \gamma(y_2/2)(by_2) = (\gamma b y_2^2)/2.$$

Eq. (6.6): $(\gamma b y_1^2)/2 - (\gamma b y_2^2)/2 = \rho Q[Q/(by_2) - Q/(by_1)]$

$$\rightarrow(\gamma/\rho)(y_1^2 - y_2^2) = 2[Q^2/(by_1)^2](y_1)^2(1/y_2 - 1/y_1) = 2V_1^2(y_1^2)(y_1 - y_2)/(y_1 y_2)$$
$$\rightarrow g(y_1 - y_2)(y_1 + y_2) = 2V_1^2(y_1/y_2)(y_1 - y_2)$$
$$\rightarrow(y_1 + y_2)y_2 = 2[V_1^2/(gy_1)]y_1^2 = 2y_1^2F_{r,1}^2$$
$$\rightarrow y_2^2 + y_1 y_2 - 2y_1^2F_{r,1}^2 = 0$$
$$\rightarrow y_2 = \frac{-y_1 + \sqrt{y_1^2 - 4 * 1 * \left(-2y_1^2F_{r,1}^2\right)}}{2 * 1} = \frac{-y_1 + y_1\sqrt{1 + 8F_{r,1}^2}}{2} = \frac{-1 + \sqrt{1 + 8F_{r,1}^2}}{2}y_1$$

$\rightarrow$dividing y_1 by the two sides, one can get Eq. (6.9): $\dfrac{y_2}{y_1} = \dfrac{-1 + \sqrt{1 + 8F_{r,1}^2}}{2}$.

6.4.2 Choking

Choking occurs as a result of an obstruction in an open channel that restricts the subcritical or supercritical flow and leads to a critical flow regime (Sturm, 2010). That is, at the choking section, the water depth is forced to become the corresponding critical water depth. Contraction

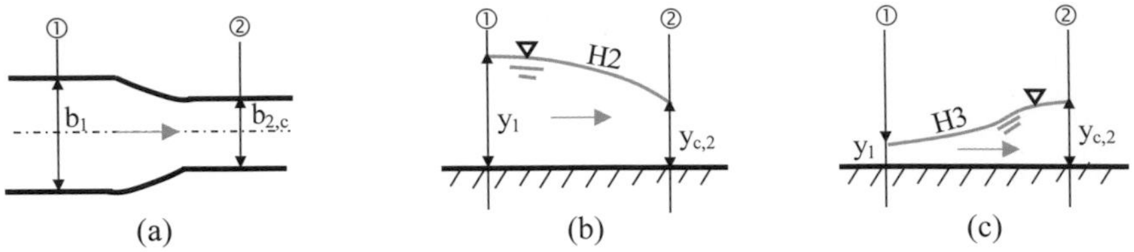

Figure 6.11 Choking in a channel due to channel contraction. (a) Map or bird's-eye view; (b) side view with approaching subcritical flow; and (c) side view with approaching supercritical flow.

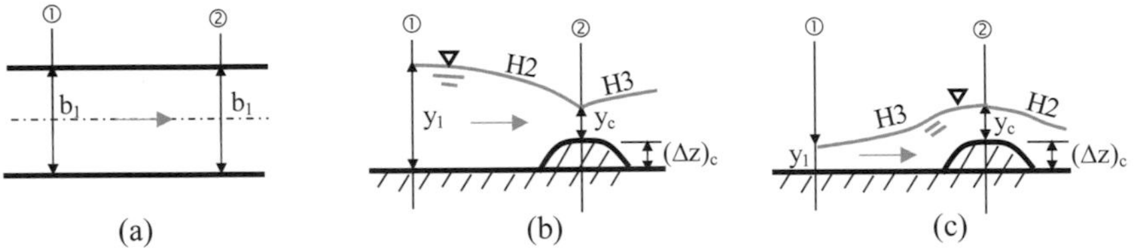

Figure 6.12 Choking in a channel due to gradual channel-bed rise. (a) Map or bird's-eye view; (b) side view with approaching subcritical flow; and (c) side view with approaching supercritical flow.

(Figure 6.11) and gradual bed rise (Figure 6.12) are two typical obstructions, as illustrated for a generic horizontal rectangular channel. Neglecting any energy loss, one can analyze choking by applying the energy equation (Eq. (2.21)) between sections ① and ②. Let $b_{2,c}$ be the channel width at section ② and $(\Delta z)_c$ be the height of the gradual bed rise when choking just occurs. If $b_2 > b_{2,c}$ or $\Delta z < (\Delta z)_c$, the approaching flow will not be retarded, with a water depth at section ②, y_2, smaller than the depth at section ①, y_1; otherwise, the approaching flow will be retarded until the flow has sufficient energy to be able to pass through the contraction or bed rise, choking at section ② (i.e., y_2 becomes the corresponding critical depth). In terms of specific energy curve for constant flow rate as shown in Figure 2.17, when the approaching flow is subcritical, the energy increase is achieved by increasing y_1, whereas when the approaching flow is supercritical, the energy increase is achieved by decreasing y_1.

Example 6.21 The rectangular channel in Figure 6.11 conveys an approaching water flow with a depth of 2.2 m and a velocity of 2.5 m s^{-1}. If $b_1 = 5$ m, determine $b_{2,c}$ when choking just occurs.

Solution

$y_1 = 2.2\,\text{m}$, $V_1 = 2.5\,\text{m}\,\text{s}^{-1}$, $b_1 = 5\,\text{m}$, $b_{2,c} = ?$

Total flow rate: $Q = (2.5\,\text{m}\,\text{s}^{-1}) * [(5\,\text{m}) * (2.2\,\text{m})] = 27.5\,\text{m}^3\,\text{s}^{-1}$.

Choking → critical depth at section ②: $y_2 = y_{c,2} = \left(\frac{(27.5/b_{2,c})^2}{9.81}\right)^{\frac{1}{3}}\,\text{m} = 4.256 b_{2,c}^{-\frac{2}{3}}\,\text{m}$

→ critical velocity at section ②:

$$V_2 = V_{c,2} = \frac{27.5}{b_{2,c}\left(4.256b_{2,c}^{-\frac{2}{3}}\right)}\,\text{ms}^{-1} = 6.4615b_{2,c}^{-\frac{1}{3}}\,\text{ms}^{-1}.$$

Horizontal channel $\rightarrow z_1 = z_2$.

When chocking just occurs, the energy equation between sections ① and ② with negligible energy loss is:

$$2.2\,\text{m} + \frac{\left(2.5\,\text{ms}^{-1}\right)^2}{2*(9.81\,\text{ms}^{-2})} = 4.256b_{2,c}^{-\frac{2}{3}} + \frac{\left(6.4615b_{2,c}^{-\frac{1}{3}}\,\text{ms}^{-1}\right)^2}{2*(9.81\,\text{ms}^{-2})} \rightarrow b_{2,c} = 4.04\,\text{m}.$$

Example 6.22 The contraction in a rectangular channel conveys an approaching water flow with a depth of 7.5 ft and a velocity of 0.8 ft sec^{-1}. The width at the pre-contraction section (i.e., section ①) is $b_1 = 15$ ft. Determine the water depth at the post-contraction section (i.e., section ②) if its width is:

(1) $b_2 = 1.0$ ft

(2) $b_2 = 2.0$ ft.

Solution

$y_1 = 7.5$ ft, $V_1 = 0.8$ ft sec^{-1}, $b_1 = 15$ ft, $y_2 = ?$

Eq. (2.46): $F_{r,1} = (0.8\,\text{ft sec}^{-1})/[(32.2\,\text{ft sec}^{-2})*(7.5\,\text{ft})]^{1/2} = 0.05 < 1$

$\rightarrow$the approaching flow is subcritical.

Determine the choking width at section ②, $b_{2,c}$, for such an approaching flow:

Total flow rate: $Q = (0.8\,\text{ft sec}^{-1})*[(15\,\text{ft})*(7.5\,\text{ft})] = 90.0$ cfs.

Critical depth at section ②: $y_{c,2} = \left(\frac{(90.0/b_{2,c})^2}{32.2}\right)^{\frac{1}{3}}\,\text{ft} = 6.3126b_{2,c}^{-\frac{2}{3}}\,\text{ft}.$

Critical velocity at section ②: $V_{c,2} = \frac{90.0}{b_{2,c}\left(6.3126b_{2,c}^{-\frac{2}{3}}\right)}\,\text{ft sec}^{-1} = 14.2572b_{2,c}^{-\frac{1}{3}}\,\text{ft sec}^{-1}.$

Horizontal channel $\rightarrow z_1 = z_2$.

Under choking conditions, the energy equation between sections ① and ② with negligible energy loss is:

$$7.5\text{ft} + \frac{\left(0.8\,\text{ft sec}^{-1}\right)^2}{2*(32.2\,\text{ft sec}^{-2})} = 6.3126b_{2,c}^{-\frac{2}{3}} + \frac{\left(14.2572b_{2,c}^{-\frac{1}{3}}\,\text{ft sec}^{-1}\right)^2}{2*(32.2\,\text{ft sec}^{-2})} \rightarrow b_{2,c} = 1.42\,\text{ft}.$$

(1) $b_2 = 1.0\,\text{ft} < b_{2,c} = 1.42\,\text{ft} \rightarrow$ the approaching flow is retarded, $y_2 = y_{c,2} = \left[\dfrac{(90.0\,\text{cfs}/1.0\,\text{ft})^2}{32.2\,\text{ft sec}^{-2}}\right]^{\frac{1}{3}} = 6.31\,\text{ft}$; $V_{c,2} = (90.0\,\text{cfs})/[(1.0\,\text{ft}) * (6.31\,\text{ft})] = 14.26\,\text{ft sec}^{-1}$.

Approaching flow is subcritical $\rightarrow y_1$ needs to be increased.

At the new value of y_1: $V_1 = (90.0\,\text{ft}^3\text{sec}^{-1})/[(15\,\text{ft})y_1] = 6.0/y_1\,\text{ft sec}^{-1}$.

The energy equation between sections ① and ② with negligible energy loss is:

$$y_1 + \frac{(6.0/y_1\,\text{ft sec}^{-1})^2}{2 * (32.2\,\text{ft sec}^{-2})} = 6.31\,\text{ft} + \frac{(14.26\,\text{ft sec}^{-1})^2}{2 * (32.2\,\text{ft sec}^{-2})} \rightarrow y_1^3 - 9.4676 y_1^2 + 0.559 = 0$$

$\rightarrow$ trial-and-error or Excel Solver: $y_1 = 9.46\,\text{ft}$

$\rightarrow V_1 = 6.0/9.46\,\text{ft sec}^{-1} = 0.63\,\text{ft sec}^{-1}$.

That is, if $b_2 = 1.0\,\text{ft}$, the approaching flow will become deeper with a lower velocity.

(2) $b_2 = 2.0\,\text{ft} > b_{2,c} = 1.42\,\text{ft} \rightarrow$ the approaching flow is not retarded, $V_2 = (90.0\,\text{cfs})/[(2.0\,\text{ft})y_2] = 45.0/y_2\,\text{ft sec}^{-1}$.

The energy equation between sections ① and ② with negligible energy loss is:

$$7.5\,\text{ft} + \frac{(0.8\,\text{ft sec}^{-1})^2}{2 * (32.2\,\text{ft sec}^{-2})} = y_2 + \frac{(45.0/y_2\,\text{ft sec}^{-1})^2}{2 * (32.2\,\text{ft sec}^{-2})} \rightarrow y_2^3 - 7.5099 y_2^2 + 31.4441 = 0$$

$\rightarrow$ trial-and-error or Excel Solver: $y_2 = 2.51\,\text{ft}$

$\rightarrow V_2 = 45.0/2.51\,\text{ft sec}^{-1} = 17.93\,\text{ft sec}^{-1}$.

That is, if $b_2 = 2.0\,\text{ft}$, the flow at section ② has a depth of $y_2 = 2.51\,\text{ft}$.

Example 6.23 The contraction in a rectangular channel conveys an approaching water flow with a depth of 0.5 ft and a velocity of 12.0 ft sec^{-1}. If the width at the pre-contraction section (i.e., section ①) is $b_1 = 15\,\text{ft}$, determine the water depth at the post-contraction section (i.e., section ②) if its width is $b_2 = 3.5\,\text{ft}$.

Solution

$y_1 = 0.5\,\text{ft}$, $V_1 = 12.0\,\text{ft sec}^{-1}$, $b_1 = 15\,\text{ft}$, $b_2 = 3.5\,\text{ft}$, $y_2 = ?$

Eq. (2.46): $F_{r,1} = (12.0\,\text{ft sec}^{-1})/[(32.2\,\text{ft sec}^{-2}) * (0.5\,\text{ft})]^{1/2} = 3.0 > 1$

$\rightarrow$the approaching flow is supercritical.

Determine the choking width at section ②, $b_{2,c}$, for such an approaching flow:

Total flow rate: $Q = (12.0\,\text{ft sec}^{-1}) * [(15\,\text{ft}) * (0.5\,\text{ft})] = 90.0\,\text{cfs}$.

Critical depth at section ②: $y_{c,2} = \left(\dfrac{(90.0/b_{2,c})^2}{32.2}\right)^{\frac{1}{3}}\,\text{ft} = 6.3126 b_{2,c}^{-\frac{2}{3}}\,\text{ft}$.

Critical velocity at section ②: $V_{c,2} = \dfrac{90.0}{b_{2,c}\left(6.3126 b_{2,c}^{-\frac{2}{3}}\right)}\,\text{ft sec}^{-1} = 14.2572 b_{2,c}^{-\frac{1}{3}}\,\text{ft sec}^{-1}$.

Horizontal channel $\rightarrow z_1 = z_2$.

Under choking conditions, the energy equation between sections ① and ② with negligible energy loss is:

$$0.5\,\text{ft} + \frac{(12.0\,\text{ft sec}^{-1})^2}{2*(32.2\,\text{ft sec}^{-2})} = 6.3126 b_{2,c}^{-\frac{2}{3}} + \frac{\left(14.2572 b_{2,c}^{-\frac{1}{3}}\,\text{ft sec}^{-1}\right)^2}{2*(32.2\,\text{ft sec}^{-2})} \to b_{2,c} = 6.44\,\text{ft}.$$

<u>Determine y_2 for $b_2 = 3.5$ ft:</u>

$b_2 = 3.5\,\text{ft} < b_{2,c} = 6.44\,\text{ft} \to$ the approaching flow is retarded,

$$y_2 = y_{c,2} = \left[\frac{(90.0\,\text{cfs}/3.5\,\text{ft})^2}{32.2\,\text{ft sec}^{-2}}\right]^{\frac{1}{3}} = 2.74\,\text{ft};\ V_{c,2} = (90.0\,\text{cfs})/[(3.5\,\text{ft})*(2.74\,\text{ft})] = 9.38\,\text{ft sec}^{-1}.$$

Approaching flow is supercritical $\to y_1$ needs to be decreased.

At the new value of y_1: $V_1 = (90.0\,\text{ft}^3\,\text{sec}^{-1})/[(15\,\text{ft})y_1] = 6.0/y_1\,\text{ft sec}^{-1}$.

The energy equation between sections ① and ② with negligible energy loss is:

$$y_1 + \frac{(6.0/y_1\,\text{ft sec}^{-1})^2}{2*(32.2\,\text{ft sec}^{-2})} = 2.74\,\text{ft} + \frac{(9.38\,\text{ft sec}^{-1})^2}{2*(32.2\,\text{ft sec}^{-2})} \to y_1^3 - 4.1062 y_1^2 + 0.5590 = 0$$

$\to$trial-and-error or Excel Solver : $y_1 = 0.39$ ft

$\to V_1 = 6.0/0.39\,\text{ft sec}^{-1} = 15.38\,\text{ft sec}^{-1}$.

That is, if $b_2 = 3.5$ ft, the approaching flow will become shallower with a higher velocity.

Example 6.24 The rectangular channel in Figure 6.12 conveys an approaching water flow with a depth of 2.2 m and a velocity of $2.5\,\text{m s}^{-1}$. Determine $(\Delta z)_c$, at which choking just occurs.

Solution

$y_1 = 2.2\,\text{m}$, $V_1 = 2.5\,\text{m s}^{-1}$, $(\Delta z)_c = ?$

Unit-width flow rate: $q = (2.5\,\text{m s}^{-1})*(2.2\,\text{m}) = 5.5\,\text{m}^2\text{s}^{-1}$.

Critical depth: $y_c = [(5.5\,\text{m}^2\text{s}^{-1})^2/(9.81\,\text{m s}^{-2})]^{1/3} = 1.46\,\text{m}$.

Critical velocity: $V_c = (5.5\,\text{m}^2\text{s}^{-1})/(1.46\,\text{m}) = 3.77\,\text{m s}^{-1}$.

At section ②: $y_2 = y_c = 1.46\,\text{m}$, $V_2 = V_c = 3.77\,\text{m s}^{-1}$.

The energy equation between sections ① and ② with negligible energy loss is:

$$2.2\,\text{m} + (2.5\,\text{m s}^{-1})^2/[2*(9.81\,\text{m s}^{-2})] = (\Delta z)_c + 1.46\,\text{m} + (3.77\,\text{m s}^{-1})^2/[2*(9.81\,\text{m s}^{-2})]$$

$\to (\Delta z)_c = 0.33\,\text{m}$.

Example 6.25 The gradual bed rise in a rectangular channel conveys an approaching water flow with a depth of 6.5 ft and a velocity of 7.5 ft sec^{-1}. Determine the flow depth over the bed rise if its height is:

(1) $\Delta z = 1.0$ ft
(2) $\Delta z = 1.5$ ft.

Solution

$y_1 = 6.5$ ft, $V_1 = 7.5$ ft sec^{-1}, $y_2 = ?$

Eq. (2.46): $F_{r,1} = (7.5\ ft\ sec^{-1})/[(32.2\ ft\ sec^{-2}) * (6.5\ ft)]^{1/2} = 0.52 < 1$

$\rightarrow$the approaching flow is subcritical.

Determine the choking rise height, $(\Delta z)_c$, for such an approaching flow:

Unit-width flow rate: $q = (7.5\ ft\ sec^{-1}) * (6.5\ ft) = 48.75\ ft^2\ sec^{-1}$.

Critical depth: $y_c = [(48.75\ ft^2\ sec^{-1})^2/(32.2\ ft\ sec^{-2})]^{1/3} = 4.19$ ft.

Critical velocity: $V_c = (48.75\ ft^2\ sec^{-1})/(4.19\ ft) = 11.63\ ft\ sec^{-1}$.

At section ②: $y_2 = y_c = 4.19$ ft, $V_2 = V_c = 11.63\ ft\ sec^{-1}$.

Under choking conditions, the energy equation between sections ① and ② with negligible energy loss is:

$6.5\ ft + (7.5\ ft\ sec^{-1})^2/[2 * (32.2\ ft\ sec^{-2})] = (\Delta z)_c + 4.19\ ft + (11.63\ ft\ sec^{-1})^2/[2 * (32.2\ ft\ sec^{-2})] \rightarrow (\Delta z)_c = 1.08$ ft.

(1) $\Delta z = 1.0\ ft < (\Delta z)_c = 1.08\ ft \rightarrow$ the approaching flow is not retarded, $V_2 = 48.75/y_2\ ft\ sec^{-1}$.
The energy equation between sections ① and ② with negligible energy loss is:
$6.5\ ft + (7.5\ ft\ sec^{-1})^2/[2 * (32.2\ ft\ sec^{-2})] = 1.0\ ft + y_2 + (48.75/y_2\ ft\ sec^{-1})^2/[2 * (32.2\ ftsec^{-2})] \rightarrow y_2^3 - 6.3734 y_2^2 + 36.9031 = 0 \rightarrow$ trial-and-error or Excel Solver: $y_2 = 3.75\ ft \rightarrow V_2 = 48.75/3.75\ ft\ sec^{-1} = 13.0\ ft\ sec^{-1}$.
That is, if $\Delta z = 1.0$ ft, the flow at section ② has a depth smaller than $y_1 = 6.5$ ft.

(2) $\Delta z = 1.5\ ft > (\Delta z)_c = 1.08\ ft \rightarrow$ the approaching flow is retarded, $y_2 = y_c = 4.19$ ft, $V_2 = V_c = 11.63\ ft\ sec^{-1}$.
Approaching flow is subcritical $\rightarrow y_1$ needs to be increased.
At the new value of y_1: $V_1 = 48.75/y_1\ ft\ sec^{-1}$.
The energy equation between sections ① and ② with negligible energy loss is:
$y_1 + (48.75/y_1\ ft\ sec^{-1})^2/[2 * (32.2\ ft\ sec^{-2})] = 1.5\ ft + 4.19\ ft + (11.63\ ft\ sec^{-1})^2/[2 * (32.2ftsec^{-2})] \rightarrow y_1^3 - 7.7903 y_1^2 + 36.9031 = 0 \rightarrow$ trial-and-error or Excel Solver: $y_1 = 2.69\ ft \rightarrow V_1 = 48.75/2.69\ ft\ sec^{-1} = 18.12\ ft\ sec^{-1}$.
That is, if $\Delta z = 1.5$ ft, the approaching flow will become shallower and faster.

Example 6.26 The gradual bed rise in a rectangular channel conveys an approaching water flow with a depth of 2.5 ft and a velocity of 19.5 ft sec^{-1}. Determine the flow depth over the bed rise if its height is $\Delta z = 3.5$ ft.

Solution

$y_1 = 2.5$ ft, $V_1 = 19.5$ ft sec^{-1}, $\Delta z = 3.5$ ft, $y_2 = ?$

Eq. (2.46): $F_{r,1} = (19.5\ ft\ sec^{-1})/[(32.2\ ft\ sec^{-2}) * (2.5\ ft)]^{1/2} = 2.17 > 1$

$\rightarrow$the approaching flow is supercritical.

Determine the choking rise height, $(\Delta z)_c$, for such an approaching flow:

Unit-width flow rate: $q = (19.5\ ft\ sec^{-1}) * (2.5\ ft) = 48.75\ ft^2\ sec^{-1}$.

Critical depth: $y_c = [(48.75\ ft^2\ sec^{-1})^2/(32.2\ ft\ sec^{-2})]^{1/3} = 4.19$ ft.

Critical velocity: $V_c = (48.75\ ft^2\ sec^{-1})/(4.19\ ft) = 11.63\ ft\ sec^{-1}$.

At section ②: $y_2 = y_c = 4.19$ ft, $V_2 = V_c = 11.63$ ft sec^{-1}.

Under choking conditions, the energy equation between sections ① and ② with negligible energy loss is:

$2.5\ ft + (19.5\ ft\ sec^{-1})^2/[2 * (32.2\ ft\ sec^{-2})] = (\Delta z)_c + 4.19\ ft + (11.63\ ft\ sec^{-1})^2/[2 * (32.2\ ft\ sec^{-2})] \rightarrow (\Delta z)_c = 2.11$ ft.

Determine y_2 for $\Delta z = 3.5$ ft:

$\Delta z = 3.5\ ft > (\Delta z)_c = 2.11$ ft$\rightarrow$ the approaching flow is retarded, $y_2 = y_c = 4.19$ ft, $V_2 = V_c = 11.63$ ft sec^{-1}.

Approaching flow is supercritical $\rightarrow y_1$ needs to be decreased.

At the new value of y_1: $V_1 = 48.75/y_1$ ft sec^{-1}.

The energy equation between sections ① and ② with negligible energy loss is:

$y_1 + (48.75/y_1\ ft\ sec^{-1})^2/[2 * (32.2\ ft\ sec^{-2})] = 3.5\ ft + 4.19\ ft + (11.63\ ft\ sec^{-1})^2/[2 * (32.2\ ft\ sec^{-2})] \rightarrow y_1^3 - 9.7903 y_1^2 + 36.9031 = 0 \rightarrow$ trial-and-error or Excel Solver: $y_1 = 2.21$ ft$\rightarrow V_1 = 48.75/2.21$ ft $sec^{-1} = 22.06$ ft sec^{-1}.

That is, if $\Delta z = 3.5$ ft, the approaching flow will become shallower and faster.

6.4.3 Supercritical Contraction and Expansion

Supercritical contraction and expansion (Figure 6.13) are common in open channels for water supply and flood diversion. These parts of channels usually have rectangular cross sections that are based on the equations in Figure 6.13 to minimize the transmission of standing waves downstream (Sturm, 2010). The contraction is between sections ② and ③, and the expansion is between sections ③ and ④. Hydraulic jump may occur in the approaching flow between sections ① and ②, followed by a critical flow (i.e., choking) at section ③. From sections ③ to ④, energy loss is negligible. If, on the one hand, no hydraulic jump occurs, the energy loss between sections ① and ② is computed by applying the energy equation (Eq. (2.21)) between sections ① and ②, and in turn

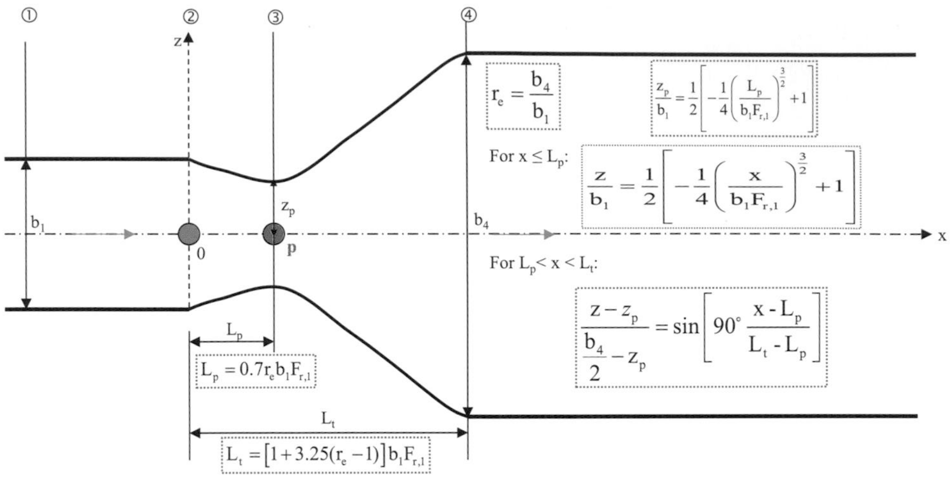

Figure 6.13 **Map view of design dimensions necessary for supercritical contraction and expansion.**

the computed energy loss is used in the energy equation between sections ① and ② to compute the water depth at section ②. On the other hand, if a hydraulic jump occurs, the water depth at section ② is determined by applying the linear momentum equation (Eq. (2.32)) between sections ① and ②, and the computed depth is then used in the energy equation to compute the energy loss between sections ① and ②. Regardless, the water depth at section ④ is computed by applying the energy equation between sections ③ and ④ with no energy loss.

It is obvious that no hydraulic jump occurs if the approaching flow is subcritical. However, when the approaching flow is supercritical, how can we know whether a hydraulic jump will occur? First, compute the critical water depth at section ③, $y_{c,3}$, using Eq. (2.38). At this section, the channel width is equal to $2z_p$. Second, compute the water depth at section ②, y_2, by applying the linear momentum equation between sections ① and ②. Finally, compare y_2 and $y_{c,3}$: if $y_2 \le y_{c,3}$, a hydraulic jump will occur; otherwise, no hydraulic jump occurs.

Example 6.27 A 2-ft-wide rectangular channel is used to supply water to a treatment plant. To minimize the transmission of standing waves, a supercritical contraction followed by a 3.8-ft-wide supercritical expansion is designed as the inlet waterway of the plant. If the approaching flow has a depth of 1.5 ft and a velocity of 10.5 ft sec^{-1}, determine the:

(1) Geometries of the inlet waterway
(2) Energy head loss between sections ① and ②
(3) Water depth at section ④.

Solution

$b_1 = 2\,\text{ft}$, $y_1 = 1.5\,\text{ft}$, $V_1 = 10.5\,\text{ft sec}^{-1}$, $b_4 = 3.8\,\text{ft}$

Eq. (2.46): $F_{r,1} = (10.5\,\text{ft sec}^{-1})/[(32.2\,\text{ft sec}^{-2}) * (1.5\,\text{ft})]^{1/2} = 1.51 > 1 \rightarrow$ supercritical flow.

Total discharge: $Q = (10.5\,\text{ft sec}^{-1}) * [(2\,\text{ft}) * (1.5\,\text{ft})] = 31.5\,\text{cfs}$.

$r_e = (3.8\,\text{ft})/(2\,\text{ft}) = 1.9$.

(1) In reference to Figure 6.13,
$L_p = 0.7 * 1.9 * (2\,\text{ft}) * 1.51 = 4.02\,\text{ft}$
$L_t = [1 + 3.25 * (1.9 - 1)] * (2\,\text{ft}) * 1.51 = 11.85\,\text{ft}$
$z_p/b_1 = 1/2 * [-1/4 * ((4.02\,\text{ft})/(2\,\text{ft} * 1.51))^{3/2} + 1] = 0.31 \rightarrow z_p = 0.31 * (2\,\text{ft}) = 0.62\,\text{ft}$.
For $x < L_p$ (supercritical contraction):

$$z/b_1 = 1/2 * [-1/4 * (x/(2\,\text{ft} * 1.51))^{3/2} + 1] = -0.02382x^{3/2} + 0.5$$
$$\rightarrow z = (-0.02382x^{3/2} + 0.5) * (2\,\text{ft}) = -0.04764x^{3/2} + 1.0\,\text{ft}.$$

For $L_p < x < L_t$ (supercritical expansion):

$$(z - 0.62)/(3.8/2 - 0.62) = \sin[90° * (x - 4.02)/(11.85 - 4.02)]$$
$$\rightarrow z = 0.62 + 1.28 \sin[11.4943°(x - 4.02)]\,\text{ft}.$$

In Figure 6.14, the computed pairs of x, z, and $-z$ are shown in columns A to C, with plotted supercritical contraction and expansion. For instance, for $x = 0.50\,\text{ft}$, $z = -0.04764 * (0.50)^{3/2} + 1.0\,\text{ft} = 0.98\,\text{ft}$, whereas for $x = 4.50\,\text{ft}$, $z = 0.62 + 1.28 \sin[11.4943°(4.50 - 4.02)]\,\text{ft} = 0.74\,\text{ft}$.

(2) The channel width at section ③: $b_3 = 2z_p = 2 * (0.62\,\text{ft}) = 1.24\,\text{ft}$.
Eq. (2.38): $y_{c,3} = [(31.5\,\text{cfs}/1.24\,\text{ft})^2/(32.2\,\text{ft sec}^{-2})]^{1/3} = 2.72\,\text{ft}$

$$\rightarrow V_{c,3} = (31.5\,\text{cfs})/[(1.24\,\text{ft}) * (2.72\,\text{ft})] = 9.34\,\text{ft sec}^{-1}$$

$y_1 = 1.5\,\text{ft} < y_{c,3} = 2.72\,\text{ft} \rightarrow$ hydraulic jump may occur between sections ① and ②.
Eq. (6.9): $y_2/y_1 = 1/2 * [-1 + (1 + 8 * 1.51^2)^{1/2}] = 1.69 \rightarrow y_2 = 1.69 * (1.5\,\text{ft}) = 2.54\,\text{ft} < y_{c,3} = 2.72\,\text{ft}$, hydraulic jump will occur $\rightarrow V_2 = (31.5\,\text{cfs})/[(2\,\text{ft}) * (2.54\,\text{ft})] = 6.20\,\text{ft sec}^{-1}$.
Horizontal channel: $z_1 = z_2$.
The energy equation (Eq. (2.21)) between sections ① and ② is:

$$1.5\,\text{ft} + (10.5\,\text{ft sec}^{-1})^2/(2 * 32.2\,\text{ft sec}^{-2}) = 2.54\,\text{ft} + (6.20\,\text{ft sec}^{-1})^2/(2 * 32.2\,\text{ft sec}^{-2}) + h_L$$

$\rightarrow h_L = 0.08\,\text{ft}$ (energy head loss).

(3) At section ④: $V_4 = (31.5\,\text{cfs})/[(3.8\,\text{ft})y_4] = 8.2895/y_4\,\text{ft sec}^{-1}$.
The energy equation (Eq. (2.21)) between sections ③ and ④ is:

$$2.72\,\text{ft} + (9.34\,\text{ft sec}^{-1})^2/(2 * 32.2\,\text{ft sec}^{-2}) = y_4 + (8.2895/y_4\,\text{ft sec}^{-1})^2/(2 * 32.2\,\text{ft sec}^{-2})$$
$$\rightarrow y_4{}^3 - 4.0746y_4{}^2 + 1.0671 = 0 \rightarrow y_4 = 4.01\,\text{ft}.$$

	A	B	C	D	E	F	G	H	I
1	b_1 =	2	ft		$F_{r,1}$ =	1.51			
2	y_1 =	1.5	ft		r_e =	1.9			
3	V_1 =	10.5	ft sec^{-1}		L_p =	4.02	ft		
4	b_4 =	3.8	ft		L_t =	11.85	ft		
5					z_p =	0.62	ft		
6	x (ft)	z (ft)	-z (ft)						
7	0.00	1.00	-1.00						
8	0.50	0.98	-0.98						
9	1.00	0.95	-0.95						
10	1.50	0.91	-0.91						
11	2.00	0.87	-0.87						
12	2.50	0.81	-0.81						
13	3.00	0.75	-0.75						
14	3.50	0.69	-0.69						
15	4.02	0.62	-0.62						
16	4.50	0.74	-0.74						
17	5.00	0.87	-0.87						
18	5.50	0.99	-0.99						
19	6.00	1.12	-1.12						
20	6.50	1.23	-1.23						
21	7.00	1.34	-1.34						
22	7.50	1.44	-1.44						
23	8.00	1.54	-1.54						
24	8.50	1.62	-1.62						
25	9.00	1.70	-1.70						
26	9.50	1.76	-1.76						
27	10.00	1.81	-1.81						
28	10.50	1.85	-1.85						
29	11.00	1.88	-1.88						
30	11.50	1.90	-1.90						
31	11.85	1.90	-1.90						

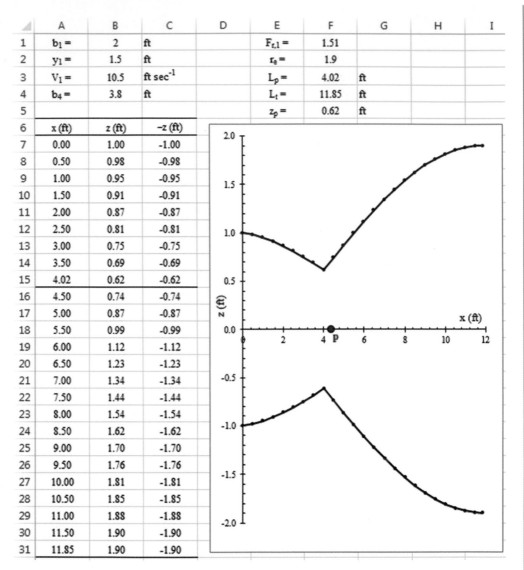

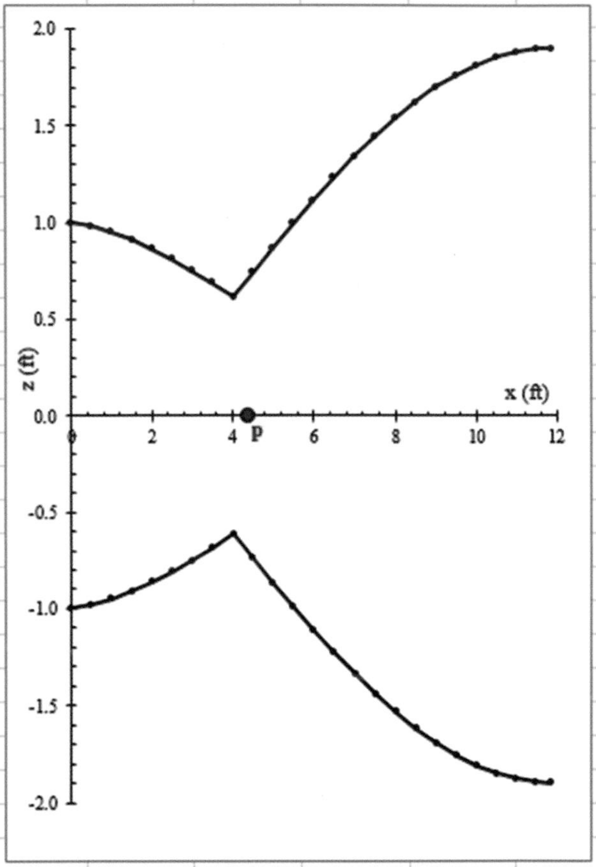

Figure 6.14 **The supercritical contraction and expansion as determined in part (1) of Example 6.27.**

6.5 Flow Measurement

A weir is a barrier built across a river that alters its flow. One purpose of a weir is as an open channel flow-measuring device. When built for this purpose, weirs are constructed to obstruct flows so that the relationship between water depth above a weir and discharge through the weir is unique

(Roberson *et al.*, 1998; Sturm, 2010). In this regard, sharp-crested rectangular, sharp-crested triangular, and broad-crested rectangular weirs as well as long-throated flumes are commonly used. Measured along flow direction, the length of a sharp-crested weir is minimal relative to the water depth above the weir and can be neglected, whereas the length of a broad-crested weir is relatively large and must be incorporated into computations.

6.5.1 Sharp-Crested Rectangular Weir

A sharp-crested rectangular weir can be installed either across the entire or partial width of a channel cross section (Figure 6.15). If the entire cross section is occupied by the weir, the width of the weir is same as that of the cross section (i.e., $b_w = b$) and thus the approaching flow is not contracted. In contrast, if the weir only occupies a portion of the cross section, the width of the weir is less than that of the cross section (i.e., $b_w < b$) and thus the approaching flow is contracted. The weir can be installed from the left or right channel bank or in the middle of the cross section, while the remaining portion of the cross section is encroached so that water can only flow through the weir.

The flow through a sharp-crested rectangular weir has a nappe-shaped HGL (Figure 6.15a). Throughout the water depth above weir crest, H, the point velocity is computed as:

$$\underbrace{v}_{\text{point velocity}} = \sqrt{2\underbrace{g}_{\text{gravitational acceleration}}\underbrace{h}_{\text{energy head above the point}}} \tag{6.14}$$

Integrating the point flow rate, vdA, throughout H and considering head losses due to contractions, one can derive the weir discharge equation expressed as (Kindsvater and Carter, 1959):

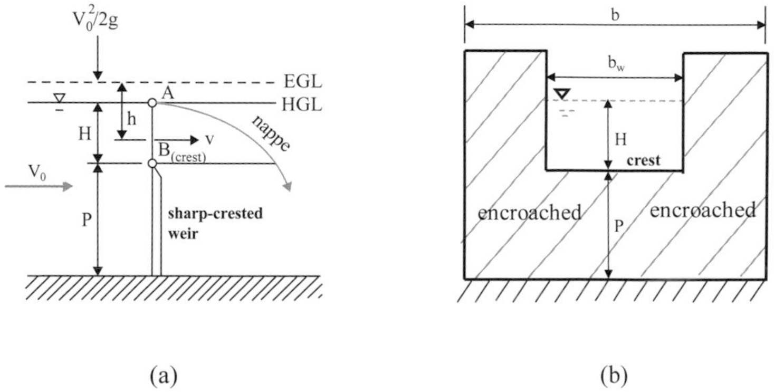

Figure 6.15 **Cross-sectional views of a sharp-crested rectangular weir.** (a) Longitudinal cross section; and (b) transverse cross section.

$$\text{weir discharge} \longrightarrow Q = \underset{\text{weir coefficient (Eq. (6.16))}}{k}\sqrt{\underset{\text{gravitational acceleration}}{2g}}\,(\underset{\text{weir width}}{b_w} - 0.1\underset{\text{number of encroached sides (= 0, 1, and 2 for no, one-side, and two-side encroachment, respectively}}{N}\underset{\text{approaching depth above weir crest}}{H})\,H^{\frac{3}{2}} \tag{6.15}$$

Finally, for $\frac{H}{P} \leq 10$, which is usually the case, the weir coefficient can be computed as:

$$\text{weir coefficient} \longrightarrow k = 0.40 + 0.05\frac{\overset{\text{approaching depth above weir crest}}{H}}{\underset{\text{weir height}}{P}} \tag{6.16}$$

Example 6.28 A 2-ft-high, 3.5-ft-wide sharp-crested rectangular weir is used to measure the discharge in a rectangular channel. If the approaching depth above the weir crest is 8.5 ft, determine the discharge through the weir if it has:

(1) No encroachment
(2) One-side encroachment
(3) Two-side encroachments.

Solution

$P = 2\,\text{ft}$, $b_w = 3.5\,\text{ft}$, $H = 8.5\,\text{ft}$

Eq. (6.16): $k = 0.40 + 0.05 * (8.5\,\text{ft})/(2\,\text{ft}) = 0.6125$

(1) No encroachment $\rightarrow N = 0$.

$$\text{Eq. (6.15): } Q = 0.6125 * (2 * 32.2\,\text{ft sec}^{-2})^{1/2} * [(3.5\,\text{ft}) - 0.1 * 0 * (8.5\,\text{ft})] * (8.5\,\text{ft})^{3/2} = 426.3\,\text{cfs}.$$

(2) One-side encroachment $\rightarrow N = 1$.

$$\text{Eq. (6.15): } mathrmQ = 0.6125 * (2 * 32.2\,\text{ft sec}^{-2})^{1/2} * [(3.5\,\text{ft}) - 0.1 * 1 * (8.5\,\text{ft})] * (8.5\,\text{ft})^{3/2} = 322.8\,\text{cfs}.$$

(3) Two-side encroachments $\rightarrow N = 2$.

$$\text{Eq. (6.15): } Q = 0.6125 * (2 * 32.2\,\text{ft sec}^{-2})^{1/2} * [(3.5\,\text{ft}) - 0.1 * 2 * (8.5\,\text{ft})] * (8.5\,\text{ft})^{3/2} = 219.3\,\text{cfs}.$$

In summary, for a given approaching depth, a sharp-crested rectangle weir with two-side encroachments has a smaller discharge capacity than that with one-side encroachment, which in turn has a smaller discharge capacity than that with no encroachment. This is because a two-side encroachment can cause greater flow contractions and thus larger head losses than a one-side encroachment. Flow is not contracted for scenarios with no encroachment.

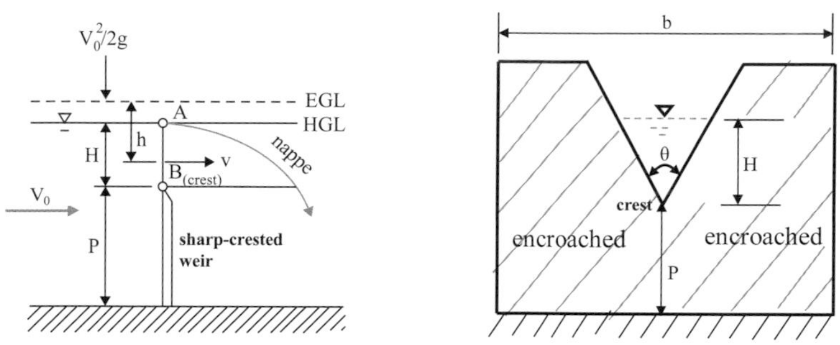

(a) (b)

Figure 6.16 **Cross-sectional views of a sharp-crested triangular weir.** (a) Longitudinal cross section; and (b) transverse cross section.

6.5.2 Sharp-Crested Triangular Weir

A sharp-crested triangular weir is also called a V-notch. As illustrated in Figure 6.16, throughout the water depth above weir crest, H, the point velocity is computed by Eq. (6.14). Integrating the point flow rate, vdA, throughout H and considering head losses due to contractions, one can derive the weir discharge equation expressed as (Kindsvater and Carter, 1959):

$$Q = \frac{8}{15} k \sqrt{2g} \tan(\theta)\, H^{\frac{5}{2}} \tag{6.17}$$

where Q = weir discharge; k = weir coefficient (= 0.58 ~ 0.60); g = gravitational acceleration; θ = weir angle; H = approaching depth above weir crest.

Example 6.29 A 2-ft-high sharp-crested triangular weir is used to measure the discharge in a rectangular channel. If the approaching depth above the weir crest is 8.5 ft, determine the discharge through the weir if it has a weir angle of:

(1) $\theta = 30°$
(2) $\theta = 60°$
(3) $\theta = 85°$.

Solution

$P = 2\,\text{ft}$, $b = 10\,\text{ft}$, $H = 8.5\,\text{ft}$, assume a weir coefficient of $k = 0.59$

(1) $\theta = 30°$

Eq. (6.17): $Q = 8/15 * 0.59 * (2 * 32.2\,\text{ft sec}^{-2})^{1/2} * \tan(30°) * (8.5\,\text{ft})^{5/2} = 307.1\,\text{cfs}$.

(2) $\theta = 60°$

Eq. (6.17): $Q = 8/15 * 0.59 * (2 * 32.2\,\text{ft sec}^{-2})^{1/2} * \tan(60°) * (8.5\,\text{ft})^{5/2} = 921.3\,\text{cfs}$.

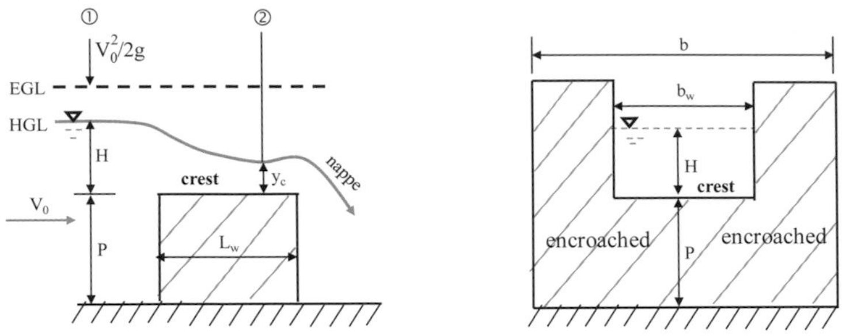

(a) (b)

Figure 6.17 **Cross sections views of a broad-crested weir.** (a) Longitudinal cross section; and (b) transverse cross section.

(3) $\theta = 85°$

Eq. (6.17): $Q = 8/15 * 0.59 * (2 * 32.2\,\text{ft sec}^{-2})^{1/2} * \tan(85°) * (8.5\,\text{ft})^{5/2} = 6079.8\,\text{cfs}.$

In summary, for a given approaching depth, a triangular weir with a larger angle has a larger discharge capacity than that with a smaller angle.

6.5.3 Broad-Crested Weir

As illustrated in Figure 6.17, the approaching flow dips over a broad-crested weir to its critical depth at a point on the weir. The occurrence of such a flow condition requires that the ratio of water depth above weir crest to weir length (i.e., H/L_w) is between 0.08 and 0.33 (Roberson *et al.*, 1998). The energy losses upstream of the critical-depth section are insignificant. Applying the energy equation (Eq. (2.21)) between sections ① and ② and rearranging, one can derive the weir discharge equation expressed as (Hager and Schwalt, 1994; Salmasi *et al.*, 2011):

$$Q = \frac{2\sqrt{2g}}{3\sqrt{3}} k b_w H^{\frac{3}{2}} \tag{6.18}$$

(Q: weir discharge; g: gravitational acceleration; k: weir coefficient (Figure 6.18); b_w: weir width; H: approaching depth above weir crest)

Example 6.30 A 2-ft-high, 6-ft-wide broad-crested weir is used to measure the discharge in a rectangular channel. If the approaching depth above the weir crest is 8.5 ft, determine the discharge through the weir.

Solution

$P = 2\,\text{ft}$, $b_w = 6\,\text{ft}$, $H = 8.5\,\text{ft}$

Figure 6.18: $(8.5\,\text{ft})/[(2\,\text{ft}) + (8.5\,\text{ft})] = 0.79 \rightarrow k = 1.035.$

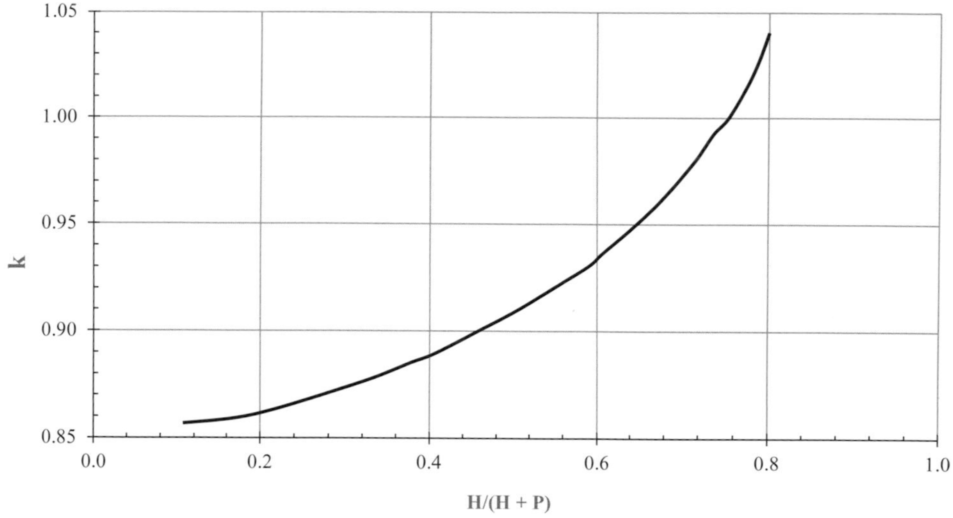

Figure 6.18 **Determination of broad-crested weir coefficient, k.** Nomogram for k versus the ratio of water depth above weir crest, H, to the summation of H and weir height, P.

Eq. (6.18): $Q = \frac{2\sqrt{2 * (32.2\,\text{ft sec}^{-2})}}{3\sqrt{3}} * 1.035 * (6\,\text{ft}) * (8.5\,\text{ft})^{\frac{3}{2}} = 475.3\,\text{cfs}.$

6.5.4 Parshall Flume

A Parshall flume (Figure 6.19), also called a long-throated flume, is designed to have a choking section (i.e., throat), at which the channel gradually rises in bed elevation and contracts in width. That is, a Parshall flume is a segment of a supercritical contraction-expansion channel with a gradual bed rise at its throat. As discussed in Sections 6.4.2 and 6.4.3, a critical flow occurs at the throat with the water depth easily determined using Eq. (2.38). Parshall flumes are applicable for subcritical approaching flows with a Froude number of less than 0.5 to avoid surface waves and instability (Sturm, 2010).

Consider a Parshall flume installed in a channel. The bird's-eye view of this configuration is shown in Figure 6.13, and the side view is as shown in Figure 6.19. The flume is the segment between sections ② and ④. Applying the energy equation (Eq. (2.21)) between sections ① and ③ and rearranging, one can derive the weir discharge equation expressed as (USBR, 2003):

$$Q = \frac{2\sqrt{2g}}{3\sqrt{3}} k b_w H^{\varepsilon} \tag{6.19}$$

Parshall flume discharge → Q; gravitational acceleration → g; exponent (Table 6.2) → ε; approaching depth above flume crest → H; flume coefficient (Table 6.2) → k; throat width → b_w

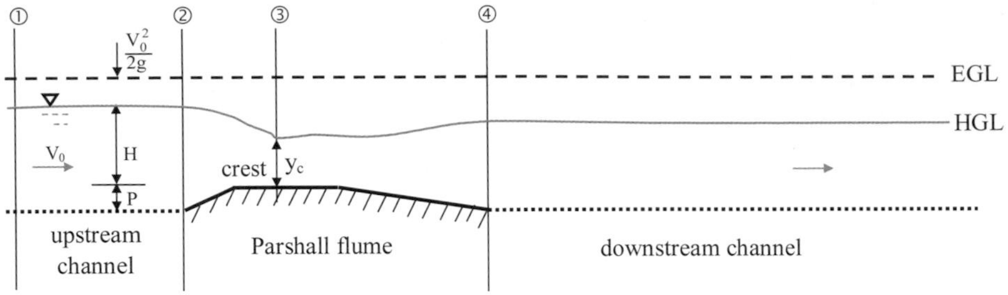

Figure 6.19 **Cross-sectional view of a Parshall flume.** Side view as installed in a flow-measuring channel, with the bird's-eye view shown in Figure 6.13.

Table 6.2 Parshall flume coefficient and exponent in Eq. (6.19).[1]

Throat width, b_w (ft)	Coefficient k	Exponent ε	Throat width, b_w (ft)	Coefficient k	Exponent E
1/12	1.313	1.55	6	1.295	1.59
2/12	1.313	1.55	7	1.295	1.60
3/12	1.285	1.55	8	1.295	1.61
6/12	1.334	1.58	10	1.275	1.60
9/12	1.325	1.53	12	1.261	1.60
1	1.279	1.55	15	1.248	1.60
1.5	1.295	1.54	20	1.234	1.60
2	1.295	1.55	25	1.226	1.60
3	1.295	1.57	30	1.221	1.60
4	1.295	1.58	40	1.214	1.60
5	1.295	1.59	50	1.210	1.60

[1] Reproduced from Table 8.6 in USBR (2003).

Example 6.31 A Parshall flume (bed rise of 2 ft and throat width of 6 ft) is used to measure the discharge in a rectangular channel. If the approaching depth above the flume crest is 8.5 ft, determine the discharge through the flume.

Solution

$P = 2\,\text{ft},\ b_w = 6\,\text{ft},\ H = 8.5\,\text{ft}$

Table 6.2: $b_w = 6\,\text{ft} \rightarrow k = 1.295,\ \varepsilon = 1.59.$

Eq. (6.19): $Q = \dfrac{2\sqrt{2 * (32.2\,\text{ft sec}^{-2})}}{3\sqrt{3}} * 1.295 * (6\,\text{ft}) * (8.5\,\text{ft})^{1.59} = 721.1\,\text{cfs}.$

6.5.5 Ultrasonic Open Channel Flow Meter

An ultrasonic open channel flow meter (USOPF) consists of a wall-mounted host and a probe (Figure 6.20). The wall-mounted host has a display and an integral keypad for programming. The

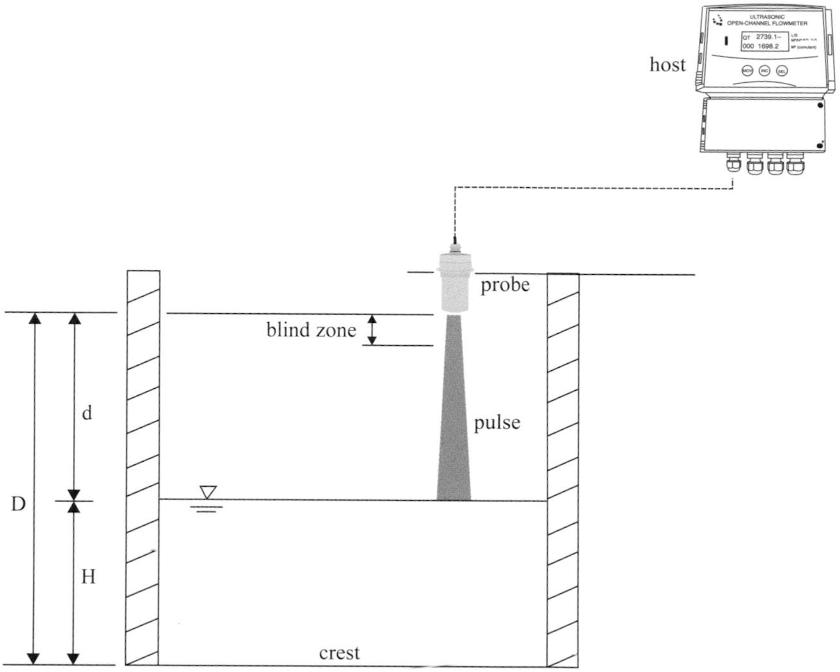

Figure 6.20 **Illustration of the principle of an ultrasonic open channel flow meter.** (*Source:* Engematic, 2018, reproduce with permission of nardini@engematic.com.br)

probe must be mounted directly above the surface to be monitored (Engematic, 2018). Both the host and the probe are made of leak-proof plastic. The USOPF is used jointly with a weir and/or flume structure to continuously measure the water depth above its crest, and the discharge is then computed using the corresponding formula discussed above. The probe is mounted on the top of either a weir or flume (e.g., Figure 6.21), and ultrasonic pulses are transmitted by the probe to the water surface, from which they are reflected back and received by the probe. Note that the probe installation height should be sufficiently large so that the water surface is always below the "blind" zone. The host measures the time between pulse transmission and reception, and calculates the distance between the sensor bottom and the monitored water surface, d, as the product of the measured time and the sound speed in air, which can be computed as:

sound speed in air [m s^{-1}] → c; air temperature [°C] → T_a

$$c = 331.3 + 0.606T_a \tag{6.20}$$

The water depth above the crest is then determined as:

water depth above crest → H; probe installation height above crest → D; distance between sensor bottom and water surface → d; sound speed in air → c; time between pulse transmission and reception → t

$$H = D - d = D - ct \tag{6.21}$$

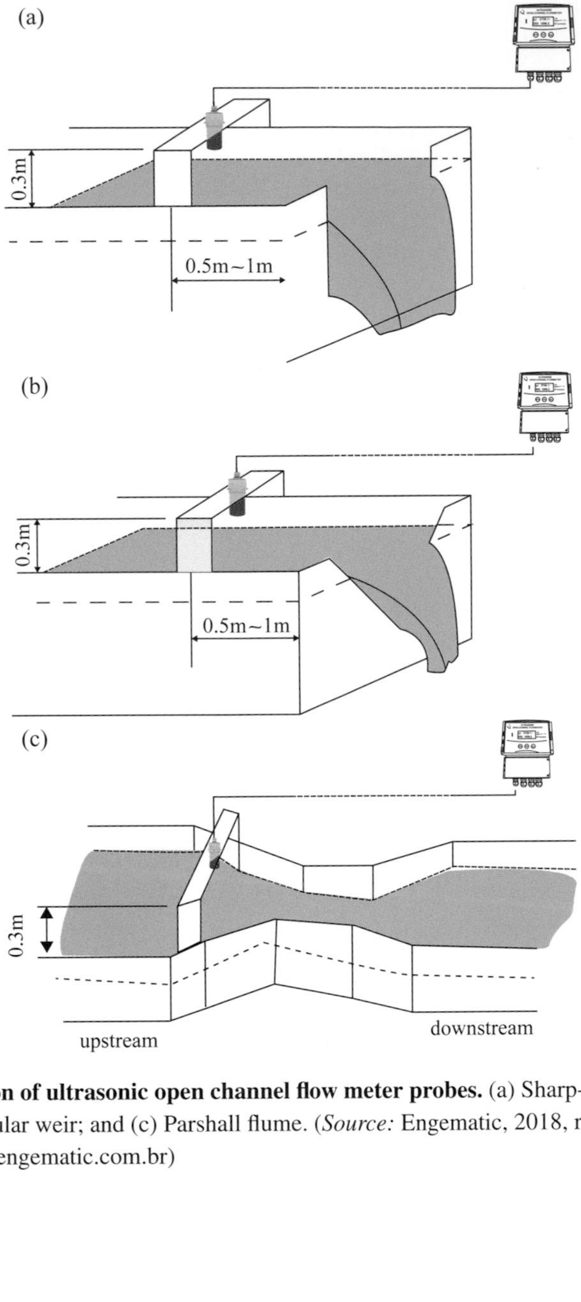

Figure 6.21 **Installation of ultrasonic open channel flow meter probes.** (a) Sharp-crested rectangular weir; (b) sharp-crested triangular weir; and (c) Parshall flume. (*Source:* Engematic, 2018, reproduced with permission of nardini@engematic.com.br)

Example 6.32 A USOPF and a Parshall flume (bed rise of 0.6 m and throat width of 1.8 m) are jointly used to continuously measure discharge in a rectangular channel. The installation height of the USOPF probe is 5 m above the flume crest. On a day with air temperature of 25°C, if the time between pulse transmission and reception is measured to be 0.01s by the host, determine the discharge through the flume.

Solution

$P = 0.6\,\text{m}$, $b_w = 1.8\,\text{m}$, $D = 5\,\text{m}$, $t = 0.01\,\text{s}$, $T_a = 25°\text{C}$

Eq. (6.20): sound speed $c = 331.3 + 0.606 * (25°\text{C})\,\text{m s}^{-1} = 346.45\,\text{m s}^{-1}$.

Pulse transmission distance: $d = (346.45\,\text{m s}^{-1}) * (0.01\,\text{s}) = 3.47\,\text{m}$.

Eq. (6.21): $H = 5\,\text{m} - 3.47\,\text{m} = 1.53\,\text{m}$.

Table 6.2: $b_w = 1.8\,\text{m} = 5.9\,\text{ft} \rightarrow k = 1.295$, $\varepsilon = 1.59$.

Eq. (6.19): $Q = \dfrac{2\sqrt{2 * (9.81\,\text{m s}^{-2})}}{3\sqrt{3}} * 1.295 * (1.8\,\text{m}) * (1.53\,\text{m})^{1.59} = 7.81\,\text{m}^3\,\text{s}^{-1}$.

6.6 Channel Design and Analysis

The engineering of a water resources system begins with two related processes, namely channel design and analysis. Channel design is the process during which cross-sectional shape, bed slope, and construction and lining materials are selected, whereas analysis is the process during which the dimensions (e.g., width and depth) of channel cross section are determined. The chosen dimensions are subject to a required minimum conveyance capacity (i.e., design peak discharge) and/or an allowable maximum velocity. In practice, although design is undertaken first and analysis second, these two processes need to be iterated until a satisfactory solution is found. Usually, the channel bed slope follows the overall natural topographic gradient, although some adjustments may need to be made by excavation and/or filling. The cross-sectional shape may be rectangular, circular, triangular, or trapezoidal, depending on whether the lining material is non-erosive (e.g., concrete-lined), erosive (e.g., unlined), or semi-erosive (e.g., grass- and riprap-lined).

6.6.1 Overall Guidelines

Many types of channels are designed and analyzed for uniform flow (i.e., by assuming that EGL slope, S, HGL slope, S_1, and channel bed slope, S_0 are the same). Non-erosive channels are expected (but not mandatory) to be designed for supercritical uniform flow with $F_r > 1.4$, whereas erosive channels must be designed for subcritical uniform flow with $F_r < 0.6$ (Roberson *et al.*, 1998). Semi-erosive channels also must be designed for subcritical uniform flow with $F_r < 1.0$ (Sturm, 2010). No channel should be designed for critical flow because it is unstable and can easily be switched to either supercritical or subcritical flow by a small disturbance. In addition, channels should be designed to have a mean velocity of $V \geq 0.7\,\text{m s}^{-1}$ (or $2\,\text{ft sec}^{-1}$) to prevent sedimentation and thus loss of conveyance capacity. Further, if a channel does not have a closed cross section or is not to be covered, the channel depth should be the normal depth plus a freeboard. The normal depth is computed using the Manning formula, as discussed in Section 6.2. The freeboard, which is added to account for uncertainties in design, analysis, construction, and operation of the channel, can be estimated using the formulas proposed the US Bureau of Reclamation (USBR) (Roberson *et al.*, 1998) and presented in Table 6.3.

Table 6.3 Channel freeboard proposed by USBR.[1]

Conveyance capacity (cfs)	Freeboard (ft)	Conveyance capacity ($m^3 s^{-1}$)	Freeboard (m)
< 20	≥ 1.0	< 0.6	≥ 0.3
$20 \sim 3000$	$1.2\sqrt{y_n}$	$0.6 \sim 85$	$0.7\sqrt{y_n}$
> 3000	$1.6\sqrt{y_n}$	> 85	$0.9\sqrt{y_n}$

[1] y_n: normal depth.

Table 6.4 Design parameters and constraints of concrete-lined channel.

Shape	Design parameter	Constraint
Circular	Diameter, D	Normal depth, $y_n \leq D$
Rectangular	Channel width, b; normal depth, y_n; freeboard	No overspill along entire channel
Triangular	Side slope, Z; normal depth, y_n; freeboard	No overspill along entire channel
Trapezoidal	Bottom width, b; side slope, Z; normal depth, y_n; freeboard	No overspill along entire channel

6.6.2 Concrete-Lined Channel

A concrete-lined channel can be designed to have a circular, rectangular, triangular, or trapezoidal shape, whose design parameters and constraints are summarized in Table 6.4. For a circular shape, it is assumed that the channel is just full but not yet pressurized and thus the Manning formula is still applicable. The design and analysis process of a circular channel is shown in Figure 6.22. For a rectangular shape (Figure 6.23), the primary design parameter is channel width, which is subject to the condition that the channel bank top elevation should be higher than the surrounding ground surface along the entire channel (i.e., no overspill occurs). For a triangular shape (Figure 6.24), the primary design parameter is side slope, which is subject to the condition of no overspill. For a trapezoidal shape (Figure 6.25), the primary design parameters are bottom width and side slope, also subject to the criteria of no overspill.

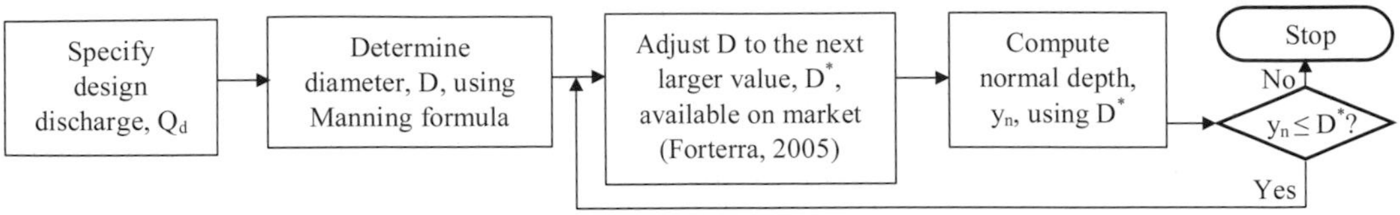

Figure 6.22 Flowchart for design and analysis of a concrete-lined circular channel.

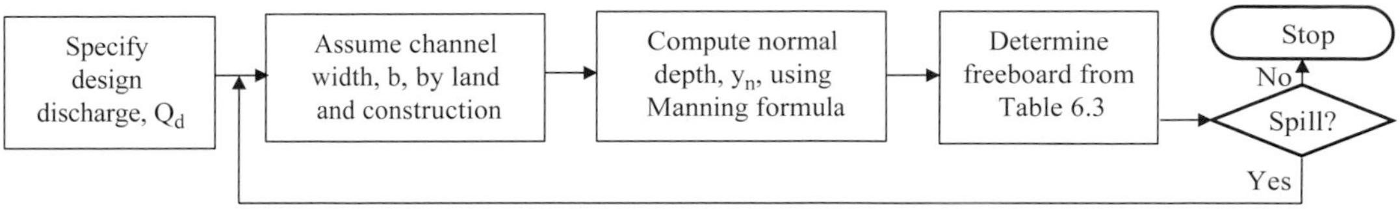

Figure 6.23 **Flowchart for design and analysis of a concrete-lined rectangular channel.**

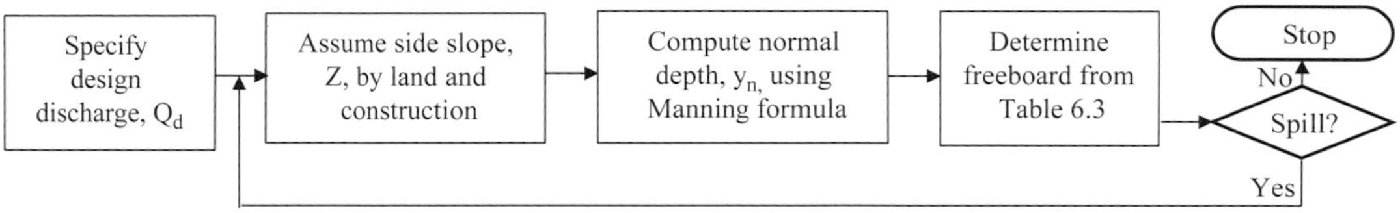

Figure 6.24 **Flowchart for design and analysis of a concrete-lined triangular channel.**

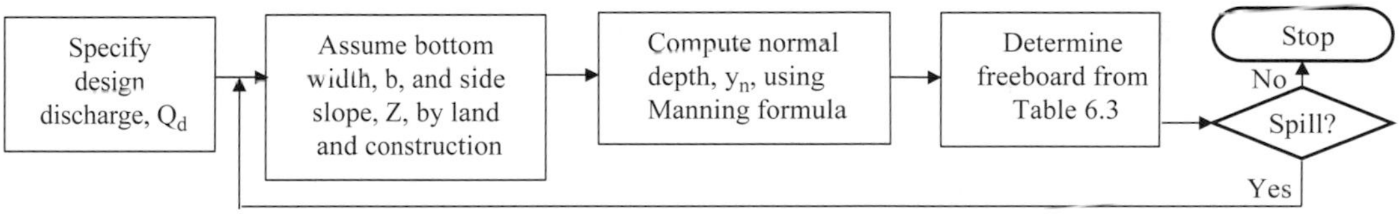

Figure 6.25 **Flowchart of design and analysis of a concrete-lined trapezoidal channel.**

Example 6.33 As shown in Figure 6.26, a concrete circular storm sewer (Manning's $n = 0.012$ and bed slope $S_0 = 0.002$) needs to carry 3–15 cfs. Determine:

(1) Its diameter
(2) Whether sediment would deposit in the sewer.

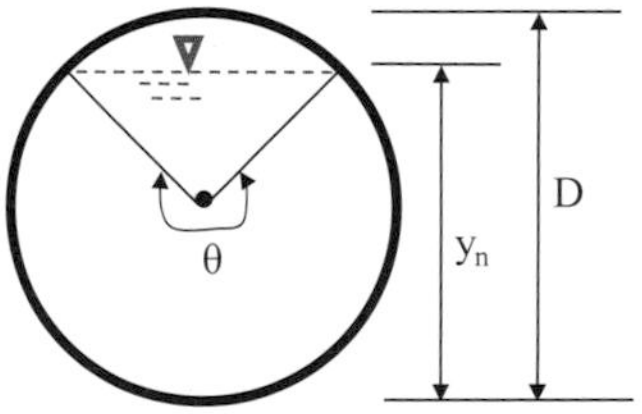

Figure 6.26 **Cross section of a circular sewer.**

Solution

$n = 0.012$, $S_0 = 0.002$, $Q_d = 15$ cfs, $Q_{min} = 3$ cfs

(1) Follow Figure 6.22.

Determine the diameter:

Flow area: $A = (\pi/4)D^2$ ft^2.

Wetted perimeter: $P_{wet} = \pi D$ ft.

Hydraulic radius: $R = [(\pi/4)D^2\ ft^2]/(\pi D\ ft) = D/4\ ft$.
Mean velocity: $V = (15\ cfs)/[(\pi/4)D^2\ ft^2] = 60/(\pi D^2)\ ft\ sec^{-1}$.
Manning formula (Eq. (2.26)): $60/(\pi D^2)\ ft\ sec^{-1} = 1.486/0.012 * (D/4\ ft)^{2/3} * 0.002^{1/2}$

$$\rightarrow D^{8/3} = 8.6900 \rightarrow D = 2.25\ ft = 27\ in.$$

Adjust D to the next larger value available on market:
From Forterra (2005) (www.forterrabp.com), $D^* = 27\ in = 2.25\ ft$.
Compute normal depth using D^*:
In reference to Figure 6.26 and Table 2.4, at the normal depth, one can calculate:
Flow area: $A = [(2.25\ ft)^2/8](\theta - \sin\theta) = 0.6328(\theta - \sin\theta)\ ft^2$.
Wetted perimeter: $P_{wet} = [(2.25\ ft)/2]\theta = 1.125\theta\ ft$.
Hydraulic radius: $R = [0.6328(\theta - \sin\theta)\ ft^2]/(1.125\theta\ ft) = 0.5625(1 - \sin\theta/\theta)\ ft$.
Mean velocity: $V = (15\ cfs)/[0.6328(\theta - \sin\theta)\ ft^2] = 23.7042/(\theta - \sin\theta)\ ft\ sec^{-1}$.
Manning formula (Eq. (2.26)):

$$23.7042/(\theta - \sin\theta)\ ft\ sec^{-1} = 1.486/0.012 * [0.5625(1 - \sin\theta/\theta)\ ft]^{2/3} * 0.002^{1/2}$$

$\rightarrow(\theta - \sin\theta)^{5/3}\theta^{-2/3} = 6.2814 \rightarrow$ trial-and-error or Excel Solver, $\theta = 4.5273$ rad
$\rightarrow$ Table 2.4: $y_n = (2.25\ ft)/2 * [1 - \cos(4.5273/2\ rad)] = 1.84\ ft$.
Check if $D^* = 27\ in = 2.25\ ft$ is acceptable:
$y_n < D^* \rightarrow$ okay.
Check Froude number at $y_n = 1.84$ ft:
$A = 0.6328 * [4.5273 - \sin(4.5273\ rad)]\ ft^2 = 3.49\ ft^2$.
$V = (15\ cfs)/(3.49\ ft^2) = 4.30\ ft\ sec^{-1}$.
Table 2.4, top width: $B = (2.25\ ft) * \sin[(4.5273\ rad)/2] = 1.73\ ft$.
Hydraulic depth: $d = (3.49\ ft^2)/(1.73\ ft) = 2.02\ ft$.
Eq. (2.46): $F_r = (4.30\ ft\ sec^{-1})/[(32.2\ ft\ sec^{-2}) * (2.02\ ft)]^{1/2} = 0.53$.
$Fr < 1 \rightarrow$ subcritical flow.

(2) Determine the velocity at $Q_{min} = 3$ cfs.
Flow area and hydraulic radius are computed as in (1).
Mean velocity: $V = (3\ cfs)/[0.6328(\theta - \sin\theta)\ ft^2] = 4.7408/(\theta - \sin\theta)\ ft\ sec^{-1}$.
Manning formula (Eq. (2.26)):

$$4.7408/(\theta - \sin\theta)\ ft\ sec^{-1} = 1.486/0.012 * [0.5625(1 - \sin\theta/\theta)\ ft]^{2/3} * 0.002^{1/2}$$

$\rightarrow(\theta - \sin\theta)^{5/3}\theta^{-2/3} = 1.2563 \rightarrow$ trial-and-error or Excel Solver, $\theta = 2.3327$ rad
$\rightarrow A = 0.6328 * [2.3327 - \sin(2.3327\ rad)]\ ft^2 = 1.0183\ ft^2$
$\rightarrow V = (3\ cfs)/(1.0183\ ft^2) = 2.95\ ft\ sec^{-1} > 2\ ft\ sec^{-1} \rightarrow$ no sedimentation.

In summary, the channel has a diameter of 27 in without having a sedimentation problem, but it cannot be designed to have a supercritical flow.

Example 6.34 As shown in Figure 6.27, a concrete-lined rectangular channel (Manning's n = 0.012 and bed slope $S_0 = 0.002$) needs to carry 3–15 cfs. Determine the channel width and depth.

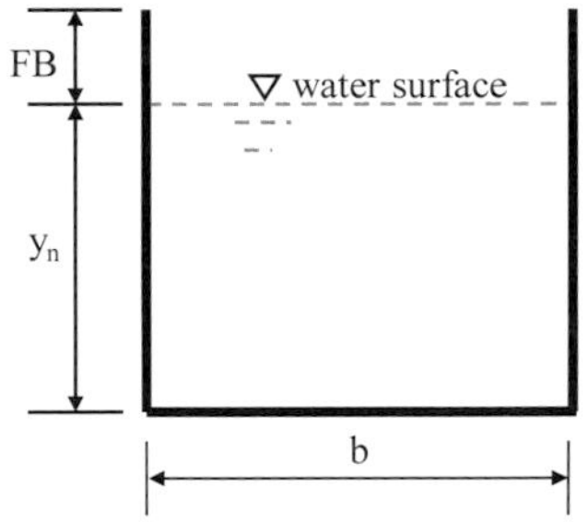

Figure 6.27 **Cross section of a rectangular channel.**

Solution

$n = 0.012$, $S_0 = 0.002$, $Q_d = 15$ cfs, $Q_{min} = 3$ cfs.

Follow Figure 6.23.

Assume a width in terms of available land:

Say $b = 5$ ft.

Compute the normal depth:

Flow area: $A = 5y_n\ \text{ft}^2$.

Wetted perimeter: $P_{wet} = 5 + 2y_n$ ft.

Hydraulic radius: $R = (5y_n)/(5 + 2y_n)$ ft.

Mean velocity: $V = (15\ \text{cfs})/(5y_n\ \text{ft}^2) = 3/y_n\ \text{ft sec}^{-1}$.

Manning formula (Eq. (2.26)): $3/y_n\ \text{ft sec}^{-1} = 1.486/0.012 * [(5y_n)/(5 + 2y_n)\ \text{ft}]^{2/3} * 0.002^{1/2}$

$$\rightarrow (5 + 2y_n)^{2/3} y_n^{-5/3} = 5.3977 \rightarrow \text{trial-and-error or Excel Solver, } y_n = 0.77\ \text{ft}$$

$$\rightarrow Q_d < 20\ \text{cfs, Table 6.3: FB} = 1.0\ \text{ft} \rightarrow \text{channel depth} = 0.77\ \text{ft} + 1.0\ \text{ft} = 1.77\ \text{ft}.$$

Check overspill and Froude number at $y_n = 0.77$ ft:

Compare the channel bank top elevation to the surrounding ground surface for overspill.

Hydraulic depth: $d = y_n = 0.77$ ft.

$V = 3/(5 * 0.77)\ \text{ft sec}^{-1} = 0.78\ \text{ft sec}^{-1}$.

Eq. (2.46): $F_r = (0.78\ \text{ft sec}^{-1})/[(32.2\ \text{ft sec}^{-2}) * (0.77\ \text{ft})]^{1/2} = 0.16$.

$F_r < 1 \rightarrow$ subcritical flow.

Check sedimentation at $Q_{min} = 3$ cfs:

Mean velocity: $V = (3\ \text{cfs})/(5y_n\ \text{ft}^2) = 0.6/y_n\ \text{ft sec}^{-1}$.

Manning formula: $0.6/y_n\ \text{ft sec}^{-1} = 1.486/0.012 * [(5y_n)/(5 + 2y_n)\ \text{ft}]^{2/3} * 0.002^{1/2}$

$$\rightarrow (5 + 2y_n)^{2/3} y_n^{-5/3} = 26.9887 \rightarrow \text{trial-and-error or Excel Solver, } y_n = 0.27\ \text{ft}.$$

$V = 0.6/0.27\ \text{ft sec}^{-1} = 2.22\ \text{ft sec}^{-1} > 2\ \text{ft sec}^{-1} \rightarrow$ no sedimentation.

In summary, the channel can have a width of 5 ft and a depth of 1.77 ft without having a sedimentation problem, but it cannot be designed to have a supercritical flow.

Example 6.35 As shown in Figure 6.28, a concrete-lined triangular channel (Manning's $n = 0.012$ and bed slope $S_0 = 0.002$) needs to carry 3–15 cfs. Determine the channel side slope.

Solution

$n = 0.012$, $S_0 = 0.002$, $Q_d = 15\ \text{cfs}$, $Q_{min} = 3\ \text{cfs}$

Follow Figure 6.24.

Assume a side slope in terms of soil properties:

Say $Z = 0.5$.

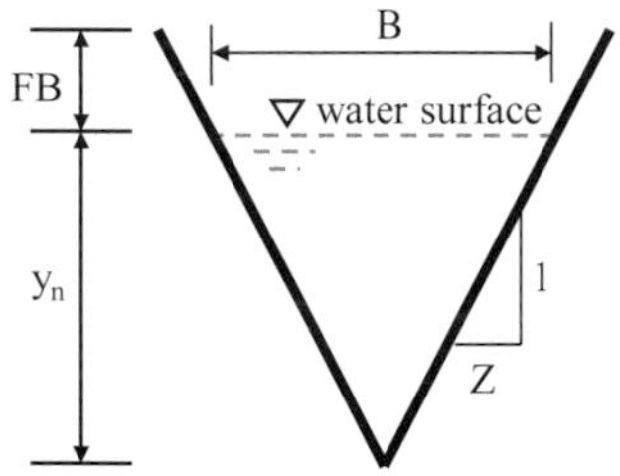

Figure 6.28 **Cross section of a triangular channel.**

Compute the normal depth:

Top width: $B = 2 * 0.5y_n\ \text{ft} = y_n\ \text{ft}$.
Flow area: $A = {}^1\!/_2 * (y_n)y_n = 0.5{y_n}^2\ \text{ft}^2$.
Wetted perimeter: $P_{wet} = 2 * (1 + 0.5^2)^{1/2} y_n\ \text{ft} = 2.2361y_n\ \text{ft}$.
Hydraulic radius: $R = (0.5{y_n}^2)/(2.2361y_n)\ \text{ft} = 0.2236y_n\ \text{ft}$.
Mean velocity: $V = (15\ \text{cfs})/(0.5{y_n}^2\ \text{ft}^2) = 30/{y_n}^2\ \text{ft sec}^{-1}$.
Manning formula (Eq. (2.26)): $30/{y_n}^2\ \text{ft sec}^{-1} = 1.486/0.012 * (0.2236y_n\ \text{ft})^{2/3} * 0.002^{1/2}$

$\rightarrow {y_n}^{8/3} = 14.7046 \rightarrow y_n = 2.74\ \text{ft}$

$\rightarrow Q_d < 20\ \text{cfs}$, Table 6.3: $FB = 1.0\ \text{ft} \rightarrow \text{channel depth} = 2.74\ \text{ft} + 1.0\ \text{ft} = 3.74\ \text{ft}$.

Check overspill and Froude number at $y_n = 2.74$ ft:

Compare the channel bank top elevation to the surrounding ground surface for overspill.
$B = 2.74\ \text{ft}$.
$A = 0.5 * (2.74\ \text{ft})^2 = 3.75\ \text{ft}^2$.
Hydraulic depth: $d = (3.75\ \text{ft}^2)/(2.74\ \text{ft}) = 1.37\ \text{ft}$.

$V = 15/3.75\,\text{ft sec}^{-1} = 4.0\,\text{ft sec}^{-1}$.

Eq. (2.46): $F_r = (4.0\,\text{ft sec}^{-1})/[(32.2\,\text{ft sec}^{-2}) * (1.37\,\text{ft})]^{1/2} = 0.60$.

$F_r < 1 \rightarrow$ subcritical flow.

Check sedimentation at $Q_{min} = 3\,\text{cfs}$:

Mean velocity: $V = (3\,\text{cfs})/(0.5{y_n}^2\,\text{ft}^2) = 6/{y_n}^2\,\text{ft sec}^{-1}$.

Manning formula: $6/{y_n}^2\,\text{ft sec}^{-1} = 1.486/0.012 * (0.2236y_n\,\text{ft})^{2/3} * 0.002^{1/2}$

$$\rightarrow {y_n}^{8/3} = 2.9409 \rightarrow y_n = 1.50\,\text{ft}.$$

$V = 30/1.50^2\,\text{ft sec}^{-1} = 13.33\,\text{ft sec}^{-1} > 2\,\text{ft sec}^{-1} \rightarrow$ no sedimentation.

In summary, the channel can have a side slope of 0.5 and a depth of 3.74 ft without having a sedimentation problem, but it cannot be designed to have a supercritical flow.

Example 6.36 As shown in Figure 6.29, a concrete-lined trapezoidal channel (Manning's $n = 0.012$ and bed slope $S_0 = 0.002$) needs to carry 3–15 cfs. Determine the channel side slope, bottom width, and depth.

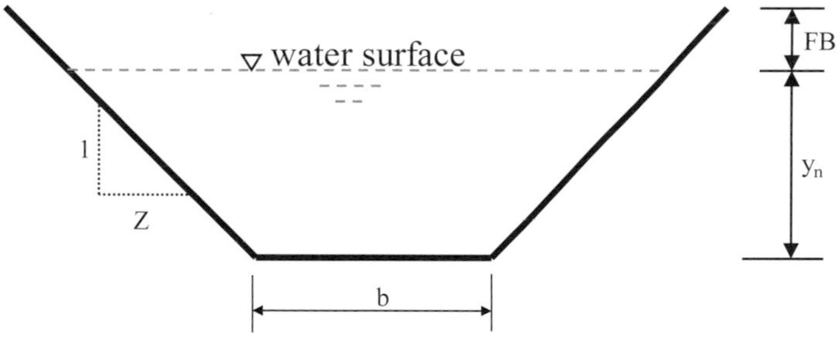

Figure 6.29 **Cross section of a trapezoidal channel.**

Solution

$n = 0.012$, $S_0 = 0.002$, $Q_d = 15\,\text{cfs}$, $Q_{min} = 3\,\text{cfs}$

Follow Figure 6.25.

Assume a side slope in terms of soil properties:

Say $Z = 1.5$.

Assume a bottom width in terms of available land:

Say $b = 1$ ft.

Compute the normal depth:

Top width: $B = 1 + 2 * 1.5y_n\,\text{ft} = 1 + 3.0y_n\,\text{ft}$.

Flow area: $A = (1 + 1.5y_n)y_n\,\text{ft}^2$.

Wetted perimeter: $P_{wet} = 1 + 2 * (1 + 1.5^2)^{1/2} y_n \text{ ft} = 1 + 3.6056 y_n \text{ ft}$.
Hydraulic radius: $R = [(1 + 1.5y_n)y_n]/[1 + 3.6056y_n] \text{ ft}$.
Mean velocity: $V = 15/[(1 + 1.5y_n)y_n] \text{ ft sec}^{-1}$.
Manning formula (Eq. (2.26)):

$$15/[(1 + 1.5y_n)y_n] \text{ ft sec}^{-1} = 1.486/0.012 * ([(1 + 1.5y_n)y_n]/[1 + 3.6056y_n])^{2/3} * 0.002^{1/2}$$
$$\rightarrow [(1 + 1.5y_n)y_n]^{5/3}(1 + 3.6056y_n)^{-2/3} = 2.7086$$

$\rightarrow$ trial-and-error or Excel Solver, $y_n = 1.25$ ft
$\rightarrow Q_d < 20$ cfs, Table 6.3: FB = 1.0 ft $\rightarrow$ channel depth = 1.25 ft + 1.0 ft = 2.25 ft.

Check overspill and Froude number at $y_n = 1.25$ ft:

Compare the channel bank top elevation to the surrounding ground surface for overspill.
$B = 1 + 3.0 * 1.25 \text{ ft} = 4.75 \text{ ft}$.
$A = (1 + 1.5 * 1.25) * 1.25 \text{ ft}^2 = 3.59 \text{ ft}^2$.
Hydraulic depth: $d = (3.59 \text{ ft}^2)/(4.75 \text{ ft}) = 0.76 \text{ ft}$.
$V = 15/3.59 \text{ ft sec}^{-1} = 4.18 \text{ ft sec}^{-1}$.
Eq. (2.46): $F_r = (4.18 \text{ ft sec}^{-1})/[(32.2 \text{ ft sec}^{-2}) * (0.76 \text{ ft})]^{1/2} = 0.85$.
$F_r < 1 \rightarrow$ subcritical flow.

Check sedimentation at $Q_{min} = 3$ cfs:

Mean velocity: $V = 3/[(1 + 1.5y_n)\, y_n] \text{ ft sec}^{-1}$.
Manning formula:

$$3/[(1 + 1.5y_n)y_n] \text{ ft sec}^{-1} = 1.486/0.012 * ([(1 + 1.5y_n)y_n]/[1 + 3.6056y_n] \text{ ft})^{2/3} * 0.002^{1/2}$$
$$\rightarrow [(1 + 1.5y_n)y_n]^{5/3}(1 + 3.6056y_n)^{-2/3} = 0.5417$$

$\rightarrow$ trial-and-error or Excel Solver, $y_n = 0.58$ ft.
$V = 3/[(1 + 1.5 * 0.58) * 0.58] \text{ ft sec}^{-1} = 2.77 \text{ ft sec}^{-1} > 2 \text{ ft sec}^{-1} \rightarrow$ no sedimentation.

In summary, the channel can have a side slope of 1.5, bottom width of 1 ft, and depth of 2.25 ft without having a sedimentation problem, but it cannot be designed to have a supercritical flow.

A concrete-lined channel can also be designed to have the most efficient cross section, where the channel conveyance capacity is maximized for fixed flow area, bed slope, and material (i.e., Manning's n). In this regard, based on the Manning formula, to maximize the conveyance capacity, the hydraulic radius needs to be maximized. Differentiating hydraulic radius with respect to normal depth and letting the derivative be equal to zero, one can derive the most efficient cross sections for circular, rectangular, and trapezoidal shapes (Table 6.5). It is not possible to design the most efficient cross section for a triangular shape because its hydraulic radius cannot be maximized. Note that for trapezoidal shape, the most efficient cross section is different depending on whether channel side slope is a variable or a constant.

Table 6.5 The most efficient cross sections.

Shape	Side slope	Diameter/width	Normal depth
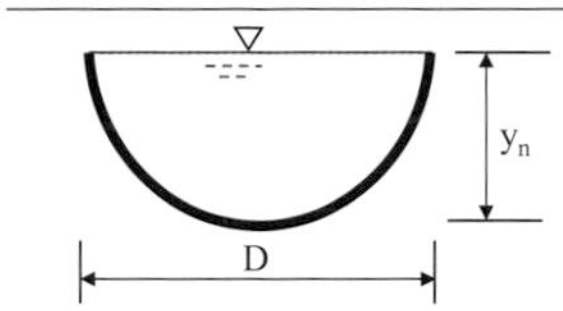		D	$y_n = \frac{D}{2}$
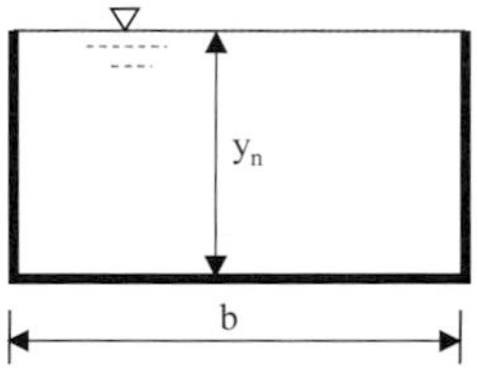		b	$y_n = \frac{b}{2}$
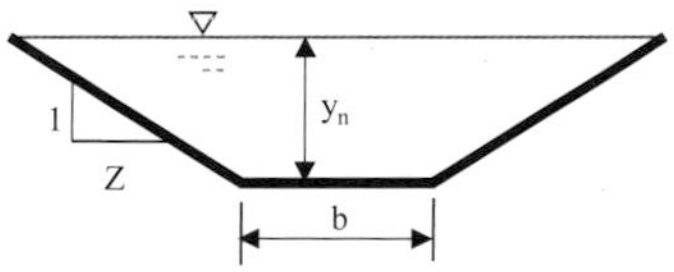	Variable: $Z = \frac{1}{\sqrt{3}}$	b	$y_n = \frac{\sqrt{3}}{2}b$
	Constant: Z	b	$y_n = \sqrt{\frac{A}{2\sqrt{1+Z^2} - Z}}$

Example 6.37 Redesign the circular channel in Example 6.33 to have the most efficient cross section. What is the percent increase of the conveyance capacity?

Solution

Example 6.33 $\rightarrow n = 0.012$, $S_0 = 0.002$, $A = 3.49\,\text{ft}^2$

Table 6.5: most efficient cross section $\rightarrow y_n = D/2$.

Determine the diameter:

Flow area: $A = 1/2 * [(\pi/4)D^2]\,\text{ft}^2 = (\pi/8)D^2\,\text{ft}^2 = 3.49\,\text{ft}^2 \rightarrow D = 2.98\,\text{ft}$.

Adjust D to the next larger value available on market:

From Forterra (2005) (www.forterrabp.com), $D^* = 36\,\text{in} = 3\,\text{ft} \rightarrow y_n = (3\,\text{ft})/2 = 1.5\,\text{ft}$.

Calculate the new conveyance capacity:

$A = (\pi/8) * (3\,\text{ft})^2 = 3.53\,\text{ft}^2$.
$P_{wet} = (1/2) * [\pi(3\,\text{ft})] = 4.71\,\text{ft}$.
$R = (3.53\,\text{ft}^2)/(4.71\,\text{ft}) = 0.75\,\text{ft}$.
$V = Q/3.53\,\text{ft sec}^{-1}$.
Manning formula: $Q/3.53\,\text{ft sec}^{-1} = 1.486/0.012 * (0.75\,\text{ft})^{2/3} * 0.002^{1/2} \rightarrow Q = 16.14\,\text{cfs}$
$\rightarrow$percent increase $= (16.14\,\text{cfs} - 15\,\text{cfs})/(15\,\text{cfs}) * 100\% = 7.6\%$.

Example 6.38 Redesign the rectangular channel in Example 6.34 to have the most efficient cross section. What is the percent increase of the conveyance capacity?

Solution

Example 6.34 $\rightarrow$n = 0.012, $S_0 = 0.002$, $A = 5 * 0.77\ \text{ft}^2 = 3.85\ \text{ft}^2$.

Table 6.5: most efficient cross section $\rightarrow y_n = b/2$.

Determine the channel width:

Flow area: $A = b(b/2)\ \text{ft}^2 = b^2/2\ \text{ft}^2 = 3.85\ \text{ft}^2 \rightarrow b = 2.77\ \text{ft} \rightarrow y_n = (2.77\ \text{ft})/2 = 1.39\ \text{ft}$.

Calculate the new conveyance capacity:

$P_{wet} = 2.77\ \text{ft} + 2 * (1.39\ \text{ft}) = 5.55\ \text{ft}$.
$R = (3.85\ \text{ft}^2)/(5.55\ \text{ft}) = 0.69\ \text{ft}$.
$V = Q/3.85\ \text{ft sec}^{-1}$.
Manning formula: $Q/3.85\ \text{ft sec}^{-1} = 1.486/0.012 * (0.69\ \text{ft})^{2/3} * 0.002^{1/2} \rightarrow Q = 16.65\ \text{cfs}$
$\rightarrow$percent increase $= (16.65\ \text{cfs} - 15\ \text{cfs})/(15\ \text{cfs}) * 100\% = 11.0\%$.

Example 6.39 Redesign the trapezoidal channel in Example 6.36 to have the most efficient cross section. What is the percent increase of the conveyance capacity? Answer these questions if side slope Z:

(1) Can be varied
(2) Is fixed.

Solution

Example 6.36 $\rightarrow$n = 0.012, $S_0 = 0.002$, $A = 3.59\ \text{ft}^2$

(1) Table 6.5: most efficient cross section $\rightarrow Z = \frac{1}{\sqrt{3}}$, $y_n = \frac{\sqrt{3}}{2}b$.

Determine the channel width:

$$\text{Flow area: } A = \left(b + \frac{1}{\sqrt{3}} * \frac{\sqrt{3}}{2}b\right) * \left(\frac{\sqrt{3}}{2}b\right) = \frac{3\sqrt{3}}{4}b^2 = 3.59\ \text{ft}^2 \rightarrow b = 1.66\ \text{ft}$$

$$\rightarrow y_n = \frac{\sqrt{3}}{2} * (1.66\ \text{ft}) = 1.44\ \text{ft}.$$

Calculate the new conveyance capacity:

$$P_{wet} = 1.66\ \text{ft} + 2 * \sqrt{1 + \left(\frac{1}{\sqrt{3}}\right)^2} * (1.44\ \text{ft}) = 4.99\ \text{ft}.$$

$R = (3.59\ \text{ft}^2)/(4.99\ \text{ft}) = 0.72\ \text{ft}$.
$V = Q/3.59\ \text{ft sec}^{-1}$.

Manning formula: $Q/3.59\,\text{ft sec}^{-1} = 1.486/0.012 * (0.72\,\text{ft})^{2/3} * 0.002^{1/2} \rightarrow Q = 15.97\,\text{cfs}$
$\rightarrow$percent increase $= (15.97\,\text{cfs} - 15\,\text{cfs})/(15\,\text{cfs}) * 100\% = 6.5\%$.

(2) Example 6.36 $\rightarrow Z = 1.5$.

Table 6.5: most efficient cross section $\rightarrow y_n = \sqrt{\dfrac{3.59\,\text{ft}^2}{2 * \sqrt{1+1.5^2} - 1.5}} = 1.31$ ft.

Determine the channel width:

Flow area: $A = (b + 1.5 * 1.31) * 1.31 = 3.59\,\text{ft}^2 \rightarrow b = 0.78$ ft.

Calculate the new conveyance capacity:

$P_{wet} = 0.78\,\text{ft} + 2 * \sqrt{1+1.5^2} * (1.31\,\text{ft}) = 5.50\,\text{ft}$.
$R = (3.59\,\text{ft}^2)/(5.50\,\text{ft}) = 0.65\,\text{ft}$.
$V = Q/3.59\,\text{ft sec}^{-1}$.
Manning formula: $Q/3.59\,\text{ft sec}^{-1} = 1.486/0.012 * (0.65\,\text{ft})^{2/3} * 0.002^{1/2} \rightarrow Q = 14.92\,\text{cfs}$
$\rightarrow$percent increase $= (14.92\,\text{cfs} - 15\,\text{cfs})/(15\,\text{cfs}) * 100\% = -0.5\%$.

Example 6.40 Show that the most efficient cross section of rectangular channel satisfies $y_n = b/2$.

Solution

Hydraulic radius: $R = A/(b + 2y_n) = A/(A/y_n + 2y_n) = Ay_n/(A + 2y_n{}^2)$.

$$\frac{dR}{dy_n} = \frac{A\left(A + 2y_n^2\right) - Ay_n\left(4y_n\right)}{\left(A + 2y_n^2\right)^2} = \frac{A^2 - 2Ay_n^2}{\left(A + 2y_n^2\right)^2} = 0 \rightarrow A = 2y_n{}^2 \rightarrow by_n = 2y_n{}^2 \rightarrow y_n = b/2.$$

6.6.3 Earth Channel

Earth channels can only have a trapezoidal cross section because of channel bank stability issues. Earth channels can be designed using the permissible velocity (V_{max}) or tractive force (τ_{max}) method (Roberson *et al.*, 1998; USDA-NRCS, 2007a). Figure 6.30 shows how to implement these two methods. To design an earth channel, the data on topography, soil type, and vegetative condition need to be collected and used to determine bed slope, Manning's n (Table 2.5), and side slope (Table 6.6). For the permissible velocity method, the value of V_{max} can be determined from Table 6.7. If a solution to the equations of flow area and hydraulic radius corresponding to V_{max} exists, the resulting bottom width, b, and normal depth, y_n, will be the design parameters; otherwise, a ratio of b to y_n is assumed until the corresponding mean velocity is not greater than V_{max}. However, for the permissible tractive force method, the value of τ_{max} is determined from Figure 6.31 or 6.32 and used to compute y_n. If the Manning formula with the computed y_n can give a meaningful value of b, the y_n and b can be treated as the design parameters; otherwise, a ratio of b to y_n is assumed until the corresponding tractive force is not greater than τ_{max}. Regardless of the methods, a freeboard needs to be determined from Table 6.3.

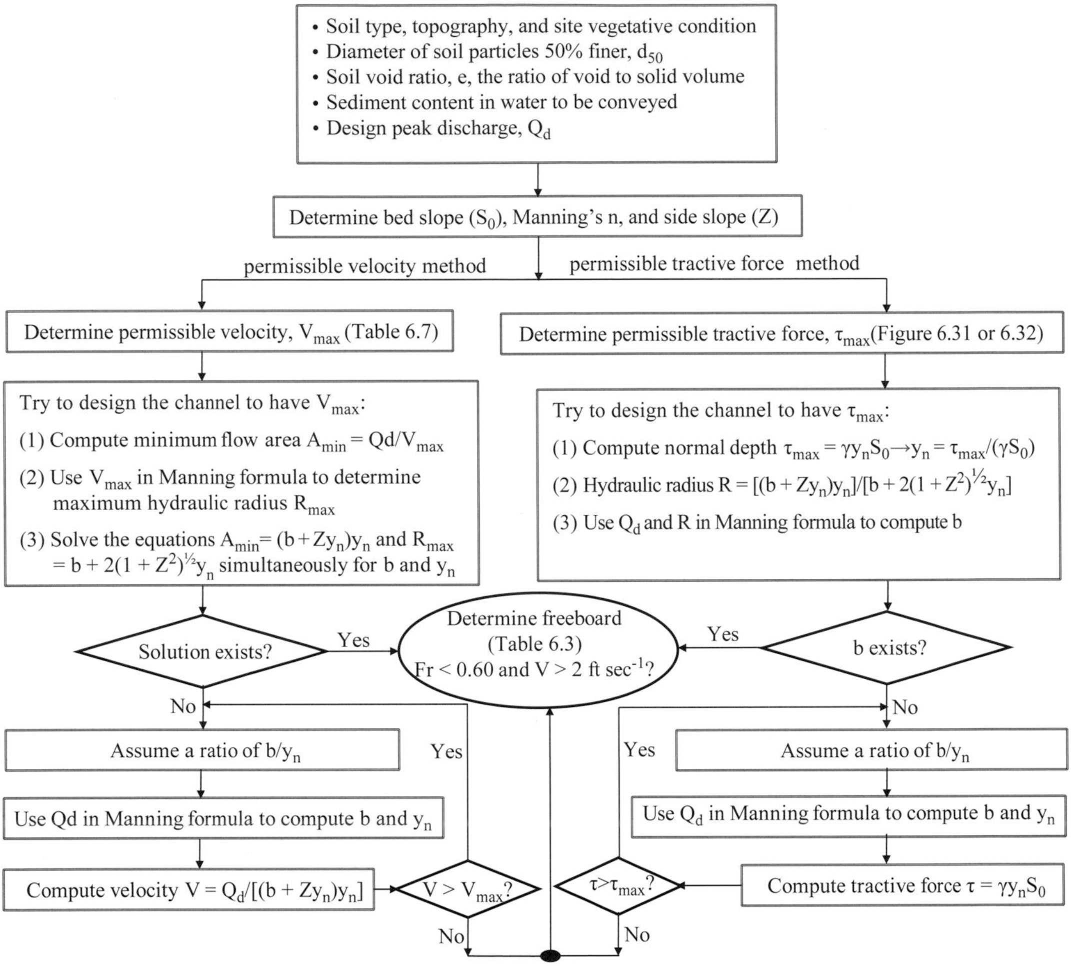

Figure 6.30 **Steps involved in the design of an earth channel.** Steps for the permissible velocity method are in the left panel and those for the permissible tractive force method are in the right panel. Steps in the central part of the flowchart apply to both methods.

Table 6.6 Side slope for earth channel.[1]

Soil type	Side slope, Z
Rock	0.25
Stiff clay	1
Stone lining	1
Firm clay	1.5
Gravelly loam	1.5
Sandy loam	3

[1] *Source:* USDA-NRCS (2007).

Table 6.7 Permissible velocity, V_{max}, for earth channel design.[1]

Soil type	Clear water without detritus		Water transporting colloidal silts		Water with noncolloidal silts, sands, gravels, or rock fragments	
	ft sec^{-1}	m s^{-1}	ft sec^{-1}	m s^{-1}	ft sec^{-1}	m s^{-1}
Fine sand (noncolloidal)	1.5	0.46	2.5	0.76	1.5	0.46
Sandy loam (noncolloidal)	1.75	0.53	2.5	0.76	2.0	0.61
Silt loam (noncolloidal)	2.0	0.61	3.0	0.91	2.0	0.61
Alluvial silt (noncolloidal)	2.0	0.61	3.5	1.07	2.0	0.61
Ordinary firm loam	2.5	0.76	3.5	1.07	2.25	0.69
Volcanic ash	2.5	0.76	3.5	1.07	2.0	0.61
Stiff clay (very colloidal)	3.75	1.14	5.0	1.52	3.0	0.91
Alluvial silt (colloidal)	3.75	1.14	5.0	1.52	3.0	0.91
Shales and hardpans	6.0	1.83	6.0	1.83	5.0	1.52
Fine gravel	2.5	0.76	5.0	1.52	3.75	1.14
Graded, loam to cobbles (noncolloidal)	3.75	1.14	5.0	1.52	5.0	1.52
Graded silt to cobbles (colloidal)	4.0	1.22	5.5	1.68	5.0	1.52
Coarse gravel (noncolloidal)	4.0	1.22	6.0	1.83	6.5	1.98
Cobbles and shingles	5.0	1.52	5.5	1.68	6.5	1.98

[1] *Source:* USDA-NRCS (2007).

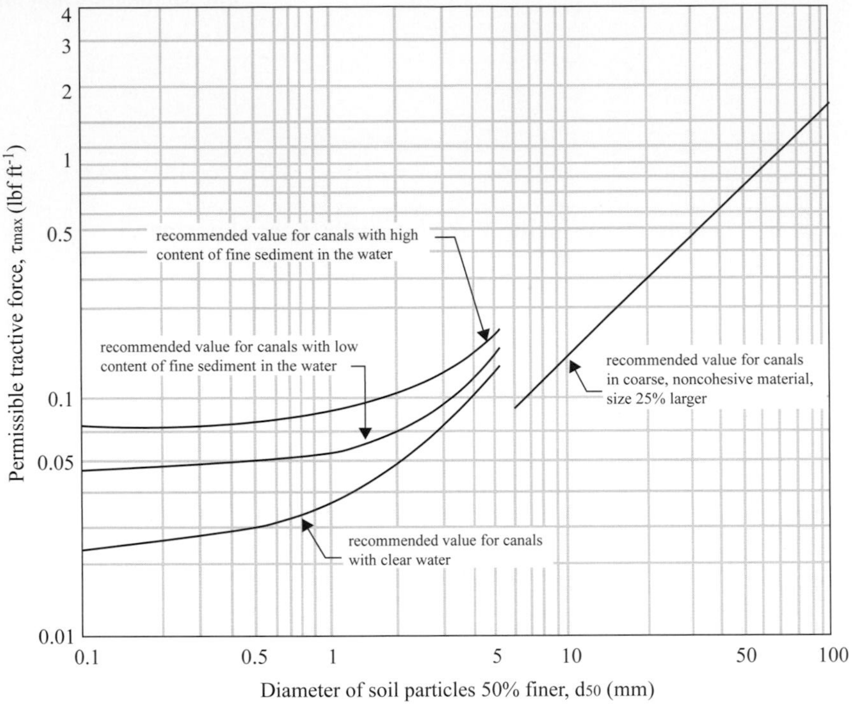

Figure 6.31 Tractive force and average particle diameter in the design of an earth channel. Nomogram shows permissible tractive force versus average particle diameter,d_{50}. (*Source:* USDA-NRCS, 2007)

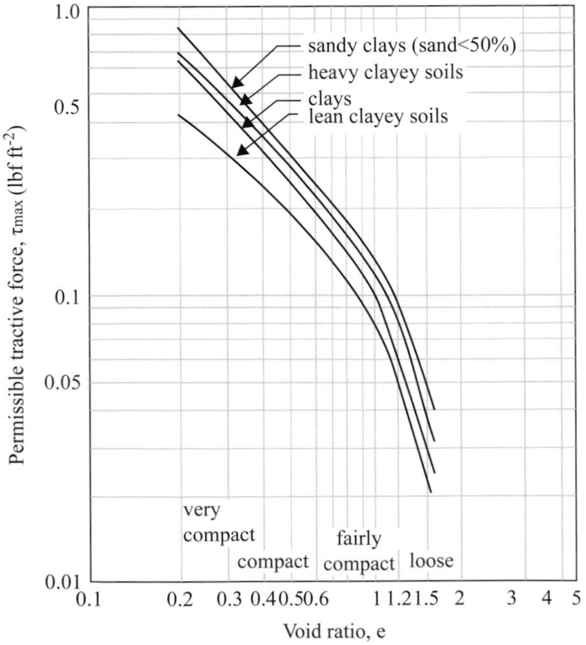

Figure 6.32 Tractive force and void ratio in the design of an earth channel. Nomogram shows permissible tractive force versus soil void ratio, e. (*Source:* USDA-NRCS, 2007)

Example 6.41 A stiff-clay channel is to be constructed to convey clear water at 150 cfs. Based on the site topography, the channel bed slope is determined to be 0.002. If the average soil particle diameter is 1 mm, design the channel cross section using the method of:

(1) Permissible velocity
(2) Permissible tractive force.

Solution

$Q_d = 150\,\text{cfs}$, $S_0 = 0.002$, $d_{50} = 1\,\text{mm}$

Table 2.5: earth artificial channel with some vegetation $\rightarrow n = 0.025$.

Table 6.6: stiff clay $\rightarrow$ side slope $Z = 1$.

(1) Table 6.7: clear water, stiff clay $\rightarrow V_{max} = 3.75\,\text{ft sec}^{-1}$.
Follow the left panel of Figure 6.30.

Try to design the channel to have V_{max}:

$A_{min} = (150\,\text{cfs})/(3.75\,\text{ft sec}^{-1}) = 40\,\text{ft}^2$.
Manning formula: $3.75\,\text{ft sec}^{-1} = 1.486/0.025 * (R_{max})^{2/3} * 0.002^{1/2} \rightarrow R_{max} = 1.68\,\text{ft}$.
Solve the following two equations simultaneously:
$40\,\text{ft}^2 = (b + 1 * y_n)y_n = (b + y_n)y_n$
$1.68\,\text{ft} = (40\,\text{ft}^2)/[b + 2 * (1 + 1^2)^{1/2}\, y_n] = 40/(b + 2.8284y_n) \rightarrow b = 23.8095 - 2.8284y_n$.
Substituting: $40 = (23.8095 - 2.8284y_n + y_n)y_n \rightarrow 1.8284{y_n}^2 - 23.8095y_n + 40 = 0$.
Solving by quadratic formula: $y_n = 11.04\,\text{ft} \rightarrow b = -7.42\,\text{ft}$ (meaningless, neglect)
$y_n = 1.98\,\text{ft} \rightarrow b = 18.21\,\text{ft}$

Check whether the overall guidelines are satisfied:

Flow area: $A = [(18.21\,\text{ft}) + 1 * (1.98\,\text{ft})] * (1.98\,\text{ft}) = 39.98\,\text{ft}^2$.
Top width: $B = (18.21\,\text{ft}) + 2 * 1 * (1.98\,\text{ft}) = 22.17\,\text{ft}$.
Hydraulic depth: $d = (39.98\,\text{ft}^2)/(22.17\,\text{ft}) = 1.80\,\text{ft}$.
Mean velocity: $V = (150\,\text{cfs})/(39.98\,\text{ft}^2) = 3.75\,\text{ft sec}^{-1}$
$V > 2\,\text{ft sec}^{-1} \rightarrow$ no sedimentation.
Froude number: $F_r = (3.75\,\text{ft sec}^{-1})/[(32.2\,\text{ft sec}^{-2}) * (1.80\,\text{ft})]^{1/2} = 0.49$
$F_r < 0.60 \rightarrow$ okay.

Determine the freeboard:

Table 6.3: $FB = 1.2 * (1.98\,\text{ft})^{1/2} = 1.69\,\text{ft} \rightarrow$ channel depth $= (1.98 + 1.69)\,\text{ft} = 3.67\,\text{ft}$.

In summary, the channel can be designed to have V_{max}. Its cross section has a bottom width of 18.21 ft and a channel depth of 3.67 ft. For such a design, the overall guidelines are satisfied.

(2) Figure 6.31: clear water, $d_{50} = 1\,\text{mm} \rightarrow \tau_{max} = 0.038\,\text{lbf ft}^{-2}$
Follow the right panel of Figure 6.30.

Try to design the channel to have τ_{max}:

$y_n = (0.038\,\text{lbf ft}^{-2})/[(62.4\,\text{lbf ft}^{-3}) * 0.002] = 0.30\,\text{ft}$.

Manning formula: $150/[(b + 1 * 0.30) * 0.30]\,\text{ft sec}^{-1} = 1.486/0.025 * ([(b + 1 * 0.30) * 0.30]/[b + 2 * (1 + 1^2)^{1/2} * 0.30])^{2/3} * 0.002^{1/2}$

$\rightarrow (b + 0.30)^{5/3}(b + 0.8485)^{-2/3} = 419.7223$

$\rightarrow$ trial-and-error or Excel Solver: $b = 419.79\,\text{ft}$.

Although the channel can be designed to τ_{max}, it may be a too wide and too shallow. One suggestion is to try different ratios of b/y_n.

Assume $b/y_n = 10$:

$b = 10y_n$.

Flow area: $A = (10y_n + 1y_n)y_n = 11{y_n}^2\,\text{ft}^2$.

Wetted perimeter: $P_{wet} = 10y_n + 2 * (1 + 1^2)^{1/2}y_n = 12.8284y_n\,\text{ft}$.

Hydraulic radius: $R = (11{y_n}^2\,\text{ft}^2)/(12.8284y_n\,\text{ft}) = 0.8575y_n\,\text{ft}$.

Mean velocity: $V = 150/(11{y_n}^2)\,\text{ft sec}^{-1} = 13.6364/{y_n}^2\,\text{ft sec}^{-1}$.

Manning formula: $13.6364/{y_n}^2\,\text{ft sec}^{-1} = 1.486/0.025 * (0.8575y_n)^{2/3} * 0.002^{1/2}$

$$\rightarrow y_n^{8/3} = 304.9192 \rightarrow y_n = 8.54\,\text{ft} \rightarrow b = 10 * (8.54\,\text{ft}) = 85.4\,\text{ft}.$$

Check the tractive force:

$\tau = (62.4\,\text{lbf ft}^{-3}) * (8.54\,\text{ft}) * 0.002 = 1.07\,\text{lbf ft}^{-2}$.

$\tau > \tau_{max} \rightarrow$ assume a much larger ratio of b/y_n.

Assume $b/y_n = 100$:

$b = 100y_n$.

Flow area: $A = (100y_n + 1y_n)y_n = 101{y_n}^2\,\text{ft}^2$.

Wetted perimeter: $P_{wet} = 100y_n + 2 * (1 + 1^2)^{1/2}y_n = 102.8284y_n\,\text{ft}$.

Hydraulic radius: $R = (101{y_n}^2\,\text{ft}^2)/(102.8284y_n\,\text{ft}) = 0.9822y_n\,\text{ft}$.

Mean velocity: $V = 150/(101{y_n}^2)\,\text{ft sec}^{-1} = 1.4851/{y_n}^2\,\text{ft sec}^{-1}$.

Manning formula: $1.4851/{y_n}^2\,\text{ft sec}^{-1} = 1.486/0.025 * (0.9822y_n)^{2/3} * 0.002^{1/2}$

$$\rightarrow {y_n}^{8/3} = 33.2078 \rightarrow y_n = 3.72\,\text{ft} \rightarrow b = 100 * (3.72\,\text{ft}) = 372.0\,\text{ft}.$$

Check the tractive force:

$\tau = (62.4\,\text{lbf ft}^{-3}) * (3.72\,\text{ft}) * 0.002 = 0.46\,\text{lbf ft}^{-2}$.

$\tau > \tau_{max} \rightarrow$ assume a much larger ratio of b/y_n.

It seems that the design to have τ_{max} is feasible.

Check whether the overall guidelines are satisfied:

Flow area: $A = [(419.79\,\text{ft}) + 1 * (0.30\,\text{ft})] * (0.30\,\text{ft}) = 126.03\,\text{ft}^2$

Top width: $B = (419.79\,\text{ft}) + 2 * 1 * (0.30\,\text{ft}) = 420.39\,\text{ft}$

Hydraulic depth: $d = (126.03\,\text{ft}^2)/(420.39\,\text{ft}) = 0.30\,\text{ft}$

Mean velocity: $V = (150\,\text{cfs})/(126.03\,\text{ft}^2) = 1.19\,\text{ft sec}^{-1}$

$V < 2\,\text{ft sec}^{-1} \rightarrow$ sedimentation would be an issue

Froude number: $F_r = (1.19\,\text{ft sec}^{-1})/[(32.2\,\text{ft sec}^{-2}) * (0.30\,\text{ft})]^{1/2} = 0.38$

$F_r < 0.60 \rightarrow$ okay.

Determine the freeboard:

Table 6.3: $FB = 1.2 * (0.30\,\text{ft})^{1/2} = 0.66\,\text{ft} \rightarrow$ channel depth $= (0.30 + 0.66)\,\text{ft} = 0.96\,\text{ft}$.

In summary, the channel can be designed to have τ_{max}. Its cross section has a bottom width of 419.79 ft and a channel depth of 0.96 ft. For such a design, the flow is subcritical but sedimentation would be a problem.

6.6.4 Grass-Lined Channel

Grass-lined channels are eco-environmentally sound and thus have been widely used as a best management practice (BMP) for watershed and stream restoration. They can have a cross section of trapezoidal, parabolic, or triangular shape (USDA-NRCS, 2007b), but trapezoidal shape is most common and thus is introduced here. The side slope of a grass-lined trapezoidal channel can be determined from Table 6.6. The process to design a grass-lined channel is shown in Figure 6.33. In terms of site-specific climate and soil conditions, the cover vegetation species is chosen, so that its vegetal retardance class and mechanic properties (namely upright vegetal height, h_s, vegetal stiffness, MEI, permissible vegetal shear, τ_p) can be determined from Tables 6.8 and 6.9, respectively. Relationships among these parameters are expressed in Eqs. (6.22) through (6.26).

The maximum allowable water depth is computed as:

$$y_{n,max} = \frac{\tau_p}{\gamma S_0} \tag{6.22}$$

maximum allowable water depth ($y_{n,max}$); permissible vegetal shear (Table 6.9) (τ_p); channel bed slope (S_0); specific weight of water (γ)

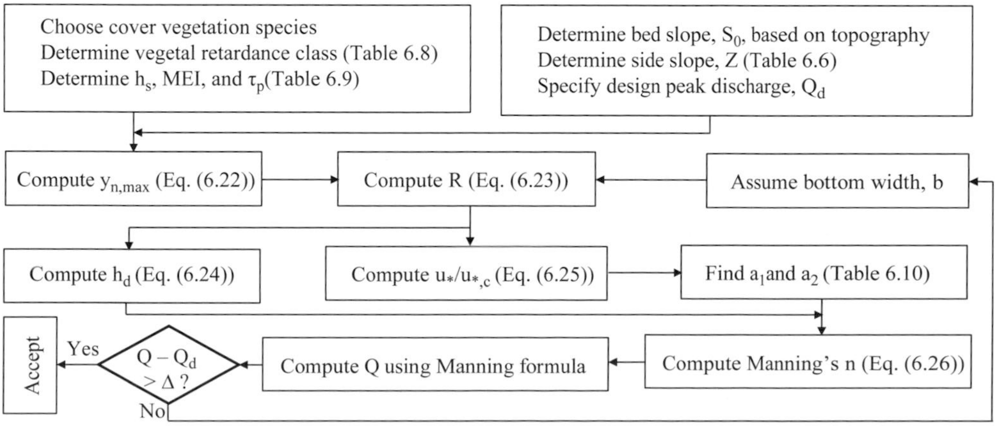

Figure 6.33 **Flowchart of the procedure to design a grass-lined channel.** Δ is a small positive value.

Table 6.8 Retardance classes of cover vegetation species for grass-lined channel.[1]

Retardance	Cover	Condition
A	Weeping lovegrass	Excellent stand, tall (average 30 in)
	Reed canarygrass or Yellow bluestem ischaemum	Excellent stand, tall (average 36 in)
B	Smooth bromegrass	Good stand, mowed (average 12 to 15 in)
	Bermudagrass	Good stand, tall (average 12 in)
	Native grass mixture (little bluestem, blue grama, and other long and short midwest grasses)	Good stand, unmowed
	Tall fescue	Good stand, unmowed (average 18 in)
	Sericea lespedeza	Good stand, not woody, tall (average 19 in)
	Grass-legume mixture—Timothy, smooth bromegrass, or orchardgrass	Good stand, uncut (average 20 in)
	Reed canarygrass	Good stand, uncut (average 12 to 15 in)
	Tall fescue, with birdsfoot trefoil or ladino clover	Good stand, uncut (average 18 in)
	Blue grama	Good stand, uncut (average 13 in)
C	Bahiagrass	Good stand, uncut (average 6 to 12 in)
	Bermudagrass	Good stand, mowed (average 6 in)
	Redtop	Good stand, headed (15 to 20 in)
	Grass-legume mixture—summer (orchardgrass, redtop, Italian ryegrass, and common lespedeza)	Good stand, uncut (6 to 8 in)
	Centipedegrass	Very dense cover (average 6 in)
	Kentucky bluegrass	Good stand, headed (6 to 12 in)
D	Bermudagrass	Good stand, cut to 2.5-in height
	Red fescue	Good stand, headed (12 to 18 in)
	Buffalograss	Good stand, uncut (3 to 6 in)
	Grass-legume mixture—fall, spring (orchardgrass, redtop, Italian ryegrass, and common lespedeza)	Good stand, uncut (4 to 5 in)
	Sericea lespedeza or Kentucky bluegrass	Good stand, cut to 2-in height. Very good stand before cutting
E	Bermudagrass	Good stand, cut to 15-in height
	Bermudagrass	Burned stubble

[1] *Source:* USDA-NRCS (2007b).

For an assumed bottom width, one can compute the hydraulic radius as:

channel side slope (Table 6.6)
assumed channel bottom width
maximum allowable water depth (Eq. (6.22))
hydraulic radius

$$R = \frac{(b + Zy_{n,max})\, y_{n,max}}{b + 2y_{n,max}\sqrt{1 + Z^2}} \tag{6.23}$$

Table 6.9 Mechanic properties of vegetal retardance classes for grass-lined channel.[1]

Retardance Class[2]	Upright vegetal height, h_s		Vegetal stiffness, MEI		Permissible vegetal shear, τ_p	
	cm	ft	N-m^2	lbf-ft^2	N m^{-2}	lbf ft^{-2}
A	91	3.0	300	725	177	3.7
B	61	2.0	20	50	100	2.1
C	20	0.66	0.5	1.2	48	1.0
D	10	0.33	0.05	0.12	29	0.6
E	4	0.13	0.005	0.012	17	0.35

[1] *Source:* USDA-NRCS (2007b).
[2] Defined in Table 6.8.

For the assumed bottom width, the deflected vegetal height can be computed as:

$$h_d = 0.14 h_s \left[\frac{1}{h_s} \left(\frac{MEI}{\gamma R S_0} \right)^{\frac{1}{4}} \right]^{1.59} \tag{6.24}$$

deflected vegetal height → h_d; upright vegetal height (Table 6.9) → h_s; vegetal stiffness (Table 6.9) → MEI; specific weight of water → γ; hydraulic radius (Eq. (6.23)) → R; channel bed slope → S_0

For the assumed bottom width, one can compute the ratio of shear velocity to its critical value as (USDA-NRCS, 2007b):

$$\frac{u_*}{u_{*,c}} = \frac{\sqrt{gRS_0}}{\min\left[0.028 + 6.33\,(MEI)^2,\, 0.23\,(MEI)^{0.106}\right]} \tag{6.25}$$

shear velocity → u_*; critical shear velocity → $u_{*,c}$; gravitational acceleration (= 9.81 m s^{-2}) → g; hydraulic radius (Eq. (6.23)) [m] → R; channel bed slope → S_0; minimum function → min; vegetal stiffness (Table 6.9) [N-m^2] → MEI

Finally, based on the computed ratio of $\frac{u_*}{u_{*,c}}$ from Eq. (6.25), one can find the two coefficients from Table 6.10. The value of Manning's n is estimated as:

$$n = \frac{K \, R^{\frac{1}{6}}}{\sqrt{8g}\left[a_1 + a_2 \log_{10}\left(\frac{R}{h_d}\right)\right]} \tag{6.26}$$

Manning's n → n; units conversion factor (= 1.486 for BG; 1.0 for SI) → K; hydraulic radius (Eq. 6.23) [ft or m] → R; gravitational acceleration (= 32.2 ft sec^{-2} or 9.81 m s^{-1}) → g; coefficients (Table 6.10) → a_1, a_2; upright vegetal height [ft or m] → h_d

In terms of the Manning's n (Eq. (6.26)), one can compute the mean velocity, V, using the Manning formula (Eq. (2.26)) and then the discharge, Q. If Q is greater than the design peak discharge, Q_d, by a small positive delta (e.g., $0.05Q_d$), the assumed b is acceptable; otherwise, assume a new value for b and repeat the above analyses.

Table 6.10 Coefficients for computing Manning's n using Eq. (6.26).[1]

$u_*/u_{*,c}$(Eq. (6.25))	a_1	a_2
≤ 1.0	0.15	1.85
(1.0, 1.5]	0.20	2.70
(1.5, 2.5]	0.28	3.08
> 2.5	0.29	3.50

[1] *Source:* USDA-NRCS (2007b).

Example 6.42 A bahiagrass-lined trapezoidal channel ($Z = 6$ and $S_0 = 0.0075$) is to be constructed to convey $4.8\,m^3\,s^{-1}$. Determine the channel bottom width and depth.

Solution

$Z = 6$, $S_0 = 0.0075$, $Q_d = 4.8\,m^3\,s$

Table 6.8: bahiagrass $\rightarrow$ retardance C.

Table 6.9: retardance C$\rightarrow h_s = 20\,cm = 0.2\,m$, MET $= 0.5\,N\text{-}m^2$, $\tau_p = 48\,N\,m^{-2}$.

Eq. (6.22): $y_{n,max} = (48\,N\,m^{-2})/[(9810\,N\,m^{-3}) * 0.0075] = 0.65\,m$.

Assume bottom width $b = 5\,m$:

Eq. (6.23): $R = [(5\,m + 6 * 0.65\,m) * (0.65\,m)]/[5\,m + 2 * (0.65\,m) * (1 + 6^2)^{1/2}] = 0.4482\,m$.

Eq. (6.24): $h_d = 0.14 * (0.2\,m) * \left[\frac{1}{0.2\,m} * \left(\frac{0.5\,N\text{ - }m^2}{(9810\,N\,m) * (0.4482\,m) * 0.0075}\right)^{\frac{1}{4}}\right]^{1.59} =$ 0.06845 m.

Eq. (6.25): $\frac{u_*}{u_{*,c}} \frac{\sqrt{(9.81\,m\,s^{-2}) * (0.4482\,m) * 0.0075}}{\min\left[0.028 + 6.33\left(0.5\,N\text{ - }m^2\right)^2, 0.23\left(0.5\,N\text{ - }m^2\right)^{0.106}\right]} = 0.8498.$

Table 6.10: $a_1 = 0.15$, $a_2 = 1.85$

Eq. (6.26): $n = \frac{1}{\sqrt{8 * (9.81\,m\,s^{-2})}} \frac{(0.4482\,m)^{\frac{1}{6}}}{\left[0.15 + 1.85 * \log_{10}\left(\frac{0.4482\,m}{0.06845\,m}\right)\right]} = 0.05950.$

Manning formula (Eq. (2.26)): $V = (1/0.05950) * (0.4482\,m)^{2/3} * 0.0075^{1/2} = 0.8524\,m\,s^{-1}$.

Flow area: $A = [(5\,m) + 6 * (0.65\,m)] * (0.65\,m) = 5.785\,m^2$.

Discharge: $Q = (0.8524\,m\,s^{-1}) * (5.785\,m^2) = 4.93\,m^3\,s^{-1}$.

$Q > Q_d \rightarrow$ the design is acceptable.

Determine freeboard (Table 6.3) and channel depth:

$FB = 0.7 * (0.65\,m)^{1/2} = 0.56\,m \rightarrow$ channel depth $= 0.65\,m + 0.56\,m = 1.21\,m$.

Check the consistency with the overall guidelines:

$V > 0.7\,m\,s^{-1} \rightarrow$ no sedimentation.

Top width: $B = 5\,m + 2 * 6 * (0.65\,m) = 12.8\,m$.

Hydraulic depth: $d = (5.785\,m^2)/(12.8\,m) = 0.4520\,m$.

Eq. (2.46): $F_r = (0.8524\,\mathrm{m\,s^{-1}})/[(9.81\,\mathrm{m\,s^{-2}}) * (0.4520\,\mathrm{m})]^{1/2} = 0.40 < 0.6 \rightarrow$ ok, subcritical.

In summary, the channel can have a bottom width of 5 m and depth of 1.21 m, and it satisfies the overall guidelines in terms of velocity and Froude number.

6.6.5 Riprap-Lined Channel

Riprap, which can be made from a variety of gravels and cobbles, is widely used as a channel lining material. Riprap-lined channels are also eco-environmentally sound and thus have been widely used as a BMP. They usually have a trapezoidal cross section, which can be designed as shown in Figure 6.34 (FHA, 2005). For an assumed riprap mean particle diameter and ratio of channel bottom width to normal depth, the critical shears and shear stresses are computed. This is iterated until the shear stresses are no larger than the corresponding critical shears. The riprap mean diameter can be assumed by referencing existing channels that have been constructed in previous projects. The freeboard can be estimated from Table 6.3. The parameters for a riprap-lined channel can be calculated using Eqs. (6.27) through (6.31).

Manning's n can be estimated as:

$$n = 0.04(d_{50})^{\frac{1}{6}} \tag{6.27}$$

Manning's n → n; d_{50} ← diameter of riprap particles finer than 50% [ft]

The critical shears are computed as:

$$\tau_{c,b} = 4d_{50} \tag{6.28}$$

critical bed shear [lbf ft^{-2}] → $\tau_{c,b}$; d_{50} ← diameter of riprap particles finer than 50% [ft]

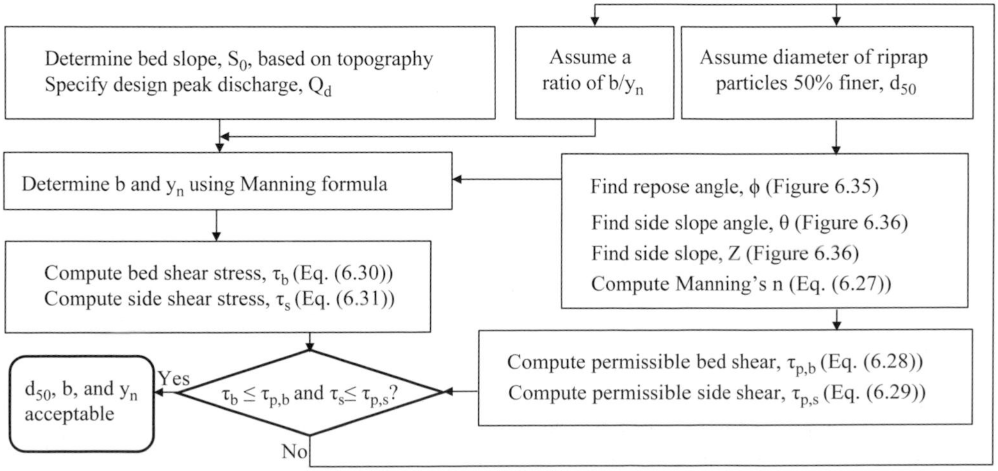

Figure 6.34 Flowchart of the procedure to design a riprap-lined channel.

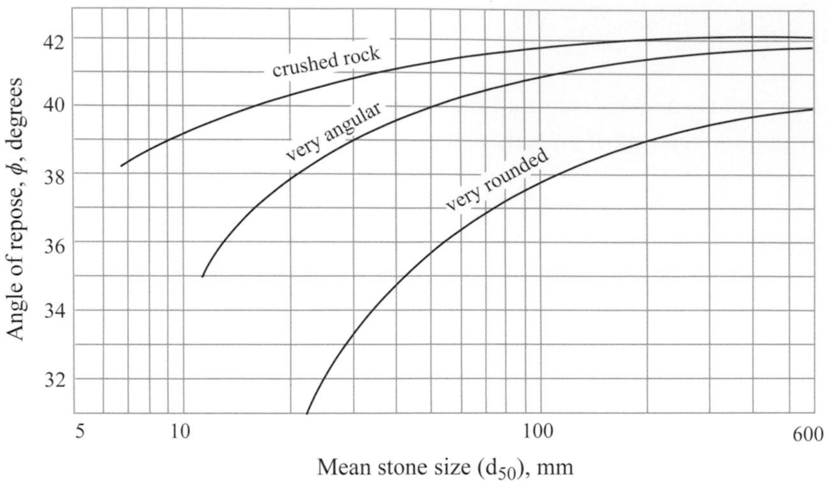

Figure 6.35 **Nomogram of repose angle of riprap particles in the design of a riprap-lined channel.** (*Source:* USDA-NRCS, 2007b)

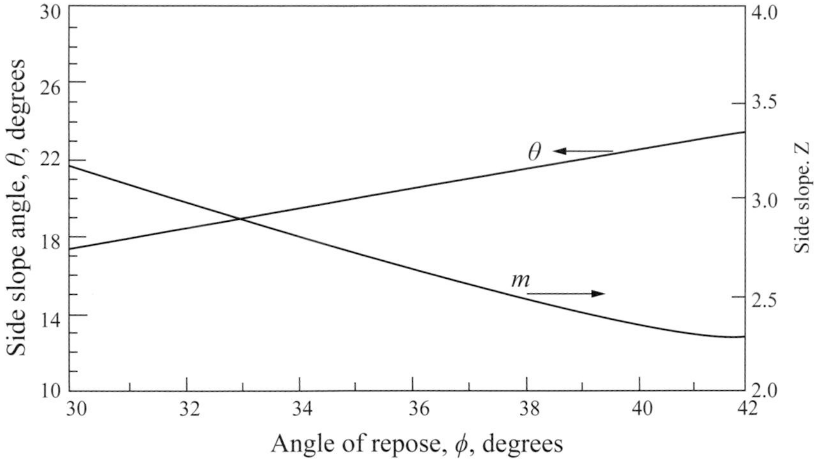

Figure 6.36 **Nomogram of side slope angle and side slope in the design of a riprap-lined channel.** (*Source:* USDA-NRCS, 2007b)

$$\tau_{c,s} = \tau_{c,b}\sqrt{1 - \frac{(\sin\theta)^2}{(\sin\phi)^2}} \tag{6.29}$$

critical side shear [lbf ft^{-2}] → $\tau_{c,s}$

critical bed shear [lbf ft^{-2}] (Eq. (6.28)) → $\tau_{c,b}$

side slope angle (Figure 6.36) → θ

repose angle (Figure 6.36) → ϕ

Finally, the shear stresses are computed as:

$$\underbrace{\tau_b}_{\text{bed shear stress [lbf ft}^{-2}]} = 1.5\underbrace{\gamma}_{\text{specific weight of water [lbf ft}^{-3}]}\underbrace{R}_{\text{hydraulic depth [ft]}}\underbrace{S_0}_{\text{channel bed slope}} \tag{6.30}$$

$$\underbrace{\tau_s}_{\text{side shear stress [lbf ft}^{-2}]} = 1.2\underbrace{\gamma}_{\text{specific weight of water [lbf ft}^{-3}]}\underbrace{R}_{\text{hydraulic depth [ft]}}\underbrace{S_0}_{\text{channel bed slopewe}} \tag{6.31}$$

Example 6.43 A riprap-lined trapezoidal channel ($S_0 = 0.001$) is to be constructed to convey 800 cfs. Crushed rock is to be used and the channel bottom width is not to exceed 15 ft. Determine the riprap size, side slope, bottom width, and channel depth.

Solution

$S_0 = 0.001$, $Q_d = 800\,\text{cfs}$, $b_{max} = 15\,\text{ft}$

Assume $d_{50} = 30\,\text{mm} = 0.09843\,\text{ft}$ and $b/y_n = 2.5$:

Figure 6.35: crushed rock $\rightarrow\phi = 41°$.

Figure 6.36: $\phi = 41° \rightarrow \theta = 23°$, $Z = 3.3$.

Eq. (6.27): $n = 0.04 * (0.09843\,\text{ft})^{1/6} = 0.0272$.

Compute the permissible shears:

Eq. (6.28): $\tau_{c,b} = 4 * (0.09843\,\text{ft}) = 0.3937\,\text{lbf ft}^{-2}$.

Eq. (6.29): $\tau_{c,s} = (0.3937\,\text{lbf ft}^{-2}) * [1 - (\sin 23°)^2/(\sin 41°)^2]^{1/2} = 0.3163\,\text{lbf ft}^{-2}$.

Determine b and y_n:

$b = 2.5y_n$.

Top width: $B = 2.5y_n + 2 * 3.3y_n = 9.1y_n\,\text{ft}$.

Flow area: $A = (2.5y_n + 9.3y_n)y_n = 5.8{y_n}^2\,\text{ft}^2$.

Wetted perimeter: $P_{wet} = 2.5y_n + 2 * (1 + 3.3^2)^{1/2}y_n = 9.3964y_n\,\text{ft}$.

Hydraulic radius: $R = (5.8{y_n}^2\,\text{ft}^2)/(9.3964y_n\,\text{ft}) = 0.6173y_n\,\text{ft}$.

Mean velocity: $V = (800\,\text{cfs})/(5.8{y_n}^2\,\text{ft}^2) = 137.931/{y_n}^2\,\text{ft sec}^{-1}$.

Manning formula: $137.931/{y_n}^2\,\text{ft sec}^{-1} = 1.486/0.0272 * (0.6173y_n\,\text{ft})^{2/3} * 0.001^{1/2}$

$$\rightarrow (y_n)^{8/3} = 110.1238 \rightarrow y_n = 5.83\,\text{ft} \rightarrow b = 2.5 * (5.83\,\text{ft}) = 14.58\,\text{ft},$$
$$B = 9.1 * (5.83\,\text{ft}) = 53.05\,\text{ft},\ A = 5.8 * (5.83\,\text{ft})^2 = 197.14\,\text{ft}^2,\ R =$$
$$0.6173 * (5.83\,\text{ft}) = 3.60\,\text{ft},\ V = 137.931/5.83^2\,\text{ft sec}^{-1} = 4.06\,\text{ft sec}^{-1}.$$

$b < bmax \rightarrow$ okay.

Compute shear stresses:

Eq. (6.30): $\tau_b = 1.5 * (62.4\,\text{lbf ft}^{-3}) * (3.60\,\text{ft}) * (0.001) = 0.3370\,\text{lbf ft}^{-2}$.

Eq. (6.31): $\tau_s = 1.2 * (62.4\,\text{lbf ft}^{-3}) * (3.60\,\text{ft}) * (0.001) = 0.2696\,\text{lbf ft}^{-2}$.

Check whether the shear stresses are less than the permissible shears:

$\tau_b < \tau_{c,b}$ and $\tau_s < \tau_{c,s} \rightarrow$ the assumed d_{50} and b/y_n are acceptable.

Check the consistency with the overall guidelines:

$V = 4.06\,\text{ft sec}^{-1} > 2\,\text{ft sec}^{-1} \rightarrow$ no sedimentation.

Hydraulic depth: $d = (197.14\,\text{ft}^2)/(53.05\,\text{ft}) = 3.72\,\text{ft}$.

$F_r = (4.06\,\text{ft sec}^{-1})/[(32.2\,\text{ft sec}^{-2}) * (3.72\,\text{ft})]^{1/2} = 0.37 < 0.6 \rightarrow$ ok, subcritical.

Determine the freeboard and channel depth:

Table 6.3: $FB = 1.2 * (5.83\,\text{ft})^{1/2} = 2.90\,\text{ft}$

$\rightarrow$channel depth $= 5.83\,\text{ft} + 2.99\,\text{ft} = 8.73\,\text{ft}$.

In summary, $d_{50} = 30\,\text{mm}$, $Z = 2.3$, $b = 14.58\,\text{ft}$, channel depth $= 8.73\,\text{ft}$. This design satisfies the overall guidelines in terms of velocity and Froude number.

PROBLEMS

6.1 Sketch the EGL and HGL of the following water system.

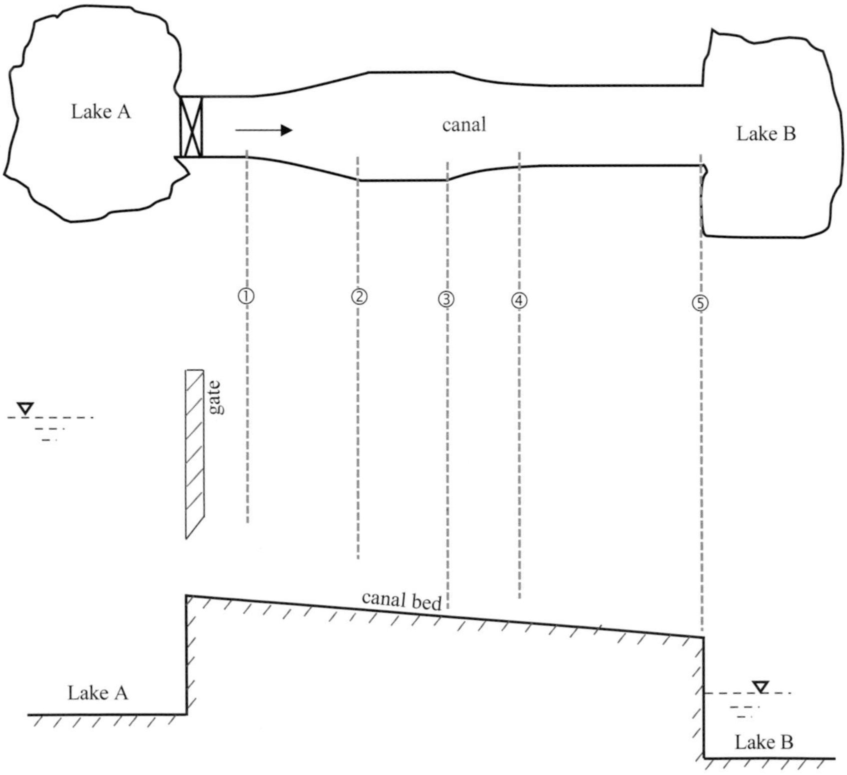

6.2 A 3-ft-wide rectangular channel (n = 0.012) is laid on a slope of 0.002. Determine the normal depth if the channel carries:
(1) 20 cfs
(2) 40 cfs
(3) 60 cfs.

6.3 A 3-ft-wide rectangular channel (n = 0.012) carries 30 cfs. Determine the normal depth if the channel is laid on a slope of:
(1) 0.001
(2) 0.003
(3) 0.005.

6.4 A triangular channel (n = 0.012) with a side slope of 1.5 is laid on a slope of 0.002. Determine the normal depth if the channel carries:
(1) 20 cfs
(2) 40 cfs
(3) 60 cfs.

6.5 A triangular channel (n = 0.012) with a side slope of 1.5 carries 30 cfs. Determine the normal depth if the channel is laid on a slope of:
(1) 0.001
(2) 0.003
(3) 0.005.

6.6 A trapezoidal channel (n = 0.028) with a bottom width of 10 m and a side slope of 3.0 is laid on a slope of 0.002. Determine the normal depth if the channel carries:
(1) $10\,m^3\,s^{-1}$
(2) $20\,m^3\,s^{-1}$
(3) $30\,m^3\,s^{-1}$.

6.7 A trapezoidal channel (n = 0.028) with a bottom width of 10 m and a side slope of 3.0 carries $25\,m^3\,s^{-1}$. Determine the normal depth if the channel is laid on a slope of:
(1) 0.001
(2) 0.003
(3) 0.005.

6.8 A 36-in-diameter circular channel (n = 0.013) is laid on a slope of 0.002. Determine the normal depth if the channel carries:
(1) 10 cfs
(2) 20 cfs
(3) 30 cfs.

6.9 A 36-in-diameter circular channel (n = 0.013) carries 10 cfs. Determine the normal depth if the channel is laid on a slope of:
(1) 0.001
(2) 0.003
(3) 0.005.

6.10 For the rectangular channel in Problem 6.2, determine its critical depth.

6.11 For the rectangular channel in Problem 6.3, determine its critical depth.

6.12 For the triangular channel in Problem 6.4, determine its critical depth.

6.13 For the triangular channel in Problem 6.5, determine its critical depth.

6.14 For the trapezoidal channel in Problem 6.6, determine its critical depth.

6.15 For the trapezoidal channel in Problem 6.7, determine its critical depth.

6.16 For the circular channel in Problem 6.8, determine its critical depth.

6.17 For the circular channel in Problem 6.9, determine its critical depth.

6.18 If a rectangular channel is controlled to have an approaching flow with a constant specific energy of 30 ft, determine its critical depth.

6.19 If a trapezoidal channel (bottom width of 5 m and side slope of 2.5) is controlled to have an approaching flow with a constant specific energy of 10 m, determine its critical depth.

6.20 Classify the slope of the rectangular channel in Problem 6.2.

6.21 Classify the slope of the rectangular channel in Problem 6.3.

6.22 Classify the slope of the triangular channel in Problem 6.4.

6.23 Classify the slope of the triangular channel in Problem 6.5.

6.24 Classify the slope of the trapezoidal channel in Problem 6.6.

6.25 Classify the slope of the trapezoidal channel in Problem 6.7.

6.26 Classify the slope of the circular channel in Problem 6.8.

6.27 Classify the slope of the circular channel in Problem 6.9.

6.28 A rectangular channel has three segments, which have the same critical depth but different normal depths. Let y_c be the critical depth and $y_{n,1}$, $y_{n,2}$, and $y_{n,3}$ be the normal depths in the upper, mid, and lower segments, respectively. The upper and lower segments have a steep slope, whereas the mid segment has a mild slope. Sketch and label all possible water surface profiles in the channel for an approaching water depth of:

(1) Larger than y_c

(2) Between $y_{n,1}$ and y_c

(3) Less than $y_{n,1}$.

6.29 A 50-m-long, 2-m-wide rectangular channel ($n = 0.013, S_0 = 0.002$) carries $3.5\,m^3\,s^{-1}$. If the channel ends in a free-fall (i.e., critical) flow, use the one-step method to compute the water surface profile by:

(1) Eq. (6.4)

(2) Eq. (6.5).

6.30 A 500-ft-long, 8-ft-wide rectangular channel ($n = 0.025$, $S_0 = 0.005$) carries 800 cfs. If the channel ends in a free-fall (i.e., critical) flow, use the multi-step method to compute the water surface profile by:

(1) Eq. (6.4)

(2) Eq. (6.5).

6.31 The figure shows a longitudinal cross section of a rectangular channel ($b = 5\,m$, $n = 0.013$) that consists of upper, mid, and lower segments. The critical depth occurs at the turning point of the upper and mid segments. If the channel carries $3.5\,m^3\,s^{-1}$, use the multi-step method by Eq. (6.4) or (6.5) to compute the water surface profile of the:

(1) Upper segment

(2) Mid segment

(3) Lower segment.

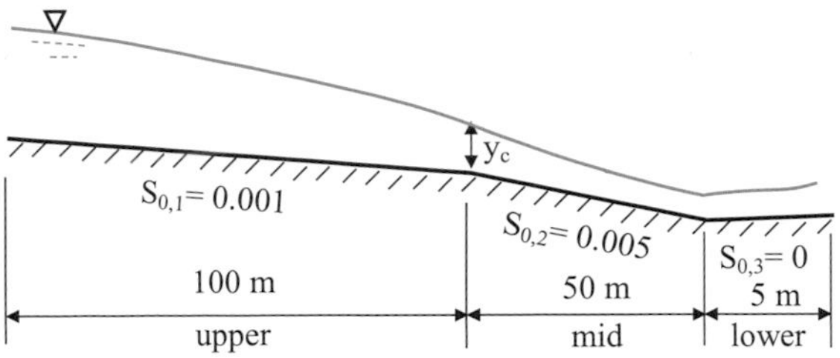

6.32 A 3-m-wide horizontal rectangular channel carries 12.9 $m^3\ s^{-1}$. If the depth at a section is 0.5 m, determine the:

(1) Conjugate depth
(2) Length of the hydraulic jump
(3) Energy dissipation by the hydraulic jump.

6.33 A horizontal triangular channel with a side slope of 0.5 carries 15 $m^3\ s^{-1}$. If the depth at a section is 0.8 m, determine the:

(1) Conjugate depth
(2) Length of the hydraulic jump
(3) Energy dissipation by the hydraulic jump.

6.34 A wide-shallow rectangular channel ($S_0 = 0.002$, $n = 0.025$) carries a unit-width flow rate of 70 $ft^2\ sec^{-1}$. A sluice gate is adjusted to produce a minimum depth of 1.0 ft just downstream of the gate. If the maximum water depth in the downstream channel is the normal depth,

(1) Determine whether a hydraulic jump will occur
(2) If so, estimate the distance from the gate to the jump.

6.35 The rectangular channel in Figure 6.11 conveys an approaching water flow with a depth of 7.5 ft and a velocity of 8.5 ft sec^{-1}. If $b_1 = 15$ ft, determine $b_{2,c}$, at which choking just occurs.

6.36 For the rectangular channel in Problem 6.35, determine the water depth at section ② if its width is:

(1) $b_2 = 5$ ft
(2) $b_2 = 10$ ft.

6.37 The rectangular channel in Figure 6.11 conveys an approaching water flow with a depth of 1.5 ft and a velocity of 10 ft sec^{-1}. If $b_1 = 15$ ft, determine:

(1) $b_{2,c}$, at which choking just occurs
(2) y_2 if $b_2 = 5$ ft
(3) y_2 if $b_2 = 9$ ft
(4) y_2 if $b_2 = 14$ ft.

6.38 The rectangular channel in Figure 6.12 conveys an approaching water flow with a depth of 7.5 ft and a velocity of 8.5 ft sec^{-1}. Determine $(\Delta z)_c$, at which choking just occurs.

6.39 For the rectangular channel in Problem 6.38, determine the water depth over the bed rise at section ② if its height is:
(1) $\Delta z = 0.5$ ft
(2) $\Delta z = 1.5$ ft
(3) $\Delta z = 2.5$ ft.

6.40 The rectangular channel in Figure 6.12 conveys an approaching water flow with a depth of 0.8 m and a velocity of 6.5 m s^{-1}. Determine:
(1) $(\Delta z)_c$, at which choking just occurs
(2) y_2 if $\Delta z = 0.5$ m
(3) y_2 if $\Delta z = 1.5$ m
(4) y_2 if $\Delta z = 2.0$ m.

6.41 A 1.5-ft-wide rectangular channel is used to supply water to a treatment plant. To minimize the transmission of standing waves, a supercritical contraction followed by a 4.5-ft-wide supercritical expansion is designed as the inlet waterway of the plant. If the approaching flow has a depth of 2.5 ft and a velocity of 5.5 ft sec^{-1}, determine the:
(1) Geometries of the inlet waterway
(2) Energy head loss between section ① and ②
(3) Water depth at section ④.

6.42 A 0.5-m-wide rectangular channel is used to supply water to a treatment plant. To minimize the transmission of standing waves, a supercritical contraction followed by a 1.0-m-wide supercritical expansion is designed as the inlet waterway of the plant. If the approaching flow has a depth of 1.2 m and a velocity of 4.5 m s^{-1}, determine the geometries of the inlet waterway and sketch the contraction and expansion.

6.43 A USOPF is used to continuously measure discharge in a rectangular channel. The installation height of the USOPF probe is 6.5 m above the crest of a flow-measuring device. On a day with air temperature of 20°C, if the time between pulse transmission and reception is measured to be 0.005 s by the host, determine the discharge if the device is a:
(1) 0.5-m-high, 1.5-m-wide sharp-crested rectangular weir (two-side encroachment)
(2) 0.5-m-high sharp-crested triangular weir with an angle of 60°
(3) 0.5-m-high, 1.5-m-wide broad-created weir
(4) Parshall flume with a bed rise of 0.3 m and a throat width of 2.0 m.

6.44 A concrete-lined channel is to be constructed to carry 5 to 20 cfs. The channel has a Manning's $n = 0.013$ and is to be laid on a longitudinal slope of 0.003. Design the channel if its shape is:
(1) Circular
(2) Triangular
(3) Rectangular
(4) Trapezoidal.

6.45 A concrete-lined channel is to be constructed to carry 0.5 to 1.5 m^3 s^{-1}. The channel has a Manning's $n = 0.015$ and is to be laid on a longitudinal slope of 0.005. Design the channel if its shape is:
(1) Circular

(2) Triangular
(3) Rectangular
(4) Trapezoidal.

6.46 Redesign the channel in Problem 6.44 to have the most efficient cross sections. What are the percent increases of the conveyance capacities?

6.47 Redesign the channel in Problem 6.45 to have most the efficient cross sections. What are the percent increases of the conveyance capacities?

6.48 A stiff-clay channel is to be constructed to convey water with colloidal silts at 200 cfs. Based on the site topography, the channel bed slope is determined to be 0.005. If the soils are compacted to have a void ratio of 0.3, design the channel cross section using the method of:
(1) Permissible velocity
(2) Permissible tractive force.

6.49 Redesign the channel in Problem 6.48 if it conveys clear water.

6.50 Redesign the channel in Problem 6.48 if it conveys water with noncolloidal sands.

6.51 A fine-gravel channel is to be constructed to convey clear water at 3.5 $m^3\,s^{-1}$. The channel is to be laid on a longitudinal slope of 0.003. If the soil particles have a mean diameter of 80 mm, design the channel cross section using the method of:
(1) Permissible velocity
(2) Permissible tractive force.

6.52 A blue grama-lined trapezoidal channel ($Z = 3.5$ and $S_0 = 0.002$) is to be constructed to convey 6300 cfs. Determine the channel bottom width and depth.

6.53 A red fescue-lined trapezoidal channel ($Z = 2.5$ and $S_0 = 0.005$) is to be constructed to convey 5.5 $m^3\,s^{-1}$. Determine the channel bottom width and depth.

6.54 A riprap-lined trapezoidal channel ($S_0 = 0.002$) is to be constructed to convey 500 cfs. Crushed rock is to be used and the channel bottom width is not to exceed 30 ft. Determine the riprap size, side slope, bottom width, and channel depth.

6.55 A riprap-lined trapezoidal channel ($S_0 = 0.002$) is to be constructed to convey 15 $m^3\,s^{-1}$. Very angular rock is to be used and the channel bottom width is not to exceed 5 m. Determine the riprap size, side slope, bottom width, and channel depth.

7 Pressurized Pipe Flow

In water resources engineering systems, one important infrastructure for conveying water is pipes that are connected in various manners. A pipe is a single segment of conduit. It is made of one type of material but can have either a constant or variant diameter. A pipeline consists of two or more pipes that are hydraulically connected at joints but do not form a loop, whereas a pipe network consists of three or more pipes that are hydraulically connected at joints and form one or more loops. This chapter introduces the application of the continuity equation (Eq. (2.18)) and energy equation (Eq. (2.21)) to formulate governing equations for a single pipe, a pipeline, and a pipe network. The approach taken is primarily through worked examples. First, the chapter introduces the principles of hydraulic and energy grade lines along a pipe or pipeline, and how to sketch them. Second, the chapter discusses viscous sublayer theory and the classification of smooth versus rough pipes. Third, the chapter discusses the governing equations of single pipes, pipelines, and pipe networks, and how to solve them using Excel Solver. Finally, the chapter analyzes flow through an orifice, which can be treated as a single pipe with a very short length, and methods of flow measurement.

7.1 Hydraulic and Energy Grade Lines

At a given location along a pressurized pipe (Figure 7.1), a piezometric tube can be installed to measure the hydrostatic pressure head, whereas a pitot tube can be installed with its inlet facing straightly against the flow direction to measure the total energy head (Finnemore and Franzini, 2002). If a series of piezometric tubes are installed along the pipe, a line formed by connecting the water surfaces in the tubes represents the hydraulic grade line (HGL), while if a series of pitot tubes are installed along the pipe, a line formed by connecting the water surfaces

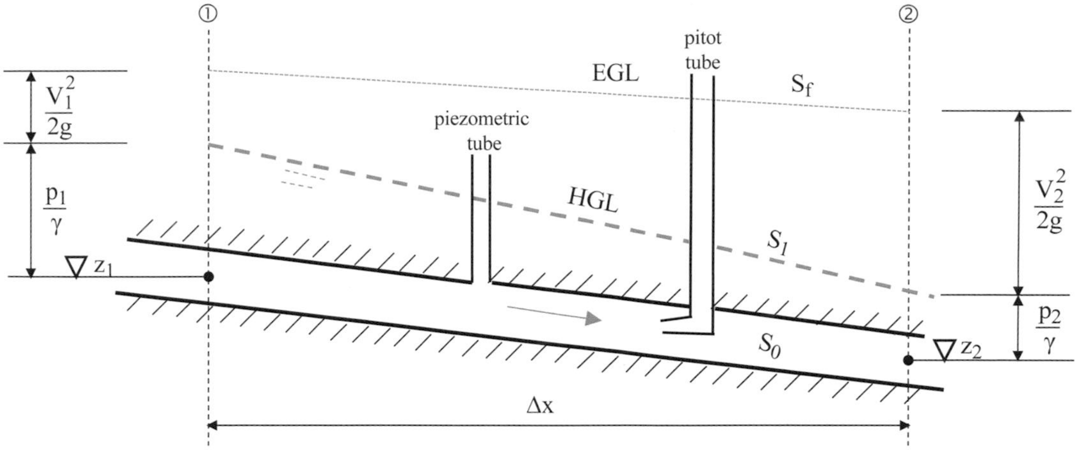

Figure 7.1 Longitudinal cross section of a pressurized pipe. Diagram shows the hydraulic grade line (HGL) and energy grade line (EGL).

in the tubes represents the energy grade line (EGL), as illustrated in Figure 7.1. The EGL defines how total energy head varies along the flow direction in the pipe and is governed by Eq. (2.19). Herein, the total energy head at any location is the summation of the pipe centerline elevation, the hydrostatic pressure head at the centerline, and the mean velocity head of the pipe cross section.

In a single pressurized pipe, friction loss can be computed using the Darcy–Weisbach equation (Eq. (2.23)) or Hazen–Williams (i.e., H–W) equation (Eq. (2.29)), but it cannot be computed using the Manning formula (Eq. (2.26)) because this formula is applicable for open channel flow only. Along a pressurized pipeline, minor losses can be computed using the formulas and coefficients presented in Table 2.3. For either a single pipe or pipeline, the HGL is governed by the equation expressed as:

pipe centerline elevation (z); hydrostatic pressure at pipe centerline (p); specific weight of water (γ); HGL slope (S_f); distance along flow direction (x)

$$\frac{d\left(z + \frac{p}{\gamma}\right)}{dx} = -S_f \tag{7.1}$$

For a pipeline, in terms of energy equation (Eq. (2.21)), the HGL and EGL can be conceptually sketched. It is convenient to first sketch the EGL and then the HGL, both starting from the furthest upstream point of the pipeline. Sketching does not have to be drawn to a scale, but it must be conceptually correct. For instance, sketching the EGL must illustrate its dips due to minor (e.g., entrance, contraction, expansion, fitting, and exit) losses and its inclinations due to friction losses, and sketching HGL must illustrate its ups and downs due to varying velocity heads. Without the addition of external energy (e.g., by a pump), along the flow direction downward, it is impossible for the EGL to rise or to have a positive slope because energy cannot be created by itself; it is possible, however, for the HGL to rise or to have a positive slope if the pipe diameter increases, causing the velocity to continually decrease, and vice versa.

Example 7.1 Sketch the EGL and HGL along a pressurized box conduit that conveys water from Pond A to Pond B (Figure 7.2).

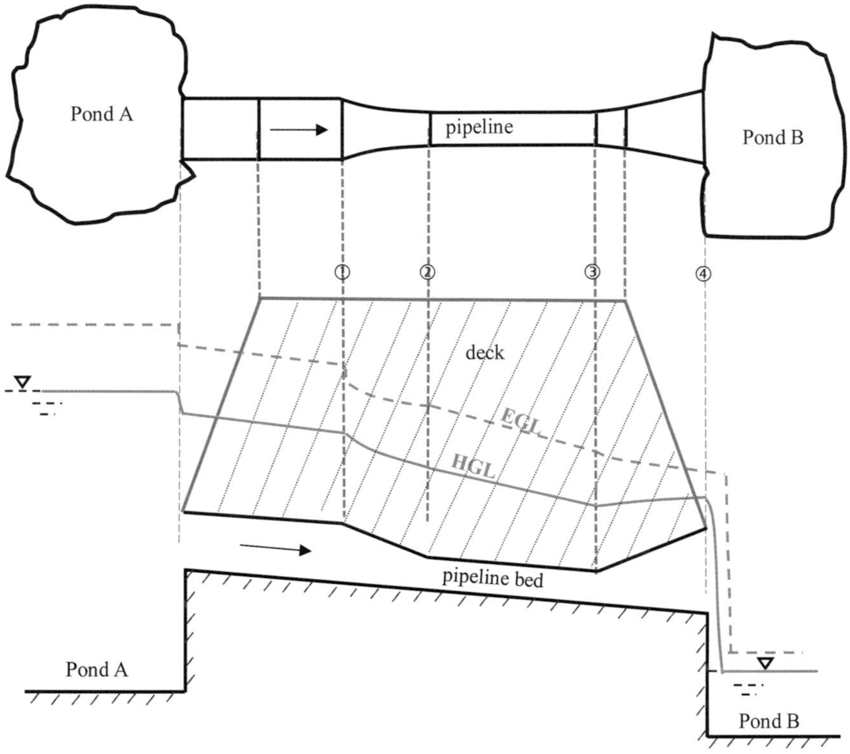

Figure 7.2 Schematic diagram illustrating two ponds connected by a pressurized conduit used in Example 7.1. The top panel shows the map (i.e., bird's-eye) view and the bottom panel shows a cross-sectional (i.e., side) view of the system.

Solution

First, let us sketch the EGL (blue thin dash line in Figure 7.2). In Pond A, the EGL is at the pond water surface if the approaching velocity is negligible but is above the pond water surface if the velocity is non-negligible. At the entrance of the conduit, due to entrance loss, the EGL has a sudden drop. From the entrance to section ①, due to friction loss, the EGL is linearly inclined downstream. From sections ① to ②, due to contraction and friction losses, the EGL is curved with an overall downward gradient. From sections ② to ③, due to friction loss, the EGL is linearly inclined downstream. From sections ③ to ④, due to expansion and friction losses, the EGL is curved with an overall downward gradient. At section ④, the EGL plunges down. In Pond B, the EGL is at the pond water surface if the leaving velocity is negligible but is above the pond water surface if the velocity is non-negligible.

Second, let us sketch the HGL (blue thin solid line in Figure 7.2). Note that at any given location along the conduit, the difference between the EGL and HGL is equal to the velocity head. In Pond A, the HGL is at the pond water surface, sharply dropping by the entrance velocity head. After the entrance, the HGL gradually declines and is parallel to the EGL. From sections ① to ②, the HGL is curved, with an overall increasing vertical distance away from the EGL because the velocity tends to increase along the flow direction. From sections ② to ③, because the conduit cross-sectional area is invariant and thus the velocity head is constant, the HGL is parallel to the EGL. From sections ③ to ④, the HGL is curved, with an overall decreasing vertical distance away from the EGL because the velocity tends to decrease along the downstream direction. From section ④ to the water surface of Pond B, the HGL plunges down with a nappe shape. In Pond B, the HGL is at the pond water surface.

7.2 Smooth Versus Rough Pipes

When using the Moody diagram (Figure 2.11) to determine the Darcy friction factor, f, one needs to classify a pipe of interest as either smooth or rough. As illustrated in Figure 7.3, the classification depends on the relative magnitude of the pipe surface roughness height, k_s, and the viscous sublayer thickness, δ_υ. If $k_s > \delta_\upsilon$, the pipe is rough, whereas if $k_s < \delta_\upsilon$, the pipe is smooth. The viscous sublayer is a very thin layer of water near the pipe wall, within which shear stress is dominant and velocity is very small (Hwang and Houghtalen, 1996; Finnemore and Franzini, 2002). The viscous sublayer theory states that : (1) the velocity within the sublayer linearly increases with distance away from, and is equal to zero at, the pipe wall surface; (2) the sublayer thickness, δ_υ, tends to decrease with an increase in flow turbulence and thus an increase in Reynolds number, R_e (Eq. (2.25)); and (3) shear stress at the pipe wall surface tends to increase as the flow turbulance increases. Thus, for a given pipe, it is probably a smooth pipe if it carries a laminar flow or turbulent flows with a relatively small R_e, but it may become a rough pipe if the flow is turbulent with a R_e large enough to make $\delta_\upsilon < k_s$.

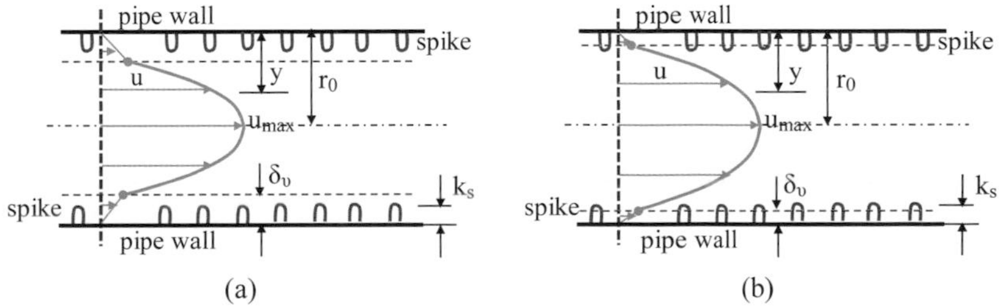

Figure 7.3 **Longitudinal cross section of pressured pipe indicating viscous sublayer and water velocity.** Velocity distribution is shown with light blue solid curve and viscous sublayer is indicated with red dash line. Diagram for: (a) smooth pipe; and (b) rough pipe. k_s: pipe wall surface roughness height; δ_υ: viscous sublayer thickness; u: point velocity at distance y from the pipe wall surface; u_{max}: maximum velocity at pipe centerline; and r_0: pipe radius.

Several relationships have been formulated to describe the viscous sublayer theory. Rewriting the shear stress at the pipe wall surface in units of velocity, one can define shear velocity (also called friction velocity) as (Schlichting and Gersten, 1999):

shear velocity

$$u_* = \sqrt{\frac{\tau_0}{\rho}} \tag{7.2}$$

shear stress at pipe wall surface

density of water

Shear stress at the pipe wall surface is computed as (Finnemore and Franzini, 2002):

shear stress at pipe wall surface

energy grade line slope

$$\tau_0 = \gamma R S_f \tag{7.3}$$

specific weight of water hydraulic radius (Table 2.4)

The viscous sublayer thickness can be computed as (Schlichting and Gersten, 1999; Finnemore and Franzini, 2002):

kinematic viscosity of water

viscous sublayer thickness

$$\delta_v = \begin{cases} \dfrac{70\nu}{u_*} & \text{for } R_e < 3000 \\[2mm] \dfrac{5\nu}{u_*} & \text{for } R_e > 3000 \end{cases} \tag{7.4}$$

shear velocity (Eq. (7.2))

Reynolds number (Eq. (2.25))

For a pressurized smooth pipe, as shown in Figure 7.3a, the velocity distribution can be approximated as one of two forms. The first form (Eq. (7.5)) assumes a linear distribution within the viscous sublayer and a logarithmic distribution from the sublayer to the pipe centerline. The second form (Eq. (7.6)) assumes a power functional distribution from the pipe wall surface to the centerline without differentiating within and outside the viscous sublayer. The equations for these two forms can be written as (Hwang and Houghtalen, 1996; Schlichting and Gersten, 1999):

point velocity

kinematic viscosity of water

distance away from pipe wall surface

shear velocity (Eq. (7.2))

$$\frac{u}{u_*} = \begin{cases} \dfrac{u_*}{\nu} y & \text{for} \quad 0 \le y < \dfrac{5\nu}{u_*} \\[3mm] \dfrac{u_*}{15\nu}\left[(5.75\log_{10}(20)+0.5)\,y - (28.75\log_{10}(20)-72.5)\,\dfrac{\nu}{u_*}\right] & \text{for} \quad \dfrac{5\nu}{u_*} \le y \le \dfrac{20\nu}{u_*} \\[3mm] 5.75\log_{10}\left(\dfrac{u_*}{\nu} y\right) + 5.5 & \text{for} \quad \dfrac{20\nu}{u_*} < y < \dfrac{10^5\nu}{u_*} \end{cases}$$

$$\tag{7.5}$$

point velocity

shape factor related to pipe material, viscosity of water, and flow turbulence

distance away from pipe wall surface

$$\frac{u}{u_{max}} = \left(\frac{y}{r_0}\right)^m \tag{7.6}$$

maximum velocity at pipe centerline

pipe radius

For a pressurized rough pipe, as shown in Figure 7.3b, the velocity distribution can be approximated as a linear distribution from the pipe wall surface to its roughness height and a logarithmic distribution beyond. The distributions can be described as (Hwang and Houghtalen, 1996; Schlichting and Gersten, 1999):

$$\frac{u}{u_*} = \begin{cases} \dfrac{8.5}{k_s} y & \text{for } y < k_s \\ 5.75 \log_{10}\left(\dfrac{y}{k_s}\right) + 8.5 & \text{for } y \geq k_s \end{cases} \tag{7.7}$$

(u: point velocity; u_*: shear velocity (Eq. (7.2)); y: distance away from pipe wall surface; k_s: pipe wall surface roughness height)

Example 7.2 A 1.2-in-diameter cast iron pipe ($k_s = 0.002$ in, $f = 0.017$) carries 0.05 cfs water 60°F. The pipe is pressurized.

(1) Determine the viscous sublayer thickness.
(2) Classify the pipe.
(3) Determine and sketch the velocity distribution.

Solution

$D = 1.2\text{ in} = 0.1\text{ ft}$, $k_s = 0.002$ in, $f = 0.017$, $Q = 0.05$ cfs, $g = 32.2\text{ ft sec}^{-2}$

Table 2.1: 60°F water $\rightarrow \rho = 1.94\text{ slug ft}^{-3}$, $\gamma = 62.4\text{ lbf ft}^{-3}$, $\upsilon = 1.22 \times 10^{-5}\text{ ft}^2\text{ sec}^{-1}$.

Flow area: $A = (\pi/4)^*(0.1\text{ ft})^2 = 0.007854\text{ ft}^2$.

Hydraulic radius: $R = (0.1\text{ ft})/4 = 0.025\text{ ft} = 0.025^*12\text{ in} = 0.3\text{ in}$.

Mean velocity: $V = (0.05\text{ cfs})/(0.007854\text{ ft}^2) = 6.3662\text{ ft sec}^{-1}$.

Eq. (2.25): $R_e = \left[(6.3662\text{ ft sec}^{-1})^*(4^*0.025\text{ ft})\right]/(1.22 \times 10^{-5}\text{ ft}^2\text{ sec}^{-1}) = 52{,}182$.

(1) Darcy–Weisbach equation (Eq. (2.23))

$$\rightarrow S_f = h_f/L = 0.017/(4^*0.025\text{ ft})^*(6.3662\text{ ft sec}^{-1})^2/(2^*32.2\text{ ft sec}^{-2}) = 0.1070$$

Eq. (7.3): $\tau_0 = (62.4\text{ lbf ft}^{-3})^*(0.025\text{ ft})^*(0.1070) = 0.1669\text{ lbf ft}^{-2}$.

Eq. (7.2): $u_* = [(0.1669\text{ lbf ft}^{-2})/(1.94\text{ slug ft}^{-3})]^{1/2} = 0.2933\text{ ft sec}^{-1}$.

Eq. (7.4): $R_e > 3000 \rightarrow \delta_\upsilon = (5^*1.22 \times 10^{-5}\text{ ft}^2\text{ sec}^{-1})/(0.2933\text{ ft sec}^{-1}) = 2.0798 \times 10^{-4}$ ft.

(2) $k_s = 0.002$ in, $\delta_\upsilon = 2.0798 \times 10^{-4}\text{ ft} = 0.002496\text{ in} \rightarrow k_s < \delta_\upsilon \rightarrow$ smooth pipe.

(3) Smooth pipe $\rightarrow$ use Eq. (7.5) to compute the velocities for selected distances away from the pipe wall surface, as summarized and plotted in Figure 7.4. The sample computations are as follows:

$$5\upsilon/u^* = 5^*(1.22 \times 10^{-5}\text{ ft}^2\text{ sec}^{-1})/(0.2933\text{ ft sec}^{-1}) = 2.0798 \times 10^{-4}\text{ ft}$$
$$20\upsilon/u_* = 20^*(1.22 \times 10^{-5}\text{ ft}^2\text{ sec}^{-1})/(0.2933\text{ ft sec}^{-1}) = 8.3191 \times 10^{-3}\text{ ft}$$
$$10^5\upsilon/u_* = 10^{5*}(1.22 \times 10^{-5}\text{ ft}^2\text{ sec}^{-1})/(0.2933\text{ ft sec}^{-1}) = 4.1596\text{ ft}.$$

	A	B	C	D	E	F	G
1	u* =	0.29330	ft sec^{-1}		5υ/u* =	2.0798E-04	ft
2	υ =	1.22E-05	ft^2 sec^{-1}		20υ/u* =	8.3191E-04	ft
3	r$_0$ =	0.05	ft		10^{5}υ/u* =	4.1596	ft
4	δ$_υ$ =	2.0798E-04	ft				
5							
6	y (ft)	u/u*	u (ft sec^{-1})				
7	0	0.00	0.00				
8	0.0001	2.40	0.71				
9	0.0002	4.81	1.41				
10	0.0002080	5.00	1.47				
11	0.00035	6.82	2.00				
12	0.00055	9.37	2.75				
13	0.00075	11.93	3.50				
14	0.0008319	12.98	3.81				
15	0.002	15.17	4.45				
16	0.005	17.46	5.12				
17	0.01	19.19	5.63				
18	0.02	20.92	6.14				
19	0.025	21.48	6.30				
20	0.03	21.93	6.43				
21	0.035	22.32	6.55				
22	0.04	22.65	6.64				
23	0.045	22.95	6.73				
24	0.05	23.21	6.81				

point velocity → u; shear velocity → u_* (Eq. 7-2); y: distance away from pipe wall surface; ν: kinematic viscosity of water

$$\frac{u}{u_*}=\begin{cases}\frac{u_*}{\nu}y & \text{for } 0\le y<\frac{5\nu}{u_*}\\ \frac{u_*}{15\nu}\left[(5.75\log_{10}(20)+0.5)y-(28.75\log_{10}(20)-72.5)\frac{\nu}{u_*}\right] & \text{for } \frac{5\nu}{u_*}\le y\le\frac{20\nu}{u_*}\\ 5.75\log_{10}\left(\frac{u_*}{\nu}y\right)+5.5 & \text{for } \frac{20\nu}{u_*}<y<\frac{10^5\nu}{u_*}\end{cases} \quad (7\text{-}5)$$

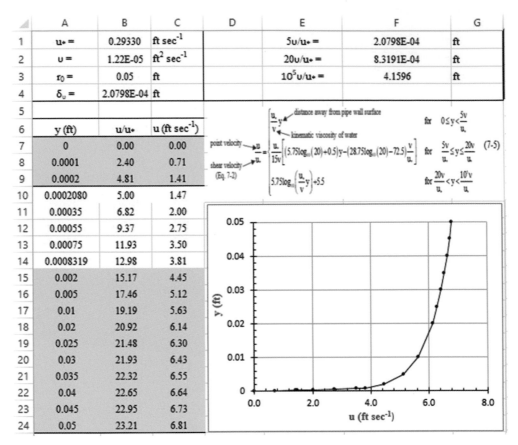

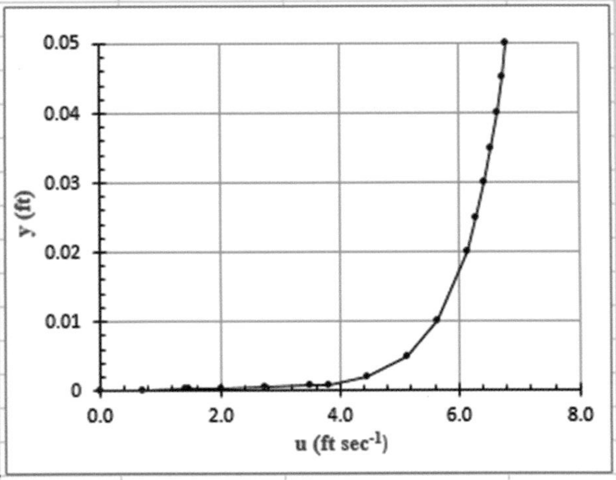

Figure 7.4 **Excel spreadsheet for computing velocities in the pipe for part (3) of Example 7.2.**

For y = 0.0002 ft:

$$0 \le y < 5\upsilon/u_* \rightarrow u/u_* = (0.2933\ \text{ft sec}^{-1})/(1.22\times10^{-5}\ \text{ft}^2\ \text{sec}^{-1})^*(0.0002\ \text{ft}) = 4.81$$
$$\rightarrow u = 1.12^*(0.2933\ \text{ft sec}^{-1}) = 1.41\ \text{ft sec}^{-1}.$$

For y = 0.00055 ft:

$$5\upsilon/u_* \le y < 20\upsilon/u_*$$
$$\rightarrow u/u_* = (0.2933\ \text{ft sec}^{-1})/(15^*1.22\times10^{-5}\ \text{ft}^2\ \text{sec}^{-1})^*[(5.75\log_{10}(20)+0.5)^*(0.00055\ \text{ft}) - (28.75\log_{10}(20)-72.5)^*(1.22\times10^{-5}\ \text{ft}^2\ \text{sec}^{-1})/(0.2933\ \text{ft sec}^{-1})] = 9.37$$
$$\rightarrow u = 9.37^*(0.2933\ \text{ft sec}^{-1}) = 2.75\ \text{ft sec}^{-1}.$$

For y = 0.005 ft:

$$20\upsilon/u_* < y < 10^5\upsilon/u_*$$
$$\rightarrow u/u_* = 5.75^* \log_{10}[(0.2933\ \text{ft sec}^{-1})/(1.22\times10^{-5}\ \text{ft}^2\ \text{sec}^{-1})^*(0.005\ \text{ft})] + 5.5$$
$$= 17.46$$
$$\rightarrow u = 17.46^*(0.2933\ \text{ft sec}^{-1}) = 5.12\ \text{ft sec}^{-1}.$$

Similar computations can be completed for other selected values of y. The results are shown in cells A6:C24, and plotted in Figure 7.4. Plotting C7:C24 along the x-axis versus A7:A24 along the y-axis yields the velocity distribution from the pipe wall surface to the centerline. As expected, the maximum velocity of 6.81 ft sec^{-1} occurs at the centerline.

Example 7.3 Repeat Example 7.2 for an increased flow rate of 0.15 cfs.

Solution

$D = 1.2\ \text{in} = 0.1\ \text{ft},\ k_s = 0.002\ \text{in},\ f = 0.017,\ Q = 0.15\ \text{cfs},\ g = 32.2\ \text{ft sec}^{-2}$

Table 2.1: 60°F water $\rightarrow \rho = 1.94\ \text{slug ft}^{-3},\ \gamma = 62.4\ \text{lbf ft}^{-3},\ \upsilon = 1.22\times10^{-5}\ \text{ft}^2\ \text{sec}^{-1}$.

Flow area: $A = (\pi/4)^*(0.1\ \text{ft})^2 = 0.007854\ \text{ft}^2$.

Hydraulic radius: $R = (0.1\ \text{ft})/4 = 0.025\ \text{ft} = 0.025^*12\ \text{in} = 0.3\ \text{in}$.

Mean velocity: $V = (0.15\ \text{cfs})/(0.007854\ \text{ft}^2) = 19.0985\ \text{ft sec}^{-1}$.

Eq. (2.25): $R_e = \left[(19.0985\ \text{ft sec}^{-1})^*(4^*0.025\ \text{ft})\right]/(1.22\times10^{-5}\ \text{ft}^2\ \text{sec}^{-1}) = 1{,}565{,}451$.

(1) Darcy–Weisbach equation (Eq. (2.23))

$$\rightarrow S_f = h_f/L = 0.017/(4^*0.025\ \text{ft})^*(19.0985\ \text{ft sec}^{-1})^2/(2^*32.2\ \text{ft sec}^{-2}) = 0.9629.$$

Eq. (7.3): $\tau_0 = (62.4\ \text{lbf ft}^{-3})^*(0.025\ \text{ft})^*(0.9629) = 1.5021\ \text{lbf ft}^{-2}$.

Eq. (7.2): $u_* = [(1.5021\ \text{lbf ft}^{-2})/(1.94\ \text{slug ft}^{-3})]^{1/2} = 0.8799\ \text{ft sec}^{-1}$.

Eq. (7.4): $R_e > 3000 \rightarrow \delta_\upsilon = (5^*1.22\times10^{-5}\ \text{ft}^2\ \text{sec}^{-1})/(0.8799\ \text{ft sec}^{-1}) = 6.9326\times10^{-5}\ \text{ft}$.

In comparison with Example 7.2, the viscous sublayer thickness decreases as a result of an increase in the flow rate.

(2) $k_s = 0.002\ \text{in},\ \delta_\upsilon = 6.9326\times10^{-5}\ \text{ft} = 0.0008319\ \text{in} \rightarrow k_s > \delta_\upsilon \rightarrow$ rough pipe.

(3) Rough pipe $\rightarrow$ use Eq. (7.7) to compute the velocities for selected distances away from the pipe wall surface, as summarized and plotted in Figure 7.5. The sample computations are as follows.

For y = 0.0001 ft:

$$y < k_s$$
$$\rightarrow u/u_* = 8.5/(0.002/12\ \text{ft})^*(0.0001\ \text{ft}) = 5.10$$
$$\rightarrow u = 16.99^*(0.8799\ \text{ft sec}^{-1}) = 14.95\ \text{ft sec}^{-1}.$$

For y = 0.005 ft:

$$y \geq k_s$$

$$\rightarrow u/u_* = 5.75^* \log_{10}[(0.005\,\text{ft})/(0.002/12\,\text{ft})] + 8.5 = 16.99$$

$$\rightarrow u = 16.99^*(0.8799\,\text{ft sec}^{-1}) = 14.95\,\text{ft sec}^{-1}.$$

Similar computations can be completed for other selected values of y. The results are shown in cells A6:C24, and plotted in Figure 7.5. Plotting C7:C24 along the x-axis versus A7:A24 along the y-axis yields the velocity distribution from the pipe wall surface to the centerline. As expected, in comparison with Example 7.2, as a result of an increase in the flow rate, the velocity at a given point is increased. The maximum velocity is increased by $[(20.01\,\text{ft sec}^{-1}) - (6.81\,\text{ft sec}^{-1})]/[(0.15\,\text{cfs}) - (0.05\,\text{cfs})] = 132\,\text{ft sec}^{-1}$ per unit cfs flow rate increase.

	A	B	C	D	E	F	G
1	u_* =	0.8799	ft sec^{-1}		k_s =	1.6667E-04	ft
2	υ =	1.22E-05	$ft^2\ sec^{-1}$				
3	r_0 =	0.05	ft				
4	δ_υ =	6.9326E-05	ft				
5							
6	y (ft)	u/u_*	u (ft sec^{-1})				
7	0	0.00	0.00				
8	0.00005	2.55	2.24				
9	0.0001	5.10	4.49				
10	0.00016667	8.50	7.48				
11	0.0005	11.24	9.89				
12	0.001	12.97	11.42				
13	0.0025	15.26	13.43				
14	0.004	16.44	14.46				
15	0.005	16.99	14.95				
16	0.01	18.72	16.48				
17	0.015	19.74	17.37				
18	0.02	20.46	18.00				
19	0.025	21.01	18.49				
20	0.03	21.47	18.89				
21	0.035	21.85	19.23				
22	0.04	22.19	19.52				
23	0.045	22.48	19.78				
24	0.05	22.74	20.01				

$$\frac{u}{u_*} = \begin{cases} \dfrac{8.5}{k_s}y & \text{for } y < k_s \\ 5.75\log_{10}\left(\dfrac{y}{k_s}\right) + 8.5 & \text{for } y \geq k_s \end{cases} \qquad (7\text{-}7)$$

point velocity → u; shear velocity (Eq. 7-2) → u_*; distance away from pipe wall surface → y; pipe wall surface roughness height → k_s

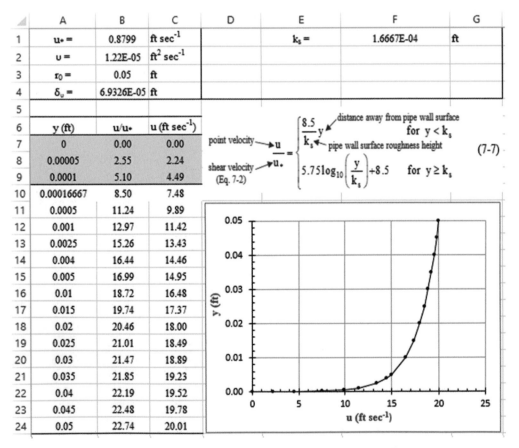

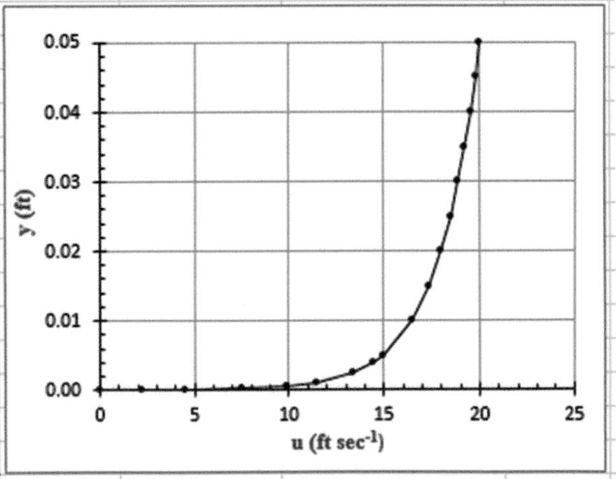

Figure 7.5 **Excel spreadsheet for computing velocities in the pipe in part (3) of Example 7.3.**

Example 7.4 The measured point velocities in a pressurized 1.2-in-diameter smooth pipe are given in cells A4:B10 in Figure 7.6. If the distribution of the velocities follows Eq. (7.6), determine the:

(1) Shape factor
(2) Mean velocity.

Solution

$D = 1.2\,\text{in} = 0.1\,\text{ft},\ r_0 = D/2 = 0.6\,\text{in} = 0.05\,\text{ft}$

(1) Taking the log of both sides of Eq. (7.6), one has $\log_{10}\left(\frac{u}{u_{max}}\right) = m\log_{10}\left(\frac{y}{r_0}\right)$. Thus, the slope of the $\log_{10}\left(\frac{u}{u_{max}}\right) \sim \log_{10}\left(\frac{y}{r_0}\right)$ regression line will be the optimal value of m. From the measured data, the maximum velocity is $u_{max} = 1.45\,\text{ft sec}^{-1}$ at the centerline. As shown in Figure 7.6, cells D4:D10 and E4:E10 compute y/r_0 and u/u_{max}, respectively, and cells G4:G10 and H4:H10 compute $\log_{10}(y/r_0)$ and $\log_{10}(u/u_{max})$, respectively. By plotting H4:H10 versus G4:G10 as a scatter plot and then adding a trend line with a zero intercept, one can determine $m = 0.3069$. The computed and measured values are very close (see cells J4:J10 versus cells B4:B10 and the right-bottom graph in Figure 7.6).

	A	B	C	D	E	F	G	H	I	J
1	r_0 =	0.6	in							
2										
3	y (in)	u (ft sec^{-1})		y/r_0	u/u_{max}		$\log_{10}(y/r_0)$	$\log_{10}(u/u_{max})$		Computed u (ft sec^{-1})
4	0.00001	0.05		0.000	0.034		-4.778	-1.462		0.05
5	0.1	0.80		0.167	0.552		-0.778	-0.258		0.84
6	0.2	1.00		0.333	0.690		-0.477	-0.161		1.03
7	0.3	1.20		0.500	0.828		-0.301	-0.082		1.17
8	0.4	1.30		0.667	0.897		-0.176	-0.047		1.28
9	0.5	1.35		0.833	0.931		-0.079	-0.031		1.37
10	0.6	1.45		1.000	1.000		0.000	0.000		1.45

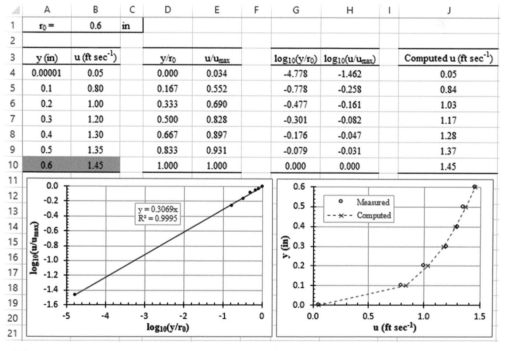

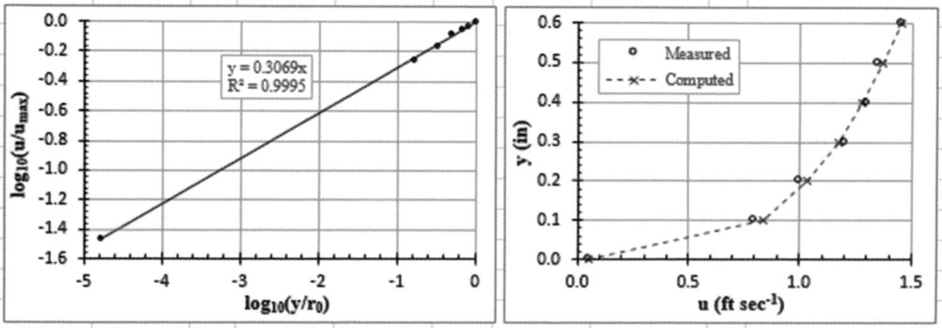

Figure 7.6 **Excel spreadsheet for determining the shape factor in Example 7.4.**

(2) The cross-sectional area: $A = \pi/4^*(0.1\ \text{ft})^2 = 0.0078540\ \text{ft}^2$.

The point velocity is: $\dfrac{u}{1.45\ \text{ft sec}^{-1}} = \left(\dfrac{y}{0.6/12\ \text{ft}}\right)^{0.3069} \rightarrow u = 3.63625\ y^{0.3069}\ \text{ft sec}^{-1}$.

For a dy increase of the distance from the pipe wall surface, the flow area between $(r_0 - y)$ and $(r_0 - y - dy)$ is: $dA = \pi(r_0 - y)^2 - \pi(r_0 - y - dy)^2 = 2\pi(r_0 - y)dy - (dy)^2$.

Given small dy, $(dy)^2 \approx 0 \rightarrow dA \approx 2\pi(r_0 - y)dy$.

The total discharge is:

$$Q = \oint_A u dA = \int_0^{r_0} \left(3.63625y^{0.3069}\right)\left[2\pi\left(r_0 - y\right)dy\right] = \int_0^{0.05} \left(3.63625y^{0.3069}\right)\left[2\pi\left(0.05 - y\right)dy\right]$$

$$= 22.84723 \int_0^{0.05} \left(0.05y^{0.3069} - y^{1.3069}\right)dy = 22.84723\left(\frac{0.05}{1.3069}y^{1.3069} - \frac{1}{2.3069}y^{2.3069}\right)\Big|_0^{0.05}$$

$$= 0.00756\ \text{cfs}.$$

The mean velocity is: $V = Q/A = (0.00756\ \text{cfs})/(0.0078540\ \text{ft}^2) = 0.963\ \text{ft sec}^{-1}$.

In practice, formulas have been used to compute the Darcy friction factor, f, because they are handy for computer applications and can minimize the subjectivity of reading values from the Moody diagram (e.g., Figure 2.11). These formulas were derived from regressing f on the pipe wall roughness height, pipe diameter, and/or Reynolds number in accordance with the Moody diagram. Thus, an f value computed using any of these formulas should not be interpreted to be more accurate than that determined from the Moody diagram.

For a pressurized smooth pipe, f can be computed using the Prandtl formula (Eq. (7.8)), Colebrook formula I (Eq. (7.9)), or Blasius formula (Eq. (7.10)). The formulas can be expressed as (Finnemore and Franzini, 2002):

Reynolds number (Eq. (2.25))

$$\frac{1}{\sqrt{f}} = 2\log_{10}\left(\frac{R_e\sqrt{f}}{2.51}\right) \tag{7.8}$$

Darcy friction factor

Reynolds number (Eq. (2.25))

$$\frac{1}{\sqrt{f}} = 1.8\log_{10}\left(\frac{R_e}{6.9}\right) \tag{7.9}$$

Darcy friction factor

Darcy friction factor

$$f = \frac{0.316}{R_e^{0.25}} \quad \text{for } 3000 \le R_e \le 10^5 \tag{7.10}$$

Reynolds number (Eq. (2.25))

For a pressurized rough pipe, f can be computed using the von Kármán formula (Eq. (7.11)), Colebrook formula II (Eq. (7.12)), or Haaland formula (Eq. (7.13)). The formulas can be expressed as (Finnemore and Franzini, 2002):

$$\frac{1}{\sqrt{f}} = 2\log_{10}\left(\frac{3.7D}{k_s}\right) \tag{7.11}$$

Darcy friction factor; pipe wall surface roughness height; pipe diameter

$$\frac{1}{\sqrt{f}} = -2\log_{10}\left(\frac{k_s}{3.7D} + \frac{2.51}{R_e\sqrt{f}}\right) \tag{7.12}$$

Darcy friction factor; pipe wall surface roughness height; pipe diameter; Reynolds number (Eq. (2.25))

$$\frac{1}{\sqrt{f}} = -1.8\log_{10}\left[\left(\frac{k_s}{3.7D}\right)^{1.11} + \frac{6.9}{R_e}\right] \tag{7.13}$$

Darcy friction factor; pipe wall surface roughness height; pipe diameter; Reynolds number (Eq. (2.25))

Example 7.5 Determine the Darcy friction factor for the pipe in Example 7.2 using the:

(1) Prandtl formula
(2) Colebrook formula I
(3) Blasius formula.

Solution

$R_e = 58,182$

(1) Eq. (7.8): $\frac{1}{\sqrt{f}} = 2\log_{10}\left(\frac{58,182\sqrt{f}}{2.51}\right) = 2\log_{10}\left(23,180\sqrt{f}\right)$.

Solving the equation by trial-and-error or Excel Solver, one can get $f = 0.02020$.

(2) Eq. (7.9): $\frac{1}{\sqrt{f}} = 1.8\log_{10}\left(\frac{58,182}{6.9}\right) = 7.0667 \rightarrow f = 0.02002$.

(3) Eq. (7.10): $f = \frac{0.316}{58,182^{0.25}} = 0.02035$.

Summary: the f values computed by the three formulas are comparable, but they are much larger than the value from the Moody diagram, which is 0.0092.

Example 7.6 Determine the Darcy friction factor for the pipe in Example 7.3 using the:

(1) von Kármán formula
(2) Colebrook formula II
(3) Haaland formula.

Solution

$D = 1.2\,\text{in} = 0.1\,\text{ft},\ k_s = 0.01024\,\text{in} = 0.0008533\,\text{ft},\ R_e = 1,565,451$

(1) Eq. (7.11): $\frac{1}{\sqrt{f}} = 2\log_{10}\left(\frac{3.7{*}0.1\,\text{ft}}{0.0008533\,\text{ft}}\right) = 5.2742 \rightarrow f = 0.03595.$

(2) Eq. (7.12): $\frac{1}{\sqrt{f}} = -2\log_{10}\left(\frac{0.0008533\,\text{ft}}{3.7{*}0.1\,\text{ft}} + \frac{2.51}{1,565,451\sqrt{f}}\right) = -2\log_{10}\left(0.002306 + \frac{1}{623,685.6574\sqrt{f}}\right)$. Solving the equation by trial-and-error or Excel Solver, one can get $f = 0.03599$.

(3) Eq. (7.13): $\frac{1}{\sqrt{f}} = -1.8\log_{10}\left[\left(\frac{0.0008533\,\text{ft}}{3.7{*}0.1\,\text{ft}}\right)^{1.11} + \frac{6.9}{1,565,451}\right] = 5.2660 \rightarrow f = 0.03606.$

Summary: the f values computed by the three formulas are comparable, but they are much larger than the value from the Moody diagram, which is 0.0092.

7.3 Pipes and Pipelines

A water system with only one pipe is rare in the real world. Such a system can be analyzed by simply using an energy equation. In most cases, pipes with different diameters and/or made of different materials are physically joined in series (with no loop) to construct a pipeline between two water bodies or between one water body and a water user. A water system with one or more pipelines needs to be analyzed by jointly using continuity and energy equations. The typical samples are two-, three, and multiple-reservoir systems.

Example 7.7 A 1000-ft-long circular pipe ($f = 0.012$) is used to convey water 60°F at 2.5 cfs from one reservoir to another with an elevation drop of 5 ft. If the pipe is pressurized and has entrance and exit loss coefficients of 0.12 and 1.0, respectively, determine its diameter.

Solution

$L = 1000\,\text{ft},\ f = 0.012,\ Q = 2.5\,\text{cfs},\ \Delta z = -5\,\text{ft},\ k_e = 0.12,\ k_{ex} = 1.0$

Table 2.1: 60°F water $\rightarrow \rho = 1.94\,\text{slug}\,\text{ft}^{-3},\ \gamma = 62.4\,\text{lbf}\,\text{ft}^{-3},\ \upsilon = 1.22\times10^{-5}\,\text{ft}^2\,\text{sec}^{-1}$.

Flow area: $A = \pi/4D^2\,\text{ft}^2$.

Mean velocity: $V = (2.5\,\text{cfs})/(\pi/4D^2\,\text{ft}^2) = 3.1831/D^2\,\text{ft}\,\text{sec}^{-1}$.

Applying the energy equation (Eq. (2.21)) between the surfaces of the two reservoirs, one has:

$$z_1 + 0 + 0 = z_2 + 0 + 0 + h_L \rightarrow h_L = z_1 - z_2 = 5\,\text{ft}.$$

Table 2.3: $h_e = 0.12V^2/(2{*}32.2\,\text{ft}\,\text{sec}^{-2}) = 0.001863V^2\,\text{ft}$.

$h_{ex} = 1.0V^2/(2{*}32.2\,\text{ft}\,\text{sec}^{-2}) = 0.01553V^2\,\text{ft}$.

Eq. (2.23): $h_f = 0.012^* [(1000\,ft)/D]^* [V^2/(2^*32.2\,ft\,sec^{-2})] = 0.1863V^2/D\,ft.$

Eq. (2.22): $h_L = (0.1863V^2/D + 0.001863V^2 + 0.01553V^2)\,ft = (0.1863/D + 0.01739)V^2\,ft$
$= (0.1863/D + 0.01739)(3.1831/D^2)^2\,ft = (1.8876/D + 0.1762)/D^4\,ft.$

Let $(1.8876/D + 0.1762)/D^4\,ft = 5\,ft \to 5D^5 - 0.1762D - 1.8876 = 0.$

Solving the equation by trial-and-error or Excel Solver, one can get $D = 0.84\,ft = 10.0\,in.$

Example 7.8 A pipeline connects two reservoirs (Figure 7.7), designated Res. A and Res. B for description purposes. Determine the flow rate. Use the Darcy–Weisbach equation to compute the friction losses.

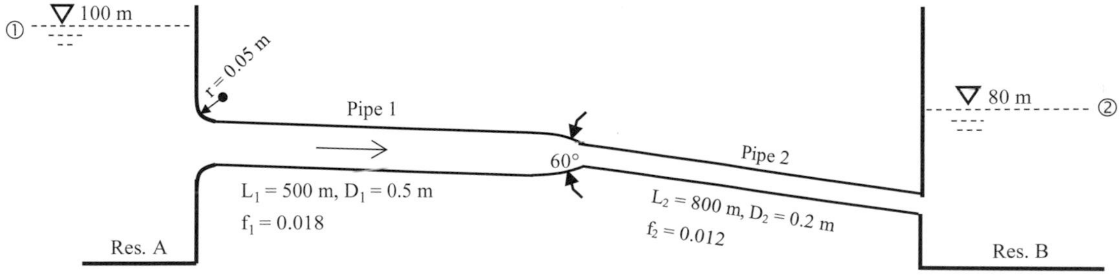

Figure 7.7 **The pipeline connecting two reservoirs used in Example 7.8.**

Solution

The energy equation (Eq. (2.21)) between ① and ②:

$$100\,m + 0 + 0 = 80\,m + 0 + 0 + h_L \to h_L = 20\,m.$$

Continuity equation (Eq. (2.17)): $Q_1 = Q_2 = Q.$

Flow areas: $A_1 = \pi/4^*(0.5\,m)^2 = \pi/16\,m^2$, $A_2 = \pi/4^*(0.2\,m)^2 = \pi/100\,m^2.$

Flow velocities: $V_1 = Q/(\pi/16\,m^2) = 5.0930Q\,m\,s^{-1}$

$$V_2 = Q/(\pi/100\,m^2) = 31.8310Q\,m\,s^{-1}.$$

Table 2.3: $r/D_1 = (0.05\,m)/(0.5\,m) = 0.1 \to$ entrance loss coefficient $k_e = 0.12$

$$\to h_e = 0.12^*(5.0930Q\,m\,s^{-1})^2/(2^*9.81\,m\,s^{-2}) = 0.1587Q^2\,m$$

$$D_2/D_1 = (0.2\,m)/(0.5\,m) = 0.4,\ \theta = 60° \to \text{contraction loss coefficient } k_c = 0.07$$

$$\to h_c = 0.07^*(31.8310Q\,m\,s^{-1})^2/(2^*9.81\,m\,s^{-2}) = 3.6149Q^2\,m$$

submerged exit $\to k_{ex} = 1.0$

$$\to h_{ex} = 1.0^*(31.8310Q\,m\,s^{-1})^2/(2^*9.81\,m\,s^{-2}) = 51.6418Q^2\,m.$$

Darcy–Weisbach equation (Eq. (2.23)):

$$h_{f,1} = 0.018^*[(500\,m)/(0.5\,m)]^* [(5.0930Q\,m\,s^{-1})^2/(2^*9.81\,m\,s^{-2})] = 23.7969Q^2\,m$$

$$h_{f,2} = 0.012^*[(800\,\text{m})/(0.2\,\text{m})]^*[(31.8310Q\,\text{m}\,\text{s}^{-1})^2/(2^*9.81\,\text{m}\,\text{s}^{-2})] = 2478.8075Q^2\,\text{m}.$$

Eq. (2.22): $h_L = h_e + h_{f,1} + h_c + h_{f,2} + h_{ex}$

$$= (0.1587Q^2 + 23.7969Q^2 + 3.6149Q^2 + 2478.8075Q^2 + 51.6418Q^2)\,\text{m}$$
$$= 2558.0198Q^2\,\text{m} = 20\,\text{m}.$$

Solving this equation, one can get $Q = 0.088\,\text{m}^3\text{s}^{-1}$.

Example 7.9 Assume the pipeline in Example 7.8 carries water at $0.25\,\text{m}^3\,\text{s}^{-1}$. Let D_1 and D_2 be the diameters of pipe 1 and 2, respectively. If $D_2/D_1 = 0.2$, determine D_1 and D_2. Consider an entrance loss coefficient of $k_e = 0.10$, a contraction loss coefficient of $k_c = 0.08$, and an exit loss coefficient of $k_{ex} = 1.0$. Use the Darcy–Weisbach equation to compute the friction losses.

Solution

The energy equation (Eq. (2.21)) between ① and ②:

$$100\,\text{m} + 0 + 0 = 80\,\text{m} + 0 + 0 + h_L \rightarrow h_L = 20\,\text{m}.$$

Continuity equation (Eq. (2.17)): $Q_1 = Q_2 = 0.25\,\text{m}^3\,\text{s}^{-1}$.

Flow areas: $A_1 = \pi/4{D_1}^2\,\text{m}^2$, $A_2 = \pi/4{D_2}^2\,\text{m}^2$.

Flow velocities: $V_1 = \left(0.25\,\text{m}^3\,\text{s}^{-1}\right) / \left(\pi/4{D_1}^2\,\text{m}^2\right) = 1/\left(\pi {D_1}^2\right)\,\text{m}\,\text{s}^{-1}$

$$V_2 = \left(0.25\,\text{m}^3\,\text{s}^{-1}\right) / \left(\pi/4{D_2}^2\,\text{m}^2\right) = 1/\left(\pi {D_2}^2\right)\,\text{m}\,\text{s}^{-1}.$$

Table 2.3: $h_e = 0.10^*\left[1/\left(\pi {D_1}^2\right)\,\text{m}\,\text{s}^{-1}\right]^2 / \left(2^*9.81\,\text{m}\,\text{s}^{-2}\right) = 0.00051642/{D_1}^4\,\text{m}$

$$h_c = 0.08^*\left[1/\left(\pi {D_2}^2\right)\,\text{m}\,\text{s}^{-1}\right]^2 / \left(2^*9.81\,\text{m}\,\text{s}^{-2}\right) = 0.00041313/{D_2}^4\,\text{m}$$
$$h_{ex} = 1.0^*\left[1/\left(\pi {D_2}^2\right)\,\text{m}\,\text{s}^{-1}\right]^2 / \left(2^*9.81\,\text{m}\,\text{s}^{-2}\right) = 0.0051642/{D_2}^4\,\text{m}.$$

Darcy–Weisbach equation (Eq. (2.23)):

$$h_{f,1} = 0.018^*\left[(500\,\text{m})/D_1\right]^*\left[\left(1/\left(\pi {D_1}^2\right)\,\text{m}\,\text{s}^{-1}\right)^2 / \left(2^*9.81\,\text{m}\,\text{s}^{-2}\right)\right] = 0.046478/{D_1}^4\,\text{m}$$
$$h_{f,2} = 0.012^*\left[(800\,\text{m})/D_2\right]^*\left[\left(1/\left(\pi {D_2}^2\right)\,\text{m}\,\text{s}^{-1}\right)^2 / \left(2^*9.81\,\text{m}\,\text{s}^{-2}\right)\right] = 0.049576/{D_2}^4\,\text{m}.$$

Eq. (2.22): $h_L = h_e + h_{f,1} + h_c + h_{f,2} + h_{ex} = (0.00051642/{D_1}^4 + 0.046478/{D_1}^4 + 0.00041313/{D_2}^4 + 0.049576/{D_2}^4 + 0.0051642/{D_2}^4)\,\text{m} = \left(0.04699442/{D_1}^4 + 0.05515333/{D_2}^4\right)\,\text{m}$.

$$D_2/D_1 = 0.2 \rightarrow D_1 = 5D_2 \rightarrow h_L = \left(0.04699442/(5D_2)^4 + 0.05515333/{D_2}^4\right)\,\text{m}$$
$$= 0.055229/{D_2}^4\,\text{m}.$$

Let $0.055229/{D_2}^4\,\text{m} = 20\,\text{m} \rightarrow {D_2}^4 = 0.0027614 \rightarrow D_2 = 0.23\,\text{m} \rightarrow D_1 = 1.15\,\text{m}$.

Example 7.10 Three reservoirs, designated Res. A, Res. B, and Res. C for description purposes, are connected by three pipes (Figure 7.8). Determine the flow rate and direction in each of the pipes. Neglect all minor losses. Use the Darcy–Weisbach equation to compute the friction losses.

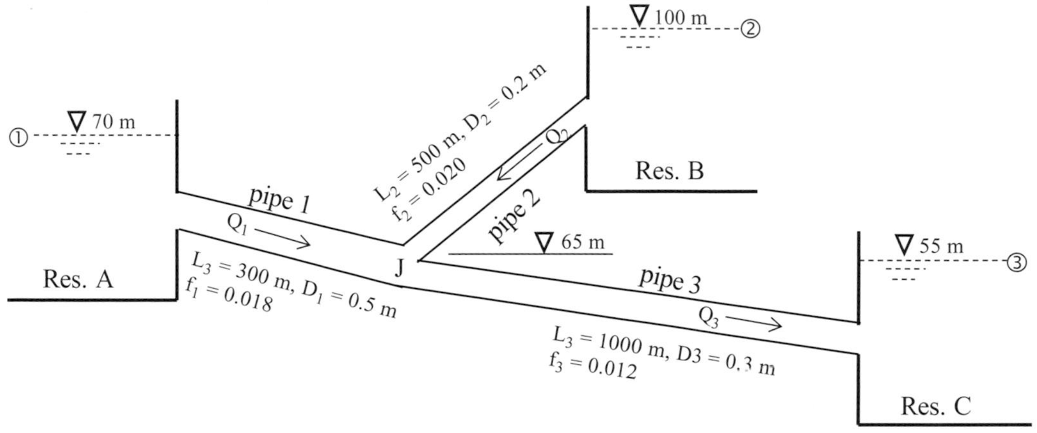

Figure 7.8 **The three-reservoir system used in Example 7.10.**

Solution

When analyzing such a system, one needs to assume the flow direction in each pipe and substitute |V|V for V^2 in the Darcy–Weisbach equation (Eq. (2.23)). If the assumed flow direction is correct, the computed Q or V is positive and there will be no difference between using |V|V and V^2. In contrast, if the assumed flow direction is wrong, the computed Q or V is negative and thus the loss term (always positive) should be on the left side of the energy equation (Eq. (2.21)). This is realized by using |V|V instead of V^2.

The assumed flow directions in the three pipes are depicted in Figure 7.8.

Flow areas: $A_1 = \pi/4^* \, (0.5\,\text{m})^2 = 0.19635\,\text{m}^2$

$$A_2 = \pi/4^* \, (0.2\,\text{m})^2 = 0.031416\,\text{m}^2$$
$$A_3 = \pi/4^* \, (0.3\,\text{m})^2 = 0.070686\,\text{m}^2.$$

Mean velocities: $V_1 = Q_1/\left(0.19635\,\text{m}^2\right) = 5.09295 Q_1\,\text{m s}^{-1}$

$$V_2 = Q_2/\left(0.031416\,\text{m}^2\right) = 31.83091 Q_2\,\text{m s}^{-1}$$
$$V_3 = Q_3/\left(0.070686\,\text{m}^2\right) = 14.14707 Q_3\,\text{m s}^{-1}.$$

Eq. (2.23): $h_{f,1} = 0.018^* \, [(300\,\text{m}) / (0.5\,\text{m})] \, {}^*[\left|5.09295 Q_1\,\text{m s}^{-1}\right| {}^* \left(5.09295 Q_1\,\text{m s}^{-1}\right) / (2^*9.81\,\text{m s}^{-2})] = 14.27788\,|Q_1|\,Q_1\,\text{m}$

$$h_{f,2} = 0.020^* \, [(500\,\text{m}) / (0.2\,\text{m})] \, {}^*[\left|31.83091 Q_2\,\text{m s}^{-1}\right| {}^*(31.83091 Q_2\,\text{m s}^{-1}) / \left(2^*9.81\,\text{m s}^{-2}\right)]$$
$$= 2582.07653\,|Q_2|\,Q_2\,\text{m}$$

$h_{f,3} = 0.012^* [(1000\,m) / (0.3\,m)] * [|14.14707Q_3\,m\,s^{-1}| * (14.14707Q_3\,m\,s^{-1}) / (2^*9.81\,m\,s^{-2})]$
$= 408.03178\,|Q_3|\,Q_3\,m.$

- For the entire system (i.e., between ①, ②, and ③), the continuity equation is: $Q_1 + Q_2 - Q_3 = 0$.
- Note that the pressure head at junction J computed from Res. A should be the same as that computed from Res. B.

  The energy equation between ① and junction J:

  $$70\,m + 0 + 0 = 65\,m + p_J/\gamma + \left(5.09295Q_1\,m\,s^{-1}\right)^2 / \left(2^*9.81\,m\,s^{-2}\right) + 14.27788\,|Q_1|\,Q_1\,m$$
  $$\rightarrow p_J/\gamma = 5 - 1.32203{Q_1}^2 - 14.27788\,|Q_1|\,Q_1\,m.$$

  The energy equation between ② and junction J:

  $$100\,m + 0 + 0 = 65\,m + p_J/\gamma + \left(31.83091Q_2\,m\,s^{-1}\right)^2 / \left(2^*9.81\,m\,s^{-2}\right) + 2582.07653\,|Q_2|\,Q_2\,m$$
  $$\rightarrow p_J/\gamma = 35 - 51.64153{Q_2}^2 - 2582.07653\,|Q_2|\,Q_2\,m.$$

  Thus: $5 - 1.32203{Q_1}^2 - 14.27788\,|Q_1|\,Q_1 = 35 - 51.64153{Q_2}^2 - 2582.07653\,|Q_2|\,Q_2$

  $$\rightarrow 14.27788\,|Q_1|\,Q_1 + 1.32203{Q_1}^2 - 2582.07653\,|Q_2|\,Q_2 - 51.64153{Q_2}^2 + 30 = 0.$$

- Similarly, the pressure head at junction J computed from Res. A should be the same as that computed from Res. C.

  The energy equation between junction J and ③:

  $$65\,m + p_J/\gamma + \left(14.14707Q_3\,m\,s^{-1}\right)^2 / \left(2^*9.81\,m\,s^{-2}\right) = 55\,m + 0 + 0 + 408.03178\,|Q_3|\,Q_3\,m$$
  $$\rightarrow p_J/\gamma = -10 - 10.20080{Q_3}^2 + 408.03178\,|Q_3|\,Q_3\,m.$$

  Thus: $5 - 1.32203{Q_1}^2 - 14.27788\,|Q_1|\,Q_1 = -10 - 10.20080{Q_3}^2 + 408.03178\,|Q_3|\,Q_3$

  $$\rightarrow 14.27788\,|Q_1|\,Q_1 + 1.32203{Q_1}^2 + 408.03178\,|Q_3|\,Q_3 - 10.20080{Q_3}^2 - 15 = 0.$$

- Find the solution using Excel Solver (Figure 7.9): set the continuity equation for the entire system as the objective function (in cell B5), subject to the other two equations (in cells B6:B7); and make the objective function equal to zero by changing Q_1, Q_2, and Q_3 (in cells B1:B3). Remember to uncheck the Make Unconstrained variables Non-Negative box to allow possible negative values for Q_1, Q_2, and Q_3.

  The initial values should satisfy the two constraints (i.e., be in the feasible region). For simplicity, if the assumed flow directions are correct, let $Q_1 = 0.0\,m^3\,s^{-1}$, $Q_2 = 0.107\,m^3\,s^{-1}$, and $Q_3 = 0.194\,m^3\,s^{-1}$. These initial values can be estimated by satisfying one constraint at a time: fixing $Q_1 = 0.0$ to solve for Q_2 and Q_3. While these values do not satisfy the continuity equation, they will be rectified when enforcing the objective function to be equal to zero.

  Figure 7.9a shows the setup of Solver with the initial values, and Figure 7.9b shows the solution: $Q_1 = 0.086\,m^3\,s^{-1}$, $Q_2 = 0.107\,m^3\,s^{-1}$, and $Q_3 = 0.193\,m^3\,s^{-1}$.

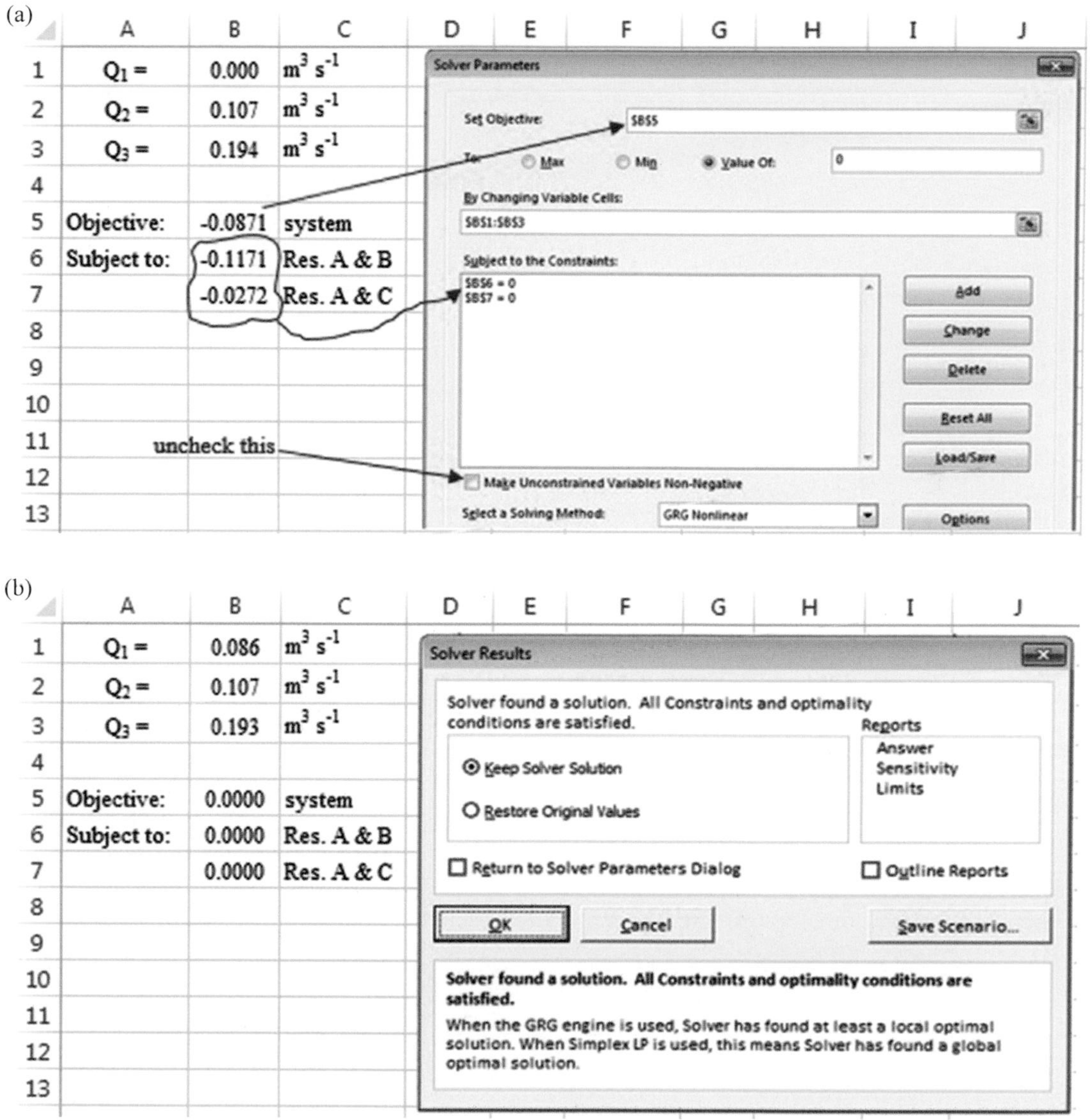

Figure 7.9 **Spreadsheet used for the three-reservoir system in Example 7.10.** (a) Setting up Excel Solver; and (b) the solution to the three-reservoir system.

Example 7.11 Four reservoirs, designated Res. A, Res. B, Res. C, and Res. D for description purposes, are connected by four pipes (Figure 7.10). Determine the flow rate and direction in each of the pipes. Neglect all minor losses. Use the Darcy–Weisbach equation to compute the friction losses.

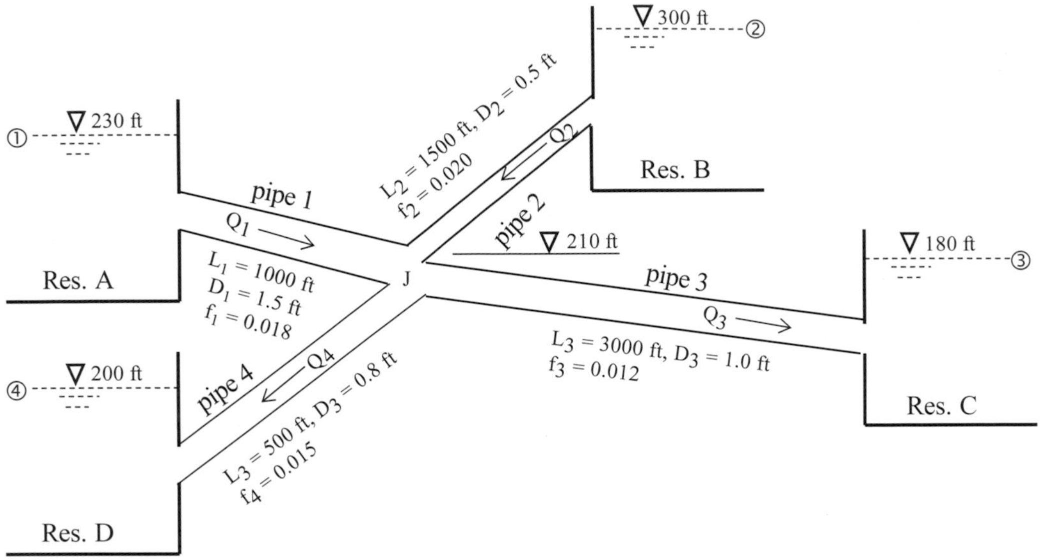

Figure 7.10 **The four-reservoir system used in Example 7.11.**

Solution

When analyzing such a system, one needs to assume the flow direction in each pipe and substitute $|V|V$ for V^2 in the Darcy–Weisbach equation (Eq. (2.23)). If the assumed flow direction is correct, the computed Q or V is positive and there will be no difference between using $|V|V$ and V^2. In contrast, if the assumed flow direction is wrong, the computed Q or V is negative and thus the loss term (always positive) should be on the left side of the energy equation (Eq. (2.21)). This is realized by using $|V|V$ instead of V^2.

The assumed flow directions in the four pipes are depicted in Figure 7.10.

Flow areas: $A_1 = \pi/4^* \, (1.5\,\text{ft})^2 = 1.76715\,\text{ft}^2$

$$A_2 = \pi/4^* \, (0.5\,\text{ft})^2 = 0.19635\,\text{ft}^2$$

$$A_3 = \pi/4^* \, (1.0\,\text{ft})^2 = 0.78540\,\text{ft}^2$$

$$A_3 = \pi/4^* \, (0.8\,\text{ft})^2 = 0.50266\,\text{ft}^2.$$

Mean velocities: $V_1 = Q_1/\left(1.76715\,\text{ft}^2\right) = 0.56588 Q_1\,\text{ft sec}^{-1}$

$$V_2 = Q_2/\left(0.19635\,\text{ft}^2\right) = 5.09295 Q_2\,\text{ft sec}^{-1}$$

$$V_3 = Q_3/\left(0.78540\,\text{ft}^2\right) = 1.27324 Q_3\,\text{ft sec}^{-1}$$

$$V_4 = Q_4/\left(0.50266\,\text{ft}^2\right) = 1.98942 Q_4\,\text{ft sec}^{-1}.$$

Eq. (2.23): $h_{f,1} = 0.018^{*}\ [(1000\,\text{ft})\ /\ (1.5\,\text{ft})]\ ^{*}[\left|0.56588Q_1\,\text{ft sec}^{-1}\right|\ ^{*}(0.56588Q_1\,\text{ft sec}^{-1})/\left(2^{*}32.2\,\text{ft sec}^{-2}\right)] = 0.059668\,|Q_1|\,Q_1\,\text{ft}$

$h_{f,2} = 0.020^{*}\ [(1500\,\text{ft})\ /\ (0.5\,\text{ft})]\ ^{*}[\left|5.09295Q_2\,\text{ft sec}^{-1}\right|\ ^{*}(5.09295Q_2\,\text{ft sec}^{-1})/\left(2^{*}32.2\,\text{ft sec}^{-2}\right)]$
$= 24.16597\,|Q_2|\,Q_2\text{ft}$

$h_{f,3} = 0.012^{*}\ [(3000\,\text{ft})\ /\ (1.0\,\text{ft})]\ ^{*}[\left|1.27324Q_3\,\text{ft sec}^{-1}\right|\ ^{*}(1.27324Q_3\,\text{ft sec}^{-1})/\left(2^{*}32.2\,\text{ft sec}^{-2}\right)]$
$= 0.90623\,|Q_3|\,Q_3\text{ft}$

$h_{f,4} = 0.015^{*}\ [(500\,\text{ft})\ /\ (0.8\,\text{ft})]\ ^{*}[\left|1.98942Q_4\,\text{ft sec}^{-1}\right|\ ^{*}(1.98942Q_4\,\text{ft sec}^{-1})/\left(2^{*}32.2\,\text{ft sec}^{-2}\right)]$
$= 0.57615\,|Q_4|\,Q_4\text{ft}.$

- For the entire system (i.e., between sections ①, ②, ③, and ④), the continuity equation is: $Q_1 + Q_2 - Q_3 - Q_4 = 0$.
- Note that the pressure head at junction J computed from Res. A should be the same as that computed from Res. B.
  The energy equation between ① and junction J:

$$230\,\text{ft} + 0 + 0 = 210\,\text{ft} + p_J/\gamma + \left(0.56588Q_1\,\text{ft sec}^{-1}\right)^2/\left(2^{*}32.2\,\text{ft sec}^{-2}\right)$$
$$+\,0.059668\,|Q_1|\,Q_1\text{ft} \rightarrow p_J/\gamma = 20 - 0.0049724Q_1{}^2 - 0.059668\,|Q_1|\,Q_1\text{ft}.$$

  The energy equation between ② and junction J:

$$300\,\text{ft} + 0 + 0 = 210\,\text{ft} + p_J/\gamma + \left(5.09295Q_2\,\text{ft sec}^{-1}\right)^2/\left(2^{*}32.2\,\text{ft sec}^{-2}\right)$$
$$+\,24.16597\,|Q_2|\,Q_2\text{ft} \rightarrow p_J/\gamma = 90 - 0.40277Q_2{}^2 - 24.16597\,|Q_2|\,Q_2\text{ft}.$$

  Thus: $20 - 0.0049724Q_1{}^2 - 0.059668\,|Q_1|\,Q_1 = 90 - 0.40277Q_2{}^2 - 24.16597\,|Q_2|\,Q_2$

$$\rightarrow 0.059668\,|Q_1|\,Q_1 + 0.0049724Q_1{}^2 - 24.16597\,|Q_2|\,Q_2 - 0.40277Q_2{}^2 + 70 = 0.$$

- Similarly, the pressure head at junction J computed from Res. A should be the same as that computed from Res. C.
  The energy equation between junction J and ③:

$$210\,\text{ft} + p_J/\gamma + \left(1.27324Q_3\,\text{ft sec}^{-1}\right)^2/\left(2^{*}32.2\,\text{ft sec}^{-2}\right) = 180\,\text{ft} + 0 + 0 + 0.90623\,|Q_3|\,Q_3\text{ft}$$
$$\rightarrow p_J/\gamma = -30 - 0.025173Q_3{}^2 + 0.90623\,|Q_3|\,Q_3\text{ft}.$$

  Thus: $20 - 0.0049724Q_1{}^2 - 0.059668\,|Q_1|\,Q_1 = -30 - 0.025173Q_3{}^2 + 0.90623\,|Q_3|\,Q_3$

$$\rightarrow 0.059668\,|Q_1|\,Q_1 + 0.004972Q_1{}^2 + 0.90623\,|Q_3|\,Q_3 - 0.025173Q_3{}^2 - 50 = 0.$$

- Similarly, the pressure head at junction J computed from Res. A should be the same as that computed from Res. D.

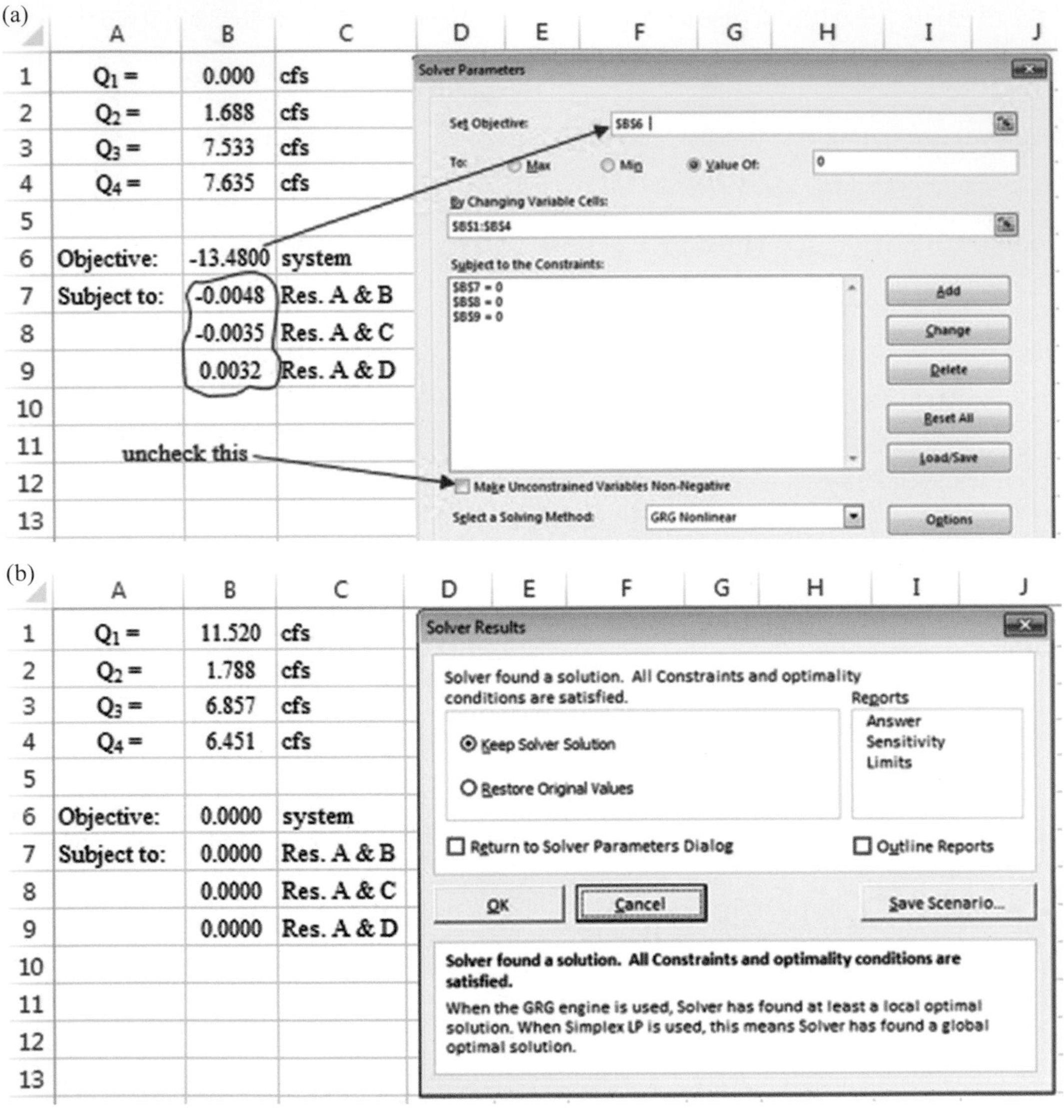

Figure 7.11 **Spreadsheet used for the four-reservoir system in Example 7.11.** (a) Setting up Excel Solver; and (b) the solution to the four-reservoir system.

The energy equation between junction J and ④:

$$210\,\text{ft} + p_J/\gamma + \left(1.98942Q_4\,\text{ft sec}^{-1}\right)^2 / \left(2^*32.2\,\text{ft sec}^{-2}\right) = 200\,\text{ft} + 0 + 0 + 0.57615\,|Q_4|\,Q_4\text{ft}$$

$$\rightarrow p_J/\gamma = -10 - 0.061456Q_4^2 + 0.57615\,|Q_4|\,Q_4\text{ft}.$$

Thus: $20 - 0.0049724Q_1^2 - 0.059668|Q_1|Q_1 = -10 - 0.061456Q_4^2 + 0.57615|Q_4|Q_4$

$$\rightarrow 0.059668|Q_1|Q_1 + 0.004972Q_1^2 + 0.57615|Q_4|Q_4 - 0.061456Q_4^2 - 30 = 0.$$

- Find the solution using Excel Solver (Figure 7.11): set the continuity equation for the entire system as the objective function (in cell B6), subject to the other three equations (in cells B7:B9); and make the objective function equal to zero by changing Q_1, Q_2, Q_3, and Q_4 (in cells B1:B4). Remember to uncheck the Make Unconstrained variables Non-Negative box to allow possible negative values for Q_1, Q_2, Q_3, and Q_4.

  The initial values should satisfy the three constraints (i.e., be in the feasible region). For simplicity, if the assumed flow directions are correct, let $Q_1 = 0.0$ cfs, the three constraints give $Q_2 = 1.688$ cfs, $Q_3 = 7.533$ cfs, and $Q_4 = 7.635$ cfs. Although these values do not satisfy the continuity equation, they will be rectified when enforcing the objective function to be equal to zero.

  Figure 7.11a shows the setup of Solver with the initial values, and Figure 7.11b shows the solution: $Q_1 = 11.520$ cfs, $Q_2 = 1.788$ cfs, $Q_3 = 6.857$ cfs, and $Q_4 = 6.451$ cfs.

7.4 Pipe Networks

A pipe network consists of loops that are formed by a number of pipes physically joined in series and/or in parallel. Such a configuration is very common in water supply, sanitary, and stormwater systems. The simplest pipe network may have only one loop, whereas a real pipe network (e.g., the water supply system in a city) can have hundreds or even thousands of loops. Regardless of its complexity, a pipe network can be analyzed by jointly applying continuity and energy equations to the entire system as well as to each loop. In this regard, the flow direction in each pipe is assumed and the equations are solved simultaneously. For a given pipe, if the assumed flow direction is correct, the computed flow rate in this pipe will be positive, otherwise it is negative. In order to make the loss term in an energy equation correct, one needs to substitute $|V|V$ for V^2 in the Darcy–Weisbach equation (Eq. (2.23)) and in computation of minor losses (e.g., Table 2.3), or $|V|^{0.85}V$ for $V^{1.85}$ in the Hazen–Williams formula (Eq. (2.30)). It is a must that each pipe is included in a continuity equation and in an energy equation. The continuity and energy equations can be solved using Excel Solver by setting the energy equation for the entire system as the objective function and making it equal to zero by adjusting the flow rates in the pipes, subject to the continuity and energy equations for all loops. The initial flow rates must satisfy the constraints (i.e., be in the feasible region) and can be estimated by manipulating the constraints. In practice, one of the computer software packages (e.g., EPANET) described in Appendix V can be used to solve the continuity and energy equations for pipe networks with a large number of loops. However, to demonstrate the hydraulic concepts and principles of pipe networks, the examples in this section are worked manually with the equations solved using Excel Solver.

Example 7.12 If the one-loop pipe network shown in Figure 7.12 carries a total flow rate of 0.5 cfs, determine the flow rate in each pipe and the overall head loss of the network. Use the Darcy–Weisbach equation to compute the friction losses and neglect minor losses.

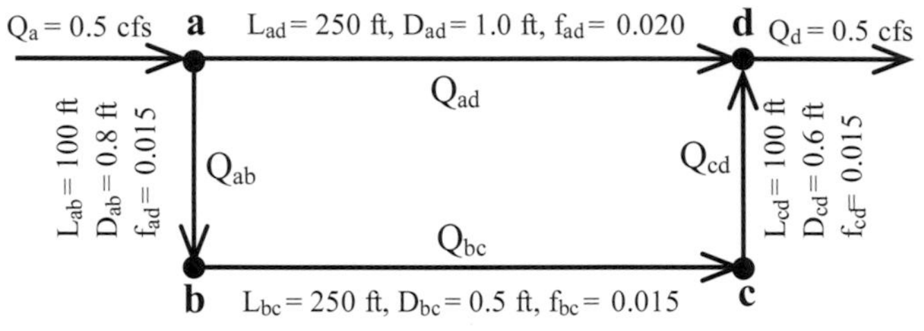

Figure 7.12 **The one-loop pipe network used in Example 7.12.**

Solution

The assumed flow direction in each pipe is shown in Figure 7.12.

$$\begin{aligned}
\text{Flow areas: } A_{ab} &= \pi/4^*(0.8\,\text{ft})^2 = 0.50266\,\text{ft}^2\\
A_{bc} &= \pi/4^*(0.5\,\text{ft})^2 = 0.19635\,\text{ft}^2\\
A_{cd} &= \pi/4^*(0.6\,\text{ft})^2 = 0.28274\,\text{ft}^2\\
A_{ad} &= \pi/4^*(1.0\,\text{ft})^2 = 0.78540\,\text{ft}^2.
\end{aligned}$$

$$\begin{aligned}
\text{Mean velocities: } V_{ab} &= Q_{ab}/\left(0.50266\,\text{ft}^2\right) = 1.98942Q_{ab}\,\text{ft sec}^{-1}\\
V_{bc} &= Q_{bc}/\left(0.19635\,\text{ft}^2\right) = 5.09295Q_{bc}\,\text{ft sec}^{-1}\\
V_{cd} &= Q_{cd}/\left(0.28274\,\text{ft}^2\right) = 3.53682Q_{cd}\,\text{ft sec}^{-1}\\
V_{ad} &= Q_{ad}/\left(0.78540\,\text{ft}^2\right) = 1.27324Q_{ad}\,\text{ft sec}^{-1}.
\end{aligned}$$

- From joints a to d, the head drop along pipeline abcd is the same as that along pipe ad, which is the energy equation for the entire pipe network:

$$\rightarrow h_{f,ad} = h_{f,ab} + h_{f,bc} + h_{f,cd}.$$

Darcy–Weisbach equation (Eq. (2.23)):

$$\begin{aligned}
h_{f,ab} &= 0.015^*\,[(100\,\text{ft})\,/\,(0.8\,\text{ft})]\,^*\\
&\quad \left[\left|1.98942Q_{ab}\,\text{ft sec}^{-1}\right|^*\left(1.98942Q_{ab}\,\text{ft sec}^{-1}\right)\right]/(2^*32.2\,\text{ft sec}^{-2})\\
&= 0.11523\,|Q_{ab}|\,Q_{ab}\,\text{ft}
\end{aligned}$$

$$h_{f,bc} = 0.015^* \left[(250\,\text{ft}) / (0.5\,\text{ft})\right]^* \left[\left|5.09295Q_{bc}\,\text{ft sec}^{-1}\right|^* \left(5.09295Q_{bc}\,\text{ft sec}^{-1}\right)\right] / (2^* 32.2\,\text{ft sec}^{-2})$$

$$= 3.02075\,|Q_{bc}|\,Q_{bc}\,\text{ft}$$

$$h_{f,cd} = 0.015^* \left[(100\,\text{ft}) / (0.6\,\text{ft})\right]^* \left[\left|3.53682Q_{cd}\,\text{ft sec}^{-1}\right|^* \left(3.53682Q_{cd}\,\text{ft sec}^{-1}\right)\right] / (2^* 32.2\,\text{ft sec}^{-2})$$

$$= 0.48560\,|Q_{cd}|\,Q_{cd}\,\text{ft}$$

$$h_{f,ad} = 0.020^* \left[(250\,\text{ft}) / (1.0\,\text{ft})\right] {}^* \left[big\left|1.27324Q_{ad}\,\text{ft sec}^{-1}\right|^* \left(1.27324Q_{ad}\,\text{ft sec}^{-1}\right)\right] / (2^* 32.2\,\text{ft sec}^{-2})$$

$$= 0.12587\,|Q_{ad}|\,Q_{ad}\,\text{ft.}$$

Thus, $0.12587\,|Q_{ad}|\,Q_{ad} = 0.11523\,|Q_{ab}|\,Q_{ab} + 3.02075\,|Q_{bc}|\,Q_{bc} + 0.48560\,|Q_{cd}|\,Q_{cd}$

$$\rightarrow 0.11523\,|Q_{ab}|\,Q_{ab} + 3.02075\,|Q_{bc}|\,Q_{bc} + 0.48560\,|Q_{cd}|\,Q_{cd} - 0.12587\,|Q_{ad}|\,Q_{ad} = 0$$

- The continuity at the joints:

  joint a: $0.5 = Q_{ab} + Q_{ad} \rightarrow Q_{ab} + Q_{ad} - 0.5 = 0$

  joint b: $Q_{ab} = Q_{bc} \rightarrow Q_{ab} - Q_{bc} = 0$

  joint c: $Q_{bc} = Q_{cd} \rightarrow Q_{bc} - Q_{cd} = 0$.

- Find the solution using Excel Solver (Figure 7.13): set the energy equation for the entire system as the objective function (in cell B6), subject to the other three equations (in cells B7:B9); and make the objective function equal to zero by changing Q_1, Q_2, Q_3, and Q_4 (in cells B1:B4). Remember to uncheck the Make Unconstrained variables Non-Negative box to allow possible negative values for Q_1, Q_2, Q_3, and Q_4.

  The initial values should satisfy the three constraints (i.e., be in the feasible region). For simplicity, if the assumed flow directions are correct, let $Q_{ab} = 0.0$ cfs, the three constraints give $Q_{ad} = 0.5$ cfs, $Q_{bc} = 0.0$ cfs, and $Q_{cd} = 0.0$ cfs. Although these values do not satisfy the energy equation, they will be rectified when enforcing the objective function to be equal to zero.

  Figure 7.13a shows the setup of Solver with the initial values, and Figure 7.13b shows the solution: $Q_{ab} = Q_{bc} = Q_{cd} = 0.079$ cfs, and $Q_{ad} = 0.421$ cfs.

- Substituting the computed flow rates into the above equation of $h_{f,ad}$, one can compute the overall head loss as:

$$h_{f,ad} = 0.12587^*\,|0.421\,\text{cfs}|\,(0.421\,\text{cfs}) = 0.022\,\text{ft}.$$

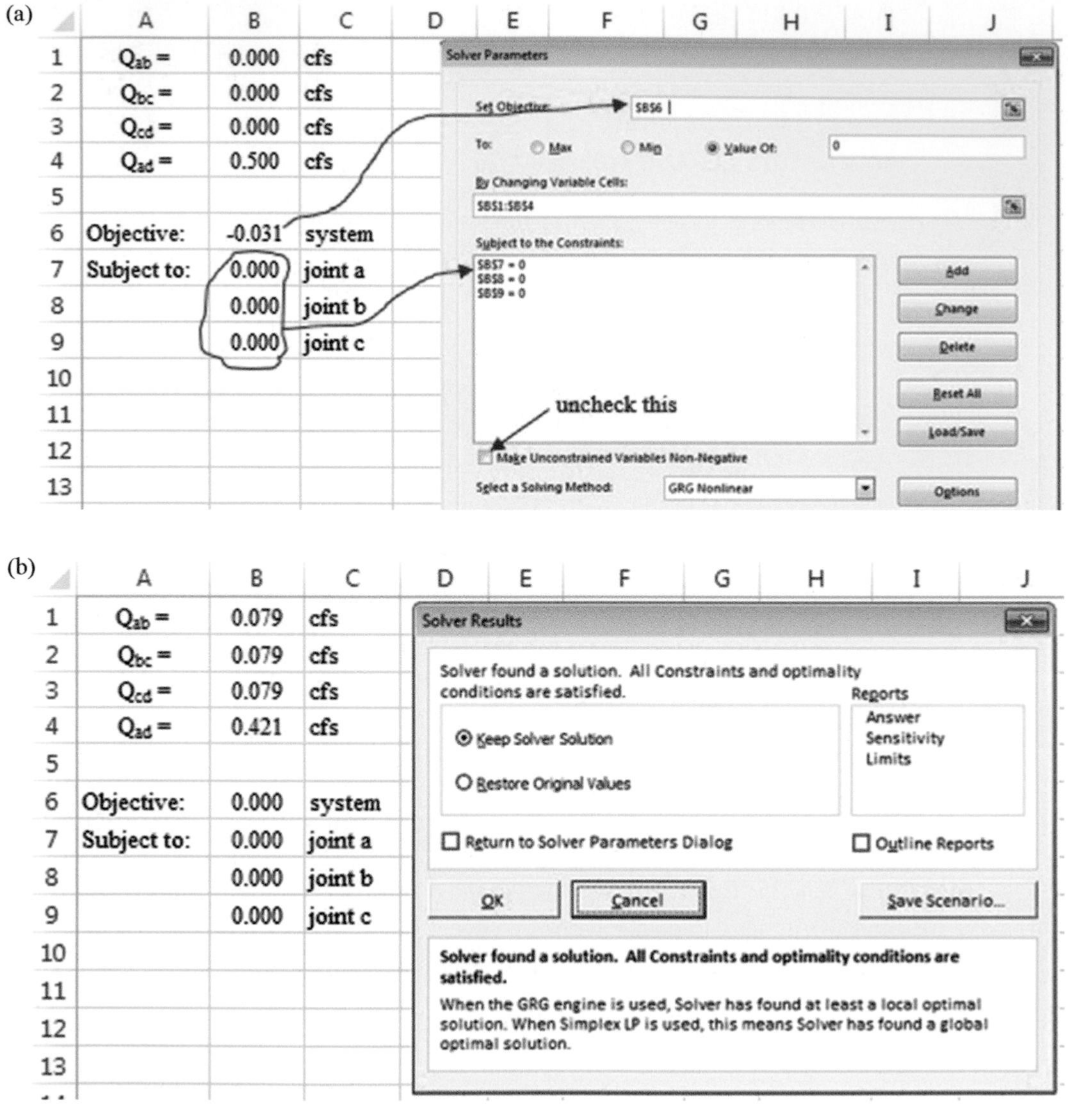

Figure 7.13 **Spreadsheet for the one-loop pipe network in Example 7.12.** (a) Setting up Excel Solver; and (b) the solution to the one-loop pipe network.

Example 7.13 If the two-loop pipe network shown in Figure 7.14 carries a total flow rate of 0.5 cfs, of which 0.4 and 0.1 cfs are diverted out of the network at joints d and g, respectively, determine the flow rate in each pipe and the overall head loss of the network. Use the Darcy–Weisbach equation to compute the friction losses and neglect minor losses.

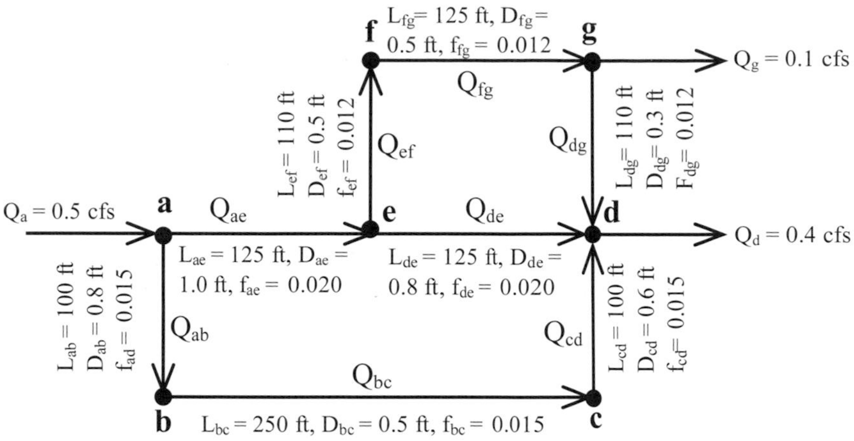

Figure 7.14 **The two-loop pipe network used in Example 7.13.**

Solution

The assumed flow direction in each pipe is shown in Figure 7.14.

$$\text{Flow areas: } A_{ab} = \pi/4^*(0.8\,\text{ft})^2 = 0.50266\,\text{ft}^2$$
$$A_{bc} = \pi/4^*(0.5\,\text{ft})^2 = 0.19635\,\text{ft}^2$$
$$A_{cd} = \pi/4^*(0.6\,\text{ft})^2 = 0.28274\,\text{ft}^2$$
$$A_{ae} = \pi/4^*(1.0\,\text{ft})^2 = 0.78540\,\text{ft}^2$$
$$A_{de} = \pi/4^*(0.8\,\text{ft})^2 = 0.50266\,\text{ft}^2$$
$$A_{ef} = \pi/4^*(0.5\,\text{ft})^2 = 0.19635\,\text{ft}^2$$
$$A_{fg} = \pi/4^*(0.5\,\text{ft})^2 = 0.19635\,\text{ft}^2$$
$$A_{dg} = \pi/4^*(0.3\,\text{ft})^2 = 0.070686\,\text{ft}^2.$$

$$\text{Mean velocities: } V_{ab} = Q_{ab}/\left(0.50266\,\text{ft}^2\right) = 1.98942 Q_{ab}\,\text{ft sec}^{-1}$$
$$V_{bc} = Q_{bc}/\left(0.19635\,\text{ft}^2\right) = 5.09295 Q_{bc}\,\text{ft sec}^{-1}$$
$$V_{cd} = Q_{cd}/\left(0.28274\,\text{ft}^2\right) = 3.53682 Q_{cd}\,\text{ft sec}^{-1}$$
$$V_{ae} = Q_{ae}/\left(0.78540\,\text{ft}^2\right) = 1.27324 Q_{ae}\,\text{ft sec}^{-1}$$
$$V_{de} = Q_{de}/\left(0.50266\,\text{ft}^2\right) = 1.98942 Q_{de}\,\text{ft sec}^{-1}$$
$$V_{ef} = Q_{ef}/\left(0.19635\,\text{ft}^2\right) = 5.09295 Q_{ef}\,\text{ft sec}^{-1}$$

$$V_{fg} = Q_{fg}/\left(0.19635\,\text{ft}^2\right) = 5.09295 Q_{fg}\ \text{ft sec}^{-1}$$
$$V_{dg} = Q_{dg}/\left(0.070686\,\text{ft}^2\right) = 14.14707 Q_{dg}\ \text{ft sec}^{-1}.$$

- From joints a to d, the total energy loss along pipeline abcd is the same as that along pipeline aefgd, which is the energy equation for the entire pipe network

  $$\rightarrow |Q_{ab}|h_{f,ab} + |Q_{bc}|h_{f,bc} + |Q_{cd}|h_{f,cd} = |Q_{ae}|h_{f,ae} + |Q_{de}|h_{f,de} + |Q_{ef}|h_{f,ef} + |Q_{fg}|h_{f,fg} + |Q_{dg}|h_{f,dg}.$$

  Darcy–Weisbach equation (Eq. (2.23)):

$$\begin{aligned}
h_{f,ab} &= 0.015^* \left[(100\,\text{ft}) / (0.8\,\text{ft})\right]^* \\
&\quad \left[\left|1.98942 Q_{ab}\ \text{ft sec}^{-1}\right|^* \left(1.98942 Q_{ab}\ \text{ft sec}^{-1}\right)\right] / (2^* 32.2\,\text{ft sec}^{-2}) \\
&= 0.11523\,|Q_{ab}|\,Q_{ab}\ \text{ft} \\
h_{f,bc} &= 0.015^* \left[(250\,\text{ft}) / (0.5\,\text{ft})\right]^* \\
&\quad \left[\left|5.09295 Q_{bc}\ \text{ft sec}^{-1}\right|^* \left(5.09295 Q_{bc}\ \text{ft sec}^{-1}\right)\right] / (2^* 32.2\,\text{ft sec}^{-2}) \\
&= 3.02075\,|Q_{bc}|\,Q_{bc}\ \text{ft} \\
h_{f,cd} &= 0.015^* \left[(100\,\text{ft}) / (0.6\,\text{ft})\right]^* \\
&\quad \left[\left|3.53682 Q_{cd}\ \text{ft sec}^{-1}\right|^* \left(3.53682 Q_{cd}\ \text{ft sec}^{-1}\right)\right] / (2^* 32.2\,\text{ft sec}^{-2}) \\
&= 0.48560\,|Q_{cd}|\,Q_{cd}\ \text{ft} \\
h_{f,ae} &= 0.020^* \left[(125\,\text{ft}) / (1.0\,\text{ft})\right]^* \\
&\quad \left[\left|1.27324 Q_{ae}\ \text{ft sec}^{-1}\right|^* \left(1.27324 Q_{ae}\ \text{ft sec}^{-1}\right)\right] / (2^* 32.2\,\text{ft sec}^{-2}) \\
&= 0.062933\,|Q_{ae}|\,Q_{ae}\ \text{ft} \\
h_{f,de} &= 0.020^* \left[(125\,\text{ft}) / (0.8\,\text{ft})\right]^* \\
&\quad \left[\left|1.98942 Q_{de}\ \text{ft sec}^{-1}\right|^* \left(1.98942 Q_{de}\ \text{ft sec}^{-1}\right)\right] / (2^* 32.2\,\text{ft sec}^{-2}) \\
&= 0.19205\,|Q_{de}|\,Q_{de}\ \text{ft} \\
h_{f,ef} &= 0.012^* \left[(110\,\text{ft}) / (0.5\,\text{ft})\right]^* \\
&\quad \left[\left|5.09295 Q_{ef}\ \text{ft sec}^{-1}\right|^* \left(5.09295 Q_{ef}\ \text{ft sec}^{-1}\right)\right] / (2^* 32.2\,\text{ft sec}^{-2}) \\
&= 1.06330\,|Q_{ef}|\,Q_{ef}\ \text{ft} \\
h_{f,fg} &= 0.012^* \left[(125\,\text{ft}) / (0.5\,\text{ft})\right]^* \\
&\quad \left[\left|5.09295 Q_{fg}\ \text{ft sec}^{-1}\right|^* \left(5.09295 Q_{fg}\ \text{ft sec}^{-1}\right)\right] / (2^* 32.2\,\text{ft sec}^{-2}) \\
&= 1.20830\,|Q_{fg}|\,Q_{fg}\ \text{ft} \\
h_{f,dg} &= 0.012^* \left[(110\,\text{ft}) / (0.3\,\text{ft})\right]^*
\end{aligned}$$

$$\left[\left|14.14707Q_{dg}\,\text{ft sec}^{-1}\right| * \left(14.14707Q_{dg}\,\text{ft sec}^{-1}\right)\right] / (2^*32.2\text{ft sec}^{-2})$$
$$= 13.67413\left|Q_{dg}\right| Q_{dg}\text{ ft.}$$

Thus, $|Q_{ab}|(0.11523|Q_{ab}|Q_{ab}) + |Q_{bc}|(3.02075|Q_{bc}|Q_{bc}) + |Q_{cd}|(0.48560|Q_{cd}|Q_{cd}) = |Q_{ae}|(0.062933|Q_{ae}|Q_{ae}) + |Q_{de}|(0.19205|Q_{de}|Q_{de}) + |Q_{ef}|(1.06330|Q_{ef}|Q_{ef}) + |Q_{fg}|(1.20830|Q_{fg}|Q_{fg}) + |Q_{dg}|(13.67413|Q_{dg}|Q_{dg})$

$$\rightarrow 0.11523Q_{ab}^3 + 3.02075Q_{bc}^3 + 0.48560Q_{cd}^3 - 0.062933Q_{ae}^3 - 0.19205Q_{de}^3 - 1.06330Q_{ef}^3 - 1.20830Q_{fg}^3 - 13.67413Q_{dg}^3 = 0.$$

- The continuity at the joints:

  joint a: $0.5 = Q_{ab} + Q_{ae} \quad \rightarrow Q_{ab} + Q_{ae} - 0.5 = 0$

  joint b: $Q_{ab} = Q_{bc} \quad \rightarrow Q_{ab} - Q_{bc} = 0$

  joint c: $Q_{bc} = Q_{cd} \quad \rightarrow Q_{bc} - Q_{cd} = 0$

  joint d: $Q_{cd} + Q_{de} + Q_{dg} = 0.4 \rightarrow Q_{cd} + Q_{de} + Q_{dg} - 0.4 = 0$

  joint e: $Q_{ae} = Q_{de} + Q_{ef} \quad \rightarrow Q_{ae} - Q_{de} - Q_{ef} = 0$

  joint f: $Q_{ef} = Q_{fg} \quad \rightarrow Q_{ef} - Q_{fg} = 0$

  joint g: $Q_{fg} = Q_{dg} + 0.1 \quad \rightarrow Q_{fg} - Q_{dg} - 0.1 = 0.$

- Find the solution using Excel Solver (Figure 7.15): set the energy equation for the entire system as the objective function (in cell B10), subject to the other seven equations (in cells B11:B17); and make the objective function equal to zero by changing flow rates (in cells B1:B8). Remember to uncheck the Make Unconstrained variables Non-Negative box to allow possible negative flow rates.

  The initial values should satisfy the seven constraints (i.e., be in the feasible region). For simplicity, if the assumed flow directions are correct, let $Q_{ab} = 0.0$ cfs and $Q_{ef} = 0.1$ cfs, the seven constraints give $Q_{bc} = 0.0$ cfs, $Q_{cd} = 0.0$ cfs, $Q_{ae} = 0.5$ cfs, $Q_{de} = 0.4$ cfs, $Q_{fg} = 0.1$ cfs, and $Q_{dg} = 0.0$ cfs. Although these values do not satisfy the energy equation, they will be rectified when enforcing the objective function to be equal to zero.

  Figure 7.15a shows the setup of Solver with the initial values, and Figure 7.15b shows the solution: $Q_{ab} = Q_{bc} = Q_{cd} = 0.134$ cfs, $Q_{ae} = 0.366$ cfs, $Q_{de} = 0.273$ cfs, $Q_{ef} = Q_{fg} = 0.093$ cfs, and $Q_{dg} = -0.007$ cfs. Note that the computed flow rate in pipe dg is negative, indicating that the actual flow direction is opposite to the assumed one.

- Substituting $Q_{ab} = Q_{bc} = Q_{cd} = 0.134$ cfs into the above equations of $h_{f,ab}$, $h_{f,bc}$, and $h_{f,cd}$, one can compute the overall head loss between joints a and d as:

$$h_{f,ad} = 0.11523^* |0.134\text{ cfs}| * (0.134\text{ cfs}) + 3.02075^* |0.134\text{ cfs}| * (0.134\text{ cfs})$$
$$+ 0.48560^* |0.134\text{ cfs}| * (0.134\text{ cfs})$$
$$= 0.065\text{ ft.}$$

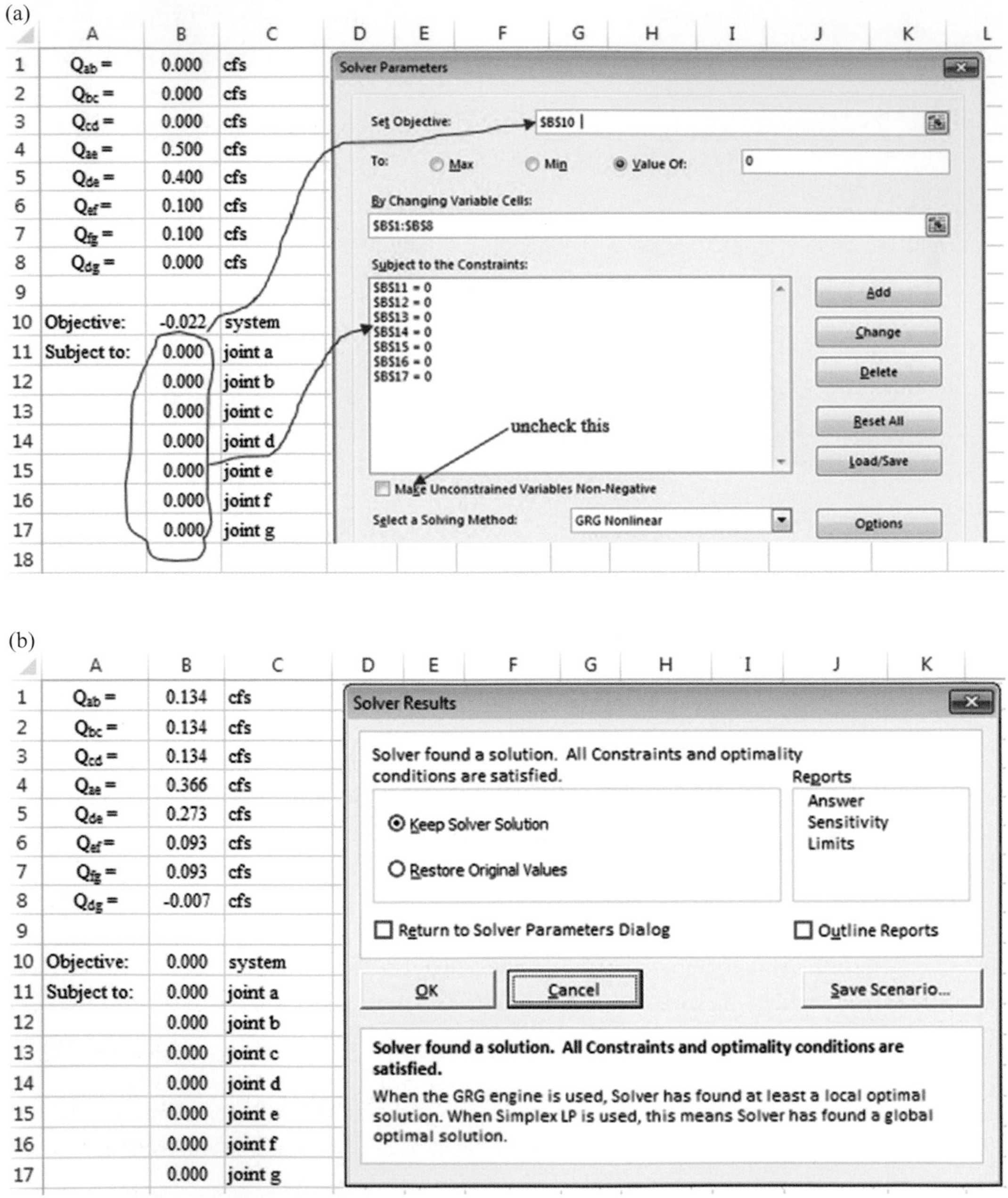

Figure 7.15 **Spreadsheet for the two-loop pipe network in Example 7.13.** (a) Setting up Excel Solver; and (b) the solution to the two-loop pipe network.

Example 7.14 For the pipe network shown in Figure 7.16 with dimensions and roughness heights given in Table 7.1, determine the flow rate and head loss in each pipe. Neglect minor losses and assume a water temperature of 20°C. Use the Darcy–Weisbach equation to compute the friction losses with the Darcy friction factors computed using one of the formulas (Eqs. (7.8) through (7.13)) for a smooth or rough pipe.

Table 7.1 Characteristics of the pipes in Figure 7.16.

Pipe	Length, L (m)	Diameter, D (m)	Roughness height, k_s (m)
ab	735	0.45	0.00026
bc	610	0.35	0.00026
ac	485	0.45	0.00026
bd	580	0.30	0.00026
cd	550	0.30	0.00026
bf	610	0.45	0.00023
ef	520	0.40	0.00026
de	400	0.25	0.00023
fg	550	0.40	0.00026
eg	370	0.60	0.00026

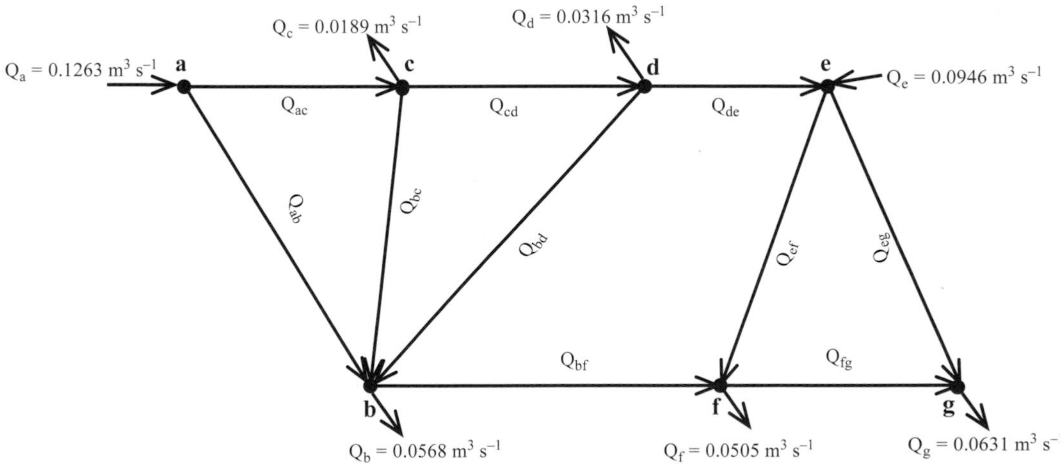

Figure 7.16 The four-loop pipe network used in Example 7.14.

Solution

The assumed flow direction in each pipe is shown in Figure 7.16.

$$\begin{aligned}
\text{Flow areas: } A_{ab} &= \pi/4^*(0.45\,\text{m})^2 = 0.15904\,\text{m}^2\\
A_{bc} &= \pi/4^*(0.35\,\text{m})^2 = 0.096211\,\text{m}^2\\
A_{ac} &= \pi/4^*(0.45\,\text{m})^2 = 0.15904\,\text{m}^2\\
A_{bd} &= \pi/4^*(0.30\,\text{m})^2 = 0.070686\,\text{m}^2\\
A_{cd} &= \pi/4^*(0.30\,\text{m})^2 = 0.070686\,\text{m}^2\\
A_{bf} &= \pi/4^*(0.45\,\text{m})^2 = 0.15904\,\text{m}^2\\
A_{ef} &= \pi/4^*(0.40\,\text{m})^2 = 0.12566\,\text{m}^2\\
A_{de} &= \pi/4^*(0.25\,\text{m})^2 = 0.049087\,\text{m}^2\\
A_{fg} &= \pi/4^*(0.40\,\text{m})^2 = 0.12566\,\text{m}^2\\
A_{eg} &= \pi/4^*(0.60\,\text{m})^2 = 0.28274\,\text{m}^2.
\end{aligned}$$

$$\begin{aligned}
\text{Mean velocities: } V_{ab} &= Q_{ab}/\left(0.15904\,\text{m}^2\right) = 6.28773Q_{ab}\,\text{m}\,\text{s}^{-1}\\
V_{bc} &= Q_{bc}/\left(0.096211\,\text{m}^2\right) = 10.39382Q_{bc}\,\text{m}\,\text{s}^{-1}\\
V_{ac} &= Q_{ac}/\left(0.15904\,\text{m}^2\right) = 6.28773Q_{ac}\,\text{m}\,\text{s}^{-1}\\
V_{bd} &= Q_{bd}/\left(0.070686\,\text{m}^2\right) = 14.14707Q_{bd}\,\text{m}\,\text{s}^{-1}\\
V_{cd} &= Q_{cd}/\left(0.070686\,\text{m}^2\right) = 14.14707Q_{cd}\,\text{m}\,\text{s}^{-1}\\
V_{bf} &= Q_{bf}/\left(0.15904\,\text{m}^2\right) = 6.28773Q_{bf}\,\text{m}\,\text{s}^{-1}\\
V_{ef} &= Q_{ef}/\left(0.12566\,\text{m}^2\right) = 7.95798Q_{ef}\,\text{m}\,\text{s}^{-1}\\
V_{de} &= Q_{de}/\left(0.049087\,\text{m}^2\right) = 20.37199Q_{de}\,\text{m}\,\text{s}^{-1}\\
V_{fg} &= Q_{fg}/\left(0.12566\,\text{m}^2\right) = 7.95798Q_{fg}\,\text{m}\,\text{s}^{-1}\\
V_{eg} &= Q_{eg}/\left(0.28274\,\text{m}^2\right) = 3.53682Q_{eg}\,\text{m}\,\text{s}^{-1}.
\end{aligned}$$

- First, assume rough pipes and estimate the Darcy friction factors using the von Kármán formula (Eq. (7.11)):

  $$\text{pipe ab} \rightarrow \frac{1}{\sqrt{f_{ab}}} = 2\log_{10}\left(\frac{3.7^*0.45\,\text{m}}{0.00026\,\text{m}}\right) = 7.61288 \rightarrow f_{ab} = 0.0173$$

  $$\text{pipe bc} \rightarrow \frac{1}{\sqrt{f_{bc}}} = 2\log_{10}\left(\frac{3.7^*0.35\,\text{m}}{0.00026\,\text{m}}\right) = 7.39459 \rightarrow f_{bc} = 0.0183$$

$$\text{pipe ac} \rightarrow \frac{1}{\sqrt{f_{ac}}} = 2\log_{10}\left(\frac{3.7^*0.45\,\text{m}}{0.00026\,\text{m}}\right) = 7.61288 \rightarrow f_{ac} = 0.0173$$

$$\text{pipe bd} \rightarrow \frac{1}{\sqrt{f_{bd}}} = 2\log_{10}\left(\frac{3.7^*0.30\,\text{m}}{0.00026\,\text{m}}\right) = 7.26070 \rightarrow f_{bd} = 0.0190$$

$$\text{pipe cd} \rightarrow \frac{1}{\sqrt{f_{cd}}} = 2\log_{10}\left(\frac{3.7^*0.30\,\text{m}}{0.00026\,\text{m}}\right) = 7.26070 \rightarrow f_{cd} = 0.0190$$

$$\text{pipe bf} \rightarrow \frac{1}{\sqrt{f_{bf}}} = 2\log_{10}\left(\frac{3.7^*0.45\,\text{m}}{0.00023\,\text{m}}\right) = 7.71937 \rightarrow f_{bf} = 0.0168$$

$$\text{pipe ef} \rightarrow \frac{1}{\sqrt{f_{ef}}} = 2\log_{10}\left(\frac{3.7^*0.40\,\text{m}}{0.00026\,\text{m}}\right) = 7.51058 \rightarrow f_{ef} = 0.0177$$

$$\text{pipe de} \rightarrow \frac{1}{\sqrt{f_{de}}} = 2\log_{10}\left(\frac{3.7^*0.25\,\text{m}}{0.00023\,\text{m}}\right) = 7.20883 \rightarrow f_{de} = 0.0192$$

$$\text{pipe fg} \rightarrow \frac{1}{\sqrt{f_{fg}}} - 2\log_{10}\left(\frac{3.7^*0.40\,\text{m}}{0.00026\,\text{m}}\right) = 7.51058 \rightarrow f_{fg} = 0.0177$$

$$\text{pipe eg} \rightarrow \frac{1}{\sqrt{f_{eg}}} = 2\log_{10}\left(\frac{3.7^*0.60\,\text{m}}{0.00026\,\text{m}}\right) = 7.86276 \rightarrow f_{eg} = 0.0162.$$

- From joints a to g, the total energy loss along pipeline abfg is the same as that along pipeline acdeg, which is the energy equation for the entire pipe network:

$$\rightarrow |Q_{ab}|h_{f,ab} + |Q_{bf}|h_{f,bf} + |Q_{fg}|h_{f,fg} = |Q_{ac}|h_{f,ac} + |Q_{cd}|h_{f,cd} + |Q_{de}|h_{f,de} + |Q_{eg}|h_{f,eg}.$$

Darcy–Weisbach equation (Eq. (2.23)):

$$\begin{aligned}
h_{f,ab} &= 0.0173^* \left[(735\,\text{m}) / (0.45\,\text{m})\right]^* \\
&\quad \left[\left|6.28773Q_{ab}\,\text{m}\,\text{s}^{-1}\right|^* \left(6.28773Q_{ab}\,\text{m}\,\text{s}^{-1}\right)\right] / (2^*9.81\text{m}\,\text{s}^{-2}) \\
&= 56.93898\,|Q_{ab}|\,Q_{ab}\,\text{m} \\
h_{f,bf} &= 0.0168^* \left[(610\,\text{m}) / (0.45\,\text{m})\right]^* \\
&\quad \left[\left|6.28773Q_{bf}\,\text{m}\,\text{s}^{-1}\right|^* \left(6.28773Q_{bf}\,\text{m}\,\text{s}^{-1}\right)\right] / (2^*9.81\text{ms}^{-2}) \\
&= 45.88972\,|Q_{bf}|\,Q_{bf}\,\text{m} \\
h_{f,fg} &= 0.0177^* \left[(550\,\text{m}) / (0.40\,\text{m})\right]^* \\
&\quad \left[\left|7.95798Q_{fg}\,\text{m}\,\text{s}^{-1}\right|^* \left(7.95798Q_{fg}\,\text{m}\,\text{s}^{-1}\right)\right] / (2^*9.81\,\text{ms}^{-2}) \\
&= 78.55659\,\left|Q_{fg}\right|\,Q_{fg}\,\text{m} \\
h_{f,ac} &= 0.0173^* \left[(485\,\text{m}) / (0.45\,\text{m})\right]^* \\
&\quad \left[\left|6.28773Q_{ac}\,\text{m}\,\text{s}^{-1}\right|^* \left(6.28773Q_{ac}\,\text{m}\,\text{s}^{-1}\right)\right] / (2^*9.81\text{m}\,\text{s}^{-2}) \\
&= 37.57198\,|Q_{ac}|\,Q_{ac}\,\text{m} \\
h_{f,cd} &= 0.0190^* \left[(550\,\text{m}) / (0.30\,\text{m})\right]^* \\
&\quad \left[\left|14.14707Q_{cd}\,\text{m}\,\text{s}^{-1}\right|^* (14.14707Q_{cd}\,\text{m}\,\text{s}^{-1})\right] / \left(2^*9.81\,\text{m}\,\text{s}^{-2}\right)
\end{aligned}$$

$$= 355.32768\,|Q_{cd}|\,Q_{cd}\ \text{m}$$

$$h_{f,de} = 0.0192^{*}\,[(400\,\text{m})\,/\,(0.25\,\text{m})]^{*}$$
$$\left[\left|20.37199Q_{de}\,\text{m}\,\text{s}^{-1}\right|^{*}(20.37199Q_{de}\,\text{m}\,\text{s}^{-1})\right]/\left(2^{*}9.81\,\text{m}\,\text{s}^{-2}\right)$$
$$= 649.81408\,|Q_{de}|\,Q_{de}\ \text{m}$$

$$h_{f,eg} = 0.0162^{*}\,[(370\,\text{m})\,/\,(0.60\,\text{m})]^{*}$$
$$\left[\left|3.53682Q_{eg}\,\text{m}\,\text{s}^{-1}\right|^{*}\left(3.53682Q_{eg}\,\text{m}\,\text{s}^{-1}\right)\right]/(2^{*}9.81\text{m}\,\text{s}^{-2})$$
$$= 6.36931\,\left|Q_{eg}\right|\,Q_{eg}\ \text{m}.$$

Thus, $|Q_{ab}|(56.93898\,|Q_{ab}|\,Q_{ab}) + |Q_{bf}|\,(45.88972\,|Q_{bf}|\,Q_{bf}) + \left|Q_{fg}\right|(78.55659\left|Q_{fg}\right|Q_{fg}) =$
$|Q_{ac}|\,(37.57198\,|Q_{ac}|\,Q_{ac}) + |Q_{cd}|\,(355.32768\,|Q_{cd}|\,Q_{cd})$
$+\,|Q_{de}|\,(649.81408\,|Q_{de}|\,Q_{de}) + \left|Q_{eg}\right|\,(6.36931\left|Q_{eg}\right|Q_{eg})$

$$\rightarrow 56.93898Q_{ab}{}^{3} + 45.88972Q_{bf}{}^{3} + 78.55659Q_{fg}{}^{3} - 37.57198Q_{ac}{}^{3} - 355.32768Q_{cd}{}^{3}$$
$$- 649.81408Q_{de}{}^{3} - 6.36931Q_{eg}{}^{3} = 0.$$

- The continuity at the joints:

  joint a: $0.1263 = Q_{ab} + Q_{ac} \rightarrow Q_{ab} + Q_{ac} - 0.1263 = 0$

  joint b: $Q_{ab} + Q_{bc} + Q_{bd} = Q_{bf} + 0.0568 \rightarrow Q_{ab} + Q_{bc} + Q_{bd} - Q_{bf} - 0.0568 = 0$

  joint c: $Q_{ac} = Q_{bc} + Q_{cd} + 0.0189 \rightarrow Q_{ac} - Q_{bc} - Q_{cd} - 0.0189 = 0$

  joint d: $Q_{cd} = Q_{bd} + Q_{de} + 0.0316 \rightarrow Q_{cd} - Q_{bd} - Q_{de} - 0.0316 = 0$

  joint e: $Q_{de} + 0.0946 = Q_{ef} + Q_{eg} \rightarrow Q_{de} - Q_{ef} - Q_{eg} + 0.0946 = 0$

  joint f: $Q_{bf} + Q_{ef} = Q_{fg} + 0.0505 \rightarrow Q_{bf} + Q_{ef} - Q_{fg} - 0.0505 = 0$

  joint g: $Q_{eg} + Q_{fg} = 0.0631 \rightarrow Q_{eg} + Q_{fg} - 0.0631 = 0.$

- To account for the energy losses in pipes bc and bd, for loop bcdb, the total energy loss along pipe cb is the same as that along pipeline cdb:

$$\rightarrow |Q_{bc}|h_{f,bc} = |Q_{cd}|h_{f,cd} + |Q_{bd}|h_{f,bd}.$$

Darcy–Weisbach equation (Eq. (2.23)):

$$h_{f,bc} = 0.0183^{*}\,[(610\,\text{m})\,/\,(0.35\,\text{m})]^{*}$$
$$\left[|10.39382Q_{bc}\,\text{m}\,\text{s}^{-1}|^{*}\left(10.39382Q_{bc}\,\text{m}\,\text{s}^{-1}\right)\right]/(2^{*}9.81\text{m}\,\text{s}^{-2})$$
$$= 175.61607\,|Q_{bc}|\,Q_{bc}\ \text{m}$$

$$h_{f,bd} = 0.0190^{*}\,[(580\,\text{m})\,/\,(0.30\,\text{m})]^{*}$$
$$\left[\left|14.14707Q_{bd}\,\text{m}\,\text{s}^{-1}\right|^{*}\left(14.14707Q_{bd}\,\text{m}\,\text{s}^{-1}\right)\right]/(2^{*}9.81\text{m}\,\text{s}^{-2})$$
$$= 374.70919\,|Q_{bd}|\,Q_{bd}\ \text{m}.$$

Thus, $|Q_{bc}|(175.61607\,|Q_{bc}|\,Q_{bc}) = |Q_{cd}|\,(355.32768\,|Q_{cd}|\,Q_{cd}) + |Q_{bd}|\,(374.70919\,|Q_{bd}|\,Q_{bd})$

$$\rightarrow 175.61607Q_{bc}^{\ 3} - 355.32768Q_{cd}^{\ 3} - 374.70919Q_{bd}^{\ 3} = 0.$$

- To account for the energy loss in pipe ef, for loop bdefb, the total energy loss along pipeline dbf is the same as that along pipeline def:

$$\rightarrow |Q_{bd}|h_{f,bd} + |Q_{bf}|h_{f,bf} = |Q_{de}|h_{f,de} + |Q_{ef}|h_{f,ef}$$

Darcy–Weisbach equation (Eq. (2.23)):

$$h_{f,ef} = 0.0177^{*}\,[(520\,\text{m})\,/\,(0.40\,\text{m})]$$

$$^{*}\,\left[|7.95798Q_{ef}\,\text{m}\,\text{s}^{-1}|^{*}\,(7.95798Q_{ef}\,\text{m}\,\text{s}^{-1})\right]/(2^{*}9.81\,\text{ms}^{-2})$$

$$= 74.27169\,|Q_{ef}|\,Q_{ef}\,\text{m}.$$

Thus, $|Q_{bd}|(374.70919\,|Q_{bd}|\,Q_{bd}) + |Q_{bf}|\,(45.88972\,|Q_{bf}|\,Q_{bf}) = |Q_{de}|\,(649.81408\,|Q_{de}|\,Q_{de}) +$ $|Q_{ef}|\,(74.27169\,|Q_{ef}|\,Q_{ef})$

$$\rightarrow 374.70919Q_{bd}^{\ 3} + 45.88972Q_{bf}^{\ 3} - 649.81408Q_{de}^{\ 3} - 74.27169Q_{ef}^{\ 3} = 0.$$

- Find the solution using Excel Solver (Figure 7.17): set the energy equation for the entire system as the objective function (in cell B12), subject to the other nine equations (in cells B13:B21); and make the objective function equal to zero by changing flow rates (in cells B1:B10). Remember to uncheck the Make Unconstrained variables Non-Negative box to allow possible negative flow rates.

The initial values should satisfy the nine constraints (i.e., be in the feasible region). For simplicity, if the assumed flow directions are correct, let $Q_{ab} = 0.0\,\text{m}^3\,\text{s}^{-1}$, $Q_{bc} = 0.0568\,\text{m}^3\,\text{s}^{-1}$, $Q_{ac} = 0.1263\,\text{m}^3\,\text{s}^{-1}$, $Q_{bd} = 0.0\,\text{m}^3\,\text{s}^{-1}$, $Q_{cd} = 0.0506\,\text{m}^3\,\text{s}^{-1}$, $Q_{bf} = 0.0\,\text{m}^3\,\text{s}^{-1}$, $Q_{ef} = 0.0505\,\text{m}^3\,\text{s}^{-1}$, $Q_{de} = 0.0190\,\text{m}^3\,\text{s}^{-1}$, $Q_{fg} = 0.0\,\text{m}^3\,\text{s}^{-1}$, and $Q_{eg} = 0.0631\,\text{m}^3\,\text{s}^{-1}$. Note that these values do not exactly satisfy the energy equations for loops bcdb and bdefb, but the errors are sufficiently close to zero for finding the solution. Although these values do not satisfy the energy equation for the entire pipe network, they will be rectified when enforcing the objective function to be equal to zero.

Figure 7.17a shows the setup of Solver with the initial values, and Figure 7.17b shows the solution: $Q_{ab} = 0.0637\,\text{m}^3\,\text{s}^{-1}$, $Q_{bc} = 0.0240\,\text{m}^3\,\text{s}^{-1}$, $Q_{ac} = 0.0626\,\text{m}^3\,\text{s}^{-1}$, $Q_{bd} = -0.0091\,\text{m}^3\,\text{s}^{-1}$, $Q_{cd} = 0.0197\,\text{m}^3\,\text{s}^{-1}$, $Q_{bf} = 0.0218\,\text{m}^3\,\text{s}^{-1}$, $Q_{ef} = 0.0139\,\text{m}^3\,\text{s}^{-1}$, $Q_{de} = -0.0028\,\text{m}^3\,\text{s}^{-1}$, $Q_{fg} = -0.0149\,\text{m}^3\,\text{s}^{-1}$, and $Q_{eg} = 0.0780\,\text{m}^3\,\text{s}^{-1}$. Note that the computed flow rates in pipes bd, de, and fg are negative, indicating that the actual flow directions in these three pipes are opposite to the assumed ones.

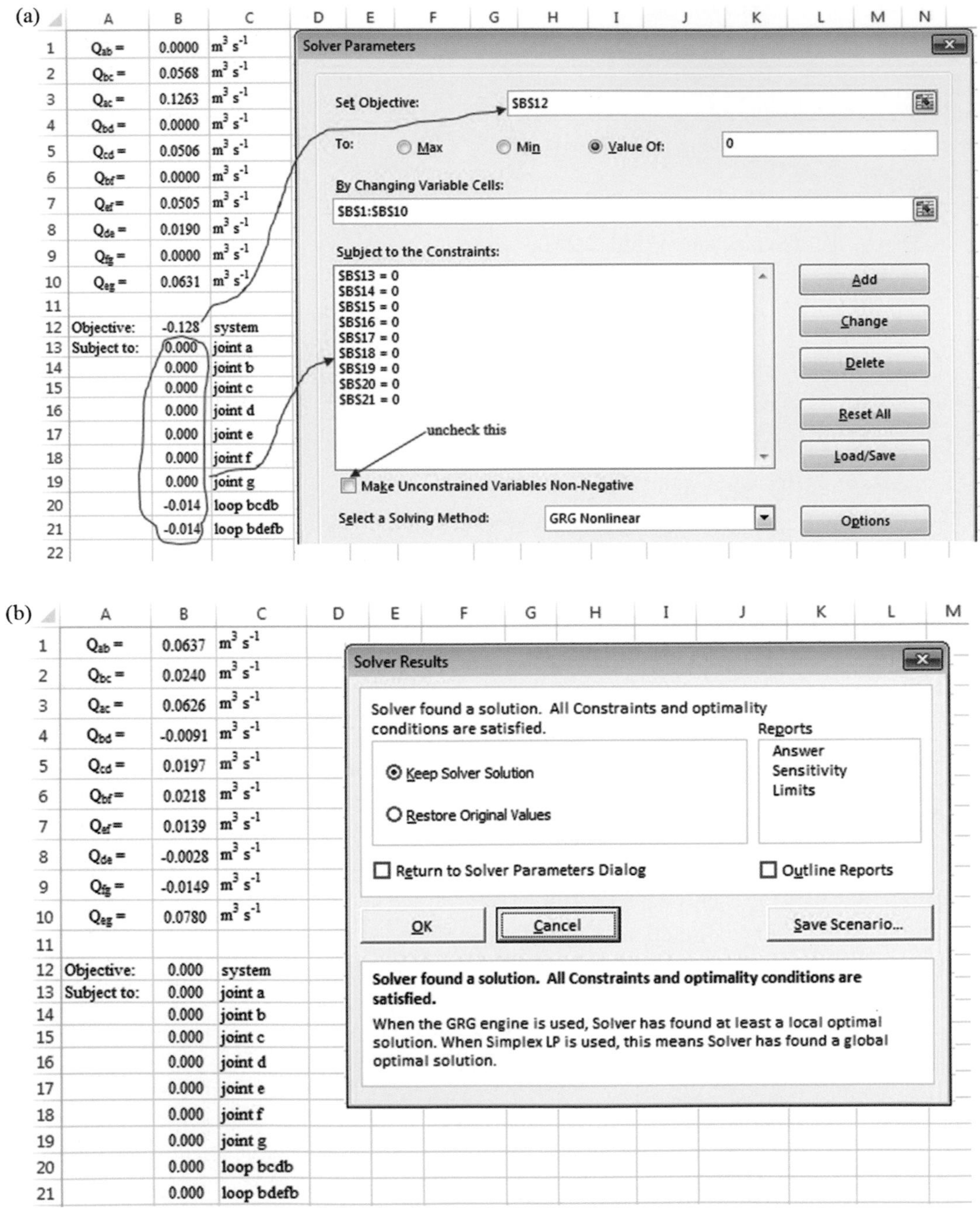

Figure 7.17 **Spreadsheet for the four-loop pipe network in Example 7.14.** (a) Setting up Excel Solver; and (b) the solution to the four-loop pipe network.

- Check if the assumption of rough pipes is correct:

  Substituting the computed flow rates into the above equations to compute the friction losses in the pipes, one can get:

$$
\begin{aligned}
h_{f,ab} &= 56.93898^* \left|0.0637\,\mathrm{m}^3\,\mathrm{s}^{-1}\right| * \left(0.0637\,\mathrm{m}^3\,\mathrm{s}^{-1}\right) = 0.231\,\mathrm{m} \\
h_{f,bc} &= 175.61607^* \left|0.0240\,\mathrm{m}^3\,\mathrm{s}^{-1}\right| * \left(0.0240\,\mathrm{m}^3\,\mathrm{s}^{-1}\right) = 0.101\,\mathrm{m} \\
h_{f,ac} &= 37.57198^* \left|0.0626\,\mathrm{m}^3\,\mathrm{s}^{-1}\right| * \left(0.0626\,\mathrm{m}^3\,\mathrm{s}^{-1}\right) = 0.147\,\mathrm{m} \\
h_{f,bd} &= 374.70919^* \left|-0.0091\,\mathrm{m}^3\,\mathrm{s}^{-1}\right| * \left(-0.0091\,\mathrm{m}^3\,\mathrm{s}^{-1}\right) = -0.031\,\mathrm{m} \\
h_{f,cd} &= 355.32768^* \left|0.0197\,\mathrm{m}^3\,\mathrm{s}^{-1}\right| * \left(0.0197\,\mathrm{m}^3\,\mathrm{s}^{-1}\right) = 0.138\,\mathrm{m} \\
h_{f,bf} &= 45.88972^* \left|0.0218\,\mathrm{m}^3\,\mathrm{s}^{-1}\right| * \left(0.0218\,\mathrm{m}^3\,\mathrm{s}^{-1}\right) = 0.022\,\mathrm{m} \\
h_{f,ef} &= 74.27169^* \left|0.0139\,\mathrm{m}^3\,\mathrm{s}^{-1}\right| * \left(0.0139\,\mathrm{m}^3\,\mathrm{s}^{-1}\right) = 0.014\,\mathrm{m} \\
h_{f,de} &= 649.81408^* \left|-0.0028\,\mathrm{m}^3\,\mathrm{s}^{-1}\right| * \left(-0.0028\,\mathrm{m}^3\,\mathrm{s}^{-1}\right) = -0.005\,\mathrm{m} \\
h_{f,fg} &= 78.55659^* \left|-0.0149\,\mathrm{m}^3\,\mathrm{s}^{-1}\right| * \left(-0.0149\,\mathrm{m}^3\,\mathrm{s}^{-1}\right) = -0.017\,\mathrm{m} \\
h_{f,eg} &= 6.36931^* \left|0.0780\,\mathrm{m}^3\,\mathrm{s}^{-1}\right| * \left(0.0780\,\mathrm{m}^3\,\mathrm{s}^{-1}\right) = 0.039\,\mathrm{m}.
\end{aligned}
$$

  Table 2.1: 20°C water $\rightarrow \rho = 998\,\mathrm{kg\,m^{-3}}$, $\gamma = 9790\,\mathrm{N\,m^{-3}}$, $\upsilon = 10^{-6}\,\mathrm{m^2\,s^{-1}}$.

  Pressurized pipe $\rightarrow$ hydraulic radius is equal to one-fourth of the diameter $\rightarrow$ shear stress at the pipe wall surface by Eq. (7.3):

$$
\begin{aligned}
\tau_{0,ab} &= \left(9790\,\mathrm{N\,m^{-3}}\right) * (0.45\,\mathrm{m}/4) * [(0.231\,\mathrm{m}) / (735\,\mathrm{m})] = 0.346\,\mathrm{N\,m^{-2}} \\
\tau_{0,bc} &= \left(9790\,\mathrm{N\,m^{-3}}\right) * (0.35\,\mathrm{m}/4) * [(0.101\,\mathrm{m}) / (610\,\mathrm{m})] = 0.142\,\mathrm{N\,m^{-2}} \\
\tau_{0,ac} &= \left(9790\,\mathrm{N\,m^{-3}}\right) * (0.45\,\mathrm{m}/4) * [(0.147\,\mathrm{m}) / (485\,\mathrm{m})] = 0.334\,\mathrm{N\,m^{-2}} \\
\tau_{0,bd} &= \left(9790\,\mathrm{N\,m^{-3}}\right) * (0.30\,\mathrm{m}/4) * [(0.031\,\mathrm{m}) / (580\,\mathrm{m})] = 0.040\,\mathrm{N\,m^{-2}} \\
\tau_{0,cd} &= \left(9790\,\mathrm{N\,m^{-3}}\right) * (0.30\,\mathrm{m}/4) * [(0.138\,\mathrm{m}) / (550\,\mathrm{m})] = 0.184\,\mathrm{N\,m^{-2}} \\
\tau_{0,bf} &= \left(9790\,\mathrm{N\,m^{-3}}\right) * (0.45\,\mathrm{m}/4) * [(0.022\,\mathrm{m}) / (610\,\mathrm{m})] = 0.039\,\mathrm{N\,m^{-2}} \\
\tau_{0,ef} &= \left(9790\,\mathrm{N\,m^{-3}}\right) * (0.40\,\mathrm{m}/4) * [(0.014\,\mathrm{m}) / (520\,\mathrm{m})] = 0.027\,\mathrm{N\,m^{-2}} \\
\tau_{0,de} &= \left(9790\,\mathrm{N\,m^{-3}}\right) * (0.25\,\mathrm{m}/4) * [(0.005\,\mathrm{m}) / (400\,\mathrm{m})] = 0.008\,\mathrm{N\,m^{-2}} \\
\tau_{0,fg} &= \left(9790\,\mathrm{N\,m^{-3}}\right) * (0.40\,\mathrm{m}/4) * [(0.017\,\mathrm{m}) / (550\,\mathrm{m})] = 0.031\,\mathrm{N\,m^{-2}} \\
\tau_{0,eg} &= \left(9790\,\mathrm{N\,m^{-3}}\right) * (0.60\,\mathrm{m}/4) * [(0.039\,\mathrm{m}) / (370\,\mathrm{m})] = 0.154\,\mathrm{N\,m^{-2}}.
\end{aligned}
$$

  Eq. (7.2): shear stresses at the wall surfaces of the pipes:

  pipe ab: $u_{*,ab} = \left[\left(0.346\,\mathrm{N\,m^{-2}}\right) / \left(998\,\mathrm{kg\,m^{-3}}\right)\right]^{1/2} = 0.0186\,\mathrm{m\,s^{-1}}$

  pipe bc: $u_{*,bc} = \left[\left(0.142\,\mathrm{N\,m^{-2}}\right) / \left(998\,\mathrm{kg\,m^{-3}}\right)\right]^{1/2} = 0.0119\,\mathrm{m\,s^{-1}}$

  pipe ac: $u_{*,ac} = \left[\left(0.334\,\mathrm{N\,m^{-2}}\right) / \left(998\,\mathrm{kg\,m^{-3}}\right)\right]^{1/2} = 0.0183\,\mathrm{m\,s^{-1}}$

pipe bd: $u_{*,bd} = \left[\left(0.040\,\mathrm{N\,m^{-2}}\right) / \left(998\,\mathrm{kg\,m^{-3}}\right)\right]^{1/2} = 0.0063\,\mathrm{m\,s^{-1}}$

pipe cd: $u_{*,cd} = \left[\left(0.184\,\mathrm{N\,m^{-2}}\right) / \left(998\,\mathrm{kg\,m^{-3}}\right)\right]^{1/2} = 0.0136\,\mathrm{m\,s^{-1}}$

pipe bf: $u_{*,bf} = \left[\left(0.039\,\mathrm{N\,m^{-2}}\right) / \left(998\,\mathrm{kg\,m^{-3}}\right)\right]^{1/2} = 0.0063\,\mathrm{m\,s^{-1}}$

pipe ef: $u_{*,ef} = \left[\left(0.027\,\mathrm{N\,m^{-2}}\right) / \left(998\,\mathrm{kg\,m^{-3}}\right)\right]^{1/2} = 0.0052\,\mathrm{m\,s^{-1}}$

pipe de: $u_{*,de} = \left[\left(0.008\,\mathrm{N\,m^{-2}}\right) / \left(998\,\mathrm{kg\,m^{-3}}\right)\right]^{1/2} = 0.0027\,\mathrm{m\,s^{-1}}$

pipe fg: $u_{*,fg} = \left[\left(0.031\,\mathrm{N\,m^{-2}}\right) / \left(998\,\mathrm{kg\,m^{-3}}\right)\right]^{1/2} = 0.0056\,\mathrm{m\,s^{-1}}$

pipe eg: $u_{*,eg} = \left[\left(0.154\,\mathrm{N\,m^{-2}}\right) / \left(998\,\mathrm{kg\,m^{-3}}\right)\right]^{1/2} = 0.0124\,\mathrm{m\,s^{-1}}$.

Eq. (2.25): Reynolds numbers of the pipe flows:

pipe ab: $R_{e,ab} = \left(6.28773{*}0.0637\,\mathrm{m\,s^{-1}}\right) * (0.45\,\mathrm{m}) / \left(10^{-6}\,\mathrm{m^2\,s^{-1}}\right) = 180,238$

pipe bc: $R_{e,bc} = \left(10.39382{*}0.0240\,\mathrm{m\,s^{-1}}\right) * (0.35\,\mathrm{m}) / \left(10^{-6}\,\mathrm{m^2\,s^{-1}}\right) = 87,308$

pipe ac: $R_{e,ac} = \left(6.28773{*}0.0626\,\mathrm{m\,s^{-1}}\right) * (0.45\,\mathrm{m}) / \left(10^{-6}\,\mathrm{m^2\,s^{-1}}\right) = 177,125$

pipe bd: $R_{e,bd} = \left(14.14707{*}0.0091\,\mathrm{m\,s^{-1}}\right) * (0.30\,\mathrm{m}) / \left(10^{-6}\,\mathrm{m^2\,s^{-1}}\right) = 38,622$

pipe cd: $R_{e,cd} = \left(14.14707{*}0.0197\,\mathrm{m\,s^{-1}}\right) * (0.30\,\mathrm{m}) / \left(10^{-6}\,\mathrm{m^2\,s^{-1}}\right) = 83,609$

pipe bf: $R_{e,bf} = \left(6.28773{*}0.0218\,\mathrm{m\,s^{-1}}\right) * (0.45\,\mathrm{m}) / \left(10^{-6}\,\mathrm{m^2\,s^{-1}}\right) = 61,683$

pipe ef: $R_{e,ef} = \left(7.95798{*}0.0139\,\mathrm{m\,s^{-1}}\right) * (0.40\,\mathrm{m}) / \left(10^{-6}\,\mathrm{m^2\,s^{-1}}\right) = 44,246$

pipe de: $R_{e,de} = \left(20.37199{*}0.0028\,\mathrm{m\,s^{-1}}\right) * (0.25\,\mathrm{m}) / \left(10^{-6}\,\mathrm{m^2\,s^{-1}}\right) = 14,260$

pipe fg: $R_{e,bf} = \left(7.95798{*}0.0149\,\mathrm{m\,s^{-1}}\right) * (0.40\,\mathrm{m}) / \left(10^{-6}\,\mathrm{m^2\,s^{-1}}\right) = 47,430$

pipe eg: $R_{e,eg} = \left(3.53682{*}0.0780\,\mathrm{m\,s^{-1}}\right) * (0.60\,\mathrm{m}) / \left(10^{-6}\,\mathrm{m^2\,s^{-1}}\right) = 165,523$.

Eq. (7.4): Reynolds numbers $> 3000 \rightarrow$ use the second part of the equation to compute the viscous sublayer thicknesses:

pipe ab: $\delta_{\upsilon,ab} = 5^* \left(10^{-6}\,\mathrm{m^2\,s^{-1}}\right) / \left(0.0186\,\mathrm{m\,s^{-1}}\right) = 0.0002688\,\mathrm{m} > 0.00026\,\mathrm{m} = k_{s,ab}$

pipe bc: $\delta_{\upsilon,bc} = 5^* \left(10^{-6}\,\mathrm{m^2\,s^{-1}}\right) / \left(0.0119\,\mathrm{m\,s^{-1}}\right) = 0.0004202\,\mathrm{m} > 0.00026\,\mathrm{m} = k_{s,bc}$

pipe ac: $\delta_{\upsilon,ac} = 5^* \left(10^{-6}\,\mathrm{m^2\,s^{-1}}\right) / \left(0.0183\,\mathrm{m\,s^{-1}}\right) = 0.0002732\,\mathrm{m} > 0.00026\,\mathrm{m} = k_{s,ac}$

pipe bd: $\delta_{\upsilon,bd} = 5^* \left(10^{-6}\,\mathrm{m^2\,s^{-1}}\right) / \left(0.0063\,\mathrm{m\,s^{-1}}\right) = 0.0111111\,\mathrm{m} > 0.00026\,\mathrm{m} = k_{s,bd}$

pipe cd: $\delta_{\upsilon,cd} = 5^* \left(10^{-6}\,\mathrm{m^2\,s^{-1}}\right) / \left(0.0136\mathrm{ms^{-1}}\right) = 0.0003676\,\mathrm{m} > 0.00026\,\mathrm{m} = k_{s,cd}$

pipe bf: $\delta_{\upsilon,bf} = 5^* \left(10^{-6}\,\mathrm{m^2\,s^{-1}}\right) / \left(0.0063\,\mathrm{m\,s^{-1}}\right) = 0.0007937\,\mathrm{m} > 0.00026\,\mathrm{m} = k_{s,bf}$

pipe ef: $\delta_{\upsilon,\mathrm{ef}} = 5^* \left(10^{-6}\,\mathrm{m^2\,s^{-1}}\right) / \left(0.0052\mathrm{ms}^{-1}\right) = 0.0009615\,\mathrm{m} > 0.00026\,\mathrm{m} = k_{s,\mathrm{ef}}$

pipe de: $\delta_{\upsilon,\mathrm{de}} = 5^* \left(10^{-6}\,\mathrm{m^2\,s^{-1}}\right) / \left(0.0027\,\mathrm{m\,s^{-1}}\right) = 0.025259\,\mathrm{m} > 0.00026\,\mathrm{m} = k_{s,\mathrm{de}}$

pipe fg: $\delta_{\upsilon,\mathrm{fg}} = 5^* \left(10^{-6}\,\mathrm{m^2\,s^{-1}}\right) / \left(0.0056\,\mathrm{m\,s^{-1}}\right) = 0.012500\,\mathrm{m} > 0.00026\,\mathrm{m} = k_{s,\mathrm{fg}}$

pipe eg: $\delta_{\upsilon,\mathrm{eg}} = 5^* \left(10^{-6}\,\mathrm{m^2\,s^{-1}}\right) / \left(0.0124\,\mathrm{m\,s^{-1}}\right) = 0.0004032\,\mathrm{m} > 0.00026\,\mathrm{m} = k_{s,\mathrm{eg}}.$

Thus, the assumption of rough pipes is not correct. All pipes are obviously smooth because the viscous sublayer thicknesses are larger than the corresponding roughness heights. The Darcy friction factors should be recalculated using one of the formulas (Eqs. (7.8) to (7.10)) for a smooth pipe and then the recalculated factors are used to repeat the above calculations until the assumed Reynolds numbers are close to the corresponding computed values. However, to be concise, the recalculations are not included here. Readers are strongly encouraged to practice the calculations by assuming smooth pipes.

Example 7.15 For the pipe network shown in Figure 7.18 with dimensions given in Table 7.2, determine the flow rate and head loss in each pipe. Neglect minor losses. Use the Hazen–Williams formula to compute the friction losses with an H–W coefficient of $C_{hw} = 100$ for all pipes.

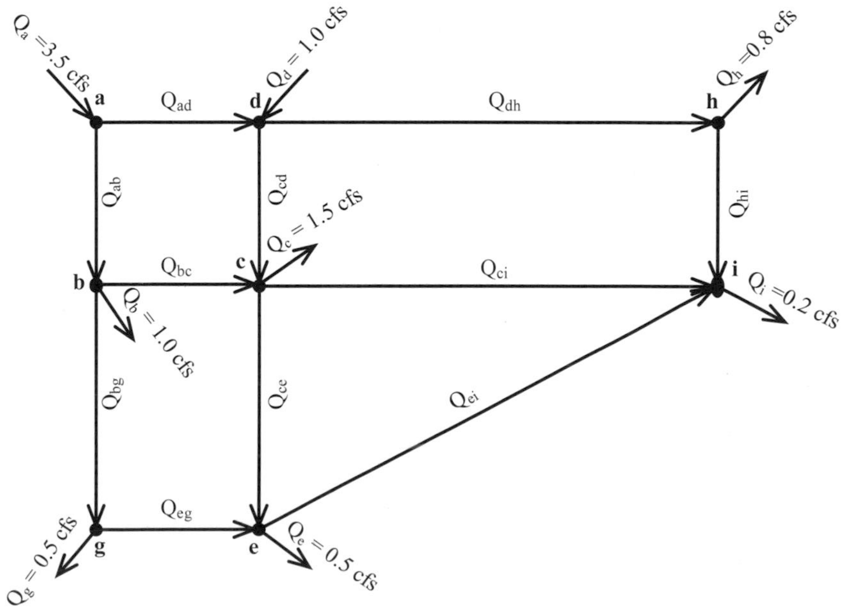

Figure 7.18 **The four-loop pipe network used in Example 7.15.**

Table 7.2 Dimensions of the pipes in Figure 7.18.

Pipe	Length, L (ft)	Diameter, D (ft)
ab	1000	2.0
bc	1000	0.5
cd	1000	2.0
ad	1000	2.0
bg	1500	1.0
ce	1500	1.0
eg	1000	0.5
ci	2800	0.2
hi	1000	2.0
dh	2800	3.0
ei	2850	0.2

Solution

The assumed flow direction in each pipe is shown in Figure 7.18.

$$\text{Flow areas: } A_{ab} = \pi/4^* \,(2.0\,\text{ft})^2 = 3.14159\,\text{ft}^2$$
$$A_{bc} = \pi/4^* \,(0.5\,\text{ft})^2 = 0.19635\,\text{ft}^2$$
$$A_{cd} = \pi/4^* \,(2.0\,\text{ft})^2 = 3.14159\,\text{ft}^2$$
$$A_{ad} = \pi/4^* \,(2.0\,\text{ft})^2 = 3.14159\,\text{ft}^2$$
$$A_{bg} = \pi/4^* \,(1.0\,\text{ft})^2 = 0.78540\,\text{ft}^2$$
$$A_{ce} = \pi/4^* \,(1.0\,\text{ft})^2 = 0.78540\,\text{ft}^2$$
$$A_{eg} = \pi/4^* \,(0.5\,\text{ft})^2 = 0.19635\,\text{ft}^2$$
$$A_{ci} = \pi/4^* \,(0.2\,\text{ft})^2 = 0.03142\,\text{ft}^2$$
$$A_{hi} = \pi/4^* \,(2.0\,\text{ft})^2 = 3.14159\,\text{ft}^2$$
$$A_{dh} = \pi/4^* \,(3.0\,\text{ft})^2 = 7.06858\,\text{ft}^2$$
$$A_{ei} = \pi/4^* \,(0.2\,\text{ft})^2 = 0.03142\,\text{ft}^2.$$

$$\text{Mean velocities: } V_{ab} = Q_{ab}/\left(3.14159\,\text{ft}^2\right) = 0.31831 Q_{ab}\,\text{ft sec}^{-1}$$
$$V_{bc} = Q_{bc}/\left(0.19635\,\text{ft}^2\right) = 5.09295 Q_{bc}\,\text{ft sec}^{-1}$$
$$V_{cd} = Q_{cd}/\left(3.14159\,\text{ft}^2\right) = 0.31831 Q_{cd}\,\text{ft sec}^{-1}$$
$$V_{ad} = Q_{ad}/\left(3.14159\,\text{ft}^2\right) = 0.31831 Q_{ad}\,\text{ft sec}^{-1}$$
$$V_{bg} = Q_{bg}/\left(0.78540\,\text{ft}^2\right) = 1.27324 Q_{bg}\,\text{ft sec}^{-1}$$
$$V_{ce} = Q_{ce}/\left(0.78540\,\text{ft}^2\right) = 1.27324 Q_{ce}\,\text{ft sec}^{-1}$$

$$V_{eg} = Q_{eg}/\left(0.19635\,ft^2\right) = 5.09295Q_{eg}\,ft\,sec^{-1}$$
$$V_{ci} = Q_{ci}/\left(0.03142\,ft^2\right) = 31.82686Q_{ci}\,ft\,sec^{-1}$$
$$V_{hi} = Q_{hi}/\left(3.14159\,ft^2\right) = 0.31831Q_{hi}\,ft\,sec^{-1}$$
$$V_{dh} = Q_{dh}/\left(7.06858\,ft^2\right) = 0.14147Q_{dh}\,ft\,sec^{-1}$$
$$V_{ei} = Q_{ei}/\left(0.03142\,ft^2\right) = 31.82686Q_{ei}\,ft\,sec^{-1}.$$

- From joints a to i, the total energy loss along pipeline abgei is the same as that along pipeline adhi, which is the energy equation for the entire pipe network:

$$\rightarrow |Q_{ab}|\,h_{f,ab} + |Q_{bg}|\,h_{f,bg} + |Q_{eg}|\,h_{f,eg} + |Q_{ei}|\,h_{f,ei} = |Q_{ad}|\,h_{f,ad} + |Q_{dh}|\,h_{f,dh} + |Q_{hi}|\,h_{f,hi}.$$

Hazen–Williams formula (Eq. (2.30)):

$$h_{f,ab} = (1000\,ft) * \left|0.31831Q_{ab}\,ft\,sec^{-1}\right|^{0.85}\left(0.31831Q_{ab}\,ft\,sec^{-1}\right)/[1.318*100*(2.0\,ft/4)^{0.63}]^{1.85}$$
$$= 0.03231\,|Q_{ab}|^{0.85}\,Q_{ab}\,ft$$

$$h_{f,bg} = (1500\,ft) * \left|1.27324Q_{bg}\,ft\,sec^{-1}\right|^{0.85}\left(1.27324Q_{bg}\,ft\,sec^{-1}\right)/[1.318*100*(1.0\,ft/4)^{0.63}]^{1.85}$$
$$= 1.41264\,|Q_{bg}|^{0.85}\,Q_{bg}\,ft$$

$$h_{f,eg} = (1000\,ft) * \left|5.09295Q_{eg}\,ft\,sec^{-1}\right|^{0.85}\left(5.09295Q_{eg}\,ft\,sec^{-1}\right)/[1.318*100*(0.5\,ft/4)^{0.63}]^{1.85}$$
$$= 27.45373\,|Q_{eg}|^{0.85}\,Q_{eg}\,ft$$

$$h_{f,ei} = (2850\,ft) * \left|31.82686Q_{ei}\,ft\,sec^{-1}\right|^{0.85}\left(31.82686Q_{ei}\,ft\,sec^{-1}\right)/[1.318*100*(0.2\,ft/4)^{0.63}]^{1.85}$$
$$= 6753.36666\,|Q_{ei}|^{0.85}\,Q_{ei}\,ft$$

$$h_{f,ad} = (1000\,ft) * \left|0.31831Q_{ad}\,ft\,sec^{-1}\right|^{0.85}\left(0.31831Q_{ad}\,ft\,sec^{-1}\right)/[1.318*100*(2.0\,ft/4)^{0.63}]^{1.85}$$
$$= 0.03231\,|Q_{ad}|^{0.85}\,Q_{ad}\,ft$$

$$h_{f,dh} = (2800\,ft) * \left|0.14147Q_{dh}\,ft\,sec^{-1}\right|^{0.85}\left(0.14147Q_{dh}\,ft\,sec^{-1}\right)/[1.318*100*(3.0\,ft/4)^{0.63}]^{1.85}$$
$$= 0.01258\,|Q_{dh}|^{0.85}\,Q_{dh}\,ft$$

$$h_{f,hi} = (1000\,ft) * \left|0.31831Q_{hi}\,ft\,sec^{-1}\right|^{0.85}\left(0.31831Q_{hi}\,ft\,sec^{-1}\right)/[1.318*100*(2.0\,ft/4)^{0.63}]^{1.85}$$
$$= 0.03231\,|Q_{hi}|^{0.85}\,Q_{hi}\,ft.$$

Thus, $|Q_{ab}|(0.03231\,|Q_{ab}|^{0.85}\,Q_{ab}) + |Q_{bg}|\,(1.41264\,|Q_{bg}|^{0.85}\,Q_{bg})$
$+\,|Q_{eg}|\,(27.45373\,|Q_{eg}|^{0.85}\,Q_{eg}) + |Q_{ei}|\,(6753.36666\,|Q_{ei}|^{0.85}\,Q_{ei}) =$
$|Q_{ad}|\,(0.03231\,|Q_{ad}|^{0.85}\,Q_{ad}) + |Q_{dh}|\,(0.01258\,|Q_{dh}|^{0.85}\,Q_{dh}) + |Q_{hi}|\,(0.03231\,|Q_{hi}|^{0.85}\,Q_{hi})$

$$\rightarrow 0.03231\,|Q_{ab}|^{1.85}\,Q_{ab} + 1.41264\,|Q_{bg}|^{1.85}\,Q_{bg} + 27.45373\,|Q_{eg}|^{1.85}\,Q_{eg} + 6753.36666\,|Q_{ei}|^{1.85}\,Q_{ei}$$
$$-\,0.03231\,|Q_{ad}|^{1.85}\,Q_{ad} - 0.01258\,|Q_{dh}|^{1.85}\,Q_{dh}$$
$$-\,0.03231\,|Q_{hi}|^{1.85}\,Q_{hi} = 0.$$

- The continuity at the joints:

  joint a: $3.5 = Q_{ab} + Q_{ad} \rightarrow Q_{ab} + Q_{ad} - 3.5 = 0$

  joint b: $Q_{ab} = Q_{bc} + Q_{bg} + 1.0 \rightarrow Q_{ab} - Q_{bc} - Q_{bg} - 1.0 = 0$

  joint c: $Q_{bc} + Q_{cd} = Q_{ce} + Q_{ci} + 1.5 \rightarrow Q_{bc} + Q_{cd} - Q_{ce} - Q_{ci} - 1.5 = 0$

  joint d: $Q_{ad} + 1.0 = Q_{cd} + Q_{dh} \rightarrow Q_{ad} - Q_{cd} - Q_{dh} + 1.0 = 0$

  joint e: $Q_{ce} + Q_{eg} = Q_{ei} + 0.5 \rightarrow Q_{ce} + Q_{eg} - Q_{ei} - 0.5 = 0$

  joint g: $Q_{bg} = Q_{eg} + 0.5 \rightarrow Q_{bg} - Q_{eg} - 0.5 = 0$

  joint h: $Q_{dh} = Q_{hi} + 0.8 \rightarrow Q_{dh} - Q_{hi} - 0.8 = 0$

  joint i: $Q_{ci} + Q_{hi} + Q_{ei} = 0.2 \rightarrow Q_{ci} + Q_{hi} + Q_{ei} - 0.2 = 0.$

- To account for the energy losses in pipes bc and cd, for loop abcda, the total energy loss along pipeline abc is the same as that along pipeline adc:

$$\rightarrow |Q_{ab}|\,h_{f,ab} + |Q_{bc}|\,h_{f,bc} = |Q_{ad}|\,h_{f,ad} + |Q_{cd}|\,h_{f,cd}.$$

Hazen–Williams formula (Eq. (2.30)):

$$h_{f,bc} = (1000\,\text{ft}) * \left|5.09295Q_{bc}\,\text{ft sec}^{-1}\right|^{0.85}\left(5.09295Q_{bc}\,\text{ft sec}^{-1}\right)/[1.318*100*(0.5\text{ft}/4)^{0.63}]^{1.85}$$
$$= 27.45373\,|Q_{bc}|^{0.85}\,Q_{bc}\,\text{ft}$$

$$h_{f,cd} = (1000\,\text{ft}) * \left|0.31831Q_{cd}\,\text{ft sec}^{-1}\right|^{0.85}\left(0.31831Q_{cd}\,\text{ft sec}^{-1}\right)/[1.318*100*(2.0\text{ft}/4)^{0.63}]^{1.85}$$
$$= 0.03231\,|Q_{cd}|^{0.85}\,Q_{cd}\,\text{ft}$$

Thus, $|Q_{ab}|\,(0.03231\,|Q_{ab}|^{0.85}\,Q_{ab}) + |Q_{bc}|\left(27.45373\,|Q_{bc}|^{0.85}\,Q_{bc}\right) =$
$|Q_{ad}|\left(0.03231\,|Q_{ad}|^{0.85}\,Q_{ad}\right) + |Q_{cd}|\left(0.03231\,|Q_{cd}|^{0.85}\,Q_{cd}\right)$

$$\rightarrow 0.03231\,|Q_{ab}|^{1.85}\,Q_{ab} + 27.45373\,|Q_{bc}|^{1.85}\,Q_{bc} - 0.03231\,|Q_{ad}|^{1.85}\,Q_{ad} - 0.03231\,|Q_{cd}|^{1.85}\,Q_{cd} = 0.$$

- To account for the energy losses in pipes ce and ci, for loop ceic, the total energy loss along pipeline cei is the same as that along pipe ci:

$$\rightarrow |Q_{ce}|\, h_{f,ce} + |Q_{ei}|\, h_{f,ei} = |Q_{ci}|\, h_{f,ci}.$$

Hazen–Williams formula (Eq. (2.30)):

$$h_{f,ce} = (1500\,ft) * \left|1.27324Q_{ce}\,ft\,sec^{-1}\right|^{0.85} \left(1.27324Q_{ce}\,ft\,sec^{-1}\right) / [1.318^{*}100^{*}(1.0\,ft/4)^{0.63}]^{1.85}$$
$$= 1.41264\,|Q_{ce}|^{0.85}\,Q_{ce}\,ft$$

$$h_{f,ci} = (2800\,ft) * \left|31.82686Q_{ci}\,ft\,sec^{-1}\right|^{0.85} \left(31.82686Q_{ci}\,ft\,sec^{-1}\right) / [1.318^{*}100^{*}(0.2\,ft/4)^{0.63}]^{1.85}$$
$$= 6634.88654\,|Q_{ci}|^{0.85}\,Q_{ci}\,ft$$

Thus, $|Q_{ce}|\,(1.41264\,|Q_{ce}|^{0.85}\,Q_{ce}) + |Q_{ei}|\,(6753.36666\,|Q_{ei}|^{0.85}\,Q_{ei})$
$= |Q_{ci}|\,(6634.88654\,|Q_{ci}|^{0.85}\,Q_{ci})$

$$\rightarrow 1.41264\,|Q_{ce}|^{1.85}\,Q_{ce} + 6753.36666\,|Q_{ei}|^{1.85}\,Q_{ei} - 6634.88654\,|Q_{ci}|^{1.85}\,Q_{ci} = 0.$$

- Find the solution using Excel Solver (Figure 7.19): set the energy equation for the entire system as the objective function (in cell B13), subject to the other ten equations (in cells B14:B23); and make the objective function equal to zero by changing flow rates (in cells B1:B11). Remember to uncheck the Make Unconstrained variables Non-Negative box to allow possible negative flow rates.

  The initial values should satisfy the nine constraints (i.e., be in the feasible region). In order to guess a set of initial values, if the assumed flow directions are correct, let $Q_{ab} = 1.75$ cfs, $Q_{bc} = 0.0$ cfs, $Q_{cd} = 0.0$ cfs, $Q_{ad} = 1.75$ cfs, $Q_{bg} = 0.75$ cfs, $Q_{ce} = -0.5$ cfs, $Q_{eg} = 0.25$ cfs, $Q_{ci} = -1.0$ cfs, $Q_{hi} = 1.95$ cfs, $Q_{dh} = 2.75$ cfs, and $Q_{ei} = -0.75$ cfs to satisfy the eight continuity equations at the joints and the energy equation for loop abcda. In Solver, make the energy equation (cell B23) for loop ceic correct by changing cells B1:B11, subject to B14:B22. This results in the initial values: $Q_{ab} = 1.7503$ cfs, $Q_{bc} = 0.0081$ cfs, $Q_{cd} = 0.1675$ cfs, $Q_{ad} = 1.7497$ cfs, $Q_{bg} = 0.7423$ cfs, $Q_{ce} = -0.5309$ cfs, $Q_{eg} = 0.2423$ cfs, $Q_{ci} = -0.7935$ cfs, $Q_{hi} = 1.7821$ cfs, $Q_{dh} = 2.5821$ cfs, and $Q_{ei} = -0.7886$ cfs. Although these values do not satisfy the energy equation for the entire system, they will be rectified when enforcing the objective function to be equal to zero.

  Figure 7.19a shows the setup of Solver with the initial values, while Figure 7.19b shows the solution: $Q_{ab} = 1.9242$ cfs, $Q_{bc} = 0.0080$ cfs, $Q_{cd} = 1.4358$ cfs, $Q_{ad} = 1.5758$ cfs, $Q_{bg} = 0.9161$ cfs, $Q_{ce} = 0.0141$ cfs, $Q_{eg} = 0.4161$ cfs, $Q_{ci} = -0.0703$ cfs, $Q_{hi} = 0.3401$ cfs, $Q_{dh} = 1.1401$ cfs, and $Q_{ei} = -0.0698$ cfs. Note that the computed flow rates in pipes ci and ei are negative, indicating that the actual flow directions in these two pipes are opposite to the assumed ones.

- Substituting the computed flow rates into the above equations to compute the friction losses in the pipes, one can get:

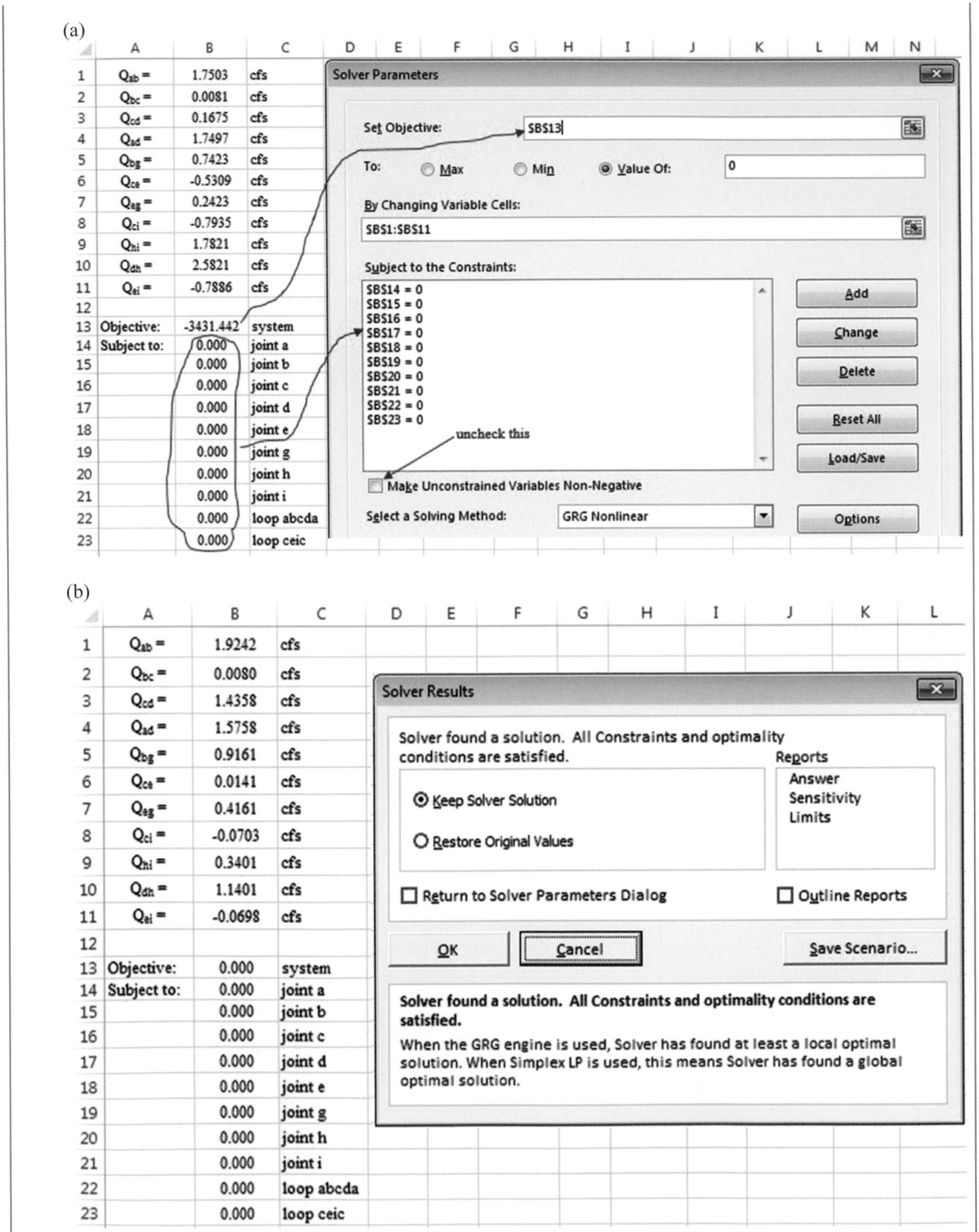

Figure 7.19 **Spreadsheet for the four-loop pipe network in Example 7.15.** (a) Setting up Excel Solver; and (b) the solution to the four-loop pipe network.

pipe ab: $h_{f,ab} = 0.03231^* \, |1.9242\,\text{cfs}|^{0.85} * (1.9242\,\text{cfs}) = 0.108\,\text{ft}$

pipe bc: $h_{f,bc} = 27.45373^* \, |0.0080\,\text{cfs}|^{0.85} * (0.0080\,\text{cfs}) = 0.004\,\text{ft}$

pipe cd: $h_{f,cd} = 0.03231^* \, |1.4358\,\text{cfs}|^{0.85} * (1.4358\,\text{cfs}) = 0.063\,\text{ft}$

pipe ad: $h_{f,ad} = 0.03231^* \, |1.5758\,\text{cfs}|^{0.85} * (1.5758\,\text{cfs}) = 0.075\,\text{ft}$

pipe bg: $h_{f,bg} = 1.41264^* \, |0.9161\,\text{cfs}|^{0.85} * (0.9161\,\text{cfs}) = 1.201\,\text{ft}$

pipe ce: $h_{f,ce} = 1.41264^* \, |0.0141\,\text{cfs}|^{0.85} * (0.0141\,\text{cfs}) = 0.001\,\text{ft}$

pipe eg: $h_{f,eg} = 27.45373^* \, |0.4161\,\text{cfs}|^{0.85} * (0.4161\,\text{cfs}) = 5.422\,\text{ft}$

pipe ci: $h_{f,ci} = 6634.88654^* \, |-0.0703\,\text{cfs}|^{0.85} * (-0.0703\,\text{cfs}) = -48.779\,\text{ft}$

pipe hi: $h_{f,hi} = 0.03231^* \, |0.3401\,\text{cfs}|^{0.85} * (0.3401\,\text{cfs}) = 0.004\,\text{ft}$

pipe dh: $h_{f,dh} = 0.01258^* \, |1.1401\,\text{cfs}|^{0.85} * (1.1401\,\text{cfs}) = 0.016\,\text{ft}$

pipe ei: $h_{f,ei} = 6753.36666^* \, |-0.0698\,\text{cfs}|^{0.85} * (-0.0698\,\text{cfs}) = -49.083\,\text{ft}.$

7.5 Orifices

In the field of hydraulics, an orifice is an opening through which water passes with an associated pressure. Conceptually, it can be considered as a pipe with a very short length in the flow (i.e., longitudinal) direction relative to the dimensions of the opening in the transverse direction. An orifice is usually used as a flow measurement and/or control device. As illustrated in Figure 7.20, an orifice can have either a sharp or rounded edge and may be unsubmerged or submerged, depending on the downstream water level. Regardless, the area of jet flow out of the orifice tends to decrease to a minimum at the *vena contracta* and then starts to increase again. For an unsubmerged orifice, the head on the orifice is the water depth above the centerline of the orifice opening, whereas for a submerged orifice, the head on the orifice is the water depth above the downstream water level. Taking into account the contraction and friction effects, based on the energy equation, one can derive (see Example 7.16) the formula to compute the discharge through an orifice, expressed as:

cross-sectional area of orifice opening

orifice discharge

head on orifice (defined in Figure 7.20)

orifice discharge coefficient (Table 7.3)

gravitational acceleration

$$Q = C_{od} A_0 \sqrt{2gh} \tag{7.14}$$

In Eq. (7.14), the orifice discharge coefficient, C_{od}, is the product of the area contraction coefficient, the ratio of jet flow area at the vena contracta to the orifice opening area (i.e., A/A_0 in Figure 7.20), and the velocity coefficient (defined by Eq. 2.15) due to replacing point velocity by mean velocity in energy equation. Common C_{od} values are listed in Table 7.3.

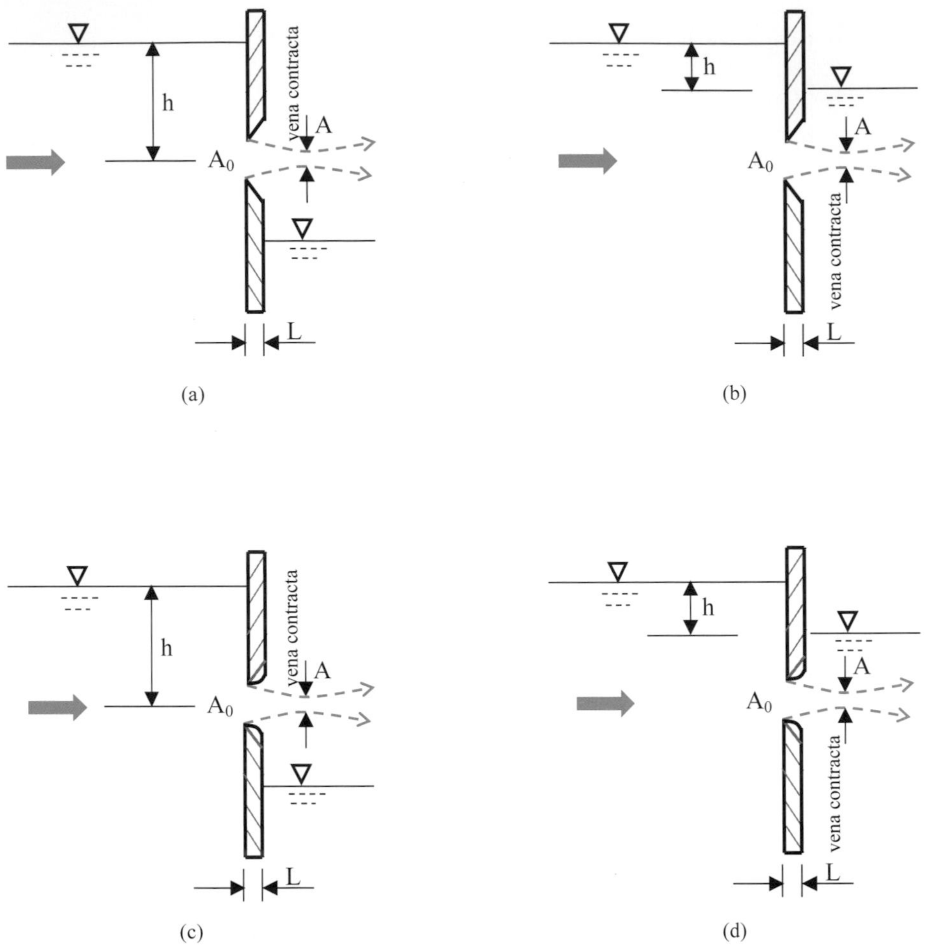

Figure 7.20 **Illustrations of various orifice flows.** Diagram for: (a) sharp-edged free-discharge orifice; (b) sharp-edged submerged orifice; (c) round-edged free-discharge orifice; and (d) round-edged submerged orifice. h: head on the orifice; A_0: cross-sectional area of the orifice opening; A: area of jet flow at the vena contracta; and L: length of the orifice in the flow direction.

Example 7.16 Derive Eq. (7.14).

Solution

In Figure 7.20, choose the orifice centerline as the datum. Given that the orifice length is very small, the friction and minor losses are minimal and can be neglected. Thus, the energy equation (Eq. (2.21)) becomes the Bernoulli equation.

If the orifice is unsubmerged (Figure 7.20a,c), the Bernoulli equation between the upstream water surface and the vena contracta can be written as:

Table 7.3 Orifice discharge coefficient, C_{od}.[1]

L/P_{wet}[2]	Sharp-edged opening	Rounded at bottom only	Rounded at bottom and one side	Rounded at bottom and two sides	Round-edged opening
0.02	0.61	0.63	0.68	0.77	0.95
0.04	0.62	0.64	0.68	0.77	0.94
0.06	0.63	0.65	0.69	0.76	0.94
0.08	0.65	0.66	0.69	0.74	0.93
0.10	0.66	0.67	0.69	0.73	0.93
0.12	0.67	0.68	0.70	0.72	0.93
0.14	0.69	0.69	0.71	0.72	0.92
0.16	0.71	0.70	0.72	0.72	0.92
0.18	0.72	0.71	0.73	0.72	0.92
0.20	0.74	0.73	0.74	0.73	0.92
0.22	0.75	0.74	0.75	0.75	0.91
0.24	0.77	0.75	0.76	0.78	0.91
0.26	0.78	0.76	0.77	0.81	0.91
0.28	0.78	0.76	0.78	0.82	0.91
0.30	0.79	0.77	0.79	0.83	0.91
0.35	0.79	0.78	0.80	0.84	0.90
0.40	0.80	0.79	0.80	0.84	0.90
0.60	0.80	0.80	0.81	0.84	0.90
0.80	0.80	0.80	0.81	0.85	0.90
1.00	0.80	0.81	0.82	0.85	0.90

[1] *Source:* Rossmiller (2014).
[2] L: orifice length in the flow direction; P_{wet}: perimeter of the orifice opening.

$$h + 0 + 0 = 0 + 0 + \alpha_2 V^2/(2g) \rightarrow V = \frac{\sqrt{2gh}}{\alpha_2}$$

$$Q = VA \rightarrow Q = \frac{\sqrt{2gh}}{\alpha_2} A = \frac{A_0\sqrt{2gh}}{\alpha_2}\left(\frac{A}{A_0}\right) = \left(\frac{A}{A_0}\frac{1}{\alpha_2}\right) A_0\sqrt{2gh} = C_{od} A_0 \sqrt{2gh}.$$

If the orifice is submerged (Figure 7.20b,d), the Bernoulli equation between the upstream water surface and the *vena contracta* can be written as:

$$z_1 + 0 + 0 = 0 + z_2 + \alpha_2 V^2/(2g) \rightarrow V = \frac{\sqrt{2g(z_1 - z_2)}}{\alpha_2} = \frac{\sqrt{2gh}}{\alpha_2}$$

$$Q = VA \rightarrow Q = \frac{\sqrt{2gh}}{\alpha_2} A = \frac{A_0\sqrt{2gh}}{\alpha_2}\left(\frac{A}{A_0}\right) = \left(\frac{A}{A_0}\frac{1}{\alpha_2}\right) A_0\sqrt{2gh} = C_{od} A_0 \sqrt{2gh}.$$

Thus, Eq. (7.14) holds for both unsubmerged and submerged orifice, but the head on the orifice is defined differently, as illustrated in Figure 7.20.

Example 7.17 If a sharp-edged orifice has a diameter of 0.5 ft, a longitudinal length of 0.1 ft, and a head of 5.6 ft, determine its discharge.

Solution

$D_0 = 0.5\,\text{ft},\ L = 0.1\,\text{ft},\ h = 5.6\,\text{ft}$

Cross-sectional area: $A_0 = \pi/4^*\ (0.5\,\text{ft})^2 = 0.19635\,\text{ft}^2$.

Cross-sectional perimeter: $P_{wet} = \pi^*\ (0.5\,\text{ft}) = 1.57080\,\text{ft}$.

Table 7.3: $L/P_{wet} = (0.1\,\text{ft})\ /\ (1.57080\,\text{ft}) = 0.06$, sharp-edged $\rightarrow C_{od} = 0.63$.

Eq. (7.14): $Q = 0.63^* \left(0.19635\,\text{ft}^2\right) * \left[\left(2^* \left(32.2\,\text{ft sec}^{-2}\right) * (5.6\,\text{ft})\right]^{1/2} = 2.35\,\text{cfs}.$

Example 7.18 If a round-edged orifice has a diameter of 0.5 ft, a longitudinal length of 0.1 ft, and a head of 5.6 ft, determine its discharge.

Solution

$D_0 = 0.5\,\text{ft},\ L = 0.1\,\text{ft},\ h = 5.6\,\text{ft}$

Cross-sectional area: $A_0 = \pi/4^*\ (0.5\,\text{ft})^2 = 0.19635\,\text{ft}^2$.

Cross-sectional perimeter: $P_{wet} = \pi^*\ (0.5\,\text{ft}) = 1.57080\,\text{ft}$.

Table 7.3: $L/P_{wet} = (0.1\,\text{ft})\ /\ (1.57080\,\text{ft}) = 0.06$, round-edged $\rightarrow C_{od} = 0.94$.

Eq. (7.14): $Q = 0.94^* \left(0.19635\,\text{ft}^2\right) * \left[\left(2^* \left(32.2\,\text{ft sec}^{-2}\right) * (5.6\,\text{ft})\right]^{1/2} = 3.51\,\text{cfs}.$

Example 7.19 A 0.05-m-wide, 0.1-m-high rectangular orifice has a longitudinal length of 0.06 m and a head of 0.3 m, determine its discharge if the orifice has a:

(1) Rounded bottom only
(2) Rounded bottom and one side
(3) Rounded bottom and two sides
(4) Round-edged opening.

Solution

$a = 0.05\ m$, $b = 0.1\ m$, $L = 0.06\ m$, $h = 0.3\ m$

Cross-sectional area: $A_0 = ab = (0.05\ m) * (0.1\ m) = 0.005\ m^2$.

Cross-sectional perimeter: $P_{wet} = 2\,(a + b) = 2^*\,(0.05\ m + 0.1\ m) = 0.3\ m$.

$L/P_{wet} = (0.06\ m)\,/\,(0.3\ m) = 0.2$.

(1) Table 7.3: $L/P_{wet} = 0.2$, rounded bottom only $\rightarrow C_{od} = 0.73$

Eq. (7.14): $Q = 0.73^*\left(0.005\ m^2\right) {}^*\left[\left(2^*\left(9.81\ m\,s^{-2}\right) * (0.3\ m)\right]^{1/2} = 0.0089\ m^3\,s^{-1}$.

(2) Table 7.3: $L/P_{wet} = 0.2$, rounded bottom and one side $\rightarrow C_{od} = 0.74$

Eq. (7.14): $Q = 0.74^*\left(0.005\ m^2\right) {}^*\left[\left(2^*\left(9.81\ m\,s^{-2}\right) * (0.3\ m)\right]^{1/2} = 0.0090\ m^3\,s^{-1}$.

(3) Table 7.3: $L/P_{wet} = 0.2$, rounded bottom and two sides $\rightarrow C_{od} = 0.73$

Eq. (7.14): $Q = 0.73^*\left(0.005\ m^2\right) {}^*\left[\left(2^*\left(9.81\ m\,s^{-2}\right) * (0.3\ m)\right]^{1/2} = 0.0089\ m^3\,s^{-1}$.

(4) Table 7.3: $L/P_{wet} = 0.2$, round-edged opening $\rightarrow C_{od} = 0.92$

Eq. (7.14): $Q = 0.92^*\left(0.005\ m^2\right) {}^*\left[\left(2^*\left(9.81\ m\,s^{-2}\right) * (0.3\ m)\right]^{1/2} = 0.011\ m^3\,s^{-1}$.

7.6 Discharge Measurements

Accurate measurement of the discharge through a pipeline and/or pipe network is always needed in practice. For instance, a water supply manager needs to determine water usage in order to bill customers; and a wastewater manager needs to determine treatment load and cost. Also, such measurements can help detect possible water or wastewater leaks. The measuring devices can be classified into three categories, namely obstruction meters, non-intrusive ultrasonic flowmeters, and area–velocity flowmeters.

7.6.1 Obstruction Meters

The rationale of obstruction meters is to reduce pressure while increasing velocity at a restricted cross-sectional area (Finnemore and Franzini, 2002), where the relationship between discharge and pressure drop can be determined by combining the continuity and Bernoulli equations. Herein, the Bernoulli equation is a special case of the energy equation when energy losses are minimal and can be neglected. Orifice, flow nozzle, venturi tube, and rotameter are most commonly used obstruction meters (Figure 7.21). The former three meters have a constant area of obstruction and the pressure drop changes with flow rate. In contrast, a rotameter has a constant pressure drop with a variable area. Also, a rotameter is only applicable for vertical flows, as shown in Figure 7.21d. The accuracy of a rotameter is lower than that of the other three obstruction meters.

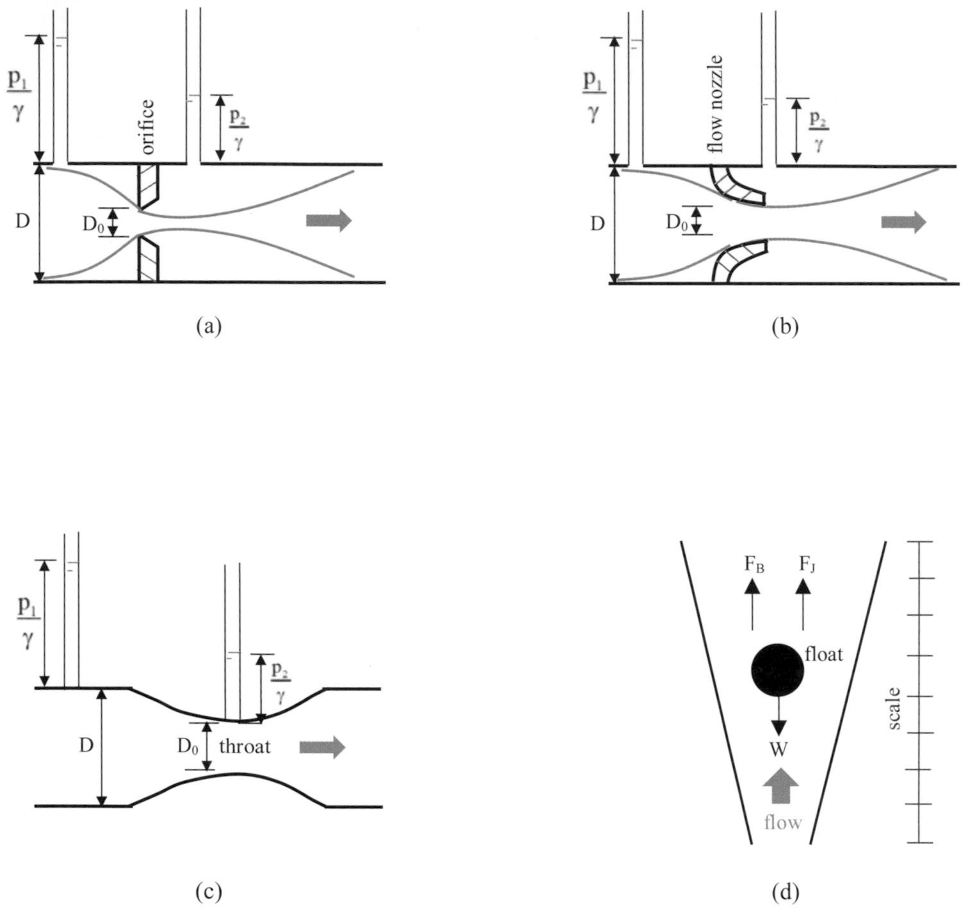

Figure 7.21 **Types of obstruction meters.** Diagram of: (a) orifice meter; (b) flow nozzle; (c) venturi tube; and (d) rotameter. F_B: buoyance force; F_J: jet flow force; and W: weight of the float.

For an orifice meter (Figure 7.21a), a flow nozzle (Figure 7.21b), and a venturi tube (Figure 7.21c), discharge can be computed using Eq. (7.14) with $h = \frac{p_1}{\gamma} - \frac{p_2}{\gamma}$ and $A_0 = \frac{\pi}{4} D_0^2$. The discharge coefficient of an orifice meter can be determined from Table 7.3, whereas the discharge coefficient of a flow nozzle is between 0.984 and 0.995 and the discharge coefficient of venturi tube is almost a constant of 0.975 (Finnemore and Franzini, 2002). For a rotameter (Figure 7.21d), the position of the float at which the upward buoyance and jet forces balance gravity is read from a scale which shows the flow rate. The scale is calibrated and provided by the manufacture of the rotameter.

7.6.2 Non-Intrusive Ultrasonic Flowmeters

Non-intrusive ultrasonic flowmeters are based on the transit-time measurement principle (Shenitech, 2007). A typical transit-time flow measurement system utilizes two transducers that function

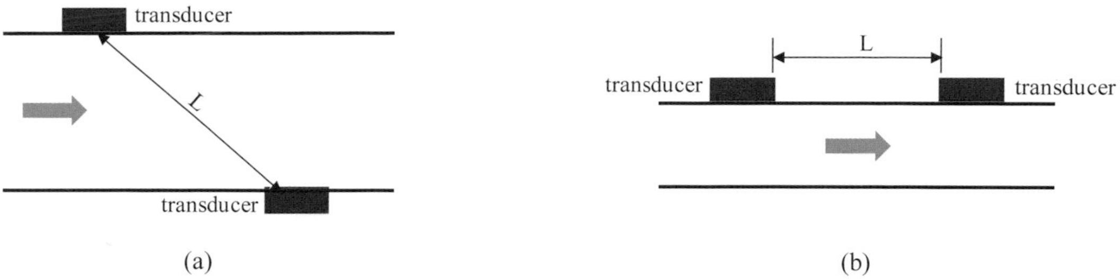

Figure 7.22 **Transducer placement on pipes.** Typical installation of for a: (a) large pipe; and (b) small pipe.

as both an ultrasonic transmitter and a receiver. The transducers are clamped onto the outside of a closed pipe at a specific distance from each other. The flowmeter operates by alternately transmitting and receiving a coded burst of sound energy between the two transducers and measuring the transit time it takes for sound to travel between the two transducers. The difference in the transit time measured is directly related to the velocity of the liquid in the pipe. For large-size pipes, the transducers are installed on the opposite sides (e.g., Figure 7.22a), whereas for medium- and small-size pipes, the transducers are installed on the same sides (e.g., Figure 7.22b).

There are various different products available on market with different prices, accuracies, and application conditions. Among them, STUF-300FxB, manufactured by Shenitech LLC (www.shenitech.com), is suitable for all commonly used full-flow (i.e., pressurized) pipes that carry pure liquids and liquids with minor suspended particles. The flow direction can be bi-directional. The STUF-300FxB meter has a high accuracy: normally 1% of velocity and possibly 0.5% with in-situ calibration. The surrounding air temperature can range from −10 to 70°C and the relative humidity can be as high as 85%. Another type of non-intrusive ultrasonic flowmeter is FloPro XCi, manufactured by Aqua Technology Group (www.aquatechnologygroup.com). This product can be used to measure partially filled (i.e., non-pressurized) pipes, which is usually necessary because pipes do not run completely full at all the times and because it is often either difficult to achieve full flow under all conditions or not economically feasible to make the needed piping changes (Aqua, 2018).

7.6.3 Area–Velocity Flowmeters

Area–velocity flowmeters are good for both full-flow and partially filled pipes. Among the products available on market, a good representative is Isco 2150, manufactured by Teledyna Isco Inc. (www.teledyneisco.com/en-us). It can measure pipes with a shallow water depth of 25 mm (or 1 in) as well as operating under pressurized conditions (Isco, 2018). It is mounted at the bottom of a pipe and uses Doppler technology to directly measure average velocity in the flow stream. An integral pressure transducer measures the liquid depth to determine the flow area. Discharge is then calculated by multiplying the area of the flow stream by its average velocity. It has a velocity accuracy of $\pm 0.03\,\mathrm{m\,s^{-1}}$ ($\pm 0.1\,\mathrm{ft\,sec^{-1}}$) and an operating air temperature range of 0 to 71°C.

PROBLEMS

7.1 Sketch the EGL and HGL along the following pressurized circular pipes conveying water from Tank A to B.

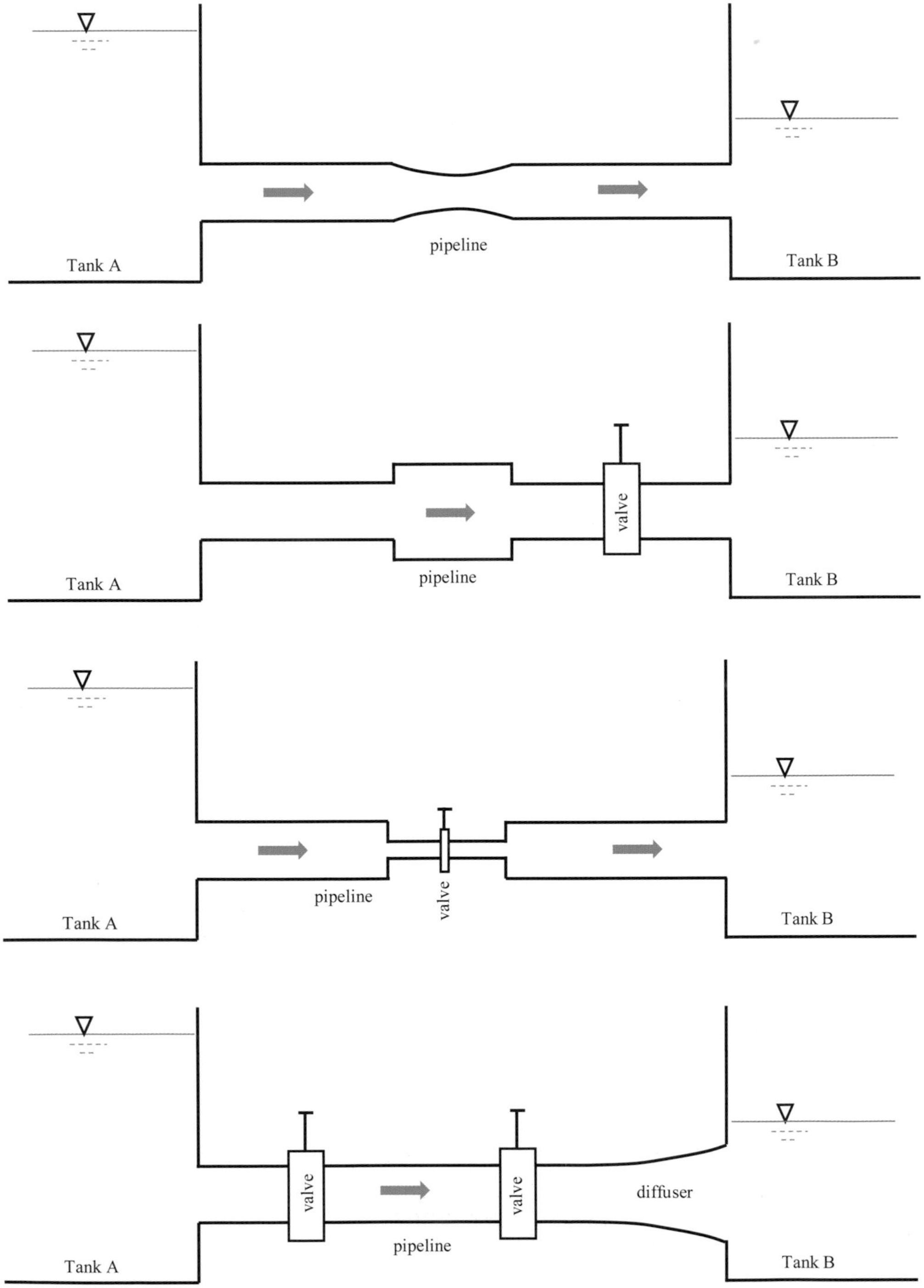

7.2 A 6-in-diameter steel pipe ($k_s = 0.001$ in, $f = 0.015$) carries 1.2 cfs water at 60°F. The pipe is pressurized.

(1) Determine the viscous sublayer thickness.
(2) Classify the pipe.
(3) Determine and sketch the velocity distribution.
(4) Determine the ratio of the mean to maximum velocity.

7.3 Redo Problem 7.2 if the flow rate is reduced to 0.2 cfs.

7.4 Redo Problem 7.2 if the flow rate is reduced to 0.6 cfs.

7.5 Redo Problem 7.2 if the flow rate is increased to 1.8 cfs.

7.6 Redo Problem 7.2 if the flow rate is increased to 4.0 cfs.

7.7 A 15-cm-diameter steel pipe ($k_s = 0.025$ mm, $f = 0.015$) carries 35 L s^{-1} water at 15°C. The pipe is pressurized.

(1) Determine the viscous sublayer thickness.
(2) Classify the pipe.
(3) Determine and sketch the velocity distribution.
(4) Determine the ratio of the mean to maximum velocity.

7.8 Redo Problem 7.7 if the flow rate is reduced to 10 L s^{-1}.

7.9 Redo Problem 7.7 if the flow rate is reduced to 20 L s^{-1}.

7.10 Redo Problem 7.7 if the flow rate is increased to 100 L s^{-1}.

7.11 Redo Problem 7.7 if the flow rate is increased to 150 L s^{-1}.

7.12 The measured point velocities in a pressurized 8-cm-diameter smooth pipe are given in the following table. If the distribution of the velocities follows Eq. (7.6), determine the:

(1) Shape factor
(2) Mean velocity.

y (cm)	0.001	0.1	0.3	0.5	0.8	1.0	1.5	2.5	3.0	3.5	4.0
u (m s^{-1})	0.02	0.1	0.2	0.3	0.4	0.5	0.6	0.8	0.9	0.95	1.0

7.13 The measured point velocities in a pressurized 2-in-diameter smooth pipe are given in the following table. If the distribution of the velocities follows Eq. (7.6), determine the:

(1) Shape factor
(2) Mean velocity.

y (in)	0.09	0.1	0.2	0.3	0.4	0.5	0.6	0.7	0.8	0.9	1.0
u (ft sec^{-1})	0.01	0.03	0.08	0.2	0.4	0.6	0.8	1.2	1.5	1.8	2.5

7.14 For the pipe of Problem 7.3:

(1) Determine which of Eqs. (7.8) through (7.13) can be used to compute the Darcy friction factor
(2) Use the equations to compute the Darcy friction factor
(3) Compare the values in (2) with the corresponding value from the Moody diagram (Figure 2.11) and discuss possible reasons for any discrepancies.

7.15 For the pipe of Problem 7.6:

(1) Determine which of Eqs. (7.8) through (7.13) can be used to compute the Darcy friction factor.

(2) Use the equations to compute the Darcy friction factor.

(3) Compare the values in (2) with the corresponding value from the Moody diagram (Figure 2.11) and discuss possible reasons for any discrepancies.

7.16 For the pipe of Problem 7.8:

(1) Determine which of Eqs. (7.8) through (7.13) can be used to compute the Darcy friction factor.

(2) Use the equations to compute the Darcy friction factor.

(3) Compare the values in (2) with the corresponding value from the Moody diagram (Figure 2.11) and discuss possible reasons for any discrepancies.

7.17 For the pipe of Problem 7.11:

(1) Determine which of Eqs. (7.8) through (7.13) can be used to compute the Darcy friction factor.

(2) Use the equations to compute the Darcy friction factor.

(3) Compare the values in (2) with the corresponding value from the Moody diagram (Figure 2.11) and discuss possible reasons for any discrepancies.

7.18 Redo Example 7.7 if the friction loss is computed using the Hazen–Williams formula. Assume an H–W coefficient of $C_{hw} = 100$ for the pipe.

7.19 Redo Example 7.8 if the friction losses are computed using the Hazen–Williams formula. Assume an H–W coefficient of $C_{hw} = 100$ for both pipes.

7.20 Redo Example 7.9 if the friction losses are computed using the Hazen–Williams formula. Assume an H–W coefficient of $C_{hw} = 100$ for all pipes.

7.21 Redo Example 7.10 if the friction losses are computed using the Hazen–Williams formula. Assume an H–W coefficient of $C_{hw} = 100$ for all pipes.

7.22 Redo Example 7.11 if the friction losses are computed using the Hazen–Williams formula. Assume an H–W coefficient of $C_{hw} = 100$ for all pipes.

7.23 A pipeline connects Lakes A and B, as shown in the following figure. Determine the discharge when the global valve is fully open. Use the Darcy–Weisbach equation to compute the friction losses.

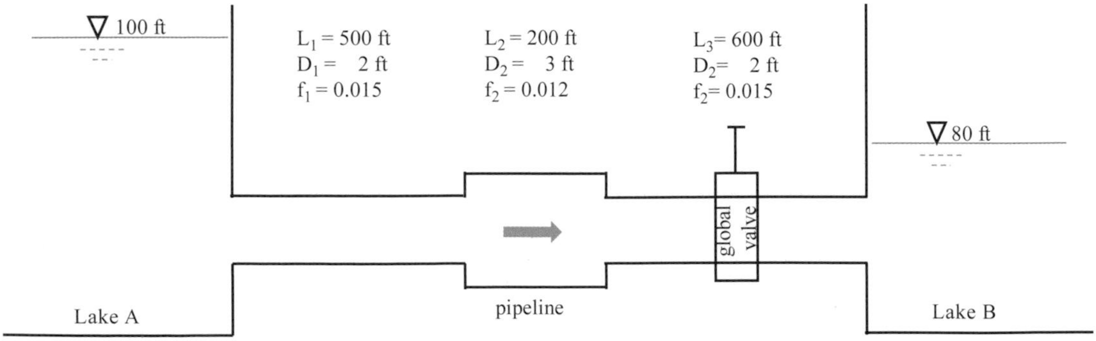

7.24 A pipeline connects Lakes A and B, as shown in the following figure. Determine the discharge when the global valve is fully open. Use the Darcy–Weisbach equation to compute the friction losses.

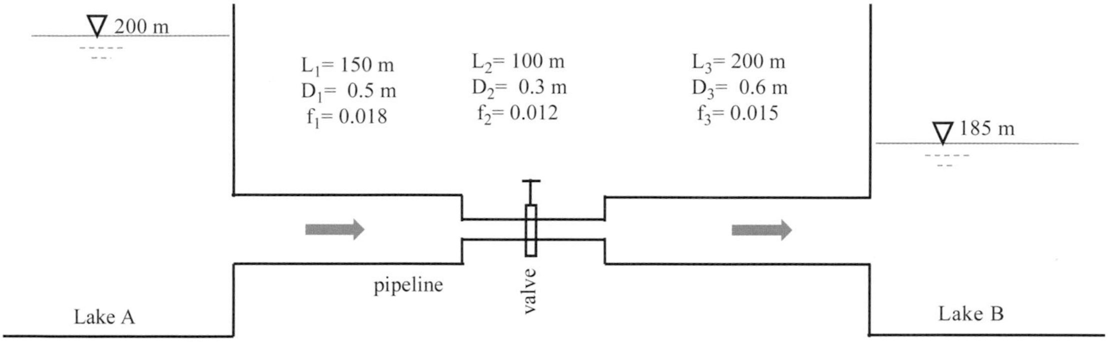

7.25 Redo Problem 7.23 if the friction losses are computed using the Hazen–Williams formula. Assume an H–W coefficient of $C_{hw} = 100$ for all pipes.

7.26 Redo Problem 7.24 if the friction losses are computed using the Hazen–Williams formula. Assume an H–W coefficient of $C_{hw} = 100$ for all pipes.

7.27 Three reservoirs, designed Res. A, Res. B, and Res. C for description purposes, are connected by three pipes, as shown in the following figure. Determine the flow rate and direction in each pipe. Use the Darcy–Weisbach equation to compute the friction losses and neglect all minor losses.

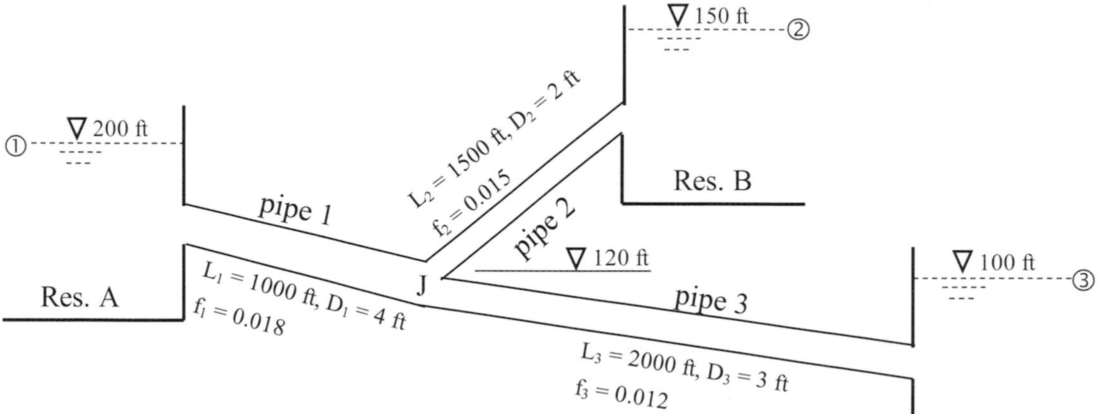

7.28 Redo Problem 7.27 if the friction losses are computed using the Hazen–Williams formula. Assume an H–W coefficient of $C_{hw} = 100$ for all pipes.

7.29 Four reservoirs, designated Res. A, Res. B, Res. C, and Res. D for description purposes, are connected by four pipes, as shown in the following figure. Determined the flow rate and direction in each pipe. Use the Darcy–Weisbach equation to compute the friction losses and neglect all minor losses.

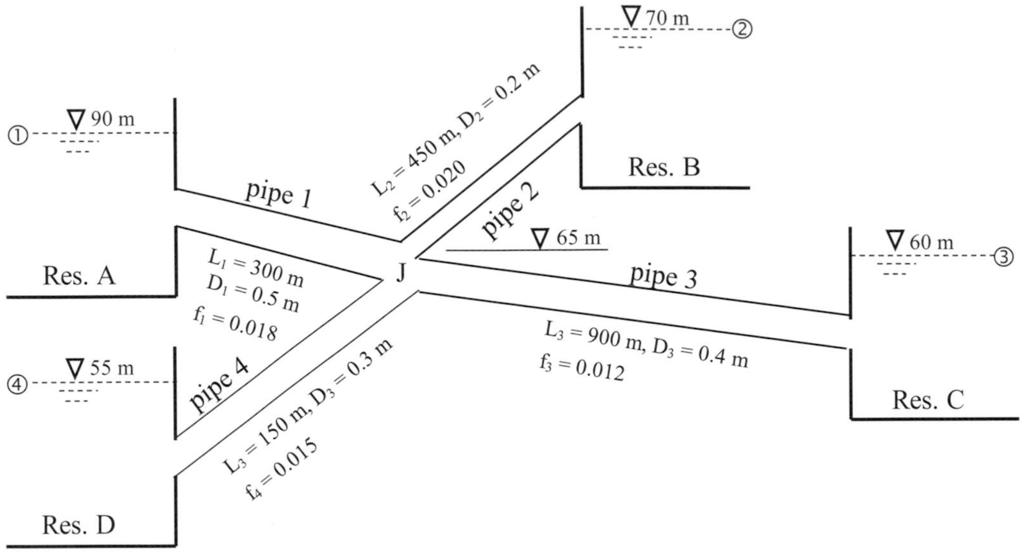

7.30 Redo Problem 7.29 if the friction losses are computed using the Hazen–Williams formula. Assume an H–W coefficient of $C_{hw} = 100$ for all pipes.

7.31 For the one-loop pipe network shown in the following figure, determine the discharge in each pipe and the overall head loss of the network. Use the Darcy–Weisback equation to compute the friction losses and neglect minor losses.

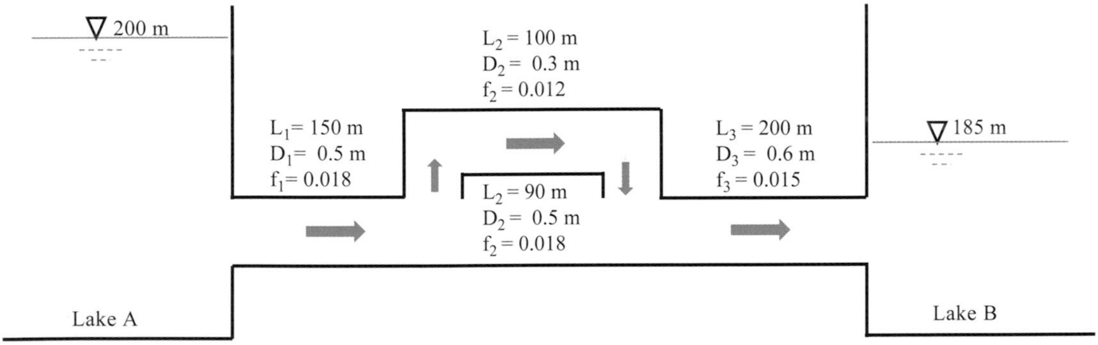

7.32 Redo Example 7.12 if the friction losses are computed using the Hazen–Williams formula. Assume an H–W coefficient of $C_{hw} = 100$ for all pipes.

7.33 Redo Example 7.13 if the friction losses are computed using the Hazen–Williams formula. Assume an H–W coefficient of $C_{hw} = 100$ for all pipes.

7.34 Redo Example 7.14 if the friction losses are computed using the Hazen–Williams formula. Assume an H–W coefficient of $C_{hw} = 100$ for all pipes.

7.35 Redo Example 7.14 if the friction losses are computed using the Darcy–Weisbach equation with f estimated using one of the formulas (Eqs. (7.8) through (7.10)) for a smooth pipe. Also, check if the pipes are smooth.

7.36 For the pipe network shown in the following figure, determine the discharge and head loss in each pipe. Use the Darcy–Weisbach equation to compute the friction losses and neglect minor losses.

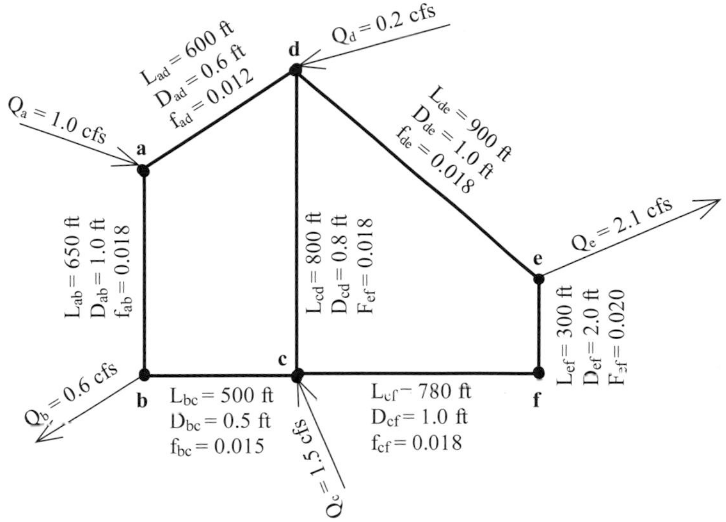

7.37 Redo Problem 7.36 if the friction losses are computed using the Hazen–Williams formula. Assume an H–W coefficient of $C_{hw} = 100$ for all pipes.

7.38 For the pipe network shown in the following figure, determine the discharge and head loss in each pipe. Use the Darcy–Weisbach equation to compute the friction losses and neglect minor losses.

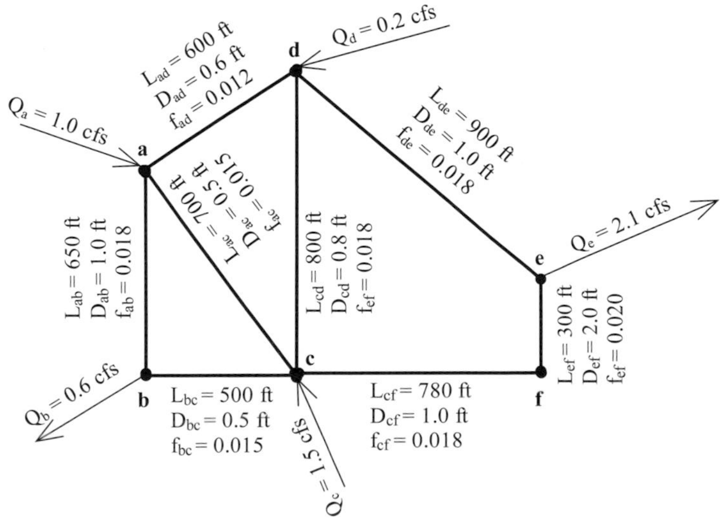

7.39 Redo Problem 7.38 if the friction losses are computed using the Hazen–Williams formula. Assume an H–W coefficient of $C_{hw} = 100$ for all pipes.

7.40 For the pipe network shown in the following figure, determine the discharge and head loss in each pipe. Use the Darcy–Weisbach equation to compute the friction losses and neglect minor losses.

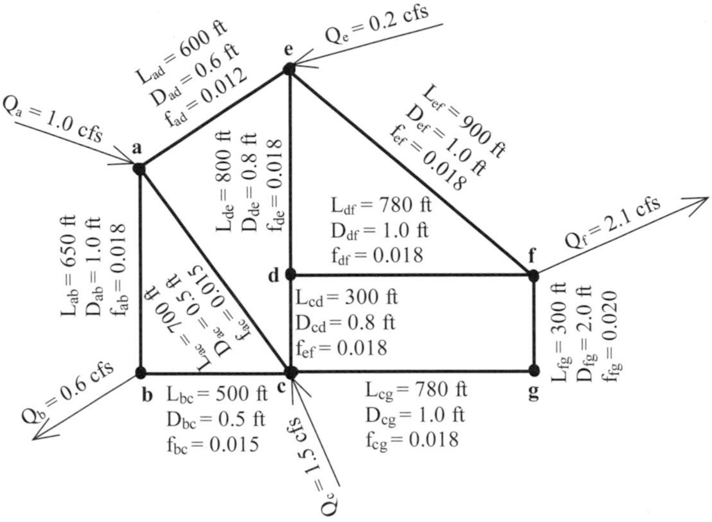

7.41 Redo Problem 7.40 if the friction losses are computed using the Hazen–Williams formula. Assume an H–W coefficient of $C_{hw} = 100$ for all pipes.

7.42 For the pipe network shown in the following figure with dimensions given in the table, determine the flow rate and head loss in each pipe. Use the Darcy–Weisbach equation to compute the friction losses and neglect minor losses.

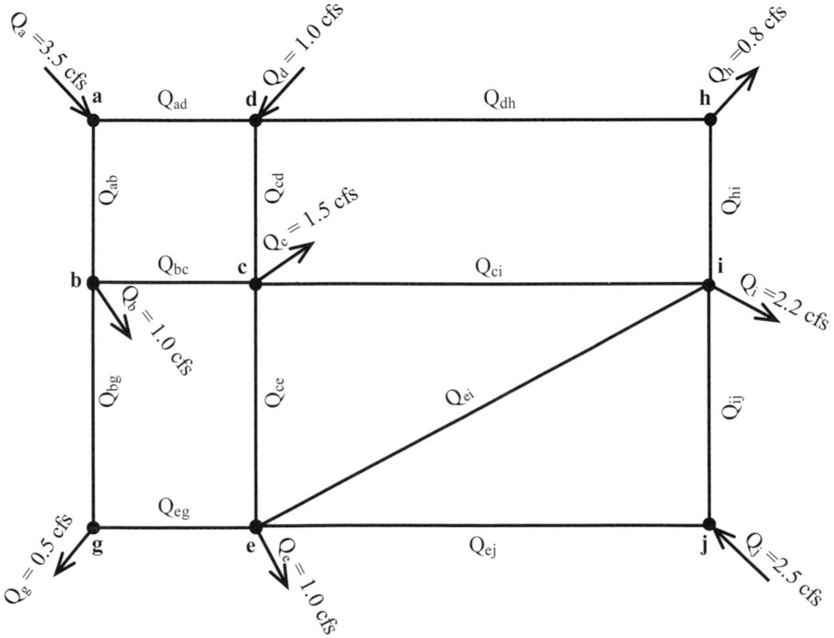

Pipe	Length, L (ft)	Diameter, D (ft)	Darcy friction factor, f
ab	1000	2.0	0.0175
bc	1000	0.5	0.0220
cd	1000	2.0	0.0170
ad	1000	2.0	0.0178
bg	1500	1.0	0.0195
ce	1500	1.0	0.0200
eg	1000	0.5	0.0215
ci	2800	1.2	0.0185
hi	1000	2.0	0.0197
dh	2800	3.0	0.0165
ei	2850	1.2	0.0180
ej	2800	1.2	0.0188
ij	1500	2.0	0.0168

7.43 For the orifice in Example 7.17, if the head is reduced to 2.5 ft, determine its discharge.

7.44 For the orifice in Example 7.18, if the head is increased to 6.5 ft, determine its discharge.

7.45 For in orifice in Example 7.19, if the head is increased to 0.8 m, determine its discharges.

7.46 The orifice in Example 7.17 is used to measure the discharge in a pipe, as illustrated in Figure 7.21a. If the piezometric readings are $p_1/\gamma = 2.2\,\text{ft}$ and $p_2/\gamma = 1.5\,\text{ft}$, determine the discharge.

7.47 The orifice in Example 7.18 is used to measure the discharge in a pipe, as illustrated in Figure 7.21a. If the piezometric readings are $p_1/\gamma = 2.2\,\text{ft}$ and $p_2/\gamma = 1.5\,\text{ft}$, determine the discharge.

7.48 The orifice in Example 7.19 is used to measure the discharge in a pipe, as illustrated in Figure 7.21a. If the piezometric readings are $p_1/\gamma = 2.2\,\text{ft}$ and $p_2/\gamma = 1.5\,\text{ft}$, determine the discharge.

7.49 A 12-in-diameter flow nozzle is used to measure the discharge in a pipe, as illustrated in Figure 7.21b. If the piezometric readings are $p_1/\gamma = 2.2\,\text{ft}$ and $p_2/\gamma = 1.5\,\text{ft}$, determine the discharge.

7.50 A venturi tube (throat diameter of 12 in) is used to measure the discharge in a pipe, as illustrated in Figure 7.21c. If the piezometric readings are $p_1/\gamma = 2.2\,\text{ft}$ and $p_2/\gamma = 1.5\,\text{ft}$, determine the discharge.

7.51 A venturi tube (throat diameter of 20 cm) is used to measure the discharge in a pipe, as illustrated in Figure 7.21c. If the piezometric readings are $p_1/\gamma = 0.5\,\text{m}$ and $p_2/\gamma = 0.2\,\text{m}$, determine the discharge.

7.52 Find a rotameter from the internet and plot its relationship between float reading and flow rate.

7.53 Find a non-intrusive ultrasonic flowmeter for full-flow pipes from the internet and describe its features and application conditions.

7.54 Find a non-intrusive ultrasonic flowmeter for partially filled pipes from the internet and describe its features and application conditions.

7.55 Find an area–velocity flowmeter from the internet and describe its features and application conditions.

8 Hydraulic Structures

Besides primary hydraulic structures (e.g., channels and pipes), water resources engineering systems may also include various appurtenant hydraulic structures, such as culverts, bridges, risers, storm sewers, spillways, and stilling basins. This chapter introduces the principles of hydraulics as they relate to these appurtenant hydraulic structures. For each of these structures, the relevant hydraulic analysis methods and computations for sizing the structure to convey a required discharge are explained and demonstrated using worked examples.

8.1 Culverts

A culvert is a structure that allows water to flow under an obstruction (e.g., road, railroad, or trail) (Figure 8.1).

8.1.1 Culvert Components

A culvert consists of four parts, namely inlet, barrel, outlet, and deck. Water flows through the inlet into the barrel and exits at the outlet. Most commonly, the barrel has a circular or rectangular (i.e., box) cross section (Figure 8.2), though it may have an arch, pipe arch, low-profile arch, high-profile arch, elliptical, semi-circular, or ConSpan-shaped cross section (USACE, 2010). The barrel can be made of concrete, metal, or plastic. Along a culvert, the invert of a particular cross section is the lowest point on the inside of the section. So, the inlet and outlet inverts are the lowest points on the insides of the corresponding cross sections. Their elevations are termed the inlet invert elevation and outlet invert elevation, respectively. The upstream water depth above the inlet invert is defined as the headwater depth (usually symbolized as HW), whereas the downstream water depth above the outlet invert is defined as the tail water depth (usually symbolized as TW).

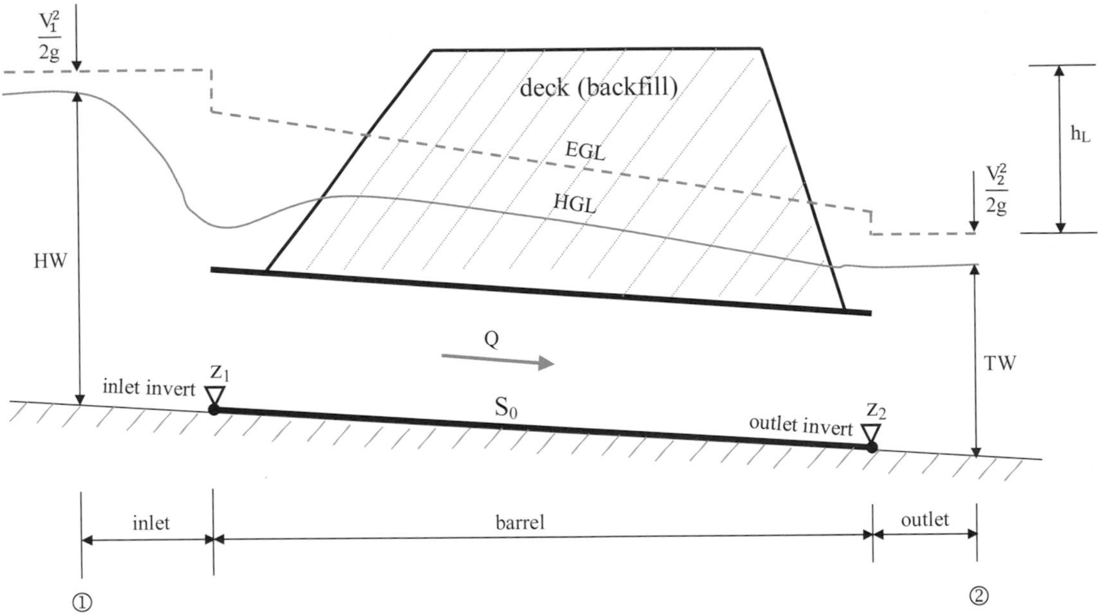

Figure 8.1 Longitudinal cross section of a culvert. Diagram shows inlet, barrel, outlet, and deck. HW: headwater depth; TW: tail water depth; and h_L: total head loss.

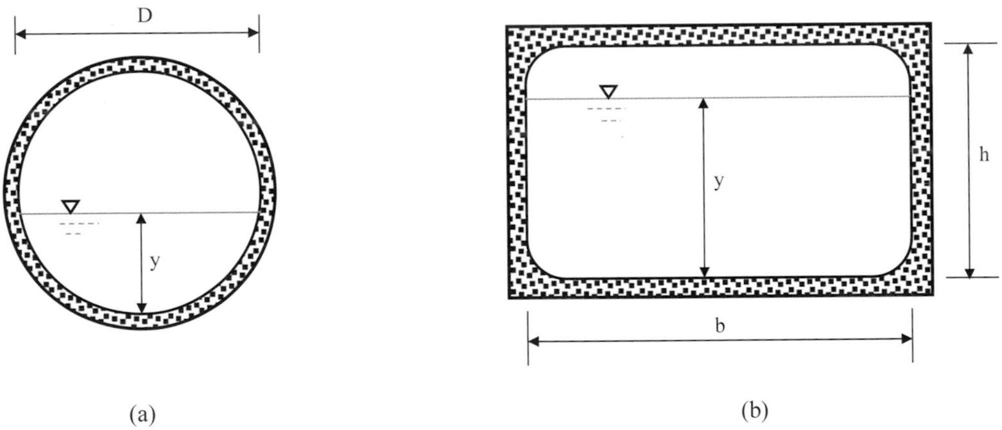

Figure 8.2 Cross sections of culvert barrels. (a) Circular barrel; and (b) box-shaped barrel.

In practice, a culvert can have one barrel or multiple barrels, depending on the required conveyance capacity and transportation standard. For multiple barrels, they can either be different or identical in terms of shape as well as inlet and outlet invert elevations (Figure 8.3). One purpose of using multiple barrels is to limit the maximum allowable size of a single barrel, carry a large total discharge, and/or control flow rates at varying water surface elevations. Another purpose of using multiple barrels is to make the deck wide enough for transportation. The deck is the volume of material above and/or surrounding the barrel(s) and consists of backfill that forms the obstruction and functions as part of a road, railroad, or trail. The only difference between a culvert with a

Figure 8.3 **Illustrations of culverts with multiple barrels at a channel.** Diagrams show cross-sectional views of culverts with: (a) two identical circular barrels; and (b) two identical box barrels and one circular barrel.

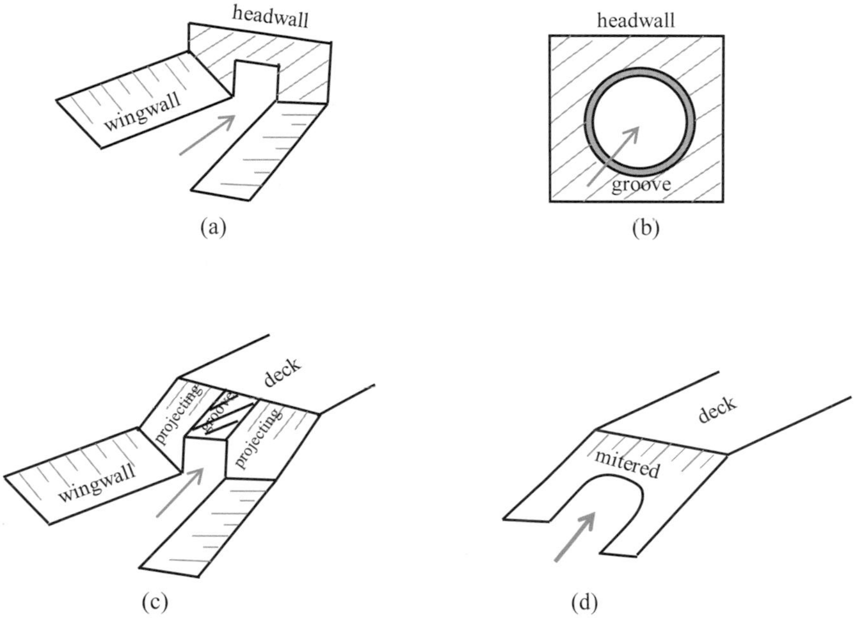

Figure 8.4 **Sketches of common culvert inlet types.** (a) Square edge with headwall; (b) groove end with headwall; (c) groove end with projecting of fill; and (d) mitered to slope. The diagrams are sketched as frontal views looking downstream.

single barrel and a culvert with multiple barrels is that for the former culvert, the total discharge is conveyed by only one barrel, whereas for the latter culvert, the total discharge is split among multiple barrels. Because the analysis methods are the same regardless of the number of barrels, the discussion hereinafter focuses on culverts with one barrel only.

Various culvert inlets (i.e., barrel entrances) have been used, such as square edge with headwall, groove end with headwall, groove end with projecting of fill, mitered to slope, beveled ring, headwall with chamfers, wingwall flares, chamfers with skewed headwall (Normann *et al.*, 1985; USACE, 2010). Figure 8.4 illustrates the four most common types of inlets. Illustrations of other types are available in USACE (2010). The selection of an entrance type depends on the shape and material of the barrel (Table 8.1). In terms of hydraulics, for a given culvert, it can either be under an inlet- or an outlet-control flow condition, depending on the headwater depth (HW) and tail water depth (TW) as well as the relative magnitude of the ratio of HW to TW (Figure 8.1). Under an inlet-control flow condition, the discharge through the barrel is equal to the flow rate flowing into the inlet, whereas under an outlet-control flow condition, the discharge through the barrel is equal to the flow rate flowing out of the outlet.

The US Department of Transportation (USDoT) Federal Highway Administration (FHA) published a series of nomographs (Herr and Bossy, 1965), which allows the inlet-control (see Section 8.1.2) headwater depth to be computed for different types of culverts operating under a wide range of flow conditions. Each of these nomographs is identified by a chart number and has from two to four separate scales representing different culvert entrance types (Table 8.1). In the 1980s, these nomographs were expressed in three mathematical equations (see Section 8.1.2) for the convenience of computations using computers. One of these equations is for the condition that the inlet is submerged, whereas the other two equations are for the condition that the inlet is unsubmerged (Normann *et al.*, 1985). The two equations for the unsubmerged condition are designated Form 1 and 2. For a particular combination of chart number and scale number, one of the two forms is applicable. The constants of the equations for circular and box culverts are also presented in Table 8.1. Moreover, the chart numbers, scale numbers, and constants of the equations for other shapes of barrels and other inlet types can be found in existing publications (e.g., Normann *et al.*, 1985; Mays, 2005). Given that subjective errors are difficult to avoid from looking up the nomographs and that it is more convenient and efficient to use the equations, the nomographs are not presented in this book.

8.1.2 Inlet-Control Conditions

For a culvert, if an open channel flow occurs just downstream of its entrance and maintains itself in most portion of the barrel, the culvert has an inlet-control condition. The discharge through the barrel is equal to the maximum flow rate that the inlet allows; it is also smaller than the maximum flow rate that the outlet allows. Such a condition may exist for situations in which the inlet and outlet are either unsubmerged or submerged (Figure 8.5).

If the inlet of a culvert barrel is submerged, which is the case when $Q/(AD^{0.5}) \geq 4.0$, where Q is the discharge through the barrel, A is the cross-sectional area of the barrel, and D is the interior height of the barrel, it functions as an orifice. The relationship between HW and Q can be expressed as (Normann *et al.*, 1985):

$$\frac{HW}{D} = C\left(\frac{Q}{AD^{0.5}}\right)^2 + Y + Z \tag{8.1}$$

barrel discharge [cfs]
headwater depth [ft]
barrel interior height [ft]
constant (Table 8.1)
barrel cross-sectional area [ft^2]
correction factor of barrel slope S_0 ($= -0.5S_0$ in general; $+0.7S_0$ for mitered inlet)
constant (Table 8.1)

If the inlet of a culvert barrel is unsubmerged, which is the case when $Q/(AD^{0.5}) \leq 3.5$, it functions as a weir. The relationship between HW and Q has two forms, designated Form 1 and Form 2, for description purposes. The choice of one form over another depends on the barrel shape and material (i.e., signified by a chart number), as shown in Table 8.1. These two forms can be expressed as (Normann *et al.*, 1985):

Table 8.1 Constants of the inlet-control equations for circular and rectangular culverts.[1]

Chart #[2]	Shape (material)	Scale #	Inlet type description	Inlet submerged[3]		Inlet unsubmerged[4]		
				C	Y	Eq. Form	K	M
1	Circular (concrete)	1	Square edge w/headwall	0.0398	0.67	1	0.0098	2.0
		2	Groove w/headwall	0.0292	0.74		0.0078	2.0
		3	Groove w/projecting	0.0317	0.69		0.0045	2.0
2	Circular (CMP)	1	Headwall	0.0379	0.69	1	0.0078	2.0
		2	Mitered to slope	0.0463	0.75		0.0210	1.33
		3	Projecting	0.0553	0.54		0.0340	1.50
3	Circular	A	Beveled ring, 45° bevels	0.0300	0.74	1	0.0018	2.50
		B	Beveled ring, 33.7° bevels	0.0243	0.83		0.0018	2.50
8	Rectangular box	1	30° to 75° wingwall flares	0.0385	0.81	1	0.026	1.0
		2	90° and 15° wingwall flares	0.4000	0.80		0.061	0.75
		3	0° wingwall flares	0.0423	0.82		0.061	0.75
9	Rectangular box	1	45° wingwall flare d = 0.0430	0.0309	0.80	2	0.510	0.667
		2	18° to 33.7° wingwall flare d = 0.0830	0.0249	0.83		0.486	0.667
10	Rectangular box	1	90° headwall w/ 3/4″ chamfers	0.0375	0.79	2	0.515	0.667
		2	90° headwall w/ 45° bevels	0.0314	0.82		0.495	0.667
		3	90° headwall w/ 33.7° bevels	0.0252	0.865		0.486	0.667
11	Rectangular box	1	3/4″ chamfers, 45° skewed headwall	0.0402	0.73	2	0.522	0.667
		2	3/4″ chamfers, 30° skewed headwall	0.0425	0.705		0.533	0.667
		3	3/4″ chamfers, 15° skewed headwall	0.04505	0.68		0.545	0.667
		4	45° bevels, 10° to 45° skewed headwall	0.0327	0.75		0.498	0.667

Table 8.1 **(Continued)**

Chart #[2]	Shape (material)	Scale #	Inlet type description	Inlet submerged[3]		Inlet unsubmerged[4]		
				C	Y	Eq. Form	K	M
12	Rectangular box, 3/4″ chamfers	1	45° non-offset wingwall flares	0.0339	0.805	2	0.497	0.667
		2	18.4° non-offset wingwall flares	0.0361	0.806		0.493	0.667
		3	30° non-offset wingwall flares	0.0386	0.71		0.495	0.667
13	Rectangular box, top bevels	1	45° wingwall flares – offset	0.0302	0.835	2	0.495	0.667
		2	33.7° wingwall flares – offset	0.0252	0.881		0.493	0.667
		3	18.4° wingwall flares – offset	0.0227	0.887		0.497	0.667
16–19	CM box	1	90° headwall	0.0379	0.69	1	0.0083	2.0
		2	Thick wall projecting	0.0419	0.64		0.0145	1.75
		3	Thin wall projecting	0.0496	0.57		0.0340	1.5
57	Rectangular	1	Tapered inlet throat	0.0179	0.97	2	0.475	0.667
58	Rectangular (concrete)	1	Side tapered – less favorable edges	0.0466	0.85	2	0.56	0.667
		2	Side tapered – more favorable edges	0.0978	0.87		0.56	0.667
59	Rectangular (concrete)	1	Side tapered – less favorable edges	0.0466	0.65	2	0.50	0.667
		2	Side tapered – more favorable edges	0.0378	0.71		0.50	0.667

[1] Reproduced from Normann *et al.* (1985).
[2] The chart numbers of 29, 30, 34, 35, 36, 40–42, 55, and 56 are for other barrel shapes.
[3] The constants for Eq. (8.1).
[4] The constants for Form 1 (Eq. (8.2)) and Form 2 (Eq. (8.3)).

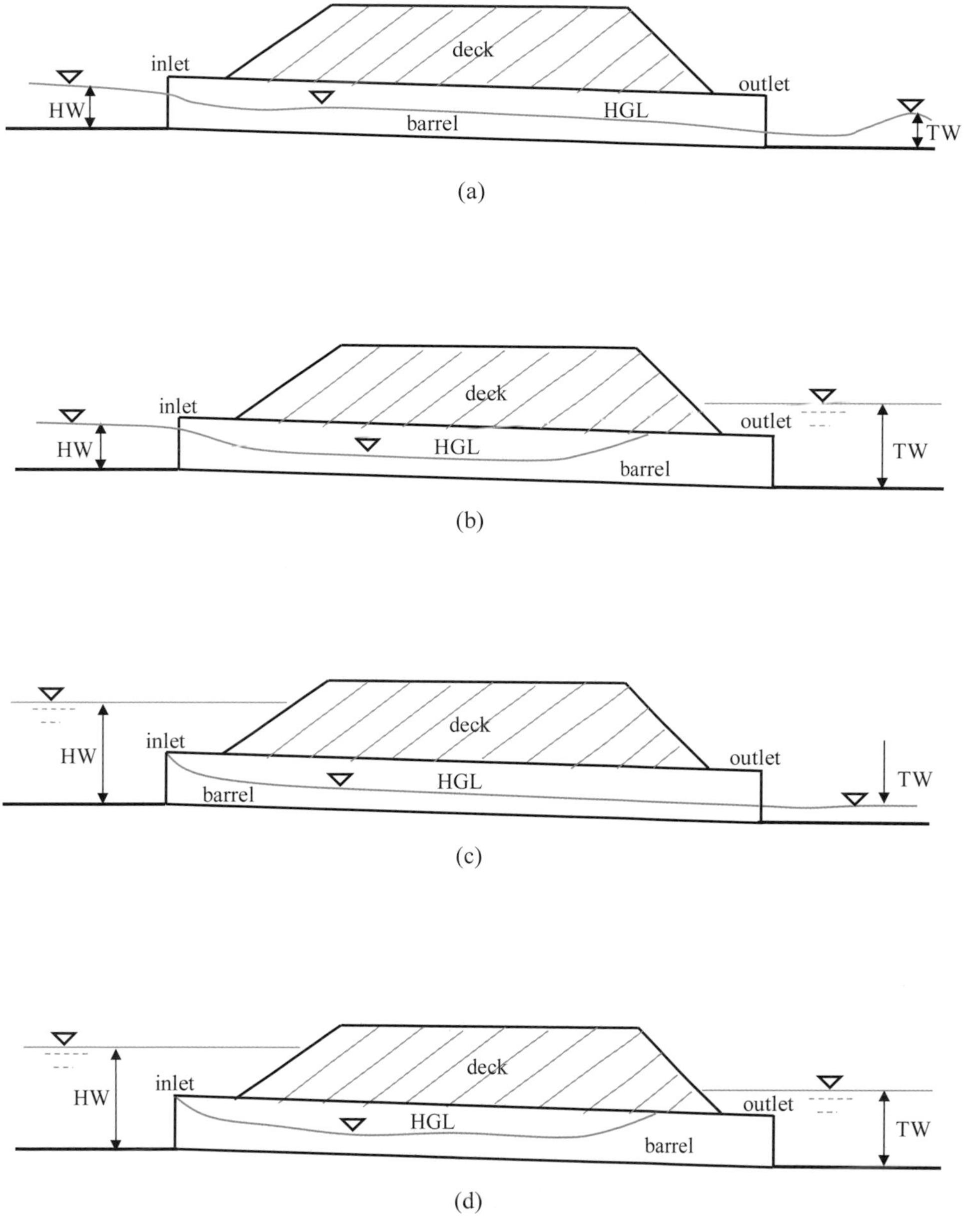

Figure 8.5 **Longitudinal cross sections of culverts with various inlet-control flow conditions.** (a) Unsubmerged inlet and outlet; (b) unsubmerged inlet and submerged outlet; (c) submerged inlet and unsubmerged outlet; and (d) submerged inlet and outlet. (*Source:* reproduced Normann *et al.*, 1985)

$$\text{Form 1}: \frac{HW}{D} = \frac{H_c}{D} + K\left(\frac{Q}{AD^{0.5}}\right)^M + Z \tag{8.2}$$

specific energy head at critical depth [ft] → H_c
headwater depth [ft] → HW
barrel interior height [ft] → D
barrel discharge [cfs] → Q
constant (Table 8.1) → K
constant (Table 8.1) → M
barrel cross-sectional area [ft^2] → A
correction factor of barrel slope S_0 (= $-0.5S_0$ in general; +0.7S_0 for mitered inlet) → Z

$$\text{Form 2}: \frac{HW}{D} = K\left(\frac{Q}{AD^{0.5}}\right)^M + Z \tag{8.3}$$

headwater depth [ft] → HW
barrel interior height [ft] → D
barrel discharge [cfs] → Q
constant (Table 8.1) → K
constant (Table 8.1) → M
barrel cross-sectional area [ft^2] → A
correction factor of barrel slope S_0 (= $-0.5S_0$ in general; +0.7S_0 for mitered inlet) → Z

If $3.5 < Q/(AD^{0.5}) < 4.0$, HW/D is the linear interpolation of either between Eq. (8.1) and (8.2) or between Eq. (8.1) and (8.3). In this regard, $Q/(AD^{0.5}) = 4.0$ is used in Eq. (8.1) and $Q/(AD^{0.5}) = 3.5$ is used in Eq. (8.2) or (8.3) to compute the corresponding values of HW/D.

Example 8.1 A circular concrete culvert with a 2-ft-diameter barrel carries 25 cfs. It is laid on a slope of 0.005. For an inlet-control condition, determine the HW if the inlet is:

(1) Square edge with headwall
(2) Groove end with headwall
(3) Groove end with projecting of fill.

Solution

$D = 2\,\text{ft}$, $S_0 = 0.005$, $Q = 25\,\text{cfs}$

Cross-sectional area: $A = \pi/4 * (2\,\text{ft})^2 = 3.14159\,\text{ft}^2$.

Table 8.1: circular concrete culvert → chart# 1.

$Q/(AD^{0.5}) = (25\,\text{cfs})/[(3.14159\,\text{ft}^2) * (2\,\text{ft})^{0.5}] = 5.62698 > 4.0 \rightarrow$ inlet submerged (Eq. (8.1)).

(1) Table 8.1: square edge w/headwall → scale# 1→C = 0.0398, Y = 0.67

$$Z = -0.5 * (0.005) = -0.0025$$
$$HW/D = 0.0398 * 5.62698^2 + 0.67 - 0.0025 = 1.92768$$
$$\rightarrow HW = 1.92768 * (2\,\text{ft}) = 3.86\,\text{ft}.$$

(2) Table 8.1: groove edge w/headwall → scale# 2→C = 0.0292, Y = 0.74

$$Z = -0.5 * (0.005) = -0.0025$$
$$HW/D = 0.0292 * 5.62698^2 + 0.74 - 0.0025 = 1.66206$$
$$\rightarrow HW = 1.66206 * (2\,\text{ft}) = 3.32\,\text{ft}.$$

(3) Table 8.1: groove edge w/projecting $\rightarrow$ scale# 3$\rightarrow$C = 0.0317, Y = 0.69

$$Z = -0.5 * (0.005) = -0.0025$$
$$HW/D = 0.0317 * 5.62698^2 + 0.69 - 0.0025 = 1.69121$$
$$\rightarrow HW = 1.69121 * (2\,ft) = 3.38\,ft.$$

Example 8.2 A circular CMP culvert (see Table 8.1) with a 1-ft-diameter barrel carries 1.8 cfs. It is laid on a slope of 0.01. The culvert has an inlet of mitered to slope. For an inlet-control condition, determine the HW.

Solution

$D = 1\,ft$, $S_0 = 0.01$, $Q = 1.8\,cfs$, $g = 32.2\,ft\,sec^{-2}$

Cross-sectional area: $A = \pi/4 * (1\,ft)^2 = 0.78540\,ft^2$.

Table 8.1: circular CMP culvert $\rightarrow$ chart# 2; mitered to slope inlet $\rightarrow$ scale# 2.

Mitered inlet$\rightarrow Z = 0.7 * (0.01) = 0.007$.

$Q/(AD^{0.5}) = (1.8\,cfs)/[(0.78540\,ft^2) * (1\,ft)^{0.5}] = 2.29183 < 3.5 \rightarrow$ inlet unsubmerged.

Table 8.1: chart# 2, scale# 2, unsubmerged $\rightarrow$ Form 1, K = 0.0210, M = 1.33.

- Determine the critical depth y_c and velocity V_c:
  Table 2.4: $A_c = (1\,ft)^2/8(\theta_c - \sin\theta_c) = 0.125(\theta_c - \sin\theta_c)\,ft^2$
  $B_c = (1\,ft)\sin(\theta_c/2) = \sin(\theta_c/2)\,ft.$
  Eq. (2.37): $(1.8\,cfs)^2/(32.2\,ft\,sec^{-2}) = [0.125(\theta_c - \sin\theta_c)]^3/\sin(\theta_c/2)$.

  By trial-and-error or Excel Solver, $\theta_c = 3.42688$ rad.
  Table 2.4: $3.42688\,rad = 2\cos^{-1}[(1\,ft - 2y_c)/(1\,ft)]$
  $$\rightarrow 1 - 2y_c = \cos(1.71344\,rad) = -0.14216 \rightarrow y_c = 0.57108\,ft.$$
  $$A_c = 0.125 * [3.42688 - \sin(3.42688\,rad)]\,ft^2 = 0.46354\,ft^2.$$
  $$V_c = (1.8\,cfs)/(0.46354\,ft^2) = 3.88316\,ft\,sec^{-1}.$$

- Determine the specific energy head at critical depth:
  Eq. (2.35): $H_c = 0.57108\,ft + (3.88316\,ft\,sec^{-1})^2/(2 * 32.2\,ft\,sec^{-2}) = 0.80523\,ft.$
- Form 1 (Eq. (8.2)):

$$HW/D = (0.80523\,ft)/(1\,ft) + 0.0210 * 2.29183^{1.33} + 0.007 = 0.87551$$
$$\rightarrow HW = 0.87551 * (1\,ft) = 0.88\,ft.$$

Example 8.3 A 2-ft-wide, 1.5-ft-high rectangular box culvert (3/4″ chamfers) is constructed as a road crossing over a small stream. It is laid on a slope of 0.006 and has an inlet with 45° non-offset wingwall flares. For an inlet-control condition, determine the HW if the culvert conveys:

(1) 10 cfs
(2) 30 cfs.

Solution

$b = 2\,\text{ft}$, $D = 1.5\,\text{ft}$, $S_0 = 0.006$, $g = 32.2\,\text{ft sec}^{-2}$

Cross-sectional area: $A = (2\,\text{ft}) * (1.5\,\text{ft}) = 3.0\,\text{ft}^2$.

Table 8.1: rectangular box culvert, 3/4″ chamfers → chart# 12

inlet of 45° non-offset wingwall flares→scale# 1.

$Z = -0.5 * 0.006 = -0.003.$

(1) $Q = 10\,\text{cfs}$

$Q/(AD^{0.5}) = (10\,\text{cfs})/[(3.0\,\text{ft}^2) * (1.5\,\text{ft})^{0.5}] = 2.72166 < 3.5 \rightarrow$ inlet unsubmerged.

Table 8.1: Chart# 12, scale# 1 → Form 2, $K = 0.497$, $M = 0.667$.

Form 2 (Eq. (8.3)): $HW/D = 0.497 * 2.72166^{0.667} - 0.003 = 0.96615$

$\rightarrow HW = 0.96615 * (1.5\,\text{ft}) = 1.45\,\text{ft}.$

(2) $Q = 30\,\text{cfs}$

$Q/(AD^{0.5}) = (30\,\text{cfs})/[(3.0\,\text{ft}^2) * (1.5\,\text{ft})^{0.5}] = 8.16497 > 4.0 \rightarrow$ inlet submerged.

Table 8.1: Chart# 12, scale# 1→$C = 0.0339$, $Y = 0.805$.

Eq. (8.1): $HW/D = 0.0339 * 8.16497^2 + 0.805 - 0.003 = 3.06200$

$\rightarrow HW = 3.06200 * (1.5\,\text{ft}) = 4.59\,\text{ft}.$

Example 8.4 A 4-ft-wide, 2-ft-high rectangular box culvert is constructed as a road crossing over a small stream. It is laid on a slope of 0.002 and has an inlet of 0° wingwall flare ($d = 0.0430$). For an inlet-control condition, determine the discharge if the HW is:

(1) 1.5 ft
(2) 3.5 ft.

Solution

$b = 4\,\text{ft}$, $D = 2\,\text{ft}$, $S_0 = 0.002$, $g = 32.2\,\text{ft sec}^{-2}$

Cross-sectional area: $A = (4\,\text{ft}) * (2\,\text{ft}) = 8\,\text{ft}^2$.

Table 8.1: rectangular box culvert with 0° wingwall flare inlet → chart# 8, scale# 3

$Z = -0.5 * 0.002 = -0.001$.

(1) HW = 1.5 ft < 2 ft→ assume that the inlet unsubmerged.
Chart# 8, scale# 3, unsubmerged → Form 1, K = 0.061, M = 0.75.

- Determine the critical depth y_c and velocity V_c:
Unit-width flow rate: $q = Q/(4\,\text{ft}) = 0.25Q\,\text{ft}^2\,\text{sec}^{-1}$.
Eq. (2.38): $y_c = [(0.25Q\,\text{ft}^2\,\text{sec}^{-1})^2/(32.2\,\text{ft}\,\text{sec}^{-2})]^{1/3} = 0.12474Q^{2/3}\,\text{ft}$.

$$V_c = q/y_c = (0.25Q\,\text{ft}^2\,\text{sec}^{-1})/(0.12474Q^{2/3}\,\text{ft}) = 2.00417Q^{1/3}\,\text{ft}\,\text{sec}^{-1}.$$

- Determine the specific energy head at critical depth:
Eq. (2.35): $H_c = 0.12474Q^{2/3}\,\text{ft} + (2.00417Q^{1/3}\,\text{ft}\,\text{sec}^{-1})^2/(2 * 32.2\,\text{ft}\,\text{sec}^{-2})$
$= 0.18711Q^{2/3}\,\text{ft}$.

- Form 1 (Eq. (8.2)):

$$(1.5\,\text{ft})/(2\,\text{ft}) = (0.18711Q^{2/3}\,\text{ft})/(2\,\text{ft}) + 0.061 * \{Q/[(8\,\text{ft}^2) * (2\,\text{ft})^{0.5}]\}^{0.75} - 0.001$$
$$\rightarrow 0.093555Q^{2/3} + 0.0098884Q^{0.75} - 0.751 = 0.$$

Solving the equation by trial-and-error or Excel Solver, one can get Q = 18.8 cfs.

- Check: $Q/(AD^{0.5}) = (18.8\,\text{cfs})/[(8\,\text{ft}^2) * (2\,\text{ft})^{0.5}] = 1.66 < 3.5$→ correct assumption.
Thus, the discharge is 18.8 cfs.

(2) HW = 3.5 ft > 2 ft→ assume that the inlet is submerged.
Chart# 8, scale# 3, submerged →C = 0.0423, Y = 0.82.
Eq. (8.1): $(3.5\,\text{ft})/(2\,\text{ft}) = 0.0423 * \{Q/[(8\,\text{ft}^2) * (2\,\text{ft})^{0.5}]\}^2 + 0.82 - 0.001$

$$\rightarrow 0.00033047Q^2 = 0.931 \rightarrow Q = 53.1\,\text{cfs}.$$

Check: $Q/(AD^{0.5}) = (53.1\,\text{cfs})/[(8\,\text{ft}^2) * (2\,\text{ft})^{0.5}] = 4.69 > 4.0$→ correct assumption.
Thus, the discharge is 53.1 cfs.

Example 8.5 A 2.5-ft-wide, 1.5-ft-high rectangular concrete culvert is constructed as a road crossing over a small stream. It is laid on a slope of 0.003 and has an inlet of side tapered (more favorable edges). For an inlet-control condition, determine the discharge if the HW is:

(1) 1.2 ft
(2) 1.8 ft.

Solution

$b = 2.5\,\text{ft}$, $D = 1.5\,\text{ft}$, $S_0 = 0.003$, $g = 32.2\,\text{ft}\,\text{sec}^{-2}$

Cross-sectional area: $A = (2.5\,\text{ft}) * (1.5\,\text{ft}) = 3.75\,\text{ft}^2$.

Table 8.1: rectangular concrete, side tapered (favorable edge) inlet → chart# 59, scale# 2.

$Z = -0.5 * 0.003 = -0.0015$.

(1) HW = 1.2 ft < 1.5 ft→ assume that the inlet is unsubmerged.
Chart# 59, scale# 2, unsubmerged → Form 2, K = 0.50, M = 0.667.
Eq. (8.3): $(1.2\,\text{ft})/(1.5\,\text{ft}) = 0.50 * \{Q/[(3.75\,\text{ft}^2) * (1.5\,\text{ft})^{0.5}]\}^{0.667} - 0.0015$

$$\rightarrow 0.36174Q^{0.667} = 0.8015 \rightarrow Q = 3.3\,\text{cfs}.$$

Check: $Q/(AD^{0.5}) = (3.3\,\text{cfs})/[(3.75\,\text{ft}^2) * (1.5\,\text{ft})^{0.5}] = 0.72 < 3.5 \rightarrow$ correct assumption.
Thus, the discharge is 3.3 cfs.

(2) HW = 1.8 ft > 1.5 ft→ assume that the inlet is submerged.
Chart# 59, scale# 2, submerged →C = 0.0378, Y = 0.71.
Eq. (8.1): $(1.8\,\text{ft})/(1.5\,\text{ft}) = 0.0378 * \{Q/[(3.75\,\text{ft}^2) * (1.5\,\text{ft})^{0.5}]\}^2 + 0.71 - 0.0015$

$$\rightarrow 0.001792Q^2 = 0.4915 \rightarrow Q = 16.6\,\text{cfs}.$$

Check: $Q/(AD^{0.5}) = (16.6\,\text{cfs})/[(3.75\,\text{ft}^2) * (1.5\,\text{ft})^{0.5}] = 3.61$, between 3.5 and 4.0→ wrong assumption → recalculate Q as the linear interpolation of discharges for $Q/(AD^{0.5}) = 3.5$ and 4.0.
Eq. (8.1): $Q/(AD^{0.5}) = 4.0 \rightarrow HW = (1.5\,\text{ft}) * [0.0378 * 4.0^2 + 0.71 - 0.0015] = 1.97\,\text{ft}$.
Eq. (8.3): $Q/(AD^{0.5}) = 3.5 \rightarrow HW = (1.5\,\text{ft}) * (0.50 * 3.5^{0.667} - 0.0015) = 1.73\,\text{ft}$.
Linear interpolation: $HW = 1.8\,\text{ft} \rightarrow Q/(AD^{0.5}) = 3.5 + (4.0 - 3.5)/(1.97\,\text{ft} - 1.73\text{ft}) * (1.8\,\text{ft} - 1.73\,\text{ft}) = 3.65 \rightarrow Q = 3.65 * [(3.75\,\text{ft}^2) * (1.5\,\text{ft})^{0.5}] = 16.8\,\text{cfs}$.
Thus, the discharge is 16.8 cfs.

8.1.3 Outlet-Control Conditions

For a culvert, an outlet-control condition exists if the maximum flow rate allowed by the inlet is larger than the maximum flow rate allowed by the outlet. That is, the discharge through the barrel is controlled by the outlet. Such a condition may exist for situations that the inlet and outlet are either submerged or unsubmerged (Figure 8.6). The possible outlet-control flow conditions include that: (1) both the inlet and outlet are submerged, leading to a pressurized flow throughout the barrel (Figure 8.6a); (2) both the inlet and outlet are submerged, leading to an open channel flow in the vicinity of the inlet but a pressurized flow in the remaining portion of the barrel (Figure 8.6b); (3) the inlet is submerged while the outlet is unsubmerged, leading to a pressurized flow throughout the barrel (Figure 8.6c); (4) the inlet is submerged while the outlet is unsubmerged, leading to a pressurized flow from the inlet to a location near the outlet and then becomes unpressurized (Figure 8.6d); and (5) both the inlet and outlet are unsubmerged, leading to an open channel flow existing throughout the barrel (Figure 8.6e).

If the inlet and outlet are submerged (Figure 8.6a,b,c) while the approach and leaving velocity heads are negligible, HW can be determined by applying the energy equation (Eq. (2.21)) between the headwater and tail water surface. The expression can be written as:

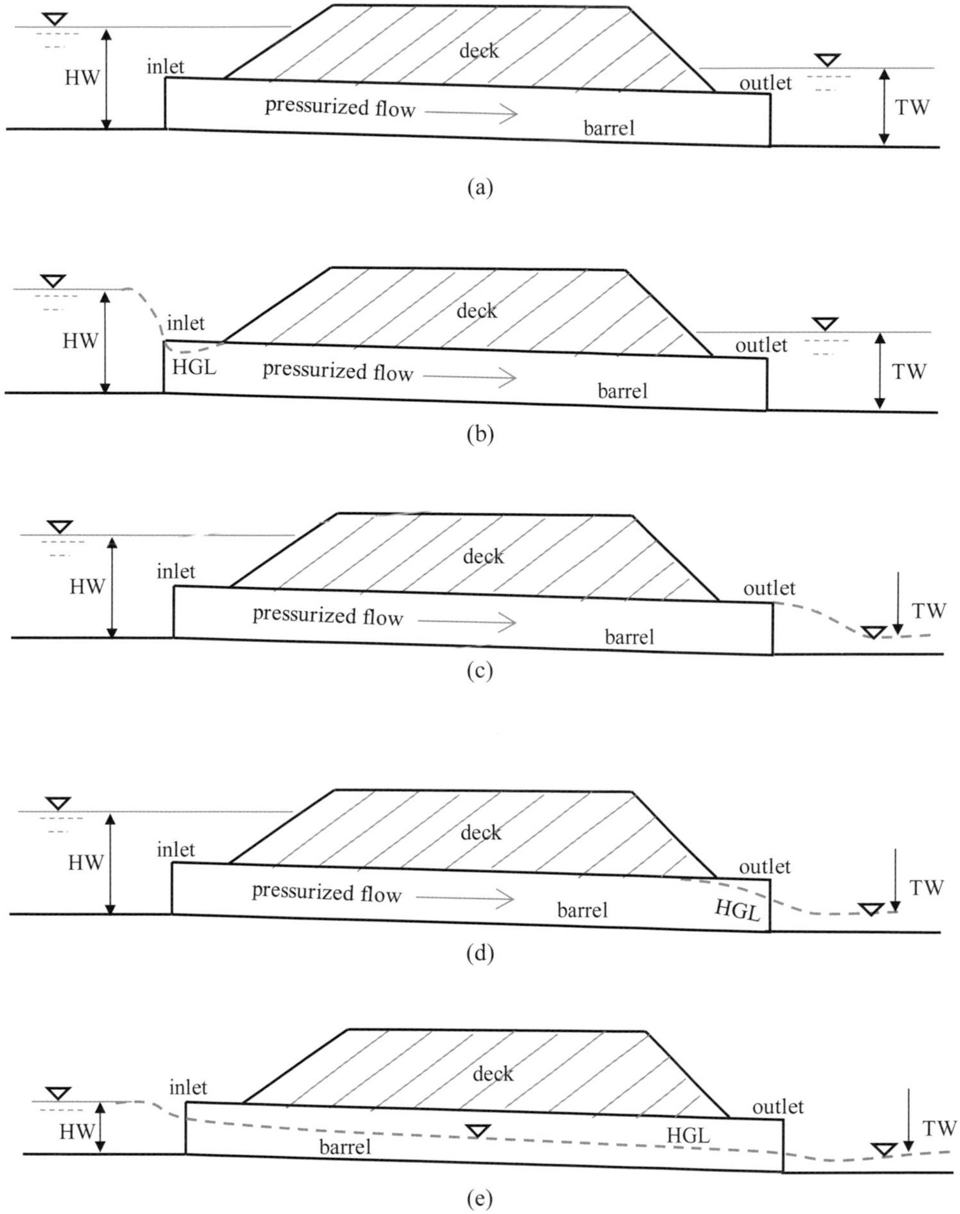

Figure 8.6 **Longitudinal cross sections of culverts with various outlet-control flow conditions.** (a) Submerged inlet and outlet with a pressurized flow throughout the barrel; (b) submerged inlet and outlet with an open channel flow in the vicinity of the inlet but pressurized flow in the remaining portion of the barrel; (c) submerged inlet and unsubmerged outlet with a pressurized flow throughout the barrel; (d) submerged inlet and unsubmerged outlet with a pressurized flow from the inlet to a location near the outlet and then open channel flow; and (e) unsubmerged inlet and outlet with open channel flow throughout the barrel. (*Source:* reproduced Normann *et al.*, 1985)

Table 8.2 Entrance loss coefficients, K_e, for common types of culvert inlets.[1]

Shape (material)	Inlet type description	K_e
Circular (concrete)	Mitered to conform to fill slope	0.7
	End section conforming to fill slope	0.5
	Projecting from fill, sq. cut end	0.5
	Square-edge headwall or headwall and wingwalls	0.5
	Rounded (radius = 1/12D) headwall or headwall and wingwalls	0.2
	Socket end of pipe (groove-end) headwall or headwall and wingwalls	0.2
	Projecting from fill, socket end (groove-end)	0.2
	Beveled edges, 33.7° or 45° bevels	0.2
	Side- or slope-tapered inlet	0.2
Circular or pipe-arch (corrugated metal)	Projecting from fill (no headwall)	0.9
	Mitered to conform to fill slope, paved or unpaved slope	0.7
	Headwall or headwall and wingwalls square-edge	0.5
	End section conforming to fill slope	0.5
	Beveled edges, 33.7° or 45° bevels	0.2
	Side- or slope-tapered inlet	0.2
Rectangular box (reinforced concrete)	Wingwalls parallel (extension of sides), square-edged at crown	0.7
	Wingwalls at 10° to 25° or 30° to 75° to barrel	0.5
	Headwall parallel to embankment (no wingwalls), square-edged on 3 edges	0.5
	Headwall parallel to embankment (no wingwalls), rounded on 3 edges to 1/12 barrel dimension	0.2
	Headwall parallel to embankment (no wingwalls), beveled edges on 3 sides	0.2
	Wingwalls at 30° to 75° to barrel, crown edge rounded to radius of 1/12 barrel dimension	0.2
	Wingwalls at 30° to 75° to barrel, beveled top edge	0.2
	Side-or slope-tapered inlet	0.2

[1] *Source:* Normann *et al.* (1985).

$$\underbrace{HW}_{\text{headwater depth}} = \overbrace{TW}^{\text{tail water depth}} + \underbrace{h_e}_{\text{entrance loss (Tables 2.3 and 8.2)}} + \underbrace{h_f}_{\text{friction loss (Eq. (2.23) or (2.29))}} + \overbrace{h_{ex}}^{\text{exit loss (Table 2.3)}} - \overbrace{L}^{\text{barrel length}}\overbrace{S_0}^{\text{barrel slope}} \tag{8.4}$$

If the outlet is unsubmerged (Figure 8.6d,e), the downstream water depth is empirically corrected to be the maximum of TW and $(D + y_c)/2$, where D is the barrel interior height and y_c is the critical water depth of the barrel, and the friction, entrance, and exit losses are computed for the situation of a pressurized barrel flow. That is, HW can be determined by rewriting Eq. (8.4) as:

$$\underbrace{HW}_{\text{headwater depth}} = \underbrace{\max}_{\text{maximum function}} \left\{ \overbrace{TW}^{\text{tail water depth}}, \frac{\overbrace{D}^{\text{barrel interior height}} + \overbrace{y_c}^{\text{critical depth}}}{2} \right\} + \underbrace{h_e}_{\text{entrance loss (Tables 2.3 and 8.2)}} + \underbrace{h_f}_{\text{friction loss (Eq. (2.23) or (2.29))}} + \overbrace{h_{ex}}^{\text{exit loss (Table 2.3)}} - \overbrace{L}^{\text{barrel length}}\overbrace{S_0}^{\text{barrel slope}} \tag{8.5}$$

Example 8.6 A culvert with a square-edged reinforced concrete box barrel (5 ft wide, 3 ft high, and 250 ft long) is constructed as a road crossing over a stream. The barrel ($f = 0.012$) is laid on a slope of 0.01 and designed to convey 200 cfs. The maximum allowable head water depth is 5 ft to prevent overtopping of the deck and flooding of adjacent structures. For a tail water depth of 3.5 ft and an outlet-control condition, determine:

(1) The headwater depth
(2) Whether the barrel is large enough and if not, its required sizes.

Solution

$b = 5\,\text{ft}, D = 3\,\text{ft}, L = 250\,\text{ft}, S_0 = 0.01, f = 0.012, Q = 200\,\text{cfs}, TW = 3.5\,\text{ft}$

Table 8.2: squared-edged box barrel $\rightarrow K_e = 0.5$.

$TW > D \rightarrow$ submerged outlet $\rightarrow K_{ex} = 1.0$.

(1) Cross-sectional area: $A = (5\,\text{ft}) * (3\,\text{ft}) = 15\,\text{ft}^2$.
Wetted perimeter: $P_{wet} = 2 * (5\,\text{ft} + 3\,\text{ft}) = 16\,\text{ft}$.
Hydraulic radius: $R = A/P_{wet} = (15\,\text{ft}^2)/(16\,\text{ft}) = 0.9375\,\text{ft}$.
Mean velocity: $V = (200\,\text{cfs})/(15\,\text{ft}^2) = 13.33333\,\text{ft sec}^{-1}$.
Entrance loss: $h_e = (0.5) * (13.33333\,\text{ft sec}^{-1})^2/(2 * 32.2\,\text{ft sec}^{-2}) = 1.38\,\text{ft}$.
Friction loss: $h_f = 0.012 * (250\,\text{ft})/(4 * 0.9375\,\text{ft}) * (13.33333\,\text{ft sec}^{-1})^2/(2 * 32.2\,\text{ft sec}^{-2}) = 2.21\,\text{ft}$.
Exit loss: $h_{ex} = 1.0 * (13.33333\,\text{ft sec}^{-1})^2/(2 * 32.2\,\text{ft sec}^{-2}) = 2.76\,\text{ft}$.
Assume $HW > D \rightarrow$ both the inlet and outlet are submerged $\rightarrow$ Eq. (8.4).
$HW = 3.5\,\text{ft} + 1.38\,\text{ft} + 2.21\,\text{ft} + 2.76\,\text{ft} - (250\,\text{ft}) * 0.01 = 7.35\,\text{ft}$.
$HW > 3\,\text{ft} \rightarrow$ the assumption is correct, $HW = 7.35\,\text{ft}$.

(2) $HW = 7.35\,\text{ft} > 5\,\text{ft} \rightarrow$ the barrel is too small and needs to be resized.
Choose new sizes: $b = 6\,\text{ft}, D = 3.5\,\text{ft}$.
Cross-sectional area: $A = (6\,\text{ft}) * (3.5\,\text{ft}) = 21\,\text{ft}^2$.
Wetted perimeter: $P_{wet} = 2 * (6\,\text{ft} + 3.5\,\text{ft}) = 19\,\text{ft}$.
Hydraulic radius: $R = A/P_{wet} = (21\,\text{ft}^2)/(19\,\text{ft}) = 1.10526\,\text{ft}$.
Mean velocity: $V = (200\,\text{cfs})/(21\,\text{ft}^2) = 9.52381\,\text{ft sec}^{-1}$.
Entrance loss: $h_e = (0.5) * (9.52381\,\text{ft sec}^{-1})^2/(2 * 32.2\,\text{ft sec}^{-2}) = 0.70\,\text{ft}$.
Friction loss: $h_f = 0.012 * (250\,\text{ft})/(4 * 1.10526\,\text{ft}) * (9.52381\,\text{ft sec}^{-1})^2/(2 * 32.2\,\text{ft sec}^{-2}) = 0.96\,\text{ft}$.
Exit loss: $h_{ex} = 1.0 * (9.52381\,\text{ft sec}^{-1})^2/(2 * 32.2\,\text{ft sec}^{-2}) = 1.41\,\text{ft}$.
Eq. (8.4): $HW = 3.5\,\text{ft} + 0.70\,\text{ft} + 0.96\,\text{ft} + 1.41\,\text{ft} - (250\,\text{ft}) * 0.01 = 4.07\,\text{ft}$.
$HW > 3.5\,\text{ft}$ and $HW < 5\,\text{ft} \rightarrow$ the new sizes are fine.
Thus, a resized barrel can be 6 ft wide and 3.5 ft high.

Example 8.7 Redo Example 8.6 if the tail water depth is lowered to 2.5 ft.

Solution

$b = 5\,ft$, $D = 3\,ft$, $L = 250\,ft$, $S_0 = 0.01$, $f = 0.012$, $Q = 200\,cfs$, $TW = 2.5\,ft$

Table 8.2: squared-edged box barrel $\rightarrow K_e = 0.5$.

$TW < D \rightarrow$ unsubmerged outlet $\rightarrow K_{ex} = 0.0$.

(1) Cross-sectional area: $A = (5\,ft) * (3\,ft) = 15\,ft^2$.
Wetted perimeter: $P_{wet} = 2 * (5\,ft + 3\,ft) = 16\,ft$.
Hydraulic radius: $R = A/P_{wet} = (15\,ft^2)/(16\,ft) = 0.9375\,ft$.
Mean velocity: $V = (200\,cfs)/(15\,ft^2) = 13.33333\,ft\,sec^{-1}$.
Entrance loss: $h_e = (0.5) * (13.33333\,ft\,sec^{-1})^2/(2 * 32.2\,ft\,sec^{-2}) = 1.38\,ft$.
Friction loss: $h_f = 0.012 * (250\,ft)/(4 * 0.9375\,ft) * (13.33333\,ft\,sec^{-1})^2/(2 * 32.2\,ft\,sec^{-2}) = 2.21\,ft$.
Exit loss: $h_{ex} = 0.0 * (13.33333\,ft\,sec^{-1})^2/(2 * 32.2\,ft\,sec^{-2}) = 0.0\,ft$.
Assume $HW > D \rightarrow$ submerged inlet but unsubmerged outlet $\rightarrow$ Eq. (8.5).

- Determine critical water depth y_c:
  Unit-width flow rate: $q = (200\,cfs)/(5\,ft) = 40\,ft^2\,sec^{-1}$.
  Eq. (2.38): $y_c = [(40\,ft^2\,sec^{-1})^2/(32.2\,ft\,sec^{-2})]^{1/3} = 3.68\,ft > D \rightarrow$ outlet pressurized $\rightarrow$ the condition is as illustrated in Figure 8.6c.
- Determine the headwater depth, HW, using Eq. (8.4):
  $HW = 2.5\,ft + 1.38\,ft + 2.21\,ft + 0.0\,ft - (250\,ft) * 0.01 = 3.59\,ft$.
  $HW > 3\,ft \rightarrow$ the assumption is correct, $HW = 4.14\,ft$.

(2) $HW = 3.59\,ft < 5\,ft \rightarrow$ the barrel is large enough.

Example 8.8 Redo Example 8.6 if the tail water depth is lowered to 2.5 ft and the design discharge is reduced to 100 cfs.

Solution

$b = 5\,ft$, $D = 3\,ft$, $L = 250\,ft$, $S_0 = 0.01$, $f = 0.012$, $Q = 100\,cfs$, $TW = 2.5\,ft$

Table 8.2: squared-edged box barrel $\rightarrow K_e = 0.5$.

$TW < D \rightarrow$ unsubmerged outlet $\rightarrow K_{ex} = 0.0$.

(1) Cross-sectional area: $A = (5\,ft) * (3\,ft) = 15\,ft^2$.
Wetted perimeter: $P_{wet} = 2 * (5\,ft + 3\,ft) = 16\,ft$.
Hydraulic radius: $R = A/P_{wet} = (15\,ft^2)/(16\,ft) = 0.9375\,ft$.
Mean velocity: $V = (200\,cfs)/(15\,ft^2) = 13.33333\,ft\,sec^{-1}$.
Entrance loss: $h_e = (0.5) * (13.33333\,ft\,sec^{-1})^2/(2 * 32.2\,ft\,sec^{-2}) = 1.38\,ft$.
Friction loss: $h_f = 0.012 * (250\,ft)/(4 * 0.9375\,ft) * (13.33333\,ft\,sec^{-1})^2/(2 * 32.2\,ft\,sec^{-2}) = 2.21\,ft$.
Exit loss: $h_{ex} = 0.0 * (13.33333\,ft\,sec^{-1})^2/(2 * 32.2\,ft\,sec^{-2}) = 0.0\,ft$.
Assume $HW > D \rightarrow$ submerged inlet but unsubmerged outlet $\rightarrow$ Eq. (8.5).

- Determine critical water depth y_c:
  Unit-width flow rate: $q = (100\,cfs)/(5\,ft) = 20\,ft^2\,sec^{-1}$.
  Eq. (2.38): $y_c = [(20\,ft^2\,sec^{-1})^2/(32.2\,ft\,sec^{-2})]^{1/3} = 2.32\,ft < D \rightarrow$ Figure 8.6d.
- Determine the headwater depth HW:
  Eq. (8.5): $HW = \max\{2.5\,ft,\ (2.32\,ft + 3\,ft)/2\} + 1.38\,ft + 2.21\,ft + 0.0\,ft - (250\,ft) * 0.01 = 3.75\,ft$.
  $HW > 3\,ft \rightarrow$ the inlet is submerged $\rightarrow$ the condition is as illustrated in Figure 8.6d.

(2) $HW = 3.75\,ft < 5\,ft \rightarrow$ the barrel is large enough.

8.1.4 Design and Operation of Culverts

In practice, engineers and managers must face two types of problems related to culverts. The first is that of culvert size; that is, a new culvert must be planned to be large enough to convey a design discharge to result in a headwater depth (HW) less than a maximum allowable value. An HW greater than this value will either cause the culvert deck to be overtopped or the structures adjacent to the backwater-influencing areas upstream of the culvert inlet to be flooded. This maximum allowable headwater depth is dictated by a sag point upstream of the deck. For instance, if a circular barrel is chosen, its required diameter needs to be determined, or if a rectangular box culvert is selected, its required interior width and height need to be determined. In contrast, the second type of problem is that of managing existing culverts. In this regard, it is imperative to develop the performance curve of a culvert of interest, which is the functional relationship of barrel discharge versus HW. Regardless of the type of problem, for a given discharge, two values of HW should be computed: one for an inlet-control flow condition and another for an outlet-control flow condition. The minimum of these two computed values is used as the HW corresponding to the discharge.

Example 8.9 Determine the size of a 100-ft-long circular concrete culvert to convey a 100-year peak discharge of 180 cfs, subject to a maximum allowable headwater depth of 7.5 ft. The culvert has one barrel ($f = 0.013$) laid on a channel bed slope of 0.002 and an inlet of square edge w/headwall. The 100-year tail water depth is 4.2 ft.

Solution

$L = 100\,ft,\ f = 0.013,\ S_0 = 0.002,\ Q = 180\,cfs,\ TW = 4.2\,ft$

Size of barrel to have maximum allowable headwater depth $\rightarrow HW = 7.5\,ft$.

- Determine the required barrel diameter D for an inlet-control condition:
  $Z = -0.5 * 0.002 = -0.001$.
  Assume $Q/(AD^{0.5}) \geq 4.0 \rightarrow$ inlet submerged $\rightarrow$ Eq. (8.1).
  Table 8.1: circular concrete (chart# 1) and square edge w/headwall (scale# 1)

$$\rightarrow C = 0.0398,\ Y = 0.67.$$

Eq. (8.1): $(7.5\,ft)/D = 0.0398 * [(180\,cfs)/(\pi/4D^2D^{0.5})]^2 + 0.67 - 0.001$

$$\rightarrow 0.669D^5 - 7.5D^4 + 2090.49108 = 0.$$

By trial-and-error or Excel Solver, $D = 4.68$ ft.
Check: cross-sectional area: $A = \pi/4 * (4.68\,\text{ft})^2 = 17.20211\,\text{ft}^2$

$$Q/(AD^{0.5}) = (180\,\text{cfs})/[(17.20211\,\text{ft}^2) * (4.68\,\text{ft})^{0.5}] = 4.84 > 4.0 \rightarrow \text{assumption correct.}$$

- Check the HW for an outlet-control condition:
  Mean velocity: $V = (180\,\text{cfs})/(17.20211\,\text{ft}^2) = 10.46383\,\text{ft sec}^{-1}$.
  Table 8.2: circular concrete, square edge w/headwall$\rightarrow K_e = 0.5$.
  $TW < D = 4.68\,\text{ft} \rightarrow$ unsubmerged outlet $\rightarrow K_{ex} = 0.0$.
  Entrance loss: $h_e = 0.5 * (10.46383\,\text{ft sec}^{-1})^2/(2 * 32.2\,\text{ft sec}^{-2}) = 0.85$ ft.
  Friction loss: $h_f = 0.013 * (100\,\text{ft})/(4.68\,\text{ft}) * (10.46383\,\text{ft sec}^{-1})^2/(2 * 32.2\,\text{ft sec}^{-2}) = 0.47$ ft.
  Exit loss: $h_{ex} = 0.0 * (10.46383\,\text{ft sec}^{-1})^2/(2 * 32.2\,\text{ft sec}^{-2}) = 0.0$ ft.
  Compute the critical depth y_c:
  Table 2.4: $A_c = (4.68\,\text{ft})^2/8(\theta_c - \sin\theta_c) = 2.7378(\theta_c - \sin\theta_c)\,\text{ft}^2$

$$B_c = (4.68\,\text{ft}) \sin(\theta_c/2) = 4.68 \sin(\theta_c/2)\,\text{ft}.$$

  Eq. (2.37): $(180\,\text{cfs})^2/(32.2\,\text{ft sec}^{-2}) = [2.7378(\theta_c - \sin\theta_c)\,\text{ft}^2]^3/[4.68 \sin(\theta_c/2)\,\text{ft}]$

$$\rightarrow (\theta_c - \sin\theta_c)^3/\sin(\theta_c/2) = 229.47207.$$

  By trial-and-error or ExcelSolver, $\theta_c = 4.57842$ rad

  Table 2.4: $4.57842\,\text{rad} = 2\cos^{-1}[(4.68\,\text{ft} - 2y_c)/(4.68\,\text{ft})]$

$$\rightarrow (4.68 - 2y_c)/4.68 = \cos(2.28921\,\text{rad}) = -0.65819 \rightarrow y_c = 3.88\,\text{ft}.$$

  $y_c < D \rightarrow$ the flow is as illustrated in Figure 8.6d and governed by Eq. (8.5):

$$HW = \max\{4.2\,\text{ft},\ (4.68\,\text{ft} + 3.88\,\text{ft})/2\} + 0.85\,\text{ft} + 0.47\,\text{ft} + 0.0\,\text{ft} - (100\,\text{ft}) * 0.002$$
$$= 5.4\,\text{ft} < 7.5\,\text{ft} \rightarrow \text{barrel with a diameter of 4.68 ft is large enough.}$$

  Summary: the diameter of the barrel is $D = 4.68$ ft.

Example 8.10 Develop the performance curve of an 80-ft-long rectangular concrete culvert, which has a 4-ft-wide, 3-ft-high barrel ($f = 0.015$) laid on a slope of 0.01 and an inlet of side tapered (more favorable edges). Assume a constant tail water depth of $TW = 0.5$ ft and a maximum allowable headwater depth of 5 ft.

Solution

$b = 4\,\text{ft}$, $D = 3\,\text{ft}$, $L = 80\,\text{ft}$, $f = 0.015$, $S_0 = 0.01$, $TW = 0.5\,\text{ft}$, $HW_{max} = 5\,\text{ft}$

Cross-sectional area: $A = (4\,\text{ft}) * (3\,\text{ft}) = 12\,\text{ft}^2$.

Wetted perimeter: $P_{wet} = 2 * (4\,\text{ft} + 3\,\text{ft}) = 14$ ft.

Hydraulic radius: $R = A/P_{wet} = (12\,\text{ft}^2)/(14\,\text{ft}) = 0.85714$ ft.

Mean velocity: $V = Q/(12\,\text{ft}^2) = Q/12\,\text{ft sec}^{-1}$.

Table 8.2: box concrete, side tapered inlet $\rightarrow K_e = 0.2$.

$TW < D \rightarrow$ unsubmerged outlet $\rightarrow K_{ex} = 0.0$

Entrance loss: $h_e = 0.2 * (Q/12\,\text{ft sec}^{-1})^2/(2 * 32.2\,\text{ft sec}^{-2}) = 2.15666 \times 10^{-5} Q^2\,\text{ft}$.

Friction loss: $h_f = 0.015 * (80\,\text{ft})/(4 * 0.85714\,\text{ft}) * (Q/12\,\text{ft sec}^{-1})^2/(2 * 32.2\,\text{ft sec}^{-2})$.
$= 3.77417 \times 10^{-5} Q^2\,\text{ft}$.

Exit loss: $h_{ex} = 0.0 * (Q/12\,\text{ft sec}^{-1})^2/(2 * 32.2\,\text{ft sec}^{-2}) = 0.0\,\text{ft}$.

Table 8.1: rectangular concrete, side tapered (favorable edges) $\rightarrow$ chart# 59, scale# 2

$$\rightarrow \text{Form2},\ K = 0.50,\ M = 0.667,\ C = 0.0378,\ Y = 0.71.$$

$Z = -0.5 * 0.01 = -0.005$.

To develop the performance curve, for an arbitrarily selected discharge, Q, compute the values of headwater depth, HW, by assuming an inlet- and outlet-control condition, respectively. The control HW (in cells O10:O35) is the maximum of these two computed values (in cells E10:E35 and M10:M35). These computations are summarized in Figure 8.7a, and the performance curve is plotted in Figure 8.7b. The sample computations are as follows:

- For $Q = 0.1$ cfs:

  Inlet-control condition

  $Q/(AD^{0.5}) = (0.1\,\text{cfs})/[(12\,\text{ft}^2) * (3\,\text{ft})^{0.5}] = 0.00481 < 3.5 \rightarrow$ Eq. (8.3).
  $HW/D = 0.50 * 0.00481^{0.667} - 0.005 = 0.009222 \rightarrow HW = 0.009222 * (3\,\text{ft}) = 0.03\,\text{ft}$.

  Outlet-control condition

  $h_e = 2.15666 \times 10^{-5} * (0.1\,\text{cfs})^2 = 2.15666 \times 10^{-7}\,\text{ft}$.
  $h_f = 3.77417 \times 10^{-5} * (0.1\,\text{cfs})^2 = 3.77417 \times 10^{-7}\,\text{ft}$.
  Unit-width flow rate: $q = (0.1\,\text{cfs})/(4\,\text{ft}) = 0.025\,\text{ft}^2\,\text{sec}^{-1}$.
  Eq. (2.38): $y_c = [(0.025\,\text{ft}^2\,\text{sec}^{-1})^2/(32.2\,\text{ft sec}^{-2})]^{1/3} = 0.02687\,\text{ft} < D \rightarrow$ Eq. (8.5).
  $HW = \max\{0.5\,\text{ft},\ (3\,\text{ft} + 0.02687\,\text{ft})/2\} + 2.15666 \times 10^{-7}\,\text{ft} + 3.77417 \times 10^{-7}\,\text{ft} + 0.0\,\text{ft} - (80\,\text{ft}) * 0.01 = 0.71\,\text{ft}$.

  Control HW

  $HW = \max\{0.03\,\text{ft},\ 0.71\,\text{ft}\} = 0.71\,\text{ft} \rightarrow$ the flow is controlled by the outlet.
  $HW < D = 3\,\text{ft} \rightarrow$ inlet unsubmerged $\rightarrow$ Figure 8.6e.

- For $Q = 31.0$ cfs:

  Inlet-control condition

  $Q/(AD^{0.5}) = (31.0\,\text{cfs})/[(12\,\text{ft}^2) * (3\,\text{ft})^{0.5}] = 1.49149 < 3.5 \rightarrow$ Eq. (8.3).
  $HW/D = 0.50 * 1.49149^{0.667} - 0.005 = 0.64779 \rightarrow HW = 0.64779 * (3\,\text{ft}) = 1.94\,\text{ft}$.

  Outlet-control condition

  $h_e = 2.15666 \times 10^{-5} * (31.0\,\text{cfs})^2 = 2.07255 \times 10^{-2}\,\text{ft}$.
  $h_f = 3.77417 \times 10^{-5} * (31.0\,\text{cfs})^2 = 3.62697 \times 10^{-2}\,\text{ft}$.
  Unit-width flow rate: $q = (31.0\,\text{cfs})/(4\,\text{ft}) = 7.750\,\text{ft}^2\,\text{sec}^{-1}$.

	A	B	C	D	E	F	G	H	I	J	K	L	M	N	O
1	b =	4	ft		A =	12	ft^2		chart# 59	C =	0.0378				
2	D =	3	ft		P_{wet} =	14	ft		scale# 2	Y =	0.71				
3	L =	80	ft		R =	0.85714	ft		(Table 8-1)	K =	0.50				
4	f =	0.02			K_e =	0.2				M =	0.667				
5	S_0 =	0.01			K_{ex} =	0.0				Z =	-0.005				
6	TW =	0.5	ft		g =	32.2	ft sec^{-2}								
7															
8			Inlet-Control						Outlet-Control						Control
9	Q (cfs)		Q/(AD$^{0.5}$)	HW/D	HW (ft)		V (ft sec^{-1})	h_e (ft)	h_f (ft)	h_{ex} (ft)	q (ft^2 sec^{-1})	y_c (ft)	HW (ft)		HW (ft)
10	0.1		0.00481	0.00922	0.03		0.00833	2.15666E-07	3.77417E-07	0.00000E+00	0.025	0.02687	0.71	outlet control (Figure 8-6e)	0.71
11	0.2		0.00962	0.01758	0.05		0.01667	8.62664E-07	1.50967E-06	0.00000E+00	0.050	0.04266	0.72		0.72
12	0.4		0.01925	0.03087	0.09		0.03333	3.45066E-06	6.03867E-06	0.00000E+00	0.100	0.06772	0.73		0.73
13	0.6		0.02887	0.04200	0.13		0.05000	7.76398E-06	1.35870E-05	0.00000E+00	0.150	0.08874	0.74		0.74
14	0.8		0.03849	0.05194	0.16		0.06667	1.38026E-05	2.41547E-05	0.00000E+00	0.200	0.10750	0.75		0.75
15	1.0		0.04811	0.06107	0.18		0.08333	2.15666E-05	3.77417E-05	0.00000E+00	0.250	0.12474	0.76		0.76
16	6.0		0.28868	0.21331	0.64		0.50000	7.76398E-04	1.35870E-03	0.00000E+00	1.500	0.41188	0.91		0.91
17	11.0		0.52924	0.32207	0.97		0.91667	2.60956E-03	4.56674E-03	0.00000E+00	2.750	0.61698	1.02		1.02
18	16.0		0.76980	0.41494	1.24		1.33333	5.52105E-03	9.66187E-03	0.00000E+00	4.000	0.79205	1.11		1.24
19	21.0		1.01036	0.49845	1.50		1.75000	9.51087E-03	1.66441E-02	0.00000E+00	5.250	0.94948	1.20	inlet control (Figure 8-5a)	1.50
20	26.0		1.25093	0.57553	1.73		2.16667	1.45790E-02	2.55134E-02	0.00000E+00	6.500	1.09477	1.29		1.73
21	31.0		1.49149	0.64779	1.94		2.58333	2.07255E-02	3.62697E-02	0.00000E+00	7.750	1.23097	1.37		1.94
22	36.0		1.73205	0.71626	2.15		3.00000	2.79503E-02	4.89132E-02	0.00000E+00	9.000	1.36001	1.46		2.15
23	41.0		1.97261	0.78162	2.34		3.41667	3.62535E-02	6.34438E-02	0.00000E+00	10.250	1.48319	1.54		2.34
24	46.0		2.21318	0.84437	2.53		3.83333	4.56349E-02	7.98614E-02	0.00000E+00	11.500	1.60145	1.63		2.53
25	50.0		2.40563	0.89294	2.68		4.16667	5.39165E-02	9.43542E-02	0.00000E+00	12.500	1.69299	1.69		2.68
26	55.0		2.64619	0.95188	2.86		4.58333	6.52390E-02	1.14169E-01	0.00000E+00	13.750	1.80406	1.78		2.86
27	60.0		2.88675	1.00906	3.03		5.00000	7.76398E-02	1.35870E-01	0.00000E+00	15.000	1.91180	1.87	inlet control (Figure 8-5c)	3.03
28	65.0		3.12731	1.06467	3.19		5.41667	9.11189E-02	1.59459E-01	0.00000E+00	16.250	2.01659	1.96		3.19
29	70.0		3.36788	1.11887	3.36		5.83333	1.05676E-01	1.84934E-01	0.00000E+00	17.500	2.11872	2.05		3.36
30	75.0		3.60844	1.18316	3.55		6.25000	1.21312E-01	2.12297E-01	0.00000E+00	18.750	2.21845	2.14		3.55
31	80.0		3.84900	1.26096	3.78		6.66667	1.38026E-01	2.41547E-01	0.00000E+00	20.000	2.31598	2.24		3.78
32	85.0		4.08956	1.33719	4.01		7.08333	1.55819E-01	2.72684E-01	0.00000E+00	21.250	2.41150	2.33		4.01
33	90.0		4.33013	1.41375	4.24		7.50000	1.74689E-01	3.05708E-01	0.00000E+00	22.500	2.50517	2.43		4.24
34	100.0		4.81125	1.58000	4.74		8.33333	2.15666E-01	3.77417E-01	0.00000E+00	25.000	2.68746	2.64		4.74
35	104.8		5.04219	1.66602	5.00		8.73333	2.36867E-01	4.14518E-01	0.00000E+00	26.200	2.77278	2.74		5.00

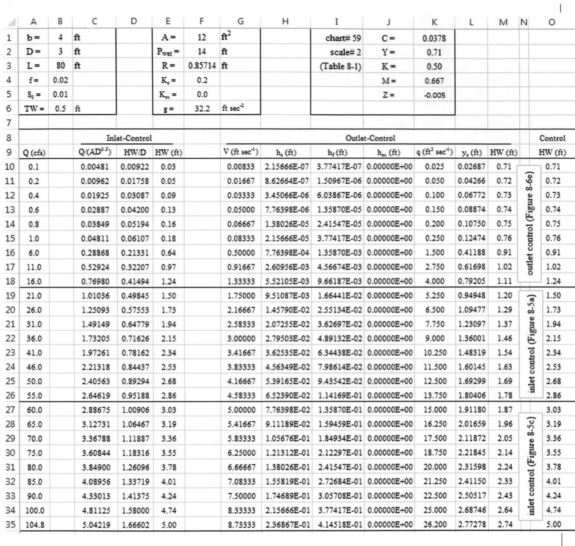

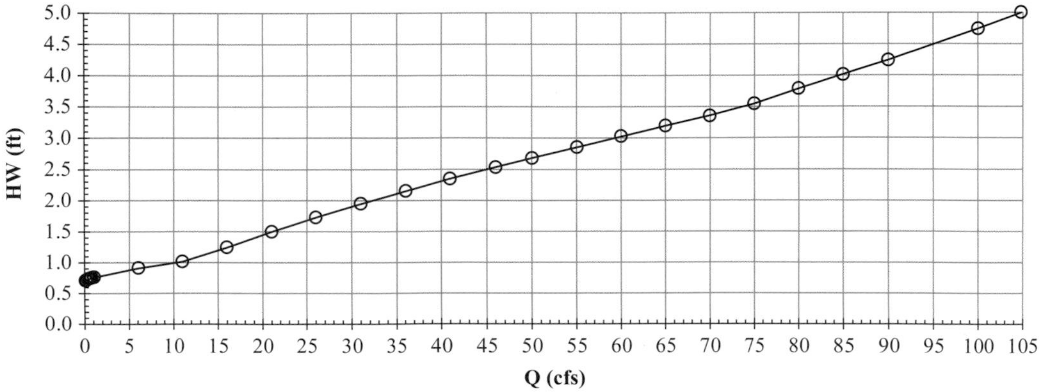

Figure 8.7 **Generation of the performance curve for the culvert in Example 8.10.** (a) Excel spreadsheet for computations; and (b) scatter plot of the performance curve.

Eq. (2.38): $y_c = [(7.750\,\text{ft}^2\,\text{sec}^{-1})^2/(32.2\,\text{ft}\,\text{sec}^{-2})]^{1/3} = 1.23097\,\text{ft} < D \rightarrow$ Eq. (8.5).
$HW = \max\{0.5\,\text{ft},\ (3\,\text{ft} + 1.23097\,\text{ft})/2\} + 2.07255 \times 10^{-2}\,\text{ft} + 3.62697 \times 10^{-2}\,\text{ft} + 0.0\,\text{ft} - (80\,\text{ft}) * 0.01 = 1.37\,\text{ft}$.

Control HW

$HW = \max\{1.94\,\text{ft},\ 1.37\,\text{ft}\} = 1.94\,\text{ft} \rightarrow$ the flow is controlled by the inlet.
$HW < D = 3\,\text{ft} \rightarrow$ inlet unsubmerged $\rightarrow$ Figure 8.5a.

- For $Q = 80.0\,\text{cfs}$:

Inlet-control condition

$Q/(AD^{0.5}) = (80.0\,\text{cfs})/[(12\,\text{ft}^2) * (3\,\text{ft})^{0.5}] = 3.8490$, which is between 3.5 and 4.0 $\rightarrow$ linear interpolation of Eqs. (8.1) and (8.3).
Eq. (8.1): $Q/(AD^{0.5}) = 4.0 \rightarrow HW/D = 0.0378 * 4.0^2 + 0.71 - 0.005 = 1.30980$.
Eq. (8.3): $Q/(AD^{0.5}) = 3.5 \rightarrow HW/D = 0.50 * 3.5^{0.667} - 0.005 = 1.14809$.
$\rightarrow$ linear interpolation, $HW/D = 1.30980 + (1.14809 - 1.30980)/(3.5 - 4.0) * (3.8490 - 4.0) = 1.26096$.
$\rightarrow HW = 1.26096 * (3\,\text{ft}) = 3.78\,\text{ft}$.

Outlet-control condition

$h_e = 2.15666 \times 10^{-5} * (80.0\,\text{cfs})^2 = 1.38026 \times 10^{-1}\,\text{ft}$.
$h_f = 3.77417 \times 10^{-5} * (80.0\,\text{cfs})^2 = 2.41547 \times 10^{-1}\,\text{ft}$.
Unit-width flow rate: $q = (80.0\,\text{cfs})/(4\,\text{ft}) = 20.0\,\text{ft}^2\,\text{sec}^{-1}$.
Eq. (2.38): $y_c = [(20.0\,\text{ft}^2\,\text{sec}^{-1})^2/(32.2\,\text{ft}\,\text{sec}^{-2})]^{1/3} = 2.31598\,\text{ft} < D \rightarrow$ Eq. (8.5).
$HW = \max\{0.5\,\text{ft},\ (3\,\text{ft} + 2.31598\,\text{ft})/2\} + 1.38026 \times 10^{-1}\,\text{ft} + 2.41547 \times 10^{-1}\,\text{ft} + 0.0\,\text{ft} - (80\,\text{ft}) * 0.01 = 2.24\,\text{ft}$.

Control HW

$HW = \max\{3.78\,\text{ft},\ 2.24\,\text{ft}\} = 3.78\,\text{ft} \rightarrow$ the flow is controlled by the inlet
$HW > D = 3\,\text{ft} \rightarrow$ inlet submerged $\rightarrow$ Figure 8.5c.

8.2 Bridges

Bridges are also widely used as a road crossing over waterways (Figure 8.8). Because the opening of a bridge cross section is usually narrower than those of both the upstream and downstream river cross sections, the flow tends to be contracted at the upstream of the bridge but expanded at the downstream, likely causing significant energy losses. The contraction and expansion can be strengthened by bridge piers, abutments, and low and high chords. As a result, water flows through a bridge opening as a rapidly varied flow, which in principle should be analyzed by firstly using linear momentum equation (Eq. (2.32)) and then energy equation (Eq. (2.21)).

Three types of flows, designated classes A, B, and C, may flow through a bridge opening. Class A flow is subcritical throughout the bridge opening, whereas class C flow is supercritical throughout the bridge opening. Class B flow can either be subcritical or supercritical but becomes a critical flow at a bridge constriction cross section. If the bridge does not exist, the water depth

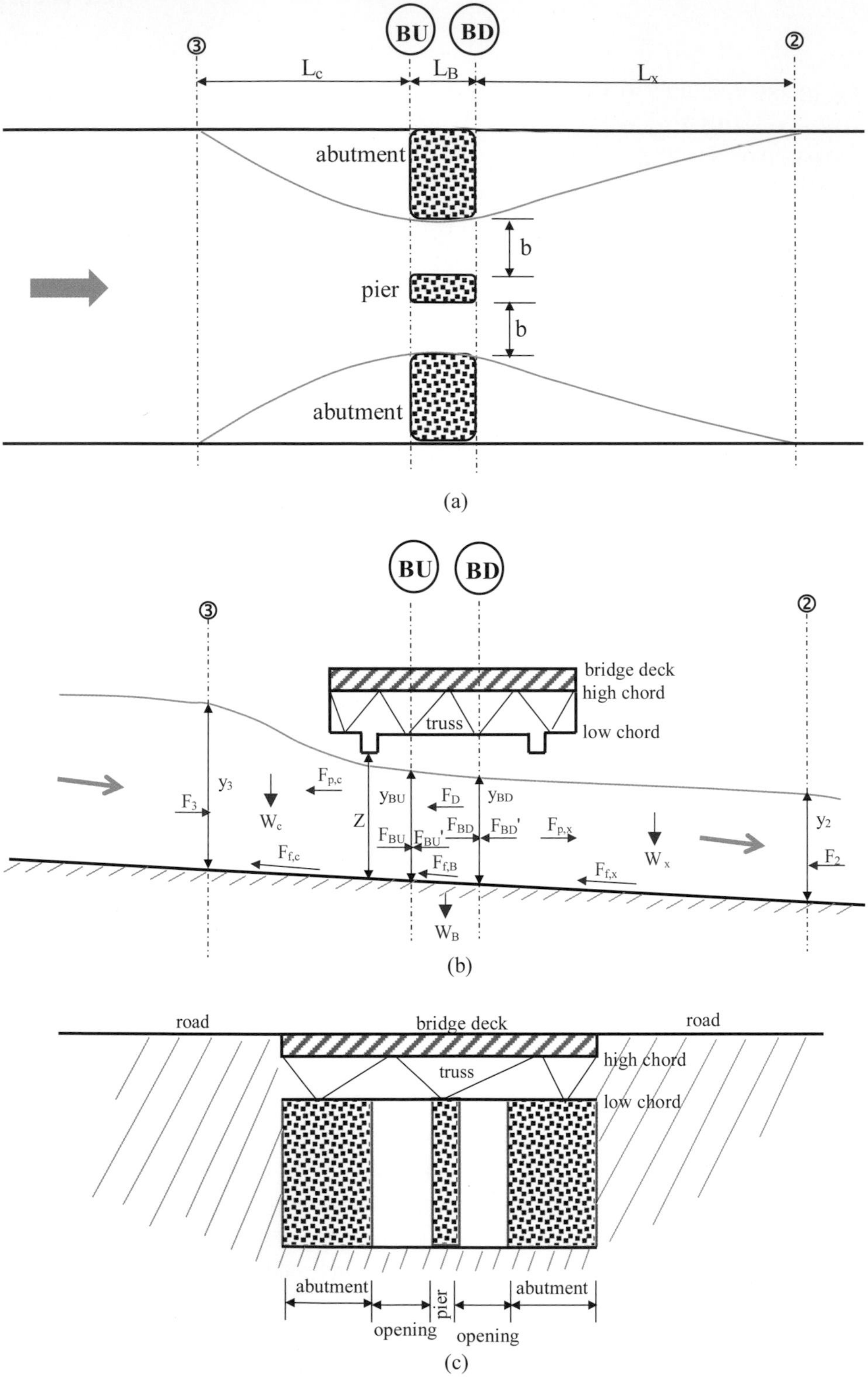

Figure 8.8 Sketches of a bridge. (a) Bird's-eye view; (b) side view; and (c) front view. The bridge has one pier and two abutments, leading to two identical openings. The red arrows signify forces acting on the corresponding control volume (see Example 8.11).

at section ② should be approximately equal to the water depth at section ③ (see Figure 8.8), that is, $y_3 \approx y_2$. In contrast, the bridge tends to increase y_3 while decreasing y_2, making $y_3 > y_2$ and a measurable drop of water surface from sections ③ to ②.

In practice, three methods have been used to determine the water surface drop. The first method is to apply linear momentum equation (Eq. (2.32)) between section ② and the downstream end of the bridge, throughout the bridge opening, and between the upstream end of the bridge and section ③ (Figure 8.8a,b). The details of this method can be found in USACE (2010). The second method is to apply energy equation (Eq. (2.21)) between sections ② and ③, with the total head loss equal to the summation of the friction loss between section ② and the downstream end of the bridge, the friction loss throughout the bridge opening, the friction loss between the upstream end of the bridge and section ③, the contraction loss from section ③ to upstream end of the bridge, and the expansion loss from downstream end of the bridge to section ②. Each of the friction losses can be computed using the Manning formula (Eq. (2.27)), and the contraction or expansion loss is computed as the multiplication of a coefficient and the velocity head (Table 2.3). A contraction loss coefficient of 0.1 and an expansion loss coefficient of 0.3 are commonly used if bridge-specific values are not available. The third method is the application of the Yarnell equation, an empirical equation that is used to predict the change in water surface from just downstream of the bridge (section ②) to just upstream of the bridge (section ③). The equation is based on approximately 2600 lab experiments in which the researchers varied piers' shape, width, length, and angle as well as flow rate. It is generally applicable for subcritical flow. The Yarnell equation can be expressed as (Yarnell, 1934):

$$\Delta z_{3-2} = 2K\left(K + \frac{10}{y_2}\frac{V_2^2}{2g} - 0.6\right)\left(\alpha + 15\alpha^4\right)\frac{V_2^2}{2g} \tag{8.6}$$

Yarnell's pier shape coefficient (Table 8.3) → K
mean velocity at section ② of Figure 8.8a,b → V_2
water surface drop from section ③ to ② of Figure 8.8a,b → Δz_{3-2}
water depth at section ② of Figure 8.8a,b → y_2
gravitational acceleration → g
obstructed area of piers and abutments divided by total unobstructed area at section ② → α

Example 8.11 A bridge crossing is constructed over a 50-ft-wide rectangular channel ($S_0 = 0.002$, $n = 0.025$). It has two openings of 10 ft wide each and one pier and two abutments, all of which have a semi-circular shape. The layout of the bridge is similar to that shown in Figure 8.8, with $L_c = 20$ ft, $L_B = 15$ ft, and $L_x = 40$ ft. For a design discharge of 200 cfs and by assuming a normal depth and no obstruction at section ②, determine the change in water surface from section ③ to ② using the:

(1) Momentum equation
(2) Energy equation
(3) Yarnell equation.

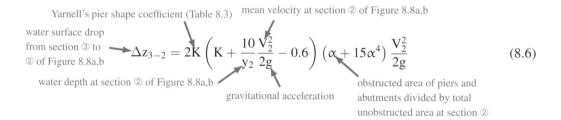

Table 8.3 Yarnell's bridge pier coefficient, K, and drag coefficient, C_D.[1]

Pier shape	K	C_D [2]
Semi-circular nose and tail	0.90	1.33
Twin-cylinder piers with connecting diaphragm	0.95	–
Twin-cylinder piers without diaphragm	1.05	1.20
90° triangular nose and tail	1.05	1.60
Square nose and tail	1.25	2.00
Ten pile trestle bent	2.50	–

[1] *Sources*: USACE (2010).

[2] Drag force: $F_D = C_D A \gamma V^2/(2g)$, where A: frontal area of pier; γ: specific weight of water; V: flow velocity; and g: gravitational acceleration (Finnemore and Franzini, 2002).

Solution

$B = 50\,\text{ft},\ b = 10\,\text{ft},\ S_0 = 0.002,\ n = 0.025,\ L_c = 20\,\text{ft},\ L_B = 15\,\text{ft},\ L_x = 40\,\text{ft},$

$$Q = 200\,\text{cfs},\ \rho = 1.94\,\text{slug ft}^{-3},\ \gamma = 62.4\,\text{lbf ft}^{-3}$$

Table 8.3: semi-circular pier and abutments $\rightarrow C_D = 1.33$.

At section ②: normal depth $\rightarrow S = S_0 = 0.002$.

Flow area: $A_2 = (50\,\text{ft})y_2 = 50y_2\,\text{ft}^2$.
Wetted perimeter: $P_{wet,2} = 50 + 2y_2\,\text{ft}$.
Hydraulic radius: $R_2 = 50y_2/(50 + 2y_2)\,\text{ft} = 25y_2/(25 + y_2)\,\text{ft}$.
Mean velocity: $V_2 = (200\,\text{cfs})/(50y_2\,\text{ft}^2) = 4/y_2\,\text{ft sec}^{-1}$.
Manning formula (Eq. (2.26)): $4/y_2\,\text{ft sec}^{-1} = 1.486/0.025 * [25y_2/(25 + y_2)\,\text{ft}]^{2/3} * 0.002^{1/2}$

$$\rightarrow 5.68190(y_2)^{5/3} = (25 + y_2)^{2/3}$$

$$\rightarrow \text{by trial-and-error or Excel Solver},\ y_2 = 1.30\,\text{ft}.$$

(1) Apply momentum equation for downstream, inside, and upstream of the bridge.

- From downstream end of the bridge to section ②

  Mean velocities: $V_{BD} = (200\,\text{cfs})/[(20\,\text{ft})y_{BD}] = 10/y_{BD}\,\text{ft sec}^{-1}$

  $$V_2 = (200\,\text{cfs})/[(50\,\text{ft}) * (1.30\,\text{ft})] = 3.07692\,\text{ft sec}^{-1}.$$

  Eq. (2.21): $h_{L,BD-2} = L_x S_0 + [y_{BD} + V_{BD}^2/(2g)] - [y_2 + V_2^2/(2g)]$

  $$= (40\,\text{ft}) * 0.002 + [y_{BD} + (10/y_{BD})^2/(2 * 32.2\,\text{ft sec}^{-2})] - [1.30\,\text{ft} + (3.07692\,\text{ft sec}^{-1})^2/(2 * 32.2\,\text{ft sec}^{-2})]$$

  $$= y_{BD} + 1.55280/y_{BD}^2 - 1.36701\,\text{ft}$$

$\rightarrow S_{BD-2} = h_{L,BD-2}/L_x = 0.025y_{BD} + 0.03882/y_{BD}{}^2 - 0.034175.$

Average flow area: $A_{BD-2} = [(20\,\text{ft})y_{BD} + (50\,\text{ft}) * (1.30\,\text{ft})]/2 = 10(y_{BD} + 3.25)\,\text{ft}^2.$

Average wetted perimeter: $P_{wet,BD-2} = [(20\,\text{ft} + 2y_{BD}) + (50\,\text{ft} + 2 * 1.30\,\text{ft})]/2$

$$= y_{BD} + 36.30\,\text{ft}.$$

Average hydraulic radius: $R_{BD-2} = \{[(20\,\text{ft})y_{BD}]/\,(20\,\text{ft}+2y_{BD}) + (50\,\text{ft}) * (1.30\,\text{ft})/(50\,\text{ft}+ 2 * 1.30\,\text{ft})\}/2 = 5y_{BD}/(10 + y_{BD}) + 0.61787\,\text{ft}.$

As shown in Figure 8.8b, the following five forces act on this control volume:

F_{BD}: hydrostatic force at downstream end of the bridge

$F_{p,x}$: reaction force from the pier and abutments

F_2: hydrostatic force at section ②

$F_{f,x}$: friction force of the channel bed and banks

W_x: weight of water of the control volume.

Eq. (2.7): $F_{BD} = (62.4\,\text{lbf ft}^{-3}) * (y_{BD}/2) * [(20\,\text{ft})y_{BD}] = 624(y_{BD})^2\text{lbf}$

$$F_{p,x} = (62.4\,\text{lbf ft}^{-3}) * (y_{BD}/2) * [(50\,\text{ft} - 20\,\text{ft})y_{BD}] = 936(y_{BD})^2\,\text{lbf}$$
$$F_2 = (62.4\,\text{lbf ft}^{-3}) * (1.30\,\text{ft}/2) * [(50\,\text{ft}) * (1.30\,\text{ft})] = 2636.4\,\text{lbf}.$$

$$F_{f,x} = \gamma R_{BD-2} S_{BD-1}(P_{wet,BD-2} L_x) = (62.4\,\text{lbf ft}^{-3})[5y_{BD}/(10 + y_{BD}) + 0.61787\,\text{ft}] * (0.025y_{BD} + 0.03882/y_{BD}{}^2 - 0.034175)*](y_{BD} + 36.30) * (40\,\text{ft})]$$
$$= 2496[5y_{BD}/(10 + y_{BD}) + 0.61787] * (0.025y_{BD} + 0.03882/y_{BD}{}^2 - 0.034175)* (y_{BD} + 36.30)\,\text{lbf}.$$

$$W_x = (62.4\,\text{lbf ft}^{-3})[(20\,\text{ft}) * y_{BD} + (50\,\text{ft}) * (1.30\,\text{ft})]/2 * (40\,\text{ft}) = 24,960(y_{BD} + 3.25)\,\text{lbf}.$$

Eq. (2.32) in the flow direction: $F_{BD} + F_{p,x} + W_x S_0 - F_{f,x} - F_2 = \rho Q(V_2 - V_{BD})$

$\rightarrow 624(y_{BD})^2 + 936(y_{BD})^2 + [24,960(y_{BD}+3.25)]*0.002 - 2496[5y_{BD}/(10+y_{BD})+0.61787]* (0.025y_{BD} + 0.03882/y_{BD}{}^2 - 0.034175) * (y_{BD} + 36.30) - 2636.4$
$= 1.94 * 200 * (3.07692 - 10/y_{BD})$

$\rightarrow 1560(y_{BD})^2 + 49.92(y_{BD} + 3.25) - 2496[5y_{BD}/(10 + y_{BD}) + 0.61787] * (0.025y_{BD} + 0.03882/y_{BD}{}^2 - 0.034175) * (y_{BD} + 36.30) + 3880/y_{BD} - 3830.24496 = 0$

$\rightarrow$by trial-and-error or Excel Solver, $y_{BD} = 1.48\,\text{ft}.$

- From upstream to downstream end of the bridge

Mean velocities: $V_{BU} = (200\,\text{cfs})/[(20\,\text{ft})y_{BU}] = 10/y_{BU}\,\text{ft sec}^{-1}$

$$V_{BD} = 10/1.48\,\text{ft sec}^{-1} = 6.75676\,\text{ft sec}^{-1}.$$

Eq. (2.21): $h_{L,BU-BD} = L_B S_0 + [y_{BU} + V_{BU}{}^2/(2g)] - [y_{BD} + V_{BD}{}^2/(2g)]$

$$= (15\,\text{ft}) * 0.002 + [y_{BU} + (10/y_{BU})^2/(2 * 32.2\,\text{ft sec}^{-2})] - [1.48\,\text{ft} + (6.75676\,\text{ft sec}^{-1})^2/(2 * 32.2\,\text{ft sec}^{-2})]$$

$$= y_{BU} + 1.55280/y_{BU}^2 - 2.15891\,\text{ft}$$

$$\rightarrow S_{BU\text{-}BD} = h_{L,BU\text{-}BD}/L_B = 0.066667y_{BU} + 0.10352/y_{BU}^2 - 0.14393.$$

Average flow area: $A_{BU\text{-}BD} = [(y_{BU} + 1.48\,\text{ft})/2] * (20\,\text{ft}) = 10(y_{BU} + 1.48)\,\text{ft}^2$.

Average wetted perimeter: $P_{wet,BU\text{-}BD} = 2 * [(10\,\text{ft} + 2y_{BU}) + (10\,\text{ft} + 2 * 1.48\,\text{ft})]/2$

$$= 2y_{BU} + 22.96\,\text{ft}.$$

Average hydraulic radius: $R_{BU\text{-}BD} = \{(20\,\text{ft})(y_{BU})/[2 * (10\,\text{ft} + 2y_{BU})] + (20\,\text{ft}) * (1.48\,\text{ft})/[2 * (10\,\text{ft} + 2 * 1.48\,\text{ft})]\}/2$

$$= 2.5y_{BU}/(5 + y_{BU}) + 0.570988\,\text{ft}.$$

As shown in Figure 8.8b, following five forces act on this control volume:

F_{BU}: hydrostatic force at upstream end of the bridge

$F_{BD'}$: hydrostatic force at downstream end of the bridge

F_D: drag force from the pier and abutments

$F_{f,B}$: friction force of the beds and sides of the openings

W_B: weight of water of the control volume.

Eq. (2.7): $F_{BU} = (62.4\,\text{lbf}\,\text{ft}^{-3}) * (y_{BU}/2) * [(20\,\text{ft})y_{BU}] = 624(y_{BU})^2\,\text{lbf}$

$$F_{BD'} = F_{BD} = 624 * (1.48)^2\,\text{lbf} = 1366.8096\,\text{lbf}.$$

$$F_D = 1.33 * [(50\,\text{ft} - 20\,\text{ft}) * y_{BU}] * (62.4\,\text{lbf}\,\text{ft}^{-3}) * (10/y_{BU}\,\text{ft}\,\text{sec}^{-1})^2/(2 * 32.2\,\text{ft}\,\text{sec}^{-2})$$
$$= 3866.08696/y_{BU}\,\text{lbf}.$$

$$F_{f,B} = \gamma R_{BU\text{-}BD} S_{BU\text{-}BD}(P_{wet,BU\text{-}BD} L_B) = (62.4\,\text{lbf}\,\text{ft}^{-3})[2.5y_{BU}/(5 + y_{BU}) + 0.570988\,\text{ft}] * (0.066667y_{BU} + 0.10352/y_{BU}^2 - 0.14393) * [(2y_{BU} + 22.96\,\text{ft}) * (15\,\text{ft})]$$
$$= 936[2.5y_{BU}/(5 + y_{BU}) + 0.570988] * (0.066667y_{BU} + 0.10352/y_{BU}^2 - 0.14393) * (2y_{BU} + 22.96)\,\text{lbf}.$$

$$W_B = (62.4\,\text{lbf}\,\text{ft}^{-3})[(20\,\text{ft}) * (y_{BU} + 1.48\,\text{ft})/2] * (15\,\text{ft}) = 9360(y_{BU} + 1.48)\,\text{lbf}.$$

Eq. (2.32) in the flow direction: $F_{BU} + W_B S_0 - F_D - F_{f,B} - F_{BD'} = \rho Q(V_{BD} - V_{BU})$
$\rightarrow 624(y_{BU})^2 + [9360(y_{BU} + 1.48)] * 0.002 - 3866.08696/y_{BU} - -936[2.5y_{BU}/(5 + y_{BU}) + 0.570988] * (0.066667y_{BU} + 0.10352/y_{BU}^2 - 0.14393) * (2y_{BU} + 22.96) - 1366.8096 = 1.94 * 200 * (6.75676 - -10/y_{BU})$
$\rightarrow 624(y_{BU})^2 + 18.72(y_{BU} + 1.48) - 3866.08696/y_{BU} - 936[2.5y_{BU}/(5 + y_{BU}) + 0.570988] * (0.066667y_{BU} + 0.10352/y_{BU}^2 - 0.14393) * (2y_{BU} + 22.96) + 3880/y_{BU} - 3988.43248 = 0$
$\rightarrow$ by trial-and-error or Excel Solver, $y_{BU} = 4.206$ ft.

- From section ③ to upstream end of the bridge

Mean velocities: $V_3 = (200\,\text{cfs})/[(50\,\text{ft})y_3] = 4/y_3\,\text{ft}\,\text{sec}^{-1}$

$$V_{BU} = 10/4.206\,\text{ft}\,\text{sec}^{-1} = 2.377556\,\text{ft}\,\text{sec}^{-1}.$$

$$\text{Eq. (2.21): } h_{L,3-BU} = L_cS_0 + [y_3 + V_3{}^2/(2g)] - [y_{BU} + V_{BU}{}^2/(2g)]$$
$$= (20\,\text{ft}) * 0.002 + [y_3 + (4/y_3)^2/(2 * 32.2\,\text{ft sec}^{-2})] - [4.206\,\text{ft} +$$
$$(2.377556\,\text{ft sec}^{-1})^2/(2 * 32.2\,\text{ft sec}^{-2})]$$
$$= y_3 + 0.24845/y_3{}^2 - 4.253776\,\text{ft}$$

$$\rightarrow S_{3-BU} = h_{L,3-BU}/L_c = 0.05y_3 + 0.012423/y_3{}^2 - 0.212689.$$

Average flow area: $A_{3-BU} = [(50\,\text{ft})y_3 + (20\,\text{ft}) * (4.206\,\text{ft})]/2 = 25(y_3 + 1.6824)\,\text{ft}^2$.
Average wetted perimeter: $P_{wet,3-BU} = [(50\,\text{ft} + 2y_3) + 2 * (10\,\text{ft} + 2 * 4.206\,\text{ft})]/.$

$$= y_3 + 43.412\,\text{ft}$$

Average hydraulic radius: $R_{3-BU} = \{[(50\,\text{ft})y_3]/(50\,\text{ft} + 2y_3) + [(20\,\text{ft}) * (4.206\text{ft})]/]2 * (10\,\text{ft} + 2 * 4.206\,\text{ft})]\}/2 = 12.5y_3/(y_3 + 25) + 1.142190\,\text{ft}$.
As shown in Figure 8.8b, the following five forces act on this control volume:
F_3: hydrostatic force at section ③
$F_{p,c}$: reaction force from the pier and abutments
$F_{BU'}$: hydrostatic force at upstream end of the bridge
$F_{f,c}$: friction force of the channel bed and banks
W_c: weight of water of the control volume.
Eq. (2.7): $F_3 = (62.4\,\text{lbf ft}^{-3}) * (y_3/2) * [(50\,\text{ft})y_3] = 1560(y_3)^2\,\text{lbf}$

$$F_{p,c} = (62.4\,\text{lbf ft}^{-3}) * (4.206\,\text{ft}/2) * [(50 - 20)\,\text{ft} * 4.206\,\text{ft}] = 16{,}558.2481\,\text{lbf}$$
$$F_{BU'} = F_{BU} = 624 * (4.206)^2\,\text{lbf} = 11{,}038.83206\,\text{lbf}.$$

$$F_{f,c} = \gamma R_{3-BU}S_{3-BU}(P_{wet,3-BU}L_c) = (62.4\,\text{lbf ft}^{-3})[12.5y_3/(y_3 + 25) + 1.42190\,\text{ft}] * (0.05y_3$$
$$+ 0.012423/y_3{}^2 - 0.212689) * [(y_3 + 43.412\,\text{ft}) * (20\,\text{ft})]$$
$$= 1248[12.5y_3/(y_3 + 25) + 1.142190] * (0.05y_3 + 0.012423/y_3{}^2 - 0.212689) * (y_3 +$$
$$43.412)\,\text{lbf}.$$
$$W_c = (62.4\,\text{lbf ft}^{-3})[(50\,\text{ft})y_3 + (20\,\text{ft}) * (4.206\,\text{ft})]/2 * (20\,\text{ft}) = 31{,}200(y_3 + 1.6824)\,\text{lbf}.$$

Eq. (2.32) in the flow direction: $F_3 + W_cS_0 - F_{p,c} - F_{f,c} - F_{BU'} = \rho Q(V_{BU} - V_3)$
$\rightarrow 1560(y_3)^2 + [31{,}200(y_3 + 1.6824)] * 0.002 - 16{,}558.2481 - 1248[12.5y_3/(y_3 + 25) + 1.142190] * (0.05y_3 + 0.012423/y_3{}^2 - 0.212689) * (y_3 + 43.412) - 11{,}038.83206 = 1.94 * 200 * (2.377556 - 4/y_3)$
$\rightarrow 1560(y_3)^2 + 62.4(y_3 + 1.6824) - 1248[12.5y_3/(y_3 + 25) + 1.142190] * (0.05y_3 + 0.012423/y_3{}^2 - 0.212689) * (y_3 + 43.412) + 1552/y_3 - 28{,}519.57189 = 0$
$\rightarrow$ by trial-and-error or Excel Solver, $y_3 = 4.18\,\text{ft}$.

- Water surface change from ③ to ②

$$\Delta z_{3-2} = [4.18\,\text{ft} + (40 + 15 + 20)\,\text{ft} * 0.002] - 1.30\,\text{ft} = 3.03\,\text{ft}.$$

(2) Apply energy equation between sections ③ and ②:

$$L = 40\,\text{ft} + 15\,\text{ft} + 20\,\text{ft} = 75\,\text{ft}$$
$$y_2 = 1.30\,\text{ft},\ V_2 = (200\,\text{cfs})/[(50\,\text{ft}) * (1.30\,\text{ft})] = 3.07692\,\text{ft sec}^{-1}$$
$$V_3 = (200\,\text{cfs})/[(50\,\text{ft})y_3] = 4/y_3\,\text{ft sec}^{-1}.$$

Compute friction loss, h_f, using the Manning formula (Eq. (2.27)):

$$A_2 = (50\,\text{ft}) * (1.30\,\text{ft}) = 65.0\,\text{ft}^2$$
$$P_{wet,2} = 50\,\text{ft} + 2 * (1.30\,\text{ft}) = 52.6\,\text{ft}$$
$$R_2 = A_2/P_{wet,2} = (65.0\,\text{ft}^2)/(52.6\,\text{ft}) = 1.23574\,\text{ft}$$
$$A_3 = (50\,\text{ft})y_3 = 50y_3\,\text{ft}^2$$
$$P_{wet,3} = 50\,\text{ft} + 2y_3 = 50 + 2y_3\,\text{ft}$$
$$R_3 = A_3/P_{wet,3} = (50y_3\text{ft}^2)/(50 + 2y_3\text{ft}) = 25y_3/(25 + y_3)\,\text{ft}$$
$$\overline{A} = \frac{A_3 + A_2}{2} = \frac{50y_3 + 65.0}{2}\,\text{ft} = 25y_3 + 32.5\,\text{ft}^2$$
$$\overline{R} = \frac{R_3 + R_2}{2} = \frac{25y_3/(25 + y_3) + 1.23574}{2}\,\text{ft} = \frac{12.5y_3}{25 + y_3} + 0.61787\,\text{ft}$$
$$\overline{V} = \frac{V_3 + V_2}{2} = \frac{4/y_3 + 3.07692}{2}\,\text{ft sec}^{-1} = \frac{2}{y_3} + 1.53846\,\text{ft sec}^{-1}.$$

$$\text{Eq. (2.27): } h_f = (75\,\text{ft}) * \left[\frac{0.025\left(\frac{2}{y_3} + 1.53846\right)}{1.486\left(\frac{12.5y_3}{25+y_3} + 0.61787\right)^{2/3}}\right]^2 = \frac{0.021228\left(\frac{2}{y_3} + 1.53846\right)^2}{\left(\frac{12.5y_3}{25+y_3} + 0.61787\right)^{4/3}}\,\text{ft}.$$

$$\text{Contraction loss: } h_c = k_c\frac{V_3^2}{2g} = 0.1 * \frac{\left(4/y_3\,\text{ft sec}^{-1}\right)^{4/3}}{2 * 32.2\,\text{ft sec}^{-2}} = \frac{0.024845}{y_3^2}\,\text{ft}.$$

$$\text{Expansion loss: } h_x = k_x\frac{V_2^2}{2g} = 0.3 * \frac{\left(3.07692\,\text{ft sec}^{-1}\right)^2}{2 * 32.2\,\text{ft sec}^{-2}} = 0.044\,\text{ft}.$$

$$\text{Eq. (2.21): } (L_c + L_B + L_x)\,S_0 + y_3 + \frac{V_3^2}{2g} = y_2 + \frac{V_2^2}{2g} + h_L$$

$$\rightarrow (20\,\text{ft} + 15\,\text{ft} + 40\,\text{ft}) * 0.002 + y_3 + \frac{(4/y_3)^2}{2 * 32.2\,\text{ft sec}^{-2}}$$
$$= 1.30\,\text{ft} + \frac{\left(3.07692\,\text{ft sec}^{-1}\right)^2}{2 * 32.2\,\text{ft sec}^{-2}} + \frac{0.021228\left(\frac{2}{y_3} + 1.53846\right)^2}{\left(\frac{12.5y_3}{25+y_3} + 0.61787\right)^{4/3}} + \frac{0.024845}{y_3^2} + 0.044\,\text{ft}$$

$$\rightarrow y_3 + \frac{0.24845}{y_3^2} = 1.34101 + \frac{0.021228\left(\frac{2}{y_3} + 1.53846\right)^2}{\left(\frac{12.5y_3}{25+y_3} + 0.61787\right)^{4/3}} + \frac{0.024845}{y_3^2}\,\text{ft}$$

→ by trial-and-error or Excel Solver, $y_3 = 1.36$ ft.

Thus, the water surface change is: $\Delta z_{3-2} = (1.36\,\text{ft} + 75\,\text{ft} * 0.002) - 1.30\,\text{ft} = 0.21\,\text{ft}$.

(3) $y_2 = 1.30\,\text{ft}$, $V_2 = (200\,\text{cfs})/[(50\,\text{ft}) * (1.30\,\text{ft})] = 3.07692\,\text{ft sec}^{-1}$
Table 8.3: semi-circular pier and abutments $\rightarrow K = 0.90$.
No obstruction at section ② $\rightarrow \alpha = (50\,\text{ft} - 20\,\text{ft})/(50\,\text{ft}) = 0.6$.
Eq. (8.6): the water surface drop is:

$$\Delta z_{3-2} = 2 * 0.90 * \left(0.90 + \frac{10}{1.30\,\text{ft}} * \frac{\left(3.07692\,\text{ft sec}^{-1}\right)^2}{2 * 32.2\,\text{ft sec}^{-2}} - 0.6\right)$$

$$* \left(0.6 + 15 * 0.6^4\right) * \frac{\left(3.07692\,\text{ft sec}^{-1}\right)^2}{2 * 32.2\,\text{ft sec}^{-2}} = 0.96\,\text{ft}.$$

Summary: The water surface change is computed differently using the three methods. In principle, the momentum equation method is more accurate than the other two methods because the flow through the bridge is rapidly varied and thus the friction may not be computed directly. The energy equation method tends to underestimate the friction loss inside the bridge openings, while the Yarnell equation is empirical and does not consider some important characteristics (e.g., length and bed slope) specific to a bridge of interest. However, as indicated by this example, the momentum equation method requires much more computational effort than the energy equation method, which in turn requires more computational efforts than the Yarnell equation method. It is recommended to use the momentum equation method if computational resources are not a restriction.

8.3 Risers

A riser is a standing (i.e., vertical) conduit that is usually used as a pond outlet control structure (e.g., Figure 8.9). Although most risers have a circular shape, rectangular risers are also constructed. To maintain expected water levels in a pond and to regulate outflow, orifices and/or weirs are included in the design of the conduit. The water flowing into the standing conduit is drained through a downward culvert. For a given water depth in the pond (H_{wdp}), the discharge through the orifices and/or weirs can be computed using the formulas presented in Chapters 6 and 7, while the outflow through the culvert can be computed using the equations presented in Section 8.1 in terms of the head water depth (HW) in the conduit, which in turn can be determined using continuity equation (Eq. (2.16)). If the head on an orifice is smaller than 1.5 times the vertical height (diameter for a circular shape) of the orifice, the orifice is assumed to be an inlet-control culvert and its discharge is computed using Eqs. (8.1), (8.2), or (8.3); otherwise, its discharge is computed as an orifice flow using Eq. (7.14). Note that the orifices and/or weirs can become submerged once the water surface inside the riser rises above their crests or bottoms with the lapse of time. For such a situation, the formula for a submerged orifice or weir should be used. In general, because the culvert is short in length and laid on a not-small slope, the discharge through the culvert can be computed by assuming an inlet-control condition. For a riser with an open top, when H_{wdp}

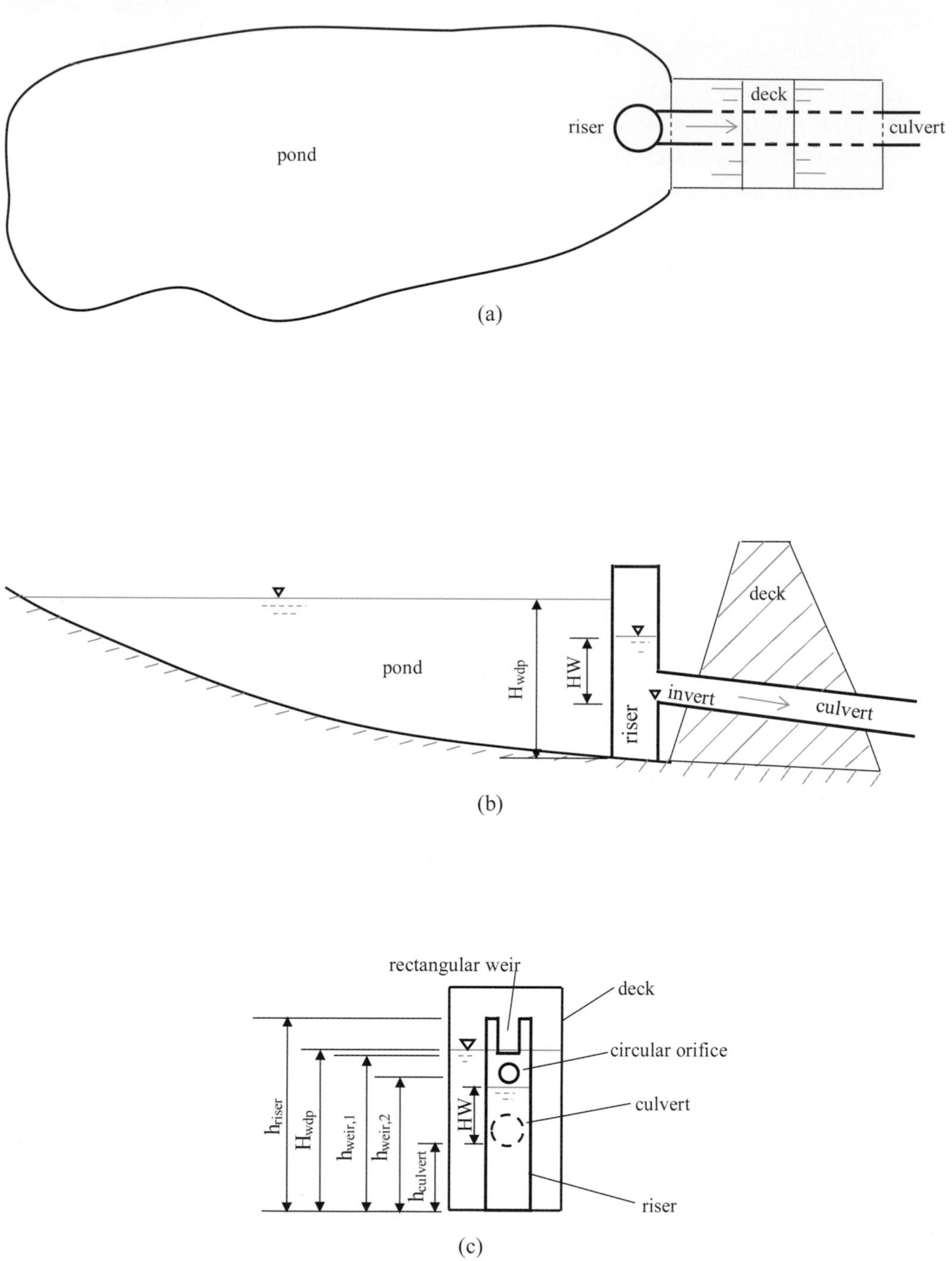

Figure 8.9 **Sketches of a pond with a riser.** The level of the pond is controlled by a riser (height of h_{riser}) with one rectangular weir (crest height of $h_{weir,1}$) and one circular orifice (crest height of $h_{weir,2}$). (a) Bird's-eye view of the pond and hydraulic structures; (b) side view of the pond and hydraulic structures; and (c) front view of the hydraulic structures. The riser is drained by a culvert (invert height of $h_{culvert}$). H_{wdp}: water depth in the pond; and HW: head water depth of the culvert.

becomes greater than its height (h_{riser}), the top opening(s) can be treated as a sharp- or broad-crested weir or an orifice, depending on the shape and submerge water depth of the opening(s), to compute the flows through them into the standing conduit.

Example 8.12 With reference to the hydraulic system shown in Figure 8.9, the riser has a diameter of $D_{riser} = 4$ ft and a height of $h_{riser} = 6$ ft. The 1.3-ft-wide rectangular weir has a crest height of $h_{weir,1} = 3.5$ ft, and the 1-ft-diameter orifice has a crest height of $h_{weir,2} = 1.5$ ft. The 1.25-ft-diameter concrete culvert has an invert height of $h_{culvert} = 0.1$ ft and is laid on a slope of $S_0 = 0.002$. For a computational time step of $\Delta t = 60$ sec, determine the flow rate into the riser (Q_{riser}), culvert head water depth (HW), and flow rate out of the culvert ($Q_{culvert}$) if the water depth in the pond (H_{wdp}) is:

(1) 1.6 ft
(2) 2.0 ft
(3) 3.0 ft
(4) 5.0 ft
(5) 6.5 ft.

Solution

$D_{riser} = 4$ ft, $h_{riser} = 6$ ft, $b_w = 1.3$ ft, $h_{weir,1} = 3.5$ ft, $D_o = 1$ ft, $h_{weir,2} = 1.5$ ft,

$D = 1.25$ ft, $h_{culvert} = 0.1$ ft, $S_0 = 0.002$, let H_{riser} be the water depth in the riser

Riser cross-sectional area: $A_{riser} = \pi/4 * (4\,\text{ft})^2 = 12.566371\,\text{ft}^2$.

Orifice cross-sectional area: $A_{orifice} = \pi/4 * (1\,\text{ft})^2 = 0.78540\,\text{ft}^2$.

Culvert cross-sectional area: $A_{culvert} = \pi/4 * (1.25\,\text{ft})^2 = 1.22718\,\text{ft}^2$.

Table 8.1: circular concrete, square edge w/headwall→$C = 0.0398$, $Y = 0.67$; Form 1 (Eq. (8.2)), $K = 0.0098$, $M = 2.0$

$Z = 0$ for the orifice; $Z = -0.5 * 0.002 = -0.001$ for the culvert.

- For $H_{wdp} \le h_{weir,2} = 1.5$ ft, no water flows into the riser and out of the culvert, that is, $Q_{riser} = 0$ cfs, $H_{riser} = 0$ ft, $Q_{culvert} = 0$ cfs.
- For $h_{weir,2} < H_{wdp} \le h_{weir,1}$, water flows into the riser through the circular orifice only, that is, $Q_{riser} = Q_{orifice} > 0$ cfs, $H_{riser} > 0$ ft, $Q_{culvert} = 0$ or > 0 cfs depending on if $H_{riser} > h_{culvert}$.

  If $H_{wdp} - h_{weir,2} \ge 1.5 * (1\,\text{ft}) = 1.5\,\text{ft}\rightarrow$ orifice flow:

  Table 7.3: sharp-edged opening, very thin wall →$C_{od} = 0.61$.

  $$\text{Eq. (7.14): } Q_{orifice} = 0.61 * (0.78540\,\text{ft}^2) * [2 * (32.2\,\text{ft sec}^{-2})(H_{wdp} - 1.5\,\text{ft} - 1/2\,\text{ft})]^{0.5}.$$
  $$= 3.84471(H_{wdp} - 2.0)^{0.5}$$

  If $H_{wdp} - h_{weir,2} < 1.5 * (1\,\text{ft}) = 1.5\,\text{ft}\rightarrow$ inlet-control culvert:

  $$\text{Table 2.4: } A_{c,o} = (1\,\text{ft})^2/8(\theta_{c,o} - \sin\theta_{c,o}) = 0.125(\theta_{c,o} - \sin\theta_{c,o})\,\text{ft}^2$$
  $$B_{c,o} = (1\,\text{ft})\sin(\theta_{c,o}/2) = \sin(\theta_{c,o}/2)\,\text{ft}$$

$$y_{c,o} = (1\,\text{ft})/2[1 - \cos(\theta_{c,o}/2)] = 0.5[1 - \cos(\theta_{c,o}/2)]\,\text{ft}$$
$$V_{c,o} = Q_{orifice}/A_{c,o}\,\text{ft sec}^{-1}.$$

Eq. (2.37): $(Q_{orifice})^2/(32.2\,\text{ft sec}^{-2}) = (A_{c,o})^3/B_{c,o} \rightarrow (Q_{orifice})^2 - 32.2(A_{c,o})^3/B_{c,o} = 0$
Eq. (8.2): $(H_{wdp} - 1.5\,\text{ft})/(1\,\text{ft}) = [y_{c,o} + V_{c,o}{}^2/(2 * 32.2\,\text{ft sec}^{-2})]/(1\,\text{ft}) + 0.0098 * \{Q_{orifice}/[(0.78540\,\text{ft}^2) * (1\,\text{ft})^{0.5}]\}^{2.0} - 0$

$$\rightarrow (y_{c,o} + 0.015528V_{c,o}{}^2) + 0.015887(Q_{orifice})^{2.0} - H_{wdp} + 1.5 = 0.$$

In Excel Solver, set objective to be Eq. (8.2) and make it become zero by changing $\theta_{c,o}$ and $Q_{orifice}$, subject to Eq. (2.37), and then solve to get $Q_{orifice}$.
If $Q_{orifice}/[(0.78540\,\text{ft}^2) * (1\,\text{ft})^{0.5}] = 1.27324Q_{orifice} \leq 3.5$, the computed $Q_{orifice}$ is fine. Otherwise, use Eq. (8.1) if $1.27324Q_{orifice} \geq 4.0$ or a linear interpolation (discussed in Section 8.1) if $3.5 < 1.27324Q_{orifice} < 4.0$ to replace Eq. (8.2), and solve in Solver to get $Q_{orifice}$.

<u>Determine HW and $Q_{culvert}$:</u>

$$H_{riser} = (Q_{riser} - Q_{culvert})(\Delta t)/A_{riser} = (Q_{riser} - Q_{culvert}) * (60\,\text{sec})/(12.566371\,\text{ft}^2)$$
$$= 4.77465(Q_{riser} - Q_{culvert})\,\text{ft}.$$
$$HW = H_{riser} - h_{culvert} = H_{riser} - 0.1\,\text{ft}.$$

Table 2.4: $A_{c,c} = (1.25\,\text{ft})^2/8(\theta_{c,c} - \sin\theta_{c,c}) = 0.19531(\theta_{c,c} - \sin\theta_{c,c})\,\text{ft}^2$
$$B_{c,c} = (1.25\,\text{ft})\sin(\theta_{c,c}/2) = 1.25\sin(\theta_{c,c}/2)\,\text{ft}$$
$$y_{c,c} = (1.25\,\text{ft})/2[1 - \cos(\theta_{c,c}/2)] = 0.625[1 - \cos(\theta_{c,c}/2)]\,\text{ft}$$
$$V_{c,c} = Q_{culvert}/A_{c,c}\,\text{ft sec}^{-1}.$$

Eq. (2.37): $(Q_{culvert})^2/(32.2\,\text{ft sec}^{-2}) = (A_{c,c})^3/B_{c,c} \rightarrow (Q_{culvert})^2 - 32.2(A_{c,c})^3/B_{c,c} = 0.$
Eq. (8.2): $HW/(1.25\,\text{ft}) = [y_{c,c} + V_{c,c}{}^2/(2 * 32.2\,\text{ft sec}^{-2})]/(1.25\,\text{ft}) + 0.0098 * \{Q_{culvert}/[(1.22718\,\text{ft}^2) * (1.25\,\text{ft})^{0.5}]\}^{2.0} - 0.001$

$$\rightarrow HW = (y_{c,c} + 0.015528V_{c,c}{}^2) + 0.0052059(Q_{culvert})^{2.0} - 0.00125.$$

In Excel Solver, set objective to be Eq. (8.2) and make it become zero by changing $\theta_{c,c}$ and $Q_{culvert}$, subject to Eq. (2.37) and $HW > 0$, solve to get $Q_{culvert}$, and then compute H_{riser} and HW.
If $Q_{culvert}/[(1.22718\,\text{ft}^2) * (1.25\,\text{ft})^{0.5}] = 0.72885Q_{culvert} \leq 3.5$, the computed $Q_{culvert}$ is fine. Otherwise, use Eq. (8.1) if $0.72885Q_{culvert} \geq 4.0$ or a linear interpolation (discussed in Section 8.1) if $3.5 < 0.72885Q_{culvert} < 4.0$ to replace Eq. (8.2), and solve in Solver to get $Q_{culvert}$.

<u>Check if the orifice is submerged:</u>

If $H_{wdp} - h_{weir,2} \geq 1.5\,\text{ft}$ and $H_{riser} > h_{weir,2} = 1.5\,\text{ft}$, recalculate $Q_{orifice}$ for submerged condition, and then recalculate $Q_{culvert}$ and HW as elaborated above.

- For $h_{weir,1} < H_{wdp} \leq h_{riser}$, water flows into the riser through both the circular orifice and the rectangular weir, that is, $Q_{riser} = Q_{orifice} + Q_{weir} > 0\,\text{cfs}$, $H_{riser} > 0\,\text{ft}$, $Q_{culvert} = 0$ or $> 0\,\text{cfs}$ depending on if $H_{riser} > h_{culvert}$.

  Determine $Q_{orifice}$:

  As elaborated above.

  Determine Q_{weir} as a sharp-crested weir:

  Eq. (6.16): weir coefficient $k = 0.40 + 0.05(H_{wdp} - 3.5\,\text{ft})/(3.5\,\text{ft}) = H_{wdp}/70 + 0.35$
  Eq. (6.15): $Q_{weir} = k(2 * 32.2\,\text{ft sec}^{-2})^{0.5} * [1.3 - 0.1 * 2(H_{wdp} - 3.5\,\text{ft})] * (H_{wdp} - 3.5)^{1.5}$

  $$= k(1.60499)(10 - H_{wdp})(H_{wdp} - 3.5)^{1.5}\,\text{cfs}.$$

  Let $Q_{riser} = Q_{orifice} + Q_{weir}$ and determine HW and $Q_{culvert}$:

  As elaborated above.

  Check if the orifice and weir are submerged:

  If $1.5\,\text{ft} < H_{riser} \leq 2.5\,\text{ft}$, recalculate $Q_{orifice}$ for submerged condition.
  If $2.5\,\text{ft} < H_{riser} \leq 6\,\text{ft}$, recalculate $Q_{orifice}$ and Q_{weir} for submerged condition.
  For either situation, recalculate $Q_{culvert}$, and HW as elaborated above.

- For $H_{wdp} > h_{riser}$, water flows into the riser through the circular orifice, the rectangular weir, and the top opening, that is, $Q_{riser} = Q_{orifice} + Q_{weir} + Q_{top} > 0\,\text{cfs}$, $H_{riser} > 0\,\text{ft}$, $Q_{culvert} = 0$ or $> 0\,\text{cfs}$ depending on if $H_{riser} > h_{culvert}$.

  Determine $Q_{orifice}$:

  As elaborated above.

  Determine Q_{weir}:

  As elaborated above.

  Determine Q_{top} as a sharp-crested weir:

  Eq. (6.16): weir coefficient $k = 0.40 + 0.05(H_{wdp} - 6\,\text{ft})/(6\,\text{ft}) = H_{wdp}/120 + 0.35$.
  Weir width: $b_w = \pi D_{riser} = 4\pi\,\text{ft} = 12.56637\,\text{ft}$, no encroachment.
  Eq. (6.15): $Q_{top} = k(2 * 32.2\,\text{ft sec}^{-2})^{0.5} * (12.56637\,\text{ft} - 0\,\text{ft}) * (H_{wdp} - 6\,\text{ft})^{1.5}$

  $$= k(100.84463)(H_{wdp} - 6)^{1.5}\,\text{cfs}.$$

  Let $Q_{riser} = Q_{orifice} + Q_{weir} + Q_{top}$ and determine HW and $Q_{culvert}$:

  As elaborated above.

  Check if the orifice, weir, and top opening are submerged:

  If $1.5\,\text{ft} < H_{riser} \leq 2.5\,\text{ft}$, recalculate $Q_{orifice}$ for submerged condition.
  If $2.5\,\text{ft} < H_{riser} \leq 6\,\text{ft}$, recalculate $Q_{orifice}$ and Q_{weir} for submerged condition.
  If $H_{riser} > 6\,\text{ft}$, recalculate $Q_{orifice}$, Q_{weir}, and Q_{top} for submerged condition.
  For either situation, recalculate $Q_{culvert}$, and HW as elaborated above.

(1) $H_{wdp} = 1.6$ ft, water flows into the riser through the orifice only.
$H_{wdp} - h_{weir,2} = 1.6\text{ ft} - 1.5\text{ ft} = 0.1\text{ ft} \leq 1.5\text{ ft} \rightarrow$ orifice as an inlet-control culvert.
Eq. (2.37): $(Q_{orifice})^2 - 32.2(A_{c,o})^3/B_{c,o} = 0$.
Eq. (8.2): $(y_{c,o} + 0.015528V_{c,o}{}^2) + 0.015887(Q_{orifice})^{2.0} - 1.6 + 1.5 = 0$

$$\rightarrow (y_{c,o} + 0.015528V_{c,o}{}^2) + 0.015887(Q_{orifice})^{2.0} - 0.1 = 0.$$

Solver (Figure 8.10a) $\rightarrow Q_{riser} = 0.03395$ cfs, HW = 0.04 ft, $Q_{culvert} = 0.00519$ cfs.
$H_{riser} = 0.14\text{ ft} < 1.5\text{ ft} \rightarrow$ orifice is unsubmerged and its head $= 1.6 - 1.5\text{ ft} = 0.1$ ft.

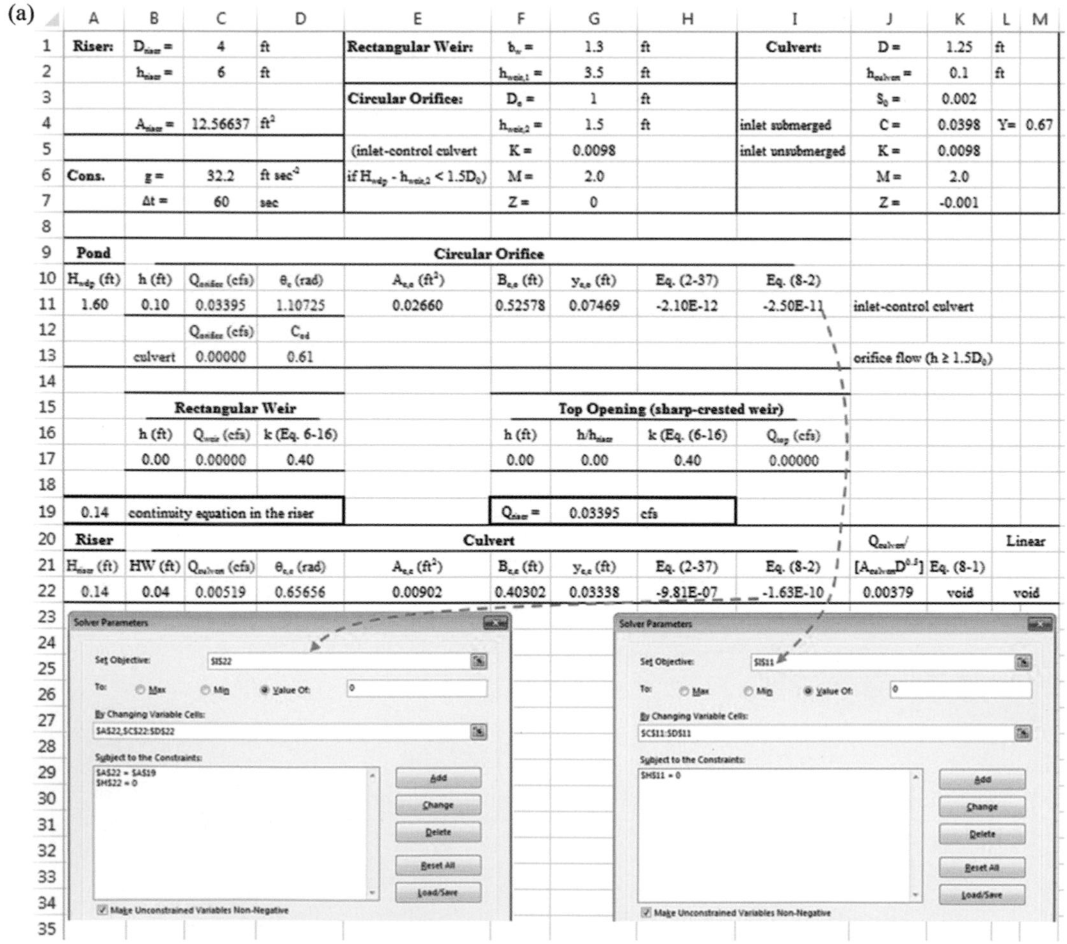

(a)
Riser: D = 4 ft; h = 6 ft; A = 12.56637 ft²
Cons. g = 32.2 ft sec⁻²; Δt = 60 sec
Rectangular Weir: b = 1.3 ft; h_weir,1 = 3.5 ft
Circular Orifice: D = 1 ft; h_weir,2 = 1.5 ft
(inlet-control culvert if H_wdp - h_weir,2 < 1.5D): K = 0.0098; M = 2.0; Z = 0
Culvert: D = 1.25 ft; h_culvert = 0.1 ft; S = 0.002
inlet submerged: C = 0.0398; Y = 0.67
inlet unsubmerged: K = 0.0098; M = 2.0; Z = -0.001
Pond
Circular Orifice
H_wdp (ft) 1.60; h (ft) 0.10; Q_orifice (cfs) 0.03395; θ (rad) 1.10725; A (ft²) 0.02660; B (ft) 0.52578; y (ft) 0.07469; Eq. (2-37) -2.10E-12; Eq. (8-2) -2.50E-11
inlet-control culvert
Q_orifice (cfs); C_d
culvert 0.00000 0.61
orifice flow (h ≥ 1.5D)
Rectangular Weir
h (ft) 0.00; Q_weir (cfs) 0.00000; k (Eq. 6-16) 0.40
Top Opening (sharp-crested weir)
h (ft) 0.00; h/h_riser 0.00; k (Eq. (6-16) 0.40; Q_top (cfs) 0.00000
0.14 continuity equation in the riser
Q_riser = 0.03395 cfs
Riser
Culvert
H_riser (ft) 0.14; HW (ft) 0.04; Q_culvert (cfs) 0.00519; θ (rad) 0.65656; A (ft²) 0.00902; B (ft) 0.40302; y (ft) 0.03338; Eq. (2-37) -9.81E-07; Eq. (8-2) -1.63E-10
Q_culvert/[A_culvert D^0.5] 0.00379; Eq. (8-1) void; Linear void
Solver Parameters
Set Objective: $I$22
To: Max Min Value Of: 0
By Changing Variable Cells: $A$22,$C$22:$D$22
Subject to the Constraints: $A$22 = $A$19; $H$22 = 0
Add Change Delete Reset All Load/Save
Make Unconstrained Variables Non-Negative
Solver Parameters
Set Objective: $I$11
To: Max Min Value Of: 0
By Changing Variable Cells: $C$11:$D$11
Subject to the Constraints: $H$11 = 0
Add Change Delete Reset All Load/Save
Make Unconstrained Variables Non-Negative

Figure 8.10 **Spreadsheet for calculating flow through the riser in Example 8.12.** Calculations are undertaken for cases in which water depth in the pond is: (a) 1.6 ft; (b) 2.0 f; (c) 3.0 ft; (d) 5.0 ft; and (e) 6.5 ft.

(b)

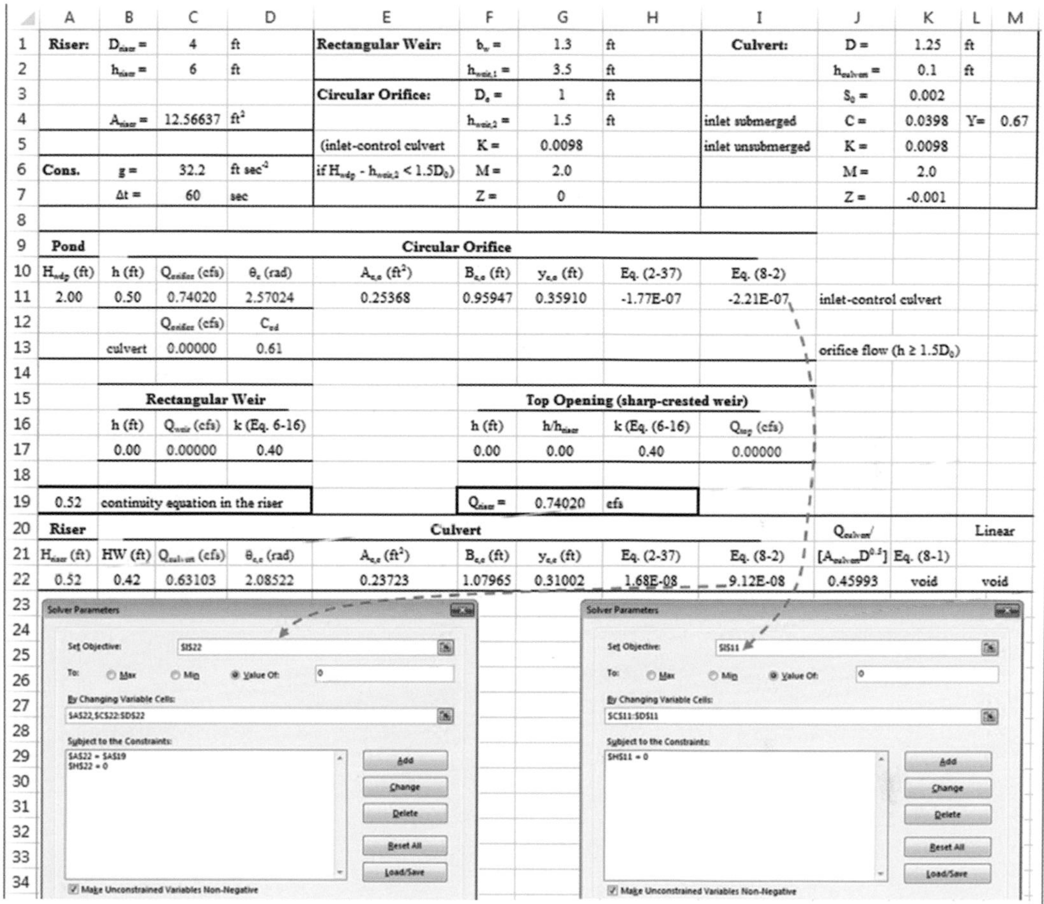

Figure 8.10 *Continued*

$Q_{orifice}$ is first computed using the Solver dialog box (*Set Objective:* I11; *To Value of*: 0; *Variable cells:* C11:D11; *Constraints:* H11 = 0) at the bottom right of Figure 8.10a and then $Q_{culvert}$ is computed using the Solver dialog box (*Set Objective:* I22; *To Value of*: 0; *Variable cells:* A22, C22:D22; *Constraints:* A22 = A19, H22 = 0) at the bottom left. The variables to be adjusted are highlighted in red color. Their initial values should satisfy Eq. (2.37).

$$Q_{culvert}/(A_{culvert}D^{0.5}) = (0.00519\,\text{cfs})/[(1.22718\,\text{ft}^2) * (1.25\,\text{ft})^{0.5}] = 0.00379 \leq 3.5 \rightarrow \text{done!}$$

(2) $H_{wdp} = 2.0$ ft, water flows into the riser through the orifice only.
$H_{wdp} - h_{weir,2} = 2.0\,\text{ft} - 1.5\,\text{ft} = 0.5\,\text{ft} \leq 1.5\,\text{ft} \rightarrow$ orifice as an inlet-control culvert.
Eq. (2.37): $(Q_{orifice})^2 - 32.2(A_{c,o})^3/B_{c,o} = 0$.

(c)

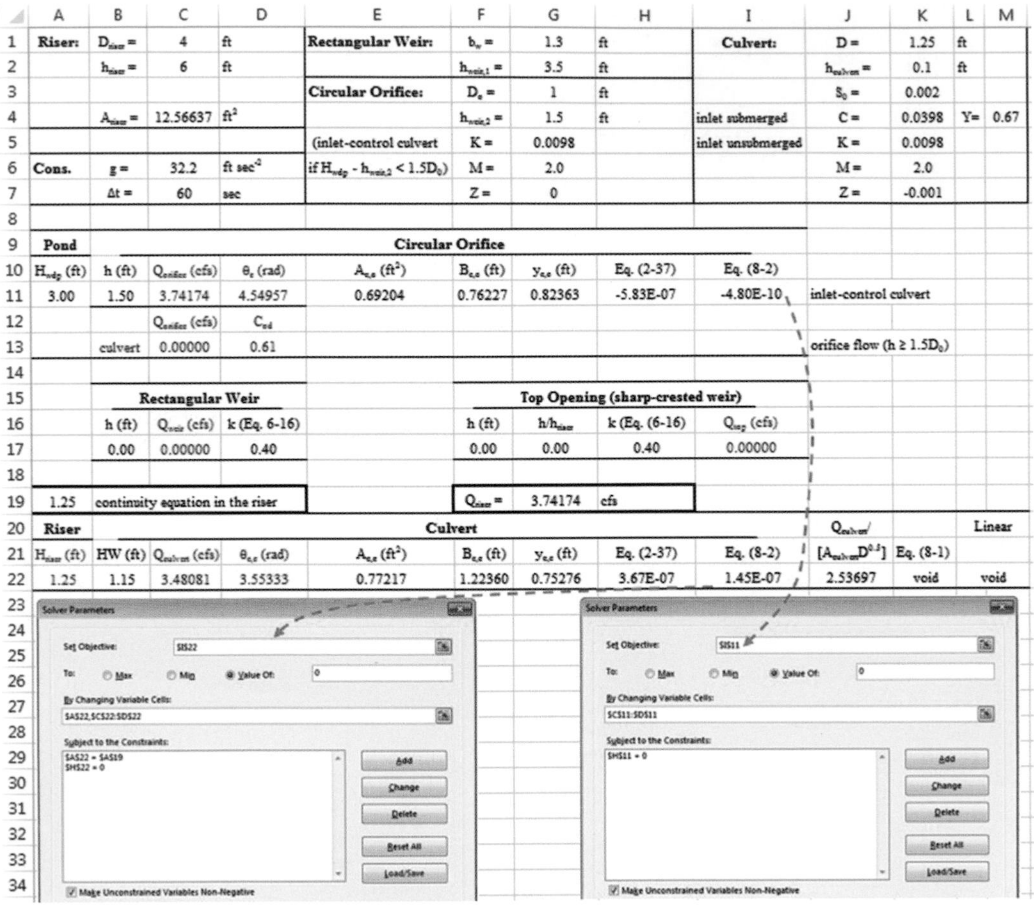

Figure 8.10 *Continued*

$$\text{Eq. (8.2)}: (y_{c,o} + 0.015528V_{c,o}{}^2) + 0.015887(Q_{orifice})^{2.0} - 2.0 + 1.5 = 0$$

$$\rightarrow (y_{c,o} + 0.015528V_{c,o}{}^2) + 0.015887(Q_{orifice})^{2.0} - 0.5 = 0.$$

Solver (Figure 8.10b) $\rightarrow Q_{riser} = 0.74020$ cfs, HW = 0.45 ft, $Q_{culvert} = 0.63103$ cfs.
$H_{riser} = 0.52$ ft < 1.5 ft→orifice is unsubmerged and its head = 2.0 – 1.5 ft = 0.5 ft.

$Q_{orifice}$ is first computed using the Solver dialog box (*Set Objective:* I11; *To Value of*: 0; *Variable cells:* C11:D11; *Constraints:* H11 = 0) at the bottom right of Figure 8.10b and then $Q_{culvert}$ is computed using the Solver dialog box (*Set Objective:* I22; *To Value of*: 0; *Variable cells:* A22, C22:D22; *Constraints:* A22 = A19, H22 = 0) at the bottom left. The variables to be adjusted are highlighted in red color. Their initial values should satisfy Eq. (2.37).

$$Q_{culvert}/(A_{culvert}D^{0.5}) = (0.63103\,\text{cfs})/[(1.22718\,\text{ft}^2) * (1.25\,\text{ft})^{0.5}] = 0.45993 \leq 3.5 \rightarrow \text{done!}$$

(d)

	A	B	C	D	E	F	G	H	I	J	K	L	M
1	**Riser:**	D_{riser} =	4	ft	**Rectangular Weir:**	b_w =	1.3	ft	**Culvert:**	D =	1.25	ft	
2		h_{riser} =	6	ft		$h_{weir,1}$ =	3.5	ft		$h_{culvert}$ =	0.1	ft	
3					**Circular Orifice:**	D_o =	1	ft		S_0 =	0.002		
4		A_{riser} =	12.56637	ft^2		$h_{weir,2}$ =	1.5	ft	inlet submerged	C =	0.0398	Y=	0.67
5					(inlet-control culvert	K =	0.0098		inlet unsubmerged	K =	0.0098		
6	**Cons.**	g =	32.2	ft sec^{-2}	if H_{wdp} - $h_{weir,2}$ < $1.5D_o$)	M =	2.0			M =	2.0		
7		Δt =	60	sec		Z =	0			Z =	-0.001		
8													
9	**Pond**				**Circular Orifice**								
10	H_{wdp} (ft)	h (ft)	$Q_{orifice}$ (cfs)	θ_c (rad)	$A_{c,o}$ (ft^2)	$B_{c,o}$ (ft)	$y_{c,o}$ (ft)	Eq. (2-37)	Eq. (8-2)				
11	5.00	1.76	0.00000	0.00000	0.00000	0.00000	0.00000	#DIV/0!	#DIV/0!	inlet-control culvert			
12			$Q_{orifice}$ (cfs)	C_{od}									
13		orifice	5.10702	0.61						orifice flow (h ≥ $1.5D_o$)			
14													
15		**Rectangular Weir**				**Top Opening (sharp-crested weir)**							
16		h (ft)	Q_{weir} (cfs)	k (Eq. 6-16)		h (ft)	h/h_{riser}	k (Eq. (6-16)	Q_{top} (cfs)				
17		1.50	6.21303	0.42		0.00	0.00	0.40	0.00000				
18													
19	3.24	continuity equation in the riser				Q_{riser} =	11.32006	cfs					
20	**Riser**				**Culvert**					$Q_{culvert}$/			Linear
21	H_{riser} (ft)	HW (ft)	$Q_{culvert}$ (cfs)	$\theta_{c,c}$ (rad)	$A_{c,c}$ (ft^2)	$B_{c,c}$ (ft)	$y_{c,c}$ (ft)	Eq. (2-37)	Eq. (8-2)	[$A_{culvert}D^{0.5}$]	Eq. (8-1)		
22	3.24	3.14	10.64240	5.45523	1.20933	0.50282	1.19720	-1.58E-12	1.09E-05	7.75666	-0.5552		void

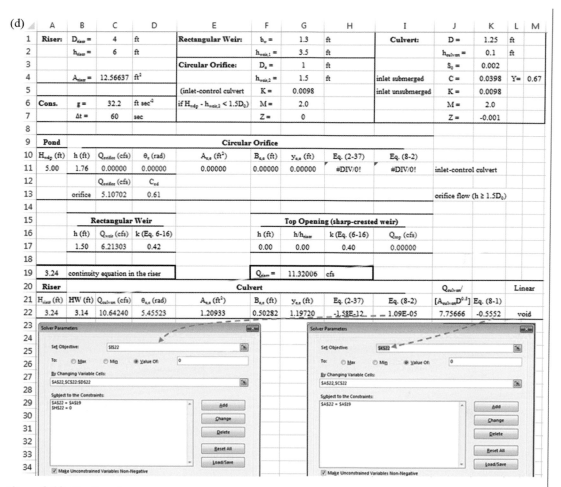

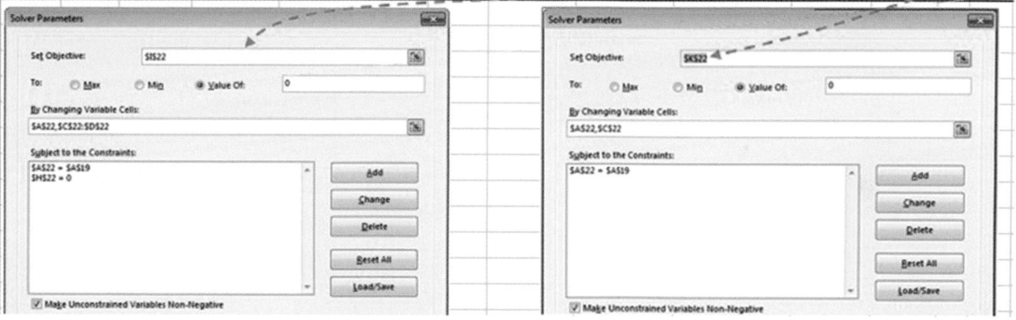

Figure 8.10 *Continued*

(3) $H_{wdp} = 3.0$ ft, water flows into the riser through the orifice only.
$H_{wdp} - h_{weir,2} = 3.0\text{ ft} - 1.5\text{ ft} = 1.5\text{ ft} \leq 1.5\text{ ft} \rightarrow$ orifice as an inlet-control culvert.
Eq. (2.37): $(Q_{orifice})^2 - 32.2(A_{c,o})^3/B_{c,o} = 0.$

$$\text{Eq. (8.2): } (y_{c,o} + 0.015528V_{c,o}{}^2) + 0.015887(Q_{orifice})^{2.0} - 3.0 + 1.5 = 0$$

$$\rightarrow (y_{c,o} + 0.015528V_{c,o}{}^2) + 0.015887(Q_{orifice})^{2.0} - 1.5 = 0.$$

Solver (Figure 8.10c) $\rightarrow Q_{riser} = 3.74174$ cfs, HW = 1.25 ft, $Q_{culvert} = 3.48081$ cfs.
$H_{riser} = 1.25\text{ ft} < 1.5\text{ ft} \rightarrow$ orifice is unsubmerged and its head = 3.0 – 1.5 ft = 1.5 ft.

(e)

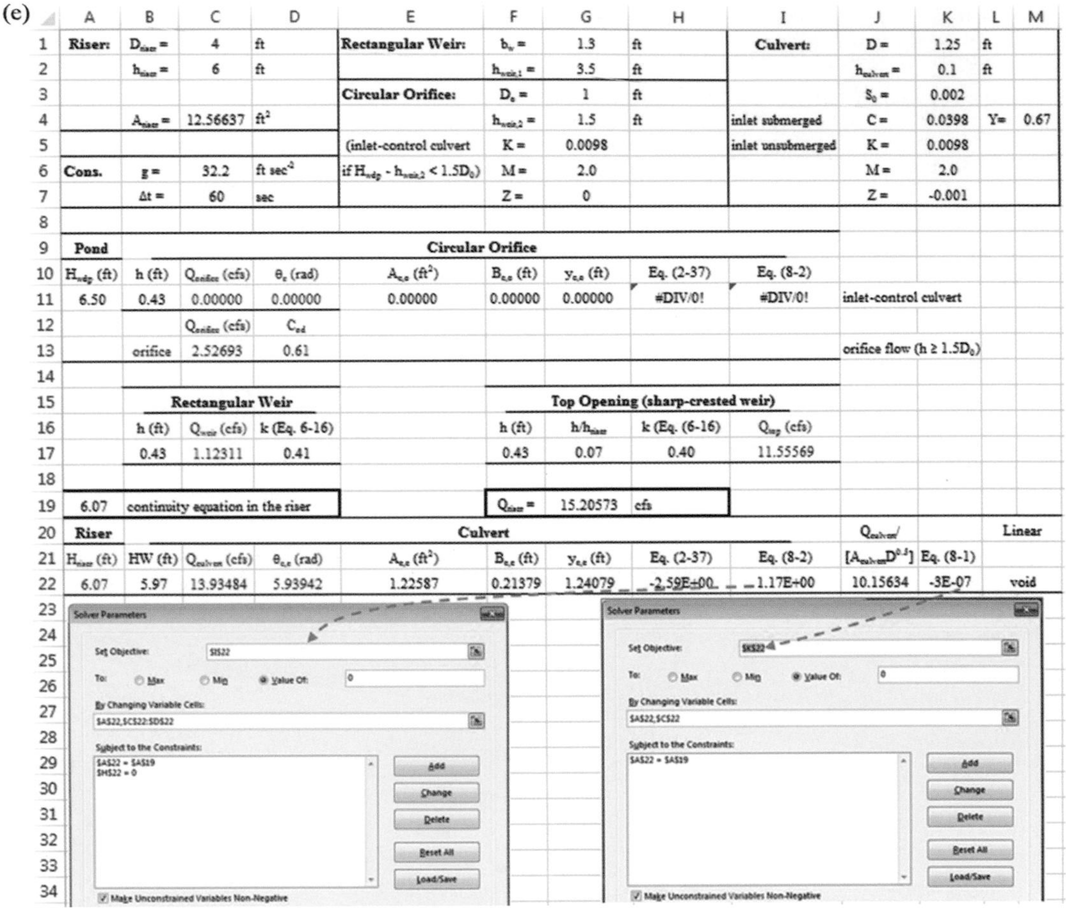

	A	B	C	D	E	F	G	H	I	J	K	L	M
1	**Riser:**	D_{riser} =	4	ft	**Rectangular Weir:**	b_w =	1.3	ft	**Culvert:**	D =	1.25	ft	
2		h_{riser} =	6	ft		$h_{weir,1}$ =	3.5	ft		$h_{culvert}$ =	0.1	ft	
3					**Circular Orifice:**	D_o =	1	ft		S_0 =	0.002		
4		A_{riser} =	12.56637	ft²		$h_{weir,2}$ =	1.5	ft	inlet submerged	C =	0.0398	Y=	0.67
5					(inlet-control culvert	K =	0.0098		inlet unsubmerged	K =	0.0098		
6	**Cons.**	g =	32.2	ft sec⁻²	if H_{wdp} - $h_{weir,2}$ < 1.5D_o)	M =	2.0			M =	2.0		
7		Δt =	60	sec		Z =	0			Z =	-0.001		
8													
9	**Pond**				**Circular Orifice**								
10	H_{wdp} (ft)	h (ft)	$Q_{orifice}$ (cfs)	θ_c (rad)	$A_{c,o}$ (ft²)	$B_{c,o}$ (ft)	$y_{c,o}$ (ft)	Eq. (2-37)	Eq. (8-2)				
11	6.50	0.43	0.00000	0.00000	0.00000	0.00000	0.00000	#DIV/0!	#DIV/0!	inlet-control culvert			
12			$Q_{orifice}$ (cfs)	C_{od}									
13		orifice	2.52693	0.61						orifice flow (h ≥ 1.5D_o)			
14													
15		**Rectangular Weir**				**Top Opening (sharp-crested weir)**							
16		h (ft)	Q_{weir} (cfs)	k (Eq. 6-16)		h (ft)	h/h_{riser}	k (Eq. (6-16)	Q_{top} (cfs)				
17		0.43	1.12311	0.41		0.43	0.07	0.40	11.55569				
18													
19	6.07	continuity equation in the riser				Q_{riser} =	15.20573	cfs					
20	**Riser**				**Culvert**					$Q_{culvert}$/			Linear
21	H_{riser} (ft)	HW (ft)	$Q_{culvert}$ (cfs)	$\theta_{c,c}$ (rad)	$A_{c,c}$ (ft²)	$B_{c,c}$ (ft)	$y_{c,c}$ (ft)	Eq. (2-37)	Eq. (8-2)	[$A_{culvert}D^{0.5}$]	Eq. (8-1)		
22	6.07	5.97	13.93484	5.93942	1.22587	0.21379	1.24079	-2.59E+00	1.17E+00	10.15634	-3E-07		void

Figure 8.10 *Continued*

$Q_{orifice}$ is first computed using the Solver dialog box (*Set Objective:* I11; *To Value of*: 0; *Variable cells:* C11:D11; *Constraints:* H11 = 0) at the bottom right of Figure 8.10c and then $Q_{culvert}$ is computed using the Solver dialog box (*Set Objective:* I22; *To Value of*: 0; *Variable cells:* A22, C22:D22; *Constraints:* A22 = A19, H22 = 0) at the bottom left. The variables to be adjusted are highlighted in red color. Their initial values should satisfy Eq. (2.37).

$$Q_{culvert}/(A_{culvert}D^{0.5}) = (3.48081\,\text{cfs})/[(1.22718\,\text{ft}^2) * (1.25\,\text{ft})^{0.5}] = 2.53697 \le 3.5 \rightarrow \text{done!}$$

(4) $H_{wdp} = 5.0$ ft, water flows into the riser through both the orifice and weir.

$H_{wdp} - h_{weir,2} = 5.0\,\text{ft} - 1.5\,\text{ft} = 3.5\,\text{ft} > 1.5\,\text{ft} \rightarrow$ orificeflow.

$H_{wdp} = 5.0\,\text{ft} > h_{weir,1} = 3.5\,\text{ft} \rightarrow$ weirflow.

Solver (Figure 8.10d) $\rightarrow Q_{riser} = 11.32006$ cfs, HW = 3.14 ft, $Q_{culvert} = 10.64242$ cfs. $H_{riser} = 3.24\,\text{ft} > 1.5\,\text{ft} \rightarrow$ orifice is submerged and its head = 5.0 – 3.24 ft = 1.76 ft.

$H_{riser} = 3.24$ ft ≤ 3.5 ft→weir is unsubmerged and its head = 5.0 − 3.5 ft = 1.5 ft.

$Q_{orifice}$ (cells C13:D13) and Q_{weir} (cells C17:D17) are computed using Eq. (6.15), while $Q_{culvert}$ is computed using the Solver dialog box (*Set Objective:* I22; *To Value of*: 0; *Variable cells:* A22, C22:D22; *Constraints:* A22 = A19, H22 = 0) at the bottom left of Figure 8.10d. Herein, Solver automatically adjusts H_{riser} and recalculates $Q_{orifice}$ until the objective and constraints are satisfied. The variables to be adjusted are highlighted in red color. Their initial values should satisfy Eq. (2.37).

$$Q_{culvert}/(A_{culvert}D^{0.5}) = (10.64242\,\text{cfs})/[(1.22718\,\text{ft}^2) * (1.25\,\text{ft})^{0.5}] = 7.75666 > 4.0$$

→ recalculate Q_{riser}, HW, and $Q_{culvert}$ using Eq. (8.1) in the Solver dialog box (*Set Objective:* K22; *To Value of*: 0; *Variable cells:* A22, C22; *Constraints:* A22 = A19) at the bottom right of Figure 8.10d.

→ $Q_{riser} = 11.32006$ cfs, HW = 3.14 ft, $Q_{culvert} = 10.64242$ cfs

→ $Q_{culvert}/(A_{culvert}D^{0.5}) = (10.64242\,\text{cfs})/[(1.22718\,\text{ft}^2) * (1.25\,\text{ft})^{0.5}] = 7.75666 > 4.0$

→ done!

Note that cell K22 cannot be made zero because the large time step of $\Delta t = 60$ sec causes large computational errors. The final results are: $Q_{riser} = 11.32005$ cfs, HW = 3.14 ft, $Q_{culvert} = 10.64242$ cfs.

(5) $H_{wdp} = 6.5$ ft, water flows into the riser through the orifice, weir, and top opening.
$H_{wdp} - h_{weir,2} = 6.5$ ft − 1.5 ft = 5.0 ft > 1.5 ft→orifice flow.
$H_{wdp} = 6.5$ ft > $h_{weir,1} = 3.5$ ft→weirflow.
$H_{wdp} = 6.5$ ft > $h_{riser} = 6$ ft→topopeningflow.

Solver (Figure 8.10e) →$Q_{riser} = 17.91614$ cfs, HW = 6.01 ft, $Q_{culvert} = 16.65699$ cfs.
$H_{riser} = 6.01$ ft > 1.5 ft→orifice is submerged and its head = 6.5 − 6.01 ft = 0.49 ft.
$H_{riser} = 6.01$ ft > 3.5 ft→weir is submerged and its head = 6.5 − 6.01 ft = 0.49 ft.
The head on the top opening = 6.5 − 6.01 ft = 0.49 ft.
$Q_{orifice}$ (cells C13:D13), Q_{weir} (cells C17:D17), and Q_{top} (cells H17:I17) are computed using Eq. (6.15), while $Q_{culvert}$ is computed using the Solver dialog box (*Set Objective:* I22; *To Value of*: 0; *Variable cells:* A22, C22:D22; *Constraints:* A22 = A19, H22 = 0) at the bottom left of Figure 8.10e. Herein, Solver automatically adjusts H_{riser} and recalculates $Q_{orifice}$ until the objective and constraints are satisfied. The variables to be adjusted are highlighted in red color. Their initial values should satisfy Eq. (2.37).

$$Q_{culvert}/(A_{culvert}D^{0.5}) = (16.65699\,\text{cfs})/[(1.22718\,\text{ft}^2) * (1.25\,\text{ft})^{0.5}] = 12.14036 > 4.0$$

→ recalculate Q_{riser}, HW, and $Q_{culvert}$ using Eq. (8.1) in the Solver dialog box (*Set Objective:* K22; *To Value of*: 0; *Variable cells:* A22, C22; *Constraints:* A22 = A19) at the bottom right of Figure 8.10e.

→ $Q_{riser} = 15.20573$ cfs, HW = 6.07 ft, $Q_{culvert} = 13.93484$ cfs

$\rightarrow Q_{culvert}/(A_{culvert}D^{0.5}) = (13.93484\,cfs)/[(1.22718\,ft^2) * (1.25\,ft)^{0.5}] = 10.15634 > 4.0$

$\rightarrow$ done!

The final results are: $Q_{riser} = 15.20573$ cfs, HW = 6.07 ft, $Q_{culvert} = 13.93484$ cfs.

8.4 Storm Sewers

A storm sewer is designed to drain runoff from impervious surfaces, such as paved roads and streets, parking lots, sidewalks, and building roofs. The runoff is collected and conveyed by the gutter along the curb of a road or street, flows through a drain inlet downward into a drainage culvert, and empties into a stream or waterbody. The hydraulics of culverts has been presented in Section 8.1, so this section focuses on analyses of gutter flow and inlet interception capacity.

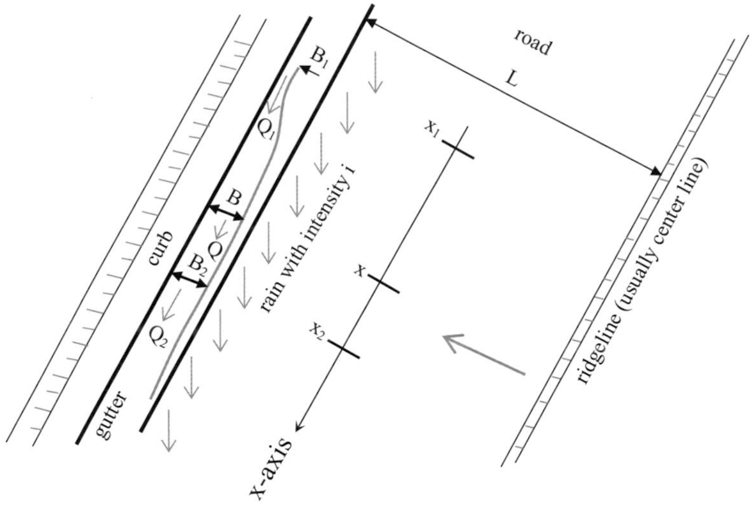

(a)

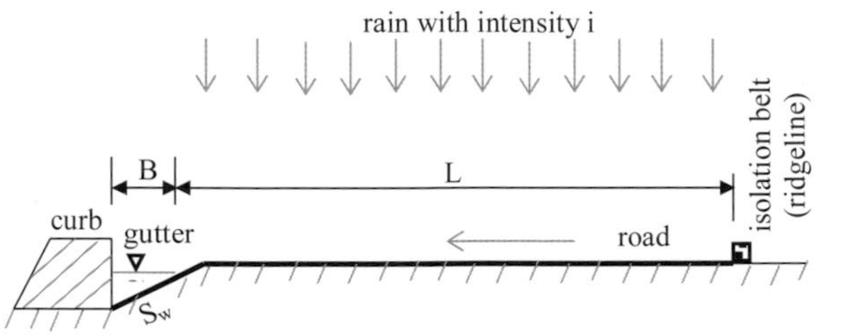

(b)

Figure 8.11 Sketches of a segment of a road with bordering gutter. (a) Bird's-eye view; and (b) transverse cross-sectional view. Notice in part (a) that the x-axis is oriented parallel to the gutter flow direction for analysis purposes.

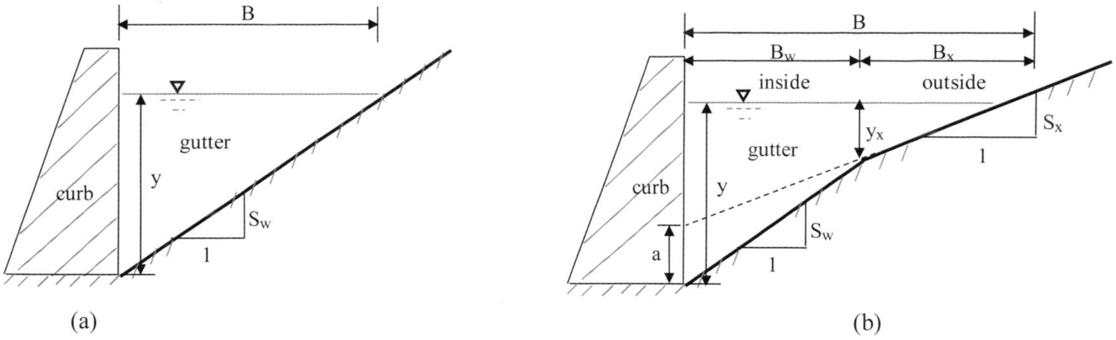

Figure 8.12 Transverse cross-sectional sketches of two types of gutters. (a) Simple gutter; and (b) composite gutter (inside plus outside). a: gutter depression; and B: spread of water.

8.4.1 Gutter Flows

A gutter is a concrete-lined triangular channel (e.g., Figure 8.11). Its side next to the curb is usually vertical, whereas the side away from the curb may have a single slope or compound slopes (Figure 8.12). That is, the gutter can have either a simple cross section, with a single side slope, or a composite cross section, with compound side slopes. The top width of water measured laterally from the curb is defined as the spread of water. Gutter flow is commonly assumed to be uniform and thus the bed slope can be used in the Manning formula (Eq. (2.26)).

For a simple gutter cross section (Figure 8.12a), the discharge can be computed using a rearrangement of the Manning formula (see Example 8.13), which can be expressed as:

$$Q = \frac{K}{n} \frac{B^{\frac{8}{3}} S_w^{\frac{5}{3}} S_0^{\frac{1}{2}}}{2^{\frac{5}{3}} \left(S_w + \sqrt{1 + S_w^2}\right)^{\frac{2}{3}}} \tag{8.7}$$

units conversion factor (= 1.486 for BG; 1.0 for SI) → K
discharge [cfs for BG, $m^3 s^{-1}$ for SI] → Q
Manning's roughness → n
spread of water [ft for BG; m for SI] → B
gutter bed slope → S_0
gutter side slope → S_w

For a compound gutter cross section (Figure 8.12b), the discharge can be computed using a rearrangement of the Manning formula (see Example 8.14), which can be expressed as:

$$Q = \frac{K}{n} \frac{\left[B^2 S_x + (B - B_x) a\right]^{\frac{5}{3}} S_0^{\frac{1}{2}}}{2^{\frac{5}{3}} \left[\left(\sqrt{1 + S_w^2} + S_x\right) B - \left(\sqrt{1 + S_w^2} - \sqrt{1 + S_x^2}\right) B_x + a\right]^{\frac{2}{3}}} \tag{8.8}$$

units conversion factor (= 1.486 for BG; 1.0 for SI) → K
discharge [cfs for BG, $m^3 s^{-1}$ for SI] → Q
Manning's roughness → n
spread of water [ft for BG; m for SI] → B
top width of water in outside cross section → B_x
gutter bed slope → S_0
side slope of inside cross section → S_w
side slope of outside cross section → S_x
gutter depression (Figure 8.12b) → a

Note that if $B \leq B_w$ (i.e., the top channel width of the inside cross section), the top water width of the outside cross section $B_x = 0$ and Eq. (8.8) will be reduced to Eq. (8.7) as shown in Example 8.15.

Example 8.13 Derive Eq. (8.7).

Solution

Reference Figure 8.12a.

Water depth in the gutter: $y = BS_w$.

Flow area: $A = 1/2\,B(BS_w) = B^2S_w/2$.

Wetted perimeter: $P_{wet} = BS_w + [B^2 + (BS_w)^2]^{1/2} = [S_w + (1 + S_w{}^2)^{1/2}]B$.

Hydraulic radius: $R = (B^2S_w/2)/\{[S_w + (1 + S_w{}^2)^{1/2}]B\} = (BS_w/2)/[S_w + (1 + S_w{}^2)^{1/2}]$.

Eq. (2.26): $V = (K/n)\{(BS_w/2)/[S_w + (1 + S_w{}^2)^{1/2}]\}^{2/3}S_0^{1/2}$.

Discharge: $Q = VA = (K/n)\{(BS_w/2)/[S_w + (1 + S_w{}^2)^{1/2}]\}^{2/3}S_0{}^{1/2}(B^2S_w/2)$

$$= (K/n)(B^{8/3})(S_w{}^{5/3})(S_0{}^{1/2})/\{(2^{5/3})[S_w + (1 + S_w{}^2)^{1/2}]^{2/3}\} \rightarrow \text{Eq. (8.7)}.$$

Example 8.14 Derive Eq. (8.8).

Solution

Reference Figure 8.12b.

Outside cross section: water depth: $y_x = B_xS_x$

flow area: $A_1 = 1/2\,B_x(B_xS_x) = (B_x{}^2S_x)/2$

wetted perimeter: $P_{wet,1} = [B_x{}^2 + (B_xS_x)^2]^{1/2} = (1 + S_x{}^2)^{1/2}B_x$.

Inside cross section: water depth: $y = BS_x + a$

$$\text{flow area: } A_2 = 1/2([B_xS_x + (BS_x + a)](B - B_x)$$
$$= 1/2[(B + B_x)S_x + a](B - B_x)$$
$$= (B^2S_x)/2 - (B_x{}^2S_x)/2 + a(B - B_x)/2$$

$$\text{wetted perimeter: } P_{wet,2} = (BS_x + a) + \{(B - B_x)^2 + [(B - B_x)S_w]^2]^{1/2}$$
$$= BS_x + a + (1 + S_w{}^2)^{1/2}B - (1 + S_w{}^2)^{1/2}B_x.$$

Entire cross section: flow area: $A = A_1 + A_2 = [B^2S_x + a(B - B_x)]/2$

$$\text{wetted perimeter: } P_{wet} = P_{wet,1} + P_{wet,2}$$
$$= [(1 + S_w{}^2)^{1/2} + S_x]B - [(1 + S_w{}^2)^{1/2} - (1 + S_x{}^2)^{1/2}]B_x + a.$$

Discharge: $Q = VA = (K/n)(A^{5/3})/(P_{wet}{}^{2/3})(S_0{}^{1/2})$

$= (K/n)\{[B^2S_x + a(B - B_x)]^{5/3}/(2^{5/3})\}/\{[(1 + S_w{}^2)^{1/2} + S_x]B - [(1 + S_w{}^2)^{1/2} - (1 + S_x{}^2)^{1/2}]B_x + a\}^{2/3}(S_0{}^{1/2}) \rightarrow Q = (K/n)[B^2S_x + a(B - B_x)]^{5/3}(S_0{}^{1/2})/(2^{5/3})/\{[(1 + S_w{}^2)^{1/2} + S_x]B - [(1 + S_w{}^2)^{1/2} - (1 + S_x{}^2)^{1/2}]B_x + a\}^{2/3} \rightarrow$ Eq. (8.8).

Example 8.15 Show that if $B_x = 0$, Eq. (8.8) will be reduced to Eq. (8.7).

Solution

If $B_x = 0$, $S_x = S_w$, $a = 0 \rightarrow$ Eq. (8.8) is simplified as:

$$Q = (K/n)(B^2S_w)^{5/3}(S_0{}^{1/2})/(2^{5/3})/\{[(1 + S_w{}^2)^{1/2} + S_w]B\}^{2/3}$$
$$\rightarrow Q = (K/n)(B^{8/3})(S_w)^{5/3}(S_0{}^{1/2})/\{(2^{5/3})[S_w + (1 + S_w{}^2)^{1/2}]^{2/3}\} \rightarrow \text{Eq. (8.7).}$$

Example 8.16 For a gutter with a composite cross section such as that shown in Figure 8.12b, its side slopes are $S_w = 0.1$ and $S_x = 0.02$. The gutter is laid on a slope of $S_0 = 0.03$ and has a Manning's n of 0.015. Determine the gutter flow rate if:

(1) $B_w = 0.7$ ft, $B_x = 0.3$ ft
(2) $B_w = 0.5$ ft, $B_x = 0.0$ ft.

Solution

$S_w = 0.1$, $S_x = 0.02$, $S_0 = 0.03$, $n = 0.015$

(1) $B_w = 0.7$ ft, $B_x = 0.3$ ft $\rightarrow B = B_w + B_x = 0.7$ ft $+ 0.3$ ft $= 1.0$ ft

Gutter depression: $a = B_w(S_w - S_x) = (0.7\text{ ft}) * (0.1 - 0.02) = 0.056$ ft.

Eq. (8.8): $Q = (1.486/0.015) * [(1.0\text{ ft})^2 * 0.02 + (1.0\text{ ft} - 0.3\text{ ft}) * (0.056\text{ ft})]^{5/3}(0.03^{1/2})/(2^{5/3})/\{[(1 + 0.1^2)^{1/2} + 0.02] * (1.0\text{ ft}) -](1 + 0.1^2)^{1/2} - (1 + 0.02^2)^{1/2}] * (0.3\text{ ft}) + 0.056\text{ ft}\}^{2/3} = 0.046$ cfs.

(2) $B_w = 0.5$ ft, $B_x = 0.0$ ft $\rightarrow B = B_w + B_x = 0.5$ ft $+ 0.0$ ft $= 0.5$ ft

Eq. (8.7): $Q = (1.486/0.015) * [(0.5\text{ ft})^{8/3}] * (0.1^{5/3}) * (0.03^{1/2})/\{(2^{5/3}) * [0.1 + (1 + 0.1^2)^{1/2}]^{2/3}\} = 0.017$ cfs.

In reality, it is a reasonable assumption that the travel time of runoff from the road or street surface into the gutter is minimal and thus can be neglected. In the following, this assumption is accepted. For a gutter with a simple cross section, during a spatiotemporally uniform storm event with an intensity i, as shown in Example 8.17, the travel time of the gutter flow from location x_1 to x_2 (Figure 8.11a) can be estimated as:

$$t_{x_1 \to x_2} = \frac{2^{\frac{8}{3}}}{3} \frac{n}{K S_0^{\frac{1}{2}}} \left(1 + \sqrt{1 + \frac{1}{S_w^2}}\right)^{\frac{2}{3}} \frac{\left(\frac{B_2}{B_1}\right)^2 - 1}{B_1^{\frac{2}{3}} \left[\left(\frac{B_2}{B_1}\right)^{\frac{8}{3}} - 1\right]} (x_2 - x_1) \tag{8.9}$$

- $t_{x_1 \to x_2}$: time of gutter flow from location x_1 to x_2 [sec]
- n: Manning's n
- K: units conversion factor (= 1.486 for BG; 1.0 for SI)
- S_0: gutter bed slope
- S_w: gutter side slope
- B_2: spread of water at location x_2 [ft for BG; m for SI]
- B_1: spread of water at location x_1 [ft for BG; m for SI]
- x_2: downstream location
- x_1: upstream location

Similarly, for a gutter with a compound cross section, if water only flows in the inside cross section (i.e., $B \leq B_w$), the travel time of the gutter flow can be estimated using Eq. (8.9). If $B > B_w$, as shown in Example 8.18, the travel time can be estimated as:

$$t_{x_1 \to x_2} = \frac{x_2 - x_1}{Q_2 - Q_1} \left\{ \frac{2}{3} S_x \left(B_2^2 - B_1^2\right) + \frac{m_1 S_x}{3 m_2} (B_2 - B_1) - \frac{m_1^2 S_x + m_2^2 B_w a}{3 m_2^2} \left(\ln \frac{m_1 + m_2 B_2}{m_1 + m_2 B_1} \right) \right\} \tag{8.10}$$

- $t_{x_1 \to x_2}$: time of gutter flow from location x_1 to x_2
- x_2: downstream location
- x_1: upstream location
- Q_2: discharge at location x_2 computed using Eq. (8.8) with B_2 [ft for BG; m for SI]
- Q_1: discharge at location x_1 computed using Eq. (8.8) with B_1 [ft for BG; m for SI]
- S_x: side slope of outside cross section
- B_2: spread of water at location x_2 [ft for BG; m for SI]
- B_1: spread of water at location x_1 [ft for BG; m for SI]
- m_2: coefficient (Eq. (8.10b))
- m_1: coefficient (Eq. (8.10a))

$$m_1 = \left(\sqrt{1 + S_w^2} - \sqrt{1 + S_x^2}\right) B_w + a \tag{8.10a}$$

- m_1: coefficient in Eq. (8.10)
- S_w: side slope of inside cross section
- S_x: side slope of outside cross section
- B_w: top channel top width of inside cross section (Figure 8.12b)
- a: gutter depression (Figure 8.12b)

$$m_2 = S_x + \sqrt{1 + S_x^2} \tag{8.10b}$$

- m_2: coefficient in Eq. (8.10)
- S_x: side slope of outside cross section

Regardless of the cross section type of a gutter, the longitudinally averaged flow velocity in the gutter is computed as:

$$\overline{V}_{x_1 \to x_2} = \frac{x_2 - x_1}{t_{x_1 \to x_2}} \tag{8.11}$$

- $\overline{V}_{x_1 \to x_2}$: longitudinally averaged velocity
- x_2: downstream location
- x_1: upstream location
- $t_{x_1 \to x_2}$: time of gutter flow from location x_1 to x_2 (Eq. (8.9) or (8.10))

Example 8.17 Derive Eq. (8.9).

Solution

Reference Figures 8.11a and 8.12a.

For an elemental length of the gutter, dx, in terms of the continuity equation, the discharge increase is equal to the runoff from the road or street surface within dx. That is:

$dQ = iLdx \rightarrow$ given $dx = Vdt$, $dQ = iLVdt$.

Example 8.13 $\rightarrow V = \frac{K}{n} \frac{B^{\frac{2}{3}} S_w^{\frac{2}{3}} S_0^{\frac{1}{2}}}{2^{\frac{2}{3}} \left[S_w + \sqrt{1+S_w^2}\right]^{\frac{2}{3}}}$.

Eq. (8.7): total derivative of $Q \rightarrow dQ = \frac{K}{n} \frac{8}{3} \frac{S_w^{\frac{5}{3}} S_0^{\frac{1}{2}}}{2^{\frac{5}{3}} \left[S_w + \sqrt{1+S_w^2}\right]^{\frac{2}{3}}} B^{\frac{5}{3}} dB$.

Substituting V and dQ back into the continuity equation and simplifying, one can get:

$$\frac{4}{3} S_w BdB = iLdt \rightarrow \frac{4}{3} S_w \int_{B_1}^{B_2} BdB = iL \int_0^{t_{x_1 \rightarrow x_2}} dt \rightarrow t_{x_1 \rightarrow x_2} = \frac{2}{3} \frac{S_w \left[B_2^2 - B_1^2\right]}{iL} = \frac{2}{3} \frac{S_w B_1^2 \left[\left(\frac{B_2}{B_1}\right)^2 - 1\right]}{iL}$$

Eq. (8.7): $Q_1 = \frac{K}{n} \frac{B_1^{\frac{8}{3}} S_w^{\frac{5}{3}} S_0^{\frac{1}{2}}}{2^{\frac{5}{3}} \left(S_w + \sqrt{1+S_w^2}\right)^{\frac{2}{3}}}$, $Q_2 = \frac{K}{n} \frac{B_2^{\frac{8}{3}} S_w^{\frac{5}{3}} S_0^{\frac{1}{2}}}{2^{\frac{5}{3}} \left(S_w + \sqrt{1+S_w^2}\right)^{\frac{2}{3}}}$

$$iL = \frac{Q_2 - Q_1}{x_2 - x_1} = \frac{K}{n} \frac{S_w^{\frac{5}{3}} S_0^{\frac{1}{2}}}{2^{\frac{5}{3}} \left(S_w + \sqrt{1+S_w^2}\right)^{\frac{2}{3}}} \frac{B_2^{\frac{8}{3}} - B_1^{\frac{8}{3}}}{x_2 - x_1} = \frac{K}{n} \frac{S_w^{\frac{5}{3}} S_0^{\frac{1}{2}}}{2^{\frac{5}{3}} \left(S_w + \sqrt{1+S_w^2}\right)^{\frac{2}{3}}} \frac{B_1^{\frac{8}{3}} \left[\left(\frac{B_2}{B_1}\right)^{\frac{8}{3}} - 1\right]}{x_2 - x_1}.$$

Substituting iL and simplifying, one can get:

$$t_{x_1 \rightarrow x_2} = \frac{2}{3} S_w B_1^2 \left[\left(\frac{B_2}{B_1}\right)^2 - 1\right] \frac{n}{K} \frac{2^{\frac{5}{3}} \left(S_w + \sqrt{1+S_w^2}\right)^{\frac{2}{3}}}{S_w^{\frac{5}{3}} S_0^{\frac{1}{2}}} \frac{x_2 - x_1}{B_1^{\frac{8}{3}} \left[\left(\frac{B_2}{B_1}\right)^{\frac{8}{3}} - 1\right]} \rightarrow \text{Eq. (8.9)}.$$

Example 8.18 Derive Eq. (8.10) if $B > B_w$.

Solution

Reference Figures 8.11a and 8.12b.

For an elemental length of the gutter, dx, in terms of the continuity equation, the discharge increase is equal to the runoff from the road or street surface within dx. That is:

$dQ = iLdx \rightarrow$ given $dx = Vdt$, $dQ = iLVdt$

Example 8.14 $\rightarrow V = \frac{Q}{A} = \frac{K}{n} \frac{\left[B^2 S_x + B_w a\right]^{\frac{2}{3}} S_0^{\frac{1}{2}}}{2^{\frac{2}{3}} \left[\left(\sqrt{1+S_w^2} + S_x\right) B - \left(\sqrt{1+S_w^2} - \sqrt{1+S_x^2}\right) B_x + a\right]^{\frac{2}{3}}}$.

Eq. (8.8): $B - B_x = B_w$, which is a constant, total derivative of Q→

$$dQ = \frac{K}{n} \frac{S_0^{\frac{1}{2}}}{2^{\frac{5}{3}}\left[\left(\sqrt{1+S_w^2}+S_x\right)B - \left(\sqrt{1+S_w^2}-\sqrt{1+S_x^2}\right)B_x + a\right]^{\frac{2}{3}}} \left\{ \frac{5}{3}\left[B^2S_x + B_w a\right]^{\frac{2}{3}}(2BS_x) \right.$$

$$+ \frac{\left[B^2S_x + B_w a\right]^{\frac{5}{3}}\left(-\frac{2}{3}\right)}{\left[\left(\sqrt{1+S_w^2}+S_x\right)B - \left(\sqrt{1+S_w^2}-\sqrt{1+S_x^2}\right)(B-B_w) + a\right]^{\frac{1}{3}}} \left[\left(\sqrt{1+S_w^2}+S_x\right)\right.$$

$$\left.\left. - \left(\sqrt{1+S_w^2}-\sqrt{1+S_x^2}\right)\right]\right\} dB$$

$$= \frac{K}{n} \frac{\left[B^2S_x + B_w a\right]^{\frac{2}{3}} S_0^{\frac{1}{2}}}{2^{\frac{2}{3}}\left[\left(\sqrt{1+S_w^2}+S_x\right)B - \left(\sqrt{1+S_w^2}-\sqrt{1+S_x^2}\right)B_x + a\right]^{\frac{2}{3}}} \left\{ \frac{5}{3}S_x B \right.$$

$$\left. - \frac{B^2S_x + B_w a}{3\left[\left(S_x+\sqrt{1+S_x^2}\right)B + \left(\sqrt{1+S_w^2}-\sqrt{1+S_x^2}\right)B_w + a\right]}\left(S_x + \sqrt{1+S_x^2}\right)\right\} dB.$$

Substituting V and dQ back into the continuity equation and simplifying, one can get:

$$\left\{ \frac{5}{3}S_x B - \frac{B^2S_x + B_w a}{3\left[\left(S_x+\sqrt{1+S_x^2}\right)B + \left(\sqrt{1+S_w^2}-\sqrt{1+S_x^2}\right)B_w + a\right]}\left(S_x + \sqrt{1+S_x^2}\right)\right\} dB$$

$$= iLdt$$

$$\rightarrow \text{let } m_1 = \left(\sqrt{1+S_w^2}-\sqrt{1+S_x^2}\right)B_w + a,\ m_2 = S_x + \sqrt{1+S_x^2}$$

$$\left\{ \frac{5}{3}S_x B - \frac{m_2 S_x}{3}\frac{B^2}{m_1+m_2B} - \frac{m_2 B_w a}{3}\frac{1}{m_1+m_2B}\right\} dB = iLdt$$

$$\rightarrow \int_{B_1}^{B_2}\left\{ \frac{5}{3}S_x B - \frac{m_2 S_x}{3}\frac{B^2}{m_1+m_2B} - \frac{m_2 B_w a}{3}\frac{1}{m_1+m_2B}\right\} dB = iL\int_0^{t_{x_1\rightarrow x_2}} dt$$

$$\rightarrow \left\{ \frac{5}{3}S_x\frac{1}{2}B^2 \right.$$

$$\left. - \frac{m_2S_x}{3}\frac{1}{m_2^3}\left[-\frac{m_2B}{2}(2m_1 - m_2B) + m_1^2\ln(m_1+m_2B)\right] - \frac{m_2B_wa}{3}\frac{1}{m_2}\ln(m_1+m_2B)\right\}\Bigg|_{B_1}^{B_2}$$

$$= iLt_{x_1\rightarrow x_2}$$

$$\rightarrow \left\{ \frac{5}{6}S_xB^2 + \frac{m_1S_x}{3m_2}B - \frac{1}{6}S_xB^2 - \frac{m_1^2S_x}{3m_2^2}\ln(m_1+m_2B) - \frac{B_wa}{3}\ln(m_1+m_2B)\right\}\Bigg|_{B_1}^{B_2} = iLt_{x_1\rightarrow x_2}$$

$$\rightarrow \left\{ \frac{2}{3}S_xB^2 + \frac{m_1S_x}{3m_2}B - \frac{m_1^2S_x + m_2^2B_wa}{3m_2^2}\ln(m_1+m_2B)\right\}\Bigg|_{B_1}^{B_2} = iLt_{x_1\rightarrow x_2}$$

$$\rightarrow \left\{\frac{2}{3}S_xB^2 + \frac{m_1S_x}{3m_2}B - \frac{m_1^2S_x + m_2^2B_wa}{3m_2^2}\ln(m_1 + m_2B)\right\}\Bigg|_{B_1}^{B_2} = iLt_{x_1\rightarrow x_2}$$

$$\rightarrow t_{x_1\rightarrow x_2} = \frac{1}{iL}\left\{\frac{2}{3}S_x\left(B_2^2 - B_1^2\right) + \frac{m_1S_x}{3m_2}(B_2 - B_1) - \frac{m_1^2S_x + m_2^2B_wa}{3m_2^2}\left(\ln\frac{m_1 + m_2B_2}{m_1 + m_2B_1}\right)\right\}.$$

Substituting $iL = \frac{Q_2 - Q_1}{x_2 - x_1}$ and rearranging, one can get:

$$t_{x_1\rightarrow x_2} = \frac{x_2 - x_1}{Q_2 - Q_1}\left\{\frac{2}{3}S_x\left(B_2^2 - B_1^2\right) + \frac{m_1S_x}{3m_2}(B_2 - B_1) - \frac{m_1^2S_x + m_2^2B_wa}{3m_2^2}\left(\ln\frac{m_1 + m_2B_2}{m_1 + m_2B_1}\right)\right\} \rightarrow \text{Eq. (8.10)}.$$

Example 8.19 For a gutter with a composite cross section such as that shown in Figure 8.12b, its side slopes are $S_w = 0.1$ and $S_x = 0.02$. The top width of the inside cross section is $B_w = 1.0$ ft. The gutter is laid on a slope of $S_0 = 0.03$ and has a Manning's n of 0.015. For two locations 200 ft apart, determine the time and longitudinally averaged velocity of gutter flow if the spreads of water at the upstream and downstream locations are:

(1) $B_1 = 0.5$ ft and $B_2 = 0.8$ ft

(2) $B_1 = 1.2$ ft and $B_2 = 2.0$ ft.

Solution

Reference to Figures 8.11 and 8.12b.

$S_w = 0.1$, $S_x = 0.02$, $B_w = 1.0$ ft, $S_0 = 0.03$, $n = 0.015$, $x_2 - x_1 = 200$ ft

(1) $B_1 \le B_w$ and $B_2 \le B_w \rightarrow$ Eqs. (8.9) and (8.11)

$$t_{x_1\rightarrow x_2} = \frac{2^{\frac{8}{3}}}{3} * \frac{0.015}{1.486 * 0.03^{\frac{1}{2}}} * \left(1 + \sqrt{1 + \frac{1}{0.1^2}}\right)^{\frac{2}{3}} * \frac{\left(\frac{0.8\text{ ft}}{0.5\text{ ft}}\right)^2 - 1}{(0.5\text{ ft})^{\frac{2}{3}} * \left[\left(\frac{0.8\text{ ft}}{0.5\text{ ft}}\right)^{\frac{8}{3}} - 1\right]} * (200\text{ ft})$$

$$= 121.13\text{ sec} = 2.02\text{ min}$$

$$\overline{V}_{x_1\rightarrow x_2} = \frac{200\text{ ft}}{121.13\text{ sec}} = 1.65\text{ ft sec}^{-1}.$$

(2) $B_1 > B_w$ and $B_2 > B_w \rightarrow$ Eqs. (8.10), (8.10a), (8.10b), and (8.11)
Gutter depression: $a = B_w(S_w - S_x) = (1.0\text{ ft}) * (0.1 - 0.02) = 0.08$ ft.

Eq. (8.8): Q_1

$$= \frac{1.486}{0.015} * \frac{\left[(1.2\,\text{ft})^2 * 0.02 + (1.0\,\text{ft}) * (0.08\,\text{ft})\right]^{\frac{5}{3}} * 0.03^{\frac{1}{2}}}{2^{\frac{5}{3}} * \left[\left(\sqrt{1+0.1^2}+0.02\right) * (1.2\,\text{ft}) - \left(\sqrt{1+0.1^2}-\sqrt{1+0.02^2}\right) * (1.2\,\text{ft} - 1.0\,\text{ft}) + (0.08\,\text{ft})\right]^{\frac{2}{3}}}$$

$$= 0.112\,\text{cfs}$$

Q_2

$$= \frac{1.486}{0.015} * \frac{\left[(2.0\,\text{ft})^2 * 0.02 + (1.0\,\text{ft}) * (0.08\,\text{ft})\right]^{\frac{5}{3}} * 0.03^{\frac{1}{2}}}{2^{\frac{5}{3}} * \left[\left(\sqrt{1+0.1^2}+0.02\right) * (2.0\,\text{ft}) - \left(\sqrt{1+0.1^2}-\sqrt{1+0.02^2}\right) * (2.0\,\text{ft} - 1.0\,\text{ft}) + (0.08\,\text{ft})\right]^{\frac{2}{3}}}$$

$$= 0.154\,\text{cfs}.$$

Eq. (8.10a): $m_1 = \left(\sqrt{1+0.1^2} - \sqrt{1+0.02^2}\right)(1.0\,\text{ft}) + 0.08\,\text{ft} = 0.084788\,\text{ft}.$
Eq. (8.10b): $m_2 = 0.02 + \sqrt{1+0.02^2} = 1.02020.$

$$\text{Eq. (8.10): } t_{x_1 \to x_2} = \frac{200\,\text{ft}}{0.154\,\text{cfs} - 0.112\,\text{cfs}} * \left\{ \begin{array}{l} \frac{2}{3} * 0.02 * \left(2.0^2 - 1.2^2\right)\,\text{ft}^2 + \frac{(0.084788\,\text{ft}) * 0.02}{3 * 1.02020} * (2.0\,\text{ft} - 1.2\,\text{ft}) - \\ \frac{(0.084788\,\text{ft})^2 * 0.02 + 1.02020^2 * (1.0\,\text{ft}) * (0.08\,\text{ft})}{3 * 1.02020^2} * \\ \ln\left(\frac{0.084788\,\text{ft} + 1.02020 * 2.0\,\text{ft}}{0.084788\,\text{ft} + 1.02020 * 1.2\,\text{ft}}\right) \end{array} \right\}$$

$$= 103.01\,\text{sec} = 1.72\,\text{min}$$

$$\text{Eq. (8.11): } \overline{V}_{x_1 \to x_2} = \frac{200\,\text{ft}}{103.01\,\text{sec}} = 1.94\,\text{ft sec}^{-1}.$$

8.4.2 Drain Inlets

A drain inlet collects overland runoff and discharges it into a drainage pipe. In practice, four types of drain inlets are used (Figure 8.13), namely grate, curb-opening, combination, and slotted. A grate inlet is an opening in the gutter covered by grates that are parallel to the flow direction, whereas a curb-opening inlet is a vertical opening in the curb covered by a top slab (Johnson and Chang, 1984; Mays, 2005). A combination inlet consists of a grate inlet and a curb-opening inlet that are placed side by side. A slotted inlet is a cut along the gutter with spacer bars to form openings. For sag locations (i.e., low or depressed areas in which water collects), because passing debris is somewhat critical (Mays, 2005), grate and slotted inlets are not appropriate due to their clogging tendencies, whereas curb-opening and combination inlets are recommended instead.

Regardless of the type of inlet, the most important hydraulic design parameter is its efficiency, which is defined as the ratio of the maximum flow that can be intercepted by the inlet (i.e., interception capacity) to the total discharge in the gutter. Mathematically, the efficiency is computed as:

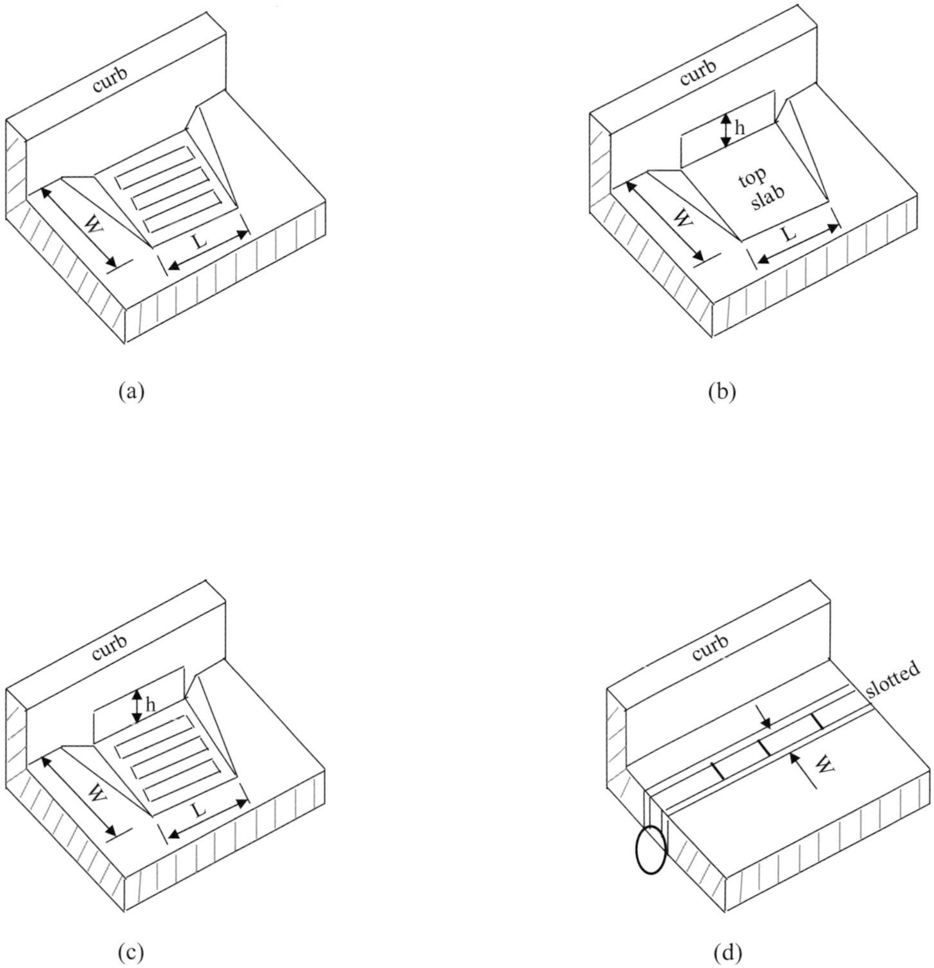

Figure 8.13 Types of drain inlets. Diagram of: (a) grate inlet; (b) curb-opening inlet; (c) combination inlet; and (d) slotted inlet. (*Source:* after Johnson and Chang, 1984)

inlet interception capacity → Q_{di}; inlet efficiency → η_{di}; total discharge in gutter → Q

$$\eta_{di} = \frac{Q_{di}}{Q} \tag{8.12}$$

The bypass fraction, essentially the fraction not captured by the inlet, is computed as:

bypass fraction → η_{bp}; inlet efficiency (Eq. (8.12)) → η_{di}

$$\eta_{bp} = 1 - \eta_{di} \tag{8.13}$$

For a given inlet, if the velocity of the gutter flow, V, is smaller than the "splash-over velocity" of the inlet, V_0, the inlet will intercept all water in the gutter with no bypass flow, that is, $\eta_{di} = 1$ and $\eta_{bp} = 0$. V_0 is the possible maximum velocity at which water can flow into the inlet. On the other hand, if $V > V_0$, some of the total discharge in the gutter will bypass the inlet, that is, $\eta_{di} < 1$ and $\eta_{bp} > 0$. When the spread of water in the gutter is larger than the inlet width (Figures 8.12 and 8.13), that is, $B > W$, the inlet will intercept both the flow over the inlet, termed frontal flow, and the flow beyond the inlet, referred to as side flow.

Grate inlets can be classified into a number of subtypes in terms of shape, spacing, and tilting angle of grate bars. The detailed classifications and their illustrations are available in Johnson and Chang (1984). Among them, the P-1-7/8 grate inlets have a bar spacing of 1-7/8 in on center, and the P-1-7/8-4 grate inlets have a bar spacing of 1-7/8 in on center and 3/8 in diameter lateral rods spaced at 4 in on center. The CV-3-3/4-4-1/4 grate inlets have a curved vane grate with 3-3/4 in longitudinal bars and 4-1/4 in transverse bar spacing on center. The 30-3-1/4 grate inlets have a 30° tilt-bar grate with 3-1/4 in on center longitudinal and lateral bar spacings, respectively. The reticuline grate inlets have a honeycomb pattern of lateral bars and longitudinal bearing bars. The values of V_0 for the common grate inlets can be determined from Figure 8.14. The interception efficiency of grate inlets is computed as:

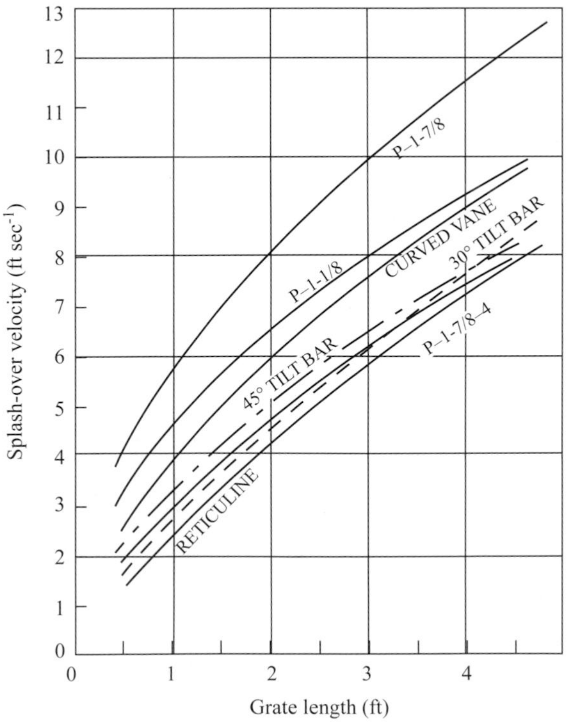

Figure 8.14 Grate inlet nomogram of splash-over velocity, V_0. (*Source:* Johnson and Chang, 1984)

(grate inlet interception efficiency; frontal-flow interception efficiency; frontal flow rate; side-flow interception efficiency; side flow rate)

$$\eta_{di} = \frac{R_f Q_f + R_s Q_s}{Q_f + Q_s} \tag{8.14}$$

R_f and R_s can be estimated as (Johnson and Chang, 1984):

(frontal-flow interception efficiency of grate inlet; mean velocity of gutter flow [ft sec^{-1}]; splash-over velocity [ft sec^{-1}] (from Figure 8.14))

$$R_f = 1 - 0.09\,(V - V_0) \tag{8.15}$$

(side-flow interception efficiency of grate inlet; mean velocity of gutter flow [ft sec^{-1}]; side slope of outside cross section (Figure 8.12); grate length [ft] (Figure 8.13))

$$R_s = \left[1 + \frac{0.15V^{1.8}}{S_x L^{2.3}}\right]^{-1} \tag{8.16}$$

Example 8.20 A gutter has a simple cross section (as shown in Figure 8.12a) with a side slope of $S_w = 0.05$ and Manning's $n = 0.016$. It is laid on a slope of $S_0 = 0.03$. A 1-ft-wide, 1.5-ft-long P-1-1/8 grate inlet is used to intercept the gutter flow. If the spread of water in the gutter is 2 ft, determine the interception efficiency and bypass fraction.

Solution

$S_w = 0.05$, $n = 0.016$, $S_0 = 0.03$, $W = 1$ ft, $L = 1.5$ ft, $B = 2$ ft

$$\text{Eq. (8.7): } Q = \frac{1.486}{0.016} * \frac{(2\,\text{ft})^{\frac{8}{3}} * 0.05^{\frac{5}{3}} * 0.03^{\frac{1}{2}}}{2^{\frac{5}{3}} * \left(0.05 + \sqrt{1 + 0.05^2}\right)^{\frac{2}{3}}} = 0.21\,\text{cfs}.$$

Gutter flow area: $A = 1/2B^2S_x = 1/2 * (2\,\text{ft})^2 * 0.05 = 0.1\,\text{ft}^2$.

Mean velocity: $V = Q/A = (0.21\,\text{cfs})/(0.1\,\text{ft}^2) = 2.1\,\text{ft sec}^{-1}$.

<u>Determine the side flow rate</u>:

Side flow area: $A_{side} = 1/2(B - W)^2 S_w = 1/2 * (2\,\text{ft} - 1\,\text{ft})^2 * 0.05 = 0.025\,\text{ft}^2$

$$P_{wet,side} = (1 + S_w{}^2)^{1/2}(B - W) = (1 + 0.05^2)^{1/2} * (2\,\text{ft} - 1\,\text{ft}) = 1.0012\,\text{ft}$$

$$R_{side} = A_{side}/P_{wet,side} = (0.025\,\text{ft}^2)/(1.0012\,\text{ft}) = 0.02497\,\text{ft}$$

$$Q_s = (1.486/0.016) * (0.025\,\text{ft}^2) * (0.02497\,\text{ft})^{2/3} * (0.03^{1/2}) = 0.0344\,\text{cfs}.$$

<u>Determine the frontal flow rate</u>:

$$Q_f = Q - Q_s = 0.21\,\text{cfs} - 0.0344\,\text{cfs} = 0.1756\,\text{cfs}.$$

Determine the interception efficiency and bypass fraction:

Figure 8.14: P-1-1/8 grate inlet, $L = 1.5\,\text{ft} \rightarrow V_0 = 5.5\,\text{ft sec}^{-1}$.

$$V \leq V_0 \rightarrow \eta_{di} = 1,\; R_f = Q_f/Q = (0.1756\,\text{cfs})/(0.21\,\text{cfs}) = 0.84,\; R_s = 1 - R_f = 0.16$$
$$\rightarrow \eta_{bp} = 1 - 1 = 0.$$

Example 8.21 A gutter has a simple cross section (as shown in Figure 8.12a) with a side slope of $S_w = 0.05$ and Manning's $n = 0.013$. It is laid on a slope of $S_0 = 0.25$. A 1-ft-wide, 1.5-ft-long P-1-1/8 grate inlet is used to intercept the gutter flow. If the spread of water in the gutter is 2 ft, determine the interception efficiency and bypass fraction.

Solution

$S_w = 0.05$, $n = 0.013$, $S_0 = 0.25$, $W = 1\,\text{ft}$, $L = 1.5\,\text{ft}$, $B = 2\,\text{ft}$

Eq. (8.7): $$Q = \frac{1.486}{0.013} * \frac{(2\,\text{ft})^{\frac{8}{3}} * 0.05^{\frac{5}{3}} * 0.25^{\frac{1}{2}}}{2^{\frac{5}{3}} * \left(0.05 + \sqrt{1 + 0.05^2}\right)^{\frac{2}{3}}} = 0.75\,\text{cfs}.$$

Gutter flow area: $A = 1/2B^2S_x = 1/2 * (2\,\text{ft})^2 * 0.05 = 0.1\,\text{ft}^2$.

Mean velocity: $V = Q/A = (0.75\,\text{cfs})/(0.1\,\text{ft}^2) = 7.5\,\text{ft sec}^{-1}$.

Determine the side flow rate:

Side flow area: $A_{side} = 1/2(B - W)^2S_w = 1/2 * (2\,\text{ft} - 1\,\text{ft})^2 * 0.05 = 0.025\,\text{ft}^2$

$$P_{wet,side} = (1 + S_w{}^2)^{1/2}(B - W) = (1 + 0.05^2)^{1/2} * (2\,\text{ft} - 1\,\text{ft}) = 1.0012\,\text{ft}$$
$$R_{side} = A_{side}/P_{wet,side} = (0.025\,\text{ft}^2)/(1.0012\,\text{ft}) = 0.02497\,\text{ft}$$
$$Q_s = (1.486/0.013) * (0.025\,\text{ft}^2) * (0.02497\,\text{ft})^{2/3} * (0.25^{1/2}) = 0.12\,\text{cfs}.$$

Determine the frontal flow rate:

$$Q_f = Q - Q_s = 0.75\,\text{cfs} - 0.12\,\text{cfs} = 0.63\,\text{cfs}.$$

Determine the interception efficiency and bypass fraction:

Figure 8.14: P-1-1/8 grate inlet, $L = 1.5\,\text{ft} \rightarrow V_0 = 5.5\,\text{ft sec}^{-1}$.

$V > V_0 \rightarrow \eta_{di} < 1,\; \eta_{bp} > 0$.

Eq. (8.15): $R_f = 1 - 0.09 * [(7.5\,\text{ft sec}^{-1}) - (5.5\,\text{ft sec}^{-1})] = 0.82$.

Eq. (8.16): $$R_s = \left[1 + \frac{0.15 * (7.5\,\text{ft sec}^{-1})^{1.8}}{0.05 * (1.5\,\text{ft})^{2.3}}\right]^{-1} = 0.022.$$

Eq. (8.14): $$\eta_{di} = \frac{0.82 * (0.63\,\text{cfs}) + 0.022 * (0.12\,\text{cfs})}{0.63\,\text{cfs} + 0.12\,\text{cfs}} = 0.69.$$

Eq. (8.13): $\eta_{bp} = 1 - 0.69 = 0.31$.

Curb-opening inlets are good for locations with a sag as well as on a small slope. They are likely to have a low interception efficiency for locations on a steep slope because flow in the gutter may not easily be diverted laterally into such an inlet.

For a curb-opening inlet on a slope, its interception efficiency can be computed as (Johnson and Chang, 1984):

$$\eta_{di} = 1 - \left(1 - \frac{L n^{0.6} S_x^{0.6}}{\Phi Q^{0.42} S_0^{0.3}}\right)^{1.8} \tag{8.17}$$

where: η_{di} = interception efficiency of curb-opening inlet on slope; L = curb-opening length [ft for BG; m for SI] (Figure 8.13); n = Manning's n; S_x = side slope of outside cross section (Figure 8.12); Φ = units conversion factor (= 0.6 for BG; 0.82 for SI); Q = discharge in gutter [cfs for BG; $m^3\ s^{-1}$ for SI]; S_0 = gutter bed slope.

A curb-opening inlet in a sag location functions as a weir when the water depth in the gutter, y, is smaller than the opening height, h, whereas it functions as an orifice when $y \geq 1.4h$. If $h < y < 1.4h$, its interception capacity can be estimated as a linear interpolation between the interception capacity for $y = h$ and that for $y = 1.4h$, otherwise its interception capacity can be estimated as (Johnson and Chang, 1984; Mays, 2005):

$$Q_{di} = \begin{cases} C_w (L + 1.8W) y^{1.5} & \text{if } y \leq h \\ 0.6hL\sqrt{2g\left(y - \frac{h}{2}\right)} & \text{if } y \geq 1.4h \end{cases} \tag{8.18}$$

where: Q_{di} = interception capacity of curb-opening inlet in sag [cfs for BG; $m^3\ s^{-1}$ for SI]; C_w = weir coefficient (= 2.3 for BG; 1.25 for SI); L = curb-opening length [ft for BG; m for SI]; W = depression lateral width [ft for BG; m for SI]; h = curb-opening height [ft for BG; m for SI]; g = gravitational acceleration (= 32.2 ft sec^{-2} for BG; 9.81 m s^{-2} for SI); y = water depth next to opening [ft for BG; m for SI].

Example 8.22 For the gutter in Example 8.20, if a 5-ft-long curb-opening inlet with an opening height of 0.3 ft is used, determine its interception efficiency and bypass fraction.

Solution

$S_w = 0.05$, $n = 0.016$, $S_0 = 0.03$, $L = 5$ ft, $h = 0.3$ ft

Example 8.20 $\rightarrow Q = 0.21$ cfs

Eq. (8.17): $S_x = S_w = 0.05 \rightarrow \eta_{di} = 1 - \left[1 - \frac{(5\ \text{ft}) * 0.016^{0.6} * 0.05^{0.6}}{0.6 * (0.21\ \text{cfs})^{0.42} * 0.03^{0.3}}\right]^{1.8} = 0.84$

Eq. (8.13): $\eta_{bp} = 1 - 0.84 = 0.16$

The curb-opening inlet is less efficient than the grate inlet.

Example 8.23 Assume that the gutter in Example 8.20 is located in a sag location. If a 1-ft-long curb-opening inlet with a 0.3-ft-high opening and a 1-ft-wide depression is used, determine its interception efficiency and bypass fraction of the inlet.

Solution

$S_w = 0.05$, $B = 2\,\text{ft}$, $W = 1\,\text{ft}$, $L = 1\,\text{ft}$, $h = 0.3\,\text{ft}$

Example 8.20 $\rightarrow Q = 0.21\,\text{cfs}$, $y = BS_w = (2\,\text{ft}) * 0.05 = 0.1\,\text{ft}$.

Eq. (8.18): $y \le h \rightarrow Q_{di} = 2.3 * (1\,\text{ft} + 1.8 * 1\,\text{ft}) * (0.1\,\text{ft})^{1.5} = 0.20\,\text{cfs}$.

Eq. (8.12): $\eta_{di} = Q_{di}/Q = (0.20\,\text{cfs})/(0.21\,\text{cfs}) = 0.95$.

Eq. (8.13): $\eta_{bp} = 1 - 0.95 = 0.05$.

For a combination inlet on a slope, its interception efficiency is the summation of the grate inlet interception efficiency (Eq. (8.14)) and the curb-opening inlet interception efficiency (Eq. (8.17)), with a upper limit of one.

For a combination inlet in a sag location, the grate inlet usually functions as an orifice, whereas the curb opening can function either as a weir or as an orifice, depending on the water depth next to the opening. Its interception capacity can be estimated as the summation of the curb-opening inlet interception capacity computed by Eq. (8.18) and the grate inlet interception. If $y \le h$ or $y \ge 1.4h$, its interception capacity is computed as:

$$Q_{di} = \begin{cases} C_w (L + 1.8W) y^{1.5} + 0.67LW\sqrt{2gy} & \text{if } y \le h \\ 0.6hL\sqrt{2g\left(y - \frac{h}{2}\right)} + 0.67LW\sqrt{2gy} & \text{if } y \ge 1.4h \end{cases} \tag{8.19}$$

Q_{di}: interception capacity of combination inlet in sag [cfs for BG; $m^3\,s^{-1}$ for SI]
C_w: weir coefficient (= 2.3 for BG; 1.25 for SI)
L: curb-opening length [ft for BG; m for SI]
W: depression lateral width [ft for BG; m for SI]
h: curb-opening height [ft for BG; m for SI]
g: gravitational acceleration (= 32.2 ft sec^{-2} for BG; 9.81 m s^{-2} for SI)
y: water depth next to opening [ft for BG; m for SI]

Example 8.24 Assume that the gutter in Example 8.20 is located in a sag location. If a 1-ft-long, 1-ft-wide combination inlet with a 0.3-ft-high curb opening is used, determine its interception efficiency and bypass fraction.

Solution

$S_w = 0.05$, $B = 2\,\text{ft}$, $W = 1\,\text{ft}$, $L = 1\,\text{ft}$, $h = 0.3\,\text{ft}$

Example 8.20 $\rightarrow Q = 0.21\,\text{cfs}$, $y = BS_w = (2\,\text{ft}) * 0.05 = 0.1\,\text{ft}$.

Eq. (8.19): $y \le h \rightarrow Q_{di} = 2.3 * (1\,\text{ft} + 1.8 * 1\,\text{ft}) * (0.1\,\text{ft})^{1.5} + 0.67 * (1\,\text{ft}) * (1\,\text{ft}) * [2 * (32.2\,\text{ftsec}^{-2}) * (0.1\,\text{ft})]^{1/2} = 1.90\,\text{cfs}$.

$Q < Q_{di} \rightarrow \eta_{di} = 1.0$, $\eta_{bp} = 0$, all gutter flow is intercepted.

For a slotted drain inlet on a slope, its interception efficiency can be computed using Eq. (8.17). As mentioned above, because of the clogging concern, slotted inlets are rarely used in sag locations.

8.5 Spillways and Stilling Basins

For dam safety, a spillway is always constructed to pass floodwater. It can be either gated or uncontrolled. Three types of spillways are widely used, namely the US Bureau Reclamation (USBR) ogee spillway, shaft spillway, and siphon spillway (Figure 8.15). Shaft and siphon spillways are cost-effective for small to moderate-sized dams, whereas a USBR ogee spillway is usually imperative for large dams. The release capacity of a shaft spillway can be analyzed using the orifice flow equations presented in Section 7.5, and the release capacity of a siphon spillway can be determined by applying the energy equation (Eq. (2.21)) between the entrance and vent cross section of the conduit (see Figure 8.15c). This section focuses on the USBR ogee spillway and its terminal structure (i.e., stilling basin).

8.5.1 USBR Ogee Spillways

The geometry of a USBR ogee spillway is defined by USBR (2014) and is shown in Figure 8.16. The spillway has a vertical upstream face, followed by two circular arches with a decreasing curvature, and then by a curved surface defined by a power function. The design water depth above the crest apex of the spillway, H_D, depends on the design peak discharge (see Chapter 4) and the storage characteristics of the reservoir formed by damming (see Chapter 3).

For an uncontrolled USBR ogee spillway, its discharge can be computed as:

$$Q = C\sqrt{2g}L\left(H + \frac{V_a^2}{2g}\right)^{\frac{3}{2}} \tag{8.20}$$

discharge of uncontrolled USBR ogee spillway → Q
discharge coefficient (Figure 8.17) → C
gravitational acceleration (= 32.2 ft sec^{-2} for BG; 9.81 m s^{-2} for SI) → g
spillway crest length (Figure 8.16b) → L
water depth above crest apex → H
approach velocity = Q/(P + H) (Figure 8.16a) → V_a

A USBR ogee spillway can be controlled by one or more gates, each of which controls one opening. The discharge through each opening can be computed as:

$$Q = C\frac{2\sqrt{2g}}{3}L\left[\left(H_1 + \frac{V_a^2}{2g}\right)^{\frac{3}{2}} - \left(H_2 + \frac{V_a^2}{2g}\right)^{\frac{3}{2}}\right] \tag{8.21}$$

discharge of gate-controlled USBR ogee spillway → Q
discharge coefficient (Figure 8.18) → C
gravitational acceleration (= 32.2 ft sec^{-2} for BG; 9.81 m s^{-2} for SI) → g
spillway crest length (Figure 8.16b) → L
water depth above crest apex (Figure 8.18) → H_1
approach velocity = Q/(P + H_1) (Figure 8.18) → V_a
water depth above lowest point of gate (Figure 8.18) → H_2

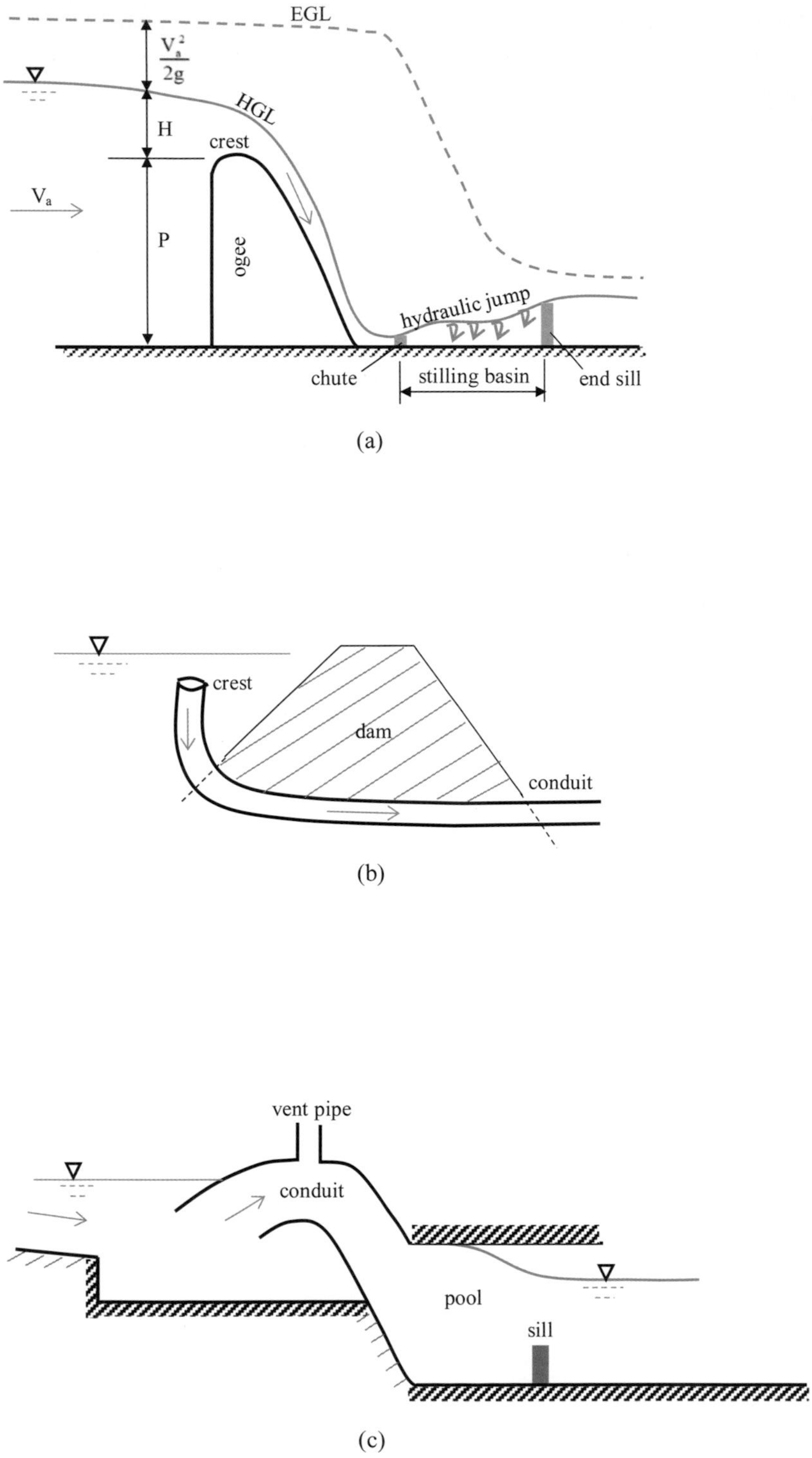

Figure 8.15 **Cross-sectional views of spillway types.** (a) Ogee spillway; (b) shaft spillway; and (c) siphon spillway.

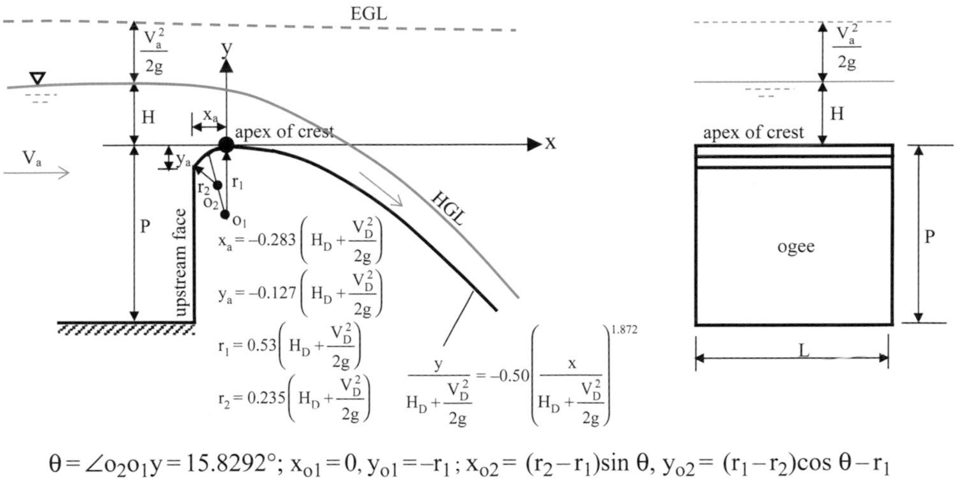

Figure 8.16 **Details of a USBR ogee spillway.** Design geometry in: (a) side view; and (b) front view. H_D: design water depth above the crest apex; and V_D: design approach flow velocity. (*Source:* after USBR, 2014)

Example 8.25 A 100-ft-high, 25-ft-long USBR ogee spillway is designed to have a crest-above water depth of 40 ft. Because the reservoir in front of the spillway is deep and wide, the approach velocity head is relatively small and can be neglected. For a flood event with a crest above water depth of 30 ft, determine the discharge through the spillway if it is:

(1) Uncontrolled
(2) Controlled by one gate being operated at an opening height of 5 ft.

Solution

$P = 100\,\text{ft},\ L = 25\,\text{ft},\ H_D = 40\,\text{ft},\ H = 30\,\text{ft}$, zero velocity head

(1) Figure 8.17a: $P/H_D = (100\,\text{ft})/(40\,\text{ft}) = 2.5 \rightarrow C_D = 0.491.$
Figure 8.17b: $H/H_D = (30\,\text{ft})/(40\,\text{ft}) = 0.75 \rightarrow C/C_D = 0.96.$
Thus, $C = 0.96 * 0.491 = 0.47136.$
Eq. (8.20): $Q = 0.47136 * \sqrt{2 * (32.2\,\text{ft sec}^{-2})} * (25\,\text{ft}) * (30\,\text{ft} + 0)^{\frac{3}{2}} = 15{,}539\,\text{cfs}.$

(2) One gate with an opening height of $d = 5$ ft.
Figure 8.18: $H_1 = H = 30\,\text{ft},\ H_2 = H_1 - d = 30\,\text{ft} - 5\,\text{ft} = 25\,\text{ft}$

$$d/H_1 = (5\,\text{ft})/(30\,\text{ft}) = 0.17 \rightarrow C = 0.703.$$

$$\text{Eq. (8.21): } Q = 0.703 * \frac{2\sqrt{2 * (32.2\,\text{ft sec}^{-2})}}{3} * (25\,\text{ft}) * \left[(30\,\text{ft} + 0\,\text{ft})^{\frac{3}{2}} - (25\,\text{ft} + 0\,\text{ft})^{\frac{3}{2}}\right]$$
$$= 3697\,\text{cfs}.$$

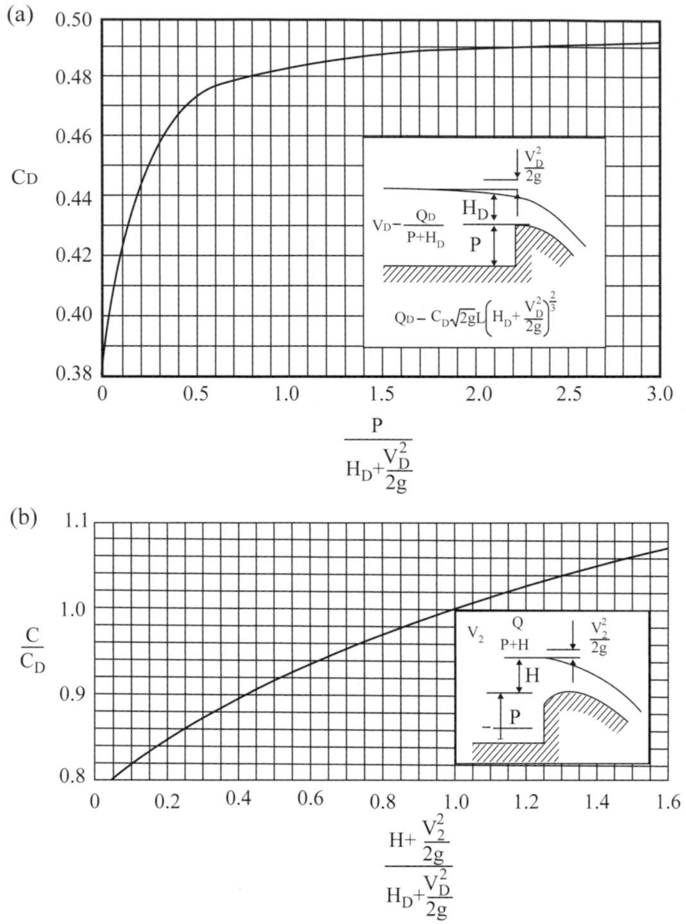

Figure 8.17 **Nomogram of discharge coefficients for USBR ogee spillway.** Diagrams are for the condition of water depth above the crest apex. (a) Design (H_D); and (b) non-design (H). (*Source:* USBR, 2014)

8.5.2 Stilling Basins

The flow through a USBR ogee spillway has a large amount of kinetic energy and cannot be directly released back into the downstream stream because of possible bed and/or bank erosion. To dissipate the energy, a stilling basin needs to be constructed at the downstream end (i.e., toe) of the spillway so that a hydraulic jump is forced to occur within the basin. As presented in Section 6.4, a hydraulic jump is very effective in dissipating energy. Given that a stilling basin commonly has a rectangular cross section, the conjugate water depths within the basin can be determined using Eq. (6.9). The USBR (2014) recommends three types of stilling basins (Figure 8.19), namely stilling basins II, III, and IV. Stilling basins II and III are applicable for the situation with an approach (i.e., pre-jump) Froude number of $F_{r,1} > 4.5$, whereas stilling basin IV is good for the situation with $F_{r,1} = 2.5$ to 4.5. The layouts and design parameters of these stilling basins are shown in Figure 8.19.

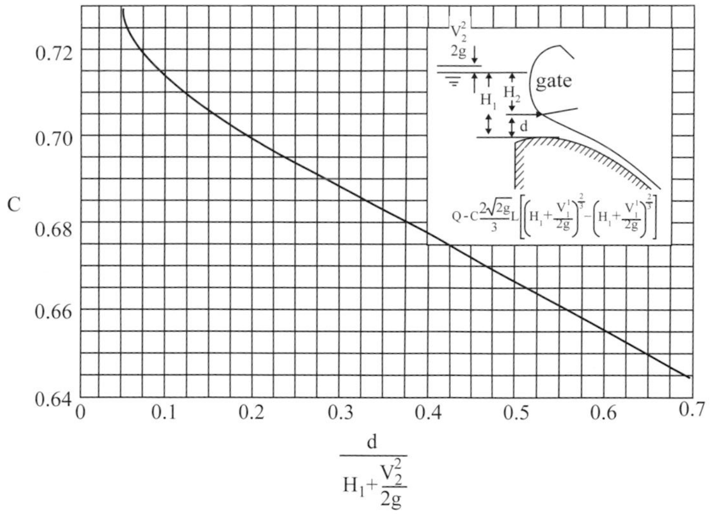

Figure 8.18 **Nomogram of discharge coefficient for gate-controlled USBR ogee spillways.** d: gate opening height; and V_a: approach flow velocity. (*Source:* USBR, 2014)

The dimensions and configurations of the stilling basins can be determined quantitatively. For stilling basin II ($F_{r,1} > 4.5$) (Figure 8.19a), its design length can be determined as (USBR, 2014):

required length of stilling basin II: L_{II}; post-jump water depth: D_2; approach Froude number at spillway toe: $F_{r,1}$

$$L_{II} = \begin{cases} D_2\left[4.0 + 0.055\left(F_{r,1} - 4.5\right)\right] & 4.5 < F_{r,1} < 10.0 \\ 4.35D_2 & F_{r,1} \geq 10.0 \end{cases} \tag{8.22}$$

For stilling basin III ($F_{r,1} > 4.5$) (Figure 8.19b), its design length and heights of baffle piers and end sill can be determined as (USBR, 2014):

required length of stilling basin III: L_{III}; post-jump water depth: D_2; approach Froude number at spillway toe: $F_{r,1}$

$$L_{III} = \begin{cases} D_2\left[2.4 + 0.073\left(F_{r,1} - 4.5\right)\right] & 4.5 < F_{r,1} < 10.0 \\ 2.8D_2 & F_{r,1} \geq 10.0 \end{cases} \tag{8.23}$$

height of baffle piers (Figure 8.19b): h_3; pre-jump water depth at spillway toe: D_1; approach Froude number at spillway toe: $F_{r,1}$

$$h_3 = D_1\left[1.3 + 0.164\left(F_{r,1} - 4.0\right)\right] \tag{8.24}$$

height of end sill (Figure 8.19b): h_4; pre-jump water depth at spillway toe: D_1; approach Froude number at spillway toe: $F_{r,1}$

$$h_4 = D_1\left[1.25 + 0.056\left(F_{r,1} - 4.0\right)\right] \tag{8.25}$$

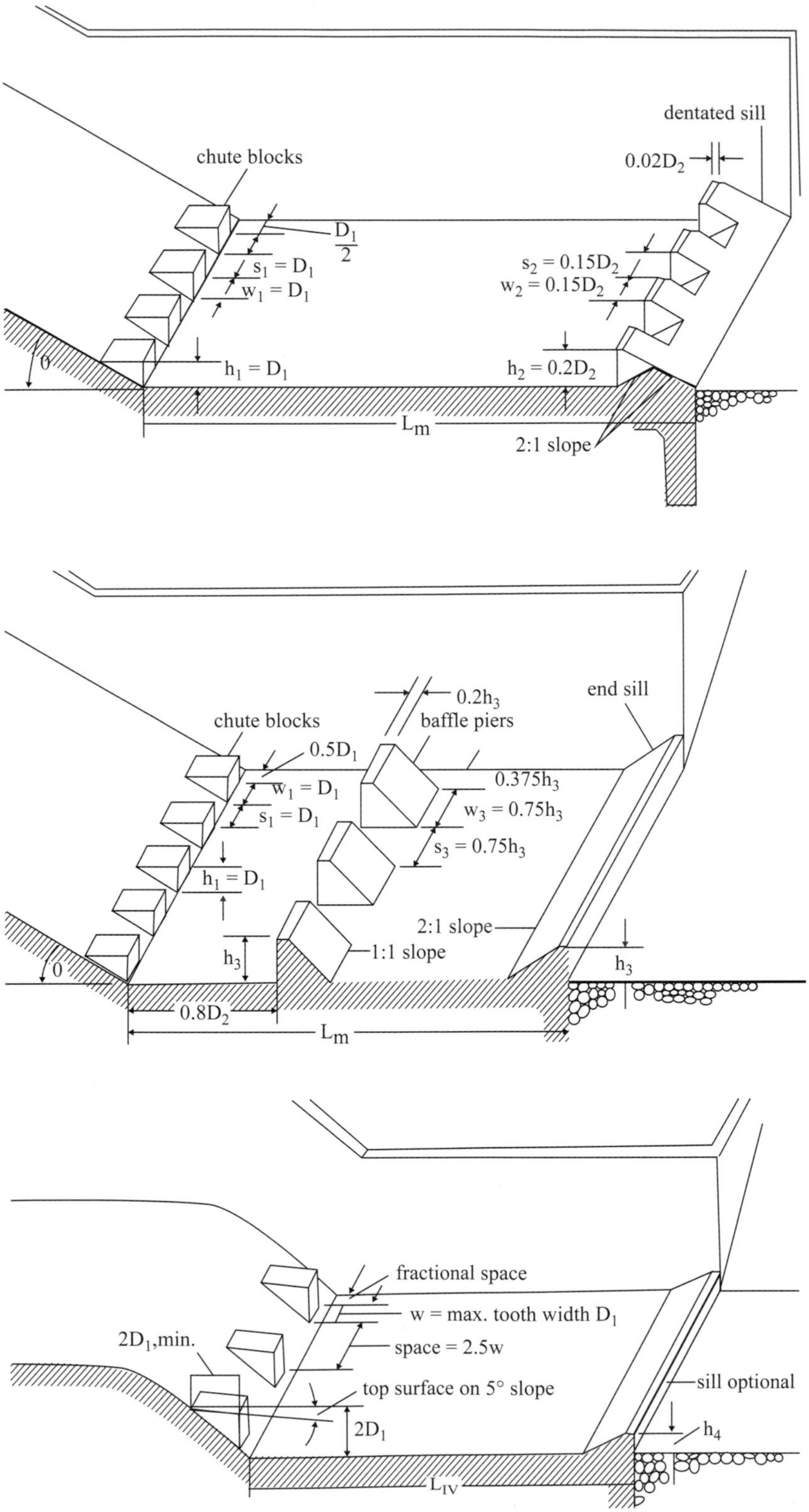

Figure 8.19 **US Bureau of Reclamation silting basin types.** Diagram of: (a) basin II; (b) basin III; and (c) basin IV. D_1 and D_2, respectively, are pre- and post-jump depths. (*Source:* USBR, 2014)

For stilling basin IV ($F_{r,1}$ = 2.5 to 4.5) (Figure 8.19c), its design length can be determined as (USBR, 2014):

post-jump water depth

approach Froude number at spillway toe

required length of stilling basin IV

$$L_{IV} = D_2\,[5.2 + 0.4\,(F_{r,1} - 2.5)] \quad (8.26)$$

Example 8.26 The design discharge through a 50-ft-long USGS ogee spillway is 15,000 cfs, which results in a water depth of 5 ft at the spillway toe. Select and design a stilling basin.

Solution

$L = 50\,\text{ft}$, $Q_D = 15{,}000\,\text{cfs}$, $D_1 = 5\,\text{ft}$

At the spillway toe:

Mean velocity: $V_1 = (15{,}000\,\text{cfs})/[(50\,\text{ft}) * (5\,\text{ft})] = 60\,\text{ft sec}^{-1}$.
Froude number: $F_{r,1} = (60\,\text{ft sec}^{-1})/[(32.2\,\text{ft sec}^{-2}) * (5\,\text{ft})]^{1/2} = 4.73$.

Determine the post-jump depth:

Eq. (6.9): $D_2 = (5\,\text{ft}) * [-1 + (1 + 8 * 4.73^2)^{1/2}]/2 = 31.04\,\text{ft}$.

Select a spillway:

$F_{r,1} > 4.5 \rightarrow$ stilling basin II or III can be selected.

Design a stilling basin II:

- Eq. (8.22): $L_{II} = (31.04\,\text{ft}) * [4.0 + 0.055 * (4.73 - 4.5)] = 125\,\text{ft}$.
- Height of chute blocks: $h_1 = 5\,\text{ft}$
  Width of chute blocks: $w_1 = 5\,\text{ft}$
  Spacing of middle chute blocks: $s_1 = 5\,\text{ft}$.
- Spacing between the wall and its adjacent chute block: $(5\,\text{ft})/2 = 2.5\,\text{ft}$
  Height of end sill: $h_2 = 0.2 * (31.04\,\text{ft}) = 6.2\,\text{ft}$
  Width of sill dents: $s_2 = 0.15 * (31.04\,\text{ft}) = 4.7\,\text{ft}$
  Width of sill bumps: $w_2 = 0.15 * (31.04\,\text{ft}) = 4.7\,\text{ft}$
  Top length of sill bumps: $0.02 * (31.04\,\text{ft}) = 0.6\,\text{ft}$.
- Number of chute blocks: $(5 + 50 - 2 * 2.5)/10 = 5$
  Number of bumps: $(50 + 4.7)/(2 * 4.7) \approx 6$
  Number of dents: $6 - 1 = 5$.

Design a stilling basin III:

- Eq. (8.23): $L_{III} = (31.04\,\text{ft}) * [2.4 + 0.073 * (4.73 - 4.5)] \approx 75\,\text{ft}$.
- Height of chute blocks: $h_1 = 5\,\text{ft}$

Width of chute blocks: $w_1 = 5$ ft

Spacing of middle chute blocks: $s_1 = 5$ ft

Spacing between the wall and its adjacent chute block: $0.5 * (5\,\text{ft}) = 2.5$ ft.

- Eq. (8.24): height of baffle piers: $h_3 = (5\,\text{ft}) * [1.3 + 0.164 * (4.73 - 4.0)] = 7.1$ ft

  Width of baffle piers: $w_3 = 0.75 * (7.1\,\text{ft}) = 5.3$ ft

  Spacing of middle baffle piers: $s_3 = 0.75 * (7.1\,\text{ft}) = 5.3$ ft

  Spacing between the wall and its adjacent baffle pier: $0.375 * (7.1\,\text{ft}) = 2.7$ ft

  Top length of baffle piers: $0.2 * (7.1\,\text{ft}) = 1.4$ ft.

- Eq. (8.25): height of end sill: $h_4 = (5\,\text{ft}) * [1.25 + 0.056 * (4.73 - 4.0)] = 6.5$ ft.

- Number of chute blocks: $(5 + 50 - 2 * 2.5)/10 = 5$

  Number of baffle piers: $(5.3 + 50 - 2 * 2.7)/10.6 \approx 5$.

Example 8.27 The design discharge through a 20-ft-long USGS ogee spillway is 1800 cfs, which results in a water depth of 2.5 ft at the spillway toe. Select and design a stilling basin.

Solution

$L = 20\,\text{ft}$, $Q_D = 1800\,\text{cfs}$, $D_1 = 2.5\,\text{ft}$

At the spillway toe:

Mean velocity: $V_1 = (1800\,\text{cfs})/[(20\,\text{ft}) * (2.5\,\text{ft})] = 36\,\text{ft sec}^{-1}$.

Froude number: $F_{r,1} = (36\,\text{ft sec}^{-1})/[(32.2\,\text{ft sec}^{-2}) * (2.5\,\text{ft})]^{1/2} = 4.01$.

Determine the post-jump depth:

Eq. (6.9): $D_2 = (2.5\,\text{ft}) * [-1 + (1 + 8 * 4.01^2)^{1/2}]/2 = 12.98$ ft.

Select a spillway:

$F_{r,1} = 2.5$ to $4.5 \rightarrow$ stilling basin IV is selected.

Design a stilling basin IV:

- Eq. (8.26): $L_{IV} = (12.98\,\text{ft}) * [5.2 + 0.4 * (4.01 - -2.5)] \approx 75$ ft.

- Height of chute blocks: $2 * (2.5\,\text{ft}) = 5$ ft

  Width of chute blocks: $w = 2.5$ ft

  Spacing of middle chute blocks: $2.5 * (2.5\,\text{ft}) = 6.25$ ft

  Top length of chute blocks: $2 * (2.5\,\text{ft}) = 5$ ft

  Spacing between the wall and its adjacent chute block: x.

- No sill.

- Number of chute blocks, N, satisfies: $2.5N + 6.25(N - 1) + 2x = 20 \rightarrow$ let $N = 3 \rightarrow x = 0$ ft.

PROBLEMS

8.1 A circular CMP culvert with a 2-ft-diameter barrel carries 30 cfs and is laid on a slope of 0.002. For an inlet-control condition, determine the HW if the culvert has an inlet with:
(1) Headwall
(2) Mitered to slope
(3) Projecting.

8.2 Redo Problem 8.1 if the flow rate is reduced to 8 cfs.

8.3 Redo Problem 8.1 if the flow rate is reduced to 16 cfs.

8.4 A 3-ft-wide, 2.5-ft-high rectangular box culvert (with an inlet of 90° headwall w/45° bevels) is constructed as a road crossing over a drainage ditch. The culvert is laid on a slope of 0.005. For an inlet-control condition, determine the HW if the culvert carries:
(1) 35 cfs
(2) 45 cfs
(3) 85 cfs.

8.5 A 6-ft-wide, 5-ft-high CM box culvert (with an inlet of thick wall projecting) is constructed as a road crossing over a stream. It is laid on a slope of 0.003. For an inlet-control condition, determine the HW if it carries:
(1) 200 cfs
(2) 250 cfs
(3) 600 cfs.

8.6 A circular concrete culvert with a 3-ft-diameter barrel is laid on a slope of 0.003. Its inlet has square edges w/headwall. For an inlet-control condition, determine the discharge when the HW is:
(1) 1.5 ft
(2) 2.5 ft
(3) 3.5 ft
(4) 4.5 ft.

8.7 A 3-ft-wide, 2-ft-high rectangular concrete culvert (with an inlet of side tapered – more favorable edges) is laid on a slope of 0.005. For an inlet-control condition, determine the discharge when the HW is:
(1) 0.5 ft
(2) 1.5 ft
(3) 2.5 ft
(4) 3.5 ft.

8.8 Redo Problem 8.7 if the inlet is changed to side tapered – less favorable edges.

8.9 A 5-ft-wide, 3-ft-high rectangular box culvert (with an inlet of 30° to 75° wingwall flares) is laid on a slope of 0.0025. For an inlet-control condition, determine the discharge when the HW is:
(1) 1.5 ft
(2) 2.5 ft

(3) 3.5 ft
(4) 4.5 ft.

8.10 Redo Problem 8.9 if the inlet is changed to 0° wingwall fares.

8.11 A 2-ft-diameter, 100-ft-long circular concrete culvert is constructed as a road crossing over a stream. The culvert barrel (f = 0.016) has an inlet of projecting from fill, square-cut end. It is laid on a slope of 0.015 and designed to convey 50 cfs. For an outlet-control condition, determine the HW if the TW is:
(1) 1.2 ft
(2) 1.8 ft
(3) 2.5 ft
(4) 3.5 ft.

8.12 A rectangular reinforced concrete box culvert (4 ft wide, 3 ft high, and 150 ft long) is laid on a slope of 0.02 and designed to convey 100 cfs. The culvert barrel has a Darcy friction factor of f = 0.015 and a slope-tapered inlet. For an outlet-control condition, determine the HW if the TW is:
(1) 1.5 ft
(2) 2.5 ft
(3) 3.8 ft
(4) 4.5 ft.

8.13 A 0.5-m-diameter, 50-m-long circular concrete culvert is constructed as a road crossing over a stream. The culvert barrel (f = 0.02) has a slope-tapered inlet. It is laid on a slope of 0.005 and designed to convey 1.5 m^3s^{-1}. For an outlet-control condition, determine the HW if the TW is:
(1) 0.2 m
(2) 0.4 m
(3) 0.8 m
(4) 1.2 m.

8.14 A rectangular reinforced concrete box culvert (1.5 m wide, 1.0 m high, and 50 m long) is laid on a slope of 0.015 and designed to convey 2.8 m^3s^{-1}. The culvert barrel has a Darcy friction factor of f = 0.018 and a slope-tapered inlet. For an outlet-control condition, determine the HW if the TW is:
(1) 0.5 m
(2) 0.8 m
(3) 1.5 m
(4) 2.0 m.

8.15 A 2-ft-diameter, 100-ft-long circular concrete culvert is laid on a slope of 0.015 and has a Darcy friction factor of 0.016. It has an inlet of mitered to conform to fill slope. If the HW and TW are 5.5 and 3.5 ft, respectively, determine the discharge through the culvert for an outlet-control condition.

8.16 Redo Problem 8.15 if the HW and TW are changed to 4.5 and 0.5 ft, respectively.

8.17 A 100-ft-long circular concrete culvert is laid on a slope of 0.015 and has a Darcy friction factor of 0.016. It has an inlet of square edge w/headwall. When the HW and TW are 5.8

and 3.2 ft, respectively, the discharge through the culvert is 200 cfs. For an outlet-control condition, determine the diameter of the culvert barrel.

8.18 A 50-m-long circular CMP culvert is laid on a slope of 0.005 and has a Darcy friction factor of 0.02. It has an inlet of mitered to slope. When the HW and TW are 2.2 and 0.2 m, respectively, the discharge through the culvert is 2.5 m^3s^{-1}. For an outlet-control condition, determine the diameter of the culvert barrel.

8.19 Size an 80-ft-long circular culvert to convey a 50-year peak discharge of 90 cfs, subject to a maximum allowable head water depth of 6.5 ft. The culvert has one barrel (f = 0.018) laid on a slope of 0.01 with an inlet of square edge w/headwall. The 50-year TW is 2.5 ft.

8.20 Size a 100-ft-long circular culvert to convey a 100-year peak discharge of 300 cfs, subject to a maximum allowable head water depth of 20 ft. The culvert has one barrel (f = 0.015) laid on a slope of 0.005 with an inlet of projecting from fill, square-cut end. The 100-year TW is 8.5 ft.

8.21 Size a 50-ft-long circular culvert to convey a 100-year peak discharge of 10 cfs, subject to a maximum allowable head water depth of 5.8 ft. The culvert has one barrel (f = 0.02) laid on a slope of 0.015 with an inlet of square edge w/headwall. The 100-year TW is 1.2 ft.

8.22 Size a 150-ft-long circular culvert to convey a 100-year peak discharge of 30 cfs, subject to a maximum allowable head water depth of 15.5 ft. The culvert has one barrel (f = 0.015) laid on a slope of 0.03 with an inlet of square edge w/headwall. The 100-year TW is 5.8 ft.

8.23 Develop the performance curve of a 100-ft-long rectangular box concrete culvert, which has a 3-ft-wide, 2-ft-high barrel (f = 0.015) laid on a slope of 0.015 with an inlet of 3/4″ chamfers, 45° skewed headwall. Assume a constant TW of 0.3 ft and a maximum allowable head water depth of 5.5 ft.

8.24 Develop the performance curve of a 50-ft-long, 2-ft-diameter circular culvert with an inlet of square edge w/headwall. The culvert barrel has a Darcy friction factor of f = 0.018 and is laid on a slope of 0.02. Assume $TW = 0.0557Q^{0.6}$, where TW, in ft, is the tail water depth; and Q, in cfs, is the discharge through the barrel. The maximum allowable head water depth is 5.5 ft.

8.25 Develop the performance curve of a 30-ft-long rectangular box concrete culvert, which has a 1.5-ft-wide, 1.0-ft-high barrel (f = 0.018) laid on a slope of 0.005 with an inlet of 3/4″ chamfers, 45° skewed headwall. Assume a constant TW of 0.2 ft and a maximum allowable head water depth of 1.5 ft.

8.26 A bridge is constructed over a 100-ft-wide rectangular channel (S_0 = 0.005, n = 0.03). It consists of two openings of 35 ft wide each, one 10-ft-wide pier, and two abutments of 10 ft wide each. The pier and abutments have a shape of square nose and tail. The layout of the bridge is as shown in Figure 8.8, with L_c = 30 ft, L_B = 60 ft, and L_x = 50 ft. For a design discharge of 500 cfs and by assuming a normal depth and no obstruction at section ②, determine the change in water surface from sections ③ to ② using the Yarnell equation.

8.27 Redo Problem 8.26 using the energy equation.

8.28 Redo Problem 8.26 using the momentum equation.

8.29 A 20-m-long bridge is to be constructed as a road crossing over a 100-m-wide rectangular channel (S_0 = 0.01, n = 0.025). The bridge will have 1-m-diameter cylindrical (no diaphragm) piers spaced 15 m apart. The channel length from the cross section (designated as ③ for description purposes) upstream of the bridge, where the flow disturbance from the piers and/or abutments is stabilized, to the upstream face of the bridge is $L_c = 25$ m. On the other hand, the channel length from the downstream face of the bridge to the cross section (designated as ② for description purposes) downstream of the bridge, where the flow disturbance from the piers and/or abutments is no longer present, is $L_x = 50$ m. For the design peak discharge, at ②, the water depth is 2.5 m and the mean velocity is 2.8 m s^{-1}.

(1) Determine the number of piers and the distance between an abutment and its adjacent pier.
(2) Sketch the bridge in bird and front views.
(3) Determine the water depth at ③ using the Yarnell equation.
(4) Determine the water depth at ③ using the energy equation.
(5) Determine the water depth at ③ using the momentum equation if the channel bed and bank friction forces are negligible.
(6) Determine the water depth at ③ using the momentum equation if the channel bed and bank friction forces are significant.

8.30 For Problem 8.29, determine the head loss caused by the piers and/or abutments when the water depth at section ③ is determined using the Yarnell and momentum equations with the channel bed and bank friction forces considered.

8.31 As shown in the following figure, a 8-ft-high, 3-ft-diameter cylindrical riser is drained by a short 1-ft-diameter circular concrete culvert (with an inlet of square edge w/headwall). The culvert is laid on a slope of 0.1. The riser is open on the top and has a 1-ft-wide sharp-crested rectangular weir 2 ft above the inlet invert of the culvert. Determine the flow rate into the riser (Q_{riser}), culvert head water depth (HW), and flow rate through the culvert ($Q_{culvert}$) if the water depth above the weir crest (H) is:

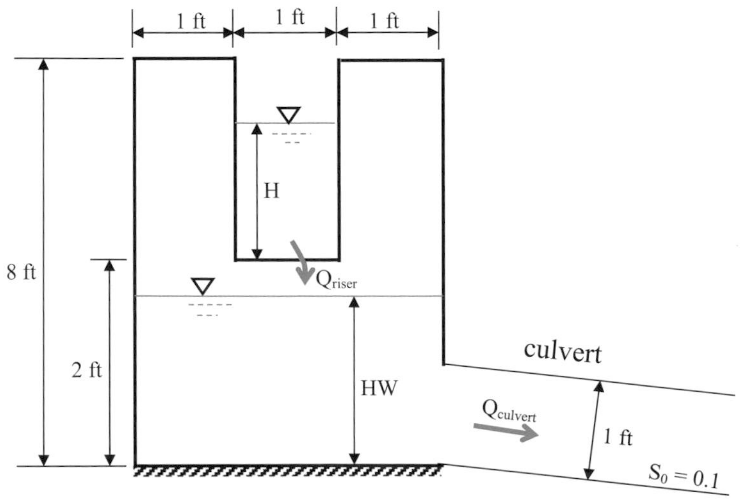

(1) 2.5 ft
(2) 3.5 ft
(3) 4.0 ft
(4) 4.5 ft
(5) 5.0 ft
(6) 5.5 ft.

8.32 Redo Problem 8.31 if the sharp-crested rectangular weir is replaced by a 30° sharp-crested triangular weir, as shown in the following figure.

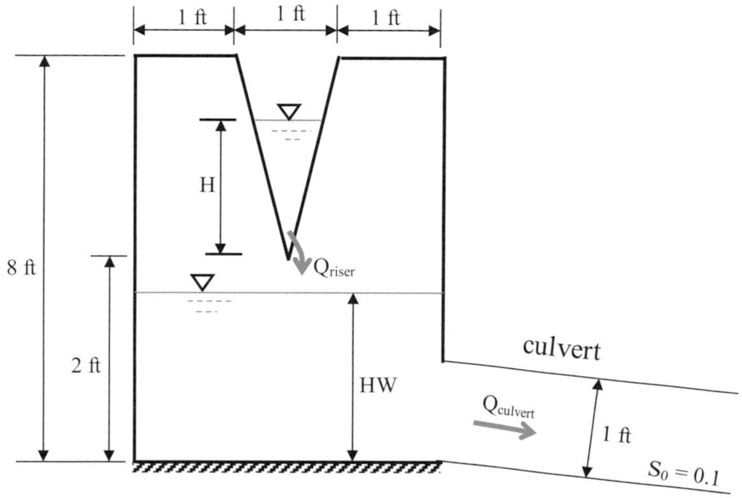

8.33 Redo Problem 8.31 if the sharp-crested rectangular weir is replaced by two orifices, as shown in the following figure. The 1-ft-diameter orifice is at 2 ft above the inlet invert of the culvert, and the 0.5-ft-diameter orifice is at 3.25 ft above the invert.

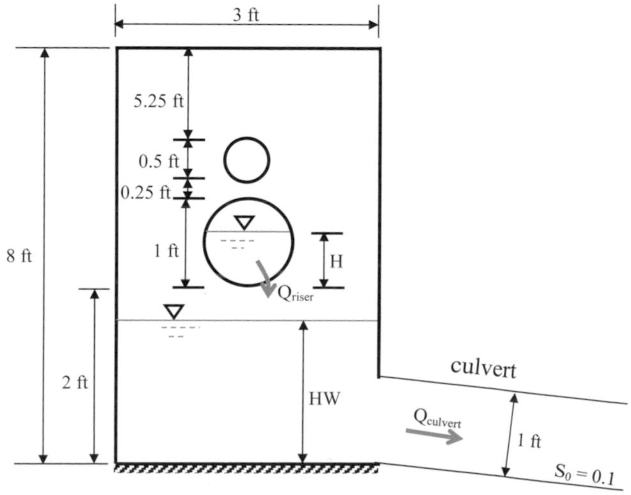

8.34 A gutter is laid on a slope of 0.01 and has a Manning's n of 0.016. The gutter has a single cross section with a side slope of $S_w = 0.2$. Determine the gutter flow rate if the spread of water is:
(1) 0.8 ft
(2) 1.5 ft
(3) 2.0 ft.

8.35 A gutter is laid on a slope of 0.005 and has a Manning's n of 0.013. The gutter has a single cross section with a side slope of $S_w = 0.1$. Determine the gutter flow rate if the spread of water is:
(1) 0.2 m
(2) 0.4 m
(3) 0.6 m.

8.36 A gutter is laid on a slope of 0.01 and has a Manning's n of 0.015. The gutter has a composite cross section with an inside side slope of $S_w = 0.15$ and an outside side slope of $S_x = 0.02$. The top width of the inside cross section is 1.5 ft. Determine the gutter flow rate if the spread of water is:
(1) 1.2 ft
(2) 1.9 ft
(3) 2.5 ft.

8.37 A gutter is laid on a slope of 0.005 and has a Manning's n of 0.012. The gutter has a composite cross section with an inside side slope of $S_w = 0.2$ and an outside side slope of $S_x = 0.01$. The top width of the inside cross section is 0.3 m. Determine the gutter flow rate if the spread of water is:
(1) 0.2 m
(2) 0.4 m
(3) 0.5 m.

8.38 A gutter is laid on a slope of 0.01 and has a Manning's n of 0.016. The gutter has a single cross section with a side slope of $S_w = 0.2$. For two locations 200 ft apart, determine the time and longitudinally averaged velocity of gutter flow if the spreads of water at the upstream and downstream locations are:
(1) $B_1 = 0.2$ ft and $B_2 = 0.6$ ft
(2) $B_1 = 0.8$ ft and $B_2 = 1.5$ ft.

8.39 A gutter is laid on a slope of 0.005 and has a Manning's n of 0.013. The gutter has a single cross section with a side slope of $S_w = 0.1$. For two locations 50 m apart, determine the time and longitudinally averaged velocity of gutter flow if the spreads of water at the upstream and downstream locations are:
(1) $B_1 = 0.2$ m and $B_2 = 0.3$ m
(2) $B_1 = 0.4$ m and $B_2 = 0.6$ m.

8.40 A gutter is laid on a slope of 0.01 and has a Manning's n of 0.015. The gutter has a composite cross section with an inside side slope of $S_w = 0.15$ and an outside side slope of $S_x = 0.02$. The top width of the inside cross section is 1.5 ft. For two locations 150 ft apart, determine the time and longitudinally averaged velocity of gutter flow if the spreads of water at the upstream and downstream locations are:

(1) $B_1 = 0.2$ ft and $B_2 = 1.2$ ft
(2) $B_1 = 1.8$ ft and $B_2 = 2.5$ ft.

8.41 A gutter is laid on a slope of 0.02 and has a Manning's n of 0.015. The gutter has a composite cross section with an inside side slope of $S_w = 0.2$ and an outside side slope of $S_x = 0.1$. The top width of the inside cross section is 0.5 ft. For two locations 80 ft apart, determine the time and longitudinally averaged velocity of gutter flow if the spreads of water at the upstream and downstream locations are:

(1) $B_1 = 0.1$ ft and $B_2 = 0.4$ ft
(2) $B_1 = 0.6$ ft and $B_2 = 0.9$ ft.

8.42 A gutter is laid on a slope of 0.005 and has a Manning's n of 0.012. The gutter has a composite cross section with an inside side slope of $S_w = 0.2$ and an outside side slope of $S_x = 0.01$. The top width of the inside cross section is 0.3 m. For two locations 30 m apart, determine the time and longitudinally averaged velocity of gutter flow if the spreads of water at the upstream and downstream locations are:

(1) $B_1 = 0.1$ m and $B_2 = 0.2$ m
(2) $B_1 = 0.4$ m and $B_2 = 0.6$ m.

8.43 A gutter is laid on a slope of 0.002 and has a Manning's n of 0.015. The gutter has a composite cross section with an inside side slope of $S_w = 0.15$ and an outside side slope of $S_x = 0.05$. The top width of the inside cross section is 0.4 m. For two locations 60 m apart, determine the time and longitudinally averaged velocity of gutter flow if the spreads of water at the upstream and downstream locations are:

(1) $B_1 = 0.1$ m and $B_2 = 0.3$ m
(2) $B_1 = 0.5$ m and $B_2 = 0.9$ m.

8.44 A gutter is laid on a slope of 0.01 and has a Manning's n of 0.016. The gutter has a single cross section with a side slope of $S_w = 0.2$. A 1-ft-wide, 1.5-ft-long P-1-1/8 grate inlet is used to drain the gutter flow. If the spread of water in the gutter is 2 ft, determine the interception efficiency and bypass fraction of the inlet.

8.45 Redo Problem 8.44 if the type of grate inlet is changed to:

(1) P-1-7/8
(2) 45° tilt bar
(3) Curved vane
(4) Reticuline.

8.46 A gutter is laid on a slope of 0.005 and has a Manning's n of 0.013. The gutter has a single cross section with a side slope of $S_w = 0.1$. A 1.2-ft-long curb-opening inlet with an opening height of 0.4 ft is used to drain the gutter flow. If the spread of water in the gutter is 1.8 ft, determine the interception efficiency and bypass fraction of the inlet.

8.47 Redo Problem 8.46 if the inlet is installed in a sag location with a 0.5-ft-wide depression.

8.48 Redo Problem 8.46 if a 1.5-ft-long, 0.8-ft-wide combination inlet with a 0.3-ft-high curb opening is installed in a sag location.

8.49 A 150-ft-high, 30-ft-long uncontrolled USBR ogee spillway is designed to have a crest-above water depth of 50 ft. Neglect the approach velocity head. Determine the:

(1) Design discharge

(2) Discharge at a crest-above water depth is 30 ft

(3) Discharge at a crest-above water depth is 65 ft.

8.50 A 50-m-high, 3-m-long uncontrolled USBR ogee spillway is designed to have a crest above water depth of 15 m. Neglect the approach velocity head. Determine the:

(1) Design discharge

(2) Discharge when the crest above water depth is 28 m.

8.51 A 90-ft-high, 20-ft-long uncontrolled USBR ogee spillway is designed to release a 100-year peak discharge of 10,000 cfs. Neglect the approach velocity head. Determine the design water depth above the spillway crest.

8.52 A 30-m-high, 5-m-long uncontrolled USBR ogee spillway is designed to release a 100-year peak discharge of $300\,m^3s^{-1}$. Neglect the approach velocity head. Determine the design water depth above the spillway crest.

8.53 A 100-ft-high, 25-ft-long gate-controlled USBR ogee spillway is designed to have a crest above water depth of 35 ft. Neglect the approach velocity head. If the maximum opening height of the gate is 10 ft, determine the:

(1) Design discharge

(2) Discharge when the crest above water depth is 20 ft with a gate opening of 5 ft

(3) Discharge when the crest above water depth is 45 ft with a gate opening of 8 ft.

8.54 The design discharge through a 25-ft-long USBR ogee spillway is 12,000 cfs, which results in a water depth of 8 ft at the spillway toe. Select and design a stilling basin.

8.55 The design discharge through a 15-ft-long USBR ogee spillway is 5250 cfs, which results in a water depth of 5 ft at the spillway toe. Select and design a stilling basin.

9 Groundwater Hydraulics

Groundwater is a very important water source. Its exploitation is part of water resources engineering and requires knowledge in the relevant hydraulics. This chapter discusses the basics of groundwater hydraulics. First, it presents how to quantify groundwater as a component of the hydrologic cycle using a water balance equation (Eq. (3.12)) for an aquifer of interest. An aquifer is the saturated zone beneath the ground surface. Second, this chapter presents aquifer properties and the governing equations of groundwater flow. Third, it examines groundwater flow induced by pumping and/or recharging. Finally, this chapter introduces well engineering concepts and procedures such as pumping and slug tests.

9.1 Groundwater in the Hydrologic Cycle

As illustrated in Figure 3.6, for an aquifer of interest, its water sources include artificial recharge, natural percolation, and inflow from other surrounding hydraulically connected aquifers. On the other side of the water budget, aquifer water can be lost to baseflow, evapotranspiration (ET), withdrawal by pumping, and outflow into other surrounding aquifers. The difference between the inflow and outflow is defined as the net inflow of the aquifer. Herein, the net inflow is positive when the inflow is greater than the outflow, but negative when the inflow is smaller than the outflow. Applying Eq. (3.12), one can write the water balance equation for the aquifer as:

$$\frac{dV_g}{dt} = Q_{ar} + Q_{pc} + Q_{nf} - Q_{bf} - Q_{et} - Q_{pp} \tag{9.1}$$

where: V_g = groundwater volume in aquifer of interest; t = time; Q_{ar} = artificial recharge rate; Q_{pc} = percolation rate; Q_{nf} = net inflow from surrounding aquifers; Q_{bf} = baseflow rate; Q_{et} = evapotranspiration rate; Q_{pp} = pumping rate

Example 9.1 In a particular year, a 20 mi^2 shallow aquifer received 0.15 in from percolation and 1.52 in net inflow from its surrounding aquifers, while it discharged 1.35 in into surface water bodies and streams. The depth to water table (i.e., groundwater surface) is large, thus direct evapotranspiration from the aquifer was negligible. Determine the groundwater volume change if withdrawal by pumping was:

(1) 0.00 in
(2) 0.25 in
(3) 0.45 in.

Solution

$A = 20\,\text{mi}^2 * \left[\left(27{,}878{,}400\,\text{ft}^2\right) / \left(1\,\text{mi}^2\right)\right] = 557{,}568{,}000\,\text{ft}^2$

Artificial recharge rate: $Q_{ar} = 0\,\text{ft}^3\,\text{yr}^{-1}$.

Percolation rate: $Q_{pc} = \left(0.15\,\text{in yr}^{-1}\right) * [(1\,\text{ft}) / (12\,\text{in})] * \left(557{,}568{,}000\,\text{ft}^2\right) = 6{,}969{,}600\,\text{ft}^3\,\text{yr}^{-1}$.

Net inflow rate: $Q_{nf} = \left(1.52\,\text{in yr}^{-1}\right) * [(1\,\text{ft}) / (12\,\text{in})] * \left(557{,}568{,}000\,\text{ft}^2\right) = 70{,}625{,}280\,\text{ft}^3\,\text{yr}^{-1}$.

Baseflow rate: $Q_{bf} = \left(1.35\,\text{in yr}^{-1}\right) * [(1\,\text{ft}) / (12\,\text{in})] * \left(557{,}568{,}000\,\text{ft}^2\right) = 62{,}726{,}400\,\text{ft}^3\,\text{yr}^{-1}$.

ET rate: $Q_{et} = \left(0.00\,\text{in yr}^{-1}\right) * [(1\,\text{ft}) / (12\,\text{in})] * \left(557{,}568{,}000\,\text{ft}^2\right) = 0\,\text{ft}^3\,\text{yr}^{-1}$.

(1) $Q_{pp} = \left(0.00\,\text{in yr}^{-1}\right) * [(1\,\text{ft}) / (12\,\text{in})] * \left(557{,}568{,}000\,\text{ft}^2\right) = 0\,\text{ft}^3\,\text{yr}^{-1}$

$$\frac{dV_g}{dt} = 0 + 6{,}969{,}600 + 70{,}625{,}280 - 62{,}726{,}400 - 0 - 0\,\text{ft}^3\,\text{yr}^{-1} = 14{,}868{,}480\,\text{ft}^3\,\text{yr}^{-1}.$$

(2) $Q_{pp} = \left(0.25\,\text{in yr}^{-1}\right) * [(1\,\text{ft}) / (12\,\text{in})] * \left(557{,}568{,}000\,\text{ft}^2\right) = 11{,}616{,}000\,\text{ft}^3\,\text{yr}^{-1}$

$$\frac{dV_g}{dt} = 0 + 6{,}969{,}600 + 70{,}625{,}280 - 62{,}726{,}400 - 0 - 11{,}616{,}000\,\text{ft}^3\,\text{yr}^{-1} = 3{,}252{,}480\,\text{ft}^3\,\text{yr}^{-1}.$$

(3) $Q_{pp} = \left(0.45\,\text{in yr}^{-1}\right) * [(1\,\text{ft}) / (12\,\text{in})] * \left(557{,}568{,}000\,\text{ft}^2\right) = 20{,}908{,}800\,\text{ft}^3\,\text{yr}^{-1}$

$$\frac{dV_g}{dt} = 0 + 6{,}969{,}600 + 70{,}625{,}280 - 62{,}726{,}400 - 0 - 20{,}908{,}800\,\text{ft}^3\,\text{yr}^{-1}$$
$$= -6{,}040{,}320\,\text{ft}^3\,\text{yr}^{-1}.$$

Example 9.2 Averaged across multiple years, the groundwater volume change in a 5 mi^2 aquifer is determined to be zero. The average annual percolation and net inflow are 0.25 and 1.32 in, respectively, and the average annual baseflow and withdrawal by pumping are 1.22 and 0.28 in, respectively. There is no artificial recharge. Determine the average annual groundwater loss to evapotranspiration.

Solution

$A = 5\,\text{mi}^2 * \left[\left(27{,}878{,}400\,\text{ft}^2\right) / \left(1\,\text{mi}^2\right)\right] = 139{,}392{,}000\,\text{ft}^2$

Zero volume change: $\frac{dV_g}{dt} = 0\,\text{ft}^3\,\text{yr}^{-1}$.

Artificial recharge rate: $Q_{ar} = 0\,\text{ft}^3\,\text{yr}^{-1}$.

Percolation rate: $Q_{pc} = \left(0.25\,\text{in yr}^{-1}\right) * [(1\,\text{ft}) / (12\,\text{in})] * \left(139{,}392{,}000\,\text{ft}^2\right) = 2{,}904{,}000\,\text{ft}^3\,\text{yr}^{-1}$.

Net inflow rate: $Q_{nf} = \left(1.32\,\text{in yr}^{-1}\right) * [(1\,\text{ft}) / (12\,\text{in})] * \left(139{,}392{,}000\,\text{ft}^2\right) = 15{,}333{,}120\,\text{ft}^3\,\text{yr}^{-1}$.

Baseflow rate: $Q_{bf} = \left(1.22\,\text{in yr}^{-1}\right) * [(1\,\text{ft}) / (12\,\text{in})] * \left(139{,}392{,}000\,\text{ft}^2\right) = 14{,}171{,}520\,\text{ft}^3\,\text{yr}^{-1}$.

Pumping rate: $Q_{pp} = \left(0.28\,\text{in yr}^{-1}\right) * [(1\,\text{ft}) / (12\,\text{in})] * \left(139{,}392{,}000\,\text{ft}^2\right) = 3{,}252{,}480\,\text{ft}^3\,\text{yr}^{-1}$.

Eq. (9.1): $0\,\text{ft}^3\,\text{yr}^{-1} = 0 + 2{,}904{,}000 + 15{,}333{,}120 - 14{,}171{,}520 - Q_{et} - 3{,}252{,}480\,\text{ft}^3\,\text{yr}^{-1}$

$\rightarrow Q_{et} = 813{,}120\,\text{ft}^3\,\text{yr}^{-1}$

$\rightarrow \left(813{,}120\,\text{ft}^3\,\text{yr}^{-1}\right) / \left(139{,}392{,}000\,\text{ft}^2\right) * [(12\,\text{in}) / (1\,\text{ft})] = 0.07\,\text{in yr}^{-1}$.

9.2 Aquifer Properties

Groundwater yields and movements depend on several properties of an aquifer. This section defines several key features and discusses some of these aquifer properties.

9.2.1 Types of Aquifers

An aquifer is a geologic formation that contains sufficient saturated permeable materials to yield a large amount of water (Todd and Mays, 2005). In contrast, a confining bed is a layer of relatively impermeable material adjacent to an aquifer. Furthermore, a confining bed can be classified as an: (1) aquiclude if its material (e.g., clay) can contain, but does not transmit, water; (2) aquitard if its material (e.g., sandy clay) cannot contain, but does transmit, water; or (3) aquifuge if its material (e.g., solid granite) cannot contain, and does not transmit, water. Regardless of its classification, the confining bed does not yield water. Similarly, an aquifer can be classified as unconfined, confined, or semiconfined (Figure 9.1). An unconfined aquifer (Figure 9.1a) has a lower (or underlying) confining bed but does not have an upper (or overlying) confining bed. A confined aquifer (Figure 9.1b) has a lower and an upper confining bed that are either an aquiclude or aquifuge. A semiconfined aquifer (Figure 9.1c,d) has a lower and an upper confining bed, at least one of which is an aquitard.

As a result of the relationships between an aquifer and its confining bed or beds, groundwater in an unconfined aquifer has a free surface on which the relative pressure is zero, so the water surface in a well penetrating the aquifer coincides with the groundwater surface in the aquifer (Figure 9.2).

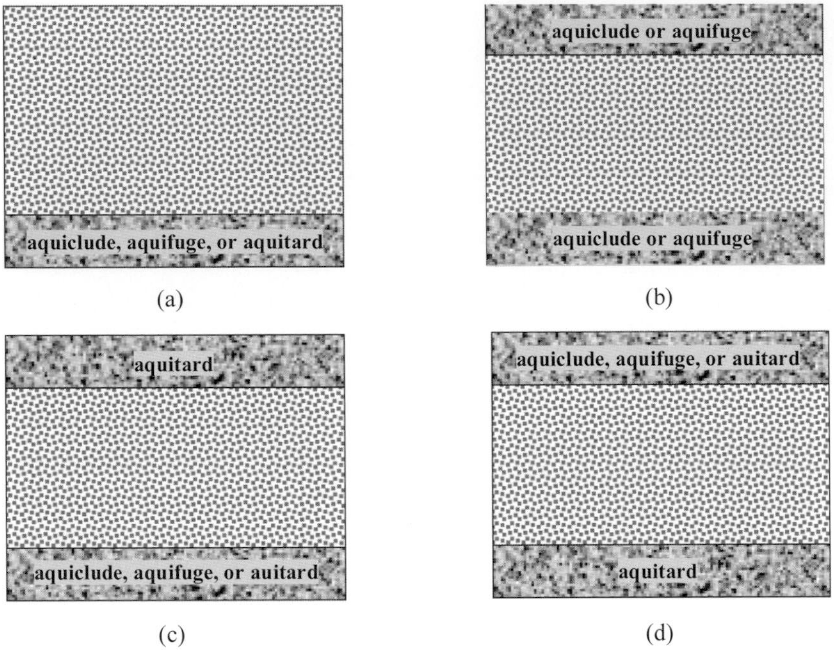

Figure 9.1 **Cross sections of aquifer types.** The aquifer is shown by the stippled pattern. (a) Unconfined aquifer; (b) confined aquifer; (c) upper semiconfined aquifer; and (d) lower semiconfined aquifer.

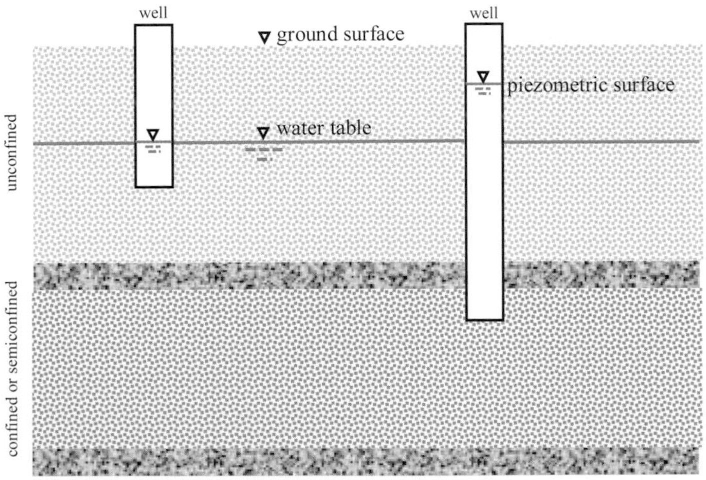

Figure 9.2 **Groundwater level in an unconfined aquifer versus that in a confined or semiconfined aquifer.**

The elevation of the groundwater surface in reference to a standard datum (i.e., benchmark) is termed the water table. The depth to water table is the vertical distance from the ground surface to the water table. In contrast, groundwater in a confined or semiconfined aquifer is subject to a pressure and does not have a free surface, so the water surface in a well penetrating the aquifer rises above the lower surface of the upper confining bed (Figure 9.2). The elevation of the water surface in the well is termed the piezometric surface. In practice, the terminology of groundwater level can be used to refer to either a water table or a piezometric surface.

In Figure 9.2, if the confining bed between the unconfined and confined aquifer is an aquiclude or aquifuge, there will be no water exchange between these two aquifers. However, if the confining bed between the unconfined and semiconfined aquifer is an aquitard, there will be water exchange between these two aquifers. The water in the unconfined aquifer will flow downward through the aquitard into the semiconfined aquifer if the water table is higher than the piezometric surface, whereas the water in the semiconfined aquifer will flow upward through the aquitard into the unconfined aquifer if the water table is lower than the piezometric surface. When the water table is equal to the piezometric surface, the water exchange between these two aquifers will reach an equilibrium condition and the net flow from one aquifer to another is zero. The exchange rate depends on the difference of the water table and piezometric surface as well as the permeability of the aquitard. The groundwater level is the hydraulic head driving groundwater flow.

9.2.2 Properties of aquifers

An aquifer can be characterized in terms of its water holding capacity, withdrawing capacity, and movement easiness (Table 9.1). The water holding capacity is described by porosity and specific retention, while the water withdrawing capacity is described by specific yield, storativity (or storage coefficient), and specific storage. The water movement easiness can be described by permeability, saturated hydraulic conductivity, and transmissivity. The typical values of saturated soil moisture and saturated hydraulic conductivity are listed in Table 3.12, and the typical values of other aquifer properties are presented in Table 9.2. These aquifer properties can be described using mathematical equations.

The aquifer properties of water holding capacity can be determined as:

$$\underbrace{n}_{\text{porosity}} = \frac{\overbrace{V_v}^{\text{volume of voids}}}{\underbrace{V_T}_{\text{total volume of aquifer soil}}} = \underbrace{\theta_{sat}}_{\text{saturated soil moisture (see Section 3.2.2)}} \tag{9.2}$$

$$\underbrace{S_r}_{\text{specific retention}} = \frac{\overbrace{V_{wr}}^{\text{maximum volume of water retained against gravity}}}{\underbrace{V_T}_{\text{total volume of aquifer soil}}} = \underbrace{\theta_{fc}}_{\text{field capacity (see Section 3.2.2)}} \tag{9.3}$$

The aquifer properties of water withdrawing capacity can be determined as:

$$\underbrace{S_y}_{\text{specific yield}} = \frac{\overbrace{V_v}^{\text{volume of voids}} - \overbrace{V_{wr}}^{\text{maximum volume of water retained against gravity}}}{\underbrace{V_T}_{\text{total volume of aquifer soil}}} = \underbrace{\theta_{sat}}_{\text{saturated soil moisture}} - \underbrace{\theta_{fc}}_{\text{field capacity}} = \frac{\overbrace{V_w}^{\text{volume of water released or absorbed}}}{\underbrace{A}_{\text{surface area of aquifer}}\underbrace{(\Delta h)}_{\text{change of water table}}} \tag{9.4}$$

Table 9.1 List of aquifer properties.

Characteristic	Property [units][1]	Symbol	Aquifer type	Brief definition
Holding capacity	Porosity [−][2]	n	All	Volumetric percentage of voids
	Specific retention [−][2]	S_r	Unconfined	Maximum volumetric percentage of water retained against gravity
Withdrawing capacity	Specific yield [−][2]	S_y	Unconfined	Maximum volumetric percentage of water drained by gravity
	Storativity (storage coefficient) [−]	S	Confined Semiconfined	Water volume released or absorbed per unit surface area of aquifer per unit change of hydraulic head normal to the surface
	Specific storage [ft^{-1} for BG; m^{-1} for SI]	S_s	All	Specific yield or storativity per unit thickness of aquifer
Movement easiness	Intrinsic permeability [ft^2 for BG; m^2 or Darcy for SI]	K	All	Size and sinuosity of connected voids; intrinsic characteristics of aquifer soil
	Saturated hydraulic conductivity [ft sec^{-1} for BG; m s^{-1} for SI]	K_{sat}	All	Easiness of water movement in aquifer; characteristics of aquifer soil and water
	Transmissivity [ft^2 sec^{-1} for BG; m^2 s^{-1} for SI]	T	All	Multiplication of saturated hydraulic conductivity and thickness of aquifer

[1] −: dimensionless; 1 Darcy ≈ 0.9869 μm^2.

[2] $n = \theta_{sat}$, $S_r = \theta_{fc}$, and $S_y = \theta_{sat} - \theta_{fc}$, where θ_{sat} is the saturated soil moisture; and θ_{fc} is the field capacity.

Table 9.2 Typical values for aquifer properties.

Material	Porosity, n	Specific retention, S_r	Specific yield, S_y	Specific storage, S_s (ft^{-1})	Permeability, κ (ft^2)
Clay	0.30 ∼ 0.42	0.25 ∼ 0.42	0.00 ∼ 0.12	$2.8 \times 10^{-4} \sim 7.8 \times 10^{-4}$	$10^{-10} \sim 10^{-9}$
Sand	0.39 ∼ 0.43	0.12 ∼ 0.20	0.23 ∼ 0.27	$0.6 \times 10^{-4} \sim 3.1 \times 10^{-4}$	$10^{-8} \sim 10^{-6}$
Sandy gravel	0.33 ∼ 0.45	0.05 ∼ 0.12	0.14 ∼ 0.38	$1.5 \times 10^{-5} \sim 3.1 \times 10^{-5}$	$10^{-4} \sim 10^{-2}$
Fissured rock	–	–	–	$0.1 \times 10^{-5} \sim 2.1 \times 10^{-5}$	$10^{-3} \sim 10^{0}$
Sound rock	–	–	–	$< 1.0 \times 10^{-6}$	$10^{-12} \sim 10^{-11}$

$$S = \frac{V_w}{A\,(\Delta h)} \tag{9.5}$$

storativity → S; volume of water released or absorbed → V_w; surface area of aquifer → A; change of hydraulic head normal to the surface → Δh

$$S_s = \begin{cases} \dfrac{S}{b} & \text{for confined or semiconfined aquifer} \\ \dfrac{S_y}{h_0} & \text{for unconfined aquifer} \end{cases} \tag{9.6}$$

specific storage → S_s; storativity (Eq. (9.5)) → S; thickness of aquifer → b; specific yield (Eq. (9.4)) → S_y; saturated thickness of unconfined aquifer → h_0

Finally, the aquifer properties of water movement easiness can be determined as:

$$\kappa = C_{cs}\, d_{50}^2 \tag{9.7}$$

intrinsic permeability → κ; dimensionless constant related to connectivity of void → C_{cs}; average diameter of aquifer soil particles → d_{50}

$$K_{sat} = \frac{g}{\nu}\,\kappa \tag{9.8}$$

saturated hydraulic conductivity → K_{sat}; gravitational acceleration (= 32.2 ft sec^{-2} for BG; 9.81 m s^{-2} for SI) → g; kinematic viscosity of water (Table 2.1) → ν; intrinsic permeability (Eq. (9.7)) → κ

$$T = \begin{cases} K_{sat}\, b & \text{for confined or semiconfined aquifer} \\ K_{sat}\, h_0 & \text{for unconfined aquifer} \end{cases} \tag{9.9}$$

transmissivity → T; saturated hydraulic conductivity → K_{sat}; thickness of aquifer → b; saturated thickness of unconfined aquifer → h_0

Example 9.3 A 100 km^2 unconfined aquifer has a specific yield of 0.2. If 3.5×10^6m^3 water is pumped by wells uniformly distributed across the aquifer area, determine the average drawdown of the water table.

Solution

$A = 100\,\text{km}^2 = 10^8\text{m}^2,\ S_y = 0.2,\ V_w = 3.5 \times 10^6\ \text{m}^3$

The total volume of aquifer impacted by pumping: $V_T = A\,(\Delta h) = 10^8\,(\Delta h)\ \text{m}^3$.

Eq. (9.4): $0.2 = (3.5 \times 10^6\text{m}^3)\ /\ [10^8\,(\Delta h)\ \text{m}^3] \rightarrow \Delta h = 0.175\,\text{m} = 17.5\,\text{cm}$, drawdown.

Example 9.4 A 30-ft-thick confined aquifer has a specific storage of $2.5 \times 10^{-4} ft^{-1}$. As a result of pumping by wells uniformly distributed across an area of 0.5 ac, the average piezometric surface beneath the area drops by 0.3 ft. Determine the total volume of groundwater pumped.

Solution

$b = 30\,ft$, $S_s = 2.5 \times 10^{-4}\,ft^{-1}$, $A = 0.5\,ac = 21{,}780\,ft^2$, $\Delta h = 0.3\,ft$

Eq. (9.6): $S = \left(2.5 \times 10^{-4}\,ft^{-1}\right) * (30\,ft) = 0.0075.$

Eq. (9.5): $\mathcal{V}_w = 0.0075 * \left(21{,}780\,ft^2\right) * (0.3\,ft) = 49.005\,ft^3.$

9.2.3 Variability of Saturated Hydraulic Conductivity

As introduced above (see Table 9.1), one measure of the movement easiness of water through an aquifer is saturated hydraulic conductivity, indicated with the symbol K_{sat}. Values for K_{sat} depend on several factors, such as soil texture and pore connectivity. Because the pore spaces in an aquifer are almost always saturated with water, for convenience, the subscript in the symbol K_{sat} will be dropped hereinafter. That is, K will be used to signify saturated hydraulic conductivity unless specifically noted.

If an aquifer has a K that does not vary from location to location in a given direction, the aquifer is considered homogenous; otherwise, it is heterogeneous. In addition, if an aquifer has a K that does not vary with direction at a given location, the aquifer is considered isotropic; otherwise, it is anisotropic. Thus, an aquifer may be: (1) homogenous isotropic; (2) homogenous anisotropic; (3) heterogeneous isotropic; or (4) heterogeneous anisotropic. Although most real aquifers are heterogeneous anisotropic, it is not uncommon for practical applications to approximate an aquifer of interest to be homogenous and isotropic. However, a layered aquifer (e.g., Figure 9.3) is obviously anisotropic and the variation of K with directions needs to be properly addressed.

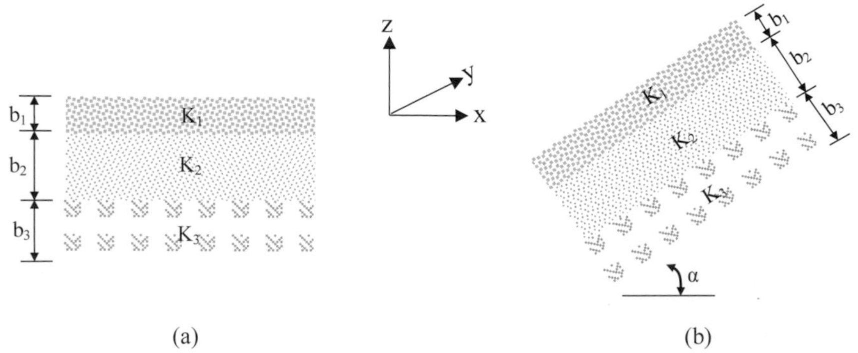

Figure 9.3 Geometric orientation of a layered aquifer. The sketches include three layers each with distinct values of saturated hydraulic conductivity, K. (a) Layers are parallel to the x-axis that lies within each planar layer; and (b) layers are inclined to the x-axis by an angle α.

For a layered aquifer such as that illustrated in Figure 9.3, each of the layers is assumed to be homogenous and isotropic but different layers have different hydraulic conductivities. The equivalent hydraulic conductivities of the aquifer as a whole in the parallel and perpendicular directions to the layers can be computed as (see Example 9.12):

total number of layers

saturated hydraulic conductivity of layer i

thickness of layer i

saturated hydraulic conductivity of aquifer in the direction of parallel to its layers

$$K_H = \frac{\sum_{i=1}^{m} (K_i b_i)}{\sum_{i=1}^{m} b_i} \tag{9.10}$$

total number of layers

thickness of layer i

saturated hydraulic conductivity of aquifer in the direction of perpendicular to its layers

$$K_V = \frac{\sum_{i=1}^{m} b_i}{\sum_{i=1}^{m} \frac{b_i}{K_i}} \tag{9.11}$$

saturated hydraulic conductivity of layer i

In the Cartesian coordinate system shown in Figure 9.3, the hydraulic conductivity in the y-axis direction is usually close to (or the same as) that in the x-axis direction, so they are not differentiated in practical applications. However, the hydraulic conductivity in the z-axis direction can be very different from that in the x-axis direction. The hydraulic conductivities in the x-, y-, and z-axis directions can be computed as:

saturated hydraulic conductivity of aquifer in the direction of parallel to its layers (Eq. (9.10))

saturated hydraulic conductivity of aquifer in the direction of perpendicular to its layers (Eq. (9.11))

saturated hydraulic conductivity of aquifer in the x-axis direction

angle between the x-axis and layers (see Figure 9.3b)

$$K_x = K_H \cos \alpha + K_v \sin \alpha \tag{9.12}$$

saturated hydraulic conductivity of aquifer in the y-axis direction

saturated hydraulic conductivity of aquifer in the x-axis direction (Eq. (9.12))

$$K_y = K_x \tag{9.13}$$

saturated hydraulic conductivity of aquifer in the direction of parallel to its layers (Eq. (9.10))

saturated hydraulic conductivity of aquifer in the direction of perpendicular to its layers (Eq. (9.11))

saturated hydraulic conductivity of aquifer in the z-axis direction

$$K_z = K_H \sin \alpha + K_v \cos \alpha \tag{9.14}$$

angle between the x-axis and layers (see Figure 9.3b)

Example 9.5 For the aquifer shown in Figure 9.3a, if $b_1 = 10\,\text{ft}$, $K_1 = 0.85\,\text{ft}\,\text{d}^{-1}$, $b_2 = 35\,\text{ft}$, $K_2 = 0.15\,\text{ft}\,\text{d}^{-1}$, $b_3 = 25\,\text{ft}$, and $K_3 = 1.55\,\text{ft}\,\text{d}^{-1}$, determine K_x, K_y, and K_z.

Solution

The aquifer layers are parallel to the x-axis direction $\rightarrow \alpha = 0°$.

Eq. (9.10): $K_H = \left[\left(0.85\,\text{ft}\,\text{d}^{-1}\right) * (10\,\text{ft}) + \left(0.15\,\text{ft}\,\text{d}^{-1}\right) * (35\,\text{ft}) + \left(1.55\,\text{ft}\,\text{d}^{-1}\right) * (25\,\text{ft})\right] / (10\,\text{ft} + 35\,\text{ft} + 25\,\text{ft}) = 0.75\,\text{ft}\,\text{d}^{-1}$.

Eq. (9.11): $K_V = (10\,\text{ft} + 35\,\text{ft} + 25\,\text{ft}) / [(10\,\text{ft}) / \left(0.85\,\text{ft}\,\text{d}^{-1}\right) + (35\,\text{ft}) / \left(0.15\,\text{ft}\,\text{d}^{-1}\right) + (25\,\text{ft}) / \left(1.55\,\text{ft}\,\text{d}^{-1}\right)] = 0.27\,\text{ft}\,\text{d}^{-1}$.

Eq. (9.12): $K_x = \left(0.75\,\text{ft}\,\text{d}^{-1}\right) * \cos(0°) + \left(0.27\,\text{ft}\,\text{d}^{-1}\right) * \sin(0°) = 0.75\,\text{ft}\,\text{d}^{-1}$.

Eq. (9.13): $K_y = K_x = 0.75\,\text{ft}\,\text{d}^{-1}$.

Eq. (9.14): $K_z = \left(0.75\,\text{ft}\,\text{d}^{-1}\right) * \sin(0°) + \left(0.27\,\text{ft}\,\text{d}^{-1}\right) * \cos(0°) = 0.27\,\text{ft}\,\text{d}^{-1}$.

Example 9.6 For the aquifer shown in Figure 9.3b, if $b_1 = 3.5\,\text{m}$, $K_1 = 0.25\,\text{m}\,\text{d}^{-1}$, $b_2 = 10.0\,\text{m}$, $K_2 = 0.05\,\text{m}\,\text{d}^{-1}$, $b_3 = 7.5\,\text{m}$, $K_3 = 2.35\,\text{m}\,\text{d}^{-1}$, and $\alpha = 15°$, determine K_x, K_y, and K_z.

Solution

The aquifer layers are inclined with $\alpha = 15°$.

Eq. (9.10): $K_H = [\left(0.25\,\text{m}\,\text{d}^{-1}\right) * (3.5\,\text{m}) + \left(0.05\,\text{m}\,\text{d}^{-1}\right) * (10.0\,\text{m}) + \left(2.35\,\text{m}\,\text{d}^{-1}\right) * (7.5\,\text{m})] / (3.5\,\text{m} + 10.0\,\text{m} + 7.5\,\text{m}) = 0.91\,\text{m}\,\text{d}^{-1}$.

Eq. (9.11): $K_V = (3.5\,\text{m} + 10.0\,\text{m} + 7.5\,\text{m}) / [(3.5\,\text{m}) / \left(0.25\,\text{m}\,\text{d}^{-1}\right) + (10.0\,\text{m}) / \left(0.05\,\text{m}\,\text{d}^{-1}\right) + (7.5\,\text{m}) / \left(2.35\,\text{m}\,\text{d}^{-1}\right)] = 0.097\,\text{m}\,\text{d}^{-1}$.

Eq. (9.12): $K_x = \left(0.91\,\text{m}\,\text{d}^{-1}\right) * \cos(15°) + \left(0.097\,\text{m}\,\text{d}^{-1}\right) * \sin(15°) = 0.90\,\text{m}\,\text{d}^{-1}$.

Eq. (9.13): $K_y = K_x = 0.90\,\text{m}\,\text{d}^{-1}$.

Eq. (9.14): $K_z = \left(0.91\,\text{m}\,\text{d}^{-1}\right) * \sin(15°) + \left(0.097\,\text{m}\,\text{d}^{-1}\right) * \cos(15°) = 0.33\,\text{m}\,\text{d}^{-1}$.

9.3 Flow in an Aquifer

This section presents the principles of groundwater movement in aquifers and discusses flow across the interface between two aquifer layers with different properties. It also presents the fluctuations of the water table, as influenced by recharge, in an unconfined aquifer surrounded by two waterbodies.

9.3.1 Darcy's Law and Governing Equation

The mean velocity in an aquifer, also called Darcy velocity, is the ratio of flow rate to the total area normal to the flow direction. However, because groundwater can only flow through pores and voids, the actual flow area is equal to the total area multiplied by the porosity. That is, the actual velocity, called pore velocity or seepage velocity, can be computed as:

$$V_\alpha = \frac{Q}{An} = \frac{V}{n} \tag{9.15}$$

(pore velocity → V_α; flow rate → Q; total area normal to flow direction → A; porosity → n; Darcy velocity → V)

Given that groundwater velocity is generally very slow, at a point in aquifer, the velocity head can be negligible in comparison with the potential and pressure heads. That is, hydraulic head, the summation of potential and pressure heads, can be used in the energy equation (Eq. (2.21)) to analyze groundwater flow in an aquifer. As illustrated in Figure 9.4, the Darcy velocity can be computed in terms of Darcy' law (Darcy, 1856), which is expressed as:

$$V = -K\frac{dh}{dL} \tag{9.16}$$

(Darcy velocity → V; hydraulic conductivity → K; hydraulic head → h; flow path length → L)

Darcy's law is valid when the Reynolds number for groundwater flow is less than one, which is true in most aquifers (Strack, 2017). Such a Reynolds number is defined as:

$$R_e = \frac{Vd_{50}}{\nu} \tag{9.17}$$

(Reynolds number → R_e; Darcy velocity → V; average diameter of aquifer soil particles → d_{50}; kinematic viscosity of groundwater → ν)

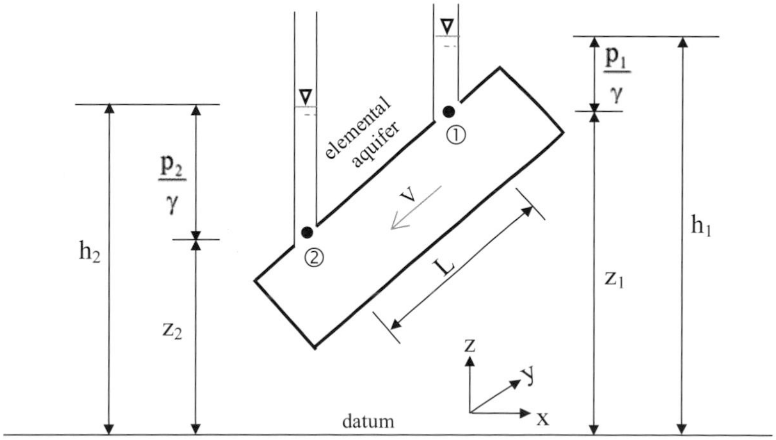

Figure 9.4 Elemental aquifer used in derivation of Darcy's law. Diagram shows the hydraulic heads, h_1 and h_2, and Darcy velocity in the aquifer.

In the Cartesian coordinate system shown in Figure 9.4, Darcy's law can be rewritten as:

$$V_x = -K_x \frac{dh}{dx} \tag{9.18}$$

Darcy velocity in x-axis direction (V_x); hydraulic conductivity in x-axis direction (K_x); hydraulic head (dh); flow path length along x-axis direction (dx)

$$V_y = -K_y \frac{dh}{dy} \tag{9.19}$$

Darcy velocity in y-axis direction (V_y); hydraulic conductivity in y-axis direction (K_y); hydraulic head (dh); flow path length along y-axis direction (dy)

$$V_z = -K_z \frac{dh}{dz} \tag{9.20}$$

Darcy velocity in z-axis direction (V_z); hydraulic conductivity in z-axis direction (K_z); hydraulic head (dh); flow path length along z-axis direction (dz)

As shown in Figure 9.5, applying mass conservation and Darcy's law (Eqs. (9.18) to (9.20)) to an elemental aquifer and simplifying, one can derive (see Example 9.13) the governing equation of groundwater flow expressed as:

$$\frac{\partial}{\partial x}\left(K_x \frac{\partial h}{\partial x}\right) + \frac{\partial}{\partial y}\left(K_y \frac{\partial h}{\partial y}\right) + \frac{\partial}{\partial z}\left(K_z \frac{\partial h}{\partial z}\right) + W_{ss} = S_s \frac{\partial h}{\partial t} \tag{9.21}$$

hydraulic conductivity in x-axis direction (K_x); hydraulic head (h); hydraulic conductivity in y-axis direction (K_y); hydraulic conductivity in z-axis direction (K_z); specific storage (Eq. (9.6)) (S_s); time (t); flow path length along x-axis direction (∂x); flow path length along y-axis direction (∂y); flow path length along z-axis direction (∂z); sink (+) or source (−) water volumetric rate per unit volume of aquifer (W_{ss})

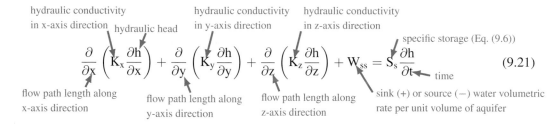

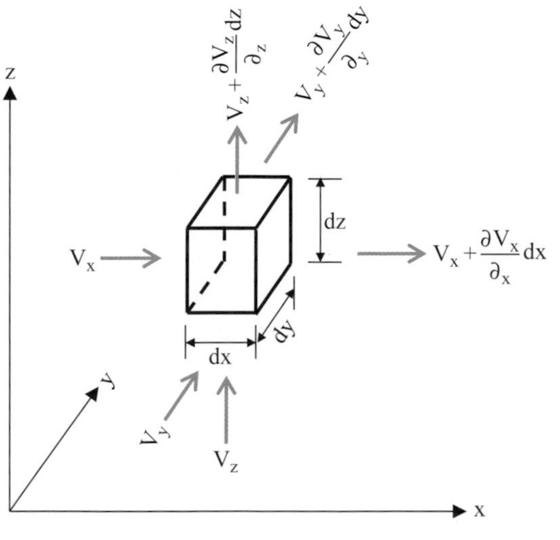

Figure 9.5 Analytical framework for mass conservation in an elemental aquifer. Influxes into and effluxes out of an elemental aquifer are shown.

In Eq. (9.21), W_{ss} represents the net rate of water per unit volume of aquifer, which is the summation of the addition (e.g., percolation and/or recharging) and the removal (e.g., pumping). If $W_{ss} > 0$, the aquifer is a sink; otherwise, it is a source.

Example 9.7 A homogeneous isotropic unconfined aquifer ($K = 0.85\,ft\,d^{-1}$, $n = 0.35$) has a horizontal confining bed. If the water table drops 1.2 ft per 20 ft horizontal distance, determine the Darcy and pore velocities as well as the travel time over a 100 ft distance.

Solution

Choose the confining bed as the datum $\rightarrow$ hydraulic head = water table.

Hydraulic gradient: $dh/dL = (-1.2\,ft)\,/\,(20\,ft) = -0.06$.

Darcy's law (Eq. (9.16)): $V = -\left(0.85\,ft\,d^{-1}\right) * (-0.06) = 0.051\,ft\,d^{-1}$.

Eq. (9.15): $V_\alpha = \left(0.051\,ft\,d^{-1}\right)/0.35 = 0.146\,ft\,d^{-1}$.

Travel time: $t = (100\,ft)\,/\left(0.146\,ft\,d^{-1}\right) \approx 685\,d$.

Example 9.8 As illustrated in Figure 9.2, an unconfined aquifer ($K_1 = 35\,ft\,d^{-1}$, $n_1 = 0.38$) overlays a semiconfined aquifer. The unconfined aquifer has a 15-ft-thick horizontal confining bed of aquitard ($K_2 = 0.5\,ft\,d^{-1}$). Assume that the aquifers and aquitard are homogeneous and isotropic. In reference to the bottom of the aquitard, the water table and piezometric surface, both of which are horizontal, are 80 and 88 ft, respectively. Determine the:

(1) Groundwater flow direction

(2) Darcy velocity of groundwater penetrating the aquitard

(3) Pore velocity of groundwater in the unconfined aquifer.

Solution

$K_1 = 35\,ft\,d^{-1}$, $n_1 = 0.38$, $z_1 = 80\,ft$, $K_2 = 0.5\,ft\,d^{-1}$,	$b_2 = 10\,ft$, $z_3 = 88\,ft$
Hydraulic head at the water table:	$h_1 = 80\,ft$
Hydraulic head at the top of the aquitard:	$h_{1,2}$
Hydraulic head at the bottom of the aquitard:	$h_3 = 88\,ft$.

(1) $h_3 > h_1 \rightarrow$ groundwater flows upward from the semiconfined to unconfined aquifer.

(2) Hydraulic gradient over the aquitard: $(dh/dL)_2 = [(h_{1,2} - 88)\,ft]\,/\,(10\,ft) = 0.1h_{1,2} - 8.8$.

Darcy's law: $V_2 = -K_2\,(dh/dL)_2 = -\left(0.5\,ft\,d^{-1}\right) * (0.1h_{1,2} - 8.8) = -0.05h_{1,2} + 4.4\,ft\,d^{-1}$.

Hydraulic gradient over the unconfined aquifer:

$$(dh/dL)_1 = [(80 - h_{1,2})\,ft]\,/\,[(80 - 10)\,ft] = (8.0 - 0.1h_{1,2})\,/7.$$

Darcy's law: $V_1 = -K_1\,(dh/dL)_1 = -\left(35\,ft\,d^{-1}\right) * [(8.0 - 0.1h_{1,2})\,/7] = 0.5h_{1,2} - 40\,ft\,d^{-1}$.

Continuity equation: $V_1 = V_2 \rightarrow 0.5h_{1,2} - 40 = -0.05h_{1,2} + 4.4 \rightarrow h_{1,2} = 80.727$ ft.

$$V_2 = -0.05 * 80.727 + 4.4\,\text{ft d}^{-1} = 0.364\,\text{ft d}^{-1}.$$

(3) $V_1 = V_2 = 0.364\,\text{ft d}^{-1}$

Eq. (9.15): $V_{\alpha,1} = V_1/n_1 = \left(0.364\,\text{ft d}^{-1}\right)/0.38 = 0.958\,\text{ft d}^{-1}$.

Example 9.9 As illustrated in Figure 9.2, an unconfined aquifer ($K_1 = 10\,\text{m d}^{-1}$) overlays a 2.0-m-thick horizontal aquitard ($K_2 = 0.65\,\text{m d}^{-1}$), which in turn overlays a semiconfined aquifer. The water table of the unconfined aquifer is 1000 m above the aquitard bed. Assume that the aquifers and aquitard are homogeneous and isotropic. If the Darcy velocity of groundwater penetrating through the aquitard from the unconfined aquifer downward into the semiconfined aquifer is measured to be $0.275\,\text{m d}^{-1}$, determine the piezometric surface of the semiconfined aquifer.

Solution

$K_1 = 10\,\text{m d}^{-1}$, $h_1 = 1000$ m, $K_2 = 0.65\,\text{m d}^{-1}$, $b_2 = 2.0$ m, $V_2 = -0.275\,\text{m d}^{-1}$

Let $h_{1,2}$ be the hydraulic head at the top of the aquitard.

Hydraulic gradient over the aquitard: $(dh/dL)_2 = [(h_{1,2} - h_2)\ \text{m}] / (2.0\,\text{m}) = 0.5h_{1,2} - 0.5h_2$.

Darcy's law: $-0.275\,\text{m d}^{-1} = -\left(0.65\,\text{m d}^{-1}\right) * (0.5h_{1,2} - 0.5h_2) \rightarrow h_2 = h_{1,2} - 11/13$ m.

Hydraulic gradient over the unconfined aquifer:

$$(dh/dL)_1 = [(1000 - h_{1,2})\ \text{m}] / (2.0\,\text{m}) = 500 - 0.5h_{1,2}.$$

Darcy's law and mass conservation for the unconfined aquifer:

$$-0.275\,\text{m d}^{-1} = -\left(10\,\text{m d}^{-1}\right) * (500 - 0.5h_{1,2}) \rightarrow h_{1,2} = 999.945\,\text{m}.$$

Thus, $h_2 = 999.945 - 11/13\,\text{m} = 999.099$ m.

Example 9.10 A 1000-m-long homogeneous isotropic confined aquifer ($K = 10\,\text{m d}^{-1}$) (Figure 9.6) has a horizontal lower bed but a linearly varying thickness. Assuming a 1-D horizontal steady flow, determine the:

(1) Flow rate per unit width of the aquifer
(2) Piezometric surface above the upper confining bed at $x = 500$ m.

Solution

$K = 10\,\text{m d}^{-1}$

(1) Hydraulic heads: $h\,(x = 0\,\text{m}) = 60\,\text{m} + 20\,\text{m} = 80\,\text{m}$

$$h\,(x = 1000\,\text{m}) = 40\,\text{m} + 30\,\text{m} = 70\,\text{m}.$$

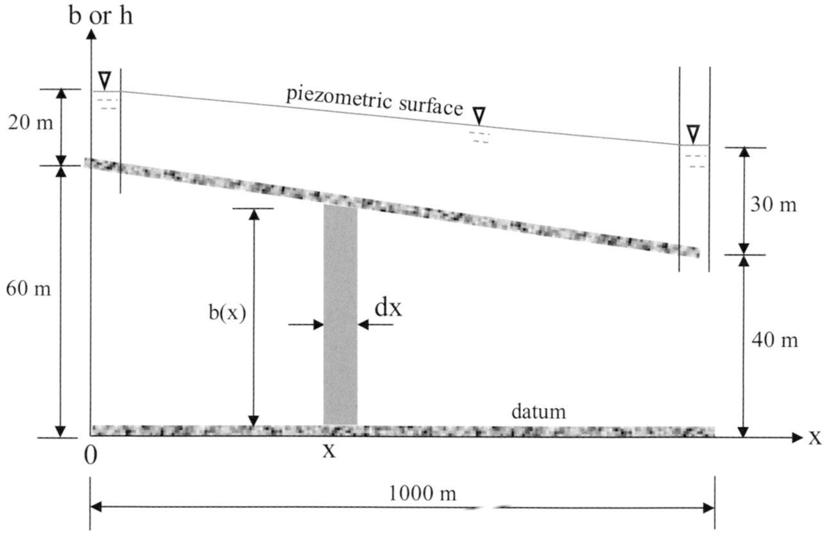

Figure 9.6 **The confined aquifer used in Example 9.10.**

Thickness: $b(x) = 60\,\text{m} + [(40\,\text{m}) - (60\,\text{m})]/(1000\,\text{m}) * x = 60 - 0.02x$ m.

Flow area per unit width: $A(x) = (60 - 0.02x\,\text{m}) * (1\,\text{m}) = 60 - 0.02x\,\text{m}^2$.

1-D horizontal flow $\rightarrow$ Eq. (9.18): $V_x = -\left(10\,\text{m}\,\text{d}^{-1}\right)(dh/dx) = -10\,(dh/dx)\ \text{m}\,\text{d}^{-1}$.

Steady flow $\rightarrow$ constant $Q = V_x A(x) = \left[-10\,(dh/dx)\ \text{m}\,\text{d}^{-1}\right]\left(60 - 0.02x\,\text{m}^2\right)$

$$= (0.2x - 600)\,(dh/dx)\ \text{m}^3\,\text{d}^{-1}$$

$$\rightarrow Q/(0.2x - 600)dx = dh \rightarrow \int_0^{1000} \frac{Q}{0.2x - 600} dx = \int_{80}^{70} dh$$

$$\rightarrow \frac{Q}{0.2} \ln(0.2x - 600)\Big|_0^{1000} = h\Big|_{80}^{70} \rightarrow \frac{Q}{0.2} \ln \frac{0.2 * 1000 - 600}{0.2 * 0 - 600} = 70 - 80$$

$$\rightarrow \frac{Q}{0.2} \ln \frac{2}{3} = -10 \rightarrow Q = 4.93\ \text{m}^3\,\text{d}^{-1}.$$

(2) At x : $4.93/(0.2x - 600)\,dx = dh \rightarrow \int_0^x \frac{4.93}{0.2x - 600} dx = \int_{80}^{h(x)} dh$

$$\rightarrow \frac{4.93}{0.2} \ln(0.2x - 600)\Big|_0^x = h\Big|_{80}^{h(x)} \rightarrow h(x) = 80 + 24.65 \ln\left(1 - \frac{x}{3000}\right)\ \text{m}.$$

$$h(x = 500\,\text{m}) = 80 + 24.65 * \ln(1 - 500/3000)\ \text{m} = 75.51\ \text{m}.$$

$$b(x = 500\,\text{m}) = 60 - 0.02 * 500\,\text{m} = 50\,\text{m}.$$

The piezometric surface above the upper bed: 75.51 m – 50 m = 25.51 m.

Example 9.11 Redo Example 9.10 if the aquifer is heterogeneous isotropic and has a spatially varying hydraulic conductivity of $K(x) = 10 + 0.005x\ \mathrm{m\,d^{-1}}$.

Solution

$K(x) = 10 + 0.005x\ \mathrm{m\,d^{-1}}$

(1) Eq. (9.18): $V_x = -(10 + 0.005x\ \mathrm{m\,d^{-1}})(dh/dx) = -(10 + 0.005x)(dh/dx)\ \mathrm{m\,d^{-1}}$.

$$Q = [-(10 + 0.005x)(dh/dx)\ \mathrm{m\,d^{-1}}](60 - 0.02x\ \mathrm{m^2})$$
$$= (0.005x + 10)(0.02x - 60)(dh/dx)\ \mathrm{m^3\,s^{-1}}$$

$$\rightarrow Q/[(0.005x + 10)(0.02x - 60)]dx = dh \rightarrow \int_0^{1000} \frac{Q}{(0.005x + 10)(0.02x - 60)}dx = \int_{80}^{70} dh$$

$$\rightarrow Q\int_0^{1000}\left[\frac{0.04}{0.02x - 60} - \frac{0.01}{0.005x + 10}\right]dx = \int_{80}^{70} dh$$

$$\rightarrow Q\left[\frac{0.04}{0.02}\ln(0.02x - 60) - \frac{0.01}{0.005}\ln(0.005x + 10)\right]\Bigg|_0^{1000} = h\Big|_{80}^{70}$$

$$\rightarrow Q\left[\frac{0.04}{0.02}\ln\left(\frac{0.02 * 1000 - 60}{0.02 * 0 - 60}\right) - \frac{0.01}{0.005}\ln\left(\frac{0.005 * 1000 + 10}{0.005 * 0 + 10}\right)\right] = 70 - 80$$

$$\rightarrow Q(-0.81093 - 0.81093) = -10 \rightarrow Q = 6.17\ \mathrm{m^3\,d^{-1}}.$$

(2) At x : $6.17/[(0.005x + 10)(0.02x - 60)]dx = dh$

$$\rightarrow \int_0^{x} \frac{6.17}{(0.005x + 10)(0.02x - 60)}dx = \int_{80}^{h(x)} dh$$

$$\rightarrow 6.17\left[\frac{0.04}{0.02}\ln(0.02x - 60) - \frac{0.01}{0.005}\ln(0.005x + 10)\right]\Bigg|_0^{x} = h\Big|_{80}^{h(x)}$$

$$\rightarrow 6.17\left[\frac{0.04}{0.02}\ln\left(\frac{0.02x - 60}{0.02 * 0 - 60}\right) - \frac{0.01}{0.005}\ln\left(\frac{0.005x + 10}{0.005 * 0 + 10}\right)\right] = h(x) - 80$$

$$\rightarrow h(x) = 80 + 12.34\left[\ln\left(1 - \frac{x}{3000}\right) - \ln\left(1 + \frac{x}{2000}\right)\right].$$

$$h(x = 500\ \mathrm{m}) = 80 + 12.34\,[\ln(1 - 500/3000) - \ln(1 + 500/2000)]\ \mathrm{m} = 75.00\ \mathrm{m}.$$

$$b(x = 500\ \mathrm{m}) = 60 - 0.02 * 500\ \mathrm{m} = 50\ \mathrm{m}.$$

The piezometric surface above the upper bed: 75.00 m − 50 m = 25.00 m.

Example 9.12 Derive Eqs. (9.10) and (9.11) by referencing Figure 9.3.

Solution

Apply Darcy's law for each layer and the aquifer as a whole; unit width.

In the horizontal direction: same hydraulic gradient for each layer and the aquifer.

For layer i: $Q_{x,i} = V_{x,i}\,(b_i \times 1) = [-K_i\,(dh/dx)]\,b_i = -K_i b_i\,(dh/dx)$

$$\text{total horizontal flow rate over the m layers: } Q_x = \sum_{i=1}^{m} Q_{x,i} = -\frac{dh}{dx}\sum_{i=1}^{m}(K_i b_i).$$

For the aquifer as a whole: $Q_x = \left(-K_H \frac{dh}{dx}\right)\sum_{i=1}^{m}(b_i \times 1) = -\frac{dh}{dx}K_H \sum_{i=1}^{m} b_i$

$$-\frac{dh}{dx}\sum_{i=1}^{m}(K_i b_i) = -\frac{dh}{dx}K_H\sum_{i=1}^{m} b_i \rightarrow K_H = \frac{\sum_{i=1}^{m}(K_i b_i)}{\sum_{i=1}^{m} b_i} \rightarrow \text{Eq. (9.10).}$$

In the vertical direction: same flow rate through each layer and the aquifer.

For layer i: $Q_z = V_{z,i}\,(1 \times 1) = [-K_i\,(\Delta h_i/b_i)] = -K_i\,(\Delta h_i/b_i) \rightarrow \Delta h_i = -\,(b_i/K_i)\,Q_z$

$$\text{total change of hydraulic head over the m layers: } \Delta h_z = \sum_{i=1}^{m}\Delta h_i = -Q_z\sum_{i=1}^{m}\frac{b_i}{K_i}.$$

For the aquifer as a whole: $Q_z = \left(-K_V \frac{\Delta h_z}{\sum_{i=1}^{m} b_i}\right)(1 \times 1) = -K_V \frac{\Delta h_z}{\sum_{i=1}^{m} b_i} \rightarrow \Delta h_z = -Q_z \frac{\sum_{i=1}^{m} b_i}{K_V}$

$$-Q_z\sum_{i=1}^{m}\frac{b_i}{K_i} = -Q_z\frac{\sum_{i=1}^{m} b_i}{K_V} \rightarrow K_V = \frac{\sum_{i=1}^{m} b_i}{\sum_{i=1}^{m}\frac{b_i}{K_i}} \rightarrow \text{Eq. (9.11).}$$

Example 9.13 Derive Eq. (9.21) by referencing Figure 9.5.

Solution

Let Q_{sink} and Q_{source} be the sink and source rates, respectively.

The total inflow into the elemental aquifer is:

$$V_x\,(dydz) + V_y\,(dxdz) + V_z\,(dxdy) + Q_{sink}.$$

The total outflow out of the elemental aquifer is:

$$\left(V_x + \frac{\partial V_x}{\partial x}dx\right)(dydz) + \left(V_y + \frac{\partial V_y}{\partial y}dy\right)(dxdz) + \left(V_z + \frac{\partial V_z}{\partial z}dz\right)(dxdy) + Q_{source}.$$

The net flux (i.e., total inflow minus total outflow) into the elemental aquifer:

$$-\frac{\partial V_x}{\partial x}dxdydz - \frac{\partial V_y}{\partial y}dxdydz - \frac{\partial V_z}{\partial z}dxdydz + Q_{sink} - Q_{source}.$$

The change rate of water volume in the elemental aquifer is:

$$\text{Eqs. (9.5) and (9.6): } S\,(dxdy)\left(\frac{\partial h}{\partial t}\right) = \frac{S}{dz}\,(dz)\,(dxdy)\left(\frac{\partial h}{\partial t}\right) = S_s\left(\frac{\partial h}{\partial t}\right)(dxdydz).$$

Based on mass conservation, the next flux is equal to the change rate:

$$-\frac{\partial V_x}{\partial x}dxdydz - \frac{\partial V_y}{\partial y}dxdydz - \frac{\partial V_z}{\partial z}dxdydz + Q_{sink} - Q_{source} = S_s\left(\frac{\partial h}{\partial t}\right)(dxdydz).$$

Dividing (dxdydz) by the two sides, one can get:

$$-\frac{\partial V_x}{\partial x} - \frac{\partial V_y}{\partial y} - \frac{\partial V_z}{\partial z} + \frac{Q_{sink} - Q_{source}}{dxdydz} = S_s\frac{\partial h}{\partial t}.$$

Substituting Eqs. (9.18) through (9.20) for the partial derivatives of velocity, one can get:

$$-\frac{\partial}{\partial x}\left(-K_x\frac{\partial h}{\partial x}\right) - \frac{\partial}{\partial y}\left(-K_y\frac{\partial h}{\partial y}\right) - \frac{\partial}{\partial z}\left(-K_z\frac{\partial h}{\partial z}\right) + \frac{Q_{sink} - Q_{source}}{dxdydz} = S_s\frac{\partial h}{\partial t}$$

$$\rightarrow \frac{\partial}{\partial x}\left(K_x\frac{\partial h}{\partial x}\right) + \frac{\partial}{\partial y}\left(K_y\frac{\partial h}{\partial y}\right) + \frac{\partial}{\partial z}\left(K_z\frac{\partial h}{\partial z}\right) + W_{ss} = S_s\frac{\partial h}{\partial t} \rightarrow \text{Eq. (9.21).}$$

9.3.2 Flow Across a Boundary

The flow direction of groundwater is conventionally measured by the counterclockwise angle from vertical. Furthermore, the direction of flow can be deflected when it reaches a boundary such as the interface between an unsaturated zone and an unconfined aquifer (Figure 9.7a) or between two layers of an aquifer with different hydraulic conductivities (Figure 9.7b).

In Figure 9.7a, points 2 and 3 have the same hydraulic head and form one equipotential line, while points 1 and 1′ have the same hydraulic head and form another equipotential line. The head difference between points 1 and 2 is equal to the length of the line connecting 1′ and 2, which is $b[\tan(\alpha)]$. In the unconfined aquifer, the length of the flow path, which is the line connecting points 1 and 3, is $[b/\cos(\alpha)]\,[\cos(90° - \alpha - \alpha_2)] = [b/\cos(\alpha)]\sin(\alpha + \alpha_2)$, and the width of the flow path is $[b/\cos(\alpha)]\,[\sin(90° - \alpha - \alpha_2)] = [b/\cos(\alpha)]\cos(\alpha + \alpha_2)$. Noting that $V_1 b = V_2\,[b/\cos(\alpha)]\cos(\alpha + \alpha_2)$ and that V_2 can be computed using Darcy's law (Eq. (9.16)), one can rearrange (see Example 9.17) the continuity equation to obtain the formula expressed as:

$$\tan(\alpha + \alpha_2) = \frac{K_2 \tan(\alpha)}{V_1} \qquad (9.22)$$

slope angle of water table (α); deflection angle at water table (Figure 9.7a) (α_2); hydraulic conductivity of the unconfined aquifer (Figure 9.7a) (K_2); percolation rate in the unsaturated zone (Figure 9.7a) (V_1)

Similarly, in Figure 9.7b, points 2 and 3 have the same hydraulic head and form one equipotential line, while points 1 and 1′ have the same hydraulic head and form another equipotential line. The head difference between points 1 and 2 is equal to the length of the line connecting points 1′ and 2, which is $b\,[\sin(\alpha_1)]$. In the upper aquifer, the length of the flow path,

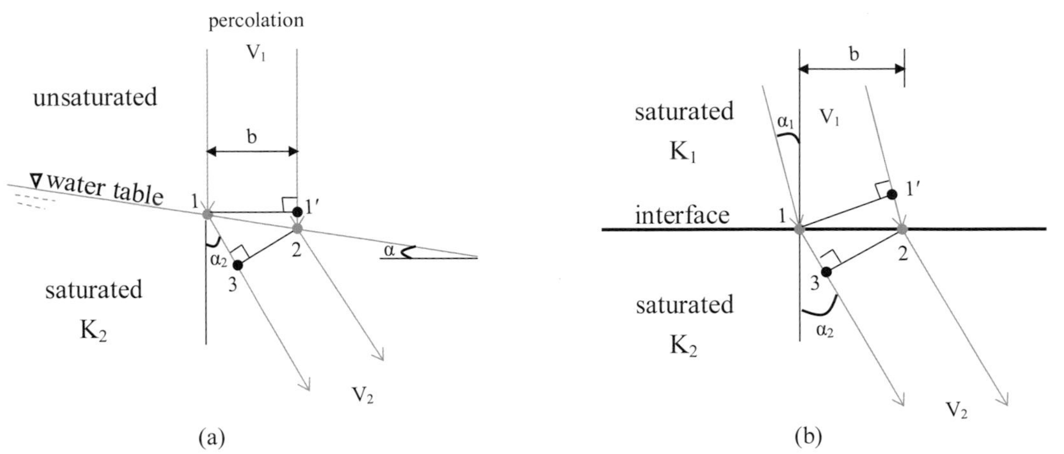

Figure 9.7 **Deflection of direction of water flow at an aquifer interface.** (a) Percolation of water downward throughout the unsaturated zone and flow direction deflected at the water table; and (b) groundwater flow deflected at the interface of two aquifers with different hydraulic conductivities.

which is the line connecting points 1′ and 2, is $b\left[\sin\left(\alpha_1\right)\right]$, and the width of the flow path is $b\left[\cos\left(\alpha_1\right)\right]$. On the other hand, in the lower aquifer, the length of flow path, which is the line connecting points 1 and 3, is $b\left[\sin\left(\alpha_2\right)\right]$, and the width of the flow path is $b\left[\cos\left(\alpha_2\right)\right]$. Noting that $V_1 b\left[\cos\left(\alpha_1\right)\right] = V_2 b\left[\cos\left(\alpha_2\right)\right]$ and that V_1 and V_2 can be computed using Darcy's law (Eq. (9.16)), one can rearrange (see Example 9.18) the continuity equation to obtain the formula expressed as:

deflection angle in lower aquifer (Figure 9.7b) — hydraulic conductivity of lower aquifer (Figure 9.7b)

$$\frac{\tan\left(\alpha_2\right)}{\tan\left(\alpha_1\right)} = \frac{K_2}{K_1} \tag{9.23}$$

deflection angle in upper aquifer (Figure 9.7b) — hydraulic conductivity of upper aquifer (Figure 9.7b)

Example 9.14 An unconfined aquifer has a hydraulic conductivity of 35 ft d^{-1} and its water table has a slope angle of 1.2°. For a percolation rate of 0.015 ft d^{-1}, determine the deflection angle when the percolation water flows across the water table.

Solution

$\alpha = 1.2°$, $K_2 = 35\,\text{ft}\,\text{d}^{-1}$, $V_1 = 0.015\,\text{ft}\,\text{d}^{-1}$

Eq. (9.22): $\tan\left(1.2° + \alpha_2\right) = \left(35\,\text{ft}\,\text{d}^{-1}\right) * \tan\left(1.2°\right) / \left(0.015\,\text{ft}\,\text{d}^{-1}\right) = 48.87637$

$$\rightarrow 1.2° + \alpha_2 = \tan^{-1}\left(48.87637\right) = 88.83° \rightarrow \alpha_2 = 87.63°.$$

Example 9.15 The deflection angle of groundwater flow in an unconfined aquifer (hydraulic conductivity of $3.5\,\mathrm{m\,d^{-1}}$) is 85°. If the confining bed of the aquifer is an aquitard with a hydraulic conductivity of $0.02\,\mathrm{m\,d^{-1}}$, determine the deflection angle when water penetrates through the aquitard.

Solution

$K_1 = 3.5\,\mathrm{m\,d^{-1}}$, $\alpha_1 = 85°$, $K_2 = 0.02\,\mathrm{m\,d^{-1}}$

Eq. (9.23): $\tan(\alpha_2) = \left(0.02\,\mathrm{m\,d^{-1}}\right) / \left(3.5\,\mathrm{m\,d^{-1}}\right) * \tan(85°) = 0.065315$

$$\rightarrow \alpha_2 = \tan^{-1}(0.065315) = 3.74°.$$

Example 9.16 An unconfined aquifer overlays and recharges a semiconfined aquifer. Let K_1, K_2, and K_3 be the hydraulic conductivities of the unconfined aquifer, aquitard, and confined aquifer, respectively. If $K_1 > K_3 > K_2$, sketch the flow path.

Solution

Eq. (9.23): $K_1 > K_2 \rightarrow \alpha_1 > \alpha_2$; $K_2 < K_3 \rightarrow \alpha_2 < \alpha_3$; $K_1 > K_3 \rightarrow \alpha_1 > \alpha_3$.

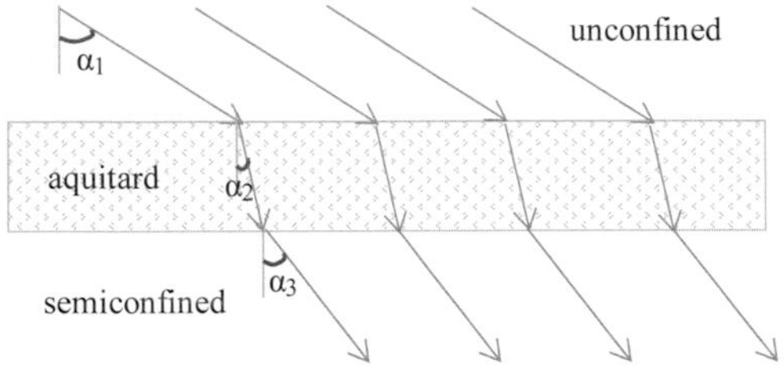

Figure 9.8 **Solution to Example 9.16.** Sketch of the flow path in the aquifers.

Example 9.17 Derive Eq. (9.22).

Solution

Reference Figure 9.7a and consider a unit-width flow.

The recharge from the unsaturated zone: $V_1 b$.

The flow rate in the unconfined aquifer: $V_2\,[b/\cos(\alpha)]\cos(\alpha + \alpha_2)$.

Continuity equation: $V_1 b = V_2\,[b/\cos(\alpha)]\cos(\alpha + \alpha_2) \rightarrow V_1\,[\cos(\alpha)] = V_2\,[\cos(\alpha + \alpha_2)]$.

Darcy's law: $V_2 = K_2\,\{[b\tan(\alpha)] / [b/\cos(\alpha)\sin(\alpha + \alpha_2)]\} = K_2 \tan(\alpha)\cos(\alpha) / \sin(\alpha + \alpha_2)$.

Substituting and simplifying: $V_1\,[\cos(\alpha)] = [K_2 \tan(\alpha)\cos(\alpha) / \sin(\alpha + \alpha_2)]\,[\cos(\alpha + \alpha_2)]$

$$\rightarrow \tan(\alpha + \alpha_2) = K_2 \tan(\alpha) / V_1 \rightarrow \text{Eq. (9.22).}$$

Example 9.18 Derive Eq. (9.23).

Solution

Reference Figure 9.7b and consider a unit-width flow.

Let $dh_{1,2}$ be the hydraulic head difference between point 1 and 2.

The flow rate in the upper aquifer: $V_1 b\,[\cos(\alpha_1)]$.

The flow rate in the lower aquifer: $V_2 b\,[\cos(\alpha_2)]$.

Continuity equation: $V_1 b\,[\cos(\alpha_1)] = V_2 b\,[\cos(\alpha_2)] \rightarrow V_1\,[\cos(\alpha_1)] = V_2\,[\cos(\alpha_2)]$.

Darcy's law: $V_1 = K_1 dh_{1,2}/\,[b \sin(\alpha_1)]\,;\; V_2 = K_2 dh_{1,2}/\,[b \sin(\alpha_2)]$.

Substituting and simplifying: $K_1 dh_{1,2}/\,[b \sin(\alpha_1)]\,[\cos(\alpha_1)] = K_2 dh_{1,2}/\,[b \sin(\alpha_2)]\,[\cos(\alpha_2)]$

$$\rightarrow \tan(\alpha_2)\,/\,\tan(\alpha_1) = K_2/K_1 \rightarrow \text{Eq. (9.23).}$$

9.3.3 Steady Water Tables Between Two Waterbodies

The water table in an unconfined aquifer that connects two waterbodies (e.g., rivers) depends on their relative water levels. If there is no percolation or recharge into the aquifer, the water in the waterbody (e.g., river 1) with a higher level will seep through the aquifer and flow into the other waterbody (e.g., river 2) with a lower level (Figure 9.9a). Otherwise, the water table may be increased to a level above which a "divide" is formed and the groundwater discharges from the aquifer into both waterbodies (Figure 9.9b). The groundwater in the aquifer on the left-hand side of the divide discharges into river 1, while the groundwater on the right-hand side discharges into river 2. By assuming that groundwater flows horizontally in an unconfined aquifer and that the groundwater discharge is proportional to the saturated aquifer thickness, which is known as the Dupuit–Forchheimer assumption (Dupuit, 1863; Forchheimer, 1886), Darcy's law can be used to analyze the seepage flow rates accurately. However, as illustrated in Figure 9.9, because the assumption is unrealistic for the water table at the upper boundary, the seepage face at the lower boundary, and the flow direction in the aquifer, the actually curved water table in the aquifer is approximated as a straight line and can only be computed with some errors. Near the lower boundary, the actual seepage face has a nappe shape and is located some distance above the water level of river 2, but it is modeled as a discharge point that is the intersect of the river water surface with the aquifer. Usually, the water table at the lower boundary has a large computational error. At the upper boundary, the actually sloped water table is simplified to be horizontal by the Dupuit–Forchheimer assumption. For simplicity, this section presents a special case that the aquifer and two waterbodies share a common horizontal confining bed as well as that the aquifer is homogenous and isotropic.

If there is no recharge (Figure 9.9a), applying Eq. (9.21) with $W_{ss} = 0$, integrating, and simplifying (see Example 9.21), one can get the formulas expressed as:

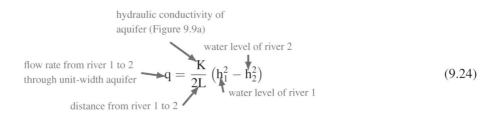

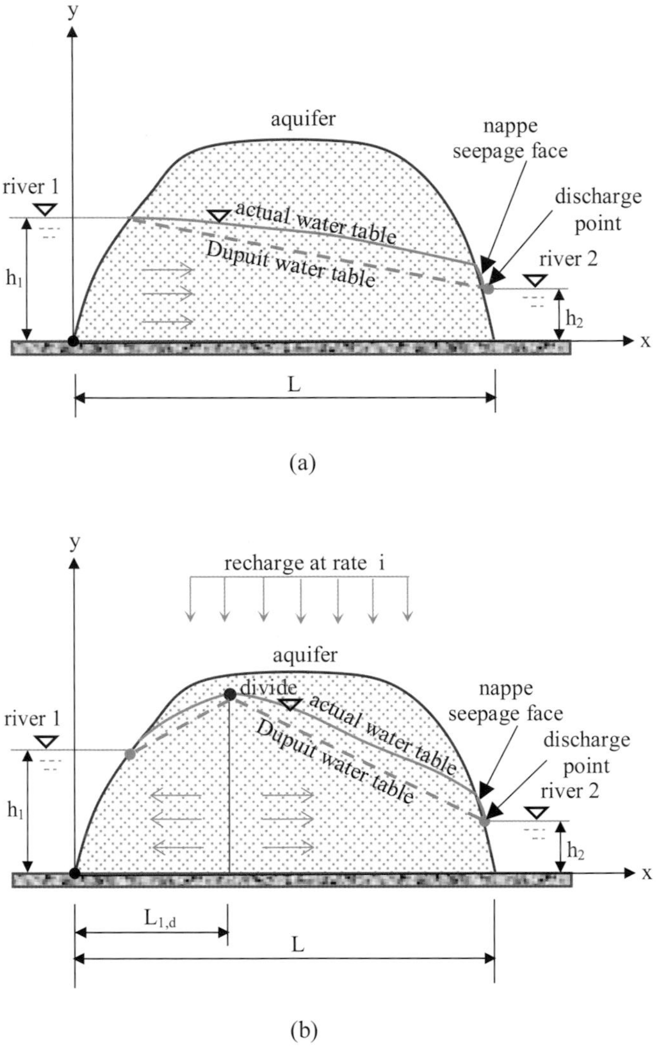

Figure 9.9 **Water table and groundwater flow in an unconfined aquifer connecting two rivers.** (a) Recharge is zero; and (b) recharge is nonzero.

$$q = \frac{K}{2L}\left(h_1^2 - h_2^2\right) \tag{9.24}$$

where: q = flow rate from river 1 to 2 through unit-width aquifer; K = hydraulic conductivity of aquifer (Figure 9.9a); L = distance from river 1 to 2; h_1 = water level of river 1; h_2 = water level of river 2.

water table at x distance from river 1; water level of river 1; flow rate from river 1 to 2 through unit-width aquifer (Eq. (9.24)); distance from river 1; hydraulic conductivity of aquifer (Figure 9.9a)

$$h_x^2 = h_1^2 - \frac{2qx}{K} \tag{9.25}$$

hydraulic gradient at x distance from river 1; flow rate from river 1 to 2 through unit-width aquifer (Eq. (9.24)); distance from river 1; hydraulic conductivity of aquifer (Figure 9.9a)

$$\frac{dh_x}{dx} = -\frac{q}{Kh_x} \tag{9.26}$$

If there is recharge (Figure 9.9b), applying Eq. (9.21) with $W_{ss} = i/L$, integrating, and simplifying (see Example 9.22), one can get the formulas expressed as:

water table at x distance from river 1; recharge rate; hydraulic conductivity of aquifer (Figure 9.9b); distance from river 1; distance from river 1 to 2; water level of river 1; water level of river 2

$$h_x^2 = \frac{i}{K}(L - x)\,x - \frac{h_1^2 - h_2^2}{L}x + h_1^2 \tag{9.27}$$

hydraulic gradient at x distance from river 1; water table at x distance from river 1 (Eq. (9.27)); recharge rate; hydraulic conductivity of aquifer (Figure 9.9b); distance from river 1 to 2; distance from river 1; water level of river 1; water level of river 2

$$\frac{dh_x}{dx} = \frac{1}{2h_x}\left[\frac{i}{K}(L - 2x) - \frac{h_1^2 - h_2^2}{L}\right] \tag{9.28}$$

distance from river 1 to "divide"; distance from river 1 to 2; hydraulic conductivity of aquifer (Figure 9.9b); recharge rate; water level of river 1; water level of river 2

$$L_{1,d} = \frac{1}{2}\left(L - \frac{K}{i}\frac{h_1^2 - h_2^2}{L}\right) \tag{9.29}$$

highest water table at "divide"; percolation rate; hydraulic conductivity of aquifer (Figure 9.9b); distance from river 1 to 2; distance from river 1 to "divide" (Eq. (9.29)); water level of river 1; water level of river 2

$$h_{max}^2 = \frac{i}{K}(L - L_{1,d})\,L_{1,d} - \frac{h_1^2 - h_2^2}{L}L_{1,d} + h_1^2 \tag{9.30}$$

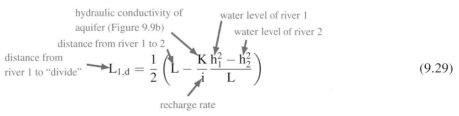

$$q_x = \frac{K}{2L}\left(h_1^2 - h_2^2\right) - \frac{i}{2}(L - 2x) \qquad (9.31)$$

hydraulic conductivity of aquifer (Figure 9.9b)
water level of river 2
recharge rate
discharge per unit-width aquifer at x distance from river 1
distance from river 1 to 2
water level of river 1

Example 9.19 In Figure 9.9a, if $h_1 = 20\,\text{ft}$, $h_2 = 15\,\text{ft}$, $K = 50\,\text{ft}\,\text{d}^{-1}$, and $L = 500\,\text{ft}$, determine the:

(1) Flow rate per unit-width aquifer from river 1 to 2
(2) Water table and its slope at $x = 250\,\text{ft}$.

Solution

There is no recharge.

(1) Eq. (9.24): $q = \left(50\,\text{ft}\,\text{d}^{-1}\right) / (2 * 500\,\text{ft}) * \left[(20\,\text{ft})^2 - (15\,\text{ft})^2\right] = 8.75\,\text{ft}^2\,\text{d}^{-1}$.

(2) Eq. (9.25): $h_x{}^2 = (20\,\text{ft})^2 - \left[2 * \left(8.75\,\text{ft}^2\,\text{d}^{-1}\right) * (250\,\text{ft})\right] / \left(50\,\text{ft}\,\text{d}^{-1}\right) = 312.5\,\text{ft}^2$

$$\rightarrow h_x = 17.68\,\text{ft}.$$

Eq. (9.26): $dh_x/dx = -\left(8.75\,\text{ft}^2\,\text{d}^{-1}\right) / \left[\left(50\,\text{ft}\,\text{d}^{-1}\right) * (17.68\,\text{ft})\right] = -0.00990$.

Example 9.20 In Figure 9.9b, if $h_1 = 5.5\,\text{m}$, $h_2 = 2.5\,\text{m}$, $K = 10\,\text{m}\,\text{d}^{-1}$, $L = 200\,\text{m}$, and $i = 0.01\,\text{m}\,\text{d}^{-1}$, determine the:

(1) Water table and its slope at $x = 100\,\text{m}$
(2) Distance from river 1 to divide and the highest water table
(3) Discharges into river 1 and 2, respectively.

Solution

There is recharge.

(1) Eq. (9.27): $h_x{}^2 = \left[\left(0.01\,\text{m}\,\text{d}^{-1}\right) / \left(10\,\text{m}\,\text{d}^{-1}\right)\right] * [(200\,\text{m}) - (100\,\text{m})] * (100\,\text{m}) - [(5.5\,\text{m})^2 - (2.5\,\text{m})^2] / (200\,\text{m}) * (100\,\text{m}) + (5.5\,\text{m})^2 = 28.25\,\text{m}^2 \rightarrow h_x = 5.32\,\text{m}$.

Eq. (9.28): $dh_x/dx = [1/(2 * 5.32\,\text{m})] * \{\left[\left(0.01\,\text{m}\,\text{d}^{-1}\right) / \left(10\,\text{m}\,\text{d}^{-1}\right)\right] * [(200\,\text{m}) - 2 * (100\,\text{m})] - \left[(5.5\,\text{m})^2 - (2.5\,\text{m})^2\right] / (200\,\text{m})\} = -0.0113$.

(2) Eq. (9.29): $L_{1,d} = 1/2 * \{200\,\text{m} - \left[\left(10\,\text{m}\,\text{d}^{-1}\right) / \left(0.01\,\text{m}\,\text{d}^{-1}\right)\right] * \left[(5.5\,\text{m})^2 - (2.5\,\text{m})^2\right] / (200\text{m})\} = 40\,\text{m}$.

Eq. (9.30): $h_{max}{}^2 = \left[\left(0.01\,\mathrm{m\,d^{-1}}\right) / \left(10\,\mathrm{m\,d^{-1}}\right)\right] * [(200\,\mathrm{m}) - (40\,\mathrm{m})] * (40\,\mathrm{m}) - [(5.5\,\mathrm{m})^2 - (2.5\,\mathrm{m})^2] / (200\,\mathrm{m}) * (40\,\mathrm{m}) + (5.5\,\mathrm{m})^2 = 31.85\,\mathrm{m}^2 \rightarrow h_{max} = 5.64\,\mathrm{m}$.

(3) Eq. (9.31): $q_{\rightarrow \mathrm{river1}} = \left[\left(10\,\mathrm{m\,d^{-1}}\right) / (2 * 200\,\mathrm{m})\right] * \left[(5.5\,\mathrm{m})^2 - (2.5\,\mathrm{m})^2\right] - (0.01\,\mathrm{m\,d^{-1}})/2 * [(200\,\mathrm{m}) - 2 * (0\,\mathrm{m})] = -0.4\,\mathrm{m^2\,d^{-1}}$.

$$q_{\rightarrow \mathrm{river2}} = \left[\left(10\,\mathrm{m\,d^{-1}}\right) / (2 * 200\,\mathrm{m})\right] * \left[(5.5\,\mathrm{m})^2 - (2.5\,\mathrm{m})^2\right] - (0.01\,\mathrm{m\,d^{-1}})/2 * [(200\,\mathrm{m}) - 2 * (200\,\mathrm{m})] = 1.6\,\mathrm{m^2\,d^{-1}}.$$

Mass balance $\rightarrow q_{\rightarrow \mathrm{river1}} = -iL_{1,d} = -\left(0.01\,\mathrm{m\,d^{-1}}\right) * (40\,\mathrm{m}) = -0.4\,\mathrm{m^2\,d^{-1}}$

$$q_{\rightarrow \mathrm{river2}} = i\,(L - L_{1,d}) = \left(0.01\,\mathrm{m\,d^{-1}}\right) * (200\,\mathrm{m} - 40\,\mathrm{m}) = 1.6\,\mathrm{m^2\,d^{-1}}$$

$$|q_{\rightarrow \mathrm{river1}}| + |q_{\rightarrow \mathrm{river2}}| = \left|-0.4\,\mathrm{m^2\,d^{-1}}\right| + \left|1.6\,\mathrm{m^2\,d^{-1}}\right| = iL = \left(0.01\,\mathrm{m\,d^{-1}}\right) * (200\,\mathrm{m}) = 2.0\,\mathrm{m^2\,d^{-1}}.$$

Example 9.21 Show Eqs. (9.24) through (9.26).

Solution

Reference to Figure 9.9a and make the Dupuit–Forchheimer assumption.

Darcy's law: $q = \left(-K\frac{dh}{dx}\right) h \rightarrow qdx = -Khdh \rightarrow q\int_0^L dx = -K\int_{h_1}^{h_2} hdh$

$$\rightarrow qL = -\frac{K}{2}\left(h_2^2 - h_1^2\right) \rightarrow q = \frac{K}{2L}\left(h_1^2 - h_2^2\right) \rightarrow \text{Eq. (9.24)}.$$

$q\int_0^x dx = -K\int_{h_1}^{h_x} hdh \rightarrow qx = -\frac{K}{2}\left(h_x^2 - h_1^2\right) \rightarrow h_x^2 = h_1^2 - \frac{2qx}{K} \rightarrow$ Eq. (9.25).

Differentiating Eq. (9.25) $\rightarrow 2h_x\frac{dh_x}{dx} = -\frac{2q}{K} \rightarrow \frac{dh_x}{dx} = -\frac{q}{Kh_x} \rightarrow$ Eq. (9.26).

Example 9.22 Show Eqs. (9.27) through (9.31).

Solution

Reference to Figure 9.9b and make the Dupuit–Forchheimer assumption.

Darcy's law: $q = \left(-K\frac{dh_x}{dx}\right) h_x \rightarrow q = -\frac{K}{2}\frac{d\left(h_x^2\right)}{dx} \rightarrow i = \frac{dq}{dx} = -\frac{K}{2}\frac{d^2\left(h_x^2\right)}{dx^2}$

$$\rightarrow \frac{d\left(h_x^2\right)}{dx} = -\frac{2i}{K}x + C_1 \rightarrow h_x^2 = -\frac{i}{K}x^2 + C_1x + C_2.$$

Boundary conditions: $x = 0$, $h_x = h_1$; $x = L$, $h_x = h_2 \rightarrow h_1{}^2 = C_2$, $h_2{}^2 = -(i/K)\,L^2 + C_1L + C_2$

$\rightarrow C_1 = -\left(h_1{}^2 - h_2{}^2\right)/L + (i/K)\,L$, $C_2 = h_1{}^2 \rightarrow h_x^2 = \frac{i}{K}(L - x)\,x - \frac{h_1^2 - h_2^2}{L}x + h_1^2 \rightarrow$ Eq. (9.27).

Differentiating Eq. (9.27) $\rightarrow 2h_x \frac{dh_x}{dx} = \frac{i}{K}(L - 2x) - \frac{h_1^2 - h_2^2}{L}$

$$\rightarrow \frac{dh_x}{dx} = \frac{1}{2h_x}\left[\frac{i}{K}(L - 2x) - \frac{h_1^2 - h_2^2}{L}\right] \rightarrow \text{Eq. (9.28).}$$

"divide" $\rightarrow \frac{dh_x}{dx} = \frac{1}{2h_x}\left[\frac{i}{K}(L - 2L_{1,d}) - \frac{h_1^2 - h_2^2}{L}\right] = 0 \rightarrow L_{1,d} = \frac{1}{2}\left(L - \frac{K}{i}\frac{h_1^2 - h_2^2}{L}\right) \rightarrow$ Eq. (9.29).

Substituting $L_{1,d}$ for x in Eq. (9.27) $\rightarrow h_{max}^2 = \frac{i}{K}(L - L_{1,d})L_{1,d} - \frac{h_1^2 - h_2^2}{L}L_{1,d} + h_1^2 \rightarrow$ Eq. (9.30).

9.4 Well Hydraulics

Wells have been widely used to withdraw or recharge groundwater since early in human history. A pumping well reduces, whereas a recharging well increases, groundwater level within its influence radius, which is the distance away from the well beyond which the groundwater is intact. Drawdown is defined as the difference between the initial and a new groundwater level. Because these two types of wells are identical except for that pumping and recharging rates have an opposite sign, this section focuses on pumping wells. Also, to develop analytical solutions, most of this section assumes a homogeneous isotropic aquifer with a uniform thickness and an initially horizontal groundwater level throughout the spatial extent of an influence radius. For such an aquifer, a fully penetrating well, which penetrates the entire aquifer thickness, induces a radial flow, which uniformly flows into or out of the well along its radial directions. The last two subsections present methods on how to account for aquifer boundaries formed by streams, rock barriers, and/or multiple wells. Also discussed will be aquifer flow induced by a partially penetrating well, which penetrates a portion of the aquifer thickness. Such a situation will induce a non-radial flow with a larger flow rate in one direction than in another direction. Moreover, an aquifer attains a steady state when its groundwater level does not vary much with time. In reality, a steady state can never be achieved. However, to simplify analyses, a steady state is usually assumed if the temporal variation of groundwater level becomes minimal after pumping or recharging for a long period of time.

9.4.1 Steady Radial Flow in a Confined Aquifer

As shown in Figure 9.10, a confined aquifer may reach a steady state after pumping for a long period of time, when the flow rate at any distance (of less than the influence radius) away from the pumping well is equal to the pumping rate. At and beyond the influence radius, the aquifer is intact with a zero drawdown and a zero-flow rate.

Applying Darcy's law to an elemental radial aquifer centered on the well and integrating, one can derive (see Example 9.23) the formula for drawdown expressed as:

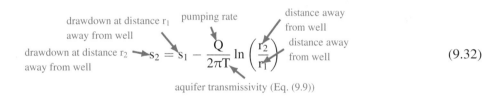

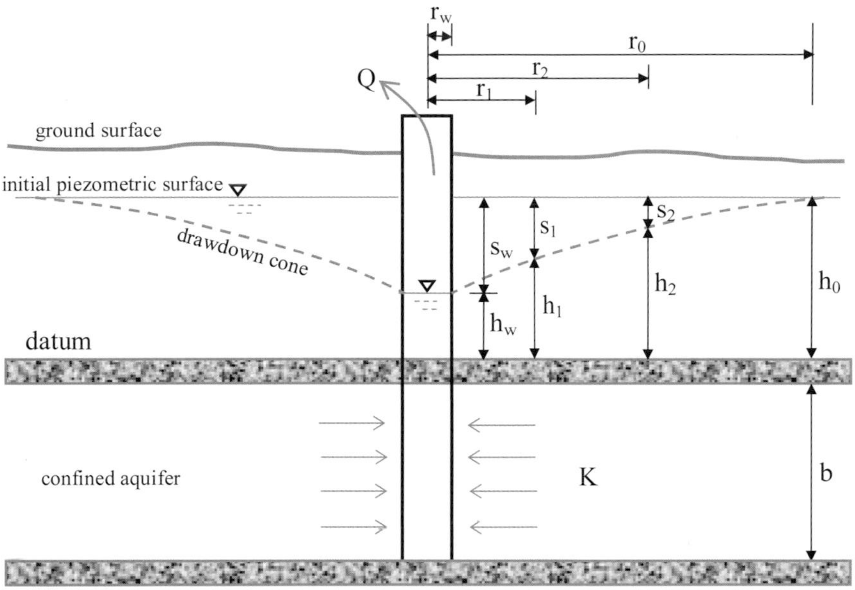

Figure 9.10 **Drawdown in a confined aquifer.** Radial flow and drawdown cone resulting from pumping. Q: pumping rate; r_w: well radius; r_0: influence radius; K: hydraulic conductivity; b: aquifer thickness; h_0: initial hydraulic head; h: hydraulic head of drawdown cone at distance r away from the well; and $s = h_0 - h$: drawdown at distance r away from the well.

$$s_2 = s_1 - \frac{Q}{2\pi T} \ln\left(\frac{r_2}{r_1}\right) \tag{9.32}$$

(s_2: drawdown at distance r_2 away from well; s_1: drawdown at distance r_1 away from well; Q: pumping rate; T: aquifer transmissivity (Eq. (9.9)); r_2, r_1: distance away from well)

Example 9.23 Derive Eq. (9.32).

Solution

Reference Figure 9.10.

Eq. (9.9): $T = Kb$.

Darcy's law (Eq. (9.16)) $\rightarrow$ radial velocity at distance r away from the well: $V_r = -K\,(dh/dr)$.

The flow area at distance r away from the well: $A = 2\pi rb$.

Continuity equation $\rightarrow$ pumping rate: $Q = -V_r A = -[-K\,(dh/dr)]\,[2\pi rb] = 2\pi Tr\,(dh/dr)$

$$\rightarrow \frac{Q}{2\pi T}\frac{1}{r}dr = dh = d(h_0 - s) = -ds$$

$$\rightarrow \frac{Q}{2\pi T}\int_{r_1}^{r_2}\frac{1}{rdr} = -\int_{s_1}^{s_2} ds \rightarrow s_2 = s_1 - \frac{Q}{2\pi T}\ln\frac{r_2}{r_1} \rightarrow \text{Eq. (9.32).}$$

Example 9.24 A 2-ft-diameter well fully penetrates a 50-ft-thick confined aquifer ($K = 30\,ft\,d^{-1}$). After pumping at $1000\,ft^3\,d^{-1}$ for a long period of time, the drawdown in the well is measured to be 1.5 ft. Determine the:

(1) Drawdown at 100 ft away from the well
(2) Influence radius
(3) Flow rate at 50 ft away from the well.

Solution

$b = 50\,ft$, $K = 30\,ft\,d^{-1}$, $Q = 1000\,ft^3\,d^{-1}$, $r_1 = r_w = 1\,ft$, $s_1 = s_w = 1.5\,ft$

Pumping for a long period of time $\rightarrow$ steady state.

Eq. (9.9): $T = \left(30\,ft\,d^{-1}\right) * (50\,ft) = 1500\,ft^2\,d^{-1}$.

(1) $r_2 = 100\,ft$

Eq. (9.32): $s_2 = (1.5\,ft) - (1000 ft^3\,d^{-1}) / \left[2\pi * (1500\,ft^2\,d^{-1})\right] * \ln\left[(100\,ft) / (1\,ft)\right] = 1.0\,ft.$

(2) $s_2 = 0$

Eq. (9.32): $0 = (1.5\,ft) - (1000\,ft^3\,d^{-1}) / \left[2\pi * (1500\,ft^2\,d^{-1})\right] * \ln\left[(r_0) / (1\,ft)\right]$

$\rightarrow r_0 = 1{,}379{,}411\,ft = 261\,mi.$

(3) $r = 50\,ft < r_0$, steady state $\rightarrow Q_r = Q = 1000\,ft^3\,d^{-1}$.

Example 9.25 A 0.5-m-diameter well fully penetrates a 30-m-thick confined aquifer. After pumping at $4500\,m^3\,d^{-1}$ for a long period of time, the drawdowns at 100 and 200 m away from the well are measured to be 3.5 and 1.5 m, respectively. Determine the:

(1) Transmissivity and hydraulic conductivity
(2) Drawdown in the well
(3) Influence radius.

Solution

$b = 30\,m$, $r_w = 0.25\,m$, $Q = 4500\,m^3\,d^{-1}$, $r_1 = 100\,m$, $s_1 = 3.5\,m$, $r_2 = 200\,m$, $s_2 = 1.5\,m$; pumping for a long period of time $\rightarrow$ steady state.

(1) Eq. (9.32): $1.5\,m = (3.5\,m) - (4500\,m^3\,d^{-1}) / (2\pi T) * \ln\left[(200\,m) / (100\,m)\right]$

$\rightarrow T = 248.22\,m^2\,d^{-1} \rightarrow K = \left(248.22\,m^2\,d^{-1}\right) / (30\,m) = 8.27\,m\,d^{-1}.$

(2) Use the drawdown closer to the well to compute s_w.

$$\text{Eq. (9.32): } 3.5\,\text{m} = s_w - (4500\,\text{m}^3\,\text{d}^{-1}) / \left[2\pi * \left(248.22\,\text{m}^2\,\text{d}^{-1}\right)\right] * \ln\left[(100\,\text{m}) / (0.25\,\text{m})\right]$$

$$\rightarrow s_w = 20.79\,\text{m}.$$

(3) Use the drawdown closer to the influence radius to compute r_0.

$$\text{Eq. (9.32): } 0 = 1.5\,\text{m} - (4500\,\text{m}^3\,\text{d}^{-1}) / \left[2\pi * \left(248.22\,\text{m}^2\,\text{d}^{-1}\right)\right] * \ln\left[(r_0) / (200\,\text{m})\right]$$

$$\rightarrow r_0 = 336.4\,\text{m}.$$

9.4.2 Steady Radial Flow in an Unconfined Aquifer

An unconfined aquifer may reach a steady state after pumping has been ongoing for a long period of time (Figure 9.11). A steady state is reached when the flow rate at any distance (of less than the influence radius) away from the pumping well is equal to the pumping rate. At and beyond the influence radius, the aquifer is intact with a zero drawdown and a zero-flow rate.

Expressions for drawdown can be derived for both cases with and without recharge. If there is no recharge, applying Darcy's law to an elemental radial aquifer centered on the well and integrating, one can derive (see Example 9.26) the formula for drawdown expressed as:

$$(h_0 - s_2)^2 = (h_0 - s_1)^2 + \frac{Q}{\pi K} \ln\left(\frac{r_2}{r_1}\right) \tag{9.33}$$

drawdown at distance r_2 away from well; drawdown at distance r_1 away from well; pumping rate; distance away from well; distance away from well; initial hydraulic head or water table; aquifer hydraulic conductivity

Similarly, if there is recharge, one can derive (see Example 9.27) the formula for drawdown expressed as:

$$Q_r = i\pi\left(r_0^2 - r^2\right) \tag{9.34}$$

flow rate at distance r away from well; recharge rate; influence radius; distance away from well

$$Q = i\pi\left(r_0^2 - r_w^2\right) \tag{9.35}$$

pumping rate; recharge rate; influence radius; well radius

$$(h_0 - s_2)^2 = (h_0 - s_1)^2 + \frac{i}{K}\left[r_0^2 \ln\left(\frac{r_2}{r_1}\right) - \frac{r_2^2 - r_1^2}{2}\right] \tag{9.36}$$

drawdown at distance r_2 away from well; drawdown at distance r_1 away from well; recharge rate; distance away from well; initial hydraulic head/water table; aquifer hydraulic conductivity; influence radius; distance away from well

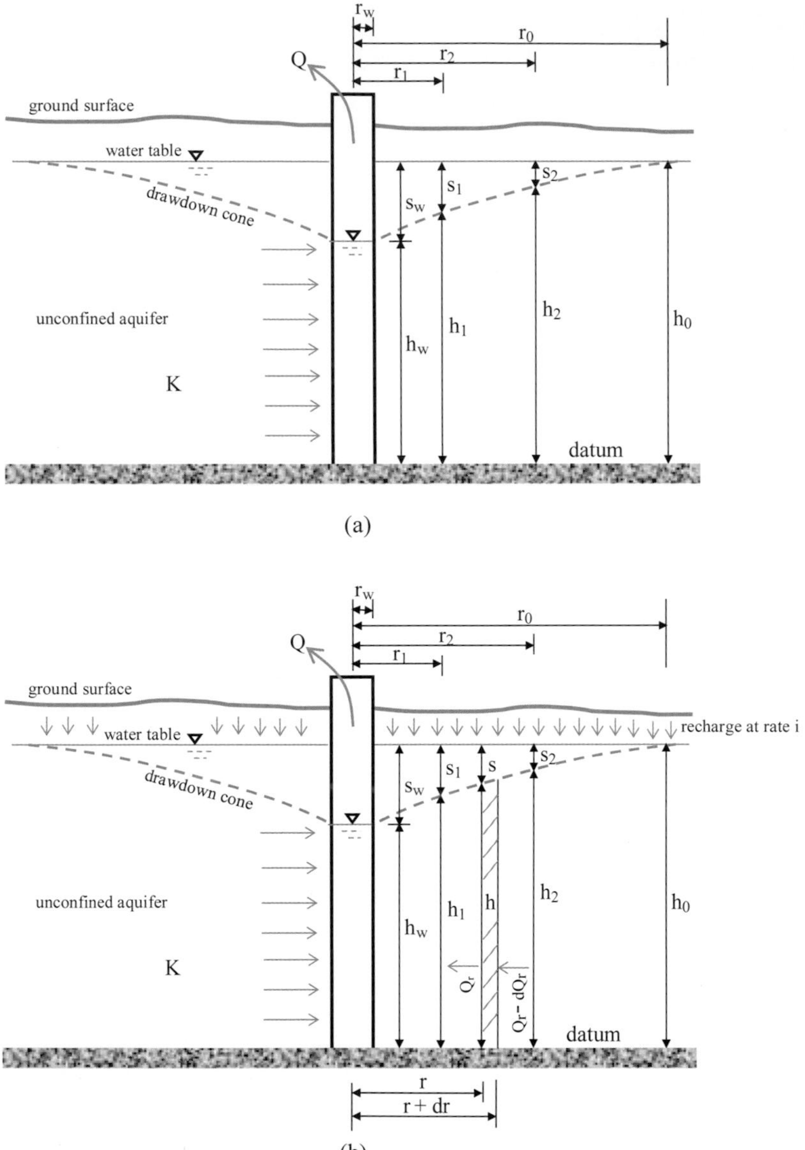

Figure 9.11 **Drawdown in an unconfined aquifer.** Radial flow and drawdown cone resulting from pumping. (a) Zero recharge; and (b) nonzero recharge. Q: pumping rate; r_w: well radius; r_0: influence radius; K: hydraulic conductivity; h_0: initial aquifer thickness or hydraulic head/water table; h: hydraulic head of drawdown cone at distance r away from the well; $s = h_0 - h$: drawdown at distance r away from the well; i: recharge rate; and Q_r: flow rate at distance r away from the well.

Example 9.26 Derive Eq. (9.33).

Solution

Reference Figure 9.11a.

Let i be the recharge rate.

For a radial elemental aquifer between distance r and (r + dr), the mass conservation is:

Darcy's law (Eq. (9.16)) $\rightarrow$ radial velocity at distance r away from the well: $V_r = -K\,(dh/dr)$.

The flow area at distance r away from the well: $A = 2\pi rh$.

Continuity equation $\rightarrow$ pumping rate: $Q = -V_r A = -\left[-K\,(dh/dr)\right]\left[2\pi rh\right] = 2\pi Krh\,(dh/dr)$

$$\rightarrow \frac{Q}{2\pi K}\frac{1}{r}dr = hdh \rightarrow \frac{Q}{2\pi K}\int_{r_1}^{r_2}\frac{1}{r}dr = \int_{h_1}^{h_2} hdh$$

$$\rightarrow \frac{Q}{2\pi K}\ln\frac{r_2}{r_1} = \frac{1}{2}\left(h_2^2 - h_1^2\right) \rightarrow h_2^2 = h_1^2 + \frac{Q}{\pi K}\ln\frac{r_2}{r_1}$$

$$\rightarrow (h_0 - s_2)^2 = (h_0 - s_1)^2 + \frac{Q}{\pi K}\ln\frac{r_2}{r_1} \rightarrow \text{Eq. (9.33).}$$

Example 9.27 Derive Eqs. (9.34) and (9.36).

Solution

Reference Figure 9.11b.

For the radial elemental aquifer between distance r and (r + dr), the flow rate change is:

$$dQ_r = i\pi r^2 - i\pi\,(r + dr)^2 \approx -2i\pi r\,(dr) \rightarrow Q_r = -i\pi r^2 + C.$$

At $r = r_0$, $Q_r = 0 \rightarrow 0 = -i\pi r_0^2 + C \rightarrow C = i\pi r_0^2$.

Thus, $Q_r = -i\pi r^2 + i\pi r_0^2 = i\pi\left(r_0^2 - r^2\right) \rightarrow$ Eq. (9.34).

In Eq. (9.34), let $r = r_w$, one can derive the pumping rate, $Q = i\pi\left(r_0^2 - r_w^2\right) \rightarrow$ Eq. (9.34).

Darcy's law (Eq. (9.16)) $\rightarrow$ radial velocity at distance r away from the well: $V_r = -K\,(dh/dr)$.

The flow area at distance r away from the well: $A = 2\pi rh$.

Continuity equation $\rightarrow Q_r = -V_r A = -\left[-K\,(dh/dr)\right]\left[2\pi rh\right] = 2\pi Krh\,(dh/dr)$

$$\rightarrow 2\pi Krh\,(dh/dr) = i\pi\left(r_0^2 - r^2\right) \rightarrow 2hdh = \frac{i}{K}\left(\frac{r_0^2}{r} - r\right)dr$$

$$\rightarrow 2\int_{h_1}^{h_2} hdh = \frac{i}{K}\int_{r_1}^{r_2}\left(\frac{r_0^2}{r} - r\right)dr \rightarrow h_2^2 - h_1^2 = \frac{i}{K}\left(r_0^2\ln\frac{r_2}{r_1} - \frac{r_2^2 - r_1^2}{2}\right)$$

$$\rightarrow (h_0 - s_2)^2 = (h_0 - s_1)^2 + \frac{i}{K}\left(r_0^2\ln\frac{r_2}{r_1} - \frac{r_2^2 - r_1^2}{2}\right) \rightarrow \text{Eq. (9.36).}$$

Example 9.28 A 2-ft-diameter well fully penetrates a 50-ft-thick unconfined aquifer ($K = 30\,\text{ft}\,\text{d}^{-1}$). There is no recharge. After pumping at $1000\,\text{ft}^3\,\text{d}^{-1}$ for a long period of time, the drawdown in the well is measured to be 1.5 ft. Determine the:

(1) Drawdown at 100 ft away from the well
(2) Influence radius
(3) Flow rate at 300 ft away from the well.

Solution

$h_0 = 50\,\text{ft},\ K = 30\,\text{ft}\,\text{d}^{-1},\ Q = 1000\,\text{ft}^3\,\text{d}^{-1},\ r_1 = r_w = 1\,\text{ft},\ s_1 = s_w = 1.5\,\text{ft}$

Pumping for a long period of time $\rightarrow$ steady state.

(1) $r_2 = 100\,\text{ft}$

$$\text{Eq. (9.33)}: (50\,\text{ft} - s_2)^2 = (50\,\text{ft} - 1.5\,\text{ft})^2 + \frac{1000\,\text{ft}^3\,\text{d}^{-1}}{\pi * (30\,\text{ft}\,\text{d}^{-1})} * \ln\frac{100\,\text{ft}}{1\,\text{ft}} = 2401.1124$$

$$\rightarrow s_2 = 1.0\,\text{ft}.$$

(2) $s_2 = 0$

$$\text{Eq. (9.33)}: (50\,\text{ft} - 0\,\text{ft})^2 = (50\,\text{ft} - 1.5\,\text{ft})^2 + \frac{1000\,\text{ft}^3\,\text{d}^{-1}}{\pi * (30\,\text{ft}\,\text{d}^{-1})} * \ln\frac{r_0}{1\,\text{ft}} = 2352.25$$
$$+10.61033 \ln(r_0)$$

$$\rightarrow r_0 = 1{,}115{,}830\,\text{ft} = 211\,\text{mi}.$$

(3) $r = 300\,\text{ft} < r_0 \rightarrow Q_r = Q = 1000\,\text{ft}^3\,\text{d}^{-1}$.

Example 9.29 A 0.5-m-diameter well fully penetrates an 80-m-thick unconfined aquifer. After pumping at $4500\,\text{m}^3\,\text{d}^{-1}$ for a long period of time, the drawdowns at 100 and 200 m away from the well are measured to be 3.5 and 1.5 m, respectively. Determine the:

(1) Hydraulic conductivity and transmissivity
(2) Drawdown in the well
(3) Influence radius.

Solution

$h_0 = 80\,\text{m},\ r_w = 0.25\,\text{m},\ Q = 4500\,\text{m}^3\,\text{d}^{-1}$

$r_1 = 100\,\text{m},\ s_1 = 3.5\,\text{m},\ r_2 = 200\,\text{m},\ s_2 = 1.5\,\text{m}$

Pumping for a long period of time $\rightarrow$ steady state.

(1) Eq. (9.33): $(80\,\text{m} - 1.5\,\text{m})^2 = (80\,\text{m} - 3.5\,\text{m})^2 + \frac{4500\,\text{m}^3\,\text{d}^{-1}}{\pi K} * \ln\frac{200\,\text{m}}{100\,\text{m}}$

$$\rightarrow K = 3.2\,\text{m}\,\text{d}^{-1} \rightarrow T = Kh_0 = \left(3.2\,\text{m}\,\text{d}^{-1}\right) * (80\,\text{m}) = 256\,\text{m}^2\,\text{d}^{-1}.$$

(2) Use the drawdown location closer to the well to compute s_w:

Eq. (9.33): $(80\,\text{m} - 3.5\,\text{m})^2 = (80\,\text{m} - s_w)^2 + \frac{4500\,\text{m}^3\,\text{d}^{-1}}{\pi * (3.2\,\text{m}\,\text{d}^{-1})} * \ln\frac{100\,\text{m}}{0.25\,\text{m}}$

$\rightarrow s_w = 23.69\,\text{m}.$

(3) Use the drawdown location closer to the influence radius to compute r_0:

Eq. (9.33): $(80\,\text{m} - 0\,\text{m})^2 = (80\,\text{m} - 1.5\,\text{m})^2 + \frac{4500\,\text{m}^3\,\text{d}^{-1}}{\pi * (3.2\,\text{m}\,\text{d}^{-1})} * \ln\frac{r_0}{200\,\text{m}}$

$$\rightarrow r_0 = 340\,\text{m}.$$

Example 9.30 A 2-ft-diameter well fully penetrates a 150-ft-thick unconfined aquifer ($K = 10\,\text{ft}\,\text{d}^{-1}$). After pumping at $150{,}000\,\text{ft}^3\,\text{d}^{-1}$ for a long period of time, the drawdown at 100 ft away from the well is measured to be 35 ft. For a recharge rate of $0.001\,\text{ft}\,\text{d}^{-1}$, determine the:

(1) Influence radius
(2) Drawdown at 300 ft away from the well
(3) Drawdown at 20 ft away from the well
(4) Flow rates at 100 and 300 ft away from the well, respectively
(5) Drawdown and flow rate at 9000 ft away from the well.

Solution

$r_w = 1\,\text{ft},\ h_0 = 150\,\text{ft},\ K = 10\,\text{ft}\,\text{d}^{-1},\ Q = 150{,}000\,\text{ft}^3\,\text{d}^{-1}$

$r_1 = 100\,\text{ft},\ s_1 = 35\,\text{ft},\ i = 0.001\,\text{ft}\,\text{d}^{-1}$

(1) Eq. (9.35): $150{,}000\,\text{ft}^3\,\text{d}^{-1} = \left(0.001\,\text{ft}\,\text{d}^{-1}\right) * \pi * \left[r_0^2 - (1\,\text{ft})^2\right] \rightarrow r_0 = 6910\,\text{ft}.$

(2) $r_2 = 300\,\text{ft} < r_0$

Eq. (9.36): $(150\,\text{ft} - s_2)^2 = (150\,\text{ft} - 35\,\text{ft})^2 + \frac{0.001\,\text{ft}\,\text{d}^{-1}}{10\,\text{ft}\,\text{d}^{-1}} * \left[(6910\text{ft})^2 \ln\left(\frac{300\,\text{ft}}{100\,\text{ft}}\right) - \frac{(300\,\text{ft})^2 - (100\,\text{ft})^2}{2}\right]$

$\rightarrow (150\,\text{ft} - s_2)^2 = 18{,}466.66494 \rightarrow s_2 = 14.11\,\text{ft}.$

(3) $r_2 = 20\,\text{ft} < r_0$

$$\text{Eq. (9.36): } (150\,\text{ft} - s_2)^2 = (150\,\text{ft} - 35\,\text{ft})^2 + \frac{0.001\,\text{ft}\,\text{d}^{-1}}{10\,\text{ft}\,\text{d}^{-1}} * \left[(6910\text{ft})^2 \ln\left(\frac{20\,\text{ft}}{100\,\text{ft}}\right) - \frac{(20\,\text{ft})^2 - (100\,\text{ft})^2}{2}\right]$$

$$\rightarrow (150\,\text{ft} - s_2)^2 = 5540.71976 \rightarrow s_2 = 75.56\,\text{ft}.$$

(4) $$\text{Eq. (9.34): } Q_{r=100\text{ft}} = (0.001\,\text{ft}\,\text{d}^{-1}) * \pi * \left[(6910\,\text{ft})^2 - (100\,\text{ft})^2\right] = 149{,}974\,\text{ft}^3\,\text{d}^{-1}$$

$$Q_{r=300\text{ft}} = (0.001\,\text{ft}\,\text{d}^{-1}) * \pi * \left[(6910\,\text{ft})^2 - (300\,\text{ft})^2\right] = 149{,}722\,\text{ft}^3\,\text{d}^{-1}.$$

(5) $r_2 = 9000\,\text{ft} > r_0 \rightarrow s_2 = 0,\ Q_{r=9000\text{ft}} = 0.$

9.4.3 Unsteady Radial Flow in a Confined Aquifer

Unsteady radial flow refers to the instantaneous conditions before reaching a steady state, under which at a given time there is the same flow rate into the pumping well from all directions. For unsteady radial flow in a confined aquifer, Theis (1935) developed an analytical solution expressed as:

$$s(r,t) = \frac{Q}{4\pi T} W(u) \tag{9.37}$$

drawdown at distance r away from well and at time t; pumping rate; well function (Eq. (9.39)); variable (Eq. (9.38)); aquifer transmissivity

$$u = \frac{Sr^2}{4Tt} \tag{9.38}$$

variable in Eq. (9.37); aquifer storativity; distance away from well; aquifer transmissivity; lapse time since start of pumping

well function in Eq. (9.37)

$$W(u) = -0.5772 - \ln(u) + u + \sum_{m=2}^{\infty}\left[(-1)^{m-1}\frac{u^m}{m(m!)}\right] \tag{9.39}$$

variable in Eq. (9.37); factorial of m, i.e., $= m(m-1)(m-2)\ldots(1)$

If $u < 0.01$, W(u) can be computed by keeping the first three terms on the right side of Eq. (9.39), with a truncation error of $u^2/[2*(2!)] = u^2/4 < 2.5 \times 10^{-5}$. For a given u and a required truncation error of ε, the minimum value of m can be determined by solving the inequality equation of $\frac{u^2}{m(m!)} \le \varepsilon$. In practice, if the aquifer properties of S and T are known, the drawdown at a given distance away from the well and at a time of interest can be directly computed using Eqs. (9.37) through (9.39). However, if S and T are unknown and need to be determined, one can fit Eq. (9.37) to a measured time series of drawdown using a least-squares regression method. In this regard, for assumed values of S and T: (1) compute values of u, W(u), and s(r,t) using the above equations; (2) compute squared errors between the measured and computed values of s(r,t); and (3) compute the summation of the squared errors and minimize it by adjusting S and T. This last step can be easily done in Excel Solver.

Example 9.31 A well fully penetrates a confined aquifer ($S = 2.0 \times 10^{-4}$, $T = 1000\,m^2\,d^{-1}$) and pumps at $5000\,m^3\,d^{-1}$. For a maximum truncation error of $\varepsilon = 10^{-4}$, determine the drawdowns at 200 m away from the well after pumping for:

(1) 1 d
(2) 2 d
(3) 5 d.

Solution

$S = 2.0 \times 10^{-4}$, $T = 1000\,m^2\,d^{-1}$, $Q = 5000\,m^3\,d^{-1}$, $r = 200\,m$

(1) $t = 1\,d$

Eq. (9.38): $u = \left[(2.0 \times 10^{-4}) * (200\,m)^2\right] / \left[4 * (1000\,m^2\,d^{-1}) * (1\,d)\right] = 0.002$

$u < 0.01 \rightarrow$ first three terms, truncation error $< 2.5 \times 10^{-5}$, which is less than $\varepsilon = 10^{-4}$

$\rightarrow$ Eq. (9.18): $W(u) = -0.5772 - \ln(0.002) + 0.002 = 5.6394.$

Eq. (9.37): $s\,(r = 200\,ft,\ t = 1\,d) = (5000\,m^3\,d^{-1}) / \left[4 * \pi * (1000\,m^2\,d^{-1})\right] * 5.6394 = 2.24\,m.$

(2) $t = 2\,d$

Eq. (9.38): $u = \left[(2.0 \times 10^{-4}) * (200\,m)^2\right] / \left[4 * (1000\,m^2\,d^{-1}) * (2\,d)\right] = 0.001$

$u < 0.01 \rightarrow$ first three terms, truncation error $< 2.5 \times 10^{-5}$, which is less than $\varepsilon = 10^{-4}$

$\rightarrow$ Eq. (9.39): $W(u) = -0.5772 - \ln(0.001) + 0.001 = 6.3316.$

Eq. (9.37): $s\,(r = 200\,ft,\ t = 2\,d) = (5000\,m^3\,d^{-1}) / \left[4 * \pi * (1000\,m^2\,d^{-1})\right] * 6.3316 = 2.52\,m.$

(3) $t = 5\,d$

Eq. (9.38): $u = \left[(2.0 \times 10^{-4}) * (200\,m)^2\right] / \left[4 * (1000\,m^2\,d^{-1}) * (5\,d)\right] = 0.0004$

$u < 0.01 \rightarrow$ first three terms, truncation error $< 2.5 \times 10^{-5}$, which is less than $\varepsilon = 10^{-4}$

$\rightarrow$ Eq. (9.39): $W(u) = -0.5772 - \ln(0.0004) + 0.0004 = 7.2473.$

Eq. (9.37): $s\,(r = 200\,ft,\ t = 5\,d) = (5000\,m^3\,d^{-1}) / \left[4 * \pi * (1000\,m^2\,d^{-1})\right] * 7.2473 = 2.88\,m.$

Example 9.32 A well fully penetrates a confined aquifer and pumps at 60 ft^3 min^{-1}. The drawdowns in a monitoring well 30 ft away from the pumping well are measured as shown in cells A6:B15 of Figure 9.12. For a maximum truncation error of $\varepsilon = 10^{-4}$, determine the aquifer properties of S and T.

	A	B	C	D	E	F	G
1	Q =	60	ft^3 min^{-1}		S =	0.00807	
2	r =	30	ft		T =	8.49	ft^2 min^{-1}
3							
4	Observed Time Series				Theis Solution		Squared Error
5	t (min)	s^m (ft)	u	W(u)	Truncation Error	s^c (ft)	$(s^m - s^c)^2$ (ft^2)
6	1	0.66	0.2138	1.1685	2.1762E-05	0.6573	0.0000
7	2	0.98	0.1069	1.7628	1.3601E-06	0.9916	0.0001
8	3	1.21	0.0713	2.1342	2.6866E-07	1.2005	0.0001
9	4	1.35	0.0534	2.4046	8.5007E-08	1.3526	0.0000
10	5	1.48	0.0428	2.6173	3.4819E-08	1.4722	0.0001
11	6	1.57	0.0356	2.7926	1.6791E-08	1.5708	0.0000
12	7	1.64	0.0305	2.9418	9.0636E-09	1.6547	0.0002
13	8	1.74	0.0267	3.0715	5.3129E-09	1.7277	0.0002
14	9	1.80	0.0238	3.1864	3.3168E-09	1.7923	0.0001
15	10	1.84	0.0214	3.2894	2.1762E-09	1.8503	0.0001
16						Summation	0.0008

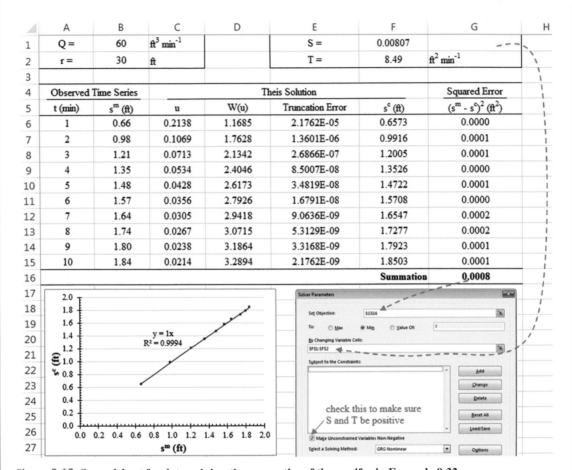

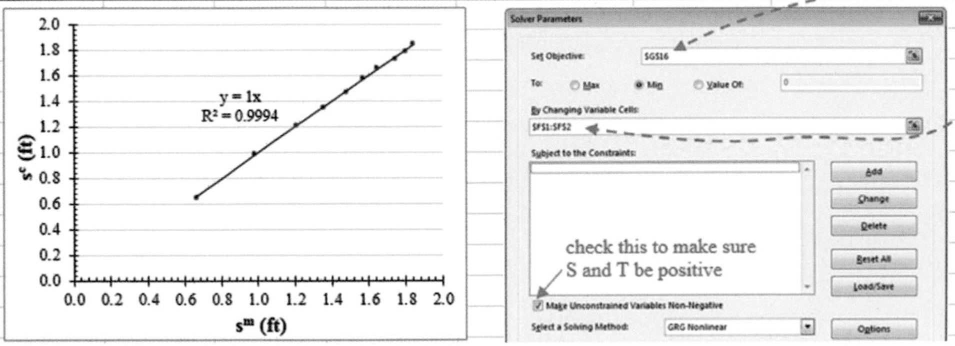

Figure 9.12 **Spreadsheet for determining the properties of the aquifer in Example 9.32.**

Solution

$Q = 60\,\text{ft}^3\,\text{min}^{-1}$, $r = 30\,\text{ft}$

Let s^m and s^c be the measured and computed values of the drawdown at time t.

The calculated values of u, W(u), and s^c are shown in cells C6:C15, D6:D15, and F6:F15 of Figure 9.12, respectively. By using m = 3, the truncation errors of W(u) (cells E6:E15) are smaller than ε. The squared errors of s^c are shown in cells G6:G15, and their summation is given in cell G16. In Solver, set objective of G16 to be minimal by changing cells F1:F2, which store S and T, respectively. Note to check the Make Unconstrained Variables Non-Negative box to make sure S and

T are positive. Also, to avoid a possible divergence, click the Options button to check Central Derivative under the GRG Nonlinear tab. The s^c versus s^m is plotted on the bottom left, and the settings of Solver are displayed on the bottom right of the figure. Both the visualization plot and statistics (coefficient of determination $R^2 = 0.9994$ and a near-one slope) indicate a good fitness of the Theis solution to the measured drawdown data. The computed aquifer properties are $S = 0.00807$ and $T = 8.49\,\mathrm{ft^2\,min^{-1}}$.

The computations are implemented in following two sequential steps:

Step 1: for instance, assume $S = 0.005$ and $T = 8.0\,\mathrm{ft^2\,min^{-1}}$:

- $t = 1\,\mathrm{min}$
  Eq. (9.38): $u = \left[0.005 * (30\,\mathrm{ft})^2\right] / \left[4 * \left(8.0\,\mathrm{ft^2\,min^{-1}}\right) * (1\,\mathrm{min})\right] = 0.1406.$
  Eq. (9.39) with $m = 3$:
  The truncation error $= 0.1406^4 / [4 * (4!)] = 4.07 \times 10^{-6} < \varepsilon = 10^{-4} \rightarrow$ fine

  $W(u) = -0.5772 - \ln(0.1406) + 0.1406 - 0.1406^2 / [2 * (2!)] + 0.1406^3 / [3 * (3!)] = 1.5203.$

  Eq. (9.37): $s^c = \left(60\,\mathrm{ft^3\,min^{-1}}\right) / \left[4 * \pi * \left(8.0\,\mathrm{ft^2\,min^{-1}}\right)\right] * 1.5203 = 0.9074\,\mathrm{ft}.$
  The squared error $= (0.9074\,\mathrm{ft} - 0.66\,\mathrm{ft})^2 = 0.0612\,\mathrm{ft^2}$.
- Similar calculations can be done for other times to get the corresponding squared errors.
- Compute the summation of the squared errors to be 1.0668 using *sum( )* in cell G16.

Step 2: minimize the summation by adjusting S and T in Solver.

9.4.4 Unsteady Radial Flow in an Unconfined Aquifer

For unsteady radial flow in an unconfined aquifer without recharge, Neuman (1975) developed an analytical solution expressed as:

$$s(r, t) = \frac{Q}{4\pi T} W(u_a, u_y, \eta) \tag{9.40}$$

drawdown at distance r away from well and at time t → $s(r, t)$
pumping rate → Q
aquifer transmissivity → T
well function (Figure 9.13) → W
earlier drawdown variable (Eq. (9.41)) → u_a
later drawdown variable (Eq. (9.42)) → u_y
anisotropy parameter (Eq. (9.43)) → η

$$u_a = \frac{Sr^2}{4Tt} \tag{9.41}$$

earlier drawdown variable in Eq. (9.40) → u_a
aquifer storativity → S
distance away from well → r
aquifer transmissivity → T
lapse time since start of pumping → t

$$u_y = \frac{S_y r^2}{4Tt} \tag{9.42}$$

where u_y: later drawdown variable in Eq. (9.40); S_y: aquifer specific yield; r: distance away from well; T: aquifer transmissivity; t: lapse time since start of pumping

$$\eta = \frac{r^2}{h_0^2}\frac{K_V}{K_H} \tag{9.43}$$

where η: anisotropy parameter in Eq. (9.40); r: distance away from well; h_0: initial hydraulic head/water table; K_V: vertical hydraulic conductivity; K_H: horizontal hydraulic conductivity

The Neuman solution suggests that the groundwater flow of an unconfined aquifer induced by a pumping well is characterized differently at the earlier versus later stage of pumping. At the earlier stage (i.e., within 4 min for sand and coarser materials, 30 min for find sand, and 170 min for silt and clay), the unconfined aquifer reacts in the same way as a confined aquifer and groundwater subject to a pressure is removed as a function of u_a (Eq. (9.41)), inducing a primarily horizontal flowing direction. However, at the later stage (i.e., after 4, 30, and 170 min for sand, fine sand, and silt and clay, respectively), groundwater subject to a near-zero relative pressure is removed as a function of u_y (Eq. (9.42)), inducing both horizontal and vertical flowing directions. During the earlier stage, the drawdown tends to increase gradually; during the later stage, however, the drawdown can increase quickly.

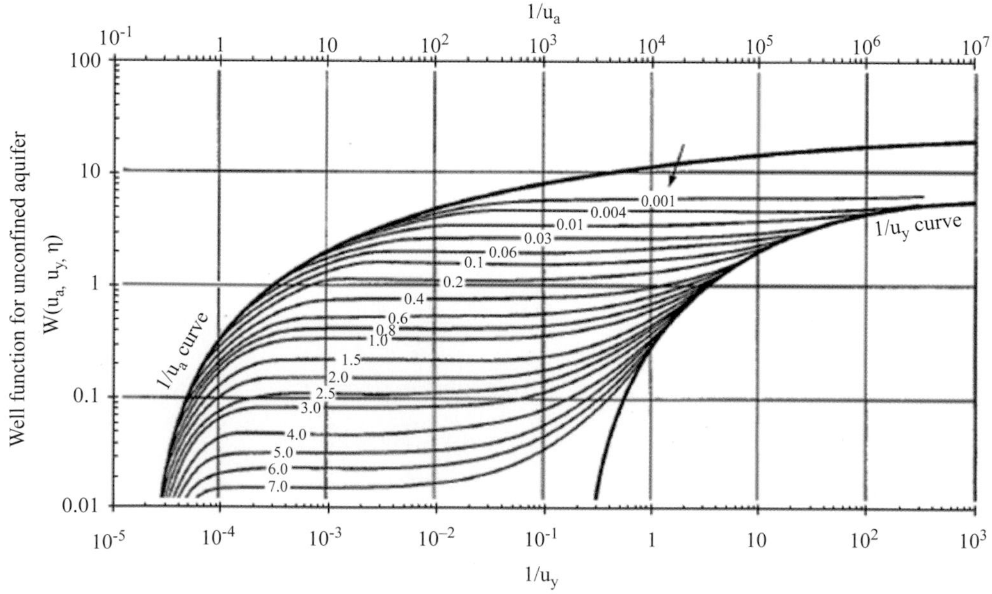

Figure 9.13 **Neuman type curves for unconfined aquifer.** The $1/u_a$ curve is for earlier stage, while the $1/u_y$ curve is for later stage. (*Source:* reproduced from Figure 4.5.3 in Todd and Mays, 2005)

In practice, on the one hand, if the aquifer properties S, T, S_y, K_H, and K_V are known, the drawdown at a given distance away from the well and at a time of interest can be directly computed using Eqs. (9.40) through (9.43). On the other hand, if the properties are unknown, they can be determined using a measured time series of drawdown by following nine steps: (1) plot $s \sim t$ in a log-log scale that is the same as the Neuman type curves shown in Figure 9.13; (2) match the $s \sim t$ curve with a Neuman type curve for earlier drawdown as much as possible; (3) read η of the best-matched Neuman type curve; (4) choose an arbitrary and convenient match point, and read the corresponding values of $W(u_a, u_y, \eta)$, $1/u_a$, s, and t; (5) compute T using Eq. (9.40), S using Eq. (9.41), and K_V/K_H using Eq. (9.43); (6) compute $K_H = T/h_0$ and then $K_V = (K_V/K_H)\,K_H$; (7) match the $s \sim t$ curve with the Neuman type curve (with the η determined in step 3 for later drawdown; (8) choose an arbitrary and convenient match point, and read the corresponding values of $1/u_y$ and t; and (9) compute S_y using Eq. (9.42).

Example 9.33 The properties of an unconfined aquifer consisting of fine sands are: $S = 0.005$, $S_y = 0.025$, $K_H = 50\,\text{m}\,\text{d}^{-1}$, $K_V = 5\,\text{m}\,\text{d}^{-1}$, and $h_0 = 60\,\text{m}$. For a pumping rate of $5000\,\text{m}^3\,\text{d}^{-1}$, determine the drawdown after pumping for 5 d at a distance away from the well of:

(1) 20 m
(2) 100 m
(3) 200 m.

Solution

$S = 0.005$, $S_y = 0.025$, $K_H = 50\,\text{m}\,\text{d}^{-1}$, $K_V = 5\,\text{m}\,\text{d}^{-1}$, $h_0 = 60\,\text{m}$

$Q = 5000\,\text{m}^3\,\text{d}^{-1}$; fine sand $\rightarrow$ the threshold time from the earlier to latter stage is 30 min.

$t = 5\,\text{d} > 30\,\text{min} \rightarrow$ later stage $\rightarrow 1/u_y$ should be used to determine $W(u_a, u_y, \eta)$.

$$T = K_H h_0 = (50\,\text{m}\,\text{d}^{-1}) * (60\,\text{m}) = 3000\,\text{m}^2\,\text{d}^{-1}.$$

(1) $r = 20\,\text{m}$

Eq. (9.42): $u_y = \left[0.025 * (20\,\text{m})^2\right] / \left[4 * (3000\,\text{m}^2\,\text{d}^{-1}) * (5\,\text{d})\right] = 1/6000$

$\rightarrow 1/u_y = 6000 = 6.0 \times 10^3.$

Eq. (9.43): $\eta = \left[(20\,\text{m})^2 / (60\,\text{m})^2\right] * \left[(5\,\text{m}\,\text{d}^{-1}) / (50\,\text{m}\,\text{d}^{-1})\right] = 0.011.$

Figure 9.13: $1/u_y = 6.0 \times 10^3 \rightarrow W(u_a, u_y, \eta) = 6.0.$

Eq. (9.40): $s(r = 20\,\text{m}, t = 5\,\text{d}) = (5000\,\text{m}^3\,\text{d}^{-1}) / \left[4 * \pi * (3000\,\text{m}^2\,\text{d}^{-1})\right] * 6.0 = 0.80\,\text{m}.$

(2) $r = 100\,\text{m}$

Eq. (9.42): $u_y = \left[0.025 * (100\,\text{m})^2\right] / \left[4 * (3000\,\text{m}^2\,\text{d}^{-1}) * (5\,\text{d})\right] = 1/240$

$\rightarrow 1/u_y = 240 = 2.4 \times 10^2.$

Eq. (9.43): $\eta = \left[(100\,\mathrm{m})^2 / (60\,\mathrm{m})^2\right] * \left[\left(5\,\mathrm{m\,d^{-1}}\right) / \left(50\,\mathrm{m\,d^{-1}}\right)\right] = 0.278.$

Figure 9.13: $1/u_y = 2.4 \times 10^2 \rightarrow W\left(u_a, u_y, \eta\right) = 5.5.$

Eq. (9.40): $s\,(r = 100\,\mathrm{m},\ t = 5\,\mathrm{d}) = \left(5000\,\mathrm{m^3\,d^{-1}}\right) / \left[4 * \pi * \left(3000\,\mathrm{m^2\,d^{-1}}\right)\right] * 5.5 = 0.73\,\mathrm{m}.$

(3) $r = 200\,\mathrm{m}$

Eq. (9.42): $u_y = \left[0.025 * (200\,\mathrm{m})^2\right] / \left[4 * \left(3000\,\mathrm{m^2\,d^{-1}}\right) * (5\,\mathrm{d})\right] = 1/60$

$\rightarrow 1/u_y = 60 = 6.0 \times 10^1.$

Eq. (9.43): $\eta = \left[(200\,\mathrm{m})^2 / (60\,\mathrm{m})^2\right] * \left[\left(5\,\mathrm{m\,d^{-1}}\right) / \left(50\,\mathrm{m\,d^{-1}}\right)\right] = 1.11.$

Figure 9.13: $1/u_y = 6.0 \times 10^1 \rightarrow W\left(u_a, u_y, \eta\right) = 4.0.$

Eq. (9.40): $s\,(r = 200\,\mathrm{m},\ t = 5\,\mathrm{d}) = \left(5000\,\mathrm{m^3\,d^{-1}}\right) / \left[4 * \pi * \left(3000\,\mathrm{m^2\,d^{-1}}\right)\right] * 4.0 = 0.53\,\mathrm{m}.$

Example 9.34 The properties of an unconfined aquifer consisting of fine sands are: $S = 0.005$, $S_y = 0.025$, $K_H = 50\,\mathrm{m\,d^{-1}}$, $K_V = 5\,\mathrm{m\,d^{-1}}$, and $h_0 = 60\,\mathrm{m}$. For a pumping rate of $5000\,\mathrm{m^3\,d^{-1}}$, determine the drawdown at 100 m away from the well after pumping for:

(1) 20 min
(2) 1 d
(3) 2 d
(4) 3 d.

Solution

$S = 0.005$, $S_y = 0.025$, $K_H = 50\,\mathrm{m\,d^{-1}}$, $K_V = 5\,\mathrm{m\,d^{-1}}$, $h_0 = 60\,\mathrm{m}$, $r = 100\,\mathrm{m}$

$Q = 5000\,\mathrm{m^3\,d^{-1}}$; find sand $\rightarrow$ the threshold time from the earlier to latter stage is 30 min.

$$T = K_H h_0 = \left(50\,\mathrm{m\,d^{-1}}\right) * (60\,\mathrm{m}) = 3000\,\mathrm{m^2\,d^{-1}}.$$

Eq. (9.43): $\eta = \left[(100\,\mathrm{m})^2 / (60\,\mathrm{m})^2\right] * \left[\left(5\,\mathrm{m\,d^{-1}}\right) / \left(50\,\mathrm{m\,d^{-1}}\right)\right] = 0.278.$

(1) $t = 20\,\mathrm{min} < 30\,\mathrm{min} \rightarrow$ earlier stage $\rightarrow 1/u_a$ should be used to determine $W\left(u_a, u_y, \eta\right)$.

Eq. (9.41): $u_a = \left[0.005 * (100\,\mathrm{m})^2\right] / \left[4 * \left(3000\,\mathrm{m^2\,d^{-1}}\right) * (20\,\mathrm{min} * 1\,\mathrm{d}/1440\,\mathrm{min})\right] = 10/3$

$\rightarrow 1/u_a = 0.3 = 3.0 \times 10^{-1}.$

Figure 9.13: $1/u_a = 3.0 \times 10^{-1} \rightarrow W\left(u_a, u_y, \eta\right) = 0.015.$

Eq. (9.40): $s\,(r = 100\,\mathrm{m},\ t = 20\,\mathrm{min}) = \left(5000\,\mathrm{m^3\,d^{-1}}\right) / \left[4 * \pi * \left(3000\,\mathrm{m^2\,d^{-1}}\right)\right] * 0.015$

$= 0.002\,\mathrm{m}.$

(2) $t = 1\,d > 30\,min \rightarrow$ later stage $\rightarrow 1/u_y$ should be used to determine $W(u_a, u_y, \eta)$.

Eq. (9.42): $u_y = \left[0.025 * (100\,m)^2\right] / \left[4 * \left(3000\,m^2\,d^{-1}\right) * (1\,d)\right] = 1/48$

$\rightarrow 1/u_y = 48 = 4.8 \times 10^1$.

Figure 9.13: $1/u_y = 4.8 \times 10^1 \rightarrow W(u_a, u_y, \eta) = 4.0$.

Eq. (9.40): $s\,(r = 100\,m,\ t = 1\,d) = \left(5000\,m^3\,d^{-1}\right) / \left[4 * \pi * \left(3000\,m^2\,d^{-1}\right)\right] * 4.0 = 0.53\,m$.

(3) $t = 2\,d > 30\,min \rightarrow$ later stage $\rightarrow 1/u_y$ should be used to determine $W(u_a, u_y, \eta)$.

Eq. (9.42): $u_y = \left[0.025 * (100\,m)^2\right] / \left[4 * \left(3000\,m^2\,d^{-1}\right) * (2\,d)\right] = 1/96$

$\rightarrow 1/u_y = 96 = 9.6 \times 10^1$.

Figure 9.13: $1/u_y = 9.6 \times 10^1 \rightarrow W(u_a, u_y, \eta) = 4.5$.

Eq. (9.40): $s\,(r = 100\,m,\ t = 2\,d) = \left(5000\,m^3\,d^{-1}\right) / \left[4 * \pi * \left(3000\,m^2\,d^{-1}\right)\right] * 4.5 = 0.60\,m$.

(4) $t = 3\,d > 30\,min \rightarrow$ later stage $\rightarrow 1/u_y$ should be used to determine $W(u_a, u_y, \eta)$.

Eq. (9.42): $u_y = \left[0.025 * (100\,m)^2\right] / \left[4 * \left(3000\,m^2\,d^{-1}\right) * (3\,d)\right] = 1/144$

$\rightarrow 1/u_y = 144 = 1.44 \times 10^2$.

Figure 9.13: $1/u_y = 1.44 \times 10^2 \rightarrow W(u_a, u_y, \eta) = 4.7$.

Eq. (9.40): $s\,(r = 100\,m,\ t = 3\,d) = \left(5000\,m^3\,d^{-1}\right) / \left[4 * \pi * \left(3000\,m^2\,d^{-1}\right)\right] * 4.7 = 0.62\,m$.

Example 9.35 A well fully penetrates a 30-ft-thick unconfined aquifer and pumps at $15\,ft^3\,min^{-1}$, resulting in a time series of drawdown at 200 ft away from the pumping well as shown in cells A1:B20 of Figure 9.14. Determine the aquifer properties of T, S, S_y, K_H, and K_V.

Solution

$h_0 = 30\,ft$, $Q = 15\,ft^3\,min^{-1}$, $r = 200\,ft$

- The $s \sim t$ curve is plotted and shown at the bottom left of the figure. To superpose the $s \sim t$ curve on the Neuman type curves, the scales of the x- and y-axes should be adjusted to be the same. In addition, the chart area of the $s \sim t$ curve should be set to have no fill (i.e., be transparent) by right-clicking the chart area and selecting Format Plot Area....
- By moving around and superposing the $s \sim t$ curve on the Neuman type curves, the earlier drawdowns (in red color) best match the type curve of $\eta = 0.4$, as shown in the graph at the top right of the figure. The arbitrary match point is shown as the solid black dot, which corresponds to $W(u_a, u_y, \eta) = 1.0$, $1/u_a = 0.42$, $s = 0.18\,ft$, and $t = 10\,min$.
  Eq. (9.40): $0.18\,ft = \left(15\,ft^3\,min^{-1}\right) / (4\pi T) * 1.0 \rightarrow T = 6.63\,ft^2\,min^{-1}$.
  Eq. (9.41): $1/0.42 = S * (200\,ft)^2 / \left[4 * \left(6.63\,ft^2\,min^{-1}\right) * (10\,min)\right] \rightarrow S = 0.016$.

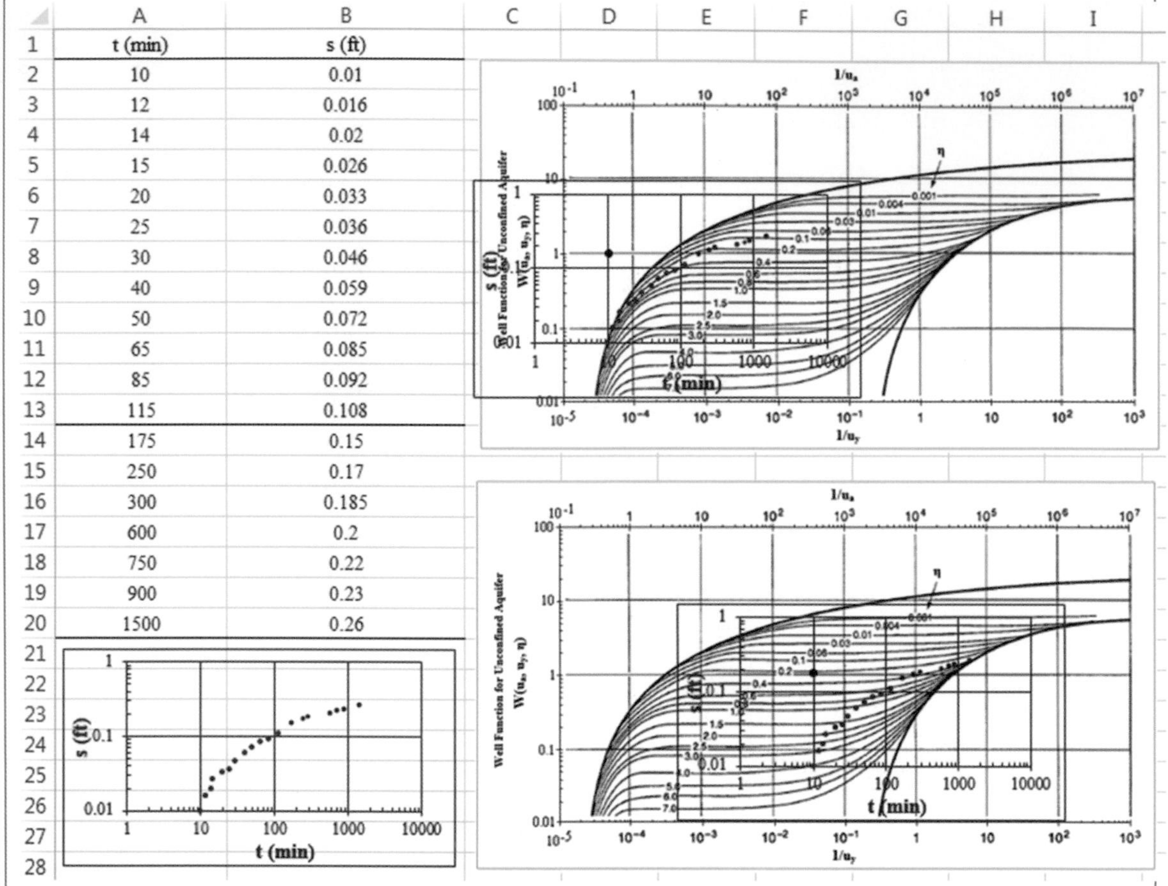

t (min)	s (ft)
10	0.01
12	0.016
14	0.02
15	0.026
20	0.033
25	0.036
30	0.046
40	0.059
50	0.072
65	0.085
85	0.092
115	0.108
175	0.15
250	0.17
300	0.185
600	0.2
750	0.22
900	0.23
1500	0.26

Figure 9.14 **Spreadsheet for determining the properties of the aquifer in Example 9.35.**

Eq. (9.43): $0.4 = (200\,\text{ft})^2 / (30\,\text{ft})^2\,(K_V/K_H) \rightarrow K_V/K_H = 0.009.$

$$K_H = T/h_0 = \left(6.63\,\text{ft}^2\,\text{min}^{-1}\right) / (30\,\text{ft}) = 0.22\,\text{ft min}^{-1}.$$

$$K_V = 0.009 * \left(0.22\,\text{ft min}^{-1}\right) = 0.0020\,\text{ft min}^{-1}.$$

- By moving around and superposing the s ~ t curve on the Neuman type curves, the later drawdowns (in blue color) best match the type curve of $\eta = 0.4$, as shown in the graph at the bottom right of the figure. The arbitrary match point is shown as the solid black dot, which corresponds to $1/u_y = 0.03$ and $t = 10\,\text{min}$.
  Eq. (9.42): $1/0.03 = S_y * (200\,\text{ft})^2 / \left[4 * \left(6.63\,\text{ft}^2\,\text{min}^{-1}\right) * (10\,\text{min})\right] \rightarrow S_y = 0.22.$

Table 9.3 Gamma function for the Bessel function in Eq. (9.45).[1]

m	Γ (m + 1)
0	1
1	1
2	2
3	6
4	24
5	120
6	720
7	5040
8	40,320
9	362,880
10	3,628,800

[1] m: integer accumulator variable.

9.4.5 Radial Flow in a Semiconfined Aquifer

For steady radial flow in a semiconfined aquifer overlain by an unconfined aquifer (e.g., Figure 9.2), De Glee (1930, 1951) derived the formula expressed as:

$$s = \frac{Q}{2\pi T} B_0 \left(\frac{r}{\sqrt{T \frac{b'}{K'}}} \right) \tag{9.44}$$

s: steady drawdown in semiconfined aquifer
Q: pumping rate
T: transmissivity of semiconfined aquifer
B_0: zero-order Bessel function (Eq. (9.45); Figure 9.15; Excel: BESSELJ( ))
r: distance away from well
b′: thickness of aquitard
K′: hydraulic conductivity of aquitard

$$B_0 \left(\frac{r}{\sqrt{T \frac{b'}{K'}}} \right) = \sum_{m=0}^{\infty} \left[\frac{(-1)^m}{m!\Gamma(m+1)} \left(\frac{r}{2\sqrt{T \frac{b'}{K'}}} \right)^{2m} \right] \tag{9.45}$$

B_0: zero-order Bessel function in Eq. (9.44) (Figure 9.15)
r: distance away from the well
T: transmissivity of semiconfined aquifer
b′: thickness of aquitard
K′: hydraulic conductivity of aquitard
Γ: gamma function (Table 9.3; Excel: GAMMA())

In Eq. (9.44), the ratio of b′/K′ is called the resistance of aquitard. In Eq. (9.45), m can be selected to achieve a required truncation error. The zero-order Bessel function can be determined from Figure 9.15 or computed in Excel using the *Besselj( )* function.

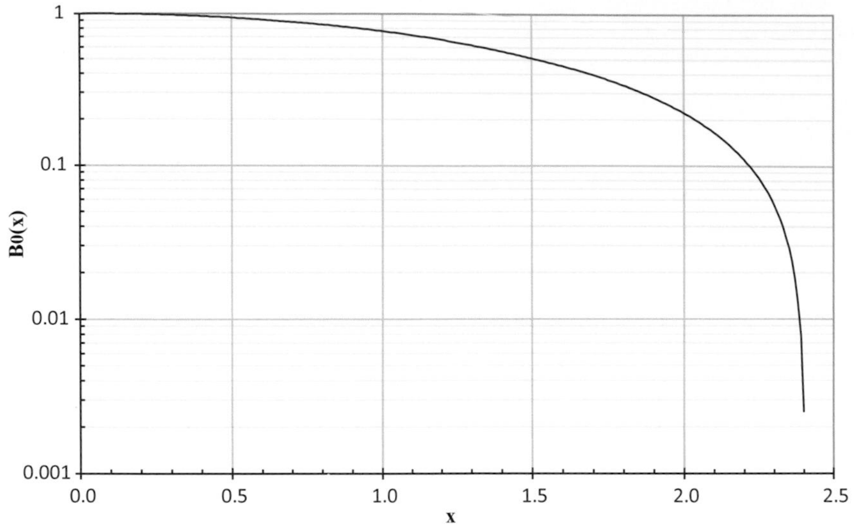

Figure 9.15 **The zero-order Bessel function, B_0 (x), versus independent variable, x.**

Example 9.36 The transmissivity of a semiconfined aquifer is $1000\,m^2\,d^{-1}$. Its aquitard bed has a thickness of 5 m and a hydraulic conductivity of $0.005\,m\,d^{-1}$. After pumping at $5000\,m^3\,d^{-1}$ for a long period of time, determine the drawdown at a distance away from the well of:

(1) 20 m
(2) 1000 m
(3) 2000 m.

Solution

$T = 1000\,m^2\,d^{-1}$, $b' = 5\,m$, $K' = 0.005\,m\,d^{-1}$, $Q = 5000\,m^3\,d^{-1}$

Pumping for a long period of time → steady state.

Let $x = \frac{r}{\sqrt{T\frac{b'}{K'}}}$

(1) $r = 20\,m$

$$x = (20\,m) / \left[\left(1000\,m^2\,d^{-1}\right) * (5\,m) / \left(0.005\,m\,d^{-1}\right)\right]^{1/2} = 0.02.$$

Figure 9.15: $B_0(x) = 0.999$.

Eq. (9.44): $s = \left(5000\,m^3\,d^{-1}\right) / \left[2 * \pi * \left(1000\,m^2\,d^{-1}\right)\right] * 0.999 = 0.80\,m.$

(2) $r = 1000\,m$

$$x = (1000\,m) / \left[\left(1000\,m^2\,d^{-1}\right) * (5\,m) / \left(0.005\,m\,d^{-1}\right)\right]^{1/2} = 1.0.$$

Figure 9.15: $B_0(x) = 0.765$.

Eq. (9.44): $s = \left(5000\,\mathrm{m^3\,d^{-1}}\right) / \left[2 * \pi * \left(1000\,\mathrm{m^2\,d^{-1}}\right)\right] * 0.765 = 0.61\,\mathrm{m}.$

(3) $r = 2000\,\mathrm{m}$

$$x = (2000\,\mathrm{m}) / \left[\left(1000\,\mathrm{m^2\,d^{-1}}\right) * (5\,\mathrm{m}) / \left(0.005\,\mathrm{m\,d^{-1}}\right)\right]^{1/2} = 2.0.$$

Figure 9.15: $B_0(x) = 0.224$.

Eq. (9.44): $s = \left(5000\,\mathrm{m^3\,d^{-1}}\right) / \left[2 * \pi * \left(1000\,\mathrm{m^2\,d^{-1}}\right)\right] * 0.224 = 0.18\,\mathrm{m}.$

Example 9.37 The aquitard bed of a semiconfined aquifer has a resistance of 10,000 d. After pumping at $4000\,\mathrm{m^3\,d^{-1}}$ for a long period of time, the drawdown at 500 m away from the well is measured to be 0.75 m. Determine the transmissivity of the semiconfined aquifer.

Solution

$b'/K' = 10{,}000\,\mathrm{d}$, $Q = 4000\,\mathrm{m^3\,d^{-1}}$, $r = 500\,\mathrm{m}$, $s = 0.75\,\mathrm{m}$

Pumping for a long period of time → steady state.

Let $x = \dfrac{r}{\sqrt{T\frac{b'}{K'}}}$

$$x = (500\,\mathrm{m}) / \left[T * (10{,}000\,\mathrm{d})\right]^{1/2} = 5T^{-1/2}$$

Eq. (9.44): $0.75\,\mathrm{m} = \left(4000\,\mathrm{m^3\,d^{-1}}\right) / (2\pi T)\, B_0\left(5T^{-1/2}\right)$.

Trial-and-error or Excel Solver → $T = 842.5\,\mathrm{m^2\,d^{-1}}$.

For unsteady radial flow in a semiconfined aquifer overlain by an unconfined aquifer (e.g., Figure 9.2), Hantush and Jacob (1955) developed an analytical solution expressed as:

$$s(r,t) = \frac{Q}{4\pi T} W\left(u, \frac{r}{\sqrt{T\frac{b'}{K'}}}\right) \quad (9.46)$$

s (r, t): drawdown at distance r away from well and at time t; Q: pumping rate; W: well function (Figure 9.16); u: variable (Eq. (9.47)); T: transmissivity of semiconfined aquifer; r: distance away from well; b′: thickness of aquitard; K′: hydraulic conductivity of aquitard

$$u = \frac{Sr^2}{4Tt} \quad (9.47)$$

u: variable in Eq. (9.46); S: storativity of semiconfined aquifer; r: distance away from well; T: transmissivity of semiconfined aquifer; t: lapse time since start of pumping

If the properties of a semiconfined aquifer (S and T) and aquitard (b′ and K′) are known, the drawdown at a given distance away from the well and at a time of interest can be directly computed using Eqs. (9.46) and (9.47). However, if the properties are unknown, they can be determined using a measured time series of drawdown by the following five steps: (1) plot $s \sim t$ in a log-log scale

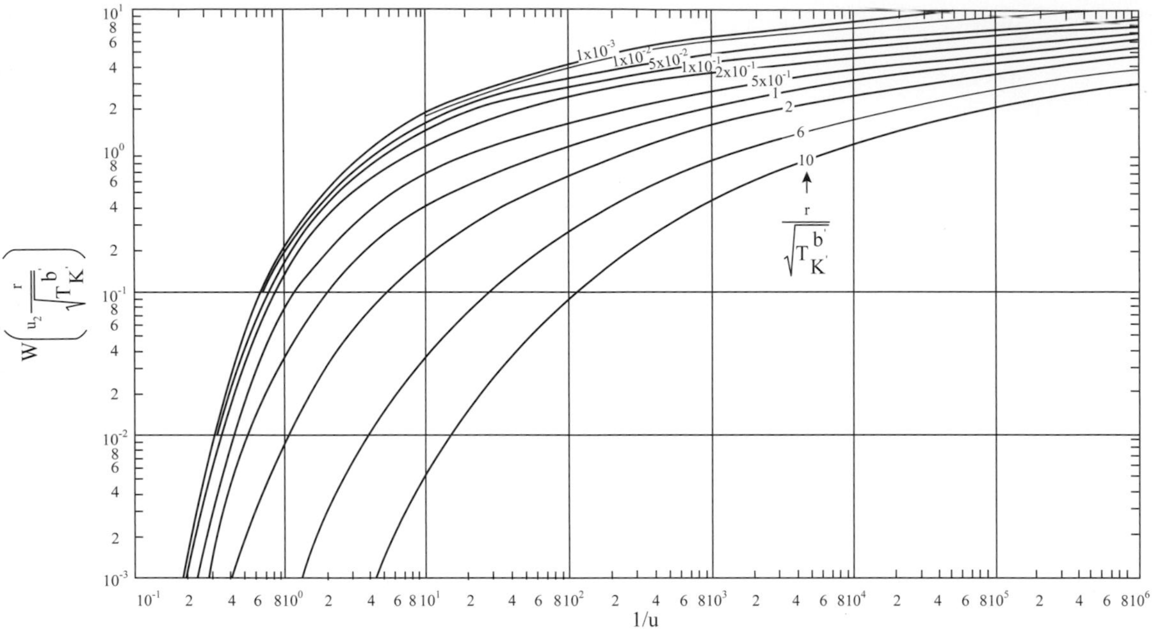

Figure 9.16 **Hantush–Jacob type curves for a semiconfined aquifer.** The y-axis is the well function defined by Eq. (9.46), and the x-axis is the variable defined by Eq. (9.47). T: transmissivity of semiconfined aquifer; b′: thickness of aquitard; K′: hydraulic conductivity of aquitard; r: distance away from pumping well; and u: variable in Eq. (9.47). (*Source:* reproduced from Figure 4.6.2 in Todd and Mays, 2005)

so as to correspond to the Hantush–Jacob type curves shown in Figure 9.16; (2) match the $s \sim t$ curve with a Hantush–Jacob type curve as closely as possible; (3) read the value of $\frac{r}{\sqrt{T\frac{b'}{K'}}}$ from the Hantush–Jacob type curve; (4) choose an arbitrary and convenient match point, and read the corresponding values of $W\left(u, \frac{r}{\sqrt{T\frac{b'}{K'}}}\right)$, 1/u, s, and t; and (5) compute T using Eq. (9.46), S using Eq. (9.47), and the aquitard resistance b′/K′.

Example 9.38 A semiconfined aquifer has a transmissivity of 10,000 $ft^2\,d^{-1}$ and a storativity of 0.005. Its aquitard bed has a thickness of 15 ft and a hydraulic conductivity of 0.0015 $ft\,d^{-1}$. After pumping at 150,000 $ft^3\,d^{-1}$ for 10 d, determine the drawdown at a distance away from the well of:

(1) 100 ft
(2) 500 ft.

Solution

$T = 10{,}000\,ft^2\,d^{-1}$, $S = 0.005$, $b' = 15\,ft$, $K' = 0.0015\,ft\,d^{-1}$

$Q = 150{,}000\,ft^3\,d^{-1}$, $t = 10\,d$

(1) r = 100 ft

$$\frac{r}{\sqrt{T\frac{b'}{K'}}} = \frac{100\,\text{ft}}{\sqrt{\left(10{,}000\,\text{ft}^2\,\text{d}^{-1}\right) * \frac{15\,\text{ft}}{0.0015\,\text{ft}\,\text{d}^{-1}}}} = 0.01.$$

$$\text{Eq. (9.47): } u = \frac{0.005 * (100\,\text{ft})^2}{4 * \left(10{,}000\,\text{ft}^2\,\text{d}^{-1}\right) * (10\,\text{d})} = 0.000125 \rightarrow 1/u = 8000 = 8.0 \times 10^3.$$

$$\text{Figure 9.16: } W\left(u, \frac{r}{\sqrt{T\frac{b'}{K'}}}\right) = 5.0.$$

Eq. (9.46): $s\,(r = 100\,\text{ft},\ t = 10\,\text{d}) = \left(150{,}000\,\text{ft}^3\,\text{d}^{-1}\right) / \left[4 * \pi * \left(10{,}000\,\text{ft}^2\,\text{d}^{-1}\right)\right] * 5.0 = 6.0\,\text{ft}$.

(2) r = 500 ft

$$\frac{r}{\sqrt{T\frac{b'}{K'}}} = \frac{500\,\text{ft}}{\sqrt{\left(10{,}000\,\text{ft}^2\,\text{d}^{-1}\right) * \frac{15\,\text{ft}}{0.0015\,\text{ft}\,\text{d}^{-1}}}} = 0.05.$$

$$\text{Eq. (9.47): } u = \frac{0.005 * (500\,\text{ft})^2}{4 * \left(10{,}000\,\text{ft}^2\,\text{d}^{-1}\right) * (10\,\text{d})} = 0.003125 \rightarrow 1/u = 320 = 3.2 \times 10^2.$$

$$\text{Figure 9.16: } W\left(u, \frac{r}{\sqrt{T\frac{b'}{K'}}}\right) = 4.0.$$

Eq. (9.46): $s\,(r = 500\,\text{ft},\ t = 10\,\text{d}) = \left(150{,}000\,\text{ft}^3\,\text{d}^{-1}\right) / \left[4 * \pi * \left(10{,}000\,\text{ft}^2\,\text{d}^{-1}\right)\right] * 4.0 = 4.8\,\text{ft}$.

Example 9.39 The properties of a semiconfined aquifer are: $S = 0.002$ and $T = 10{,}000\,\text{ft}^2\,\text{d}^{-1}$. Its aquitard has a resistance of 10,000 d. For a pumping rate of $150{,}000\,\text{ft}^3\,\text{d}^{-1}$, determine the drawdown at 100 ft away from the well after pumping for:

(1) 1 d
(2) 5 d
(3) 10 d.

Solution

$S = 0.002$, $T = 10{,}000\ \text{ft}^2\ \text{d}^{-1}$, $b'/K' = 10{,}000\ \text{d}$, $Q = 150{,}000\ \text{ft}^3\ \text{d}^{-1}$, $r = 100\ \text{ft}$

$$\frac{r}{\sqrt{T\frac{b'}{K'}}} = \frac{100\ \text{ft}}{\sqrt{\left(10{,}000\ \text{ft}^2\ \text{d}^{-1}\right) * \left(10{,}000\ \text{d}^{-1}\right)}} = 0.01.$$

(1) $t = 1\ \text{d}$

Eq. (9.47): $u = \dfrac{0.002 * (100\ \text{ft})^2}{4 * \left(10{,}000\ \text{ft}^2\ \text{d}^{-1}\right) * (1\ \text{d})} = 0.0005 \rightarrow 1/u = 2000 = 2.0 \times 10^3.$

Figure 9.16: $W\left(u, \dfrac{r}{\sqrt{T\frac{b'}{K'}}}\right) = 6.0.$

Eq. (9.46): $s\,(r = 100\ \text{ft},\ t = 1\ \text{d}) = \left(150{,}000\ \text{ft}^3\ \text{d}^{-1}\right) / \left[4 * \pi * \left(10{,}000\ \text{ft}^2\ \text{d}^{-1}\right)\right] * 6.0 = 7.2\ \text{ft}.$

(2) $t = 5\ \text{d}$

Eq. (9.47): $u = \dfrac{0.002 * (100\ \text{ft})^2}{4 * \left(10{,}000\ \text{ft}^2\ \text{d}^{-1}\right) * (5\ \text{d})} = 0.0001 \rightarrow 1/u = 10{,}000 = 1.0 \times 10^4.$

Figure 9.16: $W\left(u, \dfrac{r}{\sqrt{T\frac{b'}{K'}}}\right) = 7.3.$

Eq. (9.46): $s\,(r = 100\ \text{ft},\ t = 5\ \text{d}) = \left(150{,}000\ \text{ft}^3\ \text{d}^{-1}\right) / \left[4 * \pi * \left(10{,}000\ \text{ft}^2\ \text{d}^{-1}\right)\right] * 7.3 = 8.7\ \text{ft}.$

(3) $t = 10\ \text{d}$

Eq. (9.47): $u = \dfrac{0.002 * (100\ \text{ft})^2}{4 * \left(10{,}000\ \text{ft}^2\ \text{d}^{-1}\right) * (10\ \text{d})} = 0.00005 \rightarrow 1/u = 20{,}000 = 2.0 \times 10^4.$

Figure 9.16: $W\left(u, \dfrac{r}{\sqrt{T\frac{b'}{K'}}}\right) = 7.7.$

Eq. (9.46): $s\,(r = 100\ \text{ft},\ t = 10\ \text{d}) = \left(150{,}000\ \text{ft}^3\ \text{d}^{-1}\right) / \left[4 * \pi * \left(10{,}000\ \text{ft}^2\ \text{d}^{-1}\right)\right] * 7.7 = 9.2\ \text{ft}.$

Example 9.40 A well fully penetrates a semiconfined aquifer and pumps at 20 $ft^3\ min^{-1}$, resulting in a time series of drawdown at 100 ft away from the pumping well as shown in cells A1:B20 of Figure 9.17. Determine the properties of the aquifer and aquitard, namely T, S, and b'/K'.

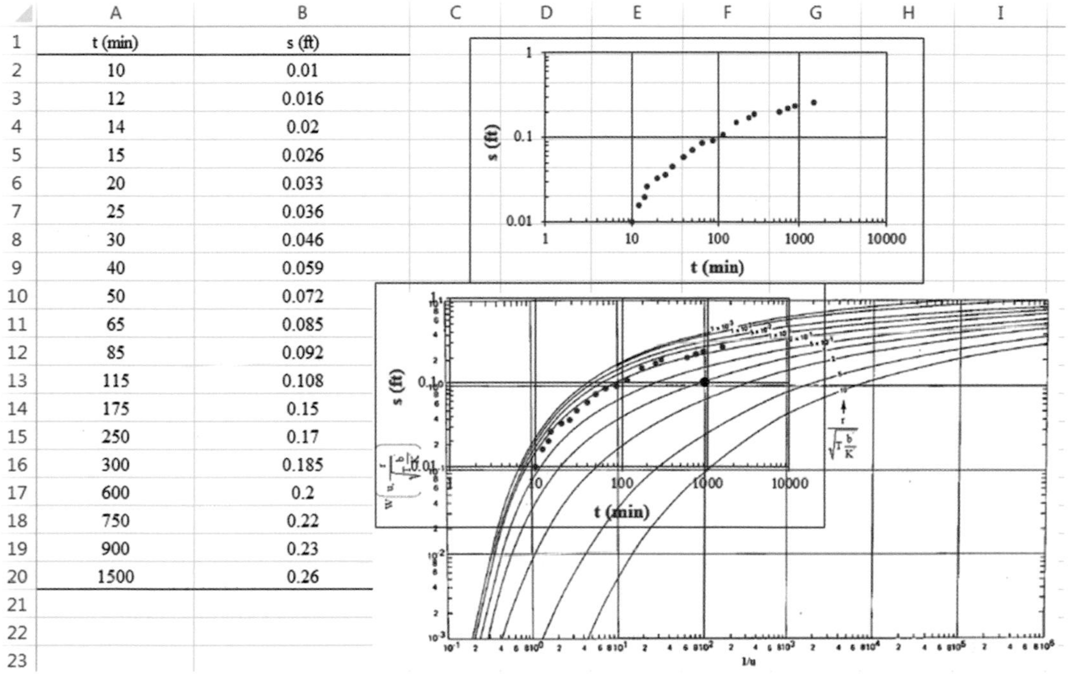

t (min)	s (ft)
10	0.01
12	0.016
14	0.02
15	0.026
20	0.033
25	0.036
30	0.046
40	0.059
50	0.072
65	0.085
85	0.092
115	0.108
175	0.15
250	0.17
300	0.185
600	0.2
750	0.22
900	0.23
1500	0.26

Figure 9.17 **Spreadsheet for determining the properties of the aquifer in Example 9.40.**

Solution

$Q = 20\,ft^3\,min^{-1}, r = 100\,ft$

- The s ~ t curve is plotted and shown at the top right of the figure. To superpose the s ~ t curve on the Hantush–Jacob type curves, the scales of x- and y-axis should be adjusted to be the same. In addition, the chart area of the s ~ t curve should be set to have no fill (i.e., be transparent) by right-clicking the chart area and selecting Format Plot Area....
- By moving around and superposing the s ~ t curve on the Hantush–Jacob type curves, the drawdowns best match the type curve for $\frac{r}{\sqrt{T\frac{b'}{K'}}} = 0.1$, as shown in the graph at the bottom right of the figure. The arbitrary match point is shown as the solid black dot, which corresponds to $W\left(u, \frac{r}{\sqrt{T\frac{b'}{K'}}}\right) = 1.0$, $1/u = 100$, $s = 0.09$ ft, and $t = 950$ min.

$$\text{Eq. (9.46): } 0.09\,\text{ft} = \left(20\,\text{ft}^3\,\text{min}^{-1}\right) / (4\pi\text{T}) * 1.0 \rightarrow \text{T} = 17.68\,\text{ft}^2\,\text{min}^{-1}.$$

$$\text{Eq. (9.47): } 1/100 = \text{S} * (100\,\text{ft})^2 / \left[4 * \left(17.68\,\text{ft}^2\,\text{min}^{-1}\right) * (950\,\text{min})\right] \rightarrow \text{S} = 0.067.$$

$$\frac{\text{r}}{\sqrt{\text{T}\frac{\text{b}'}{\text{K}'}}} = 0.1 \rightarrow \text{b}'/\text{K}' = (100\,\text{ft}/0.1)^2 / \left(17.68\,\text{ft}^2\,\text{min}^{-1}\right) = 56{,}561\,\text{min} = 39.28\,\text{d}.$$

9.4.6 Multiple Pumping Rates and Well Flow in an Aquifer with Boundaries

If multiple pumping or recharging wells are installed in the same aquifer, the overall change of groundwater level at a location can be estimated as the summation of the drawdowns induced by all pumping wells and the buildups created by all recharging wells. The drawdown induced by a pumping well can be computed using one of the formulas presented previously in this chapter, whereas the buildup created by a recharging well can be computed using the same formulas with recharge rate as a negative value. That is, the drawdowns are positive but the buildups are negative. This is called superposition theory, which can be mathematically expressed as:

overall drawdown at a location and time → $s = \sum_i s_i$ ← drawdown/buildup induced by well i at the same location and time (9.48)

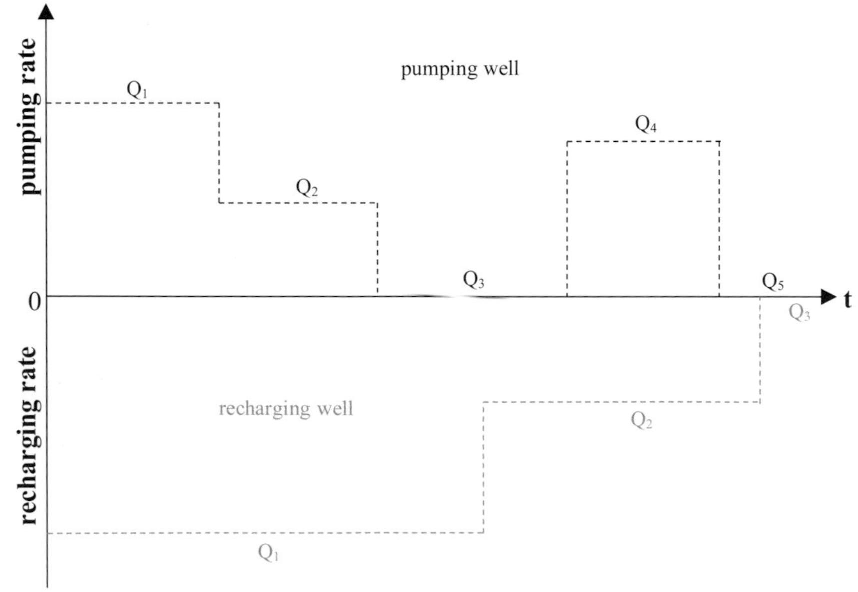

Figure 9.18 **Pumping and recharge wells in a single aquifer.** Diagram for multiple pumping rates (in black) and recharge rates (in light blue).

In reality, for a given period of time, it is common for a pumping or recharging well to run at multiple flow rates. As illustrated in Figure 9.18, the pumping well has five pumping rates, while the recharging well has three recharge rates. The pumping rate is Q_1 at the beginning and then reduced to Q_2. The pumping well stops running ($Q_3 = 0$) for some time, starts to pump at Q_4, and then stops running again. At the same time, the recharge rate is Q_1 at the beginning and then reduced to Q_2. At that point, the recharging well stops running. The overall change of groundwater level at a location can be determined in terms of the superposition theory (Eq. (9.48)). In this regard, for a time period of interest, s_1 is computed using Q_1 for the entire time period, s_2 using $(Q_2 - Q_1)$ for the time period from the beginning of Q_2 onward, and so on.

Example 9.41 A well fully penetrates a confined aquifer ($S = 2.0 \times 10^{-4}$, $T = 1000\,m^2\,d^{-1}$). It pumps at $5000\,m^3\,d^{-1}$ for 0.2 d and $2000\,m^3\,d^{-1}$ for an additional 0.8 d, and then stops running. For a maximum truncation error of $\varepsilon = 10^{-4}$, determine the overall change of groundwater level at 200 m away from the well after pumping for:

(1) 0.2 d
(2) 0.8 d
(3) 2.5 d.

Solution

$S = 2.0 \times 10^{-4}$, $T = 1000\,m^2\,d^{-1}$, $r = 200\,m$

$$t_1 = 0.2\,d,\ Q_1 = 5000\,m^3\,d^{-1},\ t_2 = 0.8\,d,\ Q_2 = 2000\,m^3\,d^{-1},\ Q_3 = 0\,m^3\,d^{-1}$$

(1) $t = 0.2\,d = t_1 \rightarrow$ the overall drawdown is induced by Q_1 only.

Eq. (9.38): $u = \left[\left(2.0 \times 10^{-4}\right) * (200\,m)^2\right] / \left[4 * \left(1000\,m^2\,d^{-1}\right) * (0.2\,d)\right] = 0.01.$

$u \le 0.01 \rightarrow$ first three terms, truncation error $< 2.5 \times 10^{-5}$, which is less than $\varepsilon = 10^{-4}$

$$\rightarrow \text{Eq. (9.39)}: W(u) = -0.5772 - \ln(0.01) + 0.01 = 4.0380.$$

Eq. (9.37): $s = \left(5000\,m^3\,d^{-1}\right) / \left[4 * \pi * \left(1000\,m^2\,d^{-1}\right)\right] * 4.0380 = 1.61\,m.$

(2) $t = 0.8\,d > t_1$ and $< t_1 + t_2 \rightarrow$ the overall change is induced by Q_1 and Q_2.

- Compute s_1 induced by Q_1 for $t = 0.8\,d$:

  Eq. (9.38): $u = \left[\left(2.0 \times 10^{-4}\right) * (200\,m)^2\right] / \left[4 * \left(1000\,m^2\,d^{-1}\right) * (0.8\,d)\right] = 0.0025.$

  $u \le 0.01 \rightarrow$ first three terms, truncation error $< 2.5 \times 10^{-5}$, which is less than $\varepsilon = 10^{-4}$

  $$\rightarrow \text{Eq. (9.39)}: W(u) = -0.5772 - \ln(0.0025) + 0.0025 = 5.4168.$$

  Eq. (9.37): $s_1 = \left(5000\,m^3\,d^{-1}\right) / \left[4 * \pi * \left(1000\,m^2\,d^{-1}\right)\right] * 5.4168 = 2.16\,m.$

- Compute s_2 induced by $Q_2 - Q_1 = -3000\,m^3\,s^{-1}$ for $t - t_1 = 0.6\,d$:

  Eq. (9.38): $u = \left[\left(2.0 \times 10^{-4}\right) * (200\,m)^2\right] / \left[4 * \left(1000\,m^2\,d^{-1}\right) * (0.6\,d)\right] = 1/300.$

$u \le 0.01 \rightarrow$ first three terms, truncation error $< 2.5 \times 10^{-5}$, which is less than $\varepsilon = 10^{-4}$

$\rightarrow$ Eq. (9.39): $W(u) = -0.5772 - \ln(1/300) + 1/300 = 5.1299$.

Eq. (9.37): $s_2 = (-3000\,m^3\,d^{-1}) / [4 * \pi * (1000\,m^2\,d^{-1})] * 5.1299 = -1.23\,m$

- Eq. (9.48): overall change $s = 2.16\,m + (-1.23\,m) = 0.93\,m$.

(3) $t = 2.5\,d > t_1 + t_2 \rightarrow$ the overall change is induced by Q_1, Q_2, and Q_3.

- Compute s_1 induced by Q_1 for $t = 2.5\,d$:

Eq. (9.38): $u = \left[(2.0 \times 10^{-4}) * (200\,m)^2\right] / \left[4 * (1000\,m^2\,d^{-1}) * (2.5\,d)\right] = 0.0008$.

$u \le 0.01 \rightarrow$ first three terms, truncation error $< 2.5 \times 10^{-5}$, which is less than $\varepsilon = 10^{-4}$

$\rightarrow$ Eq. (9.39): $W(u) = -0.5772 - \ln(0.0008) + 0.0008 = 6.5545$.

Eq. (9.37): $s_1 = (5000\,m^3\,d^{-1}) / [4 * \pi * (1000\,m^2\,d^{-1})] * 6.5545 = 2.61\,m$.

- Compute s_2 induced by $Q_2 - Q_1 = -3000\,m^3\,s^{-1}$ for $t - t_1 = 2.3\,d$:

Eq. (9.38): $u = \left[(2.0 \times 10^{-4}) * (200\,m)^2\right] / \left[4 * (1000\,m^2\,d^{-1}) * (2.3\,d)\right] = 8.69565 \times 10^{-4}$.

$u \le 0.01 \rightarrow$ first three terms, truncation error $< 2.5 \times 10^{-5}$, which is less than $\varepsilon = 10^{-4}$

$\rightarrow$ Eq. (9.39): $W(u) = -0.5772 - \ln(8.69565 \times 10^{-4}) + 8.69565 \times 10^{-4} = 6.4712$.

Eq. (9.37): $s_2 = (-3000\,m^3\,d^{-1}) / [4 * \pi * (1000\,m^2\,d^{-1})] * 6.4712 = -1.54\,m$.

- Compute s_3 induced by $Q_3 - Q_2 = -2000\,m^3\,s^{-1}$ for $t - (t_1 + t_2) = 1.5\,d$:

Eq. (9.38): $u = \left[(2.0 \times 10^{-4}) * (200\,m)^2\right] / \left[4 * (1000\,m^2\,d^{-1}) * (1.5\,d)\right] = 1/750$.

$u \le 0.01 \rightarrow$ first three terms, truncation error $< 2.5 \times 10^{-5}$, which is less than $\varepsilon = 10^{-4}$

$\rightarrow$ Eq. (9.39): $W(u) = -0.5772 - \ln(1/750) + 1/750 = 6.0442$.

Eq. (9.37): $s_3 = (-2000\,m^3\,d^{-1}) / [4 * \pi * (1000\,m^2\,d^{-1})] * 6.0442 = -0.96\,m$.

- Eq. (9.48): overall change $s = 2.61\,m + (-1.54\,m) + (-0.96\,m) = 0.11\,m$.

Example 9.42 A recharging well fully penetrates a confined aquifer ($S = 2.0 \times 10^{-4}$, $T = 1000\,m^2\,d^{-1}$). It recharges at $5000\,m^3\,d^{-1}$ for 0.2 d and $2000\,m^3\,d^{-1}$ for another 0.8 d, and then stops running. For a maximum truncation error of $\varepsilon = 10^{-4}$, determine the overall change of groundwater level at 200 m away from the well after recharging for:

(1) 0.2 d
(2) 0.8 d
(3) 2.5 d.

Solution

$S = 2.0 \times 10^{-4}$, $T = 1000\,m^2\,d^{-1}$, $r = 200\,m$

$$t_1 = 0.2\,d,\ Q_1 = -5000\,m^3\,d^{-1},\ t_2 = 0.8\,d,\ Q_2 = -2000\,m^3\,d^{-1},\ Q_3 = 0\,m^3\,d^{-1}$$

(1) $t = 0.2\,d = t_1 \rightarrow$ the overall buildup is created by Q_1 only.

Eq. (9.38): $u = \left[\left(2.0 \times 10^{-4}\right) * (200\,m)^2\right] / \left[4 * \left(1000\,m^2\,d^{-1}\right) * (0.2\,d)\right] = 0.01.$

$u \le 0.01 \rightarrow$ first three terms, truncation error $< 2.5 \times 10^{-5}$, which is less than $\varepsilon = 10^{-4}$

$$\rightarrow \text{Eq. (9.39): } W(u) = -0.5772 - \ln(0.01) + 0.01 = 4.0380.$$

Eq. (9.37): $s = \left(-5000\,m^3\,d^{-1}\right) / \left[4 * \pi * \left(1000\,m^2\,d^{-1}\right)\right] * 4.0380 = -1.61\,m.$

(2) $t = 0.8\,d > t_1$ and $< t_1 + t_2 \rightarrow$ the overall change is induced by Q_1 and Q_2.

- Compute s_1 created by Q_1 for $t = 0.8\,d$:

  Eq. (9.38): $u = \left[\left(2.0 \times 10^{-4}\right) * (200\,m)^2\right] / \left[4 * \left(1000\,m^2\,d^{-1}\right) * (0.8\,d)\right] = 0.0025.$

  $u \le 0.01 \rightarrow$ first three terms, truncation error $< 2.5 \times 10^{-5}$, which is less than $\varepsilon = 10^{-4}$

  $$\rightarrow \text{Eq. (9.39): } W(u) = -0.5772 - \ln(0.0025) + 0.0025 = 5.4168.$$

  Eq. (9.37): $s_1 = \left(-5000\,m^3\,d^{-1}\right) / \left[4 * \pi * \left(1000\,m^2\,d^{-1}\right)\right] * 5.4168 = -2.16\,m.$

- Compute s_2 created by $Q_2 - Q_1 = 3000\,m^3\,s^{-1}$ for $t - t_1 = 0.6\,d$:

  Eq. (9.38): $u = \left[\left(2.0 \times 10^{-4}\right) * (200\,m)^2\right] / \left[4 * \left(1000\,m^2\,d^{-1}\right) * (0.6\,d)\right] = 1/300.$

  $u \le 0.01 \rightarrow$ first three terms, truncation error $< 2.5 \times 10^{-5}$, which is less than $\varepsilon = 10^{-4}$

  $$\rightarrow \text{Eq. (9.39): } W(u) = -0.5772 - \ln(1/300) + 1/300 = 5.1299.$$

  Eq. (9.37): $s_2 = \left(3000\,m^3\,d^{-1}\right) / \left[4 * \pi * \left(1000\,m^2\,d^{-1}\right)\right] * 5.1299 = 1.23\,m.$

- Eq. (9.48): overall change $s = -2.16\,m + 1.23\,m = -0.93\,m.$

(3) $t = 2.5\,d > t_1 + t_2 \rightarrow$ the overall change is induced by Q_1, Q_2, and Q_3.

- Compute s_1 created by Q_1 for $t = 2.5\,d$:

  Eq. (9.38): $u = \left[\left(2.0 \times 10^{-4}\right) * (200\,m)^2\right] / \left[4 * \left(1000\,m^2\,d^{-1}\right) * (2.5\,d)\right] = 0.0008.$

  $u \le 0.01 \rightarrow$ first three terms, truncation error $< 2.5 \times 10^{-5}$, which is less than $\varepsilon = 10^{-4}$

  $$\rightarrow \text{Eq. (9.39): } W(u) = -0.5772 - \ln(0.0008) + 0.0008 = 6.5545.$$

  Eq. (9.37): $s_1 = \left(-5000\,m^3\,d^{-1}\right) / \left[4 * \pi * \left(1000\,m^2\,d^{-1}\right)\right] * 6.5545 = -2.61\,m.$

- Compute s_2 created by $Q_2 - Q_1 = 3000\,m^3\,s^{-1}$ for $t - t_1 = 2.3\,d$:

  Eq. (9.38): $u = \left[\left(2.0 \times 10^{-4}\right) * (200\,m)^2\right] / \left[4 * \left(1000\,m^2\,d^{-1}\right) * (2.3\,d)\right] = 8.69565 \times 10^{-4}.$

$u \le 0.01 \rightarrow$ first three terms, truncation error $< 2.5 \times 10^{-5}$, which is less than $\varepsilon = 10^{-4}$

$$\rightarrow \text{Eq. (9.39): } W(u) = -0.5772 - \ln(8.69565 \times 10^{-4}) + 8.69565 \times 10^{-4} = 6.4712.$$

Eq. (9.37): $s_2 = \left(3000\,\text{m}^3\,\text{d}^{-1}\right) / \left[4 * \pi * \left(1000\,\text{m}^2\,\text{d}^{-1}\right)\right] * 6.4712 = 1.54\,\text{m}.$

- Compute s_3 induced by $Q_3 - Q_2 = 2000\,\text{m}^3\,\text{s}^{-1}$ for $t - (t_1 + t_2) = 1.5\,\text{d}$:

Eq. (9.38): $u = \left[\left(2.0 \times 10^{-4}\right) * (200\,\text{m})^2\right] / \left[4 * \left(1000\,\text{m}^2\,\text{d}^{-1}\right) * (1.5\,\text{d})\right] = 1/750.$

$u \le 0.01 \rightarrow$ first three terms, truncation error $< 2.5 \times 10^{-5}$, which is less than $\varepsilon = 10^{-4}$

$$\rightarrow \text{Eq. (9.39): } W(u) = -0.5772 - \ln(1/750) + 1/750 = 6.0442.$$

Eq. (9.37): $s_3 = \left(2000\,\text{m}^3\,\text{d}^{-1}\right) / \left[4 * \pi * \left(1000\,\text{m}^2\,\text{d}^{-1}\right)\right] * 6.0442 = 0.96\,\text{m}.$

- Eq. (9.48): overall change $s = -2.61\,\text{m} + 1.54\,\text{m} + 0.96\,\text{m} = -0.11\,\text{m}.$

For the drawdown formulas presented in Sections 9.4.1 through 9.4.5, one of the profound assumptions is that the aquifer has a spatial extent larger than the influence radius, within which there are no boundaries formed by waterbodies (e.g., streams) and/or impermeable barriers (e.g., rocks). If such a boundary is present, flow in the aquifer will be non-radial, with either more water flowing into the well from the aquifer closer to a waterbody or less water from the aquifer closer to an impermeable barrier (Figure 9.19). As a result, at a given distance away from the well, the drawdown on the side closer to the boundary will be smaller or larger than that on the farther side. However, the pumping well has no influence on the aquifer beyond the boundary. Such boundaries are more common for unconfined than confined aquifers.

If a waterbody boundary is present, the drawdown at the boundary is actually zero rather than a value computed by a formula, whereas if a barrier boundary is present, the flow at the boundary is actually zero rather than a value computed by a formula. Such a discrepancy between the formula computation and the actual value is caused by the invalidity of the aforementioned assumption (i.e., that the aquifer has a spatial extent larger than the influence radius). To make this assumption valid, an *image recharging well* can be introduced to eliminate the waterbody boundary (Figure 9.19a). Similarly, an *image pumping well* can be introduced to eliminate the barrier boundary (Figure 9.19b). The distance from the real pumping well to the boundary should be the same as that from the image recharging or pumping well to the boundary. The recharge rate of the image recharging well should be equal to the negative of the pumping rate of the real well, the pumping rate of the image pumping well should be the same as the pumping rate of the real well. As a result, at the waterbody boundary, the computed drawdown from the real pumping well will be canceled by the computed buildup from the image well, whereas at the barrier boundary, the computed flow from the real well will be canceled by the computed flow from the image well. The drawdown at a location of interest can be determined by the superposition theory (Eq. (9.48)), with the drawdown induced by the real well and the buildup or drawdown induced by the image well computed by assuming no boundary.

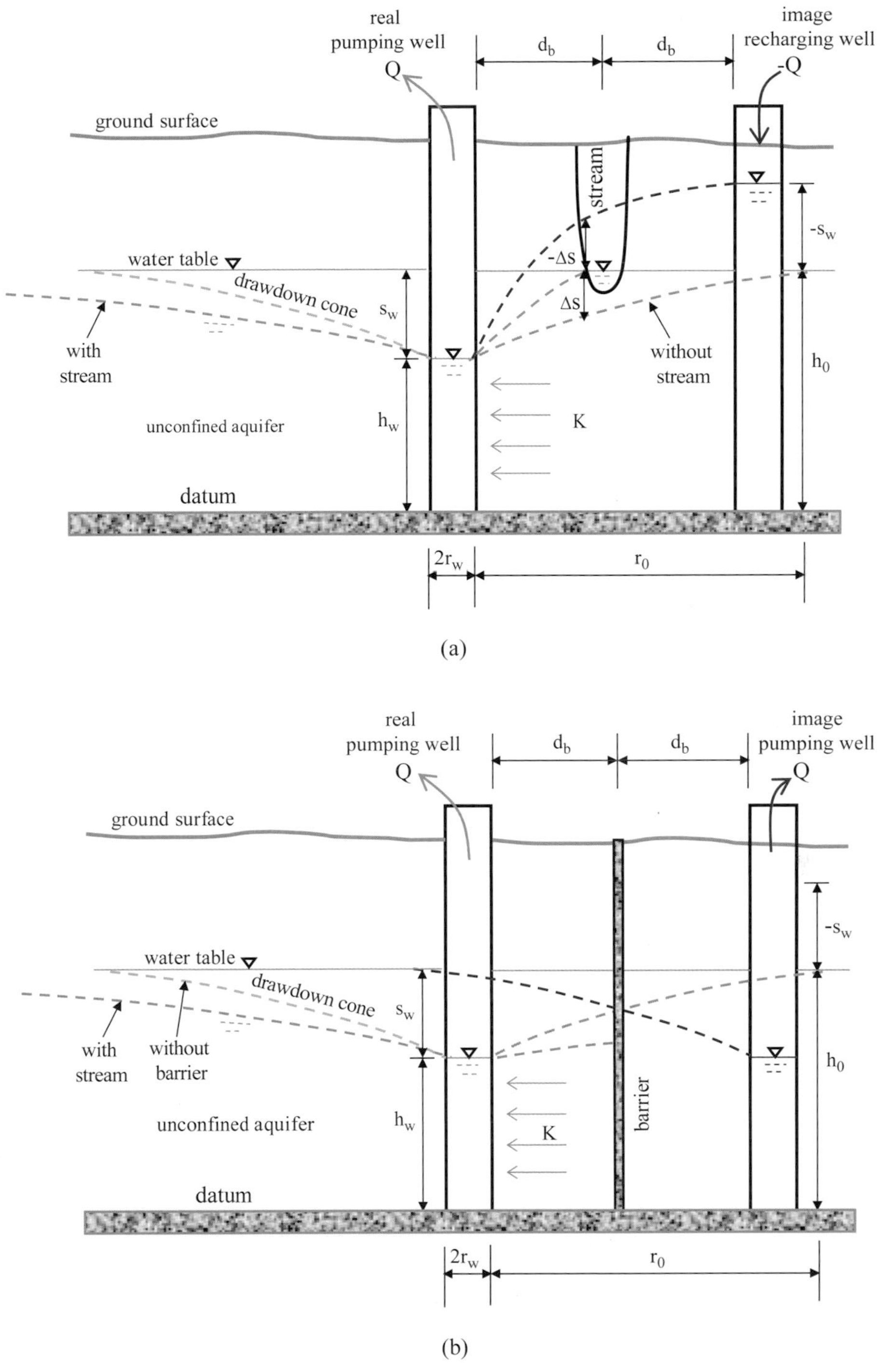

Figure 9.19 **Effect of a pumping well in an unconfined aquifer in proximity to a boundary.** Diagram for case in which a: (a) stream acts as a boundary; and (b) barrier acts as a boundary. An image recharging well can be introduced to eliminate the stream boundary, whereas an image pumping well can be introduced to eliminate the barrier boundary. Note that the distance from the pumping well to the boundary must be equal to that from the image well to the boundary. Gray dashed line: drawdown cone induced by the real well by assuming no boundary; light blue dashed line: actual drawdown cone induced by the real well with the boundary; and red dashed line: buildup or drawdown induced by the image well by assuming no boundary.

Example 9.43 Introduce image wells to eliminate the boundaries shown in Figure 9.20.

Solution

Use image recharging wells for waterbody boundaries while image pumping wells for barrier boundaries.

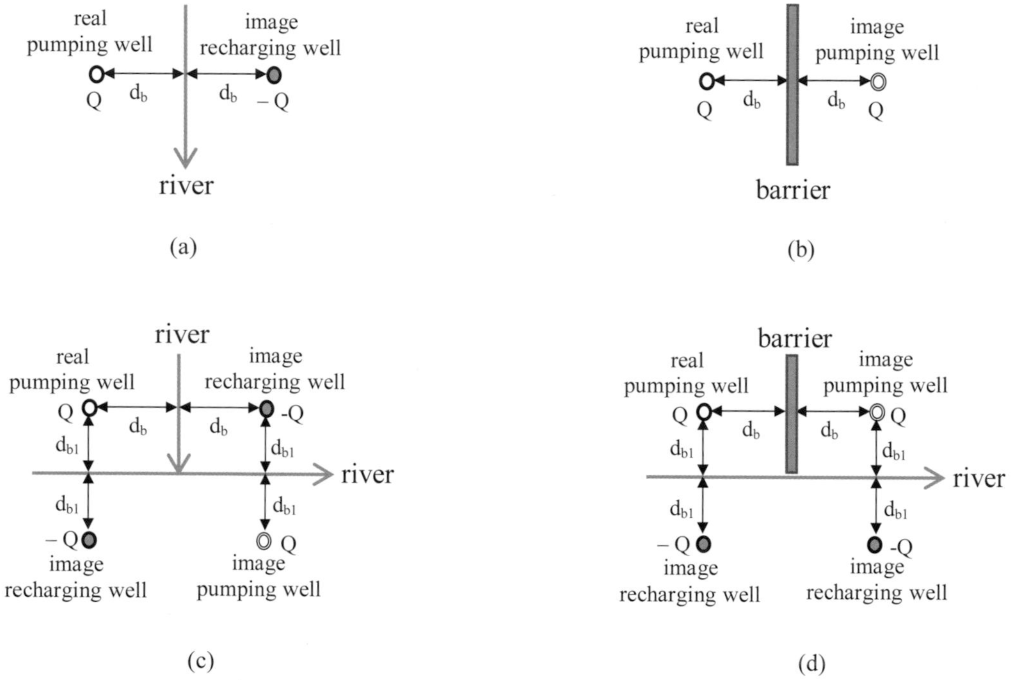

Figure 9.20 **Image wells in Example 9.43.** Diagram for eliminating the boundary of: (a) a river, (b) a barrier, (c) two rivers, and (d) one river and one barrier.

Example 9.44 The properties of an unconfined aquifer consisting of fine sands are: $S = 0.005$, $S_y = 0.025$, $K_H = 50\,\text{m}\,\text{d}^{-1}$, $K_V = 5\,\text{m}\,\text{d}^{-1}$, and $h_0 = 40\,\text{m}$. As illustrated in Figure 9.21, at 50 m away from a river, a 0.2-m-diameter well pumps at $5000\,\text{m}^3\,\text{d}^{-1}$. After pumping for 5 d, determine the drawdowns at locations of:

(1) $x = -30\,\text{m}, y = 0\,\text{m}$
(2) $x = -20\,\text{m}, y = 40\,\text{m}$
(3) $x = -100\,\text{m}, y = 0\,\text{m}$
(4) $x = 60\,\text{m}, y = 40\,\text{m}$.

Solution

$S = 0.005$, $S_y = 0.025$, $K_H = 50\,\text{m}\,\text{d}^{-1}$, $K_V = 5\,\text{m}\,\text{d}^{-1}$, $h_0 = 40\,\text{m}$

$$r_w = 0.1\,\text{m},\ x_w = -50\,\text{m},\ y_w = 0\,\text{m},\ Q_w = 5000\,\text{m}^3\,\text{d}^{-1},\ t = 5\,\text{d}$$

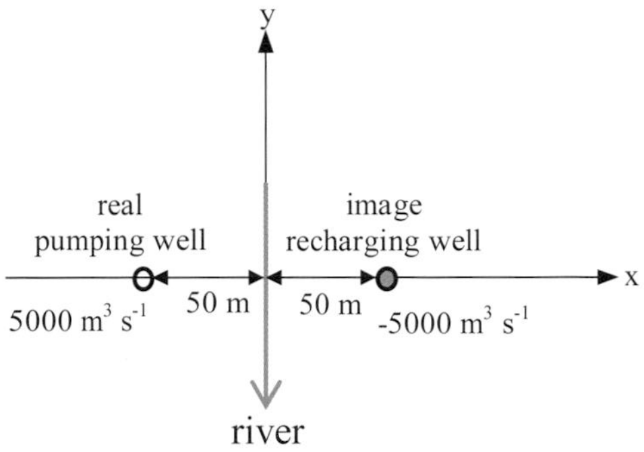

Figure 9.21 **The pumping well, river boundary, and image well used in Example 9.44.**

Find sand $\rightarrow$ the threshold time from earlier to later stage is 30 min.

$$T = K_H h_0 = \left(50\,m\,d^{-1}\right) * (40\,m) = 2000\,m^2\,d^{-1}.$$

$t = 5\,d > 30\,min \rightarrow$ later stage $\rightarrow 1/u_y$ should be used to determine $W\left(u_a, u_y, \eta\right)$.

Introduce an image recharging well to eliminate the river boundary, as shown in Figure 9.21: $r_i = 0.1\,m$, $x_i = 50\,m$, $y_i = 0\,m$, $Q_i = -5000\,m^3\,d^{-1}$.

(1) $x = -30\,m$, $y = 0\,m \rightarrow$ on the same side with the pumping well $\rightarrow$ some drawdown

- Compute the drawdown induced by the real pumping well:

  Distance from the real well to the location: $r_r = \sqrt{(-50+30)^2 + (0-0)^2}\,m = 20\,m$.

  Eq. (9.42): $u_y = \left[0.025 * (20\,m)^2\right] / \left[4 * \left(2000\,m^2\,d^{-1}\right) * (5\,d)\right] = 0.00025$

  $\rightarrow 1/u_y = 4000 = 4.0 \times 10^3$.

  Eq. (9.43): $\eta = \left[(20\,m)^2 / (40\,m)^2\right] * \left[\left(5\,m\,d^{-1}\right) / \left(50\,m\,d^{-1}\right)\right] = 0.025$.

  Figure 9.13: $1/u_y = 4.0 \times 10^3 \rightarrow W\left(u_a, u_y, \eta\right) = 6.0$.

  Eq. (9.40): $s_r = \left(5000\,m^3\,d^{-1}\right) / \left[4 * \pi * \left(2000\,m^2\,d^{-1}\right)\right] * 6.0 = 1.19\,m$.

- Compute the buildup created by the image recharging well:

  Distance from the image well to the location: $r_i = \sqrt{(50+30)^2 + (0-0)^2}\,m = 80\,m$.

  Eq. (9.42): $u_y = \left[0.025 * (80\,m)^2\right] / \left[4 * \left(2000\,m^2\,d^{-1}\right) * (5\,d)\right] = 0.004$

  $\rightarrow 1/u_y = 250 = 2.5 \times 10^2$.

  Eq. (9.43): $\eta = \left[(80\,m)^2 / (40\,m)^2\right] * \left[\left(5\,m\,d^{-1}\right) / \left(50\,m\,d^{-1}\right)\right] = 0.4$.

  Figure 9.13: $1/u_y = 2.5 \times 10^2 \rightarrow W\left(u_a, u_y, \eta\right) = 5.2$.

  Eq. (9.40): $s_i = \left(-5000\,m^3\,d^{-1}\right) / \left[4 * \pi * \left(2000\,m^2\,d^{-1}\right)\right] * 5.2 = -1.03\,m$.

- Superposition theory (Eq. (9.48)): $s = s_r + s_i = (1.19\,m) + (-1.03\,m) = 0.16\,m$.

(2) $x = -20\,m$, $y = 40\,m \rightarrow$ on the same side with the pumping well $\rightarrow$ some drawdown.

- Compute the drawdown induced by the real pumping well:

  Distance from the real well to the location: $r_r = \sqrt{(-50+20)^2 + (0-40)^2}\,m = 50\,m$.

  Eq. (9.42): $u_y = \left[0.025 * (50\,m)^2\right] / \left[4 * \left(2000\,m^2\,d^{-1}\right) * (5\,d)\right] = 0.0015625$

  $$\rightarrow 1/u_y = 640 = 6.4 \times 10^2.$$

  Eq. (9.43): $\eta = \left[(50\,m)^2 / (40\,m)^2\right] * \left[\left(5\,m\,d^{-1}\right) / \left(50\,m\,d^{-1}\right)\right] = 0.15625$.

  Figure 9.13: $1/u_y = 6.4 \times 10^2 \rightarrow W\left(u_a, u_y, \eta\right) = 5.5$.

  Eq. (9.40): $s_r = \left(5000\,m^3\,d^{-1}\right) / \left[4 * \pi * \left(2000\,m^2\,d^{-1}\right)\right] * 5.5 = 1.09\,m$.

- Compute the buildup created by the image recharging well:

  Distance from the image well to the location: $r_i = \sqrt{(50+20)^2 + (0-40)^2}\,m = 80.6226\,m$.

  Eq. (9.42): $u_y = \left[0.025 * (80.6226\,m)^2\right] / \left[4 * \left(2000\,m^2\,d^{-1}\right) * (5\,d)\right] = 0.0040625$

  $$\rightarrow 1/u_y = 246 = 2.46 \times 10^2.$$

  Eq. (9.43): $\eta = \left[(80.6226\,m)^2 / (40\,m)^2\right] * \left[\left(5\,m\,d^{-1}\right) / \left(50\,m\,d^{-1}\right)\right] = 0.4063$.

  Figure 9.13: $1/u_y = 2.46 \times 10^2 \rightarrow W\left(u_a, u_y, \eta\right) = 5.2$.

  Eq. (9.40): $s_i = \left(-5000\,m^3\,d^{-1}\right) / \left[4 * \pi * \left(2000\,m^2\,d^{-1}\right)\right] * 5.2 = -1.03\,m$.

- Superposition theory (Eq. (9.48)): $s = s_r + s_i = (1.09\,m) + (-1.03\,m) = 0.06\,m$.

(3) $x = -100\,m$, $y = 0\,m \rightarrow$ on the same side with the pumping well $\rightarrow$ some drawdown.

- Compute the drawdown induced by the real pumping well:

  Distance from the real well to the location: $r_r = \sqrt{(-50+100)^2 + (0-0)^2}\,m = 50\,m$.

  Eq. (9.42): $u_y = \left[0.025 * (50\,m)^2\right] / \left[4 * \left(2000\,m^2\,d^{-1}\right) * (5\,d)\right] = 0.0015625$

  $$\rightarrow 1/u_y = 640 = 6.4 \times 10^2.$$

  Eq. (9.43): $\eta = \left[(50\,m)^2 / (40\,m)^2\right] * \left[\left(5\,m\,d^{-1}\right) / \left(50\,m\,d^{-1}\right)\right] = 0.15625$.

  Figure 9.13: $1/u_y = 6.4 \times 10^2 \rightarrow W\left(u_a, u_y, \eta\right) = 5.5$.

  Eq. (9.40): $s_r = \left(5000\,m^3\,d^{-1}\right) / \left[4 * \pi * \left(2000\,m^2\,d^{-1}\right)\right] * 5.5 = 1.09\,m$.

- Compute the buildup created by the image recharging well:

  Distance from the image well to the location: $r_i = \sqrt{(50+100)^2 + (0-0)^2}\,m = 150\,m$.

Eq. (9.42): $u_y = \left[0.025 * (150\,m)^2\right] / \left[4 * \left(2000\,m^2\,d^{-1}\right) * (5\,d)\right] = 0.0140625$

$$\rightarrow 1/u_y = 71.1 = 7.11 \times 10^1.$$

Eq. (9.43): $\eta = \left[(150\,m)^2 / (40\,m)^2\right] * \left[\left(5\,m\,d^{-1}\right) / \left(50\,m\,d^{-1}\right)\right] = 1.40625.$

Figure 9.13: $1/u_y = 7.11 \times 10^1 \rightarrow W(u_a, u_y, \eta) = 1.8.$

Eq. (9.40): $s_i = \left(-5000\,m^3\,d^{-1}\right) / \left[4 * \pi * \left(2000\,m^2\,d^{-1}\right)\right] * 1.8 = -0.36\,m.$

- Superposition theory (Eq. (9.48)): $s = s_r + s_i = (1.09\,m) + (-0.36\,m) = 0.73\,m.$

(4) $x = 60\,m$, $y = 40\,m \rightarrow$ on other side of the river $\rightarrow$ pumping has no impact, $s = 0\,m$.

Example 9.45 Redo Example 9.44 if the river boundary is replaced by a barrier boundary.

Solution

$S = 0.005$, $S_y = 0.025$, $K_H = 50\,m\,d^{-1}$, $K_V = 5\,m\,d^{-1}$, $h_0 = 40\,m$

$$r_w = 0.1\,m,\ x_w = -50\,m,\ y_w = 0\,m,\ Q_w = 5000\,m^3\,d^{-1},\ t = 5\,d$$

Find sand $\rightarrow$ the threshold time from earlier to later stage is 30 min.

$$T = K_H h_0 = \left(50\,m\,d^{-1}\right) * (40\,m) = 2000\,m^2\,d^{-1}.$$

$t = 5\,d > 30\,min \rightarrow$ later stage $\rightarrow 1/u_y$ used to determine $W(u_a, u_y, \eta)$.

Introduce an image pumping well to eliminate the barrier boundary: $r_i = 0.1\,m$, $x_i = 50\,m$, $y_i = 0\,m$, $Q_i = 5000\,m^3\,d^{-1}$.

(1) $x = -30\,m$, $y = 0\,m \rightarrow$ on the same side with the pumping well $\rightarrow$ some drawdown.

- Compute the drawdown induced by the real pumping well:

  Distance from the real well to the location: $r_r = \sqrt{(-50+30)^2 + (0-0)^2}\,m = 20\,m.$

  Eq. (9.42): $u_y = \left[0.025 * (20\,m)^2\right] / \left[4 * \left(2000\,m^2\,d^{-1}\right) * (5\,d)\right] = 0.00025$

  $$\rightarrow 1/u_y = 4000 = 4.0 \times 10^3.$$

  Eq. (9.43): $\eta = \left[(20\,m)^2 / (40\,m)^2\right] * \left[\left(5\,m\,d^{-1}\right) / \left(50\,m\,d^{-1}\right)\right] = 0.025.$

  Figure 9.13: $1/u_y = 4.0 \times 10^3 \rightarrow W(u_a, u_y, \eta) = 6.0.$

  Eq. (9.40): $s_r = \left(5000\,m^3\,d^{-1}\right) / \left[4 * \pi * \left(2000\,m^2\,d^{-1}\right)\right] * 6.0 = 1.19\,m.$

- Compute the drawdown induced by the image pumping well:

  Distance from the image well to the location: $r_i = \sqrt{(50+30)^2 + (0-0)^2}\,m = 80\,m.$

Eq. (9.42): $u_y = \left[0.025 * (80\,m)^2\right] / \left[4 * \left(2000\,m^2\,d^{-1}\right) * (5\,d)\right] = 0.004$

$\rightarrow 1/u_y = 250 = 2.5 \times 10^2.$

Eq. (9.43): $\eta = \left[(80\,m)^2 / (40\,m)^2\right] * \left[\left(5\,m\,d^{-1}\right) / \left(50\,m\,d^{-1}\right)\right] = 0.4.$

Figure 9.13: $1/u_y = 2.5 \times 10^2 \rightarrow W\left(u_a, u_y, \eta\right) = 5.2.$

Eq. (9.40): $s_i = \left(5000\,m^3\,d^{-1}\right) / \left[4 * \pi * \left(2000\,m^2\,d^{-1}\right)\right] * 5.2 = 1.03\,m.$

- Superposition theory (Eq. (9.48)): $s = s_r + s_i = (1.19\,m) + (1.03\,m) = 2.22\,m.$

(2) $x = -20\,m$, $y = 40\,m \rightarrow$ on the same side with the pumping well $\rightarrow$ some drawdown.

- Compute the drawdown induced by the real pumping well:

Distance from the real well to the location: $r_r = \sqrt{(-50+20)^2 + (0-40)^2}\,m = 50\,m.$

Eq. (9.42): $u_y = \left[0.025 * (50\,m)^2\right] / \left[4 * \left(2000\,m^2\,d^{-1}\right) * (5\,d)\right] = 0.0015625$

$\rightarrow 1/u_y = 640 = 6.4 \times 10^2.$

Eq. (9.43): $\eta = \left[(50\,m)^2 / (40\,m)^2\right] * \left[\left(5\,m\,d^{-1}\right) / \left(50\,m\,d^{-1}\right)\right] = 0.15625.$

Figure 9.13: $1/u_y = 6.4 \times 10^2 \rightarrow W\left(u_a, u_y, \eta\right) = 5.5.$

Eq. (9.40): $s_r = \left(5000\,m^3\,d^{-1}\right) / \left[4 * \pi * \left(2000\,m^2\,d^{-1}\right)\right] * 5.5 = 1.09\,m.$

- Compute the drawdown induced by the image pumping well:

Distance from the image well to the location: $r_i = \sqrt{(50+20)^2 + (0-40)^2}\,m = 80.6226\,m.$

Eq. (9.42): $u_y = \left[0.025 * (80.6226\,m)^2\right] / \left[4 * \left(2000\,m^2\,d^{-1}\right) * (5\,d)\right] = 0.0040625$

$\rightarrow 1/u_y = 246 = 2.46 \times 10^2.$

Eq. (9.43): $\eta = \left[(80.6226\,m)^2 / (40\,m)^2\right] * \left[\left(5\,m\,d^{-1}\right) / \left(50\,m\,d^{-1}\right)\right] = 0.4063.$

Figure 9.13: $1/u_y = 2.46 \times 10^2 \rightarrow W\left(u_a, u_y, \eta\right) = 5.2.$

Eq. (9.40): $s_i = \left(5000\,m^3\,d^{-1}\right) / \left[4 * \pi * \left(2000\,m^2\,d^{-1}\right)\right] * 5.2 = 1.03\,m.$

- Superposition theory (Eq. (9.48)): $s = s_r + s_i = (1.09\,m) + (1.03\,m) = 2.12\,m.$

(3) $x = -100\,m$, $y = 0\,m \rightarrow$ on the same side with the pumping well $\rightarrow$ some drawdown.

- Compute the drawdown induced by the real pumping well:

Distance from the real well to the location: $r_r = \sqrt{(-50+100)^2 + (0-0)^2}\,m = 50\,m.$

Eq. (9.42): $u_y = \left[0.025 * (50\,m)^2\right] / \left[4 * \left(2000\,m^2\,d^{-1}\right) * (5\,d)\right] = 0.0015625$

$\rightarrow 1/u_y = 640 = 6.4 \times 10^2.$

Eq. (9.43): $\eta = \left[(50\,\text{m})^2 / (40\,\text{m})^2\right] * \left[(5\,\text{m}\,\text{d}^{-1}) / (50\,\text{m}\,\text{d}^{-1})\right] = 0.15625.$

Figure 9.13: $1/u_y = 6.4 \times 10^2 \rightarrow W(u_a, u_y, \eta) = 5.5.$

Eq. (9.40): $s_r = (5000\,\text{m}^3\,\text{d}^{-1}) / \left[4 * \pi * (2000\,\text{m}^2\,\text{d}^{-1})\right] * 5.5 = 1.09\,\text{m}.$

- Compute the drawdown induced by the image pumping well:

Distance from the image well to the location: $r_i = \sqrt{(50 + 100)^2 + (0 - 0)^2}\,\text{m} = 150\,\text{m}.$

Eq. (9.42): $u_y = \left[0.025 * (150\,\text{m})^2\right] / \left[4 * (2000\,\text{m}^2\,\text{d}^{-1}) * (5\,\text{d})\right] = 0.0140625$

$$\rightarrow 1/u_y = 71.1 = 7.11 \times 10^1.$$

Eq. (9.43): $\eta = \left[(150\,\text{m})^2 / (40\,\text{m})^2\right] * \left[(5\,\text{m}\,\text{d}^{-1}) / (50\,\text{m}\,\text{d}^{-1})\right] = 1.40625.$

Figure 9.13: $1/u_y = 7.11 \times 10^1 \rightarrow W(u_a, u_y, \eta) = 1.8.$

Eq. (9.40): $s_i = (5000\,\text{m}^3\,\text{d}^{-1}) / \left[4 * \pi * (2000\,\text{m}^2\,\text{d}^{-1})\right] * 1.8 = 0.36\,\text{m}.$

- Superposition theory (Eq. (9.48)): $s = s_r + s_i = (1.09\,\text{m}) + (0.36\,\text{m}) = 1.45\,\text{m}.$

(4) $x = 60\,\text{m}$, $y = 40\,\text{m} \rightarrow$ on other side of the barrier $\rightarrow$ pumping has no impact, $s = 0\,\text{m}.$

Example 9.46 A 2-ft-diameter well fully penetrates a 50-ft-thick unconfined aquifer ($K = 30\,\text{ft}\,\text{d}^{-1}$). After pumping at $1000\,\text{ft}^3\,\text{d}^{-1}$ for a long period of time, the drawdown in the well is measured to be 1.5 ft. If there is a boundary 200 ft away from the pumping well, determine whether the pumping is impacted by the boundary.

Solution

$h_0 = 50\,\text{ft}$, $K = 30\,\text{ft}\,\text{d}^{-1}$, $Q = 1000\,\text{ft}^3\,\text{d}^{-1}$, $r_1 = 1\,\text{ft}$, $s_1 = 1.5\,\text{ft}$, $d_b = 200\,\text{ft}$

Pumping for a long period of time $\rightarrow$ steady state.

If the influence radius $r_0 < d_b$, the pumping is not impacted; otherwise, it is impacted.

Eq. (9.33): $(50\,\text{ft} - 0\,\text{ft})^2 = (50\,\text{ft} - 1.5\,\text{ft})^2 + (1000\,\text{ft}^3\,\text{d}^{-1}) / \left[\pi * (30\,\text{ft}\,\text{d}^{-1})\right] * \ln\left[r_0 / (1\,\text{ft})\right]$

$\rightarrow \ln(r_0) = 13.9251 \rightarrow r_0 = 1116\,\text{ft} \rightarrow$ the pumping is impacted by the boundary.

Example 9.47 A 100-ft-thick confined aquifer has a transmissivity of $25{,}000\,\text{ft}^2\,\text{d}^{-1}$. Three fully penetrating wells with a diameter of 1 ft are located along a straight line and spaced at 500 ft. If the groundwater exploration permit allows a maximum drawdown of 5 ft in any of the wells after pumping for a long period of time, determine the maximum allowable pumping rates of the wells.

Solution

$b = 100\,\text{ft}$, $T = 25{,}000\,\text{ft}^2\,\text{d}^{-1}$, $d_b = 500\,\text{ft}$, $r_w = 0.5\,\text{ft}$, $s_{max} = 5\,\text{ft}$

Pumping for a long period of time $\rightarrow$ steady state.

Let r_0 be the influence radius of the wells.

Let Q_1, Q_2, and Q_3 be the pumping rates of the left-hand, middle, and right-hand wells, respectively. In terms of Eqs. (9.32) and (9.48), the overall drawdown in each of the wells can be computed as:

Left-hand well: $$s_1 = \frac{Q_1}{2\pi T}\ln\left(\frac{r_0}{r_w}\right) + \frac{Q_2}{2\pi T}\ln\left(\frac{r_0}{d_b}\right) + \frac{Q_3}{2\pi T}\ln\left(\frac{r_0}{2d_b}\right).$$

Middle well: $$s_2 = \frac{Q_1}{2\pi T}\ln\left(\frac{r_0}{d_b}\right) + \frac{Q_2}{2\pi T}\ln\left(\frac{r_0}{r_w}\right) + \frac{Q_3}{2\pi T}\ln\left(\frac{r_0}{d_b}\right).$$

Right-hand well: $$s_3 = \frac{Q_1}{2\pi T}\ln\left(\frac{r_0}{2d_b}\right) + \frac{Q_2}{2\pi T}\ln\left(\frac{r_0}{d_b}\right) + \frac{Q_3}{2\pi T}\ln\left(\frac{r_0}{r_w}\right).$$

As shown in Figure 9.22, the above equations can be solved in Excel Solver as follows:

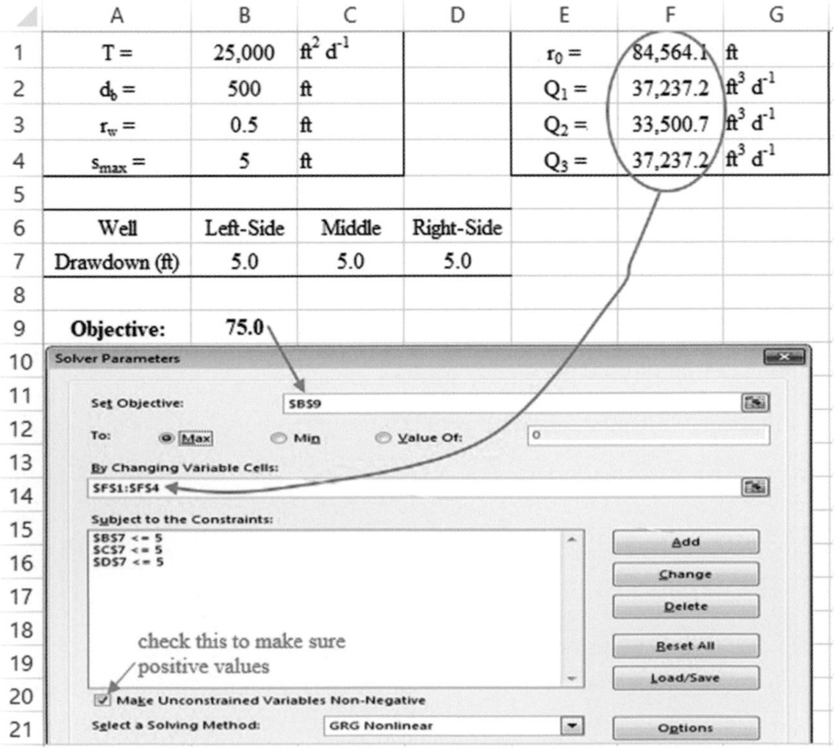

Figure 9.22 **Spreadsheet with settings of Excel Solver for Example 9.47.**

Objective (cell B9): $\max\left\{s_1{}^2 + s_2{}^2 + s_3{}^2\right\}$

By changing (cell F1:F4): $Q_1, Q_2, Q_3,$ and r_0

Subject to: $s_1 \leq 5, s_2 \leq 5, s_3 \leq 5.$

The objective is maximized so that a maximum amount of groundwater can be explored within the maximum drawdown limit of 5 ft. The results are: $Q_1 = 37{,}237\,\text{ft}^3\,\text{d}^{-1}$, $Q_2 = 33{,}500\,\text{ft}^3\,\text{d}^{-1}$, and $Q_3 = 37{,}237\,\text{ft}^3\,\text{d}^{-1}$.

9.4.7 Flows in a Partially Penetrating Well

If a well partially penetrates an aquifer, the flow is supplied by a portion of the aquifer (Figure 9.23). As a result, the drawdown at a location of interest is larger for a partially penetrating well than for a fully penetrating well (Hantush, 1961a,b; Todd and Mays, 2005). That is, the drawdown in a partially penetrating well can be computed as:

$$s_{w,p} = s_w + \Delta s_w \tag{9.49}$$

drawdown in partially penetrating well → $s_{w,p}$; drawdown in fully penetrating well → s_w; drawdown increase of partially versus fully penetrating well → Δs_w

In a confined aquifer, at a steady state, the drawdown increase, Δs_w, in a partially penetrating well (Figure 9.23a) can be estimated as (Hantush, 1961a,b):

$$\Delta s_w = \frac{Q}{2\pi T}\frac{b - h_s}{b}\ln\left[\frac{b - h_s}{b}\frac{h_s}{r_w}\right] \tag{9.50}$$

drawdown increase in partially versus fully penetrating well of confined aquifer → Δs_w; pumping rate → Q; thickness of aquifer → b; well penetration depth → h_s; transmissivity of aquifer → T; well radius → r_w

In an unconfined aquifer, at a steady state, the drawdown increase in a partially penetrating well (Figure 9.23b) can be estimated as (Hantush, 1961a,b):

$$\Delta s_w = \frac{Q}{2\pi K\left(h_0 - s_w\right)}\frac{h_0 - h_s}{h_0}\ln\left[\frac{h_0 - h_s}{h_0}\frac{h_s}{r_w}\right] \tag{9.51}$$

drawdown increase in partially versus fully penetrating well of unconfined aquifer → Δs_w; pumping rate → Q; well penetration depth → h_s; hydraulic conductivity → K; initial hydraulic head/water table → h_0; drawdown in fully penetrating well → s_w; well radius → r_w

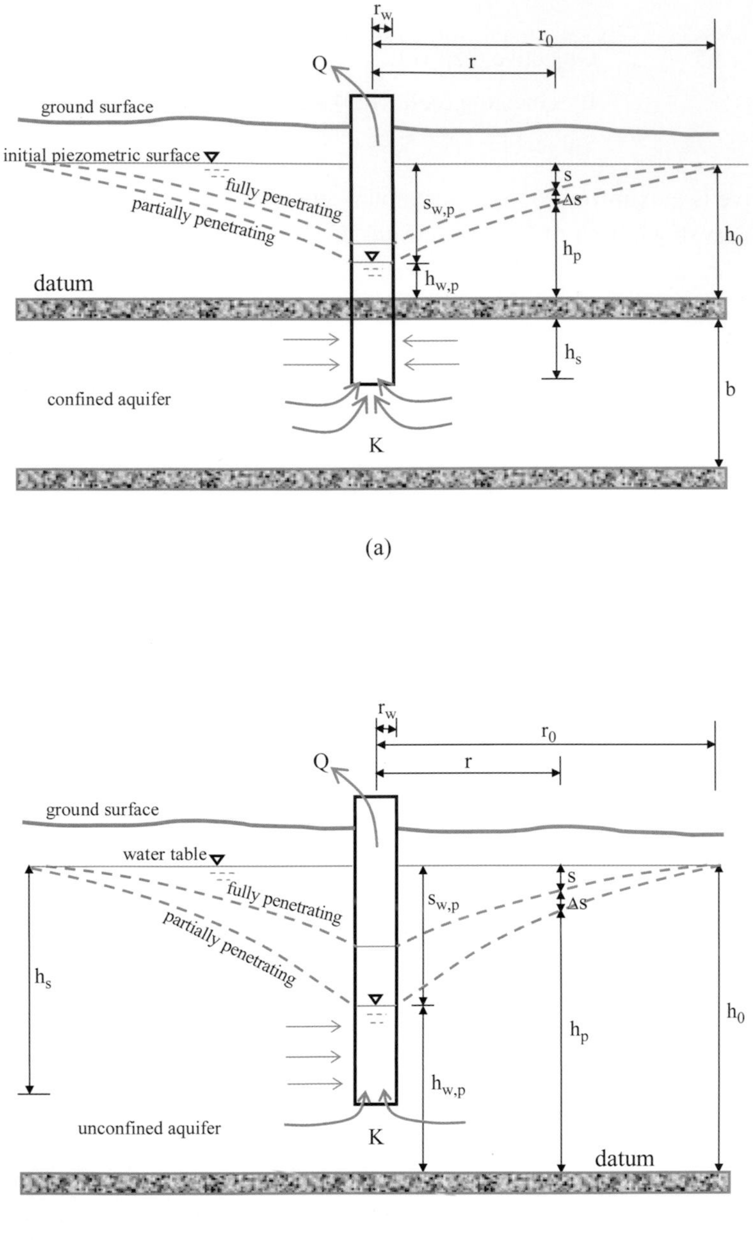

Figure 9.23 **Penetration influence on drawdown.** Diagrams show drawdown increase, Δs, from a partially penetrating well versus a fully penetrating well in: (a) a confined aquifer; and (b) an unconfined aquifer.

Example 9.48 A 2-ft-diameter well fully penetrates into a 50-ft-thick confined aquifer ($K = 30\,\text{ft}\,\text{d}^{-1}$). After pumping at $1000\,\text{ft}^3\,\text{d}^{-1}$ for a long period of time, the drawdown in the well is measured to be 1.5 ft. If the penetration depth of the well is changed to 20 ft, determine the drawdown in the well.

Solution

$r_w = 1\,\text{ft}$, $b = 50\,\text{ft}$, $K = 30\,\text{ft}\,\text{d}^{-1}$, $Q = 1000\,\text{ft}^3\,\text{d}^{-1}$, $s_w = 1.5\,\text{ft}$, $h_s = 20\,\text{ft}$

Pumping for a long period of time → steady state.

$$T = Kb = \left(30\,\text{ft}\,\text{d}^{-1}\right) * (50\,\text{ft}) = 1500\,\text{ft}^2\,\text{d}^{-1}.$$

Eq. (9.50): $\Delta s_w = \dfrac{1000\,\text{ft}^3\,\text{d}^{-1}}{2\pi * \left(1500\,\text{ft}^2\,\text{d}^{-1}\right)} * \dfrac{50\,\text{ft} - 20\,\text{ft}}{50\,\text{ft}} * \ln\left[\dfrac{50\,\text{ft} - 20\,\text{ft}}{50\,\text{ft}} * \dfrac{20\,\text{ft}}{1\,\text{ft}}\right] = 0.16\,\text{ft}.$

Eq. (9.49): $s_{w,p} = 1.5\,\text{ft} + 0.16\,\text{ft} = 1.66\,\text{ft}.$

Example 9.49 A 0.5-m-diameter well fully penetrates into an unconfined aquifer ($K = 5.5\,\text{m}\,\text{d}^{-1}$) with an initial saturated thickness of 10 m. After pumping at $25\,\text{m}^3\,\text{d}^{-1}$ for a long period of time, the drawdown in the well is measured to be 0.6 m. If the penetration depth of the well is changed to 5 m, determine the drawdown in the well.

Solution

$r_w = 0.25\,\text{m}$, $h_0 = 10\,\text{m}$, $K = 5.5\,\text{m}\,\text{d}^{-1}$, $Q = 25\,\text{m}^3\,\text{d}^{-1}$, $s_w = 0.6\,\text{m}$, $h_s = 5\,\text{m}$

Eq. (9.51): $\Delta s_w = \dfrac{25\,\text{m}^3\,\text{d}^{-1}}{2\pi * (5.5\,\text{m}\,\text{d}^{-1})\,(10\,\text{m} - 0.6\,\text{m})} * \dfrac{10\,\text{m} - 5\,\text{m}}{10\,\text{m}} * \ln\left[\dfrac{10\,\text{m} - 5\,\text{m}}{10\,\text{m}} * \dfrac{5\,\text{m}}{0.25\,\text{m}}\right] =$ 0.09 m.

Eq. (9.49): $s_{w,p} = 0.6\,\text{m} + 0.09\,\text{m} = 0.69\,\text{m}.$

9.5 Well Engineering

As an engineering practice, a pumping or slug test is commonly conducted to determine the properties of an aquifer of interest. A pumping test is a field experiment that pumps at a constant rate for a time period and records drawdowns versus time either in a monitoring well some distance away or in the pumping well. The aquifer properties are determined by analyzing the drawdown time series data, as demonstrated in Example 9.32, Example 9.35, and Example 9.40. In contrast, a slug test is a field experiment that quickly adds or removes water from a well and then records changes of hydraulic head in the well versus time. If the experiment adds water, it is a falling-head test; otherwise, it is a rising-head test. The aquifer properties can be determined by analyzing the time series data of hydraulic head change using the Bouwer–Rice method (Bouwer, 1989). The method assumes that the aquifer is homogenous and isotropic and that the initial groundwater level is invariant during the slug test. It is applicable for partially and fully penetrating wells in

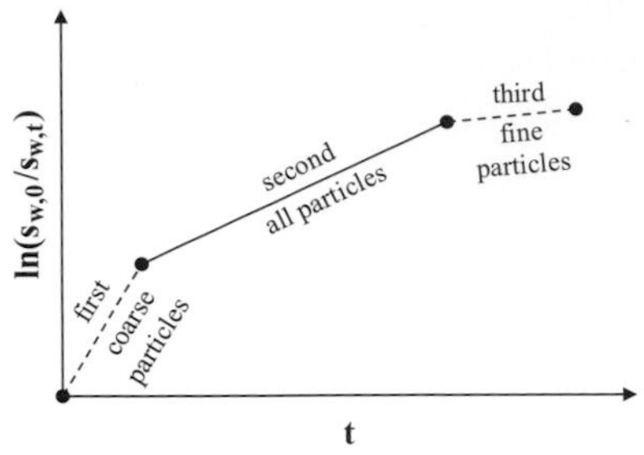

Figure 9.24 **Typical results of slug tests.** Diagram shows the three linear segments resulting from plotting measured time series. The slope of the second segment is usually used to determine the saturated hydraulic conductivity of the aquifer.

unconfined, confined, and semiconfined aquifers. To prepare a slug test, a hole with a specified radius is dug. A portion of the hole is cemented to prevent water flowing into the well, while the remaining portion is screened to allow water to flow into the well. For the balance of this section, *well* refers to the screened portion of the hole.

Let h_0 be the initial hydraulic head and h_t be the hydraulic head at time t. The horizontal (i.e., radial) hydraulic conductivity can be determined using the slope of the regression line of $\ln(h_0/h_t)$ versus t. In most situations, the regression consists of three linear segments (Figure 9.24). The first segment has a steepest slope, representing quick responses from coarse soil particles (e.g., gravels), whereas the third segment has a gentlest slope, representing slow responses from fine particles (e.g., silts). The second segment has an intermediate slope, representing overall responses from all-sized particles. In practice, the slope of the second segment is usually used to determine the hydraulic conductivity of the aquifer. Bouwer (1989) proposed a linear regression equation expressed as:

$$\ln\left(\frac{s_{w,0}}{s_{w,t}}\right) = K_H \frac{2L_s}{r_{ce}^2 \ln\left(\frac{r_e}{r_{we}}\right)} t \tag{9.52}$$

initial head displacement ($s_{w,0}$); head displacement at time t ($s_{w,t}$); horizontal hydraulic conductivity (K_H); well hole inner effective radius (Eq. (9.53)) (r_{ce}); well screen length (L_s); time (t); effective radius of the test (r_e); well screen outer effective radius (Eq. (9.54)) (r_{we})

$$r_{ce} = \sqrt{[1-(\theta_{sat}-\theta_{fc})]r_c^2 + (\theta_{sat}-\theta_{fc})r_w^2} \tag{9.53}$$

well hole inner effective radius in Eq. (9.52) (r_{ce}); saturated soil moisture (θ_{sat}); field capacity (θ_{fc}); well hole inner radius (r_c); well screen outer radius (r_w)

$$r_{we} = r_w\sqrt{\frac{K_V}{K_H}} \tag{9.54}$$

well screen outer effective radius in Eq. (9.52) (r_{we}); well screen outer radius (r_w); vertical hydraulic conductivity (K_V); horizontal hydraulic conductivity (K_H)

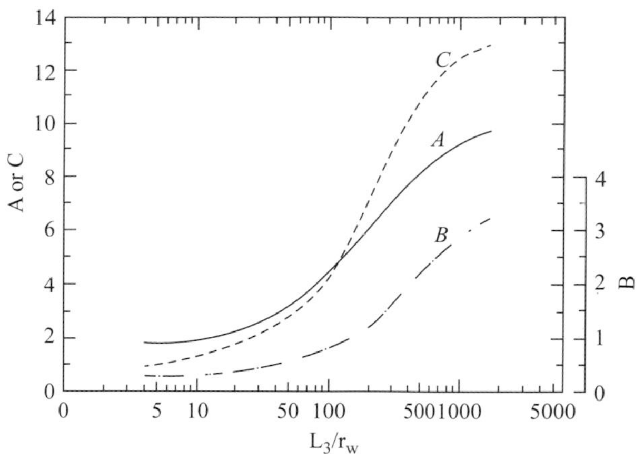

Figure 9.25 **The Bouwer–Rice method for determining coefficients in Eq. (9.55).** L_s: well screen length; and r_w: well screen outer radius. (*Sources:* Bouwer and Rice, 1976; Bouwer, 1989)

$$
\ln\left(\frac{r_e}{r_{we}}\right) =
\begin{cases}
\left[\dfrac{1.1}{\ln\left(\dfrac{h_s}{r_w}\right)} + \dfrac{A + B\ln\left(\dfrac{b - h_s}{r_w}\right)}{\dfrac{L_s}{r_w}}\right]^{-1} & \text{for partially penetrating well} \\[4em]
\left[\dfrac{1.1}{\ln\left(\dfrac{h_s}{r_w}\right)} + \dfrac{C}{\dfrac{L_s}{r_w}}\right]^{-1} & \text{for fully penetrating well}
\end{cases}
\tag{9.55}
$$

coefficient (Figure 9.25)

aquifer thickness (= h_0 for unconfined aquifer)

effective radius of the test

well penetration depth — well screen length

well screen outer effective radius

coefficient (Figure 9.25)

Example 9.50 A falling-head slug test is performed by quickly placing a steel block into a well, causing an instantaneous head rise of 2.5 ft in the well. The 25-ft-thick unconfined aquifer has a saturated soil moisture of 0.35 and a field capacity of 0.15. The well penetrates into the aquifer by 20 ft, of which 15 ft is screened. The well hole inner radius is 0.5 ft, and the well screen outer radius is 0.2 ft. The head displacement measurements are given in cells A6:B39 of Figure 9.26. If the aquifer is homogeneous and isotropic, determine the hydraulic conductivity of the aquifer using the Bouwer–Rice method.

Solution

$r_{ce} = \sqrt{[1 - (0.35 - 0.15)] * (0.5\,\text{ft})^2 + (0.35 - 0.15) * (0.2\,\text{ft})^2} = 0.45607\,\text{ft}$. As shown in cells C6:C39 of Figure 9.26, $\ln(s_{w,0}/s_{w,t})$ is calculated and plotted versus t. The slope for the second segment is determined as 0.0021921. From Figure 9.25, for $L_s/r_w = (15\,\text{ft}) / (0.2\,\text{ft}) = 75$,

	A	B	C	D	E	F	G	H
1	$b = h_0 =$	25	ft	$h_s =$	20	ft	$L_s/r_w =$	75
2	$\theta_{sat} =$	0.35		$L_s =$	15	ft	A =	3.8
3	$\theta_{fc} =$	0.15		$r_c =$	0.5	ft	B =	0.7
4	$s_{w,0} =$	2.5	ft	$r_w =$	0.2	ft	C =	3.4
5								
6	t (s)	$s_{w,t}$ (ft)	$\ln(s_{w,0}/s_{w,t})$					
7	0	2.50	0.00000					
8	2	2.42	0.03252					
9	5	2.32	0.07472					
10	10	2.29	0.08774					
11	15	2.29	0.08774					
12	20	2.08	0.18392					
13	25	1.96	0.24335					
14	30	1.93	0.25877					
15	35	1.81	0.32296					
16	40	1.80	0.32850					
17	45	1.67	0.40347					
18	50	1.58	0.45887					
19	55	1.54	0.48451					
20	60	1.38	0.59421					
21	100	1.30	0.65393					
22	150	1.20	0.73397					
23	200	1.10	0.82098					
24	250	1.00	0.91629					
25	300	0.90	1.02165		Eq. (9-53):	$r_{ce} =$	0.45607	ft
26	350	0.80	1.13943					
27	400	0.70	1.27297		Eq. (9-55):	$\ln(r_o/r_{we}) =$	3.12919	
28	450	0.60	1.42712					
29	500	0.55	1.51413		Eq. (9-52):	$K_H =$	4.756E-05	ft s^{-1}
30	550	0.50	1.60944					
31	600	0.45	1.71480					
32	650	0.40	1.83258					
33	700	0.35	1.96611					
34	750	0.32	2.05573					
35	800	0.31	2.08747					
36	850	0.30	2.12026					
37	900	0.29	2.15417					
38	950	0.28	2.18926					
39	1000	0.27	2.22562					

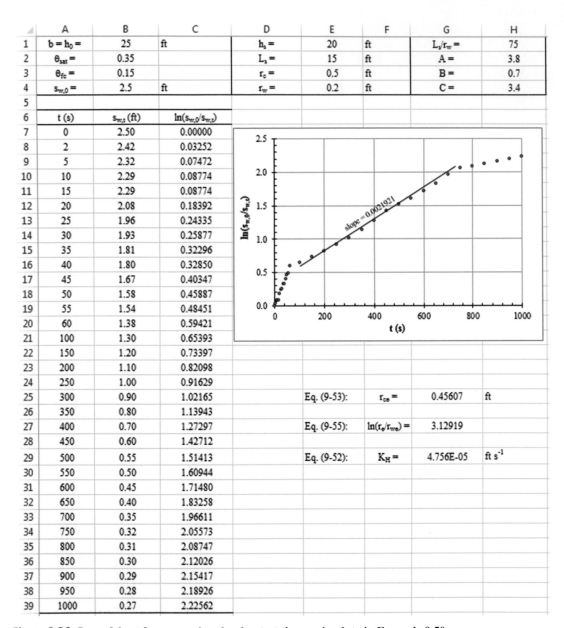

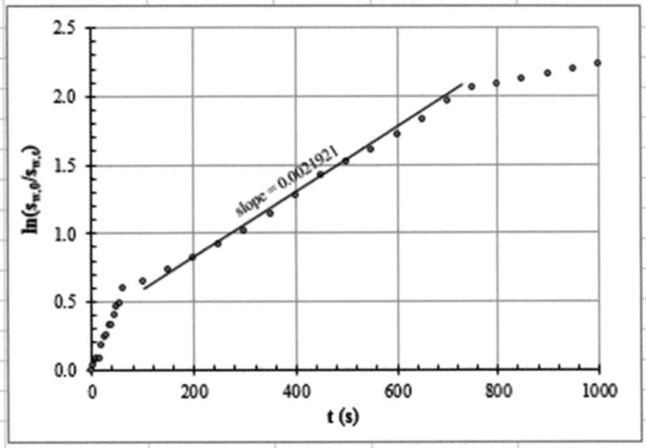

Figure 9.26 **Spreadsheet for processing the slug test time series data in Example 9.50.**

A = 3.8, B = 0.7, and C = 3.4. Using Eq. (9.5) for partially penetrating well, one has: $\ln\left(\frac{r_e}{r_{we}}\right) =$

$$\left[\frac{1.1}{\ln\left(\frac{20\,\text{ft}}{0.2\,\text{ft}}\right)} + \frac{3.8 + 0.7 * \ln\left(\frac{25\,\text{ft} - 20\,\text{ft}}{0.2\,\text{ft}}\right)}{\frac{15\,\text{ft}}{0.2\,\text{ft}}}\right]^{-1} = 3.12919.$$

The slope in Eq. (9.52) is

$$K_H \frac{2L_s}{r_{ce}^2 \ln\left(\frac{r_e}{r_{we}}\right)} = K_H \frac{2 * (15\,\text{ft})}{(0.45607\,\text{ft})^2 * 3.12919} = 46.09208K_H = 0.0021921$$

$$\rightarrow K_H = 4.756 \times 10^{-5}\text{ft}\,\text{s}^{-1} = 4.11\,\text{ft}\,\text{d}^{-1}.$$

PROBLEMS

9.1 In a particular year, a shallow aquifer receives 0.25 in percolation and 2.5 in net inflow from its surrounding aquifers, while the aquifer discharges 1.5 in into surface water bodies and streams. Because the depth to groundwater level is large, the direct evapotranspiration from the aquifer is negligible. If the withdrawal by pumping is 0.5 in, determine the groundwater storage change.

9.2 In a particular year, a shallow aquifer receives 5.5 cm percolation and discharges 30 cm into surrounding surface water bodies and streams. The evapotranspiration from the aquifer is 100 cm, while the withdrawal by pumping is 15 cm. There is no artificial recharge. If the groundwater storage change is −10 cm, determine the net inflow from the surrounding aquifers.

9.3 Averaged across multiple years, the groundwater storage change in an aquifer is determined to be zero. The average annual percolation is 0.5 in and the net inflow is 1.5 in, while the average annual baseflow and withdrawal by pumping are 1.0 and 0.35 in, respectively. There is no artificial recharge. Determine the average annual groundwater loss to evapotranspiration.

9.4 Averaged across multiple years, the groundwater storage change in an aquifer is determined to be zero. The average annual artificial recharge and percolation are 2.5 and 15 cm, respectively. The average annual baseflow is 25 cm. There is no withdrawal and the evapotranspiration is negligible. Determine the net inflow from the surrounding aquifers.

9.5 A 50 km^2 unconfined aquifer has a specific yield of 0.18. If $5.5 \times 10^6\,\text{m}^3$ water is pumped by wells uniformly distributed across the aquifer area, determine the average drawdown of the water table.

9.6 A 20 mi^2 unconfined aquifer has a specific yield of 0.18. As a result of pumping by wells uniformly distributed across the aquifer area, the water table is lowered by 1.5 ft, determine the total volume of groundwater pumped.

9.7 A 100 km^2 confined aquifer has a storativity of 0.025. If $3.5 \times 10^6\ m^3$ water is pumped by wells uniformly distributed across the aquifer area, determine the average drawdown of the piezometric surface.

9.8 A 50-ft-thick confined aquifer has a specific storage of $1.5 \times 10^{-4}\ ft^{-1}$. As a result of pumping by wells uniformly distributed across an area of 1.5 ac, the average piezometric surface beneath the area is lowered by 1.2 ft. Determine the total volume of groundwater pumped.

9.9 For the aquifer shown in Figure 9.3a, if $b_1 = 15$ ft, $K_1 = 1.85\ ft\,d^{-1}$, $b_2 = 25$ ft, $K_2 = 1.15\ ft\,d^{-1}$, $b_3 = 30$ ft, and $K_3 = 0.55\ ft\,d^{-1}$, determine K_x, K_y, and K_z.

9.10 For the aquifer shown in Figure 9.3b, $b_1 = 5$ m, $K_1 = 1.2\ m\,d^{-1}$, $b_2 = 15$ m, $K_2 = 0.85\ m\,d^{-1}$, $b_3 = 18$ m, and $K_3 = 1.5\ m\,d^{-1}$. If $\alpha = 10°$, determine K_x, K_y, and K_z.

9.11 A homogeneous isotropic unconfined aquifer ($K = 2.5\ ft\,d^{-1}$, $n = 0.3$) has an underlying horizontal confining bed. If the water table drops 5 ft per 100 ft horizontal distance, determine the Darcy velocity and pore velocity as well as the travel time over a 1000 ft distance.

9.12 A homogeneous isotropic unconfined aquifer ($K = 1.2\ m\,d^{-1}$, $n = 0.3$) has a horizontal confining bed. If the water table drops 1.5 m per 50 m horizontal distance, determine the Darcy velocity and pore velocity as well as the travel time over a 600 m distance.

9.13 The average diameter of the aquifer soil particles in Problem 9.11 is 0.001 in. If the groundwater temperature is 50°F, is Darcy's law valid? Why or why not?

9.14 The average diameter of the aquifer soil particles in Problem 9.12 is 0.15 mm. If the groundwater temperature is 10°C, is Darcy's law valid? Why or why not?

9.15 As illustrated in Figure 9.2, an unconfined aquifer ($K_1 = 45\ ft\,d^{-1}$, $n_1 = 0.35$, $d_{50} = 0.005$ in) overlays a semiconfined aquifer. The unconfined aquifer has a 50-ft-thick horizontal confining bed of aquitard ($K_2 = 0.2\ ft\,d^{-1}$, $n_2 = 0.3$). Assume that the aquifers and aquitard are homogeneous and isotropic. The groundwater temperature is 50°F. In reference to the bottom of the aquitard, the horizontal water table and piezometric surface are 100 and 98 ft, respectively. Determine the:

(1) Hydraulic head at the top of the aquitard
(2) Travel time from the water table to the top of the aquitard
(3) Travel time over the aquitard
(4) Validity of Darcy's law for the unconfined aquifer.

9.16 As illustrated in Figure 9.2, an unconfined aquifer ($K_1 = 10\ m\,d^{-1}$, $n_1 = 0.35$, $d_{50} = 0.001$ mm) overlays a semiconfined aquifer. The unconfined aquifer has a 20-m-thick horizontal confining bed of aquitard ($K_2 = 0.05\ m\,d^{-1}$, $n_2 = 0.3$). Assume that the aquifers and aquitard are homogeneous and isotropic. The groundwater temperature is 10°C. In reference to the bottom of the aquitard, the horizontal piezometric surface of the semiconfined aquifer is 35 m. If the upward Darcy velocity through the aquitard is $0.025\ m\,d^{-1}$, determine the:

(1) Water table of the unconfined aquifer
(2) Hydraulic head at the top of the aquitard
(3) Travel time from the top of the aquitard to the water table
(4) Travel time over the aquitard
(5) Validity of Darcy's law for the unconfined aquifer and aquitard.

9.17 A 200-m-long homogeneous isotropic confined aquifer ($K = 5\,\text{m}\,\text{d}^{-1}$) shown in the following figure has a horizontal lower bed and a constant thickness of 50 m. Assume a 1-D horizontal steady flow.

(1) Determine the flow rate per unit width of the aquifer.

(2) Compute and plot the longitudinal profile of piezometric surface.

(3) Determine the piezometric surface above the upper confining bed at $x = 100\,\text{m}$.

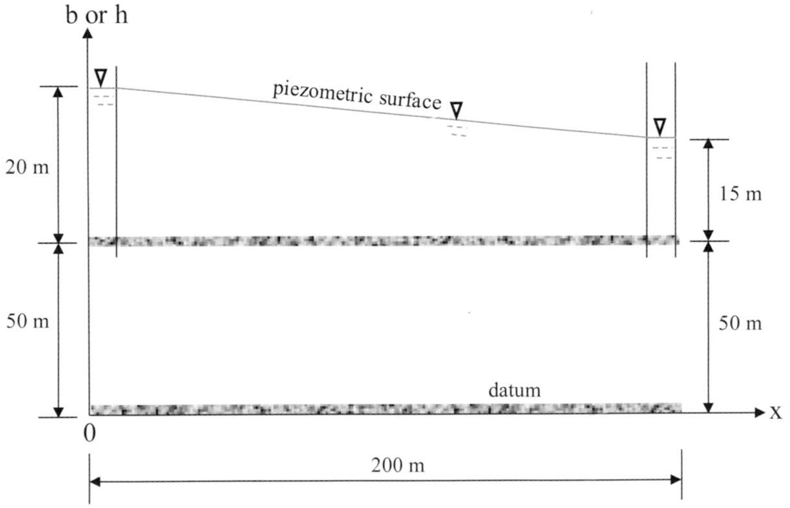

9.18 Redo Problem 9.17 if the aquifer has a spatially varying hydraulic conductivity of $K(x) = 5 + 0.01x\,\text{m}\,\text{d}^{-1}$.

9.19 A 1500-ft-long homogeneous isotropic confined aquifer ($K = 3.5\,\text{ft}\,\text{d}^{-1}$) shown in the following figure has a horizontal lower bed and a linearly varying thickness. Assume a 1-D horizontal steady flow.

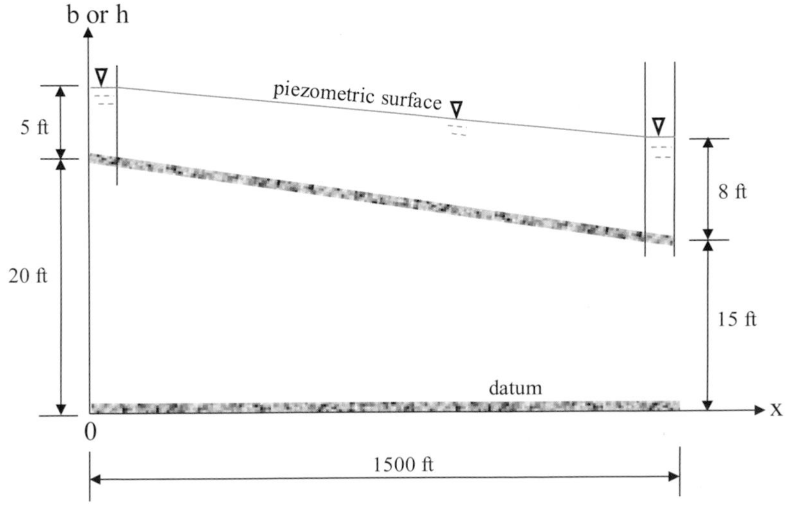

(1) Determine the flow rate per unit width of the aquifer.
(2) Compute and plot the longitudinal profile of piezometric surface.
(3) Determine the piezometric surface above the upper confining bed at $x = 750\,ft$.

9.20 Redo Problem 9.19 if the aquifer has a spatially varying hydraulic conductivity of $K(x) = 3.5 - 0.001x\ ft\,d^{-1}$.

9.21 An unconfined aquifer has a hydraulic conductivity of $50\,ft\,d^{-1}$ and its water table has a slope angle of 2.5°. For a percolation rate of $0.005\,ft\,d^{-1}$, determine the deflection angle when the percolation water flows across the water table.

9.22 An unconfined aquifer has a hydraulic conductivity of $10\,m\,d^{-1}$ and its water table has a slope angle of 1.5°. For a percolation rate of $0.001\,m\,d^{-1}$, determine the deflection angle when the percolation water flows across the water table.

9.23 The deflection angle of groundwater flow in an unconfined aquifer (hydraulic conductivity of $5.5\,m\,d^{-1}$) is 65°. If the confining bed of the aquifer is an aquitard with a hydraulic conductivity of $0.002\,m\,d^{-1}$, determine the deflection angle when water penetrates through the aquitard.

9.24 The deflection angle of groundwater flow in an unconfined aquifer (hydraulic conductivity of $25\,ft\,d^{-1}$) is 55°. If the confining bed of the aquifer is an aquitard with a hydraulic conductivity of $0.005\,ft\,d^{-1}$, determine the deflection angle when water penetrates through the aquitard.

9.25 An unconfined aquifer overlays and recharges a semiconfined aquifer. Let K_1, K_2, and K_3 be the hydraulic conductivities of the unconfined aquifer, aquitard, and confined aquifer, respectively. Sketch the flow path if:
(1) $K_1 > K_3 > K_2$
(2) $K_3 > K_1 > K_2$.

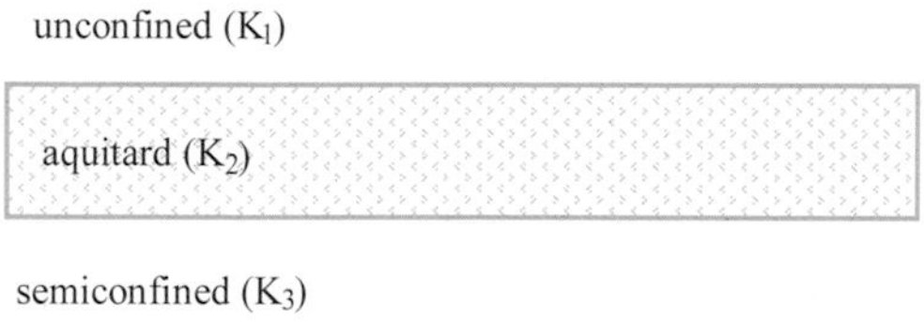

9.26 In Figure 9.9a, $h_1 = 35\,ft$, $h_2 = 20\,ft$, $K = 25\,ft\,d^{-1}$, and $L = 400\,ft$.
(1) Determine the flow rate per unit-width aquifer.
(2) Compute and plot the longitudinal profile of the water table.
(3) Determine the water table and its slope at $x = 200\,ft$.

9.27 In Figure 9.9a, $h_1 = 15\,m$, $h_2 = 10\,m$, $K = 8\,m\,d^{-1}$, and $L = 250\,m$.
(1) Determine the flow rate per unit-width aquifer.
(2) Compute and plot the longitudinal profile of the water table.
(3) Determine the water table and its slope at $x = 125\,m$.

9.28 In Figure 9.9b, $h_1 = 35\,ft$, $h_2 = 20\,ft$, $K = 25\,ft\,d^{-1}$, $L = 400\,ft$, and $i = 0.05\,ft\,d^{-1}$. Determine the:
(1) Distance from river 1 to the divide

(2) Highest water table
(3) Water table and its slope at $x = 200$ ft
(4) Discharges into river 1 and 2
(5) Total discharge into the rivers.

9.29 In Figure 9.9b, $h_1 = 15$ m, $h_2 = 10$ m, $K = 8$ m d^{-1}, $L = 250$ m, and $i = 0.02$ m d^{-1}. Determine the:
(1) Distance from river 1 to the divide
(2) Highest water table
(3) Water table and its slope at $x = 125$ m
(4) Discharges into river 1 and 2
(5) Total discharge into the rivers.

9.30 In Figure 9.9b, $h_1 = 35$ ft, $h_2 = 20$ ft, $K = 25$ ft d^{-1}, and $L = 400$ ft. Determine the recharge rate that results in a divide at:
(1) River 1
(2) The midpoint between the two rivers
(3) River 2.

9.31 In Figure 9.9b, $h_1 = 15$ m, $h_2 = 10$ m, $K = 8$ m d^{-1}, and $L = 250$ m. Determine the recharge rate that will result in a divide at:
(1) River 1
(2) The midpoint between the two rivers
(3) River 2.

9.32 In Figure 9.9b, $h_1 = 35$ ft, $K = 25$ ft d^{-1}, $L = 400$ ft, and $i = 0.05$ ft d^{-1}. Determine the distance from river 1 to the divide if:
(1) $h_2 = 25$ ft
(2) $h_2 = 30$ ft
(3) $h_2 = 35$ ft.

9.33 A 1-ft-diameter well fully penetrates a 25-ft-thick confined aquifer ($K = 20$ ft d^{-1}). After pumping at 800 ft^3 d^{-1} for a long period of time, the drawdown in the well is measured to be 3.5 ft. Determine the:
(1) Drawdown at 50 ft away from the well
(2) Influence radius
(3) Flow rate at 200 ft away from the well
(4) Flow rate at 600 ft away from the well.

9.34 A 0.5-m-diameter well fully penetrates a 20-m-thick confined aquifer ($K = 5.5$ m d^{-1}). After pumping at 100 m^3 d^{-1} for a long period of time, the drawdown in the well is measured to be 1.5 m. Determine the:
(1) Drawdown at 50 m away from the well
(2) Influence radius
(3) Flow rate at 100 m away from the well
(4) Flow rate at 3000 m away from the well
(5) Flow rate at 9500 m away from the well.

9.35 A 0.2-m-diameter well fully penetrates a 50-m-thick confined aquifer. After pumping at $500\,m^3\,d^{-1}$ for a long period of time, the drawdowns at 100 and 200 m away from the well are measured to be 4.5 and 3.5 m, respectively. Determine the:

(1) Transmissivity and hydraulic conductivity
(2) Drawdown in the well
(3) Influence radius.

9.36 A 1-ft-diameter well fully penetrates a 50-ft-thick confined aquifer. After pumping at $1000\,ft^3\,d^{-1}$ for a long period of time, the drawdowns at 200 and 500 ft away from the well are measured to be 4.5 and 2.5 ft, respectively. Determine the:

(1) Transmissivity and hydraulic conductivity
(2) Drawdown in the well
(3) Influence radius.

9.37 A 1-ft-diameter well fully penetrates a 25-ft-thick unconfined aquifer ($K = 8.0\,ft\,d^{-1}$). After pumping at $500\,ft^3\,d^{-1}$ for a long period of time, the drawdown in the well is measured to be 2.0 ft. Determine the:

(1) Drawdown at 50 ft away from the well
(2) Influence radius
(3) Flow rate at 50 ft away from the well
(4) Drawdown and flow rate at 150 ft away from the well.

9.38 A 0.3-m-diameter well fully penetrates a 10-m-thick unconfined aquifer ($K = 10.5\,m\,d^{-1}$). After pumping at $50\,m^3\,d^{-1}$ for a long period of time, the drawdown in the well is measured to be 0.5 m. Determine the:

(1) Drawdown at 50 m away from the well
(2) Influence radius
(3) Flow rate at 50 m away from the well
(4) Drawdown and flow rate at 150 m away from the well.

9.39 A 0.2-m-diameter well fully penetrates a 50-m-thick unconfined aquifer. After pumping at $500\,m^3\,d^{-1}$ for a long period of time, the drawdowns at 50 and 100 m away from the well are measured to be 3.8 and 2.5 m, respectively. Determine the:

(1) Hydraulic conductivity and transmissivity
(2) Drawdown in the well
(3) Influence radius.

9.40 A 2-ft-diameter well fully penetrates a 100-ft-thick unconfined aquifer. After pumping at $800\,ft^3\,d^{-1}$ for a long period of time, the drawdowns at 100 and 500 ft away from the well are measured to be 2.5 and 1.5 ft, respectively. Determine the:

(1) Hydraulic conductivity and transmissivity
(2) Drawdown in the well
(3) Influence radius.

9.41 A 2-ft-diameter well fully penetrates a 100-ft-thick unconfined aquifer ($K = 15.5\,ft\,d^{-1}$). After pumping at $1800\,ft^3\,d^{-1}$ for a long period of time, the drawdown at 100 ft away from the well is measured to be 3.5 ft. For a recharge rate of $0.002\,ft\,d^{-1}$, determine the:

(1) Influence radius

(2) Drawdown in the well
(3) Drawdown at 50 ft away from the well
(4) Drawdown at 250 ft away from the well
(5) Flow rates at 50 and 250 ft away from the well
(6) Drawdown and flow rate at 1000 ft away from the well.

9.42 A 0.5-m-diameter well fully penetrates an 80-m-thick unconfined aquifer ($K = 4.5\,\mathrm{m\,d^{-1}}$). After pumping at $1000\,\mathrm{m^3\,d^{-1}}$ for a long period of time, the drawdown at 50 m away from the well is measured to be 6.5 m. For a recharge rate of $0.001\,\mathrm{m\,d^{-1}}$, determine the:
(1) Influence radius
(2) Drawdown in the well
(3) Drawdown at 100 m away from the well
(4) Drawdown at 350 m away from the well
(5) Flow rates at 100 and 350 m away from the well
(6) Drawdown and flow rate at 800 m away from the well.

9.43 A well fully penetrates a confined aquifer ($S = 1.2 \times 10^{-4}$, $T = 800\,\mathrm{m^2\,d^{-1}}$) and pumps at $1000\,\mathrm{m^3\,d^{-1}}$. For a maximum truncation error of $\varepsilon = 10^{-4}$, determine the drawdown at 100 m away from the well after pumping for:
(1) 1 d
(2) 3 d
(3) 6 d.

9.44 A well fully penetrates a confined aquifer ($S = 2.5 \times 10^{-4}$, $T = 9500\,\mathrm{ft^2\,d^{-1}}$) and pumps at $5000\,\mathrm{ft^3\,d^{-1}}$. For a maximum truncation error of $\varepsilon = 10^{-4}$, determine the drawdown at 200 ft away from the well after pumping for:
(1) 1 d
(2) 3 d
(3) 6 d.

9.45 A well fully penetrates a confined aquifer and pumps at $100\,\mathrm{ft^3\,min^{-1}}$. The drawdowns in a monitoring well 50 ft away from the pumping well are measured as shown in the following table. For a maximum truncation error of $\varepsilon = 10^{-4}$, determine the aquifer properties of S and T.

t (min)	s (ft)
1	0.24
2	1.01
3	1.24
4	1.49
5	1.51
6	1.60
7	1.62
8	1.74
9	2.10
10	2.29

9.46 Redo Problem 9.45 if the pumping rate is 200 $ft^3\ min^{-1}$.

9.47 The properties of an unconfined aquifer consisting of silt and clayey soil particles are: $S = 0.001$, $S_y = 0.02$, $K_H = 0.5\ m\ d^{-1}$, $K_V = 0.02\ m\ d^{-1}$, and $h_0 = 30\ m$. For a pumping rate of 50 $m^3\ d^{-1}$, determine the drawdown after pumping for 5 d at a distance away from the well of:
(1) 10 m
(2) 50 m
(3) 100 m.

9.48 The properties of an unconfined aquifer consisting of sandy soil particles are: $S = 0.15$, $S_y = 0.25$, $K_H = 15\ ft\ d^{-1}$, $K_V = 0.05\ ft\ d^{-1}$, and $h_0 = 50\ ft$. For a pumping rate of 1000 $ft^3\ d^{-1}$, determine the drawdown after pumping for 10 d at a distance away from the well of:
(1) 50 ft
(2) 100 ft
(3) 200 ft.

9.49 The properties of an unconfined aquifer consisting of silt and clay soil particles are: $S = 0.001$, $S_y = 0.02$, $K_H = 0.5\ m\ d^{-1}$, $K_V = 0.02\ m\ d^{-1}$, and $h_0 = 30\ m$. For a pumping rate of 50 $m^3\ d^{-1}$, determine the drawdown at 50 m away from the well after pumping for:
(1) 50 min
(2) 120 min
(3) 2.5 d
(4) 10 d.

9.50 The properties of an unconfined aquifer consisting of sandy soil particles are: $S = 0.15$, $S_y = 0.25$, $K_H = 15\ ft\ d^{-1}$, $K_V = 0.05\ ft\ d^{-1}$, and $h_0 = 50\ ft$. For a pumping rate of 1000 $ft^3\ d^{-1}$, determine the drawdown at 100 ft away from the well after pumping for:
(1) 2 min
(2) 10 min
(3) 60 min
(4) 1 d.

9.51 A well fully penetrates a 50-ft-thick unconfined aquifer and pumps at 25 $ft^3\ min^{-1}$, resulting in a time series of drawdown at 100 ft away from the pumping well as shown in the following table. Determine the aquifer properties of T, S, S_y, K_H, and K_V.

t (min)	s (ft)	t (min)	s (ft)
3	0.010	85	0.327
5	0.072	115	0.344
10	0.100	175	0.400
15	0.150	250	0.429
20	0.180	300	0.448
25	0.200	600	0.520
30	0.231	750	0.580
40	0.239	900	0.600
50	0.248	1500	0.750
65	0.252		

9.52 Redo Problem 9.51 if the pumping rate is $35\ ft^3\ min^{-1}$.

9.53 The transmissivity of a semiconfined aquifer is $500\ m^2\ d^{-1}$. Its aquitard bed has a thickness of 10 m and a hydraulic conductivity of $0.001\ m\ d^{-1}$. After pumping at $2500\ m^3\ d^{-1}$ for a long period of time, determine the drawdown at a distance away from the well of:
(1) 50 m
(2) 150 m
(3) 500 m.

9.54 The transmissivity of a semiconfined aquifer is $50\ ft^2\ d^{-1}$. Its aquitard bed has a thickness of 25 ft and a hydraulic conductivity of $0.0005\ ft\ d^{-1}$. After pumping at $7500\ ft^3\ d^{-1}$ for a long period of time, determine the drawdown at a distance away from the well of:
(1) 100 ft
(2) 350 ft
(3) 500 ft.

9.55 The aquitard bed of a semiconfined aquifer has a resistance of 50,000 d. After pumping at $2500\ m^3\ d^{-1}$ for a long period of time, the drawdown at 100 m away from the well is measured to be 1.5 m. Determine the transmissivity of the semiconfined aquifer.

9.56 A semiconfined aquifer has a transmissivity of $8000\ ft^2\ d^{-1}$ and a storativity of 0.002. Its aquitard bed has a thickness of 20 ft and a hydraulic conductivity of $0.001\ ft\ d^{-1}$. After pumping at $5500\ ft^3\ d^{-1}$ for 15 d, determine the drawdown at a distance away from the well of:
(1) 100 ft
(2) 250 ft
(3) 500 ft.

9.57 A semiconfined aquifer has a transmissivity of $750\ m^2\ d^{-1}$ and a storativity of 0.01. Its aquitard bed has a thickness of 10 m and a hydraulic conductivity of $0.001\ m\ d^{-1}$. After pumping at $1000\ m^3\ d^{-1}$ for 5 d, determine the drawdown at a distance away from the well of:
(1) 50 m
(2) 150 m
(3) 350 m.

9.58 The properties of a semiconfined aquifer are: $S = 0.002$ and $T = 8000\ ft^2\ d^{-1}$. Its aquitard has a resistance of 20,000 d. For a pumping rate of $5500\ ft^3\ d^{-1}$, determine the drawdown at 100 ft away from the well after pumping for:
(1) 1 d
(2) 5 d
(3) 10 d.

9.59 The properties of a semiconfined aquifer are: $S = 0.01$ and $T = 750\ m^2\ d^{-1}$. Its aquitard has a resistance of 10,000 d. For a pumping rate of $1000\ m^3\ d^{-1}$, determine the drawdown at 150 m away from the well after pumping for:
(1) 1 d
(2) 3 d
(3) 8 d.

9.60 A well fully penetrates a semiconfined aquifer and pumps at 50 $ft^3\ min^{-1}$, resulting in a time series of drawdown at 100 ft away from the pumping well as shown in the following table. Determine the properties of the aquifer and aquitard, including T, S, and b'/K'.

t (min)	s (ft)	t (min)	s (ft)
2.5	0.017	85	0.437
7	0.088	115	0.477
14	0.149	175	0.502
15	0.169	250	0.52
20	0.18	300	0.539
25	0.232	600	0.711
30	0.274	750	0.764
40	0.311	900	0.766
50	0.327	1500	0.774
65	0.369		

9.61 Redo Problem 9.60 if the pumping rate is 30 $ft^3\ min^{-1}$.

9.62 A well fully penetrates a confined aquifer ($S = 1.5 \times 10^{-4}$, $T = 2000\ m^2\ d^{-1}$). It pumps at 3000 $m^3\ d^{-1}$ for 0.5 d and 1000 $m^3\ d^{-1}$ for a subsequent 1.5 d, and then stops pumping. For a maximum truncation error of $\varepsilon = 10^{-4}$, determine the overall change of groundwater level at 100 m away from the well after pumping for:

(1) 0.5 d
(2) 1.0 d
(3) 2.0 d
(4) 5.5 d.

9.63 A recharging well fully penetrates a confined aquifer ($S = 1.5 \times 10^{-4}$, $T = 25{,}000\ ft^2\ d^{-1}$). It recharges at 10,000 $ft^3\ d^{-1}$ for 0.5 d and 5000 $ft^3\ d^{-1}$ for a subsequent 1.5 d, and then stops recharging. For a maximum truncation error of $\varepsilon = 10^{-4}$, determine the overall change of groundwater level at 100 ft away from the well after recharging for:

(1) 0.5 d
(2) 1.0 d
(3) 2.0 d
(4) 5.5 d.

9.64 Introduce image wells to eliminate the boundaries shown in the following figures.

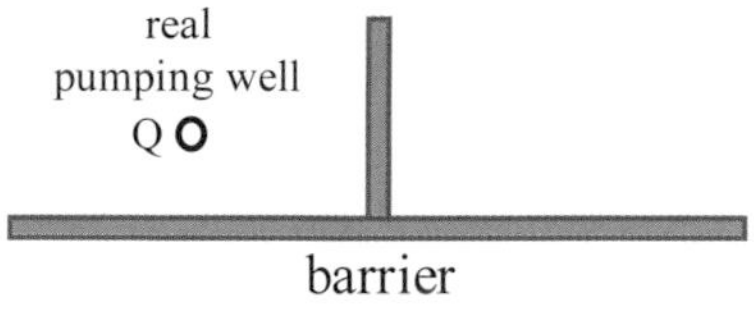

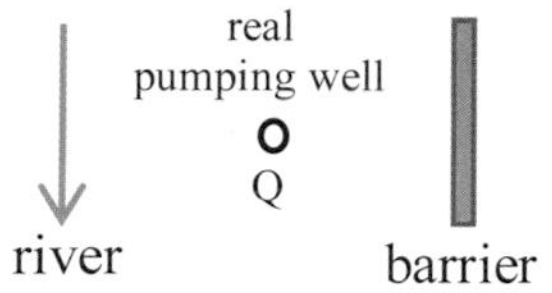

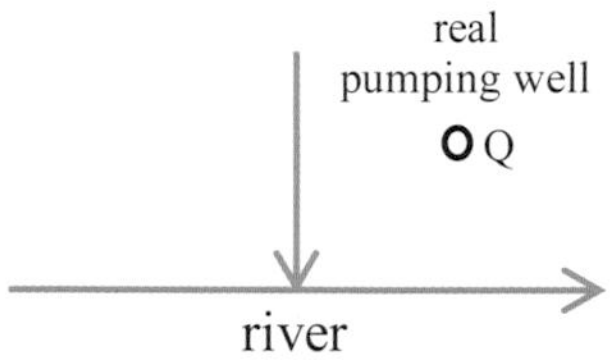

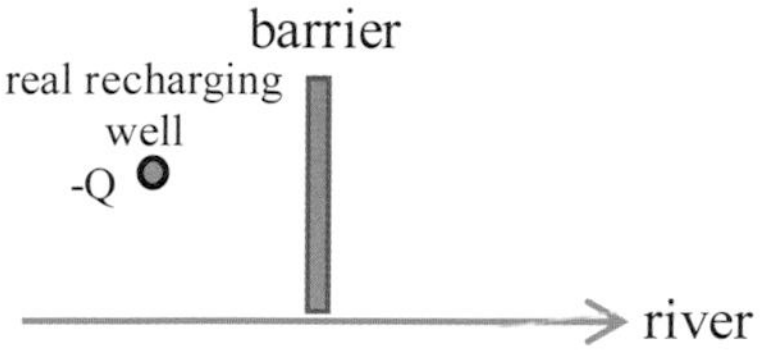

9.65 The properties of an unconfined aquifer consisting of fine-sand soil particles are: $S = 0.005$, $S_y = 0.025$, $K_H = 50\,\mathrm{m\,d^{-1}}$, $K_V = 5\,\mathrm{m\,d^{-1}}$, and $h_0 = 80\,\mathrm{m}$. As illustrated in the following figure, a 0.3-m-diameter well ($x_w = 100\,\mathrm{m}$, $y_w = 50\,\mathrm{m}$) pumps at $2500\,\mathrm{m^3\,d^{-1}}$. After pumping for 10 d, determine the drawdown at a location of:

(1) $x = 0\,\mathrm{m}$, $y = 0\,\mathrm{m}$
(2) $x = 100\,\mathrm{m}$, $y = 100\,\mathrm{m}$
(3) $x = 200\,\mathrm{m}$, $y = 50\,\mathrm{m}$
(4) $x = 200\,\mathrm{m}$, $y = 100\,\mathrm{m}$
(5) $x = -50\,\mathrm{m}$, $y = 50\,\mathrm{m}$
(6) $x = 100\,\mathrm{m}$, $y = -50\,\mathrm{m}$
(7) $x = -100\,\mathrm{m}$, $y = -100\,\mathrm{m}$.

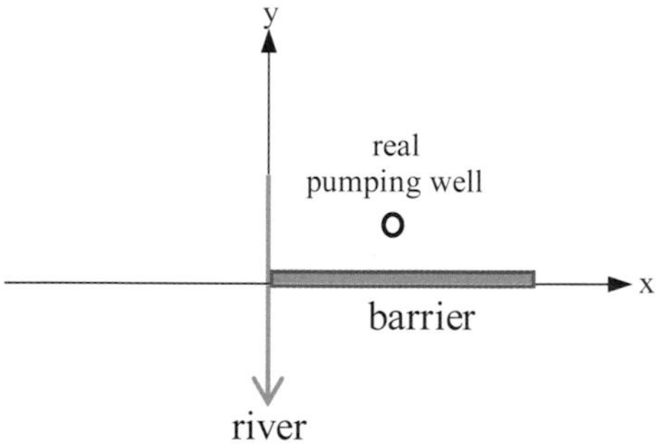

9.66 A 1-ft-diameter well fully penetrates a 30-ft-thick unconfined aquifer ($K = 25\,\mathrm{ft\,d^{-1}}$). After pumping at $500\,\mathrm{ft^3\,d^{-1}}$ for a long period of time, the drawdown in the well is measured to be 3.5 ft. If there is a boundary 300 ft away from the pumping well, determine whether pumping is impacted by the boundary.

9.67 A 25-m-thick confined aquifer has a transmissivity of $800\,m^2\,d^{-1}$. Four fully penetrating wells with a diameter of 0.5 m are located along a straight line and spaced at 800 m. If the groundwater exploration permit allows a maximum drawdown of 2.5 m in any of the wells after pumping for a long period of time, determine the maximum allowable pumping rates of the wells.

9.68 A 1-ft-diameter well fully penetrates into a 50-ft-thick confined aquifer ($K = 10\,ft\,d^{-1}$). After pumping at $2000\,ft^3\,d^{-1}$ for a long period of time, the drawdown in the well is measured as 2.5 ft. If the penetration depth of the well is changed to 30 ft, determine the drawdown in the well.

9.69 A 0.2-m-diameter well fully penetrates into an unconfined aquifer ($K = 20\,m\,d^{-1}$) with an initial saturated thickness of 50 m. After pumping at $50\,m^3\,d^{-1}$ for a long period of time, the drawdown in the well is measured to be 1.5 m. If the penetration depth of the well is changed to 25 m, determine the drawdown in the well.

9.70 A falling-head slug test is performed by quickly placing a steel block into a well, causing an instantaneous head rise of 3.44 ft in the well. The 50-ft-thick unconfined aquifer has a saturated soil moisture of 0.38 and a field capacity of 0.20. The well penetrates into the aquifer by 35 ft, of which 20 ft is screened. The well hole inner radius is 0.6 ft, and the well screen outer radius is 0.3 ft. The head displacement measurements are given in the following table. If the aquifer is homogeneous and isotropic, determine the hydraulic conductivity of the aquifer using the Bouwer–Rice method.

t (s)	$s_{w,t}$ (ft)	t (s)	$s_{w,t}$ (ft)
0	3.44	250	2.00
2	3.40	300	1.85
5	3.38	350	1.70
10	3.36	400	1.55
15	3.35	450	1.45
20	3.30	500	1.32
25	3.19	550	1.21
30	3.01	600	1.20
35	2.88	650	1.18
40	2.77	700	1.16
45	2.75	750	1.13
50	2.71	800	1.11
55	2.66	850	1.10
60	2.64	900	1.08
100	2.50	950	1.06
150	2.35	1000	1.04
200	2.20		

10 Unsteady-State Flow

In all previous chapters, the assumption of steady-state flow has underpinned analysis and computation. Although steady-state flow is relatively straightforward for practical applications, unsteady-state flow is more common in natural and manmade systems. As discussed in Section 2.5.1, for unsteady-state flow, at a given location of the conveying waterway, at least one flow condition (i.e., velocity or depth) varies with time. This chapter discusses three types of unsteady-state flows. The first type, stormwater rising and falling in a manhole, is governed by the continuity equation. Its analysis is needed to appropriately design manholes to avoid possible overspill. The second type, transient flow in a pressurized pipe or open channel, is governed by the Saint–Venant equations. Such a transient flow can be induced by a gate (valve) closing or opening as well as by a channel cross section contracting or expanding. Its analysis is very important for the design and operation of pipeline and channel systems. The third type, transient discharge in a shallow aquifer induced by recharge and/or withdrawal, is governed by the continuity equation and Darcy's law. Its analysis is needed for groundwater management.

10.1 Stormwater Rising and Falling in a Manhole

As illustrated in Figure 10.1, a manhole is a vertical access shaft (i.e., covered chamber) from the ground surface to the junction of two or more stormwater pipelines to allow cleaning, inspection, connections, and repairs (Gorse *et al.*, 2013). The chamber is usually made of concrete with a circular cross section, and has one or more inflow pipes and one or more outflow pipes to allow stormwater runoff to flow into and out of the chamber. For a design storm, the maximum water depth in the chamber should not be larger than the chamber depth (i.e., $y_{max} \leq H$). Otherwise, the water in the chamber will overspill out of the manhole to flood the adjacent areas. Thus, it is needed to determine how the water depth in the chamber varies with time during the design storm event. This can be done using the continuity equation expressed as:

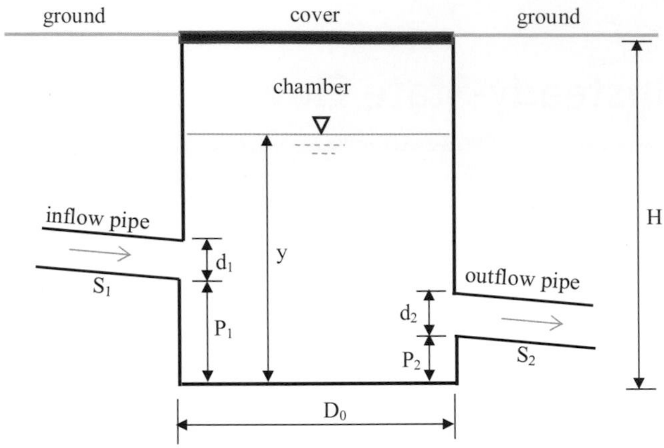

Figure 10.1 **Cross section of a stormwater manhole.** Diagram shows flow routing.

$$A_0 \frac{dy}{dt} = Q_{1,t} - Q_{2,t} \tag{10.1}$$

(cross-sectional area of chamber: A_0; water depth in chamber at time t: y; time: t; inflow rate at time t: $Q_{1,t}$; outflow rate at time t: $Q_{2,t}$)

In Eq. (10.1), $Q_{1,t}$ is usually determined from a hydrologic analysis as presented in Chapter 4, whereas $Q_{2,t}$ is zero if $y \leq P_2$ and can be determined using Eq. (8.2) if $P_2 < y \leq P_2 + d_2$ or using Eq. (8.3) if $y > P_2 + d_2$. That is, $Q_{2,t}$ can be determined by assuming that the outflow pipe is inlet-controlled, which is usually true.

A backward finite-difference format of Eq. (10.1) can be used to determine the water depth in the chamber and the outflow rate at a time of interest. It is written as:

$$y_{t+1} = y_t + \frac{\Delta t}{2A_0}\left(Q_{1,t+1} + Q_{1,t}\right) - \frac{\Delta t}{2A_0}\left(Q_{2,t+1} + Q_{2,t}\right) \tag{10.2}$$

(water depth in chamber at time t + 1: y_{t+1}; water depth in chamber at time t: y_t; computational time step: Δt; cross-sectional area of chamber: A_0; inflow rate at time t + 1: $Q_{1,t+1}$; inflow rate at time t: $Q_{1,t}$; outflow rate at time t + 1: $Q_{2,t+1}$; outflow rate at time t: $Q_{2,t}$)

If $y_{t+1} \leq P_2$, $Q_{2,t+1}$ is zero and y_{t+1} is independent of $Q_{2,t+1}$. That is:

$$Q_{2,t+1} = 0 \tag{10.3}$$

(outflow rate at time t + 1: $Q_{2,t+1}$)

If $P_2 < y_{t+1} \leq P_2 + d_2$, for an outflow pipe inlet to which the Form I equation is applicable, by Eq. (8.2) and in BG units, y_{t+1} and $Q_{2,t+1}$ are related as:

specific energy head at critical depth of time t + 1 (Eq. (10.5))
outflow rate at time t + 1
constant (Table 8.1)
water depth in chamber at time t + 1
correction factor of outflow pipe slope S_0 (= −0.5S_0 in general; +0.7S_0 for mitered inlet)
diameter of outflow pipe
constant (Table 8.1)
cross-sectional area of outflow pipe

$$\frac{y_{t+1}}{d_2} = \frac{H_{c,t+1}}{d_2} + K\left(\frac{Q_{2,t+1}}{A_2 d_2^{0.5}}\right)^M + Z \tag{10.4}$$

specific energy head at critical depth at time t + 1 in Eq. (10.4)
outflow rate at time t + 1
flow area at $y_{c,t+1}$ (Eq. (10.8))
critical water depth (Eq. (10.6)) at time t + 1
gravitational acceleration

$$H_{c,t+1} = y_{c,t+1} + \frac{1}{2g}\left(\frac{Q_{2,t+1}}{A_{c,t+1}}\right)^2 \tag{10.5}$$

diameter of outflow pipe
angle at $y_{c,t+1}$ (Table 2.4) subject to Eq. (10.7)
critical water depth at time t + 1

$$y_{c,t+1} = \frac{d_2}{2}\left[1 - \cos\left(\frac{\theta_{t+1}}{2}\right)\right] \tag{10.6}$$

outflow rate at time t + 1
flow area at $y_{c,t+1}$ (Eq. (10.8))
top width at $y_{c,t+1}$ (Eq. (10.9))
gravitational acceleration

$$\frac{(Q_{2,t+1})^2}{g} = \frac{(A_{c,t+1})^3}{B_{c,t+1}} \tag{10.7}$$

diameter of outflow pipe
angle at $y_{c,t+1}$ (Table 2.4) in Eq. (10.6)
flow area at $y_{c,t+1}$ in Eq. (10.7)

$$A_{c,t+1} = \frac{d_2^2}{8}\left[\theta_{t+1} - \sin(\theta_{t+1})\right] \tag{10.8}$$

angle at $y_{c,t+1}$ (Table 2.4) in Eq. (10.6)
top width at $y_{c,t+1}$ in Eq. (10.7)
diameter of outflow pipe

$$B_{c,t+1} = d_2\left[\sin\left(\frac{\theta_{t+1}}{2}\right)\right] \tag{10.9}$$

If $P_2 < y_{t+1} \le P_2 + d_2$, for an outflow pipe inlet to which the Form II equation is applicable, by Eq. (8.3) and in BG units, y_{t+1} and $Q_{2,t+1}$ are related as:

constant (Table 8.1)
outflow rate at time t + 1
water depth in chamber at time t + 1
correction factor of outflow pipe slope S_0 (= −0.5S_0 in general; +0.7S_0 for mitered inlet)
diameter of outflow pipe
constant (Table 8.1)
cross-sectional area of outflow pipe

$$\frac{y_{t+1}}{d_2} = K\left(\frac{Q_{2,t+1}}{A_2 d_2^{0.5}}\right)^M + Z \tag{10.10}$$

Finally, if $y_{t+1} > P_2 + d_2$, by Eq. (8.1) and in BG units, y_{t+1} and $Q_{2,t+1}$ are related as:

$$\frac{y_{t+1}}{d_2} = C\left(\frac{Q_{2,t+1}}{A_2 d_2^{0.5}}\right)^2 + Y + Z \qquad (10.11)$$

water depth in chamber at time t + 1 (y_{t+1}); outflow rate at time t + 1 ($Q_{2,t+1}$); constant (Table 8.1) (C and Y); diameter of outflow pipe (d_2); cross-sectional area of outflow pipe (A_2); correction factor of outflow pipe slope S_0 (= $-0.5S_0$ in general; $+0.7S_0$ for mitered inlet) (Z)

For an inflow hydrograph and an initial water depth in the chamber, the variations of water depth with time can be computed using the above equations in an Excel spreadsheet. If $P_2 < y_{t+1} \leq P_2 + d_2$ and the Form I equation is applicable, a trial-and-error calculation is needed to make the computed y_{t+1} using Eq. (10.2) equal to that using Eq. (10.4) by adjusting $Q_{2,t+1}$ and θ_{t+1}, subject to Eq. (10.7). In this regard, set the objective of Solver as the difference of the computed value of y_{t+1} by Eq. (10.2) from that by Eq. (10.7) to be equal to zero. The computed maximum water depth indicates whether water in the chamber would overspill out.

Example 10.1 For the manhole illustrated in Figure 10.1, $D_0 = 6\,\text{ft}$, $H = 15\,\text{ft}$, $d_2 = 2\,\text{ft}$, $P_2 = 0.5\,\text{ft}$, and $S_2 = 0$. The inlet of the outflow circular concrete pipe has a square edge with headwall. For the inflow hydrograph shown in cells A9:B29 of Figure 10.2, determine and plot the variations of water depth in the manhole chamber versus time. Assume that the chamber is initially empty.

Solution

$A_0 = \pi/4 * (6\,\text{ft})^2 = 28.27433\,\text{ft}^2$, $A_2 = \pi/4 * (2\,\text{ft})^2 = 3.14159\,\text{ft}^2$

Table 8.1: circular concrete pipe with square edge w/headwall

$\rightarrow$ chart #1, scale #1 $\rightarrow$ Form I, $K = 0.0098$, $M = 2.0$; $C = 0.0398$, $Y = 0.67$

$S_2 = 0 \rightarrow Z = 0.0$. Use a computational time step of $\Delta t = 0.2\,\text{min.} = 12\,\text{sec.}$

- $t = 0.0\,\text{min} \rightarrow Q_{1,0} = 0.00\,\text{cfs}$, $y_0 = 0.00\,\text{ft}$

$$y_0 < P_2 = 0.5\,\text{ft} \rightarrow Q_{2,0} = 0.000\,\text{cfs}.$$

<u>Step results (cells C9:D9)</u>: $y_0 = 0.00\,\text{ft}$, $Q_{2,0} = 0.000\,\text{cfs}$.

- $t = 0.2\,\text{min} \rightarrow Q_{1,1} = 0.61\,\text{cfs}$

	A	B	C	D	E	F	G	H	I	J	K	L
1	D_o =	6	ft	d_2 =	2	ft	K =	0.0098	C =	0.0398		
2	H =	15	ft	P_2 =	0.50	ft	M =	2.0	Y =	0.67		
3	A_o =	28.2743	ft^2	S_2 =	0.00		Z =	0.00				
4	g =	32.2	ft sec^{-2}	A_2 =	3.14159	ft^2	**From Table 8-1 for Chart #1 Scale #1 culvert**					
5	Δt =	0.2	min									
6												
7	Inflow Hydrograph		y_t	$Q_{2,t}$	Eq. (10-4)	θ_t	$A_{c,t}$ (Eq. 10-8)	$B_{c,t}$ (Eq. 10-9)	Eq. (10-7)	$y_{c,t}$ (Eq. 10-6)	$H_{c,t}$ (Eq. 10-5)	Objective
8	t (min)	$Q_{1,t}$ (cfs)	(ft)	(cfs)	or (10-11)	(rad)	(ft^2)	(ft)		(ft)	(ft)	
9	0.0	0.00	0.00	0.000	0.0000	0.00	0.000	0.000	#DIV/0!	0.000	#DIV/0!	0.0000
10	0.2	0.61	0.1294	0.000	0.1294	0.00	0.000	0.000	#DIV/0!	0.000	#DIV/0!	0.0000
11	0.4	1.69	0.6174	0.001	0.6174	0.11	0.000	0.111	0.000	0.002	0.617	0.0000
12	0.6	1.86	0.7904	2.734	0.7904	2.27	0.749	1.811	0.000	0.576	0.783	0.0000
13	0.8	4.90	0.9036	3.492	0.9036	2.43	0.892	1.876	0.000	0.654	0.892	0.0000
14	1.0	5.31	1.1666	5.479	1.1666	2.79	1.224	1.969	0.000	0.826	1.137	0.0000
15	1.2	7.22	1.2883	6.478	1.2883	2.94	1.374	1.990	0.000	0.901	1.247	0.0000
16	1.4	7.98	1.4575	7.924	1.4575	3.14	1.574	2.000	0.000	1.002	1.395	0.0000
17	1.6	8.26	1.4867	8.178	1.4867	3.18	1.607	2.000	0.000	1.018	1.420	0.0000
18	1.8	9.95	1.6238	9.386	1.6238	3.33	1.759	1.991	0.000	1.094	1.536	0.0000
19	2.0	9.49	1.6730	9.823	1.6730	3.38	1.812	1.985	0.000	1.121	1.577	0.0000
20	2.2	8.11	1.5221	8.488	1.5221	3.22	1.648	1.999	0.000	1.038	1.451	0.0000
21	2.4	7.74	1.4382	7.757	1.4382	3.12	1.552	2.000	0.000	0.991	1.378	0.0000
22	2.6	6.88	1.3711	7.179	1.3711	3.04	1.473	1.998	0.000	0.951	1.320	0.0000
23	2.8	4.96	1.1780	5.571	1.1780	2.81	1.239	1.972	0.000	0.833	1.147	0.0000
24	3.0	3.17	0.9342	3.708	0.9342	2.48	0.931	1.891	0.000	0.674	0.921	0.0000
25	3.2	1.81	0.7183	2.289	0.7183	2.15	0.659	1.761	0.000	0.526	0.713	0.0000
26	3.4	0.91	0.5332	1.303	0.5332	1.84	0.438	1.591	0.000	0.394	0.532	0.0000
27	3.6	0.02	0.4541	0.000	0.4541	0.00	0.000	0.000	#DIV/0!	0.000	#DIV/0!	0.0000
28	3.8	0.00	0.4584	0.000	0.4584	0.00	0.000	0.000	#DIV/0!	0.000	#DIV/0!	0.0000
29	4.0	0.00	0.4584	0.000	0.4584	0.00	0.000	0.000	#DIV/0!	0.000	#DIV/0!	0.0000
30–41												

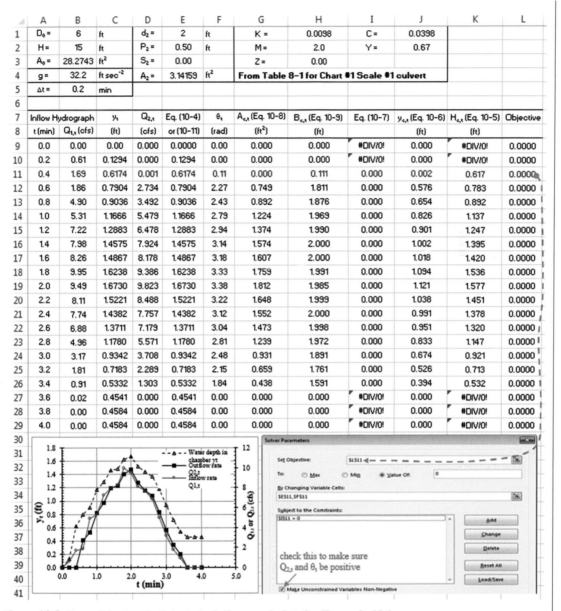

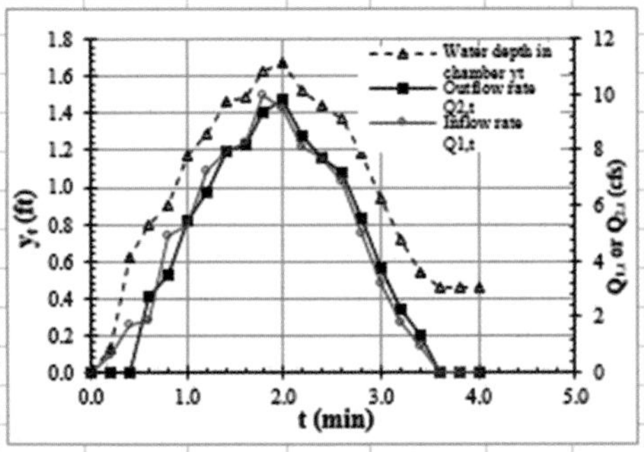

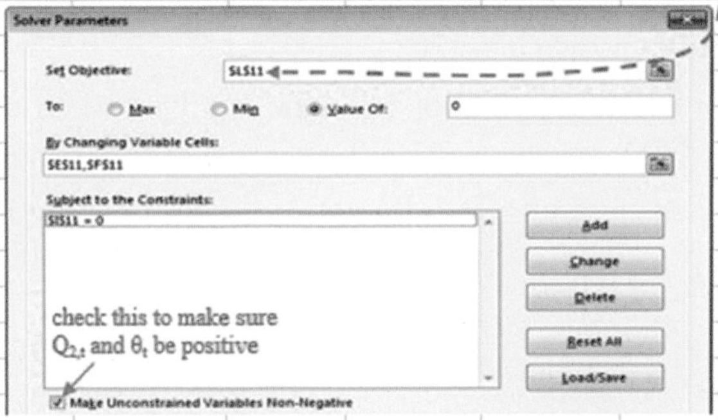

Figure 10.2 **Spreadsheet with data, calculations, and plots for Example 10.1.**

Assume $Q_{2,1} = 0.000$ cfs:

Eq. (10.2): $y_1 = 0.00\,\text{ft} + (12\,\text{sec})/\left(2 * 28.27433\,\text{ft}^2\right) * (0.61\,\text{cfs} + 0.00\,\text{cfs}) - (12\,\text{sec})/\left(2 * 28.27433\,\text{ft}^2\right) * (0.000\,\text{cfs} + 0.000\,\text{cfs}) = 0.1294\,\text{ft}.$

$$y_1 < P_2 = 0.5\,\text{ft} \rightarrow Q_{2,1} = 0.000\,\text{cfs}.$$

Step results (cells C10:D10): $y_1 = 0.1294$ ft, $Q_{2,1} = 0.000$ cfs.

- $t = 0.4$ min$\rightarrow Q_{1,2} = 1.69$ cfs

  Assume $Q_{2,2} = 0.000$ cfs:

  Eq. (10.2): $y_2 = 0.1294\,\text{ft} + (12\,\text{sec})/\left(2 * 28.27433\,\text{ft}^2\right) * (1.69\,\text{cfs} + 0.61\,\text{cfs}) - (12\,\text{sec})/\left(2 * 28.27433\,\text{ft}^2\right) * (0.000\,\text{cfs} + 0.000\,\text{cfs}) = 0.6175\,\text{ft}.$

  $$y_2 > P_2 = 0.5\,\text{ft} \rightarrow Q_{2,2} > 0.000\,\text{cfs}.$$

  Use Solver to recalculate y_2 and $Q_{2,2}$:

  Eq. (10.8): build the equation in cell G11.

  Eq. (10.9): build the equation in cell H11.

  Eq. (10.7): build the equation in cell I11.

  Eq. (10.6): build the equation in cell J11.

  Eq. (10.5): build the equation in cell K11.

  Eq. (10.4) or (10.11): *IF( )* function to build the equations depending on y_2 in cell C11.

  Objective in cell L11: C11 – E11.

  As shown at the bottom right of Figure 10.2, Solver is used to make L11 = 0 by changing D11 and F11, subject to I11 = 0.

  Step results (cells C11:D11): $y_1 = 0.6174$ ft, $Q_{2,1} = 0.001$ cfs.

- $t = 0.6$ min, 0.8 min, 1.0 min …

  Build the equations and use Solver to find the solutions, as done for $t = 0.4$ min.

  The results are shown in cells C12:D29 of Figure 10.2.

- The plot of the results is shown at the bottom left of Figure 10.2.

Example 10.2 Redo Example 10.1 for the inflow hydrograph shown in cells A9:B29 of Figure 10.3.

Solution

Following the procedure elaborated in Example 10.1, one can develop the spreadsheet shown in Figure 10.3 to undertake the calculations. The maximum water depth and outflow rate are determined to be 14.1307 ft and 49.925 cfs, respectively, both occurring at t = 3.6 min. The results are shown in cells C9:D29 and plotted at the bottom left of Figure 10.3, and the settings of the Solver dialog box are shown at the bottom right of the figure.

	A	B	C	D	E	F	G	H	I	J	K	L
1	D_0 =	6	ft	d_2 =	2	ft	K =	0.0098	C =	0.0398		
2	H =	15	ft	P_2 =	0.50	ft	M =	2.0	Y =	0.67		
3	A_0 =	28.2743	ft^2	S_2 =	0.00		Z =	0.00				
4	g =	32.2	ft sec^{-2}	A_2 =	3.14159	ft^2	**From Table 8-1 for Chart #1 Scale #1 culvert**					
5	Δt =	0.2	min									
6												
7	Inflow Hydrograph		y_t	$Q_{2,t}$	Eq. (10-4)	θ_t	$A_{c,t}$ (Eq. 10-8)	$B_{c,t}$ (Eq. 10-9)	Eq. (10-7)	$y_{c,t}$ (Eq. 10-6)	$H_{c,t}$ (Eq. 10-5)	Objective
8	t (min)	$Q_{1,t}$ (cfs)	(ft)	(cfs)	or (10-11)	(rad)	(ft^2)	(ft)		(ft)	(ft)	
9	0.0	0.00	0.00	0.000	0.0000	0.00	0.000	0.000	#DIV/0!	0.000	#DIV/0!	0.0000
10	0.2	1.00	0.2122	0.000	0.2122	0.00	0.000	0.000	#DIV/0!	0.000	#DIV/0!	0.0000
11	0.4	2.00	0.8487	0.001	0.8487	0.11	0.000	0.105	0.000	0.001	0.849	0.0000
12	0.6	5.00	1.1684	5.493	1.1684	2.79	1.227	1.970	0.000	0.827	1.138	0.0000
13	0.8	10.00	1.4740	8.067	1.4740	3.16	1.593	2.000	0.000	1.011	1.409	0.0000
14	1.0	25.00	4.6548	11.943	4.6548	3.63	2.049	1.941	0.000	1.241	1.769	0.0000
15	1.2	13.00	5.8005	20.658	5.8005	4.50	2.741	1.554	0.000	1.630	2.512	0.0000
16	1.4	10.00	4.4015	8.935	4.4015	3.27	1.704	1.996	0.000	1.067	1.494	0.0000
17	1.6	9.00	4.4644	9.769	4.4644	3.38	1.806	1.986	0.000	1.118	1.572	0.0000
18	1.8	8.00	4.3293	7.868	4.3293	3.14	1.567	2.000	0.000	0.998	1.390	0.0000
19	2.0	7.00	4.2943	7.297	4.2943	3.06	1.489	1.998	0.000	0.959	1.332	0.0000
20	2.2	6.00	4.2260	6.025	4.2260	2.88	1.307	1.982	0.000	0.868	1.198	0.0000
21	2.4	5.00	4.1869	5.159	4.1869	2.74	1.174	1.960	0.000	0.800	1.100	0.0000
22	2.6	30.00	5.9496	21.534	5.9496	4.58	2.787	1.503	0.000	1.660	2.587	0.0000
23	2.8	50.00	10.1340	38.747	10.1340	5.62	3.118	0.650	0.000	1.946	4.344	0.0000
24	3.0	50.00	13.0973	47.288	13.0973	5.84	3.134	0.443	0.000	1.975	5.510	0.0000
25	3.2	50.00	13.8420	49.202	13.8420	5.87	3.136	0.410	0.000	1.979	5.802	0.0000
26	3.4	50.00	14.0630	49.756	14.0630	5.88	3.136	0.401	0.000	1.980	5.888	0.0000
27	3.6	50.00	14.1307	49.925	14.1307	5.88	3.136	0.399	0.000	1.980	5.915	0.0000
28	3.8	0.00	7.7471	30.157	7.7471	5.23	3.050	1.004	0.000	1.865	3.383	0.0000
29	4.0	0.00	0.7806	2.672	0.7806	2.25	0.737	1.805	0.000	0.569	0.773	0.0000

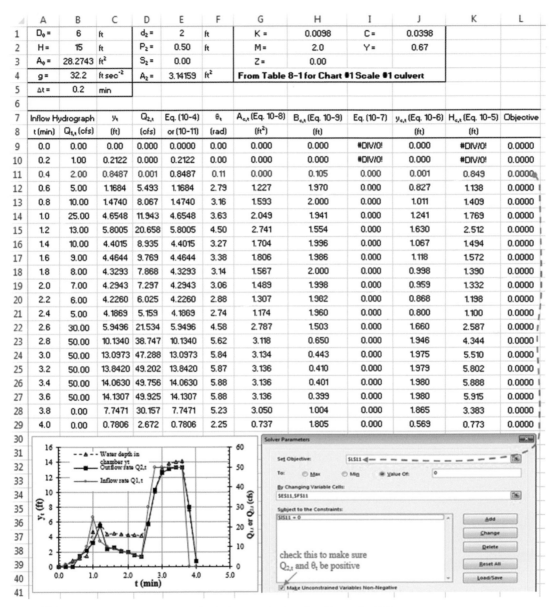

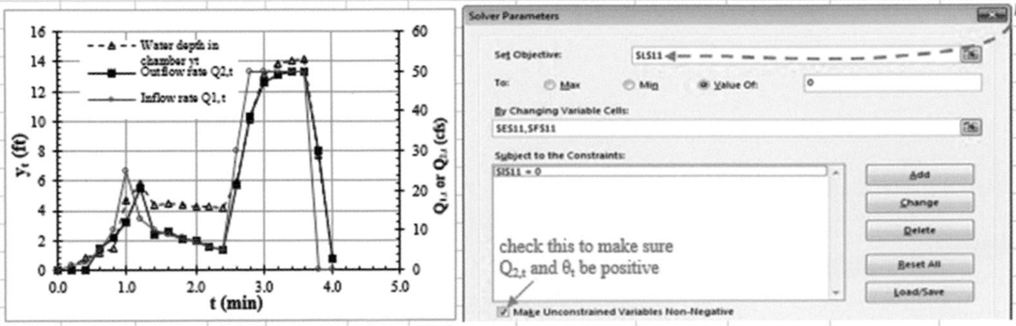

Figure 10.3 **Spreadsheet with data, calculations, and plots for Example 10.2.**

10.2 Transient Flow in a Pressurized Pipe or Open Channel

A transient flow will occur when flow is changed from one steady state to another. It is one type of unsteady-state flow and can be induced by situations such as opening or closing of valves, starting or stopping of pumps or turbines, and contracting or expanding channel cross sections. Transient flow causes changes of pressure and flow rate, propagating a surge wave back and forth in a hydraulic system until it is dissipated by friction (Roberson *et al.*, 1998). If the flow rate is suddenly reduced, the transient pressure could become large enough to result in damage to the hydraulic system and even the loss of lives and properties. In contrast, if the flow rate is suddenly increased, the transient pressure could be reduced to a value less than the saturation vapor pressure of water, likely resulting in cavitation issues and subsequent damage. For a pipeline system, control devices, such as surge tanks, air chambers, and/or pressure regulating valves, can be installed to keep the impacts of transient flow within prescribed limits (Roberson *et al.*, 1998). A surge tank is a standpipe connected to the pipeline. An associated air chamber has compressed air at the top that allows water flow into the chamber when the pressure in the pipeline increases and out of the chamber when the pressure in the pipeline decreases. A regulating valve can be used to control the pressure in the pipeline within a desired range.

Transient flow is governed by the Saint–Venant equations, which include a continuity equation (Eq. (10.12)) and a momentum equation (Eq. (10.13)). Their derivations are given in Examples 10.3 and 10.4. The equations can be expressed as:

density of water; flow area; flow velocity

$$\frac{\partial}{\partial t}(\rho A) + \frac{\partial}{\partial x}(\rho A V) = 0 \tag{10.12}$$

time; location along flow direction

force along flow direction; density of water; flow velocity

$$\lim_{\Delta x \to 0}\left(\sum \frac{F}{\Delta x}\right) = \frac{\partial}{\partial t}(\rho A V) + \frac{\partial}{\partial x}(\rho A V^2) \tag{10.13}$$

length along flow direction; time; flow area; location along flow direction

Example 10.3 Derive Eq. (10.12).

Solution

As illustrated in Figure 10.4, the left-hand section of the flow element moves at $V_{b,1}$, while the right-hand section moves at $V_{b,2}$. Using the relative velocities $V_1 - V_{b,1}$ and $V_2 - V_{b,2}$, one can analyze the element as a stationary control volume.

In terms of mass conservation: $\frac{d}{dt}\int_{x_1}^{x_2}(\rho A)\,dx = \rho_1 A_1(V_1 - V_{b,1}) - \rho_2 A_2(V_2 - V_{b,2})$.

Based on Leibnitz's rule: $\frac{d}{dt}\int_{x_1}^{x_2}(\rho A)\,dx = \int_{x_1}^{x_2}\frac{\partial}{\partial t}(\rho A)\,dx + \rho_2 A_2\frac{dx_2}{dt} - \rho_1 A_1\frac{dx_1}{dt}$.

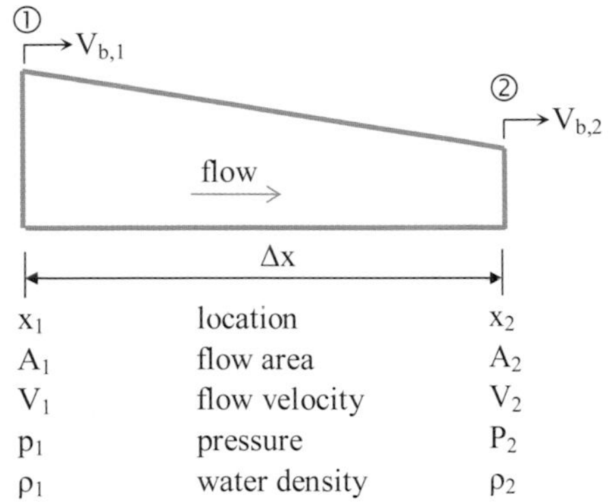

Figure 10.4 **Diagram used for Examples 10.3 and 10.4.** Illustration of mass conservation in a transient flow element at time t. $V_{b,1}$: velocity of section ①; and $V_{b,2}$: velocity of section ②.

Note: $\frac{dx_1}{dt} = V_{b,1}, \frac{dx_2}{dt} = V_{b,2} \rightarrow \frac{d}{dt}\int_{x_1}^{x_2}(\rho A)\,dx = \int_{x_1}^{x_2}\frac{\partial}{\partial t}(\rho A)\,dx + \rho_2 A_2 V_{b,2} - \rho_1 A_1 V_{b,1}.$

Substitute: $\int_{x_1}^{x_2}\frac{\partial}{\partial t}(\rho A)\,dx + \rho_2 A_2 V_{b,2} - \rho_1 A_1 V_{b,1} = \rho_1 A_1 (V_1 - V_{b,1}) - \rho_2 A_2 (V_2 - V_{b,2})$

$$\rightarrow \int_{x_1}^{x_2}\frac{\partial}{\partial t}(\rho A)\,dx + \rho_2 A_2 V_2 - \rho_1 A_1 V_1 = 0$$

$$\rightarrow \text{mean value theorem, } \frac{\partial}{\partial t}(\rho A)(x_2 - x_1) + \rho_2 A_2 V_2 - \rho_1 A_1 V_1 = 0$$

$$\rightarrow \frac{\partial}{\partial t}(\rho A)(\Delta x) + \rho_2 A_2 V_2 - \rho_1 A_1 V_1 = 0$$

$$\rightarrow \text{divide by } \Delta x, \ \frac{\partial}{\partial t}(\rho A) + \frac{\rho_2 A_2 V_2 - \rho_1 A_1 V_1}{\Delta x} = 0$$

$$\rightarrow \text{let } \Delta x \text{ approach zero, } \frac{\partial}{\partial t}(\rho A) + \frac{\partial}{\partial x}(\rho A V) = 0 \rightarrow \text{Eq. (10.12).}$$

Example 10.4 Derive Eq. (10.13).

Solution

As illustrated in Figure 10.4, the left-hand section of the flow element moves at $V_{b,1}$, while the right-hand section moves at $V_{b,2}$. Using the relative velocities $V_1 - V_{b,1}$ and $V_2 - V_{b,2}$, one can analyze the element as a stationary control volume.

In terms of momentum conservation:

$$\sum F = \frac{d}{dt}\int_{x_1}^{x_2} (\rho AV)\, dx + \rho_2 A_2 (V_2 - V_{b,2}) V_2 - \rho_1 A_1 (V_1 - V_{b,1}) V_1.$$

Based on Leibnitz's rule: $\frac{d}{dt}\int_{x_1}^{x_2} (\rho AV)\, dx = \int_{x_1}^{x_2} \frac{\partial}{\partial t}(\rho AV)\, dx + \rho_2 A_2 V_2 \frac{dx_2}{dt} - \rho_1 A_1 V_1 \frac{dx_1}{dt}.$

Note: $\frac{dx_1}{dt} = V_{b,1},\ \frac{dx_2}{dt} = V_{b,2}$

$$\rightarrow \sum F = \int_{x_1}^{x_2} \frac{\partial}{\partial t}(\rho AV)\, dx + \rho_2 A_2 V_2 V_{b,2} - \rho_1 A_1 V_1 V_{b,1} + \rho_2 A_2 (V_2 - V_{b,2}) V_2 - \rho_1 A_1 (V_1 - V_{b,1}) V_1$$

$$\rightarrow \sum F = \int_{x_1}^{x_2} \frac{\partial}{\partial t}(\rho AV)\, dx + \rho_2 A_2 V_2^2 - \rho_1 A_1 V_1^2$$

$\rightarrow$ mean value theorem, $\sum F = \frac{\partial}{\partial t}(\rho AV)(\Delta x) + \rho_2 A_2 V_2^2 - \rho_1 A_1 V_1^2$

$\rightarrow$ divide by Δx, $\sum \frac{F}{\Delta x} = \frac{\partial}{\partial t}(\rho AV) + \frac{\rho_2 A_2 V_2^2 - \rho_1 A_1 V_1^2}{\Delta x}$

$\rightarrow$ let Δx approach zero, $\lim_{\Delta x \rightarrow 0} \sum \frac{F}{\Delta x} = \frac{\partial}{\partial t}(\rho AV) + \frac{\partial}{\partial x}(\rho AV^2) \rightarrow$ Eq. (10.13).

10.2.1 Pressurized Circular Pipes

To analyze transient flow in pressurized pipes, generically, water compressibility and pipe deformation need to be considered. With an increase in pressure, water density will slightly decrease while the pipe wall thickness and/or diameter will slightly increase. In contrast, with a decrease of pressure, water density will slightly increase while the pipe wall thickness and/or diameter will slightly increase. These changes can be quantitively related as follows.

The change of water density can be computed as (Finnemore and Franzini, 2002):

$$d\rho = \frac{dp}{E_{m,w}}\rho \tag{10.14}$$

(change of water density: $d\rho$; water pressure change: dp; water density: ρ; bulk elastic modulus of water (Table 10.1): $E_{m,w}$)

The change of pipe wall thickness and/or diameter can be computed as (Timoshenko and Gere, 1961):

$$d\varepsilon = \frac{dp}{E_{m,p}}\frac{D}{2\varepsilon} \tag{10.15}$$

(change of pipe wall thickness and/or diameter: $d\varepsilon$; water pressure change: dp; pipe diameter: D; pipe wall thickness: ε; bulk elastic modulus of pipe (Table 10.1): $E_{m,p}$)

Table 10.1 Bulk elastic modulus for selected materials.

Material	Elastic modulus	
	(10^9 Pa)	(10^6 psi)
Water	2.15	0.312
Concrete	30 to 38	4.35 to 5.51
Steel	200	29
PVC	3.4	0.49

Due to the deformation, the change of pipe cross-sectional area is computed as:

$$d\Lambda = 2A\,(d\varepsilon) \tag{10.16}$$

change of pipe cross-sectional area; change of pipe wall thickness and/or diameter; pipe cross-sectional area

The pressure wave velocity (see Example 10.5 for derivation) can be computed as (Roberson *et al.*, 1998):

$$V_s = \pm\sqrt{\frac{E_{m,w}}{\rho}\,\frac{1}{1+\dfrac{E_{m,w}D}{E_{p,w}\varepsilon}}} \tag{10.17}$$

pressure wave velocity; bulk elastic modulus of water (Table 10.1); water density; pipe diameter; pipe wall thickness; bulk elastic modulus of pipe (Table 10.1)

Finally, given Eqs. (10.14) through (10.17), as shown in Examples 10.5 and 10.6, the Saint–Venant equations (Eqs. (10.12) and (10.13)) for pressurized circular pipes can be rewritten as:

$$\frac{\partial p}{\partial t} + V\frac{\partial p}{\partial x} + \rho V_s^2\frac{\partial V}{\partial x} = 0 \tag{10.18}$$

water pressure; flow velocity; pressure wave velocity (Eq. (10.17)); time; location along flow direction; water density

$$\frac{1}{\rho}\frac{\partial p}{\partial x} + \frac{\partial V}{\partial t} + V\frac{\partial V}{\partial x} + \frac{f}{2D}V\,|V| = 0 \tag{10.19}$$

water pressure; flow velocity; Darcy friction factor (see Figure 2.11); water density; length along flow direction; time; pipe diameter

Example 10.5 Derive Eqs. (10.17) and (10.18).

Solution

$$\text{Eq. (10.12)}: \frac{\partial}{\partial t}(\rho A) + \frac{\partial}{\partial x}(\rho A V) = A\frac{\partial \rho}{\partial t} + \rho\frac{\partial A}{\partial t} + \rho\frac{\partial}{\partial x}(AV) + AV\frac{\partial \rho}{\partial x}$$

$$= A\frac{\partial \rho}{\partial t} + \rho\frac{\partial A}{\partial t} + \rho A\frac{\partial V}{\partial x} + \rho V\frac{\partial A}{\partial x} + AV\frac{\partial \rho}{\partial x}$$

$$= A\left(\frac{\partial \rho}{\partial t} + V\frac{\partial \rho}{\partial x}\right) + \rho\left(\frac{\partial A}{\partial t} + V\frac{\partial A}{\partial x}\right) + \rho A\frac{\partial V}{\partial x}$$

$$= A\frac{d\rho}{dt} + \rho\frac{dA}{dt} + \rho A\frac{\partial V}{\partial x} = 0.$$

Eq. (10.14): $\frac{d\rho}{dt} = \frac{\rho}{E_{m,w}}\frac{dp}{dt}.$

Eq. (10.16) and (10.15): $\frac{dA}{dt} = 2A\frac{d\varepsilon}{dt} = 2A\frac{1}{E_{m,p}}\frac{D}{2\varepsilon}\frac{dp}{dt} = \frac{DA}{E_{m,p}\varepsilon}\frac{dp}{dt}.$

Substitute $d\rho/dt$ and dA/dt back into Eq. (10.12):

$$A\frac{d\rho}{dt} + \rho\frac{dA}{dt} + \rho A\frac{\partial V}{\partial x} = A\frac{\rho}{E_{m,w}}\frac{dp}{dt} + \rho\frac{DA}{E_{m,p}\varepsilon}\frac{dp}{dt} + \rho A\frac{\partial V}{\partial x} = \rho A\left(\frac{1}{E_{m,w}} + \frac{D}{E_{m,p}\varepsilon}\right)\frac{dp}{dt} + \rho A\frac{\partial V}{\partial x}$$

$$= \rho A\frac{1}{E_{m,w}}\left(1 + \frac{E_{m,w}D}{E_{m,p}\varepsilon}\right)\frac{dp}{dt} + \rho A\frac{\partial V}{\partial x} = A\frac{\rho}{E_{m,w}}\left(1 + \frac{E_{m,w}D}{E_{m,p}\varepsilon}\right)\frac{dp}{dt} + \rho A\frac{\partial V}{\partial x}$$

$$= A\left[\frac{E_{m,w}}{\rho}\left(1 + \frac{E_{m,w}D}{E_{m,p}\varepsilon}\right)^{-1}\right]^{-1}\frac{dp}{dt} + \rho A\frac{\partial V}{\partial x} = 0.$$

Define pressure wave velocity as: $V_s^2 = \frac{E_{m,w}}{\rho}\left(1 + \frac{E_{m,w}D}{E_{m,p}\varepsilon}\right)^{-1} \rightarrow$ Eq. (10.17).

Divide by A and multiply by V_s^2 on both sides:

$$\frac{dp}{dt} + \rho V_s^2\frac{\partial V}{\partial x} = \frac{\partial p}{\partial t} + V\frac{\partial p}{\partial x} + \rho V_s^2\frac{\partial V}{\partial x} = 0 \rightarrow \text{Eq. (10.18)}.$$

Example 10.6 Derive Eq. (10.19).

Solution

For a flow element of a pressurized circular pipe, the resultant force is:

$$\sum F = p_1 A - p_2 A - \tau_0(\pi D)(\Delta x) = -(p_2 - p_1)A - \tau_0(\pi D)(\Delta x).$$

In terms of the Darcy–Weisbach equation (Eq. (2.23)) and using V |V| to replace V^2 to allow back-and-forth flows, the friction force on the pipe wall is:

$$\tau_0(\pi D)(\Delta x) = \gamma A h_f = \gamma A\left(f\frac{\Delta x}{D}\frac{V|V|}{2g}\right) = \rho A(\Delta x)\frac{f}{2D}V|V|.$$

Substitute this relationship back into the previous equation, divide by Δx on both sides, and then let Δx approach zero:

$$\lim_{\Delta x\to 0}\sum\frac{F}{\Delta x} = -\lim_{\Delta x\to 0}\left(\frac{p_2-p_1}{\Delta x}\right)A - \rho A\frac{f}{2D}V|V| = -A\frac{\partial p}{\partial x} - \rho A\frac{f}{2D}V|V|.$$

The right side of Eq. (10.13):

$$\frac{\partial}{\partial t}(\rho AV) + \frac{\partial}{\partial x}(\rho AV^2) = \frac{\partial}{\partial t}(\rho AV) + \frac{\partial}{\partial x}(\rho AVV) = \frac{\partial}{\partial t}(\rho AV) + \rho AV\frac{\partial V}{\partial x} + V\frac{\partial}{\partial x}(\rho AV)$$

$$= \rho A\frac{\partial V}{\partial t} + V\frac{\partial}{\partial t}(\rho A) + \rho AV\frac{\partial V}{\partial x} + V\frac{\partial}{\partial x}(\rho AV) = \rho A\frac{\partial V}{\partial t} + \rho AV\frac{\partial V}{\partial x}$$

$$+ V\left[\frac{\partial}{\partial t}(\rho A) + \frac{\partial}{\partial x}(\rho AV)\right].$$

Eq. (10.12) $\rightarrow \frac{\partial}{\partial t}(\rho A) + \frac{\partial}{\partial x}(\rho AV) = 0$

$$\rightarrow \text{the right side of Eq. (10.13)} = \rho A\frac{\partial V}{\partial t} + \rho AV\frac{\partial V}{\partial x}.$$

Thus: $-A\frac{\partial p}{\partial x} - \rho A\frac{f}{2D}V|V| = \rho A\frac{\partial V}{\partial t} + \rho AV\frac{\partial V}{\partial x}$

$\rightarrow$ divide by ρA on both sides and rearrange: $\frac{1}{\rho}\frac{\partial p}{\partial x} + \frac{\partial V}{\partial t} + V\frac{\partial V}{\partial x} + \frac{f}{2D}V|V| = 0$

$\rightarrow$ Eq. (10.19).

For most pressured pipe flow conditions, $V\frac{\partial p}{\partial x}$ and $V\frac{\partial V}{\partial x}$ in Eqs. (10.18) and (10.19) are very small relative to the other terms. Neglecting these two terms and then using the method of characteristics (MOC), as shown in Example 10.7, one can derive the solution of Eqs. (10.18) and (10.19), which is the intersection between the two characteristic lines in the x–t plane (Figure 10.5). The first line has a positive slope of $|V_s|$ and is called the positive characteristic line, whereas the second line has a negative slope of $-|V_s|$ and is called the negative characteristic line.

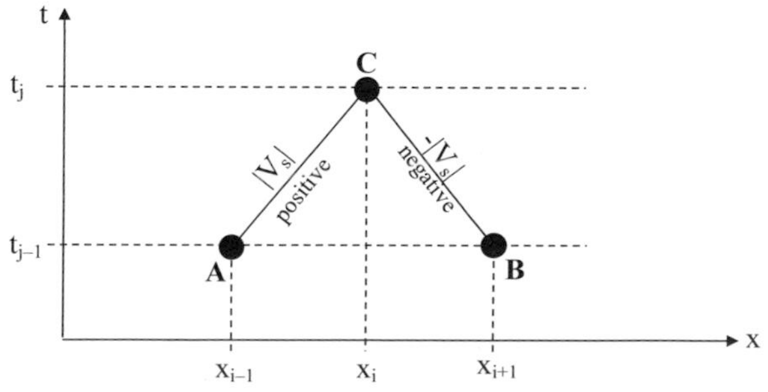

Figure 10.5 **Method of characteristics (MOC).** The MOC solution to Eqs. (10.18) and (10.19) when the two relatively small terms of $V\frac{\partial p}{\partial x}$ and $V\frac{\partial V}{\partial x}$ are neglected.

The positive characteristic line is defined as:

location along flow direction

pressure wave velocity (Eq. (10.17))

$$\frac{dx}{dt} = |V_s| \tag{10.20}$$

time

water pressure; flow velocity; Darcy friction factor (see Figure 2.11)

$$\frac{1}{\rho |V_s|}\frac{dp}{dt} + \frac{dV}{dt} + \frac{f}{2D}V|V| = 0 \tag{10.21}$$

water density; time; pipe diameter

pressure wave velocity (Eq. (10.17))

Similarly, the negative characteristic line is defined as:

location along flow direction

pressure wave velocity (Eq. (10.17))

$$\frac{dx}{dt} = -|V_s| \tag{10.22}$$

time

water pressure; flow velocity; Darcy friction factor (see Figure 2.11)

$$-\frac{1}{\rho |V_s|}\frac{dp}{dt} + \frac{dV}{dt} + \frac{f}{2D}V|V| = 0 \tag{10.23}$$

water density; time; pipe diameter

pressure wave velocity (Eq. (10.17))

In Figure 10.5, point A signifies the pressure or velocity at location x_{i-1} and time t_{j-1}, while point B signifies the pressure or velocity at location x_{i+1} and time t_{j-1}, where $i = 1, 2, \ldots$, and $j = 1, 2, \ldots$. If the pressures and velocities at points A and B are known, one can solve Eqs. (10.21) and (10.23) simultaneously to obtain the pressure and velocity at location x_i and time t_j as signified by point C. For a pipeline system of interest, the initial (i.e., time t_0) values of pressure and velocity at all locations should be known, so the corresponding values at other times can be determined. As shown in Example 10.8, by integrating Eq. (10.21) along the positive characteristic line AC and Eq. (10.23) along the negative characteristic line BC and rearranging, one can derive the solution as:

water pressure at point C; water pressure at point A and B; water density; pressure wave velocity (Eq. (10.17)); Darcy friction factor (see Figure 2.11)

$$p_C = \frac{p_A + p_B}{2} + \frac{\rho |V_s|}{2}(V_A - V_B) - \frac{\rho f |V_s|}{4D}(V_A|V_A| - V_B|V_B|)(t_j - t_{j-1}) \tag{10.24}$$

flow velocity at point A and B; pipe diameter; start and end time flow velocity at point C; flow velocity at point A and B; water pressure at point A and B; Darcy friction factor (see Figure 2.11)

$$V_C = \frac{V_A + V_B}{2} + \frac{1}{2\rho\,|V_s|}(p_A - p_B) - \frac{f}{4D}(V_A\,|V_A| + V_B\,|V_B|)\,(t_j - t_{j-1}) \tag{10.25}$$

water density; pressure wave velocity (Eq. (10.17)); pipe diameter; start and end time

To apply Eqs. (10.24) and (10.25), the pipeline is subdivided into a number of segments, each of which has two nodes Δx apart. For convergence, the computational time step should be $\Delta t \leq \Delta x/|V_s|$. The values of pressure and velocity at the first and last nodes are determined by the upper and lower boundary conditions, respectively.

Example 10.7 Derive Eqs. (10.20) through (10.23).

Solution

Neglecting $V\frac{\partial p}{\partial x}$ and $V\frac{\partial V}{\partial x}$, one has:

$$\text{Eq. (10.18)}\rightarrow\frac{\partial p}{\partial t} + \rho V_s^2\frac{\partial V}{\partial x} = 0$$

$$\text{Eq. (10.19)}\rightarrow\frac{1}{\rho}\frac{\partial p}{\partial x} + \frac{\partial V}{\partial t} + \frac{f}{2D}V\,|V| = 0.$$

Multiply the first equation by λ and then add it and the second equation together:

$$\frac{1}{\rho}\frac{\partial p}{\partial x} + \frac{\partial V}{\partial t} + \frac{f}{2D}V\,|V| + \lambda\left(\frac{\partial p}{\partial t} + \rho V_s^2\frac{\partial V}{\partial x}\right) = 0$$

$$\rightarrow\lambda\left(\frac{\partial p}{\partial t} + \frac{1}{\lambda\rho}\frac{\partial p}{\partial x}\right) + \left(\frac{\partial V}{\partial t} + \lambda\rho V_s^2\frac{\partial V}{\partial x}\right) + \frac{f}{2D}V\,|V| = 0.$$

Select λ to satisfy the relationship: $\frac{dx}{dt} = \frac{1}{\lambda\rho} = \lambda\rho V_s^2 \rightarrow \lambda = \pm\frac{1}{\rho\,|V_s|}$

$$\rightarrow\lambda\frac{dp}{dt} + \frac{dV}{dt} + \frac{f}{2D}V\,|V| = 0.$$

$$\lambda = \frac{1}{\rho\,|V_s|}\rightarrow\frac{1}{\rho\,|V_s|}\frac{dp}{dt} + \frac{dV}{dt} + \frac{f}{2D}V\,|V| = 0\rightarrow\text{Eqs. (10.20) and (10.21)}$$

$$\lambda = -\frac{1}{\rho\,|V_s|}\rightarrow -\frac{1}{\rho\,|V_s|}\frac{dp}{dt} + \frac{dV}{dt} + \frac{f}{2D}V\,|V| = 0\rightarrow\text{Eqs. (10.22) and (10.23).}$$

Example 10.8 Derive Eqs. (10.24) and (10.25).

Solution

Eq. (10.21) $\rightarrow dp + \rho|V_s|\,dV + \frac{\rho f|V_s|}{2D}V|V|\,dt = 0$

$$\rightarrow \int_A^C dp + \rho|V_s|\int_A^C dV + \frac{\rho f|V_s|}{2D}\int_A^C V|V|\,dt = 0$$

$$\rightarrow p_C - p_A + \rho|V_s|(V_C - V_A) + \frac{\rho f|V_s|}{2D}V_A|V_A|\left(t_j - t_{j-1}\right) = 0.$$

Eq. (10.23) $\rightarrow dp - \rho|V_s|\,dV - \frac{\rho f|V_s|}{2D}V|V|\,dt = 0$

$$\rightarrow \int_B^C dp - \rho|V_s|\int_B^C dV - \frac{\rho f|V_s|}{2D}\int_B^C V|V|\,dt = 0$$

$$\rightarrow p_C - p_B - \rho|V_s|(V_C - V_B) - \frac{\rho f|V_s|}{2D}V_B|V_B|\left(t_j - t_{j-1}\right) = 0.$$

Add $\rightarrow 2p_C - p_A - p_B - \rho|V_s|(V_A - V_B) + \frac{\rho f|V_s|}{2D}(V_A|V_A| - V_B|V_B|)\left(t_j - t_{j-1}\right) = 0$

$$\rightarrow 2p_C = p_A + p_B + \rho|V_s|(V_A - V_B) - \frac{\rho f|V_s|}{2D}(V_A|V_A| - V_B|V_B|)\left(t_j - t_{j-1}\right)$$

$\rightarrow$divide by 2 on both sides to get Eq. (10.24).

Subtract $\rightarrow -p_A + p_B + 2\rho|V_s|V_C - \rho|V_s|(V_A + V_B) + \frac{\rho f|V_s|}{2D}(V_A|V_A| + V_B|V_B|)\left(t_j - t_{j-1}\right) = 0$

$$\rightarrow 2\rho|V_s|V_C = \rho|V_s|(V_A + V_B) + (p_A - p_B) - \frac{\rho f|V_s|}{2D}(V_A|V_A| + V_B|V_B|)\left(t_j - t_{j-1}\right)$$

$\rightarrow$divide by $2\rho|Vs|$ on both sides to get Eq. (10.25).

Example 10.9 In Figure 10.6, the exit of the concrete circular pipe is controlled by a closed valve. The entrance loss coefficient is 0.5. If the valve is suddenly opened, determine the pressure and velocity profiles using $\Delta x = 500$ ft.

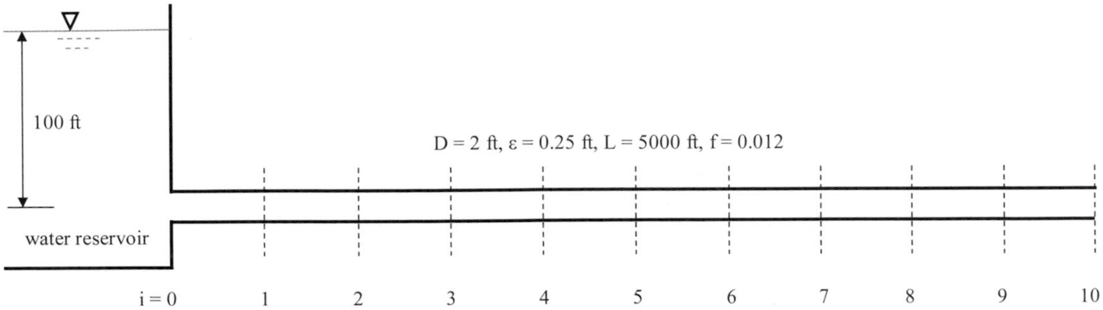

Figure 10.6 **The pipeline system of Example 10.9.**

Solution

$H_0 = 100\,\text{ft}$, $\rho = 1.94\,\text{slug ft}^{-3}$, $g = 32.2\,\text{ft sec}^{-2}$

Table 10.1: $E_{m,w} = 0.312 \times 10^6\,\text{psi} = 44.928 \times 10^6\,\text{lbf ft}^{-2}$

$$E_{m,p} = 5.0 \times 10^6\,\text{psi} = 720 \times 10^6\,\text{lbf ft}^{-2}.$$

$\Delta x = 500\,\text{ft} \rightarrow$ number of segments $= (5000\,\text{ft}) / (500\,\text{ft}) = 10$, number of nodes $= 11$.

$$\text{Eq. (10.17): } |V_s| = \sqrt{\frac{44.928 \times 10^6\,\text{lbf ft}^{-2}}{1.94\,\text{slug ft}^{-3}} * \frac{1}{1 + \frac{\left(44.928 \times 10^6\,\text{lbf ft}^{-2}\right) * (2\,\text{ft})}{\left(720 \times 10^6\,\text{lbf ft}^{-2}\right) * (0.25\,\text{ft})}}} = 3930.32\,\text{ft sec}^{-1}.$$

Computational time step: $\Delta t = \Delta x / |V_s| = (500\,\text{ft}) / \left(3930.32\,\text{ft sec}^{-1}\right) = 0.1272\,\text{sec}.$

Initial condition: cells D9:N10 in Figure 10.7, highlighted in red.

$$p_{i,0} = \rho g H_0 = \left(1.94\,\text{slug ft}^{-3}\right) * \left(32.2\,\text{ft sec}^{-2}\right) * (100\,\text{ft}) = 6246.8\,\text{lbf ft}^{-2}\ (i = 0, 1, 2, \ldots, 10).$$

At time $j = 0$, water at all nodes is not flowing $\rightarrow V_{i,0} = 0.00\,\text{ft sec}^{-1}$ $(i = 0, 1, 2, \ldots, 10)$.

Upper boundary condition: cells D11:D34 in Figure 10.7, highlighted in blue.

At time $j \geq 1$:

$$p_{0,j} = 6246.8\,\text{lbf ft}^{-2}\ (j = 1, 2, \ldots)$$

Applying Eq. (10.25) to the reservoir and Node 1, one has:

$$\begin{aligned} V_{0,j} &= \frac{0 + V_{1,j-1}}{2} + \frac{1}{2\rho |V_s|}\left(\rho g H_0 - p_{1,j-1}\right) - \frac{f}{4D}\left(0\,|0| + V_{1,j-1}\left|V_{1,j-1}\right|\right)(\Delta t) \\ &= \frac{V_{1,j-1}}{2} + \frac{1}{2 * \left(1.94\,\text{slug ft}^{-3}\right) * \left(3930.32\,\text{ft sec}^{-1}\right)} * \left(6246.8\,\text{lbf ft}^{-2} - p_{1,j-1}\right) \\ &\quad - \frac{0.012}{4 * (2\,\text{ft})} V_{1,j-1}\left|V_{1,j-1}\right| * (0.1272\,\text{sec}) \\ &= \frac{V_{1,j-1}}{2} + \left(6.55753 \times 10^{-5}\right) * \left(6246.8 - p_{1,j-1}\right) - \left(1.908 \times 10^{-4}\right) V_{1,j-1}\left|V_{1,j-1}\right|. \end{aligned}$$

Lower boundary condition: cells N11:N34 in Figure 10.7, highlighted in blue.

At time $j \geq 1$:

Node 10 is exposed to atmosphere $\rightarrow p_{10,j} = 0.00\,\text{lbf ft}^{-2}$.

Applying Eq. (10.25) to Node 9 and the ambient atmosphere, one has:

$$V_{10,j} = \frac{V_{9,j-1} + 0}{2} + \frac{1}{2\rho |V_s|}\left(p_{9,j-1} - 0\right) - \frac{f}{4D}\left(V_{9,j-1}\left|V_{9,j-1}\right| + 0\,|0|\right)(\Delta t)$$

$$= \frac{V_{9,j-1}}{2} + \frac{1}{2*\left(1.94\,\text{slug ft}^{-3}\right)*\left(3930.32\,\text{ft sec}^{-1}\right)}*\left(p_{9,j-1}\right)$$

$$- \frac{0.012}{4*(2\,\text{ft})}V_{9,j-1}\left|V_{9,j-1}\right|*(0.1272\,\text{sec})$$

$$= \frac{V_{9,j-1}}{2} + \left(6.55753\times10^{-5}\right)p_{9,j-1} - \left(1.908\times10^{-4}\right)V_{9,j-1}\left|V_{9,j-1}\right|.$$

Interior nodes ($i = 1, 2, \ldots, 9$) at time $j \geq 1$: use Eqs. (10.24) and (10.25).

The calculations can be done in an Excel spreadsheet, as shown in Figure 10.7. The profiles of pressure and velocity at the first 12 time steps are shown in Figure 10.8a,b, respectively, and the variations of pressure and velocity at Nodes 5 and 9 are plotted in Figure 10.8c.

	A	B	C	D	E	F	G	H	I	J	K	L	M	N
1	H_0 =	100	ft	D =	2	ft	Table 10-1:	$E_{m,w}$ =	4.5E+07	lbf ft^{-2}				
2	ρ =	1.94	slug ft^{-3}	ε =	0.25	ft		$E_{m,p}$ =	7.2E+08	lbf ft^{-3}				
3	g =	32.2	ft sec^{-2}	L =	5000	ft								
4	$p_{0,j}$ =	6246.8	lbf ft^{-2}	f =	0.012		Eq. (10-17):	$\lvert V_s \rvert$ =	3930.32	ft sec^{-1}				
5														
6	Δx =	500	ft	Δt =	0.1272	sec								
7									Node i =					
8	j	t_j (sec)	Variables	0	1	2	3	4	5	6	7	8	9	10
9	0	0	$p_{i,0}$ (lbf ft^{-2})	6246.80	6246.80	6246.80	6246.80	6246.80	6246.80	6246.80	6246.80	6246.80	6246.80	6246.80
10			$V_{i,0}$ (ft sec^{-1})	0.00	0.00	0.00	0.00	0.00	0.00	0.00	0.00	0.00	0.00	0.00
11	1	0.1272	$p_{i,1}$ (lbf ft^{-2})	6246.80	6246.80	6246.80	6246.80	6246.80	6246.80	6246.80	6246.80	6246.80	6246.80	0.00
12			$V_{i,1}$ (ft sec^{-1})	0.00	0.00	0.00	0.00	0.00	0.00	0.00	0.00	0.00	0.00	0.41
13	2	0.2544	$p_{i,2}$ (lbf ft^{-2})	6246.80	6246.80	6246.80	6246.80	6246.80	6246.80	6246.80	6246.80	6246.80	1561.94	0.00
14			$V_{i,2}$ (ft sec^{-1})	0.00	0.00	0.00	0.00	0.00	0.00	0.00	0.00	0.00	0.61	0.41
15	3	0.3816	$p_{i,3}$ (lbf ft^{-2})	6246.80	6246.80	6246.80	6246.80	6246.80	6246.80	6246.80	6246.80	1562.49	1561.94	0.00
16			$V_{i,3}$ (ft sec^{-1})	0.00	0.00	0.00	0.00	0.00	0.00	0.00	0.00	0.61	0.61	0.41
17	4	0.5088	$p_{i,4}$ (lbf ft^{-2})	6246.80	6246.80	6246.80	6246.80	6246.80	6246.80	6246.80	1563.04	1562.49	1561.40	0.00
18			$V_{i,4}$ (ft sec^{-1})	0.00	0.00	0.00	0.00	0.00	0.00	0.00	0.61	0.61	0.61	0.41
19	5	0.6360	$p_{i,5}$ (lbf ft^{-2})	6246.80	6246.80	6246.80	6246.80	6246.80	6246.80	1563.59	1563.04	1561.94	1561.67	0.00
20			$V_{i,5}$ (ft sec^{-1})	0.00	0.00	0.00	0.00	0.00	0.00	0.61	0.61	0.61	0.61	0.41
21	6	0.7632	$p_{i,6}$ (lbf ft^{-2})	6246.80	6246.80	6246.80	6246.80	6246.80	1564.14	1563.59	1562.49	1562.22	1561.12	0.00
22			$V_{i,6}$ (ft sec^{-1})	0.00	0.00	0.00	0.00	0.00	0.61	0.61	0.61	0.61	0.61	0.41
23	7	0.8904	$p_{i,7}$ (lbf ft^{-2})	6246.80	6246.80	6246.80	6246.80	1564.69	1564.14	1563.04	1562.77	1561.67	1561.40	0.00
24			$V_{i,7}$ (ft sec^{-1})	0.00	0.00	0.00	0.00	0.61	0.61	0.61	0.61	0.61	0.61	0.41
25	8	1.0176	$p_{i,8}$ (lbf ft^{-2})	6246.80	6246.80	6246.80	1565.24	1564.69	1563.59	1563.32	1562.22	1561.94	1560.85	0.00
26			$V_{i,8}$ (ft sec^{-1})	0.00	0.00	0.00	0.61	0.61	0.61	0.61	0.61	0.61	0.61	0.41
27	9	1.1448	$p_{i,9}$ (lbf ft^{-2})	6246.80	6246.80	1565.79	1565.24	1564.14	1563.87	1562.77	1562.49	1561.40	1561.12	0.00
28			$V_{i,9}$ (ft sec^{-1})	0.00	0.00	0.61	0.61	0.61	0.61	0.61	0.61	0.61	0.61	0.41
29	10	1.2720	$p_{i,10}$ (lbf ft^{-2})	6246.80	1566.33	1565.79	1564.69	1564.41	1563.32	1563.04	1561.94	1561.67	1560.57	0.00
30			$V_{i,10}$ (ft sec^{-1})	0.00	0.61	0.61	0.61	0.61	0.61	0.61	0.61	0.61	0.61	0.41
31	11	1.3992	$p_{i,11}$ (lbf ft^{-2})	6246.80	1566.33	1565.24	1564.96	1563.87	1563.59	1562.49	1562.22	1561.12	1560.85	0.00
32			$V_{i,11}$ (ft sec^{-1})	0.61	0.61	0.61	0.61	0.61	0.61	0.61	0.61	0.61	0.61	0.41
33	12	1.5264	$p_{i,12}$ (lbf ft^{-2})	6246.80	3905.74	1565.51	1564.41	1564.14	1563.04	1562.77	1561.67	1561.40	1560.30	0.00
34			$V_{i,12}$ (ft sec^{-1})	0.61	0.92	0.61	0.61	0.61	0.61	0.61	0.61	0.61	0.61	0.41

Figure 10.7 **Excerpt of the spreadsheet for the calculations for Example 10.9.**

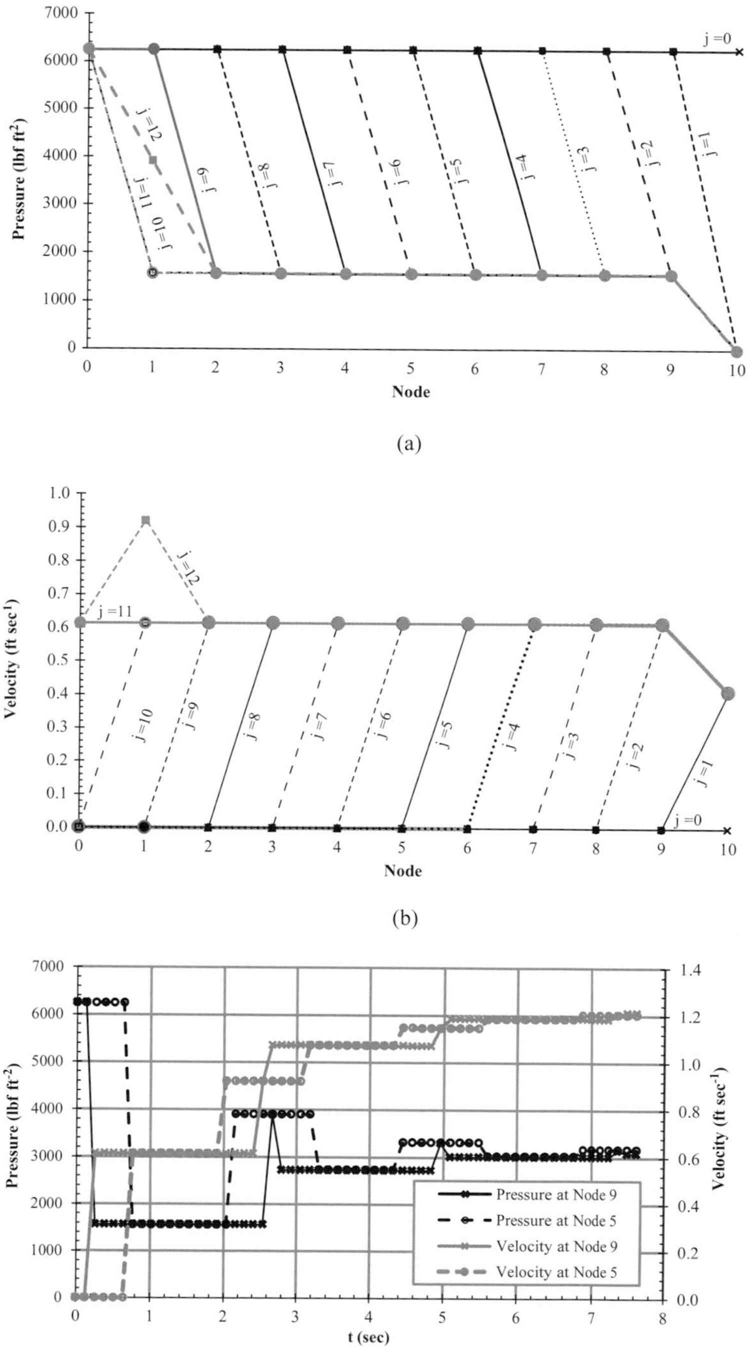

(c)

Figure 10.8 **Computed pressures and velocities for Example 10.9.** (a) Pressure profiles; (b) velocity profiles; and (c) fluctuations of pressure and velocity at Nodes 5 and 9.

At all nodes, the initial pressure and velocity are 6246.8 $lbf\,ft^{-2}$ and zero, respectively. Once the valve is suddenly opened, the pressure at Node 10 will drop to zero because it is exposed to atmosphere, while the velocity at the node is calculated to be 0.41 $ft\,sec^{-1}$. With the lapse of time until the 11th time step ($j = 11$), as expected, the pressures at the interior nodes decrease, but the velocities increase, from downstream to upstream. Starting from the 12th time step ($j = 12$), the pressures start to increase while the velocities continue to increase, from upstream to downstream. Such downstream–upstream–downstream fluctuation processes will repeat every few time steps, ultimately approaching a new steady-state flow condition, as indicated by the velocities at all nodes tending to become very similar (Figure 10.8c).

For instance, at time $j = 1$, the upper boundary condition (cells D11:D12) is $p_{0,1} = 6246.8\,lbf\,ft^{-2}$ and $V_{0,1} = \left(0.00\,ft\,sec^{-1}\right)/2 + \left(6.55753 \times 10^{-5}\right) * \left(6246.8\,lbf\,ft^{-2} - 6246.8\,lbf\,ft^{-2}\right) - \left(1.908 \times 10^{-4}\right) * \left(0.00\,ft\,sec^{-1}\right) * \left|0.00\,ft\,sec^{-1}\right| = 0.00\,ft\,sec^{-1}$, whereas the lower boundary condition (cells N11:N12) is $p_{10,1} = 0.00\,lbf\,ft^{-2}$ and $V_{10,1} = \left(0.00\,ft\,sec^{-1}\right)/2 + \left(6.55753 \times 10^{-5}\right) * \left(6246.8\,lbf\,ft^{-2}\right) - \left(1.908 \times 10^{-4}\right) * \left(0.00\,ft\,sec^{-1}\right) * \left|0.00\,ft\,sec^{-1}\right| = 0.41\,ft\,sec^{-1}$. At node $i = 9$ (cells M11:M12), $p_{9,1} = (6246.80\,lbf\,ft^{-2} + 6246.80\,lbf\,ft^{-2})/2 + \left[(1.94\,slug\,ft^{-3}) * \left(3930.32\,ft\,sec^{-1}\right)\right]/2 * \left(0.00\,ft\,sec^{-1} - 0.00\,ft\,sec^{-1}\right) - \left[\left(1.94\,slug\,ft^{-3}\right) * 0.012 * \left(3930.32\,ft\,sec^{-1}\right)\right]/(4 * 2\,ft) * \left[\left(0.00\,ft\,sec^{-1}\right)\left|0.00\,ft\,sec^{-1}\right| - \left(0.00\,ft\,sec^{-1}\right)\left|0.00\,ftsec^{-1}\right|\right] * (0.1272\,sec) = 6246.80\,lbf\,ft^{-2}$; and $V_{9,1} = \left(0.00\,ft\,sec^{-1} + 0.00\,ft\,sec^{-1}\right)/2 + [1/\left[2 * \left(1.94\,slug\,ft^{-3}\right) * \left(3930.32\,ft\,sec^{-1}\right)\right] * \left(6246.80\,lbf\,ft^{-2} - 6246.80\,lbf\,ft^{-2}\right) - 0.012/(4 * 2\,ft) * \left[\left(0.00\,ft\,sec^{-1}\right)\left|0.00\,ft\,sec^{-1}\right| + \left(0.00\,ft\,sec^{-1}\right)\left|0.00\,ft\,sec^{-1}\right|\right] * (0.1272\,sec) = 0.00\,ft\,sec^{-1}$. Similar calculations are done for nodes $i = 8, 7, \ldots, 1$. At times $j = 2, 3, \ldots$, determine the upper and lower boundary conditions and do the calculations for the interior nodes in similar ways.

Example 10.10 In Figure 10.6, the exit of the concrete circular pipe is controlled by a valve. The entrance loss coefficient is 0.5. The valve is fully open and a steady-state flow through the pipe has been established. If the valve is suddenly closed, determine the pressure and velocity profiles using $\Delta x = 500$ ft.

Solution

$H_0 = 100\,ft$, $k_e = 0.5$, $\rho = 1.94\,slug\,ft^{-3}$, $g = 32.2\,ft\,sec^{-2}$

Table 10.1: $E_{m,w} = 0.312 \times 10^6\,psi = 44.928 \times 10^6\,lbf\,ft^{-2}$

$$E_{m,p} = 5.0 \times 10^6\,psi = 720 \times 10^6\,lbf\,ft^{-2}.$$

$\Delta x = 500\,ft \rightarrow$ number of segments $= (5000\,ft)/(500\,ft) = 10$, number of nodes $= 11$.

Eq. (10.17): $|V_s| = \sqrt{\dfrac{44.928\times 10^6\ \mathrm{lbf\,ft^{-2}}}{1.94\ \mathrm{slug\,ft^{-3}}} * \dfrac{1}{1+\dfrac{\left(44.928\times 10^6\ \mathrm{lbf\,ft^{-2}}\right)*(2\ \mathrm{ft})}{\left(720\times 10^6\ \mathrm{lbf\,ft^{-2}}\right)*(0.25\ \mathrm{ft})}}} =$

$3930.32\ \mathrm{ft\,sec^{-1}}$.

Computational time step: $\Delta t = \Delta x/|V_s| = (500\ \mathrm{ft})/\left(3930.32\ \mathrm{ft\,sec^{-1}}\right) = 0.1272\ \mathrm{sec}$.

Steady-state flow:

Energy equation (Eq. (2.21)) between the reservoir surface and the pipeline exit:

$$100\ \mathrm{ft} = V_0^2/\left(2*32.2\ \mathrm{ft\,sec^{-2}}\right) + [0.012*(5000\ \mathrm{ft})/(2\ \mathrm{ft})+0.5]\,V_0^2/\left(2*32.2\ \mathrm{ft\,sec^{-2}}\right)$$
$$\rightarrow 100 = (31.5/64.4)\,V_0^2 \rightarrow V_0 = 14.30\ \mathrm{ft\,sec^{-1}}.$$

Initial condition: cells D9:N10 in Figure 10.9, highlighted in red.

At Node 0 $\rightarrow p_{0,0} = \rho g H_0 = \left(1.94\ \mathrm{slug\,ft^{-3}}\right)*\left(32.2\ \mathrm{ft\,sec^{-2}}\right)*(100\ \mathrm{ft}) = 6246.8\ \mathrm{lbf\,ft^{-2}}$.

At Node 1 $\rightarrow p_{1,0} = p_{0,0} - (\rho g)\,[1+f(\Delta x)/D+k_e]\left[V_0^2/(2g)\right]$

$$= 6246.8\ \mathrm{lbf\,ft^{-2}} - \left(1.94\ \mathrm{slug/ft^{-3}}\right)*\left(32.2\ \mathrm{ft\,sec^{-2}}\right)*[1+0.012*(500\ \mathrm{ft})/(2\ \mathrm{ft})+0.5]$$
$$*\left[\left(14.30\ \mathrm{ft\,sec^{-1}}\right)^2/\left(2*32.2\ \mathrm{ft\,sec^{-2}}\right)\right] = 5354.40\ \mathrm{lbf\,ft^{-2}}.$$

At Node 2 $\rightarrow p_{2,0} = p_{1,0} - (\rho g)\,[f(\Delta x)/D]\left[V_0^2/(2g)\right]$

$$= 5354.40\ \mathrm{lbf\,ft^{-2}} - \left(1.94\ \mathrm{slug/ft^{-3}}\right)*\left(32.2\ \mathrm{ft\,sec^{-2}}\right)*[0.012*(500\ \mathrm{ft})/(2\ \mathrm{ft})]$$
$$*\left[\left(14.30\ \mathrm{ft\,sec^{-1}}\right)^2/\left(2*32.2\ \mathrm{ft\,sec^{-2}}\right)\right] = 4759.47\ \mathrm{lbf\,ft^{-2}}.$$

At Node 3 $\rightarrow p_{3,0} = p_{2,0} - (\rho g)\,[f(\Delta x)/D]\left[V_0^2/(2g)\right]$

$$= 4759.47\ \mathrm{lbf\,ft^{-2}} - \left(1.94\ \mathrm{slug/ft^{-3}}\right)*\left(32.2\ \mathrm{ft\,sec^{-2}}\right)*[0.012*(500\ \mathrm{ft})/(2\ \mathrm{ft})]$$
$$*\left[\left(14.30\ \mathrm{ft\,sec^{-1}}\right)^2/\left(2*32.2\ \mathrm{ft\,sec^{-2}}\right)\right] = 4164.53\ \mathrm{lbf\,ft^{-2}}.$$

At Node 4 $\rightarrow p_{4,0} = p_{3,0} - (\rho g)\,[f(\Delta x)/D]\left[V_0^2/(2g)\right]$

$$= 4164.53\ \mathrm{lbf\,ft^{-2}} - \left(1.94\ \mathrm{slug/ft^{-3}}\right)*\left(32.2\ \mathrm{ft\,sec^{-2}}\right)*[0.012*(500\ \mathrm{ft})/(2\ \mathrm{ft})]$$
$$*\left[\left(14.30\ \mathrm{ft\,sec^{-1}}\right)^2/\left(2*32.2\ \mathrm{ft\,sec^{-2}}\right)\right] = 3569.60\ \mathrm{lbf\,ft^{-2}}.$$

At Node 5 $\rightarrow p_{5,0} = p_{4,0} - (\rho g)\,[f(\Delta x)/D]\left[V_0^2/(2g)\right]$

$$= 3569.60\ \mathrm{lbf\,ft^{-2}} - \left(1.94\ \mathrm{slug/ft^{-3}}\right)*\left(32.2\ \mathrm{ft\,sec^{-2}}\right)*[0.012*(500\ \mathrm{ft})/(2\ \mathrm{ft})]$$
$$*\left[\left(14.30\ \mathrm{ft\,sec^{-1}}\right)^2/\left(2*32.2\ \mathrm{ft\,sec^{-2}}\right)\right] = 2974.67\ \mathrm{lbf\,ft^{-2}}.$$

At Node 6$\rightarrow p_{6,0} = p_{5,0} - (\rho g)\,[f\,(\Delta x)\,/D]\,\left[V_0{}^2/\,(2g)\right]$

$$= 2974.67\,\text{lbf ft}^{-2} - \left(1.94\,\text{slug/ft}^{-3}\right) * \left(32.2\,\text{ft sec}^{-2}\right) * [0.012 * (500\,\text{ft})\,/(2\,\text{ft})] * \left[\left(14.30\,\text{ft sec}^{-1}\right)^2/\left(2 * 32.2\,\text{ft sec}^{-2}\right)\right] = 2379.73\,\text{lbf ft}^{-2}.$$

At Node 7$\rightarrow p_{7,0} = p_{6,0} - (\rho g)\,[f\,(\Delta x)\,/D]\,\left[V_0{}^2/\,(2g)\right]$

$$= 2379.73\,\text{lbf ft}^{-2} - \left(1.94\,\text{slug/ft}^{-3}\right) * \left(32.2\,\text{ft sec}^{-2}\right) * [0.012 * (500\,\text{ft})\,/(2\,\text{ft})] * \left[\left(14.30\,\text{ft sec}^{-1}\right)^2/\left(2 * 32.2\,\text{ft sec}^{-2}\right)\right] = 1784.80\,\text{lbf ft}^{-2}.$$

At Node 8$\rightarrow p_{8,0} = p_{7,0} - (\rho g)\,[f\,(\Delta x)\,/D]\,\left[V_0{}^2/\,(2g)\right]$

$$= 1784.80\,\text{lbf ft}^{-2} - \left(1.94\,\text{slug/ft}^{-3}\right) * \left(32.2\,\text{ft sec}^{-2}\right) * [0.012 * (500\,\text{ft})\,/(2\,\text{ft})] * \left[\left(14.30\,\text{ft sec}^{-1}\right)^2/\left(2 * 32.2\,\text{ft sec}^{-2}\right)\right] = 1189.87\,\text{lbf ft}^{-2}.$$

At Node 9$\rightarrow p_{9,0} = p_{8,0} - (\rho g)\,[f\,(\Delta x)\,/D]\,\left[V_0{}^2/\,(2g)\right]$

$$= 1189.87\,\text{lbf ft}^{-2} - \left(1.94\,\text{slug/ft}^{-3}\right) * \left(32.2\,\text{ft sec}^{-2}\right) * [0.012 * (500\,\text{ft})\,/(2\,\text{ft})] * \left[\left(14.30\,\text{ft sec}^{-1}\right)^2/\left(2 * 32.2\,\text{ft sec}^{-2}\right)\right] = 594.93\,\text{lbf ft}^{-2}.$$

At Node 10$\rightarrow p_{10,0} = p_{9,0} - (\rho g)\,[f\,(\Delta x)\,/D]\,\left[V_0{}^2/\,(2g)\right]$

$$= 594.93\,\text{lbf ft}^{-2} - \left(1.94\,\text{slug/ft}^{-3}\right) * \left(32.2\,\text{ft sec}^{-2}\right) * [0.012 * (500\,\text{ft})\,/(2\,\text{ft})] * \left[\left(14.30\,\text{ft sec}^{-1}\right)^2/\left(2 * 32.2\,\text{ft sec}^{-2}\right)\right] = 0.00\,\text{lbf ft}^{-2}.$$

Steady-state flow $\rightarrow V_{i,0} = 14.30\,\text{ft sec}^{-1}$ $(i = 0, 1, 2, \ldots, 10)$.

Upper boundary condition: cells D11:D34 in Figure 10.9, highlighted in blue.

At time $j \geq 1$:

$$p_{0,j} = 6246.8\,\text{lbf ft}^{-2}$$

Applying Eq. (10.25) to the reservoir and Node 1, one has:

$$V_{0,j} = \frac{0 + V_{1,j-1}}{2} + \frac{1}{2\rho\,|V_s|}\left(\rho g H_0 - p_{1,j-1}\right) - \frac{f}{4D}\left(0\,|0| + V_{1,j-1}\,\left|V_{1,j-1}\right|\right)(\Delta t)$$

$$= \frac{V_{1,j-1}}{2} + \frac{1}{2 * \left(1.94\,\text{slug ft}^{-3}\right) * \left(3930.32\,\text{ft sec}^{-1}\right)} * \left(6246.8\,\text{lbf ft}^{-2} - p_{1,j-1}\right) - \frac{0.012}{4 * (2\,\text{ft})} V_{1,j-1}\left|V_{1,j-1}\right| * (0.1272\,\text{sec})$$

$$= \frac{V_{1,j-1}}{2} + \left(6.55753 \times 10^{-5}\right) * \left(6246.8 - p_{1,j-1}\right) - \left(1.908 \times 10^{-4}\right) V_{1,j-1}\left|V_{1,j-1}\right|.$$

Lower boundary condition: cells N11:N34 in Figure 10.9, highlighted in blue

At time $j \geq 1$:

Water stops flowing at Node 10$\rightarrow V_{10,j} = 0.00\,\text{ft sec}^{-1}$.

Applying Eq. (10.24) to Node 9 and the ambient atmosphere, one has:

$$
\begin{aligned}
p_{10,j} &= \frac{p_{9,j-1}+0}{2} + \frac{\rho\,|V_s|}{2}\left(V_{9,j-1}-0\right) - \frac{\rho f\,|V_s|}{4D}\left(V_{9,j-1}\left|V_{9,j-1}\right| - 0\,|0|\right)(\Delta t) \\
&= \frac{p_{9,j-1}}{2} + \frac{\left(1.94\,\text{slug ft}^{-3}\right) * \left(3930.32\,\text{ft sec}^{-1}\right)}{2} * \left(V_{9,j-1}\right) \\
&\quad - \frac{\left(1.94\,\text{slug ft}^{-3}\right) * 0.012 * \left(3930.32\,\text{ft sec}^{-1}\right)}{4 * (2\,\text{ft})} V_{9,j-1}\left|V_{9,j-1}\right| * (0.1272\,\text{sec}) \\
&= \frac{p_{9,j-1}}{2} + 3812.4104 V_{9,j-1} - 1.45482 V_{9,j-1}\left|V_{9,j-1}\right|.
\end{aligned}
$$

Interior nodes ($i = 1, 2, \ldots, 9$) at time $j \geq 1$: use Eqs. (10.24) and (10.25)

The calculations can be done in an Excel spreadsheet, as shown in Figure 10.9. The profiles of pressure and velocity at the first 12 time steps are shown in Figure 10.10a,b, respectively, and the variations of pressure and velocity at Nodes 5 and 9 are plotted in Figure 10.10c.

Once the valve is suddenly closed, the velocity at Node 10 will become zero immediately while the pressure at the node is increased to $54,511.43\,\text{lbf ft}^{-2}$. At each of the other nodes, the pressure and velocity remain the same as those of the steady-state flow. With the lapse of time, the pressures at a node either increase or decrease, whereas the velocities at the same node can be positive (i.e., toward downstream) or negative (i.e., toward upstream). The pressures at some nodes vary from a negative to positive value. At the 11th time step ($j = 11$), the water in the pipeline starts to flow upstream back into the reservoir. Such downstream–upstream–downstream fluctuation processes will repeat every few time steps, ultimately approaching a new steady-state flow condition, as indicated by the velocities at all nodes tending to become very similar (Figure 10.10c).

For instance, at time $j = 1$, the upper boundary condition (cells D11:D12) is $p_{0,1} = 6246.8\,\text{lbf ft}^{-2}$ and $V_{0,1} = \left(14.30\,\text{ft sec}^{-1}\right)/2 + \left(6.55753 \times 10^{-5}\right) * \left(6246.8\,\text{lbf ft}^{-2} - 5354.20\,\text{lbf ft}^{-2}\right) - \left(1.908 \times 10^{-4}\right) * \left(14.30\,\text{ft sec}^{-1}\right) * \left|14.30\,\text{ft sec}^{-1}\right| = 7.17\,\text{ft sec}^{-1}$, whereas the lower boundary condition (cells N11:N12) is $V_{10,1} = 0.00\,\text{ft sec}^{-1}$ and $p_{10,1} = \left(594.93\,\text{lbf ft}^{-2}\right)/2 + 3812.4104 * \left(14.30\,\text{ft sec}^{-1}\right) - 1.45482 * \left(14.30\,\text{ft sec}^{-1}\right) * \left|14.30\,\text{ft sec}^{-1}\right| = 54,511.43\,\text{lbf ft}^{-2}$. At node $i = 9$ (cells M11:M12),
$p_{9,1} = \left(1189.87\,\text{lbf ft}^{-2} + 0.00\,\text{lbf ft}^{-2}\right)/2 + \left[\left(1.94\,\text{slug ft}^{-3}\right) * \left(3930.32\,\text{ft sec}^{-1}\right)\right]/2 * \left(14.30\,\text{ft sec}^{-1} - 14.30\,\text{ft sec}^{-1}\right) - \left[\left(1.94\,\text{slug ft}^{-3}\right) * 0.012 * \left(3930.32\,\text{ft sec}^{-1}\right)\right]/(4 * 2\,\text{ft}) * \left[\left(14.30\,\text{ft sec}^{-1}\right)\left|14.30\,\text{ft sec}^{-1}\right| - \left(14.30\,\text{ft sec}^{-1}\right)\left|14.30\,\text{ft sec}^{-1}\right|\right] * (0.1272\,\text{sec}) = 594.93\,\text{lbf ft}^{-2}$; and

$V_{9,1} = (14.30\,\text{ft sec}^{-1} + 14.30\,\text{ft sec}^{-1})/2 + 1/\left[2 * \left(1.94\,\text{slug ft}^{-3}\right) * \left(3930.32\,\text{ft sec}^{-1}\right)\right] * \left(1189.87\,\text{lbf ft}^{-2} - 0.00\,\text{lbf ft}^{-2}\right) - 0.012/\left(4 * 2\,\text{ft}\right) * \left[\left(14.30\,\text{ft sec}^{-1}\right)\left|14.30\,\text{ft sec}^{-1}\right| + \left(14.30\,\text{ft sec}^{-1}\right)\left|14.30\,\text{ft sec}^{-1}\right|\right] * (0.1272\,\text{sec}) = 14.30\,\text{ft sec}^{-1}$. Similar calculations can be done for nodes i = 8, 7, . . ., 1.

At other times j = 2, 3, . . ., determine the upper and lower boundary conditions and do the calculations for the interior nodes in similar ways.

	A	B	C	D	E	F	G	H	I	J	K	L	M	N
1	H_0 =	100	ft	D =	2	ft	Table 10-1:	E_{aw} =	4.5E+07	lbf ft^{-2}				
2	ρ =	1.94	slug ft^{-3}	ε =	0.25	ft		E_{ap} =	7.2E+08	lbf ft^{-3}				
3	g =	32.2	ft sec^{-2}	L =	5000	ft								
4	$p_{0,j}$ =	6246.8	lbf ft^{-2}	f =	0.012		Eq. (10-17):	$\lvert V_s \rvert$ =	3930.32	ft sec^{-1}				
5				ke =	0.5									
6	Δx =	500	ft	Δt =	0.1272	sec								
7									Node i =					
8	j	t_j (sec)	Variables	0	1	2	3	4	5	6	7	8	9	10
9	0	0	$p_{i,0}$ (lbf ft^{-2})	6246.80	5354.40	4759.47	4164.53	3569.60	2974.67	2379.73	1784.80	1189.87	594.93	0.00
10			$V_{i,0}$ (ft sec^{-1})	14.30	14.30	14.30	14.30	14.30	14.30	14.30	14.30	14.30	14.30	14.30
11	1	0.1272	$p_{i,1}$ (lbf ft^{-2})	6246.80	5503.13	4759.47	4164.53	3569.60	2974.67	2379.73	1784.80	1189.87	594.93	54511.43
12			$V_{i,1}$ (ft sec^{-1})	7.17	14.32	14.30	14.30	14.30	14.30	14.30	14.30	14.30	14.30	0.00
13	2	0.2544	$p_{i,2}$ (lbf ft^{-2})	6246.80	-21455.55	4907.39	4164.53	3569.60	2974.67	2379.73	1784.80	1189.87	82064.65	54511.47
14			$V_{i,2}$ (ft sec^{-1})	7.17	10.78	14.32	14.30	14.30	14.30	14.30	14.30	14.30	3.61	0.00
15	3	0.3816	$p_{i,3}$ (lbf ft^{-2})	6246.80	-21455.16	-21922.19	4311.65	3569.60	2974.67	2379.73	1784.80	82381.15	82064.71	54789.94
16			$V_{i,3}$ (ft sec^{-1})	7.19	10.78	10.80	14.32	14.30	14.30	14.30	14.30	3.65	3.61	0.00
17	4	0.5088	$p_{i,4}$ (lbf ft^{-2})	6246.80	-21519.56	-21921.80	-22389.35	3715.91	2974.67	2379.73	82698.04	82381.21	82482.02	54789.97
18			$V_{i,4}$ (ft sec^{-1})	7.19	10.81	10.80	10.82	14.32	14.30	14.30	3.69	3.65	3.63	0.00
19	5	0.6360	$p_{i,5}$ (lbf ft^{-2})	6246.80	-21519.18	-21986.46	-22388.96	-22857.04	3120.18	83015.31	82698.09	82798.33	82482.08	55067.87
20			$V_{i,5}$ (ft sec^{-1})	7.20	10.81	10.82	10.82	10.83	14.32	3.72	3.69	3.67	3.63	0.00
21	6	0.7632	$p_{i,6}$ (lbf ft^{-2})	6246.80	-21582.91	-21986.07	-22453.88	-22856.66	57033.05	83160.08	83115.01	82798.38	82898.53	55067.90
22			$V_{i,6}$ (ft sec^{-1})	7.20	10.83	10.82	10.84	10.83	0.31	3.74	3.70	3.67	3.65	0.00
23	7	0.8904	$p_{i,7}$ (lbf ft^{-2})	6246.80	-21582.53	-22050.07	-22453.50	57265.39	57032.91	57152.58	83259.58	83214.63	82898.58	55345.21
24			$V_{i,7}$ (ft sec^{-1})	7.22	10.83	10.85	10.84	0.34	0.31	0.29	3.72	3.69	3.65	0.00
25	8	1.0176	$p_{i,8}$ (lbf ft^{-2})	6246.80	-21645.61	-22049.69	57498.14	57265.24	57384.87	57152.44	57271.84	83358.99	83314.15	55345.25
26			$V_{i,8}$ (ft sec^{-1})	7.22	10.86	10.85	0.37	0.34	0.32	0.29	0.28	3.71	3.67	0.00
27	9	1.1448	$p_{i,9}$ (lbf ft^{-2})	6246.80	-21645.24	57731.32	57498.00	57617.58	57384.73	57504.09	57271.69	57390.82	83458.30	55621.97
28			$V_{i,9}$ (ft sec^{-1})	7.24	10.86	0.40	0.37	0.36	0.32	0.31	0.28	0.26	3.69	0.00
29	10	1.2720	$p_{i,10}$ (lbf ft^{-2})	6246.80	57964.92	57731.17	57850.70	57617.43	57736.75	57503.94	57623.03	57390.67	57509.52	55765.90
30			$V_{i,10}$ (ft sec^{-1})	7.23	0.43	0.40	0.39	0.36	0.34	0.31	0.29	0.26	0.25	0.00
31	11	1.3992	$p_{i,11}$ (lbf ft^{-2})	6246.80	57964.77	58084.25	57850.56	57969.83	57736.60	57855.65	57622.88	57741.70	57581.12	29698.45
32			$V_{i,11}$ (ft sec^{-1})	-3.18	0.43	0.42	0.39	0.37	0.34	0.32	0.29	0.28	0.24	0.00
33	12	1.5264	$p_{i,12}$ (lbf ft^{-2})	6246.80	18486.88	58084.10	58203.32	57969.68	58088.68	57855.50	57974.28	57813.29	44779.74	29698.09
34			$V_{i,12}$ (ft sec^{-1})	-3.18	-4.78	0.42	0.40	0.37	0.35	0.32	0.31	0.27	1.98	0.00

Figure 10.9 **Excerpt of the spreadsheet for the calculations for Example 10.10.**

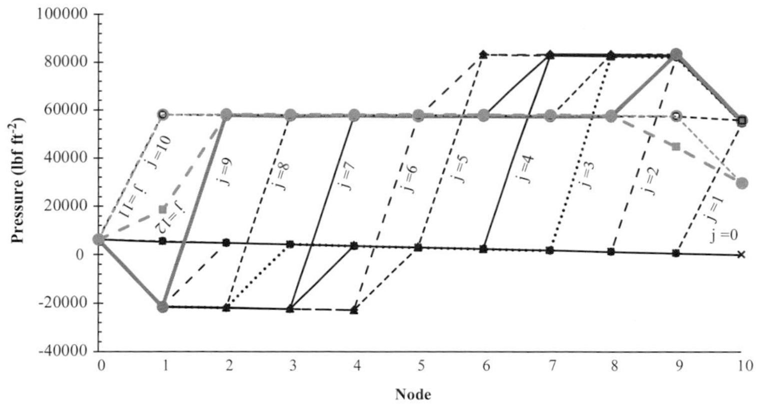

(a)

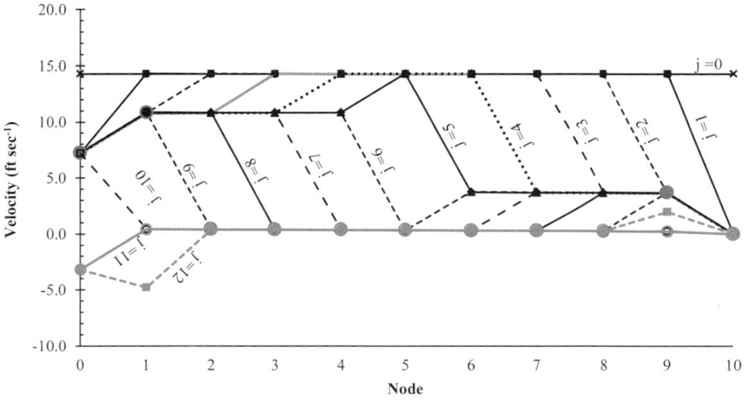

(b)

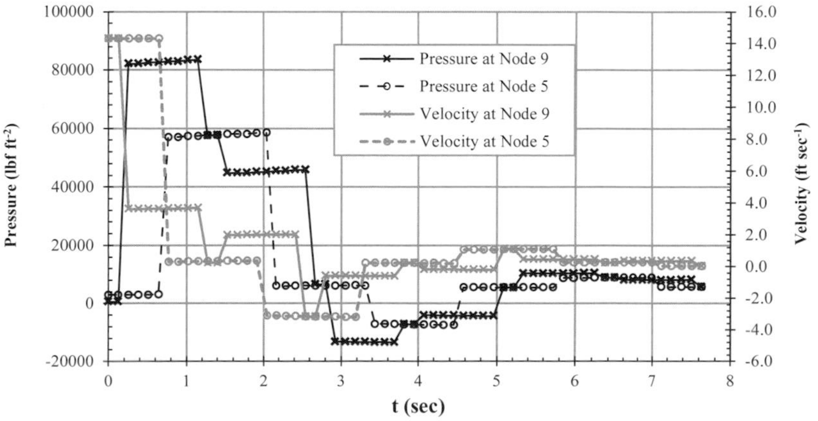

(c)

Figure 10.10 **Computed pressures and velocities for Example 10.10.** (a) Pressure profiles; (b) velocity profiles; and (c) fluctuations of pressure and velocity at Nodes 5 and 9.

Example 10.11 Redo Example 10.9 if the valve is gradually opened during the first four computational time steps. That is, at the first time step, the valve is 3/4 closed; at the second time step, the valve is 1/2 closed; at the third time step, the valve is 1/4 closed; and at the fourth time step, the valve is zero closed (i.e., fully opened). The loss coefficients at 3/4, 1/2, 1/4, and zero closed are 17, 5.6, 0.26, and 0, respectively.

Solution

$H_0 = 100\,\text{ft}$, $k_e = 0.5$, $\rho = 1.94\,\text{slug ft}^{-3}$, $g = 32.2\,\text{ft sec}^{-2}$

Table 10.1: $E_{m,w} = 0.312 \times 10^6\,\text{psi} = 44.928 \times 10^6\,\text{lbf ft}^{-2}$

$$E_{m,p} = 5.0 \times 10^6\,\text{psi} = 720 \times 10^6\,\text{lbf ft}^{-2}.$$

$\Delta x = 500\,\text{ft} \rightarrow$ number of segments $= (5000\,\text{ft}) / (500\,\text{ft}) = 10$, number of nodes $= 11$.

$$\text{Eq. (10.17): } |V_s| = \sqrt{\frac{44.928 \times 10^6\,\text{lbf ft}^{-2}}{1.94\,\text{slug ft}^{-3}} * \frac{1}{1 + \dfrac{\left(44.928 \times 10^6\,\text{lbf ft}^{-2}\right) * (2\,\text{ft})}{\left(720 \times 10^6\,\text{lbf ft}^{-2}\right) * (0.25\,\text{ft})}}} = 3930.32\,\text{ft sec}^{-1}.$$

Computational time step: $\Delta t = \Delta x / |V_s| = (500\,\text{ft}) / \left(3930.32\,\text{ft sec}^{-1}\right) = 0.1272\,\text{sec}.$

Initial condition: cells D9:N10 in Figure 10.11, highlighted in red.

$$p_{i,0} = \rho g H_0 = \left(1.94\,\text{slug ft}^{-3}\right) * \left(32.2\,\text{ft sec}^{-2}\right) * (100\,\text{ft}) = 6246.8\,\text{lbf ft}^{-2}\ (i = 0, 1, 2, \ldots, 10).$$

At time $j = 0$, water at all nodes is not flowing $\rightarrow V_{i,0} = 0.00\,\text{ft sec}^{-1}$ $(i = 0, 1, 2, \ldots, 10)$.

Upper boundary condition: cells D11:D34 in Figure 10.11, highlighted in blue.

At time $j \geq 1$:

$$p_{0,j} = 6246.8\,\text{lbf ft}^{-2}$$

Applying Eq. (10.25) to the reservoir and Node 1, one has:

$$\begin{aligned} V_{0,j} &= \frac{0 + V_{1,j-1}}{2} + \frac{1}{2\rho|V_s|}\left(\rho g H_0 - p_{1,j-1}\right) - \frac{f}{4D}\left(0\,|0| + V_{1,j-1}\left|V_{1,j-1}\right|\right)(\Delta t) \\ &= \frac{V_{1,j-1}}{2} + \frac{1}{2 * \left(1.94\,\text{slug ft}^{-3}\right) * \left(3930.32\,\text{ft sec}^{-1}\right)} * \left(6246.8\,\text{lbf ft}^{-2} - p_{1,j-1}\right) \\ &\quad - \frac{0.012}{4 * (2\,\text{ft})} V_{1,j-1}\left|V_{1,j-1}\right| * (0.1272\,\text{sec}) \\ &= \frac{V_{1,j-1}}{2} + \left(6.55753 \times 10^{-5}\right) * \left(6246.8 - p_{1,j-1}\right) - \left(1.908 \times 10^{-4}\right) V_{1,j-1}\left|V_{1,j-1}\right|. \end{aligned}$$

Lower boundary condition: cells N11:N34 in Figure 10.11, highlighted in blue.

	A	B	C	D	E	F	G	H	I	J	K	L	M	N
1	H_0 =	100	ft	D =	2	ft	Table 10-1:	$E_{m,w}$ =	4.5E+07	lbf ft^{-2}	k_{valve} =	17	3/4 closed	
2	ρ =	1.94	slug ft^{-3}	ε =	0.25	ft		$E_{m,p}$ =	7.2E+08	lbf ft^{-3}		5.6	1/2 closed	
3	g =	32.2	ft sec^{-2}	L =	5000	ft						0.26	1/4 closed	
4	$p_{0,j}$ =	6246.8	lbf ft^{-2}	f =	0.012		Eq. (10-17):	$\|V_s\|$ =	3930.32	ft sec^{-1}		0	fully open	
5														
6	Δx =	500	ft	Δt =	0.1272	sec								
7								Node i =						
8	j	t_j (sec)	Variables	0	1	2	3	4	5	6	7	8	9	10
9	0	0	$p_{i,0}$ (lbf ft^{-2})	6246.80	6246.80	6246.80	6246.80	6246.80	6246.80	6246.80	6246.80	6246.80	6246.80	6246.80
10			$V_{i,0}$ (ft sec^{-1})	0.00	0.00	0.00	0.00	0.00	0.00	0.00	0.00	0.00	0.00	0.00
11	1	0.1272	$p_{i,1}$ (lbf ft^{-2})	6246.80	6246.80	6246.80	6246.80	6246.80	6246.80	6246.80	6246.80	6246.80	6246.80	2.77
12			$V_{i,1}$ (ft sec^{-1})	0.00	0.00	0.00	0.00	0.00	0.00	0.00	0.00	0.00	0.00	0.41
13	2	0.2544	$p_{i,2}$ (lbf ft^{-2})	6246.80	6246.80	6246.80	6246.80	6246.80	6246.80	6246.80	6246.80	6246.80	1563.33	0.91
14			$V_{i,2}$ (ft sec^{-1})	0.00	0.00	0.00	0.00	0.00	0.00	0.00	0.00	0.00	0.61	0.41
15	3	0.3816	$p_{i,3}$ (lbf ft^{-2})	6246.80	6246.80	6246.80	6246.80	6246.80	6246.80	6246.80	6246.80	1563.88	1562.40	0.04
16			$V_{i,3}$ (ft sec^{-1})	0.00	0.00	0.00	0.00	0.00	0.00	0.00	0.00	0.61	0.61	0.41
17	4	0.5088	$p_{i,4}$ (lbf ft^{-2})	6246.80	6246.80	6246.80	6246.80	6246.80	6246.80	6246.80	1564.43	1562.95	1561.69	0.00
18			$V_{i,4}$ (ft sec^{-1})	0.00	0.00	0.00	0.00	0.00	0.00	0.00	0.61	0.61	0.61	0.41
19	5	0.6360	$p_{i,5}$ (lbf ft^{-2})	6246.80	6246.80	6246.80	6246.80	6246.80	6246.80	1564.97	1563.50	1562.24	1561.67	0.00
20			$V_{i,5}$ (ft sec^{-1})	0.00	0.00	0.00	0.00	0.00	0.00	0.61	0.61	0.61	0.61	0.41
21	6	0.7632	$p_{i,6}$ (lbf ft^{-2})	6246.80	6246.80	6246.80	6246.80	6246.80	1565.52	1564.05	1562.79	1562.22	1561.40	0.00
22			$V_{i,6}$ (ft sec^{-1})	0.00	0.00	0.00	0.00	0.00	0.61	0.61	0.61	0.61	0.61	0.41
23	7	0.8904	$p_{i,7}$ (lbf ft^{-2})	6246.80	6246.80	6246.80	6246.80	1566.07	1564.60	1563.34	1562.77	1561.94	1561.40	0.00
24			$V_{i,7}$ (ft sec^{-1})	0.00	0.00	0.00	0.00	0.61	0.61	0.61	0.61	0.61	0.61	0.41
25	8	1.0176	$p_{i,8}$ (lbf ft^{-2})	6246.80	6246.80	6246.80	1566.62	1565.14	1563.89	1563.32	1562.49	1561.94	1561.12	0.00
26			$V_{i,8}$ (ft sec^{-1})	0.00	0.00	0.00	0.61	0.61	0.61	0.61	0.61	0.61	0.61	0.41
27	9	1.1448	$p_{i,9}$ (lbf ft^{-2})	6246.80	6246.80	1567.17	1565.69	1564.44	1563.87	1563.04	1562.49	1561.67	1561.12	0.00
28			$V_{i,9}$ (ft sec^{-1})	0.00	0.00	0.61	0.61	0.61	0.61	0.61	0.61	0.61	0.61	0.41
29	10	1.2720	$p_{i,10}$ (lbf ft^{-2})	6246.80	1567.72	1566.24	1564.98	1564.41	1563.59	1563.04	1562.22	1561.67	1560.85	0.00
30			$V_{i,10}$ (ft sec^{-1})	0.00	0.61	0.61	0.61	0.61	0.61	0.61	0.61	0.61	0.61	0.41
31	11	1.3992	$p_{i,11}$ (lbf ft^{-2})	6246.80	1566.79	1565.53	1564.96	1564.14	1563.59	1562.77	1562.22	1561.40	1560.85	0.00
32			$V_{i,11}$ (ft sec^{-1})	0.61	0.61	0.61	0.61	0.61	0.61	0.61	0.61	0.61	0.61	0.41
33	12	1.5264	$p_{i,12}$ (lbf ft^{-2})	6246.80	3905.35	1565.51	1564.69	1564.14	1563.32	1562.77	1561.94	1561.40	1560.57	0.00
34			$V_{i,12}$ (ft sec^{-1})	0.61	0.92	0.61	0.61	0.61	0.61	0.61	0.61	0.61	0.61	0.41

Figure 10.11 **Excerpt of the spreadsheet for the calculations for Example 10.11.**

At time $j \geq 1$, applying Eq. (10.25) to Node 9 and the ambient atmosphere, one has:

$$V_{10,j} = \frac{V_{9,j-1} + 0}{2} + \frac{1}{2\rho\,|V_s|}\left(p_{9,j-1} - 0\right) - \frac{f}{4D}\left(V_{9,j-1}\left|V_{9,j-1}\right| + 0\,|0|\right)(\Delta t)$$

$$= \frac{V_{9,j-1}}{2} + \frac{1}{2 * \left(1.94\,\text{slug ft}^{-3}\right) * \left(3930.32\,\text{ft sec}^{-1}\right)} * \left(p_{9,j-1}\right)$$

$$- \frac{0.012}{4 * (2\,\text{ft})} V_{9,j-1}\left|V_{9,j-1}\right| * (0.1272\,\text{sec})$$

$$= \frac{V_{9,j-1}}{2} + \left(6.55753 \times 10^{-5}\right) p_{9,j-1} - \left(1.908 \times 10^{-4}\right) V_{9,j-1} \left|V_{9,j-1}\right|.$$

At time $j = 1$, the valve is 3/4 closed → Table 2.3: $p_{10,1} = \rho k_{valve}\left(V_{10,1}{}^2/2\right)$

$$\rightarrow p_{10,1} = \left(1.94 \text{ slug ft}^{-3}\right) * 17 * \left(V_{10,1}{}^2/2\right) = 16.49 V_{10,1}{}^2 \text{lbf ft}^{-2}.$$

At time $j = 2$, the valve is 1/2 closed → Table 2.3: $p_{10,2} = \rho k_{valve}\left(V_{10,2}{}^2/2\right)$

$$\rightarrow p_{10,2} = \left(1.94 \text{ slug ft}^{-3}\right) * 5.6 * \left(V_{10,2}{}^2/2\right) = 5.432 V_{10,2}{}^2 \text{lbf ft}^{-2}.$$

At time $j = 3$, the valve is 1/4 closed → Table 2.3: $p_{10,3} = \rho k_{valve}\left(V_{10,3}{}^2/2\right)$

$$\rightarrow p_{10,3} = \left(1.94 \text{ slug ft}^{-3}\right) * 0.26 * \left(V_{10,3}{}^2/2\right) = 0.2522 V_{10,3}{}^2 \text{lbf ft}^{-2}.$$

At time $j \geq 4$, the valve is fully opened → Table 2.3: $p_{10,j} = \rho k_{valve}\left(V_{10,j}{}^2/2\right)$

$$\rightarrow p_{10,j} = \left(1.94 \text{ slug ft}^{-3}\right) * 0 * \left(V_{10,1}{}^2/2\right) = 0.00 \text{ lbf ft}^{-2}.$$

Interior nodes ($i = 1, 2, \ldots, 9$) at time $j \geq 1$: use Eqs. (10.24) and (10.25).

The calculations can be done in an Excel spreadsheet, as shown in Figure 10.11. The profiles of pressure and velocity at the first 12 time steps are shown in Figure 10.12a,b, respectively, and the variations of pressure and velocity at Nodes 5 and 9 are plotted in Figure 10.12c.

At all nodes, the initial values of pressure and velocity are 6246.8 lbf ft^{-2} and zero, respectively. At Node 10, when the valve is partially opened, the relationship between the pressure head and velocity follows the fitting loss equation given in Table 2.3. Once the valve is fully opened, however, the pressure will drop to zero because the node is exposed to atmosphere. The velocity can be determined by applying Eq. (10.25) to Node 9 and the ambient atmosphere. With the lapse of time until the 11th time step ($j = 11$), the pressures at the interior nodes decrease, but the velocities increase, from downstream to upstream. Starting from the 12th time step ($j = 12$), the pressures start to increase while the velocities continue to increase, from upstream to downstream. Such downstream–upstream–downstream fluctuation processes will repeat every few time steps, ultimately approaching a new steady-state flow condition, as indicated by the velocities at all nodes tending to become very similar (Figure 10.12c).

For instance, at time $j = 1$, the upper boundary condition (cells D11:D12) is $p_{0,1} = 6246.8 \text{ lbf ft}^{-2}$ and $V_{0,1} = \left(0.00 \text{ ft sec}^{-1}\right)/2 + \left(6.55753 \times 10^{-5}\right) * \left(6246.8 \text{ lbf ft}^{-2} - 6246.8 \text{ lbf ft}^{-2}\right) - \left(1.908 \times 10^{-4}\right) * \left(0.00 \text{ ft sec}^{-1}\right) * \left|0.00 \text{ ft sec}^{-1}\right| = 0.00 \text{ ft sec}^{-1}$, whereas the lower boundary condition (cells N11:N12) is $V_{10,1} = \left(0.00 \text{ ft sec}^{-1}\right)/2 + \left(6.55753 \times 10^{-5}\right) * \left(6246.8 \text{ lbf ft}^{-2}\right) - \left(1.908 \times 10^{-4}\right) * \left(0.00 \text{ ft sec}^{-1}\right) * \left|0.00 \text{ ft sec}^{-1}\right| = 0.41 \text{ ft sec}^{-1}$ and $p_{10,1} = \left(1.94 \text{ slug ft}^{-3}\right) * 17 * \left[\left(0.41 \text{ ft sec}^{-1}\right)^2/2\right] = 2.77 \text{ lbf ft}^{-2}$. At node $i = 9$ (cells M11:M12),

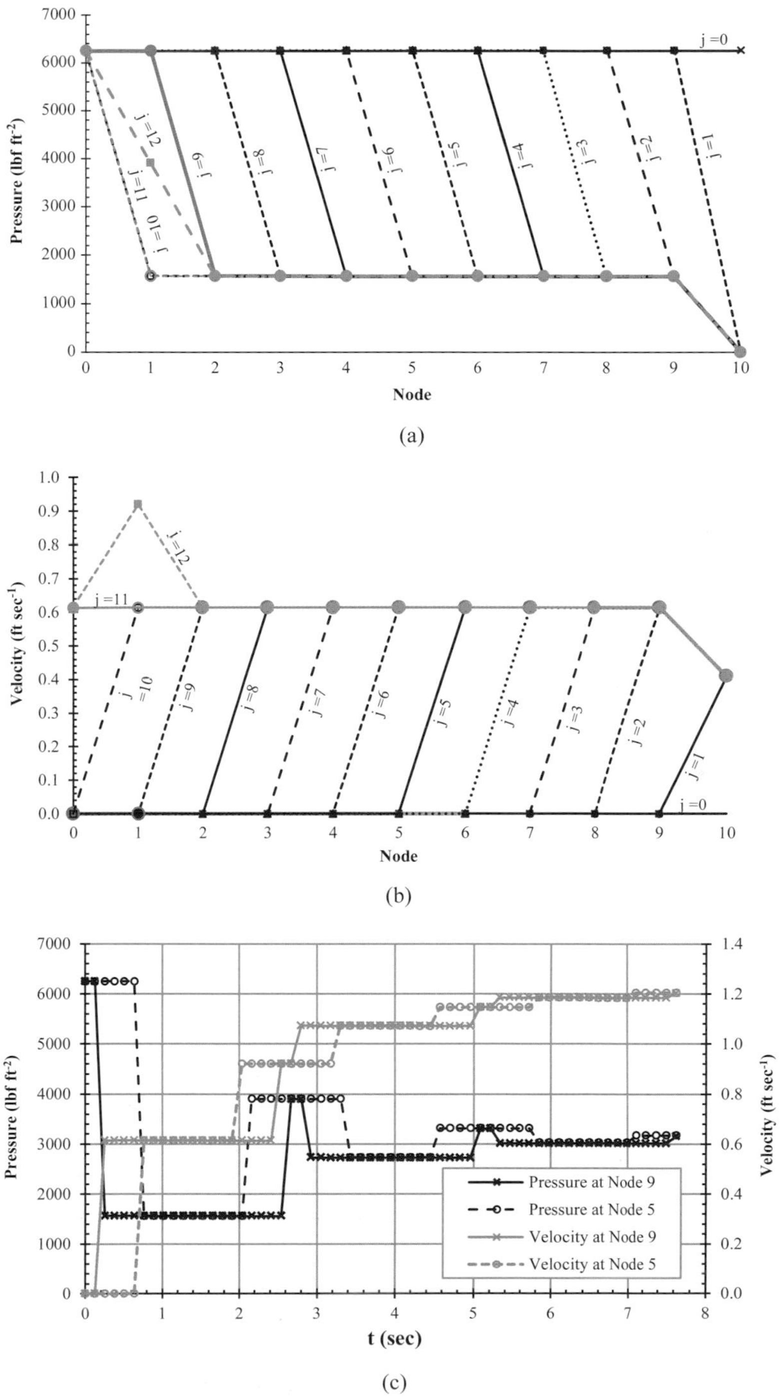

Figure 10.12 Computed pressures and velocities for Example 10.11. (a) Pressure profiles; (b) velocity profiles; and (c) fluctuations of pressure and velocity at Nodes 5 and 9.

$p_{9,1} = \left(6246.80\,\text{lbf ft}^{-2} + 6246.80\,\text{lbf ft}^{-2}\right)/2 + \left[\left(1.94\,\text{slug ft}^{-3}\right) * (3930.32\,\text{ft sec}^{-1})\right]/2*$ $(0.00\,\text{ft sec}^{-1} + 0.00\,\text{ft sec}^{-1}) - \left[\left(1.94\,\text{slug ft}^{-3}\right) * 0.012 * \left(3930.32\,\text{ft sec}^{-1}\right)\right]/(4*$ $2\,\text{ft}) * \left[(0.00\,\text{ft sec}^{-1})\,|0.00\,\text{ft sec}^{-1}| - (0.00\,\text{ft sec}^{-1})\,|0.00\,\text{ft sec}^{-1}|\right] *$ $(0.1272\,\text{sec}) = 6246.80\,\text{lbf ft}^{-2}$; and $V_{9,1} = \left(0.00\,\text{ft sec}^{-1} + 0.00\,\text{ft sec}^{-1}\right)/2$ $+[1/\left[2 * \left(1.94\,\text{slug ft}^{-3}\right) * (3930.32\,\text{ft sec}^{-1})\right] * \left(6246.80\,\text{lbf ft}^{-2} - 6246.80\,\text{lbf ft}^{-2}\right) -$ $0.012/(4 * 2\,\text{ft}) * \left[(0.00\,\text{ft sec}^{-1})\,|0.00\,\text{ft sec}^{-1}| + (0.00\,\text{ft sec}^{-1})\,|0.00\,\text{ft sec}^{-1}|\right] * (0.1272\,\text{sec}) =$ $0.00\,\text{ft sec}^{-1}$. Similar calculations can be done for nodes $i = 8, 7, \ldots, 1$. At times $j = 2, 3, \ldots$, determine the upper and lower boundary conditions and do the calculations for the interior nodes in similar ways.

Example 10.12 Redo Example 10.10 if the valve is gradually closed during the first four computational time steps. That is, at the first time step, the valve is 1/4 closed; at the second time step, the valve is 1/2 closed; at the third time step, the valve is 3/4 closed; and at the fourth time step, the valve is fully closed. The loss coefficients at zero, 1/4, 1/2, and 3/4 closed are 0, 0.26, 5.6, and 17, respectively.

Solution

$H_0 = 100\,\text{ft}$, $k_e = 0.5$, $\rho = 1.94\,\text{slug ft}^{-3}$, $g = 32.2\,\text{ft sec}^{-2}$

Table 10.1: $E_{m,w} = 0.312 \times 10^6\,\text{psi} = 44.928 \times 10^6\,\text{lbf ft}^{-2}$

$$E_{m,p} = 5.0 \times 10^6\,\text{psi} = 720 \times 10^6\,\text{lbf ft}^{-2}.$$

$\Delta x = 500\,\text{ft} \rightarrow$ number of segments $= (5000\,\text{ft})/(500\,\text{ft}) = 10$, number of nodes $= 11$.

$$\text{Eq. (10.17): } |V_s| = \sqrt{\frac{44.928 \times 10^6\,\text{lbf ft}^{-2}}{1.94\,\text{slug ft}^{-3}} * \frac{1}{1 + \frac{\left(44.928 \times 10^6\,\text{lbf ft}^{-2}\right)*(2\,\text{ft})}{\left(720 \times 10^6\,\text{lbf ft}^{-2}\right)*(0.25\,\text{ft})}}} = 3930.32\,\text{ft sec}^{-1}.$$

Computational time step: $\Delta t = \Delta x/|V_s| = (500\,\text{ft})/\left(3930.32\,\text{ft sec}^{-1}\right) = 0.1272\,\text{sec}$.

Steady-state flow:

Energy equation (Eq. (2.21)) between the reservoir surface and the pipeline exit:

$$100\,\text{ft} = V_0^2/\left(2 * 32.2\,\text{ft sec}^{-2}\right) + [0.012 * (5000\,\text{ft})/(2\,\text{ft}) + 0.5]\,V_0^2/\left(2 * 32.2\,\text{ft sec}^{-2}\right)$$
$$\rightarrow 100 = (31.5/64.4)\,V_0^2 \rightarrow V_0 = 14.30\,\text{ft sec}^{-1}.$$

Initial condition: cells D9:N10 in Figure 10.13, highlighted in red.

At Node $0 \rightarrow p_{0,0} = \rho g H_0 = \left(1.94\,\text{slug ft}^{-3}\right) * \left(32.2\,\text{ft sec}^{-2}\right) * (100\,\text{ft}) = 6246.8\,\text{lbf ft}^{-2}$.

At Node $1 \rightarrow p_{1,0} = p_{0,0} - (\rho g)\,[1 + f(\Delta x)/D + k_e]\left[V_0^2/(2g)\right]$

$$= 6246.8\,\text{lbf ft}^{-2} - \left(1.94\,\text{slug/ft}^{-3}\right) * \left(32.2\,\text{ft sec}^{-2}\right) * [1 + 0.012 * (500\,\text{ft})/(2\,\text{ft}) + 0.5]$$
$$* \left[\left(14.30\,\text{ft sec}^{-1}\right)^2/\left(2 * 32.2\,\text{ft sec}^{-2}\right)\right] = 5354.40\,\text{lbf ft}^{-2}.$$

$$\text{At Node } 2 \rightarrow p_{2,0} = p_{1,0} - (\rho g)\,[f\,(\Delta x)\,/D]\,\left[V_0{}^2/\,(2g)\right]$$

$$= 5354.40\,\text{lbf ft}^{-2} - \left(1.94\,\text{slug/ft}^{-3}\right) * \left(32.2\,\text{ft sec}^{-2}\right) * [0.012 * (500\,\text{ft})\,/(2\,\text{ft})]$$

$$* \left[\left(14.30\,\text{ft sec}^{-1}\right)^2 / \left(2 * 32.2\,\text{ft sec}^{-2}\right)\right] = 4759.47\,\text{lbf ft}^{-2}.$$

$$\text{At Node } 3 \rightarrow p_{3,0} = p_{2,0} - (\rho g)\,[f\,(\Delta x)\,/D]\,\left[V_0{}^2/\,(2g)\right]$$

$$= 4759.47\,\text{lbf ft}^{-2} - \left(1.94\,\text{slug/ft}^{-3}\right) * \left(32.2\,\text{ft sec}^{-2}\right) * [0.012 * (500\,\text{ft})\,/(2\,\text{ft})]$$

$$* \left[\left(14.30\,\text{ft sec}^{-1}\right)^2 / \left(2 * 32.2\,\text{ft sec}^{-2}\right)\right] = 4164.53\,\text{lbf ft}^{-2}.$$

$$\text{At Node } 4 \rightarrow p_{4,0} = p_{3,0} - (\rho g)\,[f\,(\Delta x)\,/D]\,\left[V_0{}^2/\,(2g)\right]$$

$$= 4164.53\,\text{lbf ft}^{-2} - \left(1.94\,\text{slug/ft}^{-3}\right) * \left(32.2\,\text{ft sec}^{-2}\right) * [0.012 * (500\,\text{ft})\,/(2\text{ft})]$$

$$* \left[\left(14.30\,\text{ft sec}^{-1}\right)^2 / \left(2 * 32.2\,\text{ft scc}^{-2}\right)\right] = 3569.60\,\text{lbf ft}^{-2}.$$

$$\text{At Node } 5 \rightarrow p_{5,0} = p_{4,0} - (\rho g)\,[f\,(\Delta x)\,/D]\,\left[V_0{}^2/\,(2g)\right]$$

$$= 3569.60\,\text{lbf ft}^{-2} - \left(1.94\,\text{slug/ft}^{-3}\right) * \left(32.2\,\text{ft sec}^{-2}\right) * [0.012 * (500\,\text{ft})\,/(2\,\text{ft})]$$

$$* [(14.30\,\text{ft sec}^{-1})^2 / (2 * 32.2\,\text{ft sec}^{-2})] = 2974.67\,\text{lbf ft}^{-2}.$$

$$\text{At Node } 6 \rightarrow p_{6,0} = p_{5,0} - (\rho g)\,[f\,(\Delta x)\,/D]\,\left[V_0{}^2/\,(2g)\right]$$

$$= 2974.67\,\text{lbf ft}^{-2} - \left(1.94\,\text{slug/ft}^{-3}\right) * \left(32.2\,\text{ft sec}^{-2}\right) * [0.012 * (500\,\text{ft})\,/(2\,\text{ft})]$$

$$* [(14.30\,\text{ft sec}^{-1})^2 / (2 * 32.2\,\text{ft sec}^{-2})] = 2379.73\,\text{lbf ft}^{-2}.$$

$$\text{At Node } 7 \rightarrow p_{7,0} = p_{6,0} - (\rho g)\,[f\,(\Delta x)\,/D]\,\left[V_0{}^2/\,(2g)\right]$$

$$= 2379.73\,\text{lbf ft}^{-2} - \left(1.94\,\text{slug/ft}^{-3}\right) * \left(32.2\,\text{ft sec}^{-2}\right) * [0.012 * (500\,\text{ft})\,/(2\,\text{ft})]$$

$$* [(14.30\,\text{ft sec}^{-1})^2 / (2 * 32.2\,\text{ft sec}^{-2})] = 1784.80\,\text{lbf ft}^{-2}.$$

$$\text{At Node } 8 \rightarrow p_{8,0} = p_{7,0} - (\rho g)\,[f\,(\Delta x)\,/D]\,\left[V_0{}^2/\,(2g)\right]$$

$$= 1784.80\,\text{lbf ft}^{-2} - \left(1.94\,\text{slug/ft}^{-3}\right) * \left(32.2\,\text{ft sec}^{-2}\right) * [0.012 * (500\,\text{ft})\,/(2\,\text{ft})]$$

$$* [(14.30\,\text{ft sec}^{-1})^2 / (2 * 32.2\,\text{ft sec}^{-2})] = 1189.87\,\text{lbf ft}^{-2}.$$

$$\text{At Node } 9 \rightarrow p_{9,0} = p_{8,0} - (\rho g)\,[f\,(\Delta x)\,/D]\,\left[V_0{}^2/\,(2g)\right]$$

$$= 1189.87\,\text{lbf ft}^{-2} - \left(1.94\,\text{slug/ft}^{-3}\right) * \left(32.2\,\text{ft sec}^{-2}\right) * [0.012 * (500\,\text{ft})\,/(2\,\text{ft})]$$

$$* [(14.30\,\text{ft sec}^{-1})^2 / (2 * 32.2\,\text{ft sec}^{-2})] = 594.93\,\text{lbf ft}^{-2}.$$

At Node 10$\rightarrow p_{10,0} = p_{9,0} - (\rho g)\,[f\,(\Delta x)\,/D]\,\left[\left(V_0{}^2/\,(2g)\right]\right.$

$$= 594.93\,\text{lbf ft}^{-2} - \left(1.94\,\text{slug/ft}^{-3}\right) * \left(32.2\,\text{ft sec}^{-2}\right) * [0.012 * (500\,\text{ft})\,/(2\text{ft})]$$

$$* \left(14.30\,\text{ft sec}^{-1}\right)^2 / \left(2 * 32.2\,\text{ft sec}^{-2}\right)] = 0.00\,\text{lbf ft}^{-2}.$$

Steady-state flows at all nodes $\rightarrow V_{i,0} = 14.30\,\text{ft sec}^{-1}$ $(i = 0, 1, 2, \ldots, 10)$.

Upper boundary condition: cells D11:D34 in Figure 10.13, highlighted in blue.

At time $j \geq 1$:

$$p_{0,j} = 6246.8\,\text{lbf ft}^{-2}$$

Applying Eq. (10.25) to the reservoir and Node 1, one has:

$$V_{0,j} = \frac{0 + V_{1,j-1}}{2} + \frac{1}{2\rho\,|V_s|}\left(\rho g H_0 - p_{1,j-1}\right) - \frac{f}{4D}\left(0\,|0| + V_{1,j-1}\left|V_{1,j-1}\right|\right)(\Delta t)$$

$$= \frac{V_{1,j-1}}{2} + \frac{1}{2 * \left(1.94\,\text{slug ft}^{-3}\right) * \left(3930.32\,\text{ft sec}^{-1}\right)} * \left(6246.8\,\text{lbf ft}^{-2} - p_{1,j-1}\right)$$

$$- \frac{0.012}{4 * (2\,\text{ft})} V_{1,j-1}\left|V_{1,j-1}\right| * (0.1272\,\text{sec})$$

$$= \frac{V_{1,j-1}}{2} + \left(6.55753 \times 10^{-5}\right) * \left(6246.8 - p_{1,j-1}\right) - \left(1.908 \times 10^{-4}\right) V_{1,j-1}\left|V_{1,j-1}\right|.$$

Lower boundary condition: cells N11:N34 in Figure 10.13, highlighted in blue.

At time $1 \leq j \leq 3$, applying Eq. (10.25) to Node 9 and the ambient atmosphere, one has:

$$V_{10,j} = \frac{V_{9,j-1} + 0}{2} + \frac{1}{2\rho\,|V_s|}\left(p_{9,j-1} - 0\right) - \frac{f}{4D}\left(V_{9,j-1}\left|V_{9,j-1}\right| + 0\,|0|\right)(\Delta t)$$

$$= \frac{V_{9,j-1}}{2} + \frac{1}{2 * \left(1.94\,\text{slug ft}^{-3}\right) * \left(3930.32\,\text{ft sec}^{-1}\right)} * \left(p_{9,j-1}\right)$$

$$- \frac{0.012}{4 * (2\,\text{ft})} V_{9,j-1}\left|V_{9,j-1}\right| * (0.1272\,\text{sec})$$

$$= \frac{V_{9,j-1}}{2} + \left(6.55753 \times 10^{-5}\right) p_{9,j-1} - \left(1.908 \times 10^{-4}\right) V_{9,j-1}\left|V_{9,j-1}\right|.$$

At time $j = 1$, the valve is 1/4 closed $\rightarrow k_{valve} = 0.26$

$$\rightarrow p_{10,1} = \left(1.94\,\text{slug ft}^{-3}\right) * 0.26 * \left(V_{10,1}{}^2/2\right) = 0.2522 V_{10,1}{}^2\text{lbf ft}^{-2}.$$

At time $j = 2$, the valve is 1/2 closed $\rightarrow k_{valve} = 5.6$

$$\rightarrow p_{10,2} = \left(1.94\,\text{slug ft}^{-3}\right) * 5.6 * \left(V_{10,2}{}^2/2\right) = 5.432 V_{10,2}{}^2\text{lbf ft}^{-2}.$$

At time $j = 3$, the valve is 3/4 closed $\rightarrow k_{valve} = 17$

$$\rightarrow p_{10,3} = \left(1.94\,\text{slug ft}^{-3}\right) * 17 * \left(V_{10,3}{}^2/2\right) = 16.49 V_{10,3}{}^2\text{lbf ft}^{-2}.$$

At time $j \geq 4$, the valve is fully closed $\rightarrow$ water stops flowing at Node 10

$$\rightarrow V_{10,j} = 0.00\,\text{ft sec}^{-1}$$

$\rightarrow$ applying Eq. (10.24) to Node 9 and the ambient atmosphere, one has:

$$p_{10,j} = \frac{p_{9,j-1}+0}{2} + \frac{\rho\,|V_s|}{2}\left(V_{9,j-1}-0\right) - \frac{\rho f\,|V_s|}{4D}\left(V_{9,j-1}\left|V_{9,j-1}\right| - 0\,|0|\right)(\Delta t)$$

$$= \frac{p_{9,j-1}}{2} + \frac{\left(1.94\,\text{slug ft}^{-3}\right) * \left(3930.32\,\text{ft sec}^{-1}\right)}{2} * \left(V_{9,j-1}\right)$$

$$- \frac{\left(1.94\,\text{slug ft}^{-3}\right) * 0.012 * \left(3930.32\,\text{ft sec}^{-1}\right)}{4 * (2\,\text{ft})} V_{9,j-1}\left|V_{9,j-1}\right| * (0.1272\,\text{sec})$$

$$= \frac{p_{9,j-1}}{2} + 3812.4104 V_{9,j-1} - 1.45482 V_{9,j-1}\left|V_{9,j-1}\right|.$$

Interior nodes ($i = 1, 2, \ldots, 9$) at time $j \geq 1$: use Eqs. (10.24) and (10.25).

The calculations can be done in an Excel spreadsheet, as shown in Figure 10.13. The profiles of pressure and velocity at the first 12 time steps are shown in Figure 10.14a,b, respectively, and the variations of pressure and velocity at Nodes 5 and 9 are plotted in Figure 10.14c.

At Node 10, when the valve is partially opened, the relationship between the pressure head and velocity follows the fitting loss equation given in Table 2.3. Once the valve is fully closed, however, the velocity will become zero. The velocity for a partial valve opening can be determined by applying Eq. (10.25) to Node 9 and the ambient atmosphere. With the lapse of time until the 12th time step ($j = 12$), the pressures at the interior nodes increase, but the velocities decrease, from downstream to upstream. Starting from the 13th time step ($j = 13$), the pressures start to decrease while the velocities increase with a reversed direction, from upstream to downstream. Such downstream–upstream–downstream fluctuation processes will repeat every few time steps, ultimately approaching a new steady-state flow condition, as indicated by the velocities at all nodes tending to become the same (Figure 10.14c).

For instance, at time $j = 1$, the upper boundary condition (cells D11:D12) is $p_{0,1} = 6246.8\,\text{lbf ft}^{-2}$ and $V_{0,1} = \left(14.30\,\text{ft sec}^{-1}\right)/2 + \left(6.55753 \times 10^{-5}\right) * \left(6246.8\,\text{lbf ft}^{-2} - 5354.20\,\text{lbf ft}^{-2}\right) - \left(1.908 \times 10^{-4}\right) * \left(14.30\,\text{ft sec}^{-1}\right) * \left|14.30\,\text{ft sec}^{-1}\right| = 7.17\,\text{ft sec}^{-1}$, whereas the lower boundary condition (cells N11:N12) is $V_{10,1} = \left(14.30\,\text{ft sec}^{-1}\right)/2 + \left(6.55753 \times 10^{-5}\right) * \left(594.93\,\text{lbf ft}^{-2}\right) - \left(1.908 \times 10^{-4}\right) * \left(14.30\,\text{ft sec}^{-1}\right) * \left|14.30\,\text{ft sec}^{-1}\right| = 7.15\,\text{ft sec}^{-1}$; and $p_{10,1} = \left(1.94\,\text{slug ft}^{-3}\right) * 0.26 * \left[\left(7.15\,\text{ft sec}^{-1}\right)^2/2\right] = 12.89\,\text{lbf ft}^{-2}$. At node $i = 9$ (cells M11:M12), $p_{9,1} = \left(1189.87\,\text{lbf ft}^{-2} + 0.00\,\text{lbf ft}^{-2}\right)/2 + \left[\left(1.94\,\text{slug ft}^{-3}\right) * \left(3930.32\,\text{ftsec}^{-1}\right)\right]/2 * \left(14.30\,\text{ft sec}^{-1} - 14.30\,\text{ft sec}^{-1}\right) - \left[\left(1.94\,\text{slug ft}^{-3}\right) * 0.012 * \left(3930.32\,\text{ft sec}^{-1}\right)\right]/(4 * 2\,\text{ft}) *$

	A	B	C	D	E	F	G	H	I	J	K	L	M	N
1	H_0 =	100	ft	D =	2	ft	Table 10-1:	$E_{m,w}$ =	4.5E+07	lbf ft^{-2}	k_{valve} =	17	3/4 closed	
2	ρ =	1.94	slug ft^{-3}	ε =	0.25	ft		$E_{m,p}$ =	7.2E+08	lbf ft^{-3}		5.6	1/2 closed	
3	g =	32.2	ft sec^{-2}	L =	5000	ft						0.26	1/4 closed	
4	$p_{0,j}$ =	6246.8	lbf ft^{-2}	f =	0.012		Eq. (10-17):	$\lvert V_s \rvert$ =	3930.32	ft sec^{-1}		0	fully open	
5				ke =	0.5									
6	Δx =	500	ft	Δt =	0.1272	sec								
7									Node i =					
8	j	t_j (sec)	Variables	0	1	2	3	4	5	6	7	8	9	10
9	0	0	$p_{i,0}$ (lbf ft^{-2})	6246.80	5354.40	4759.47	4164.53	3569.60	2974.67	2379.73	1784.80	1189.87	594.93	0.00
10			$V_{i,0}$ (ft sec^{-1})	14.30	14.30	14.30	14.30	14.30	14.30	14.30	14.30	14.30	14.30	14.30
11	1	0.1272	$p_{i,1}$ (lbf ft^{-2})	6246.80	5503.13	4759.47	4164.53	3569.60	2974.67	2379.73	1784.80	1189.87	594.93	12.89
12			$V_{i,1}$ (ft sec^{-1})	7.17	14.32	14.30	14.30	14.30	14.30	14.30	14.30	14.30	14.30	7.15
13	2	0.2544	$p_{i,2}$ (lbf ft^{-2})	6246.80	-21455.55	4907.39	4164.53	3569.60	2974.67	2379.73	1784.80	1189.87	27634.02	277.64
14			$V_{i,2}$ (ft sec^{-1})	7.17	10.78	14.32	14.30	14.30	14.30	14.30	14.30	14.30	10.75	7.15
15	3	0.3816	$p_{i,3}$ (lbf ft^{-2})	6246.80	-21455.16	-21922.19	4311.65	3569.60	2974.67	2379.73	1784.80	28099.72	27766.41	846.83
16			$V_{i,3}$ (ft sec^{-1})	7.19	10.78	10.80	14.32	14.30	14.30	14.30	14.30	10.77	10.73	7.17
17	4	0.5088	$p_{i,4}$ (lbf ft^{-2})	6246.80	-21519.56	-21921.80	-22389.35	3715.91	2974.67	2379.73	28565.94	28231.57	28115.47	54641.32
18			$V_{i,4}$ (ft sec^{-1})	7.19	10.81	10.80	10.82	14.32	14.30	14.30	10.79	10.75	10.72	0.00
19	5	0.6360	$p_{i,5}$ (lbf ft^{-2})	6246.80	-21519.18	-21986.46	-22388.96	-22857.04	3120.18	29032.70	28697.25	28579.72	82258.95	54770.44
20			$V_{i,5}$ (ft sec^{-1})	7.20	10.81	10.82	10.82	10.83	14.32	10.80	10.77	10.74	3.62	0.00
21	6	0.7632	$p_{i,6}$ (lbf ft^{-2})	6246.80	-21582.91	-21986.07	-22453.88	-22856.66	3200.05	29308.18	29044.50	82575.01	82452.10	54919.21
22			$V_{i,6}$ (ft sec^{-1})	7.20	10.83	10.82	10.84	10.83	7.37	10.80	10.76	3.66	3.63	0.00
23	7	0.8904	$p_{i,7}$ (lbf ft^{-2})	6246.80	-21582.53	-22050.07	-22453.50	3511.28	3330.48	3301.16	83035.58	82767.72	82675.40	55047.71
24			$V_{i,7}$ (ft sec^{-1})	7.22	10.83	10.85	10.84	7.39	7.35	7.34	3.71	3.67	3.64	0.00
25	8	1.0176	$p_{i,8}$ (lbf ft^{-2})	6246.80	-21645.61	-22049.69	3823.32	3641.33	3611.22	56999.28	56965.64	83135.17	82867.62	55196.51
26			$V_{i,8}$ (ft sec^{-1})	7.22	10.86	10.85	7.41	7.37	7.36	0.29	0.28	3.70	3.65	0.00
27	9	1.1448	$p_{i,9}$ (lbf ft^{-2})	6246.80	-21645.24	4136.19	3953.01	3922.11	57231.16	57197.09	57118.61	57084.80	83234.68	55324.40
28			$V_{i,9}$ (ft sec^{-1})	7.24	10.86	7.43	7.39	7.38	0.32	0.31	0.28	0.26	3.68	0.00
29	10	1.2720	$p_{i,10}$ (lbf ft^{-2})	6246.80	4449.88	4265.51	4233.81	57463.47	57428.95	57350.46	57316.20	57237.68	57203.67	55616.75
30			$V_{i,10}$ (ft sec^{-1})	7.23	7.45	7.41	7.40	0.35	0.34	0.31	0.29	0.26	0.25	0.00
31	11	1.3992	$p_{i,11}$ (lbf ft^{-2})	6246.80	4578.83	4546.34	57696.19	57661.22	57582.72	57548.02	57469.49	57435.04	57428.03	29541.39
32			$V_{i,11}$ (ft sec^{-1})	3.83	7.43	7.42	0.39	0.37	0.34	0.32	0.29	0.28	0.24	0.00
33	12	1.5264	$p_{i,12}$ (lbf ft^{-2})	6246.80	-8209.53	57929.34	57893.92	57815.40	57780.25	57701.71	57666.82	57659.79	44543.34	29619.57
34			$V_{i,12}$ (ft sec^{-1})	3.82	5.72	0.42	0.40	0.37	0.35	0.32	0.31	0.27	1.97	0.00

Figure 10.13 **Excerpt of the spreadsheet for the calculations for Example 10.12.**

$\left[\left(14.30\,\text{ft sec}^{-1}\right)\left|14.30\,\text{ft sec}^{-1}\right| - \left(14.30\,\text{ft sec}^{-1}\right)\left|14.30\,\text{ft sec}^{-1}\right|\right] * (0.1272\,\text{sec}) = 594.93\,\text{lbf ft}^{-2}$; and $V_{9,1} = \left(14.30\,\text{ft sec}^{-1} + 14.30\,\text{ft sec}^{-1}\right)/2 + 1/\left[2 * \left(1.94\,\text{slug ft}^{-3}\right) * \left(3930.32\,\text{ft sec}^{-1}\right)\right] * \left(1189.87\,\text{lbf ft}^{-2} - 0.00\,\text{lbf ft}^{-2}\right) - 0.012/(4 * 2\,\text{ft}) * \left[\left(14.30\,\text{ft sec}^{-1}\right)\left|14.30\,\text{ft sec}^{-1}\right| + \left(14.30\,\text{ft sec}^{-1}\right)\left|14.30\,\text{ft sec}^{-1}\right|\right] * (0.1272\,\text{sec}) = 14.30\,\text{ft sec}^{-1}$. Similar calculations can be done for nodes $i = 8, 7, \ldots, 1$. At other times $j = 2, 3, \ldots$, determine the upper and lower boundary conditions and do the calculations for the interior nodes in similar ways.

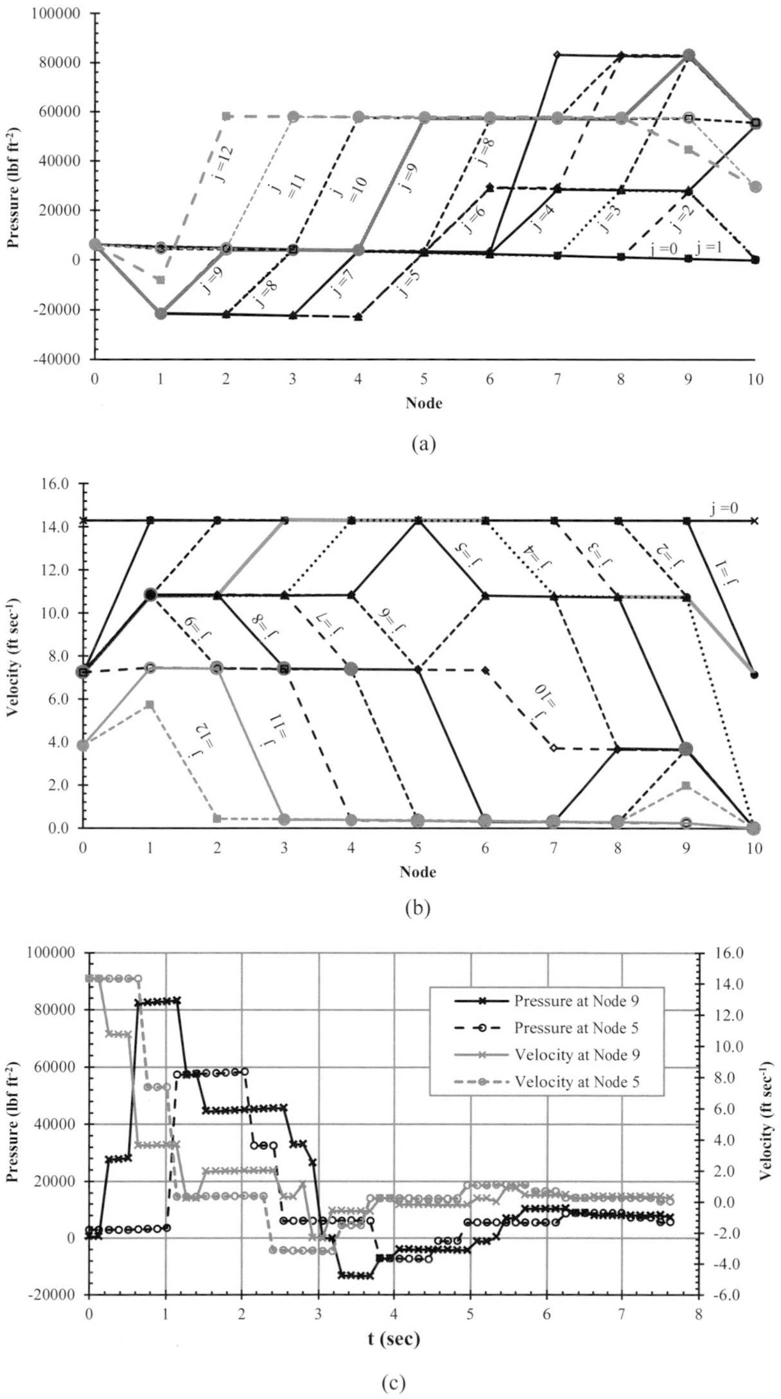

Figure 10.14 **Computed pressures and velocities for Example 10.12.** (a) Pressure profiles; (b) velocity profiles; and (c) fluctuations of pressure and velocity at Nodes 5 and 9.

10.2.2 Open Channels

In an open channel, the change of water density with pressure is minimal and thus can be neglected. As shown in Example 10.13 and Example 10.14, the Saint–Venant equations (Eqs. (10.12) and (10.13)) can be rewritten as:

water depth; flow velocity; flow area

$$\frac{\partial y}{\partial t}+V\frac{\partial y}{\partial x}+\frac{A}{B}\frac{\partial V}{\partial x}=0 \tag{10.26}$$

time; location along flow direction; top width

water depth; flow velocity; energy line gradient; gravitational acceleration

$$g\frac{\partial y}{\partial x}+\frac{\partial V}{\partial t}+V\frac{\partial V}{\partial x}+g\left(S_f-S_0\right)=0 \tag{10.27}$$

location along flow direction; time; channel bed slope

Example 10.13 Derive Eq. (10.26).

Solution

ρ is constant $\rightarrow$ Eq. (10.12): $\frac{\partial A}{\partial t}+V\frac{\partial A}{\partial x}+A\frac{\partial V}{\partial x}=0$

$$\partial A=B\left(\partial y\right)\rightarrow\frac{B\left(\partial y\right)}{\partial t}+V\frac{B\left(\partial y\right)}{\partial x}+A\frac{\partial V}{\partial x}=0\rightarrow\frac{\partial y}{\partial t}+V\frac{\partial y}{\partial x}+\frac{A}{B}\frac{\partial V}{\partial x}=0\rightarrow\text{Eq. (10.26).}$$

Example 10.14 Derive Eq. (10.27).

Solution

For a flow element of an open channel, the resultant force is:

$$\sum F=\rho g h_{c,1}A_1-\rho g h_{c,2}A_2-\left[\rho g\frac{\frac{A_1}{P_{wet,1}}+\frac{A_2}{P_{wet,2}}}{2}S_f\right]\left(\Delta x\frac{P_{wet,1}+P_{wet,2}}{2}\right)+\rho g\left(\Delta x\frac{A_1+A_2}{2}\right)S_0.$$

Divide Δx by both sides and let Δx approach zero:

$$\lim_{\Delta x\to 0}\sum\frac{F}{\Delta x}=\lim_{\Delta x\to 0}\left(\frac{\rho g h_{c,1}A_1-\rho g h_{c,2}A_2}{\Delta x}\right)-\rho g A S_f+\rho g A S_0=-\rho g\frac{\partial}{\partial x}\left(h_c A\right)+\rho g A\left(S_0-S_f\right).$$

In terms of moment theory: $h_c A=\int_0^A y\,dA$.

Leibniz's rule and mean value theorem:

$$\rightarrow\frac{\partial}{\partial x}\left(h_c A\right)=\frac{\partial}{\partial x}\int_0^A y\,d\Lambda=\int_0^A\frac{\partial y}{\partial x}d\Lambda+Y\frac{\partial A}{\partial x}-0\frac{\partial 0}{\partial x}=\frac{\partial y}{\partial x}\int_0^A d\Lambda=A\frac{\partial y}{\partial x}$$

$$\to \lim_{\Delta x \to 0} \sum \frac{F}{\Delta x} = -\rho g A \frac{\partial y}{\partial x} + \rho g A (S_0 - S_f).$$

Given that ρ is constant, the right side of Eq. (10.13):

$$\frac{\partial}{\partial t}(\rho AV) + \frac{\partial}{\partial x}(\rho AV^2) = \rho \frac{\partial}{\partial t}(AV) + \rho \frac{\partial}{\partial x}(AVV) = \rho \frac{\partial}{\partial t}(AV) + \rho AV \frac{\partial V}{\partial x} + \rho V \frac{\partial}{\partial x}(AV)$$

$$= \rho A \frac{\partial V}{\partial t} + \rho V \frac{\partial A}{\partial t} + \rho AV \frac{\partial V}{\partial x} + \rho V^2 \frac{\partial A}{\partial x} = \rho A \frac{\partial V}{\partial t} + \rho AV \frac{\partial V}{\partial x} + V \left[\frac{\partial}{\partial t}(\rho A) + \frac{\partial}{\partial x}(\rho AV) \right].$$

Eq. (10.12) $\to \frac{\partial}{\partial t}(\rho A) + \frac{\partial}{\partial x}(\rho AV) = 0$

$$\to \text{the right side of Eq. (10.13)} = \rho A \frac{\partial V}{\partial t} + \rho AV \frac{\partial V}{\partial x}.$$

Thus: $-\rho g A \frac{\partial y}{\partial x} + \rho g A (S_0 - S_f) = \rho A \frac{\partial V}{\partial t} + \rho AV \frac{\partial V}{\partial x}$

$\to$ divide by ρA on both sides and rearrange $g \frac{\partial y}{\partial x} + \frac{\partial V}{\partial t} + V \frac{\partial V}{\partial x} + g (S_f - S_0) = 0$

$\to$ Eq. (10.27).

Using the method of characteristics (MOC), as shown in Example 10.15, one can derive the solution of Eqs. (10.26) and (10.27), which is the intersection between two characteristic curves in the x–t plane (Figure 10.15). One of the curves has a positive slope and is called the positive characteristic curve, whereas the other curve has a negative slope and is called the negative characteristic curve.

The celerity of a shallow-water wave is the denominator in the formula for the Froude number (Eq. (2.46)). However, celerity is positive when the wave moves downstream, such as when induced by gate opening, but is negative when the wave moves upstream, such as when induced by gate closing (Roberson *et al.*, 1998). The magnitude of the celerity is computed as:

$$c = \sqrt{g \frac{A}{B}} \tag{10.28}$$

[c: celerity; g: gravitational acceleration; A: flow area; B: top width]

The positive characteristic curve is defined as:

$$\frac{dx}{dt} = V + c \tag{10.29}$$

[x: location along flow direction; t: time; V: flow velocity; c: celerity (Eq. (10.28))]

$$\frac{g}{c} \frac{dy}{dt} + \frac{dV}{dt} + g (S_f - S_0) = 0 \tag{10.30}$$

[g: gravitational acceleration; c: celerity (Eq. (10.28)); y: water depth; t: time; V: flow velocity; S_f: energy line gradient; S_0: channel bed slope]

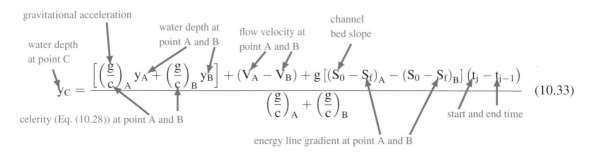

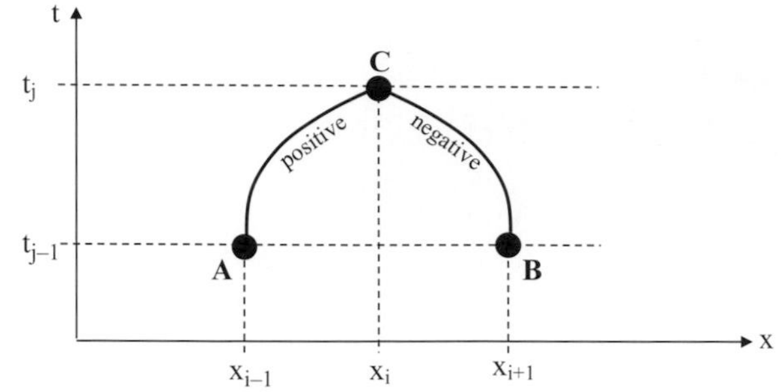

Figure 10.15 **The MOC solution to Eqs. (10.30) and (10.32).**

The negative characteristic curve is defined as:

$$\frac{dx}{dt} = V - c \tag{10.31}$$

location along flow direction; flow velocity; celerity (Eq. (10.28)); time

$$-\frac{g}{c}\frac{dy}{dt} + \frac{dV}{dt} + g\,(S_f - S_0) = 0 \tag{10.32}$$

gravitational acceleration; water depth; flow velocity; energy line gradient; celerity (Eq. (10.28)); time; channel bed slope

In Figure 10.15, point A signifies the water depth and velocity at location x_{i-1} and time t_{j-1}, while point B signifies the water depth and velocity at location x_{i+1} and time t_{j-1}, where $i = 1, 2, \ldots$, and $j = 1, 2, \ldots$. If the values of water depth and velocity at points A and B are known, one can solve Eqs (10.30) and (10.32) simultaneously to obtain the water depth and velocity at location x_i and time t_j as signified by point C. For an open channel of interest, the initial (i.e., time t_0) values of water depth and velocity at all locations should be known, so the corresponding values at other times can be determined.

As shown in Example 10.16, integrating Eq. (10.30) along the positive characteristic curve AC and Eq. (10.32) along the negative characteristic curve BC and rearranging, one can derive the solution as:

$$y_C = \frac{\left[\left(\frac{g}{c}\right)_A y_A + \left(\frac{g}{c}\right)_B y_B\right] + (V_A - V_B) + g\left[(S_0 - S_f)_A - (S_0 - S_f)_B\right](t_j - t_{j-1})}{\left(\frac{g}{c}\right)_A + \left(\frac{g}{c}\right)_B} \tag{10.33}$$

water depth at point C; gravitational acceleration; water depth at point A and B; flow velocity at point A and B; channel bed slope; celerity (Eq. (10.28)) at point A and B; start and end time; energy line gradient at point A and B

$$V_C = \frac{\left[\left(\frac{c}{g}\right)_A V_A + \left(\frac{c}{g}\right)_B V_B\right] + (y_A - y_B) + \left[c_A\left(S_0 - S_f\right)_A + c_B\left(S_0 - S_f\right)_B\right]\left(t_j - t_{j-1}\right)}{\left(\frac{c}{g}\right)_A + \left(\frac{c}{g}\right)_B} \tag{10.34}$$

celerity (Eq. (10.28)) at point A and B
flow velocity at point C
gravitational acceleration
flow velocity at point A and B
water depth at point A and B
channel bed slope
energy line gradient at point A and B
start and end time

To apply Eqs. (10.33) and (10.34), the channel is subdivided into a number of segments, each of which has two nodes Δx apart. For convergence, the computational time step should be $\Delta t \leq \Delta x/\left(V_0 + c_0\right)$, where V_0 and c_0 are the velocity and celerity, respectively, of the steady-state uniform flow in the channel. The values of water depth and velocity at the first and last nodes are defined by the upper and lower boundary conditions, respectively.

Example 10.15 Derive Eqs. (10.29) through (10.32).

Solution

$$\text{Eq. (10.26)} \rightarrow \frac{\partial y}{\partial t} + V\frac{\partial y}{\partial x} + \frac{A}{B}\frac{\partial V}{\partial x} = 0$$

$$\text{Eq. (10.27)} \rightarrow g\frac{\partial y}{\partial x} + \frac{\partial V}{\partial t} + V\frac{\partial V}{\partial x} + g\left(S_f - S_0\right) = 0.$$

Multiply the first equation by λ and then add it and the second equation together:

$$\lambda\left(\frac{\partial y}{\partial t} + V\frac{\partial y}{\partial x} + \frac{A}{B}\frac{\partial V}{\partial x}\right) + \left[g\frac{\partial y}{\partial x} + \frac{\partial V}{\partial t} + V\frac{\partial V}{\partial x} + g\left(S_f - S_0\right)\right] = 0$$

$$\rightarrow \lambda\left[\frac{\partial y}{\partial t} + \left(V + \frac{g}{\lambda}\right)\frac{\partial y}{\partial x}\right] + \left[\frac{\partial V}{\partial t} + \left(V + \lambda\frac{A}{B}\right)\frac{\partial V}{\partial x}\right] = g\left(S_0 - S_f\right).$$

Select λ to satisfy the relationship: $\frac{dx}{dt} = V + \frac{g}{\lambda} = V + \lambda\frac{A}{B} \rightarrow \lambda = \pm\sqrt{g\frac{B}{A}}$

$$\rightarrow \lambda\frac{dy}{dt} + \frac{dV}{dt} + g\left(S_f - S_0\right) = 0 \rightarrow \frac{\lambda}{g}\frac{dy}{dt} + \frac{1}{g}\frac{dV}{dt} + \left(S_f - S_0\right) = 0.$$

Eq. (10.28) $\rightarrow c = g/|\lambda|$

$$\lambda = \sqrt{g\frac{B}{A}} \rightarrow \frac{dx}{dt} = V + c;\ \frac{g}{c}\frac{dy}{dt} + \frac{dV}{dt} + g\left(S_f - S_0\right) = 0 \rightarrow \text{Eqs. (10.29) and (10.30)}$$

$$\lambda = -\sqrt{g\frac{B}{A}} \rightarrow \frac{dx}{dt} = V - c;\ -\frac{g}{c}\frac{dy}{dt} + \frac{dV}{dt} + g\left(S_f - S_0\right) = 0 \rightarrow \text{Eqs. (10.31) and (10.32).}$$

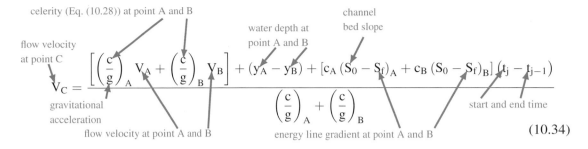

Example 10.16 Derive Eqs. (10.33) and (10.34).

Solution

Eq. (10.30) $\rightarrow \frac{g}{c}dy + dV + g\left(S_f - S_0\right)dt = 0$

$$\rightarrow \int_A^C \frac{g}{c}dy + \int_A^C dV + \int_A^C g\left(S_f - S_0\right)dt = 0$$

$$\rightarrow \left(\frac{g}{c}\right)_A \left(y_C - y_A\right) + \left(V_C - V_A\right) + g\left(S_f - S_0\right)_A\left(t_j - t_{j-1}\right) = 0.$$

Eq. (10.32) $\rightarrow -\frac{g}{c}dy + dV + g\left(S_f - S_0\right)dt = 0$

$$\rightarrow -\int_B^C \frac{g}{c}dy + \int_B^C dV + \int_B^C g\left(S_f - S_0\right)dt = 0$$

$$\rightarrow -\left(\frac{g}{c}\right)_B \left(y_C - y_B\right) + \left(V_C - V_B\right) + g\left(S_f - S_0\right)_B\left(t_j - t_{j-1}\right) = 0.$$

Subtract $\rightarrow \left[\left(\frac{g}{c}\right)_A + \left(\frac{g}{c}\right)_B\right]y_C - \left(\frac{g}{c}\right)_A y_A - \left(\frac{g}{c}\right)_B y_B - V_A + V_B$
$+g\left[\left(S_f - S_0\right)_A - \left(S_f - S_0\right)_B\right]\left(t_j - t_{j-1}\right) = 0$

$$\rightarrow \left[\left(\frac{g}{c}\right)_A + \left(\frac{g}{c}\right)_B\right]y_C = \left[\left(\frac{g}{c}\right)_A y_A + \left(\frac{g}{c}\right)_B y_B\right] + \left(V_A - V_B\right) + g\left[\left(S_0 - S_f\right)_A - \left(S_0 - S_f\right)_B\right]\left(t_j - t_{j-1}\right)$$

$\rightarrow$divide by $\left[\left(\frac{g}{c}\right)_A + \left(\frac{g}{c}\right)_B\right]$ on both sides to get Eq. (10.33).

Divide by $\left(\frac{g}{c}\right)_A \rightarrow \left(y_C - y_A\right) + \left(\frac{c}{g}\right)_A\left(V_C - V_A\right) + c_A\left(S_f - S_0\right)_A\left(t_j - t_{j-1}\right) = 0.$

Divide by $\left(\frac{g}{c}\right)_B \rightarrow -\left(y_C - y_B\right) + \left(\frac{c}{g}\right)_B\left(V_C - V_B\right) + c_B\left(S_f - S_0\right)_B\left(t_j - t_{j-1}\right) = 0.$

Add $\rightarrow -\left(y_A - y_B\right) + \left[\left(\frac{c}{g}\right)_A + \left(\frac{c}{g}\right)_B\right]V_C - \left[\left(\frac{c}{g}\right)_A V_A + \left(\frac{c}{g}\right)_B V_B\right] -$
$\left[c_A\left(S_0 - S_f\right)_A + c_B\left(S_0 - S_f\right)_B\right]\left(t_j - t_{j-1}\right) = 0$

$$\rightarrow \left[\left(\frac{c}{g}\right)_A + \left(\frac{c}{g}\right)_B\right]V_C = \left[\left(\frac{c}{g}\right)_A V_A + \left(\frac{c}{g}\right)_B V_B\right] + \left(y_A - y_B\right) + \left[c_A\left(S_0 - S_f\right)_A + c_B\left(S_0 - S_f\right)_B\right]\left(t_j - t_{j-1}\right)$$

$\rightarrow$divide by $\left[\left(\frac{c}{g}\right)_A + \left(\frac{c}{g}\right)_B\right]$ on both sides to get Eq. (10.34).

Example 10.17 A 50-m-long, 5-m-wide rectangular channel with a bed slope of 0.0005 and a Manning's n of 0.012 diverts water from a shallow fresh lake to irrigate a large field. The channel is designed for a steady-state uniform flow with a normal depth of 1.224 m. The lake water depth above the channel upstream-end invert is the same as the normal depth. At its downstream end, the channel is controlled by a sluice gate. If the gate is suddenly closed, determine the water depth and velocity profiles using $\Delta x = 5\,m$.

Solution

$L = 50\,m$, $b = 5\,m$, $S_0 = 0.0005$, $n = 0.012$, $y_0 = 1.224\,m$, $\Delta x = 5\,m$

Steady-state uniform flow in the channel:

$A_0 = (5\,m) * (1.224\,m) = 6.120\,m^2$
$B_0 = 5\,m$
$P_0 = 5\,m + 2 * (1.224\,m) = 7.448\,m$
$R_0 = A_0/P_0 = (6.120\,m^2) / (7.448\,m) = 0.822\,m$
Manning formula: $V_0 = (1/0.012) * (0.822\,m)^{2/3} * 0.0005^{1/2} = 1.634\,m\,s^{-1}$
$Q_0 = (1.634\,m\,s^{-1}) * (6.120\,m^2) = 10.0\,m^3 s^{-1}$
$c_0 = \left[(9.81\,m\,s^{-2}) * (6.120\,m^2) / (5\,m)\right]^{1/2} = 3.465\,m\,s^{-1}$.
$\Delta x = 5\,m \rightarrow$ 10 segments and 11 nodes, that is, $i = 0, 1, 2, \ldots, 10$.

$\Delta t = \Delta x / (V_0 + c_0) = (5\,m) / (1.634\,m\,s^{-1} + 3.465\,m\,s^{-1}) = 0.981\,s.$

Initial condition: cells D9:N13 in Figure 10.16, highlighted in red.

For $i = 0, 1, 2, \ldots, 10$:

$$y_{i,0} = y_0 = 1.224\,m$$
$$V_{i,0} = V_0 = 1.634\,m\,s^{-1}$$
$$c_{i,0} = c_0 = 3.465\,m\,s^{-1}$$
$$R_{i,0} = R_0 = 0.822\,m$$
$$S_{i,0} = \left(nV_{i,0}/R_{i,0}{}^{2/3}\right)^2 = \left[0.012 * (1.634\,m\,s^{-1}) / (0.822\,m)^{2/3}\right]^2 = 0.0005.$$

Upper boundary condition: cells D14:D48 in Figure 10.16, highlighted in blue.

At time $j \geq 1$:
$y_{0,j} = 1.224\,m$
Applying Eq. (10.34) to the lake and Node 1, one has:

$$V_{0,j} = \frac{\left[\left(\frac{c}{g}\right)_{1,j-1} V_{1,j-1}\right] + (y_0 - y_{1,j-1}) + \left[c_{1,j-1}(S_0 - S_f)_{1,j-1}\right](\Delta t)}{\left(\frac{c}{g}\right)_{1,j-1}}$$

$$A_{0,j} = 5y_{0,j}m^2$$
$$B_{0,j} = 5\,m$$
$$P_{0,j} = 5 + 2y_{0,j}m$$
$$R_{0,j} = A_{0,j}/P_{0,j} = (5y_{0,j}m^2) / (5 + 2y_{0,j}m) = 5y_{0,j}/(5 + 2y_{0,j})\ m$$
$$c_{0,j} = (gA_{0,j}/B_{0,j})^{1/2} = \left[g(5y_{0,j}m^2) / (5\,m)\right]^{1/2} = (gy_{0,j})^{1/2}\ m\,s^{-1}.$$

Lower boundary condition: cells N14:N48 in Figure 10.16, highlighted in blue.

At time $j \geq 1$:

$$V_{10,j} = 0.0\,\mathrm{m\,s^{-1}}$$

Applying Eq. (10.33) to Node 1 and the atmosphere, one has:

$$y_{10,j} = \frac{\left[\left(\frac{g}{c}\right)_{9,j-1} y_{9,j-1}\right] + V_{9,j-1} + g\left(S_0 - S_f\right)_{9,j-1}(\Delta t)}{\left(\frac{g}{c}\right)_{9,j-1}}$$

$$A_{10,j} = 5y_{10,j}\mathrm{m}^2$$
$$B_{10,j} = 5\,\mathrm{m}$$
$$P_{10,j} = 5 + 2y_{10,j}\mathrm{m}$$
$$R_{10,j} = A_{10,j}/P_{10,j} = \left(5y_{10,j}\mathrm{m}^2\right)/\left(5 + 2y_{10,j}\mathrm{m}\right) = 5y_{10,j}/\left(5 + 2y_{10,j}\right)\,\mathrm{m}$$
$$c_{10,j} = \left(gA_{10,j}/B_{10,j}\right)^{1/2} = \left[g\left(5y_{10,j}\mathrm{m}^2\right)/(5\,\mathrm{m})\right]^{1/2} = \left(gy_{10,j}\right)^{1/2}\,\mathrm{m\,s^{-1}}.$$

Interior nodes ($i = 1, 2, \ldots, 9$) at time $j \geq 1$: use Eqs. (10.33) and (10.34).

The calculations can be done in an Excel spreadsheet, as shown in Figure 10.16. The profiles of water depth and velocity at 1.96 s ($j = 2$) and 12.75 s ($j = 13$) after the gate is suddenly closed are shown in Figure 10.17.

At Node 10, when the gate is suddenly closed, the velocity will become zero while the water depth can be determined by applying Eq. (10.33) to Node 9 and the ambient atmosphere. As shown in Figure 10.17, with the lapse of time until the 11th time step ($j = 11$), the water depths at the interior nodes increase, but the velocities decrease, from downstream to upstream. Starting from the 12th time step ($j = 12$), the water depths start to decrease while the velocities increase with a reversed direction, from upstream to downstream. Such downstream–upstream–downstream fluctuation processes will repeat every few time steps, ultimately approaching a new steady-state flow condition, as indicated by the velocities at all nodes that tend become very similar (Figure 10.17a).

For instance, at time $j = 1$, the upper boundary condition (cells D14:D18) is $y_{0,1} = 1.224\,\mathrm{m}$; $V_{0,1} = \left[\left(3.465\,\mathrm{m\,s^{-1}}\right)/\left(9.81\,\mathrm{m\,s^{-2}}\right) * \left(1.634\,\mathrm{m\,s^{-1}}\right) + (1.224\,\mathrm{m} - 1.224\,\mathrm{m}) + \left(3.465\,\mathrm{m\,s^{-1}}\right) * (0.0005 - 0.0005) * (0.981\,\mathrm{s})\right] / \left[\left(3.465\,\mathrm{m\,s^{-1}}\right)/\left(9.81\,\mathrm{m\,s^{-2}}\right)\right] = 1.634\,\mathrm{m\,s^{-1}}$;
$A_{0,1} = (5\,\mathrm{m}) * (1.224\,\mathrm{m}) = 6.120\,\mathrm{m}^2$; $B_{0,1} = 5\,\mathrm{m}$; $P_{0,1} = 5\,\mathrm{m} + 2 * (1.224\,\mathrm{m}) = 7.448\,\mathrm{m}$;
$R_{0,1} = \left(6.120\,\mathrm{m}^2\right)/(7.448\,\mathrm{m}) = 0.822\,\mathrm{m}$; $c_{0,1} = \left[\left(9.81\mathrm{m\,s^{-2}}\right) * (1.224\,\mathrm{m})\right]^{1/2} = 3.465\,\mathrm{m\,s^{-1}}$;
and $S_{f,0,1} = \left[\left(0.012 * 1.634\,\mathrm{m\,s^{-1}}\right)/(0.822\,\mathrm{m})^{2/3}\right]^2 = 0.0005$. In contrast,

	A	B	C	D	E	F	G	H	I	J	K	L	M	N
1	b =	5	m	Q_0 =	10.00	$m^3 s^{-1}$	g =	9.81	$m s^{-2}$					
2	Z =	0		y_0 =	1.224	m	c_0 =	3.465	$m s^{-1}$					
3	n =	0.012		A_0 =	6.120	m^2	V_0 =	1.635	$m s^{-1}$					
4	L =	50	m	B_0 =	5.000	m	K =	1						
5	S_0 =	0.0005		R_0 =	0.822	m								
6	Δx =	5	m	Δt =	0.981	sec								
7									Node i =					
8	j	t_j (sec)	Variables	0	1	2	3	4	5	6	7	8	9	10
9	0	0.000	$y_{i,0}$ (m)	1.224	1.224	1.224	1.224	1.224	1.224	1.224	1.224	1.224	1.224	1.224
10			$V_{i,0}$ ($m s^{-1}$)	1.634	1.634	1.634	1.634	1.634	1.634	1.634	1.634	1.634	1.634	1.634
11			$c_{i,0}$ ($m s^{-1}$)	3.465	3.465	3.465	3.465	3.465	3.465	3.465	3.465	3.465	3.465	3.465
12			$R_{i,0}$ (m)	0.822	0.822	0.822	0.822	0.822	0.822	0.822	0.822	0.822	0.822	0.822
13			$S_{f,i,0}$	0.00050	0.00050	0.00050	0.00050	0.00050	0.00050	0.00050	0.00050	0.00050	0.00050	0.00050
14	1	0.981	$y_{i,1}$ (lbf ft^{-2})	1.224	1.224	1.224	1.224	1.224	1.224	1.224	1.224	1.224	1.224	1.801
15			$V_{i,1}$ (ft sec^{-1})	1.634	1.634	1.634	1.634	1.634	1.634	1.634	1.634	1.634	1.634	0.000
16			$c_{i,1}$ ($m s^{-1}$)	3.465	3.465	3.465	3.465	3.465	3.465	3.465	3.465	3.465	3.465	4.204
17			$R_{i,1}$ (m)	0.822	0.822	0.822	0.822	0.822	0.822	0.822	0.822	0.822	0.822	1.047
18			$S_{f,i,1}$	0.00050	0.00050	0.00050	0.00050	0.00050	0.00050	0.00050	0.00050	0.00050	0.00050	0.00000
19	2	1.961	$y_{i,2}$ (lbf ft^{-2})	1.224	1.224	1.224	1.224	1.224	1.224	1.224	1.224	1.224	1.800	1.801
20			$V_{i,2}$ (ft sec^{-1})	1.634	1.634	1.634	1.634	1.634	1.634	1.634	1.634	1.634	0.003	0.000
21			$c_{i,2}$ ($m s^{-1}$)	3.465	3.465	3.465	3.465	3.465	3.465	3.465	3.465	3.465	4.202	4.204
22			$R_{i,2}$ (m)	0.822	0.822	0.822	0.822	0.822	0.822	0.822	0.822	0.822	1.047	1.047
23			$S_{f,i,2}$	0.00050	0.00050	0.00050	0.00050	0.00050	0.00050	0.00050	0.00050	0.00050	0.00000	0.00000
24	3	2.942	$y_{i,3}$ (lbf ft^{-2})	1.224	1.224	1.224	1.224	1.224	1.224	1.224	1.224	1.799	1.800	1.803
25			$V_{i,3}$ (ft sec^{-1})	1.634	1.634	1.634	1.634	1.634	1.634	1.634	1.634	0.005	0.003	0.000
26			$c_{i,3}$ ($m s^{-1}$)	3.465	3.465	3.465	3.465	3.465	3.465	3.465	3.465	4.201	4.202	4.206
27			$R_{i,3}$ (m)	0.822	0.822	0.822	0.822	0.822	0.822	0.822	0.822	1.046	1.047	1.048
28			$S_{f,i,3}$	0.00050	0.00050	0.00050	0.00050	0.00050	0.00050	0.00050	0.00050	0.00000	0.00000	0.00000
29	4	3.922	$y_{i,4}$ (lbf ft^{-2})	1.224	1.224	1.224	1.224	1.224	1.224	1.224	1.798	1.799	1.803	1.803
30			$V_{i,4}$ (ft sec^{-1})	1.634	1.634	1.634	1.634	1.634	1.634	1.634	0.008	0.005	0.003	0.000
31			$c_{i,4}$ ($m s^{-1}$)	3.465	3.465	3.465	3.465	3.465	3.465	3.465	4.200	4.201	4.205	4.206
32			$R_{i,4}$ (m)	0.822	0.822	0.822	0.822	0.822	0.822	0.822	1.046	1.046	1.047	1.048
33			$S_{f,i,4}$	0.00050	0.00050	0.00050	0.00050	0.00050	0.00050	0.00050	0.00000	0.00000	0.00000	0.00000
34	5	4.903	$y_{i,5}$ (lbf ft^{-2})	1.224	1.224	1.224	1.224	1.224	1.224	1.797	1.798	1.802	1.803	1.806
35			$V_{i,5}$ (ft sec^{-1})	1.634	1.634	1.634	1.634	1.634	1.634	0.011	0.008	0.005	0.003	0.000
36			$c_{i,5}$ ($m s^{-1}$)	3.465	3.465	3.465	3.465	3.465	3.465	4.199	4.200	4.204	4.205	4.209
37			$R_{i,5}$ (m)	0.822	0.822	0.822	0.822	0.822	0.822	1.046	1.046	1.047	1.047	1.048
38			$S_{f,i,5}$	0.00050	0.00050	0.00050	0.00050	0.00050	0.00050	0.00000	0.00000	0.00000	0.00000	0.00000
39	6	5.883	$y_{i,6}$ (lbf ft^{-2})	1.224	1.224	1.224	1.224	1.224	1.797	1.797	1.801	1.802	1.805	1.806
40			$V_{i,6}$ (ft sec^{-1})	1.634	1.634	1.634	1.634	1.634	0.013	0.011	0.008	0.005	0.003	0.000
41			$c_{i,6}$ ($m s^{-1}$)	3.465	3.465	3.465	3.465	3.465	4.198	4.199	4.203	4.204	4.208	4.209
42			$R_{i,6}$ (m)	0.822	0.822	0.822	0.822	0.822	1.045	1.046	1.047	1.047	1.048	1.048
43			$S_{f,i,6}$	0.00050	0.00050	0.00050	0.00050	0.00050	0.00000	0.00000	0.00000	0.00000	0.00000	0.00000
44	7	6.864	$y_{i,7}$ (lbf ft^{-2})	1.224	1.224	1.224	1.224	1.796	1.797	1.800	1.801	1.804	1.805	1.808
45			$V_{i,7}$ (ft sec^{-1})	1.634	1.634	1.634	1.634	0.016	0.013	0.011	0.008	0.005	0.003	0.000
46			$c_{i,7}$ ($m s^{-1}$)	3.465	3.465	3.465	3.465	4.197	4.198	4.202	4.203	4.207	4.208	4.211
47			$R_{i,7}$ (m)	0.822	0.822	0.822	0.822	1.045	1.045	1.046	1.047	1.048	1.048	1.049
48			$S_{f,i,7}$	0.00050	0.00050	0.00050	0.00050	0.00000	0.00000	0.00000	0.00000	0.00000	0.00000	0.00000

Figure 10.16 **Excerpt of the spreadsheet for the calculations for Example 10.17.**

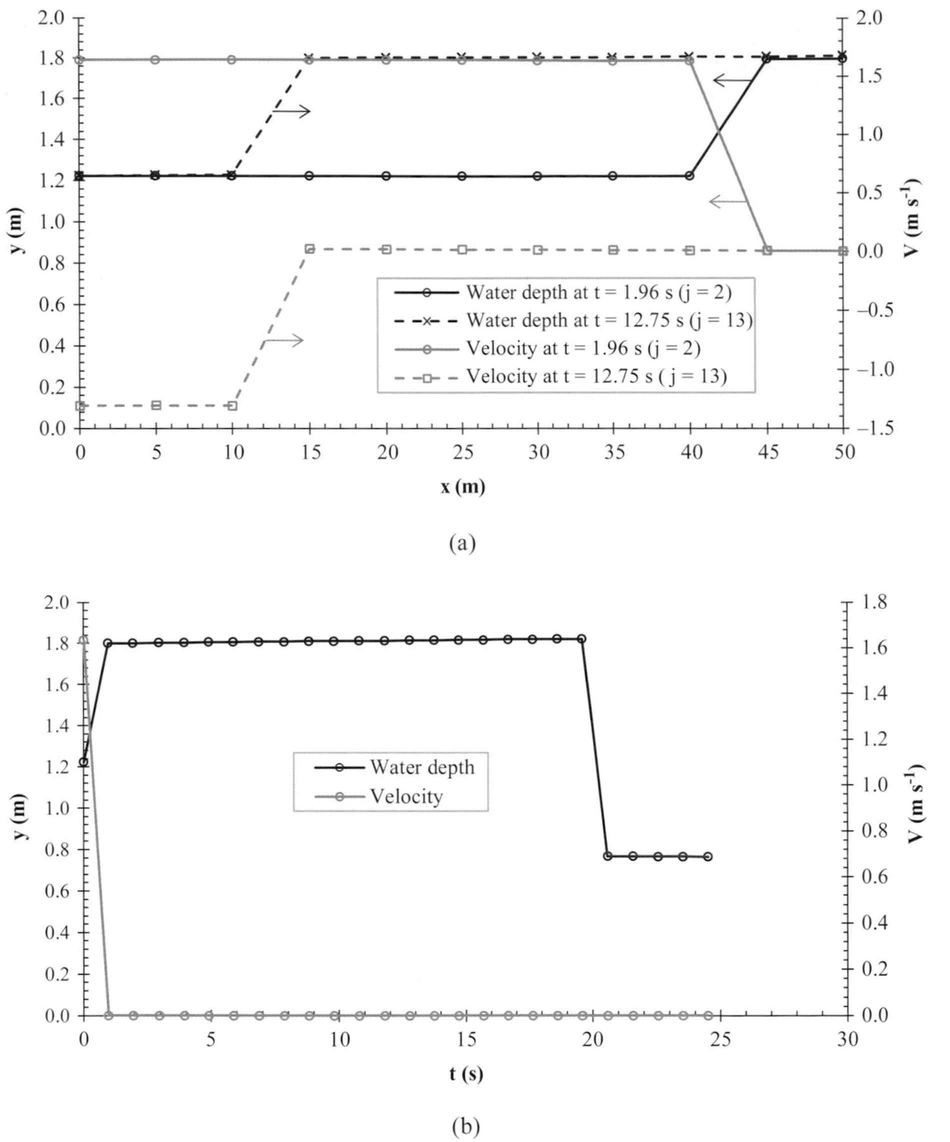

Figure 10.17 Computed water depths and velocities for Example 10.17. Profiles after the gate is suddenly closed at: (a) two selected times; and (b) Node 10.

the lower boundary condition (cells N14:N18) is $V_{10,1} = 0.0\,\mathrm{m\,s^{-1}}$; $y_{10,1} = \left[\left(9.81\,\mathrm{m\,s^{-2}}\right)/\left(3.465\,\mathrm{m\,s^{-1}}\right) * (1.224\,\mathrm{m}) + 1.634\,\mathrm{m\,s^{-1}} + \left(9.81\,\mathrm{m\,s^{-2}}\right) * (0.0005 - 0.0005) * (0.981\,\mathrm{s})\right] / \left[\left(9.81\,\mathrm{m\,s^{-2}}\right)/\left(3.465\,\mathrm{m\,s^{-1}}\right)\right] = 1.801\,\mathrm{m}$; $A_{10,1} = (5\,\mathrm{m}) * (1.801\,\mathrm{m}) = 9.005\,\mathrm{m^2}$; $B_{10,1} = 5\,\mathrm{m}$; $P_{10,1} = 5\,\mathrm{m} + 2 * (1.801\,\mathrm{m}) = 8.602\,\mathrm{m}$; $R_{10,1} = \left(9.005\,\mathrm{m^2}\right)/(8.602\,\mathrm{m}) = 1.047\,\mathrm{m}$; $c_{10,1} = \left[\left(9.81\,\mathrm{m\,s^{-2}}\right) * (1.801\,\mathrm{m})\right]^{1/2} = 4.204\,\mathrm{m\,s^{-1}}$; and $S_{f,10,1} = \left[\left(0.012 * 0.0\,\mathrm{m\,s^{-1}}\right)/(1.238\,\mathrm{m})^{2/3}\right]^2$

= 0.0. At node i = 9 (cells M14:M18), $y_{9,1} = \{[(9.81\,m\,s^{-2}) / (3.465\,m\,s^{-1}) * (1.224\,m) + [(9.81\,m\,s^{-2}) / (3.465\,m\,s^{-1}) * (1.224\,m)] + (1.634\,m\,s^{-1} - 1.634\,m\,s^{-1}) + (9.81\,m\,s^{-2}) * [(0.0005 - 0.0005) - (0.0005 - 0.0005)] * (0.981\,s)]\} / [(9.81\,m\,s^{-2}) / (3.465\,m\,s^{-1}) + (9.81\,m\,s^{-2}) / (3.465\,m\,s^{-1})] = 1.224\,m$; $V_{9,1} = \{[(3.465\,m\,s^{-1}) / (9.81\,m\,s^{-2}) * (1.634\,m\,s^{-1}) + (3.465\,m\,s^{-1}) / (9.81\,m\,s^{-2}) * (1.634\,m\,s^{-1})] + (1.224\,m - 1.224\,m) + [(3.465\,m\,s^{-1}) * (0.0005 - 0.0005) + (3.465\,m\,s^{-1}) * (0.0005 - 0.0005)] * (0.981\,s)\} / [(3.465\,m\,s^{-1}) / (9.81\,m\,s^{-2}) + (3.465\,m\,s^{-1}) / (9.81\,m\,s^{-2})] = 1.634\,m\,s^{-1}$; $A_{9,1} = (5\,m) * (1.224\,m) = 6.120\,m^2$; $B_{9,1} = 5\,m$; $P_{9,1} = 5\,m + 2 * (1.224\ m) = 7.448\,m$; $R_{9,1} = (6.120\,m^2)/(7.448\,m) = 0.822\,m$; $c_{9,1} = [(9.81\,m\,s^{-2}) * (1.224\,m)]^{1/2} = 3.465\,m\,s^{-1}$; and $S_{f,9,1} = \left[(0.012 * 1.634\,m\,s^{-1}) / (0.822\,m)^{2/3}\right]^2 = 0.0005$. Similar calculations can be done for nodes i = 8, 7, . . . , 1. At other times j = 2, 3, . . ., determine the upper and lower boundary conditions and do the calculations for the interior nodes in similar ways.

Example 10.18 A 50-m-long, 5-m-wide rectangular channel with a bed slope of 0.0005 and a Manning's n of 0.012 diverts water from a shallow fresh lake to irrigate a large field. It is controlled by a sluice gate at the downstream end. Once established, the steady-state uniform flow in the channel will have a normal depth of 1.224 m. The lake water depth above the channel upstream-end invert is the same as the normal depth. At the beginning, the gate is fully closed. If the gate is suddenly opened, determine the water depth and velocity profiles using $\Delta x = 5\,m$.

Solution

$L = 50\,m$, $b = 5\,m$, $S_0 = 0.0005$, $n = 0.012$, $y_0 = 1.224\,m$, $\Delta x = 5\,m$

Steady-state uniform flow in the channel:

$A_0 = (5\,m) * (1.224\,m) = 6.120\,m^2$

$B_0 = 5\,m$

$P_0 = 5\,m + 2 * (1.224\,m) = 7.448\,m$

$R_0 = A_0/P_0 = (6.120\,m^2) / (7.448\,m) = 0.822\,m$

Manning formula: $V_0 = (1/0.012) * (0.822\,m)^{2/3} * 0.0005^{1/2} = 1.634\,m\,s^{-1}$

$Q_0 = (1.634\,m\,s^{-1}) * (6.120\,m^2) = 10.0\,m^3 s^{-1}$

$c_0 = [(9.81\,m\,s^{-2}) * (6.120\,m^2) / (5\,m)]^{1/2} = 3.465\,m\,s^{-1}$.

$\Delta x = 5\,m \rightarrow 10$ segments and 11 nodes, that is, i = 0, 1, 2, . . . , 10.

$\Delta t = \Delta x / (V_0 + c_0) = (5\,m) / (1.634\,m\,s^{-1} + 3.465\,m\,s^{-1}) = 0.981\,s$.

Initial condition: cells D9:N13 in Figure 10.18, highlighted in red.

For i = 0, 1, 2, . . . , 10:

$y_{i,0} = y_0 = 1.224\,m$

Gate is fully closed $\rightarrow V_{i,0} = 0.0\,m\,s^{-1}$

	A	B	C	D	E	F	G	H	I	J	K	L	M	N
1	b =	5	m	Q_0 =	10.00	m^3 s^{-1}	g =	9.81	m s^{-2}					
2	Z =	0		y_0 =	1.224	m	c_0 =	3.465	m s^{-1}					
3	n =	0.012		A_0 =	6.120	m^2	V_0 =	1.635	m s^{-1}					
4	L =	50	m	B_0 =	5.000	m	K =	1						
5	S_0 =	0.0005		R_0 =	0.822	m								
6	Δx =	5	m	Δt =	0.981	sec								
7								Node i =						
8	j	t_j (sec)	Variables	0	1	2	3	4	5	6	7	8	9	10
9	0	0.000	$y_{i,0}$ (m)	1.224	1.224	1.224	1.224	1.224	1.224	1.224	1.224	1.224	1.224	1.224
10			$V_{i,0}$ (m s^{-1})	0.000	0.000	0.000	0.000	0.000	0.000	0.000	0.000	0.000	0.000	0.000
11			$c_{i,0}$ (m s^{-1})	3.465	3.465	3.465	3.465	3.465	3.465	3.465	3.465	3.465	3.465	3.465
12			$R_{i,0}$ (m)	0.822	0.822	0.822	0.822	0.822	0.822	0.822	0.822	0.822	0.822	0.822
13			$S_{fi,0}$	0.00E+00	0.00E+00	0.00E+00	0.00E+00	0.00E+00	0.00E+00	0.00E+00	0.00E+00	0.00E+00	0.00E+00	0.00E+00
14	1	0.981	$y_{i,1}$ (lbf ft^{-2})	1.224	1.224	1.224	1.224	1.224	1.224	1.224	1.224	1.224	1.224	1.226
15			$V_{i,1}$ (ft sec^{-1})	0.005	0.005	0.005	0.005	0.005	0.005	0.005	0.005	0.005	0.005	3.470
16			$c_{i,1}$ (m s^{-1})	3.465	3.465	3.465	3.465	3.465	3.465	3.465	3.465	3.465	3.465	3.468
17			$R_{i,1}$ (m)	0.822	0.822	0.822	0.822	0.822	0.822	0.822	0.822	0.822	0.822	0.822
18			$S_{fi,1}$	4.33E-09	4.33E-09	4.33E-09	4.33E-09	4.33E-09	4.33E-09	4.33E-09	4.33E-09	4.33E-09	4.33E-09	2.25E-03
19	2	1.961	$y_{i,2}$ (lbf ft^{-2})	1.224	1.224	1.224	1.224	1.224	1.224	1.224	1.224	1.224	0.616	1.227
20			$V_{i,2}$ (ft sec^{-1})	0.010	0.010	0.010	0.010	0.010	0.010	0.010	0.010	0.010	1.730	3.475
21			$c_{i,2}$ (m s^{-1})	3.465	3.465	3.465	3.465	3.465	3.465	3.465	3.465	3.465	2.459	3.470
22			$R_{i,2}$ (m)	0.822	0.822	0.822	0.822	0.822	0.822	0.822	0.822	0.822	0.495	0.823
23			$S_{fi,2}$	1.73E-08	1.73E-08	1.73E-08	1.73E-08	1.73E-08	1.73E-08	1.73E-08	1.73E-08	1.73E-08	1.10E-03	2.25E-03
24	3	2.942	$y_{i,3}$ (lbf ft^{-2})	1.224	1.224	1.224	1.224	1.224	1.224	1.224	1.224	0.618	0.617	1.049
25			$V_{i,3}$ (ft sec^{-1})	0.014	0.014	0.014	0.014	0.014	0.014	0.014	0.014	1.730	1.733	4.183
26			$c_{i,3}$ (m s^{-1})	3.465	3.465	3.465	3.465	3.465	3.465	3.465	3.465	2.462	2.460	3.207
27			$R_{i,3}$ (m)	0.822	0.822	0.822	0.822	0.822	0.822	0.822	0.822	0.496	0.495	0.739
28			$S_{fi,3}$	3.90E-08	3.90E-08	3.90E-08	3.90E-08	3.90E-08	3.90E-08	3.90E-08	3.90E-08	1.10E-03	1.10E-03	3.77E-03
29	4	3.922	$y_{i,4}$ (lbf ft^{-2})	1.224	1.224	1.224	1.224	1.224	1.224	1.224	0.620	0.619	0.460	1.050
30			$V_{i,4}$ (ft sec^{-1})	0.019	0.019	0.019	0.019	0.019	0.019	0.019	1.730	1.733	2.352	4.187
31			$c_{i,4}$ (m s^{-1})	3.465	3.465	3.465	3.465	3.465	3.465	3.465	2.465	2.464	2.125	3.210
32			$R_{i,4}$ (m)	0.822	0.822	0.822	0.822	0.822	0.822	0.822	0.497	0.496	0.389	0.740
33			$S_{fi,4}$	6.92E-08	6.92E-08	6.92E-08	6.92E-08	6.92E-08	6.92E-08	6.92E-08	1.10E-03	1.10E-03	2.81E-03	3.78E-03
34	5	4.903	$y_{i,5}$ (lbf ft^{-2})	1.224	1.224	1.224	1.224	1.224	1.224	0.621	0.620	0.464	0.461	0.965
35			$V_{i,5}$ (ft sec^{-1})	0.024	0.024	0.024	0.024	0.024	0.024	1.731	1.733	2.345	2.355	4.455
36			$c_{i,5}$ (m s^{-1})	3.465	3.465	3.465	3.465	3.465	3.465	2.468	2.467	2.133	2.127	3.077
37			$R_{i,5}$ (m)	0.822	0.822	0.822	0.822	0.822	0.822	0.498	0.497	0.391	0.389	0.696
38			$S_{fi,5}$	1.08E-07	1.08E-07	1.08E-07	1.08E-07	1.08E-07	1.08E-07	1.09E-03	1.10E-03	2.77E-03	2.81E-03	4.63E-03
39	6	5.883	$y_{i,6}$ (lbf ft^{-2})	1.224	1.224	1.224	1.224	1.224	0.623	0.622	0.467	0.464	0.400	0.967
40			$V_{i,6}$ (ft sec^{-1})	0.029	0.029	0.029	0.029	0.029	1.731	1.734	2.338	2.348	2.615	4.459
41			$c_{i,6}$ (m s^{-1})	3.465	3.465	3.465	3.465	3.465	2.471	2.470	2.140	2.134	1.982	3.080
42			$R_{i,6}$ (m)	0.822	0.822	0.822	0.822	0.822	0.499	0.498	0.393	0.392	0.345	0.697
43			$S_{fi,6}$	1.56E-07	1.56E-07	1.56E-07	1.56E-07	1.56E-07	1.09E-03	1.10E-03	2.73E-03	2.77E-03	4.07E-03	4.63E-03
44	7	6.864	$y_{i,7}$ (lbf ft^{-2})	1.224	1.224	1.224	1.224	0.624	0.623	0.470	0.468	0.405	0.401	0.922
45			$V_{i,7}$ (ft sec^{-1})	0.034	0.034	0.034	0.034	1.732	1.734	2.331	2.341	2.602	2.617	4.562
46			$c_{i,7}$ (m s^{-1})	3.465	3.465	3.465	3.465	2.475	2.473	2.147	2.142	1.992	1.983	3.007
47			$R_{i,7}$ (m)	0.822	0.822	0.822	0.822	0.499	0.499	0.396	0.394	0.348	0.346	0.673
48			$S_{fi,7}$	2.12E-07	2.12E-07	2.12E-07	2.12E-07	1.09E-03	1.09E-03	2.69E-03	2.73E-03	3.98E-03	4.07E-03	5.08E-03

Figure 10.18 **Excerpt of the spreadsheet for the calculations for Example 10.18.**

$$c_{i,0} = c_0 = 3.465\,\mathrm{m\,s^{-1}}$$
$$R_{i,0} = R_0 = 0.822\,\mathrm{m}$$
$$S_{i,0} = \left(nV_{i,0}/R_{i,0}{}^{2/3}\right)^2 = \left[0.012 * \left(0.0\,\mathrm{m\,s^{-1}}\right)/\left(0.822\,\mathrm{m}\right)^{2/3}\right]^2 = 0.0.$$

Upper boundary condition: cells D14:D48 in Figure 10.18, highlighted in blue.

At time $j \geq 1$:

$$y_{0,j} = 1.224\,\mathrm{m}$$

Applying Eq. (10.34) to the lake and Node 1, one has:

$$V_{0,j} = \frac{\left[\left(\frac{c}{g}\right)_{1,j-1} V_{1,j-1}\right] + \left(y_0 - y_{1,j-1}\right) + \left[c_{1,j-1}\left(S_0 - S_f\right)_{1,j-1}\right](\Delta t)}{\left(\frac{c}{g}\right)_{1,j-1}}.$$

$$A_{0,j} = 5y_{0,j}\mathrm{m}^2$$
$$B_{0,j} = 5\,\mathrm{m}$$
$$P_{0,j} = 5 + 2y_{0,j}\mathrm{m}$$
$$R_{0,j} = A_{0,j}/P_{0,j} = \left(5y_{0,j}\mathrm{m}^2\right)/\left(5 + 2y_{0,j}\mathrm{m}\right) = 5y_{0,j}/\left(5 + 2y_{0,j}\right)\ \mathrm{m}$$
$$c_{0,j} = \left(gA_{0,j}/B_{0,j}\right)^{1/2} = \left[g\left(5y_{0,j}\mathrm{m}^2\right)/\left(5\,\mathrm{m}\right)\right]^{1/2} = \left(gy_{0,j}\right)^{1/2}\ \mathrm{m\,s^{-1}}$$
$$S_{0,j} = \left(nV_{0,j}/R_{0,j}{}^{2/3}\right)^2.$$

Lower boundary condition: cells N14:N48 in Figure 10.18, highlighted in blue.

At time $j \geq 1$:

Applying Eq. (10.33) to Node 9 and the ambient atmosphere, one has:

$$y_{10,j} = \frac{\left[\left(\frac{g}{c}\right)_{9,j-1} y_{9,j-1}\right] + V_{9,j-1} + g\left[\left(S_0 - S_f\right)_{9,j-1}\right](\Delta t)}{\left(\frac{g}{c}\right)_{9,j-1}}.$$

Applying Eq. (10.34) to Node 9 and the ambient atmosphere, one has:

$$V_{10,j} = \frac{\left[\left(\frac{c}{g}\right)_{9,j-1} V_{9,j-1}\right] + y_{9,j-1} + \left[c_{9,j-1}\left(S_0 - S_f\right)_{9,j-1}\right](\Delta t)}{\left(\frac{c}{g}\right)_{9,j-1}}.$$

$$A_{10,j} = 5y_{10,j}\,\mathrm{m}^2$$
$$B_{10,j} = 5\,\mathrm{m}$$
$$P_{10,j} = 5 + 2y_{10,j}\,\mathrm{m}$$
$$R_{10,j} = A_{10,j}/P_{10,j} = \left(5y_{10,j}\,\mathrm{m}^2\right)/\left(5 + 2y_{10,j}\,\mathrm{m}\right) = 5y_{10,j}/(5 + 2y_{10,j})\,\mathrm{m}$$

$$c_{10,j} = \left(gA_{10,j}/B_{10,j}\right)^{1/2} = \left[\left(9.81\,\mathrm{m\,s^{-2}}\right) * \left(5y_{10,j}\,\mathrm{m^2}\right) / (5\,\mathrm{m})\right]^{1/2} = \left(9.81y_{10,j}\right)^{1/2}\ \mathrm{m\,s^{-1}}$$

$$S_{10,j} = \left(nV_{10,j}/R_{10,j}{}^{2/3}\right)^2.$$

<u>Interior nodes ($i = 1, 2, \ldots, 9$) at time $j \geq 1$</u>: use Eqs. (10.33) and (10.34)

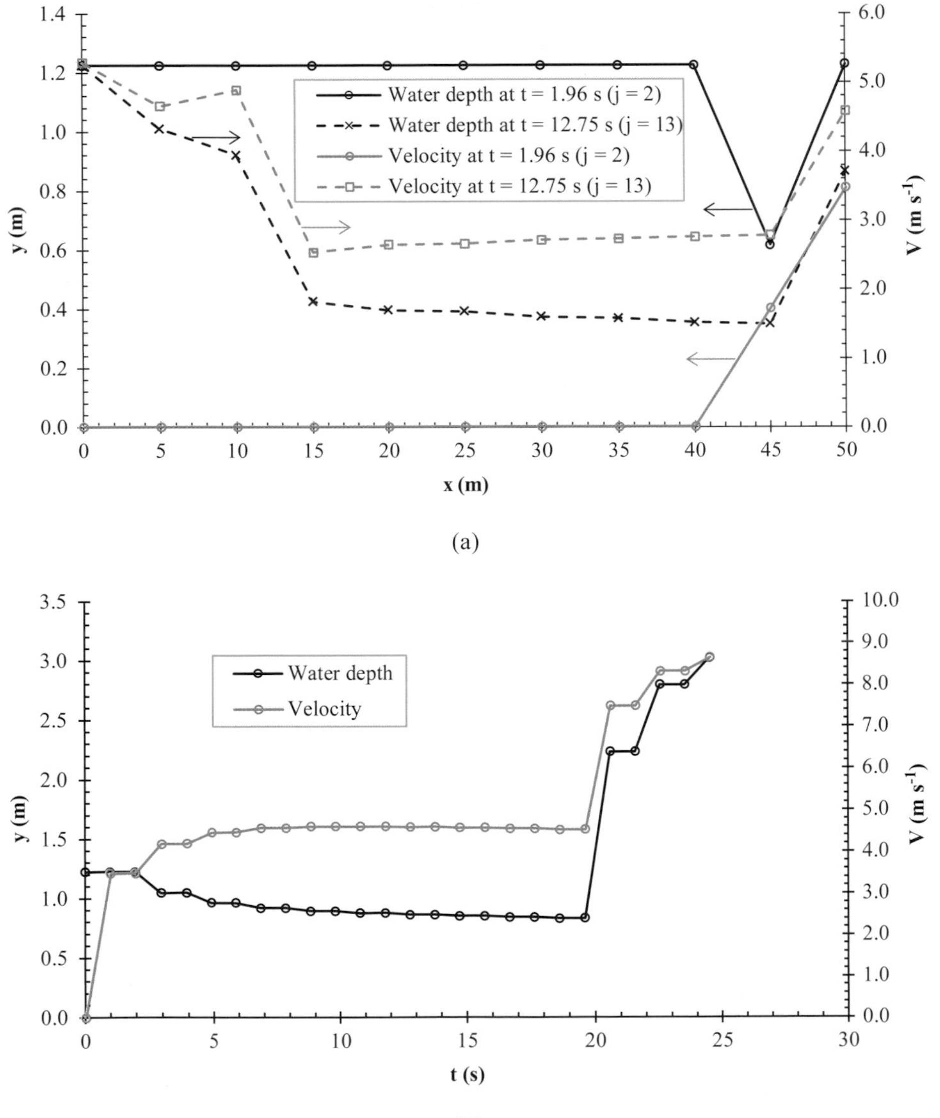

Figure 10.19 **Computed water depths and velocities for Example 10.18.** Profiles after the gate is suddenly opened at: (a) two selected times; and (2) Node 10.

The calculations can be done in an Excel spreadsheet, as shown in Figure 10.18. The profiles of water depth and velocity at 1.96 s($j = 2$) and 12.75 s ($j = 13$) after the gate is suddenly opened are shown in Figure 10.19.

At Node 10, when the gate is suddenly opened, the water depth can be determined by applying Eq. (10.33) to Node 9 and the ambient atmosphere while the velocity can be determined by applying Eq. (10.34) to Node 9 and the ambient atmosphere. As shown in Figure 10.19, with the lapse of time until the 11th time step ($j = 11$), the water depths at the interior nodes decrease, but the velocities increase, from downstream to upstream. Starting from the 12th time step ($j = 12$), the water depths start to increase while the velocities increase, from upstream to downstream. Such downstream–upstream–downstream fluctuation processes will repeat every few time steps, ultimately approaching a new steady-state flow condition, as indicated by the velocities at all nodes tending to become the same again (Figure 10.19a).

For instance, at time $j = 1$, the upper boundary condition (cells D14:D18) is $y_{0,1} = 1.224\,\text{m}$; $V_{0,1} = \left[\left(3.465\,\text{m}\,\text{s}^{-1}\right) / \left(9.81\,\text{m}\,\text{s}^{-2}\right) * \left(0.0\,\text{m}\,\text{s}^{-1}\right) + (1.224\,\text{m} - 1.224\,\text{m}) + \left(3.465\,\text{m}\,\text{s}^{-1}\right) * (0.0005 - 0.0000) * (0.981\,\text{s})\right] / \left[\left(3.465\,\text{m}\,\text{s}^{-1}\right) / \left(9.81\,\text{m}\,\text{s}^{-2}\right)\right] = 0.005\,\text{m}\,\text{s}^{-1}$;
$A_{0,1} = (5\,\text{m}) * (1.224\,\text{m}) = 6.120\,\text{m}^2$; $B_{0,1} = 5\,\text{m}$; $P_{0,1} = 5\,\text{m} + 2 * (1.224\,\text{m}) = 7.448\,\text{m}$;
$R_{0,1} = (6.120\,\text{m}^2) / (7.448\,\text{m}) = 0.822\,\text{m}$; $c_{0,1} = \left[\left(9.81\,\text{m}\,\text{s}^{-2}\right) * (1.224\,\text{m})\right]^{1/2} = 3.465\,\text{m}\,\text{s}^{-1}$;
and $S_{f,0,1} = \left[\left(0.012 * 0.005\,\text{m}\,\text{s}^{-1}\right) / (0.822\,\text{m})^{2/3}\right]^2 = 4.33 \times 10^{-9}$. In contrast, the lower boundary condition (cells N14:N18) is $y_{10,1} = \left[\left(9.81\,\text{m}\,\text{s}^{-2}\right) / \left(3.465\,\text{m}\,\text{s}^{-1}\right) * (1.224\,\text{m}) + 0.0\,\text{m}\,\text{s}^{-1} + \left(9.81\,\text{m}\,\text{s}^{-2}\right) * (0.0005 - 0.0) * (0.981\,\text{s})\right] / \left[\left(9.81\,\text{m}\,\text{s}^{-2}\right) / \left(3.465\,\text{m}\,\text{s}^{-1}\right)\right] = 1.226\,\text{m}$; $V_{10,1} = \left[\left(3.465\,\text{m}\,\text{s}^{-1}\right) / \left(9.81\,\text{m}\,\text{s}^{-2}\right) * \left(0.0\,\text{m}\,\text{s}^{-1}\right) + 1.224\,\text{m} + \left(3.465\,\text{m}\,\text{s}^{-1}\right) * (0.0005 - 0.0000) * (0.981\,\text{s})\right] / \left[\left(3.465\,\text{m}\,\text{s}^{-1}\right) / \left(9.81\,\text{m}\,\text{s}^{-2}\right)\right] = 3.470\,\text{m}\,\text{s}^{-1}$; $A_{10,1} = (5\,\text{m}) * (1.226\,\text{m}) = 6.13\,\text{m}^2$; $B_{10,1} = 5\,\text{m}$;
$P_{10,1} = 5\,\text{m} + 2 * (1.226\,\text{m}) = 7.452\,\text{m}$; $R_{10,1} = \left(6.13\,\text{m}^2\right) / (7.452\,\text{m}) = 0.822\,\text{m}$;
$c_{10,1} = \left[\left(9.81\,\text{m}\,\text{s}^{-2}\right) * (1.226\,\text{m})\right]^{1/2} = 3.468\,\text{m}\,\text{s}^{-1}$ and
$S_{f,10,1} = \left[\left(0.012 * 3.470\,\text{m}\,\text{s}^{-1}\right) / (0.822\,\text{m})^{2/3}\right]^2 = 2.25 \times 10^{-3}$. At node $i = 9$ (cells M14:M18), $y_{9,1} = \left\{\left[\left(9.81\,\text{m}\,\text{s}^{-2}\right) / \left(3.465\,\text{m}\,\text{s}^{-1}\right) * (1.224\,\text{m}) + \left[\left(9.81\,\text{m}\,\text{s}^{-2}\right) / \left(3.465\,\text{m}\,\text{s}^{-1}\right) * (1.224\,\text{m})\right] + \left(0.0\,\text{m}\,\text{s}^{-1} - 0.0\,\text{m}\,\text{s}^{-1}\right) + \left(9.81\,\text{m}\,\text{s}^{-2}\right) * [(0.0005 - 0.0) - (0.0005 - 0.0)] * (0.981\,\text{s})\right]\right\} / \left[\left(9.81\,\text{m}\,\text{s}^{-2}\right) / \left(3.465\,\text{m}\,\text{s}^{-1}\right) + \left(9.81\,\text{m}\,\text{s}^{-2}\right) / \left(3.465\,\text{m}\,\text{s}^{-1}\right)\right] = 1.224\,\text{m}$; $V_{9,1} = \left\{\left[\left(3.465\,\text{m}\,\text{s}^{-1}\right) / \left(9.81\,\text{m}\,\text{s}^{-2}\right) * \left(0.0\,\text{m}\,\text{s}^{-1}\right) + \left(3.465\,\text{m}\,\text{s}^{-1}\right) / \left(9.81\,\text{m}\,\text{s}^{-2}\right) * \left(0.0\,\text{m}\,\text{s}^{-1}\right)\right] + (1.224\,\text{m} - 1.224\,\text{m}) + \left[\left(3.465\,\text{m}\,\text{s}^{-1}\right) * (0.0005 - 0.0) + \left(3.465\,\text{m}\,\text{s}^{-1}\right) * (0.0005 - 0.0)\right] * (0.981\,\text{s})\right\} / \left[\left(3.465\,\text{m}\,\text{s}^{-1}\right) / \left(9.81\,\text{m}\,\text{s}^{-2}\right) + \left(3.465\,\text{m}\,\text{s}^{-1}\right) / \left(9.81\,\text{m}\,\text{s}^{-2}\right)\right] = 0.005\,\text{m}\,\text{s}^{-1}$;
$A_{9,1} = (5\,\text{m}) * (1.224\,\text{m}) = 6.120\,\text{m}^2$; $B_{9,1} = 5\,\text{m}$; $P_{9,1} = 5\,\text{m} + 2 * (1.224\,\text{m}) = 7.448\,\text{m}$;
$R_{9,1} = (6.120\,\text{m}^2) / (7.448\,\text{m}) = 0.822\,\text{m}$; $c_{9,1} = \left[\left(9.81\,\text{m}\,\text{s}^{-2}\right) * (1.224\,\text{m})\right]^{1/2} = 3.465\,\text{m}\,\text{s}^{-1}$; and
$S_{f,9,1} = \left[\left(0.012 * 0.005\,\text{m}\,\text{s}^{-1}\right) / (0.822\,\text{m})^{2/3}\right]^2 = 4.33 \times 10^{-9}$. Similar calculations can be done for nodes $i = 8, 7, \ldots, 1$. At other times $j = 2, 3, \ldots$, determine the upper and lower boundary conditions and do the calculations for the interior nodes in similar ways.

Example 10.19 A 50-m-long, 5-m-wide rectangular channel has a bed slope of 0.0005 and a Manning's n of 0.012. The channel, which is designed for a steady-state uniform flow with a normal depth of 1.224 m, diverts water from a shallow fresh lake to irrigate a field. The lake water depth above the channel upstream-end invert is the same as the normal depth. At the beginning, the sluice gate at the downstream end of the channel is closed. The gate is gradually opened during the first four time steps, resulting in gate openings of a = 0.1, 0.2, 0.5, and 0.8 m in sequence. At a given opening, the velocity at the channel cross section just upstream of the gate can be computed as $0.48a\sqrt{\frac{2g}{y}}$, where y is the water depth. Determine the water depth and velocity profiles using $\Delta x = 5$ m.

Solution

$L = 50\,\text{m}$, $b = 5\,\text{m}$, $S_0 = 0.0005$, $n = 0.012$, $y_0 = 1.224\,\text{m}$, $\Delta x = 5\,\text{m}$

Steady-state uniform flow in the channel:

$A_0 = (5\,\text{m}) * (1.224\,\text{m}) = 6.120\,\text{m}^2$

$B_0 = 5\,\text{m}$

$P_0 = 5\,\text{m} + 2 * (1.224\,\text{m}) = 7.448\,\text{m}$

$R_0 = A_0/P_0 = \left(6.120\,\text{m}^2\right) / (7.448\,\text{m}) = 0.822\,\text{m}$

Manning formula: $V_0 = (1/0.012) * (0.822\,\text{m})^{2/3} * 0.0005^{1/2} = 1.634\,\text{m}\,\text{s}^{-1}$

$Q_0 = \left(1.634\,\text{m}\,\text{s}^{-1}\right) * \left(6.120\,\text{m}^2\right) = 10.0\,\text{m}^3\,\text{s}^{-1}$

$c_0 = \left[\left(9.81\,\text{m}\,\text{s}^{-2}\right) * \left(6.120\,\text{m}^2\right) / (5\,\text{m})\right]^{1/2} = 3.465\,\text{m}\,\text{s}^{-1}$.

$\Delta x = 5\,\text{m} \rightarrow$ 10 segments and 11 nodes, that is, $i = 0, 1, 2, \ldots, 10$.

$\Delta t = \Delta x / (V_0 + c_0) = (5\,\text{m}) / \left(1.634\,\text{m}\,\text{s}^{-1} + 3.465\,\text{m}\,\text{s}^{-1}\right) = 0.981\,\text{s}$.

Initial condition: cells D9:N13 in Figure 10.20, highlighted in red.

For $i = 0, 1, 2, \ldots, 10$:

$y_{i,0} = y_0 = 1.224\,\text{m}$

Gate is fully closed $\rightarrow V_{i,0} = 0.0\,\text{m}\,\text{s}^{-1}$

$$c_{i,0} = c_0 = 3.465\,\text{m}\,\text{s}^{-1}$$

$$R_{i,0} = R_0 = 0.822\,\text{m}$$

$$S_{i,0} = \left(nV_{i,0}/R_{i,0}{}^{2/3}\right)^2 = \left[0.012 * \left(0.0\,\text{m}\,\text{s}^{-1}\right) / (0.822\,\text{m})^{2/3}\right]^2 = 0.0.$$

Upper boundary condition: cells D14:D48 in Figure 10.20, highlighted in blue.

At time $j \geq 1$:

$y_{0,j} = 1.224\,\text{m}$

Applying Eq. (10.34) to the lake and Node 1, one has:

$$V_{0,j} = \frac{\left[\left(\frac{c}{g}\right)_{1,j-1} V_{1,j-1}\right] + \left(y_0 - y_{1,j-1}\right) + \left[c_{1,j-1}\left(S_0 - S_f\right)_{1,j-1}\right](\Delta t)}{\left(\frac{c}{g}\right)_{1,j-1}}.$$

$$A_{0,j} = 5y_{0,j}\text{m}^2$$

$$B_{0,j} = 5\,\text{m}$$

$$P_{0,j} = 5 + 2y_{0,j}\text{m}$$

$$R_{0,j} = A_{0,j}/P_{0,j} = \left(5y_{0,j}\text{m}^2\right) / \left(5 + 2y_{0,j}\text{m}\right) = 5y_{0,j}/\left(5 + 2y_{0,j}\right)\ \text{m}$$

$$c_{0,j} = \left(gA_{0,j}/B_{0,j}\right)^{1/2} = \left[g\left(5y_{0,j}\text{m}^2\right)/(5\,\text{m})\right]^{1/2} = \left(gy_{0,j}\right)^{1/2}\ \text{m}\,\text{s}^{-1}$$

$$S_{0,j} = \left(nV_{0,j}/R_{0,j}{}^{2/3}\right)^2.$$

Lower boundary condition: cells N14:N48 in Figure 10.20, highlighted in blue.

At time $j \geq 1$:

Applying Eq. (10.33) to Node 9 and the ambient atmosphere, one has:

$$y_{10,j} = \frac{\left[\left(\frac{g}{c}\right)_{9,j-1} y_{9,j-1}\right] + V_{9,j-1} + g\left[c_{9,j-1}\left(S_0 - S_f\right)_{9,j-1}\right](\Delta t)}{\left(\frac{g}{c}\right)_{9,j-1}}.$$

$$A_{10,j} = 5y_{10,j}\text{m}^2$$

$$B_{10,j} = 5\,\text{m}$$

$$P_{10,j} = 5 + 2y_{10,j}\text{m}$$

$$R_{10,j} = A_{10,j}/P_{10,j} = \left(5y_{10,j}\text{m}^2\right) / \left(5 + 2y_{10,j}\text{m}\right) = 5y_{10,j}/\left(5 + 2y_{10,j}\right)\ \text{m}$$

$$c_{10,j} = \left(gA_{10,j}/B_{10,j}\right)^{1/2} = \left[g\left(5y_{10,j}\text{m}^2\right)/(5\,\text{m})\right]^{1/2} = \left(gy_{10,j}\right)^{1/2}\ \text{m}\,\text{s}^{-1}$$

$$S_{10,j} = \left(nV_{10,j}/R_{10,j}{}^{2/3}\right)^2.$$

At time $1 \leq j \leq 4$: $V_{10,j} = 0.48a_j\sqrt{\frac{2g}{y_{10,j}}}$.

At time $j \geq 5$: Applying Eq. (10.34) to Node 9 and the ambient atmosphere, one has:

$$V_{10,j} = \frac{\left[\left(\frac{c}{g}\right)_{9,j-1} V_{9,j-1}\right] + y_{9,j-1} + \left[c_{9,j-1}\left(S_0 - S_f\right)_{9,j-1}\right](\Delta t)}{\left(\frac{c}{g}\right)_{9,j-1}}.$$

Interior nodes ($i = 1, 2, \ldots, 9$) at time $j \geq 1$: use Eqs. (10.33) and (10.34).

The calculations can be done in an Excel spreadsheet, as shown in Figure 10.20. The profiles of water depth and velocity at 1.96 s ($j = 2$) and 12.75 s ($j = 13$) after the gate is gradually opened are shown in Figure 10.21.

At Node 10, when the gate is gradually opened, the water depth can be determined by applying Eq. (10.33) to Node 9 and the ambient atmosphere while the velocity is computed differently when the gate is partially to fully open. As shown in Figure 10.21, with the lapse of time until the 12th time step ($j = 12$), the water depths at the interior nodes decrease, but the velocities increase, from downstream to upstream. Starting from the 13th time step ($j = 13$), the water depths start to increase while the

	A	B	C	D	E	F	G	H	I	J	K	L	M	N
1	b =	5	m	Q_0 =	10.00	$m^3\ s^{-1}$	g =	9.81	$m\ s^{-2}$	opening :	0.1	m		
2	Z =	0		y_0 =	1.224	m	c_0 =	3.465	$m\ s^{-1}$		0.2	m		
3	n =	0.012		A_0 =	6.120	m^2	V_0 =	1.635	$m\ s^{-1}$		0.5	m		
4	L =	50	m	B_0 =	5.000	m	K =	1			0.8	m		
5	S_0 =	0.0005		R_0 =	0.822	m								
6	Δx =	5	m	Δt =	0.981	sec								
7									Node i =					
8	j	t_j (sec)	Variables	0	1	2	3	4	5	6	7	8	9	10
9	0	0.000	$y_{i,0}$ (m)	1.224	1.224	1.224	1.224	1.224	1.224	1.224	1.224	1.224	1.224	1.224
10			$V_{i,0}$ ($m\ s^{-1}$)	0.000	0.000	0.000	0.000	0.000	0.000	0.000	0.000	0.000	0.000	0.000
11			$c_{i,0}$ ($m\ s^{-1}$)	3.465	3.465	3.465	3.465	3.465	3.465	3.465	3.465	3.465	3.465	3.465
12			$R_{i,0}$ (m)	0.822	0.822	0.822	0.822	0.822	0.822	0.822	0.822	0.822	0.822	0.822
13			$S_{fi,0}$	0.00E+00	0.00E+00	0.00E+00	0.00E+00	0.00E+00	0.00E+00	0.00E+00	0.00E+00	0.00E+00	0.00E+00	0.00E+00
14	1	0.981	$y_{i,1}$ ($lbf\ ft^{-2}$)	1.224	1.224	1.224	1.224	1.224	1.224	1.224	1.224	1.224	1.224	1.226
15			$V_{i,1}$ ($ft\ sec^{-1}$)	0.005	0.005	0.005	0.005	0.005	0.005	0.005	0.005	0.005	0.005	0.192
16			$c_{i,1}$ ($m\ s^{-1}$)	3.465	3.465	3.465	3.465	3.465	3.465	3.465	3.465	3.465	3.465	3.468
17			$R_{i,1}$ (m)	0.822	0.822	0.822	0.822	0.822	0.822	0.822	0.822	0.822	0.822	0.822
18			$S_{fi,1}$	4.33E-09	4.33E-09	4.33E-09	4.33E-09	4.33E-09	4.33E-09	4.33E-09	4.33E-09	4.33E-09	4.33E-09	6.89E-06
19	2	1.961	$y_{i,2}$ ($lbf\ ft^{-2}$)	1.224	1.224	1.224	1.224	1.224	1.224	1.224	1.224	1.224	1.192	1.227
20			$V_{i,2}$ ($ft\ sec^{-1}$)	0.010	0.010	0.010	0.010	0.010	0.010	0.010	0.010	0.010	0.101	0.384
21			$c_{i,2}$ ($m\ s^{-1}$)	3.465	3.465	3.465	3.465	3.465	3.465	3.465	3.465	3.465	3.419	3.470
22			$R_{i,2}$ (m)	0.822	0.822	0.822	0.822	0.822	0.822	0.822	0.822	0.822	0.807	0.823
23			$S_{fi,2}$	1.73E-08	1.73E-08	1.73E-08	1.73E-08	1.73E-08	1.73E-08	1.73E-08	1.73E-08	1.73E-08	1.95E-06	2.75E-05
24	3	2.942	$y_{i,3}$ ($lbf\ ft^{-2}$)	1.224	1.224	1.224	1.224	1.224	1.224	1.224	1.224	1.192	1.160	1.229
25			$V_{i,3}$ ($ft\ sec^{-1}$)	0.014	0.014	0.014	0.014	0.014	0.014	0.014	0.014	0.106	0.197	0.959
26			$c_{i,3}$ ($m\ s^{-1}$)	3.465	3.465	3.465	3.465	3.465	3.465	3.465	3.465	3.419	3.373	3.472
27			$R_{i,3}$ (m)	0.822	0.822	0.822	0.822	0.822	0.822	0.822	0.822	0.807	0.792	0.824
28			$S_{fi,3}$	3.90E-08	3.90E-08	3.90E-08	3.90E-08	3.90E-08	3.90E-08	3.90E-08	3.90E-08	2.14E-06	7.60E-06	1.72E-04
29	4	3.922	$y_{i,4}$ ($lbf\ ft^{-2}$)	1.224	1.224	1.224	1.224	1.224	1.224	1.224	1.192	1.160	1.060	1.229
30			$V_{i,4}$ ($ft\ sec^{-1}$)	0.019	0.019	0.019	0.019	0.019	0.019	0.019	0.110	0.201	0.487	1.534
31			$c_{i,4}$ ($m\ s^{-1}$)	3.465	3.465	3.465	3.465	3.465	3.465	3.465	3.419	3.373	3.225	3.472
32			$R_{i,4}$ (m)	0.822	0.822	0.822	0.822	0.822	0.822	0.822	0.807	0.792	0.745	0.824
33			$S_{fi,4}$	6.92E-08	6.92E-08	6.92E-08	6.92E-08	6.92E-08	6.92E-08	6.92E-08	2.34E-06	7.98E-06	5.06E-05	4.39E-04
34	5	4.903	$y_{i,5}$ ($lbf\ ft^{-2}$)	1.224	1.224	1.224	1.224	1.224	1.224	1.192	1.160	1.061	0.962	1.222
35			$V_{i,5}$ ($ft\ sec^{-1}$)	0.024	0.024	0.024	0.024	0.024	0.024	0.115	0.206	0.492	0.781	0.492
36			$c_{i,5}$ ($m\ s^{-1}$)	3.465	3.465	3.465	3.465	3.465	3.465	3.419	3.373	3.226	3.072	3.462
37			$R_{i,5}$ (m)	0.822	0.822	0.822	0.822	0.822	0.822	0.807	0.792	0.745	0.695	0.821
38			$S_{fi,5}$	1.08E-07	1.08E-07	1.08E-07	1.08E-07	1.08E-07	1.08E-07	2.54E-06	8.36E-06	5.16E-05	1.43E-04	4.53E-05
39	6	5.883	$y_{i,6}$ ($lbf\ ft^{-2}$)	1.224	1.224	1.224	1.224	1.224	1.192	1.160	1.061	0.962	1.138	1.208
40			$V_{i,6}$ ($ft\ sec^{-1}$)	0.029	0.029	0.029	0.029	0.029	0.120	0.211	0.496	0.785	0.259	0.784
41			$c_{i,6}$ ($m\ s^{-1}$)	3.465	3.465	3.465	3.465	3.465	3.419	3.373	3.226	3.072	3.342	3.442
42			$R_{i,6}$ (m)	0.822	0.822	0.822	0.822	0.822	0.807	0.792	0.745	0.695	0.782	0.814
43			$S_{fi,6}$	1.56E-07	1.56E-07	1.56E-07	1.56E-07	1.56E-07	2.76E-06	8.75E-06	5.25E-05	1.44E-04	1.34E-05	1.16E-04
44	7	6.864	$y_{i,7}$ ($lbf\ ft^{-2}$)	1.224	1.224	1.224	1.224	1.192	1.160	1.061	0.962	1.138	1.078	1.228
45			$V_{i,7}$ ($ft\ sec^{-1}$)	0.034	0.034	0.034	0.034	0.125	0.216	0.501	0.789	0.264	0.419	0.265
46			$c_{i,7}$ ($m\ s^{-1}$)	3.465	3.465	3.465	3.465	3.419	3.373	3.226	3.073	3.342	3.252	3.471
47			$R_{i,7}$ (m)	0.822	0.822	0.822	0.822	0.807	0.792	0.745	0.695	0.782	0.753	0.824
48			$S_{fi,7}$	2.12E-07	2.12E-07	2.12E-07	2.12E-07	2.99E-06	9.15E-06	5.35E-05	1.46E-04	1.39E-05	3.68E-05	1.30E-05

Figure 10.20 **Excerpt of the spreadsheet for the calculations for Example 10.19.**

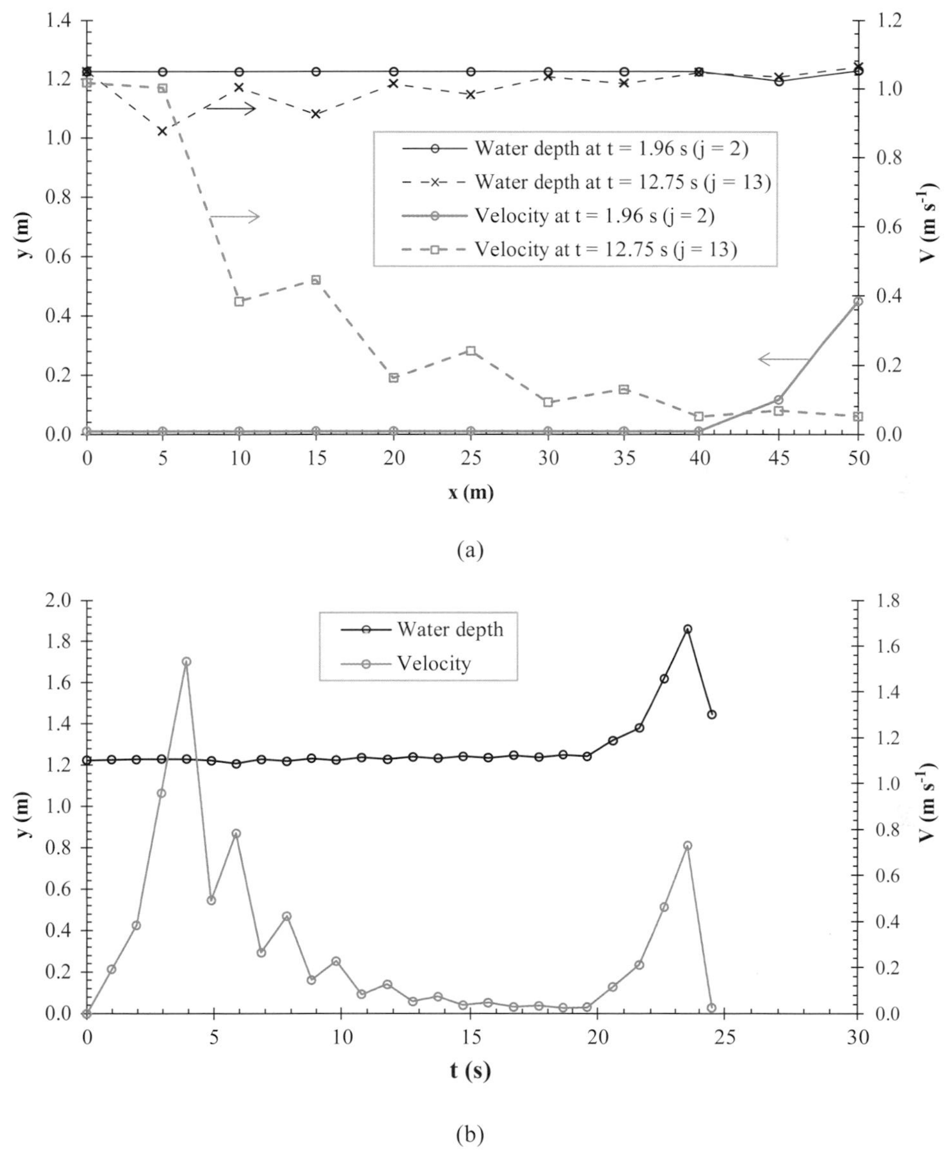

Figure 10.21 **Computed water depths and velocities for Example 10.19.** Profiles when the gate is gradually opened at: (a) two selected times; and (b) Node 10.

velocities increase, from upstream to downstream. Such downstream–upstream–downstream fluctuation processes will repeat every few time steps, ultimately approaching a new steady-state flow condition, as indicated by the velocities at all nodes tending to become very similar (Figure 10.21a).

For instance, at time j = 1, the upper boundary condition (cells D14:D18) is $y_{0,1} = 1.224\,\text{m}$; $V_{0,1} = \left[\left(3.465\,\text{m}\,\text{s}^{-1}\right) / \left(9.81\,\text{m}\,\text{s}^{-2}\right) * \left(0.0\,\text{m}\,\text{s}^{-1}\right) + (1.224\,\text{m} - 1.224\,\text{m}) + \left(3.465\,\text{m}\,\text{s}^{-1}\right) * (0.0005 - 0.0000) * (0.981\,\text{s})\right] / \left[\left(3.465\,\text{m}\,\text{s}^{-1}\right) / \left(9.81\,\text{m}\,\text{s}^{-2}\right)\right] = 0.005\,\text{m}\,\text{s}^{-1}$; $A_{0,1} =$

$(5\,\mathrm{m}) * (1.224\,\mathrm{m}) = 6.120\,\mathrm{m^2}$; $B_{0,1} = 5\,\mathrm{m}$; $P_{0,1} = 5\,\mathrm{m} + 2 * (1.224\,\mathrm{m}) = 7.448\,\mathrm{m}$; $R_{0,1} = (6.120\,\mathrm{m^2})/(7.448\,\mathrm{m}) = 0.822\,\mathrm{m}$; and $S_{f,0,1} = \left[(0.012 * 0.005\,\mathrm{m\,s^{-1}})/(0.822\,\mathrm{m})^{2/3}\right]^2 = 4.33 \times 10^{-9}$. In contrast, the lower boundary condition (cells N14:N18) is $y_{10,1} = \left[(9.81\,\mathrm{m\,s^{-2}})/(3.465\mathrm{ms^{-1}}) * (1.224\,\mathrm{m}) + 0.0\,\mathrm{m\,s^{-1}} + (9.81\,\mathrm{m\,s^{-2}}) * (0.0005 - 0.0) * (0.981\,\mathrm{s})\right] / \left[(9.81\,\mathrm{m\,s^{-2}})/(3.465\,\mathrm{m\,s^{-1}})\right] = 1.226\,\mathrm{m}$; $V_{10,1} = 0.48 * (0.1\,\mathrm{m}) * \left[2 * (9.81\,\mathrm{m\,s^{-2}})/(1.226\,\mathrm{m})\right]^{1/2} = 0.192\,\mathrm{m\,s^{-1}}$; $A_{10,1} = (5\,\mathrm{m}) * (1.226\,\mathrm{m}) = 6.13\,\mathrm{m^2}$; $B_{10,1} = 5\,\mathrm{m}$; $P_{10,1} = 5\,\mathrm{m} + 2 * (1.226\,\mathrm{m}) = 7.452\,\mathrm{m}$; $R_{10,1} = (6.13\,\mathrm{m^2})/(7.452\,\mathrm{m}) = 0.822\,\mathrm{m}$; and $S_{f,10,1} = \left[(0.012 * 0.192\,\mathrm{m\,s^{-1}})/(0.822\,\mathrm{m})^{2/3}\right]^2 = 6.89 \times 10^{-6}$. At node i = 9 (cells M14:M18), $y_{9,1} = \{[(9.81\,\mathrm{m\,s^{-2}})/(3.465\,\mathrm{m\,s^{-1}}) * (1.224\,\mathrm{m}) + [(9.81\,\mathrm{m\,s^{-2}})/(3.465\,\mathrm{m\,s^{-1}}) * (1.224\,\mathrm{m})] + (0.0\,\mathrm{m\,s^{-1}} - 0.0\,\mathrm{m\,s^{-1}}) + (9.81\,\mathrm{m\,s^{-2}}) * [(0.0005 - 0.0) - (0.0005 - 0.0)] * (0.981\,\mathrm{s})]\} / [(9.81\,\mathrm{m\,s^{-2}})/(3.465\,\mathrm{m\,s^{-1}}) + (9.81\,\mathrm{m\,s^{-2}})/(3.465\,\mathrm{m\,s^{-1}})] = 1.224\,\mathrm{m}$; $V_{9,1} = \{[(3.465\,\mathrm{m\,s^{-1}})/(9.81\,\mathrm{m\,s^{-2}}) * (0.0\,\mathrm{m\,s^{-1}}) + (3.465\,\mathrm{m\,s^{-1}})/(9.81\,\mathrm{m\,s^{-2}}) * (0.0\,\mathrm{m\,s^{-1}})] + (1.224\,\mathrm{m} - 1.224\,\mathrm{m}) + [(3.465\,\mathrm{m\,s^{-1}}) * (0.0005 - 0.0) + (3.465\,\mathrm{m\,s^{-1}}) * (0.0005 - 0.0)] * (0.981\,\mathrm{s})\} / [(3.465\,\mathrm{m\,s^{-1}})/(9.81\,\mathrm{m\,s^{-2}}) + (3.465\,\mathrm{m\,s^{-1}})/(9.81\,\mathrm{m\,s^{-2}})] = 0.005\,\mathrm{m\,s^{-1}}$; $A_{9,1} = (5\,\mathrm{m}) * (1.224\,\mathrm{m}) = 6.120\,\mathrm{m^2}$; $B_{9,1} = 5\,\mathrm{m}$; $P_{9,1} = 5\,\mathrm{m} + 2 * (1.224\,\mathrm{m}) = 7.448\,\mathrm{m}$; $R_{9,1} = (6.120\,\mathrm{m^2})/(7.448\,\mathrm{m}) = 0.822\,\mathrm{m}$; and $S_{f,9,1} = \left[(0.012 * 0.005\,\mathrm{m\,s^{-1}})/(0.822\,\mathrm{m})^{2/3}\right]^2 = 4.33 \times 10^{-9}$. Similar calculations can be done for nodes i = 8, 7, . . . , 1. At other times j = 2, 3, . . ., determine the upper and lower boundary conditions and do the calculations for the interior nodes in similar ways.

Example 10.20 A 50-m-long, 5-m-wide rectangular channel with a bed slope of 0.0005 and a Manning's n of 0.012 diverts water from a shallow fresh lake to irrigate a large field. Once established, the steady-state uniform flow in the channel will have a normal depth of 1.224 m. The lake water depth above the channel upstream-end invert is the same as the normal depth. At the beginning, the sluice gate at the end of the channel is fully open. The gate is gradually closed during the first four time steps, resulting in gate openings of a = 0.8, 0.5, 0.2, and 0.1 m in sequence. At a given opening, the velocity at the channel cross section just upstream of the gate can be computed as $0.48a\sqrt{\frac{2g}{y}}$, where y is the water depth. Once the gate is fully closed, the velocity becomes zero. Determine the water depth and velocity profiles using $\Delta x = 5\,\mathrm{m}$.

Solution

$L = 50\,\mathrm{m}$, $b = 5\,\mathrm{m}$, $S_0 = 0.0005$, $n = 0.012$, $y_0 = 1.224\,\mathrm{m}$, $\Delta x = 5\,\mathrm{m}$

Steady-state uniform flow in the channel:

$$A_0 = (5\,\mathrm{m}) * (1.224\,\mathrm{m}) = 6.120\,\mathrm{m^2}$$

$$B_0 = 5\,\mathrm{m}$$

$$P_0 = 5\,\mathrm{m} + 2 * (1.224\,\mathrm{m}) = 7.448\,\mathrm{m}$$

$$R_0 = A_0/P_0 = (6.120\,\mathrm{m^2})/(7.448\,\mathrm{m}) = 0.822\,\mathrm{m}$$

Manning formula: $V_0 = (1/0.012) * (0.822\,\text{m})^{2/3} * 0.0005^{1/2} = 1.634\,\text{m s}^{-1}$

$$Q_0 = \left(1.634\,\text{m s}^{-1}\right) * \left(6.120\,\text{m}^2\right) = 10.0\,\text{m}^3\text{s}^{-1}$$

$$c_0 = \left[\left(9.81\,\text{m s}^{-2}\right) * \left(6.120\,\text{m}^2\right) / (5\,\text{m})\right]^{1/2} = 3.465\,\text{m s}^{-1}.$$

$\Delta x = 5\,\text{m} \rightarrow 10$ segments and 11 nodes, that is, $i = 0, 1, 2, \ldots, 10$.

$\Delta t = \Delta x / (V_0 + c_0) = (5\,\text{m}) / \left(1.634\,\text{m s}^{-1} + 3.465\,\text{m s}^{-1}\right) = 0.981\,\text{s}.$

Initial condition: cells D9:N13 in Figure 10.22, highlighted in red.

For $i = 0, 1, 2, \ldots, 10$:

$$y_{i,0} = y_0 = 1.224\,\text{m}$$

Gate is fully opened $\rightarrow V_{i,0} = V_0 = 1.634\,\text{m s}^{-1}$

$c_{i,0} = c_0 = 3.465\,\text{m s}^{-1}$

$R_{i,0} = R_0 = 0.822\,\text{m}$

$S_{i,0} = \left(nV_{i,0}/R_{i,0}{}^{2/3}\right)^2 = \left[0.012 * \left(1.634\,\text{m s}^{-1}\right) / (0.822\,\text{m})^{2/3}\right]^2 = 5.00 \times 10^{-4}.$

Upper boundary condition: cells D14:D48 in Figure 10.22, highlighted in blue.

At time $j \geq 1$:

$$y_{0,j} = 1.224\,\text{m}$$

Applying Eq. (10.34) to the lake and Node 1, one has:

$$V_{0,j} = \frac{\left[\left(\frac{c}{g}\right)_{1,j-1} V_{1,j-1}\right] + \left(y_0 - y_{1,j-1}\right) + \left[c_{1,j-1}\left(S_0 - S_f\right)_{1,j-1}\right](\Delta t)}{\left(\frac{c}{g}\right)_{1,j-1}}.$$

$A_{0,j} = 5y_{0,j}\text{m}^2$

$B_{0,j} = 5\,\text{m}$

$P_{0,j} = 5 + 2y_{0,j}\text{m}$

$R_{0,j} = A_{0,j}/P_{0,j} = \left(5y_{0,j}\text{m}^2\right) / \left(5 + 2y_{0,j}\text{m}\right) = 5y_{0,j} / \left(5 + 2y_{0,j}\right)\,\text{m}$

$c_{0,j} = \left(gA_{0,j}/B_{0,j}\right)^{1/2} = \left[g\left(5y_{0,j}\text{m}^2\right) / (5\,\text{m})\right]^{1/2} = \left(gy_{0,j}\right)^{1/2}\,\text{m s}^{-1}$

$S_{0,j} = \left(nV_{0,j}/R_{0,j}{}^{2/3}\right)^2.$

Lower boundary condition: cells N14:N48 in Figure 10.22, highlighted in blue.

At time $j \geq 1$:

Applying Eq. (10.33) to Node 9 and the ambient atmosphere, one has:

$$y_{10,j} = \frac{\left[\left(\frac{g}{c}\right)_{9,j-1} y_{9,j-1}\right] + V_{9,j-1} + g\left[c_{9,j-1}\left(S_0 - S_f\right)_{9,j-1}\right](\Delta t)}{\left(\frac{g}{c}\right)_{9,j-1}}.$$

	A	B	C	D	E	F	G	H	I	J	K	L	M	N
1	b =	5	m	Q_0 =	10.00	$m^3\ s^{-1}$	g =	9.81	$m\ s^{-2}$	opening :	0.8	m		
2	Z =	0		y_0 =	1.224	m	c_0 =	3.465	$m\ s^{-1}$		0.5	m		
3	n =	0.012		A_0 =	6.120	m^2	V_0 =	1.635			0.2	m		
4	L =	50	m	B_0 =	5.000	m	K =	1			0.1	m		
5	S_0 =	0.0005		R_0 =	0.822	m								
6	Δx =	5	m	Δt =	0.981	sec								
7									Node i =					
8	j	t_j (sec)	Variables	0	1	2	3	4	5	6	7	8	9	10
9	0	0.000	$y_{i,0}$ (m)	1.224	1.224	1.224	1.224	1.224	1.224	1.224	1.224	1.224	1.224	1.224
10			$V_{i,0}$ ($m\ s^{-1}$)	1.634	1.634	1.634	1.634	1.634	1.634	1.634	1.634	1.634	1.634	1.634
11			$c_{i,0}$ ($m\ s^{-1}$)	3.465	3.465	3.465	3.465	3.465	3.465	3.465	3.465	3.465	3.465	3.465
12			$R_{i,0}$ (m)	0.822	0.822	0.822	0.822	0.822	0.822	0.822	0.822	0.822	0.822	0.822
13			$S_{fi,0}$	5.00E-04	5.00E-04	5.00E-04	5.00E-04	5.00E-04	5.00E-04	5.00E-04	5.00E-04	5.00E-04	5.00E-04	5.00E-04
14	1	0.981	$y_{i,1}$ ($lbf\ ft^{-2}$)	1.224	1.224	1.224	1.224	1.224	1.224	1.224	1.224	1.224	1.224	1.801
15			$V_{i,1}$ ($ft\ sec^{-1}$)	1.634	1.634	1.634	1.634	1.634	1.634	1.634	1.634	1.634	1.634	1.267
16			$c_{i,1}$ ($m\ s^{-1}$)	3.465	3.465	3.465	3.465	3.465	3.465	3.465	3.465	3.465	3.465	4.204
17			$R_{i,1}$ (m)	0.822	0.822	0.822	0.822	0.822	0.822	0.822	0.822	0.822	0.822	1.047
18			$S_{fi,1}$	5.00E-04	5.00E-04	5.00E-04	5.00E-04	5.00E-04	5.00E-04	5.00E-04	5.00E-04	5.00E-04	5.00E-04	2.18E-04
19	2	1.961	$y_{i,2}$ ($lbf\ ft^{-2}$)	1.224	1.224	1.224	1.224	1.224	1.224	1.224	1.224	1.224	1.555	1.801
20			$V_{i,2}$ ($ft\ sec^{-1}$)	1.634	1.634	1.634	1.634	1.634	1.634	1.634	1.634	1.634	0.696	0.792
21			$c_{i,2}$ ($m\ s^{-1}$)	3.465	3.465	3.465	3.465	3.465	3.465	3.465	3.465	3.465	3.906	4.204
22			$R_{i,2}$ (m)	0.822	0.822	0.822	0.822	0.822	0.822	0.822	0.822	0.822	0.959	1.047
23			$S_{fi,2}$	5.00E-04	5.00E-04	5.00E-04	5.00E-04	5.00E-04	5.00E-04	5.00E-04	5.00E-04	5.00E-04	7.38E-05	8.50E-05
24	3	2.942	$y_{i,3}$ ($lbf\ ft^{-2}$)	1.224	1.224	1.224	1.224	1.224	1.224	1.224	1.224	1.554	1.647	1.834
25			$V_{i,3}$ ($ft\ sec^{-1}$)	1.634	1.634	1.634	1.634	1.634	1.634	1.634	1.634	0.698	0.436	0.314
26			$c_{i,3}$ ($m\ s^{-1}$)	3.465	3.465	3.465	3.465	3.465	3.465	3.465	3.465	3.905	4.020	4.242
27			$R_{i,3}$ (m)	0.822	0.822	0.822	0.822	0.822	0.822	0.822	0.822	0.959	0.993	1.058
28			$S_{fi,3}$	5.00E-04	5.00E-04	5.00E-04	5.00E-04	5.00E-04	5.00E-04	5.00E-04	5.00E-04	7.43E-05	2.77E-05	1.32E-05
29	4	3.922	$y_{i,4}$ ($lbf\ ft^{-2}$)	1.224	1.224	1.224	1.224	1.224	1.224	1.224	1.554	1.646	1.768	1.828
30			$V_{i,4}$ ($ft\ sec^{-1}$)	1.634	1.634	1.634	1.634	1.634	1.634	1.634	0.701	0.439	0.166	0.157
31			$c_{i,4}$ ($m\ s^{-1}$)	3.465	3.465	3.465	3.465	3.465	3.465	3.465	3.904	4.019	4.165	4.234
32			$R_{i,4}$ (m)	0.822	0.822	0.822	0.822	0.822	0.822	0.822	0.958	0.993	1.036	1.056
33			$S_{fi,4}$	5.00E-04	5.00E-04	5.00E-04	5.00E-04	5.00E-04	5.00E-04	5.00E-04	7.48E-05	2.80E-05	3.78E-06	3.31E-06
34	5	4.903	$y_{i,5}$ ($lbf\ ft^{-2}$)	1.224	1.224	1.224	1.224	1.224	1.224	1.553	1.645	1.767	1.794	1.841
35			$V_{i,5}$ ($ft\ sec^{-1}$)	1.634	1.634	1.634	1.634	1.634	1.634	0.703	0.441	0.168	0.083	0.000
36			$c_{i,5}$ ($m\ s^{-1}$)	3.465	3.465	3.465	3.465	3.465	3.465	3.903	4.018	4.164	4.195	4.249
37			$R_{i,5}$ (m)	0.822	0.822	0.822	0.822	0.822	0.822	0.958	0.992	1.035	1.044	1.060
38			$S_{fi,5}$	5.00E-04	5.00E-04	5.00E-04	5.00E-04	5.00E-04	5.00E-04	7.53E-05	2.83E-05	3.90E-06	9.42E-07	0.00E+00
39	6	5.883	$y_{i,6}$ ($lbf\ ft^{-2}$)	1.224	1.224	1.224	1.224	1.224	1.552	1.644	1.766	1.793	1.840	1.831
40			$V_{i,6}$ ($ft\ sec^{-1}$)	1.634	1.634	1.634	1.634	1.634	0.705	0.444	0.171	0.086	0.003	0.000
41			$c_{i,6}$ ($m\ s^{-1}$)	3.465	3.465	3.465	3.465	3.465	3.902	4.016	4.162	4.194	4.248	4.239
42			$R_{i,6}$ (m)	0.822	0.822	0.822	0.822	0.822	0.958	0.992	1.035	1.044	1.060	1.057
43			$S_{fi,6}$	5.00E-04	5.00E-04	5.00E-04	5.00E-04	5.00E-04	7.58E-05	2.87E-05	4.02E-06	1.00E-06	8.56E-10	0.00E+00
44	7	6.864	$y_{i,7}$ ($lbf\ ft^{-2}$)	1.224	1.224	1.224	1.224	1.551	1.644	1.765	1.792	1.839	1.830	1.843
45			$V_{i,7}$ ($ft\ sec^{-1}$)	1.634	1.634	1.634	1.634	0.707	0.446	0.174	0.088	0.005	0.003	0.000
46			$c_{i,7}$ ($m\ s^{-1}$)	3.465	3.465	3.465	3.465	3.901	4.015	4.161	4.193	4.247	4.237	4.252
47			$R_{i,7}$ (m)	0.822	0.822	0.822	0.822	0.957	0.992	1.035	1.044	1.059	1.057	1.061
48			$S_{fi,7}$	5.00E-04	5.00E-04	5.00E-04	5.00E-04	7.63E-05	2.90E-05	4.14E-06	1.06E-06	3.43E-09	8.87E-10	0.00E+00

Figure 10.22 **Excerpt of the spreadsheet for the calculations for Example 10.20.**

$$A_{10,j} = 5y_{10,j}m^2$$
$$B_{10,j} = 5\,m$$
$$P_{10,j} = 5 + 2y_{10,j}m$$
$$R_{10,j} = A_{10,j}/P_{10,j} = \left(5y_{10,j}m^2\right) / \left(5 + 2y_{10,j}m\right) = 5y_{10,j}/\left(5 + 2y_{10,j}\right)\ m$$
$$c_{10,j} = \left(gA_{10,j}/B_{10,j}\right)^{1/2} = \left[g\left(5y_{10,j}m^2\right)/(5\,m)\right]^{1/1} = \left(gy_{10,j}\right)^{1/2}\ m\,s^{-1}$$
$$S_{10,j} = \left(nV_{10,j}/R_{10,j}{}^{2/3}\right)^2.$$

At time $1 \le j \le 4$: $V_{10,j} = 0.48a_j\sqrt{\frac{2g}{y_{10,j}}}$.

At time $j \ge 5$: $V_{10,j} = 0.0\,m\,s^{-1}$.

Interior nodes ($i = 1, 2, \ldots, 9$) at time $j \ge 1$: use Eqs. (10.33) and (10.34).

The calculations can be done in an Excel spreadsheet, as shown in Figure 10.22. The profiles of water depth and velocity at 1.96 s ($j = 2$) and 12.75 s ($j = 13$) after the gate is gradually opened are shown in Figure 10.23.

At Node 10, when the gate is gradually closed, the water depth can be determined by applying Eq. (10.33) to Node 9 and the ambient atmosphere while the velocity is computed differently when the gate is partially versus fully open. As shown in Figure 10.23, with the lapse of time until the 10th time step ($j = 10$), the water depths at the interior nodes increase, but the velocities decrease, from downstream to upstream. Starting from the 11th time step ($j = 11$), the water depths start to decrease while the velocities increase with a reversed direction, from upstream to downstream. Such downstream–upstream–downstream fluctuation processes will repeat every few time steps, ultimately approaching a new steady-state flow condition, as indicated by the velocities at all nodes tending to become very similar (Figure 10.23a).

For instance, at time $j = 1$, the upper boundary condition (cells D14:D18) is $y_{0,1} = 1.224\,m$; $V_{0,1} = \left[\left(3.465\,m\,s^{-1}\right)/\left(9.81\,m\,s^{-2}\right) * \left(1.634\,m\,s^{-1}\right) + (1.224\,m - 1.224\,m) + \left(3.465\,m\,s^{-1}\right) * (0.0005 - 0.0005) * (0.981\,s)\right] / \left[\left(3.465\,m\,s^{-1}\right)/\left(9.81\,m\,s^{-2}\right)\right] = 1.634\,m\,s^{-1}$; $A_{0,1} = (5\,m) * (1.224\,m) = 6.120\,m^2$; $B_{0,1} = 5\,m$; $P_{0,1} = 5\,m + 2 * (1.224\,m) = 7.448\,m$; $R_{0,1} = \left(6.120\,m^2\right)/(7.448\,m) = 0.822\,m$; and $S_{f,0,1} = \left[\left(0.012 * 1.634\,m\,s^{-1}\right)/(0.822\,m)^{2/3}\right]^2 = 5.0 \times 10^{-4}$. In contrast, the lower boundary condition (cells in N14:N18) is $y_{10,1} = \left[\left(9.81\,m\,s^{-2}\right)/\left(3.465\,m\,s^{-1}\right) * (1.224\,m) + 1.634\,m\,s^{-1} + \left(9.81\,m\,s^{-2}\right) * (0.0005 - 0.0005) * (0.981\,s)\right] / \left[\left(9.81\,m\,s^{-2}\right)/\left(3.465\,m\,s^{-1}\right)\right] = 1.801\,m$; $V_{10,1} = 0.48 * (0.8\,m) * \left[2 * \left(9.81\,m\,s^{-2}\right)/(1.801\,m)\right]^{1/2} = 1.267\,m\,s^{-1}$; $A_{10,1} = (5\,m) * (1.801\,m) = 9.005\,m^2$; $B_{10,1} = 5\,m$; $P_{10,1} = 5\,m + 2 * (1.801\,m) = 8.602\,m$; $R_{10,1} = \left(9.005\,m^2\right)/(8.602\,m) = 1.047\,m$; and $S_{f,10,1} = \left[\left(0.012 * 1.267\,m\,s^{-1}\right)/(1.047\,m)^{2/3}\right]^2 = 2.18 \times 10^{-6}$. At node $i = 9$ (cells M14:M18), $y_{9,1} = \left\{\left[\left(9.81\,m\,s^{-2}\right)/\left(3.465\,m\,s^{-1}\right) * (1.224\,m) + \left[\left(9.81\,m\,s^{-2}\right)/\left(3.465\,m\,s^{-1}\right) * (1.224\,m)\right] + \left(1.634\,m\,s^{-1} - 1.634\,m\,s^{-1}\right) + \left(9.81\,m\,s^{-2}\right) * [(0.0005 - 0.0005) - (0.0005 - 0.0005)] * (0.981\,s)\right]\right\} / \left[\left(9.81\,m\,s^{-2}\right)/\left(3.465\,m\,s^{-1}\right) + \left(9.81\,m\,s^{-2}\right)/\left(3.465\,m\,s^{-1}\right)\right] = 1.224\,m$; $V_{9,1} =$

$\{[(3.465\,\mathrm{m\,s^{-1}}) / (9.81\,\mathrm{m\,s^{-2}}) * (1.634\,\mathrm{m\,s^{-1}}) + (3.465\,\mathrm{m\,s^{-1}}) / (9.81\,\mathrm{m\,s^{-2}}) * (1.634\,\mathrm{m\,s^{-1}})] + (1.224\,\mathrm{m} - 1.224\,\mathrm{m}) + [(3.465\,\mathrm{m\,s^{-1}}) * (0.0005 - 0.0005) + (3.465\,\mathrm{m\,s^{-1}}) * (0.0005 - 0.0005)] * (0.981\,\mathrm{s})\} / [(3.465\,\mathrm{m\,s^{-1}}) / (9.81\,\mathrm{m\,s^{-2}}) + (3.465\,\mathrm{m\,s^{-1}}) / (9.81\,\mathrm{m\,s^{-2}})] = 1.634\,\mathrm{m\,s^{-1}}$; $A_{9,1} = (5\,\mathrm{m}) * (1.224\,\mathrm{m}) = 6.120\,\mathrm{m^2}$; $B_{9,1} = 5\,\mathrm{m}$; $P_{9,1} = 5\,\mathrm{m} + 2 * (1.224\,\mathrm{m}) = 7.448\,\mathrm{m}$; $R_{9,1} = (6.120\,\mathrm{m^2})/(7.448\,\mathrm{m}) = 0.822\,\mathrm{m}$; and $S_{f,9,1} = \left[(0.012 * 1.634\,\mathrm{m\,s^{-1}}) / (0.822\,\mathrm{m})^{2/3}\right]^2 = 5.0 \times 10^{-4}$. Similar calculations can be done for nodes $i = 8, 7, \ldots, 1$. At other times $j = 2, 3, \ldots$, determine the upper and lower boundary conditions and do the calculations for the interior nodes in similar ways.

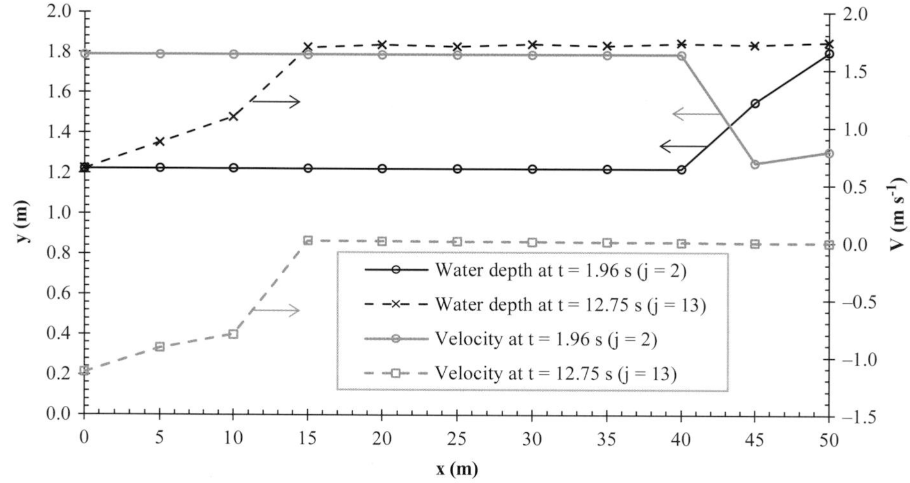

(a)

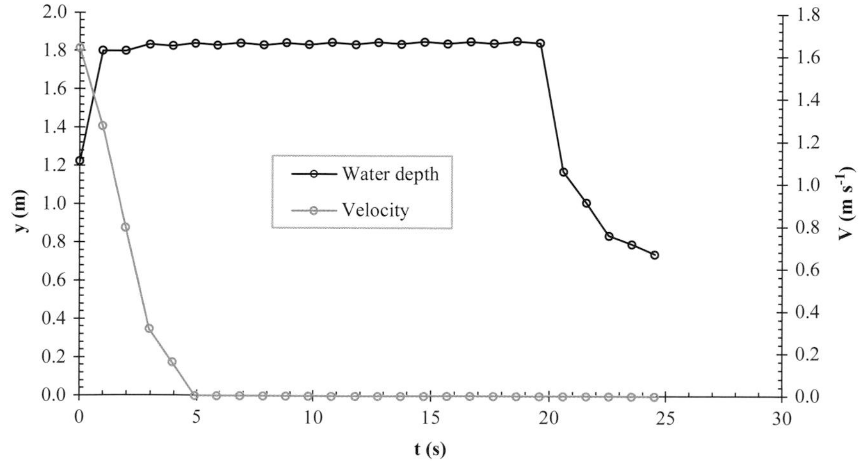

(b)

Figure 10.23 **Computed water depths and velocities for Example 10.20.** Profiles when the gate is gradually closed at: (a) two selected times; and (b) Node 10.

10.3 Transient Water Table between Two Waterbodies

As discussed in Section 9.3.3, by making the Dupuit–Forchheimer assumption (Dupuit, 1863; Forchheimer, 1886), the groundwater flow in an unconfined aquifer connecting two waterbodies (e.g., rivers) (Figure 10.24) can be considered to be 1-D horizontal. A transient flow can be induced by recharge and/or withdrawal, causing fluctuations of the water table with time. In terms of Eq. (9.21) and by dropping the subscript of hydraulic conductivity, the 1-D governing equation can be written as:

hydraulic conductivity; hydraulic head; specific storage (Eq. (9.6)); time; flow path length along x-axis direction; sink (+) or source (−) water volumetric rate per unit volume of aquifer

$$\frac{\partial}{\partial x}\left(K\frac{\partial h}{\partial x}\right) + W_{ss} = S_s\frac{\partial h}{\partial t} \tag{10.35}$$

As shown in Example 10.21, the backward finite-difference format of Eq. (10.35) can be written as:

computational time step; average hydraulic conductivity from node i − 1 to i; hydraulic head at node i − 1 and time j − 1; average hydraulic conductivity from node i to i + 1; hydraulic head at node i and time j − 1; hydraulic head at node i and time j; length of aquifer segment; specific storage at node i; hydraulic head at node i + 1 and time j − 1; sink (+) or source (−) water volumetric rate per unit volume of aquifer at node i and time j − 1

$$h_{i,j} = \frac{\Delta t}{(\Delta x)^2 S_{s,i}}\left\{K_{i-1/2}h_{i-1,j-1} - \left[K_{i-1/2} + K_{i+1/2} - \frac{(\Delta x)^2 S_{s,i}}{\Delta t}\right]h_{i,j-1} + K_{i+1/2}h_{i+1,j-1} + (\Delta x)^2 W_{ss,i,j-1}\right\} \tag{10.36}$$

Furthermore, the average hydraulic conductivities in Eq. (10.36) can be computed as the harmonic means of the corresponding nodal hydraulic conductivities as:

average hydraulic conductivity from node i − 1 to i; hydraulic conductivity at node i − 1; hydraulic conductivity at node i

$$K_{i-1/2} = \frac{2K_{i-1}K_i}{K_{i-1} + K_i} \tag{10.37}$$

average hydraulic conductivity from node i to i + 1; hydraulic conductivity at node i; hydraulic conductivity at node i + 1

$$K_{i+1/2} = \frac{2K_iK_{i+1}}{K_i + K_{i+1}} \tag{10.38}$$

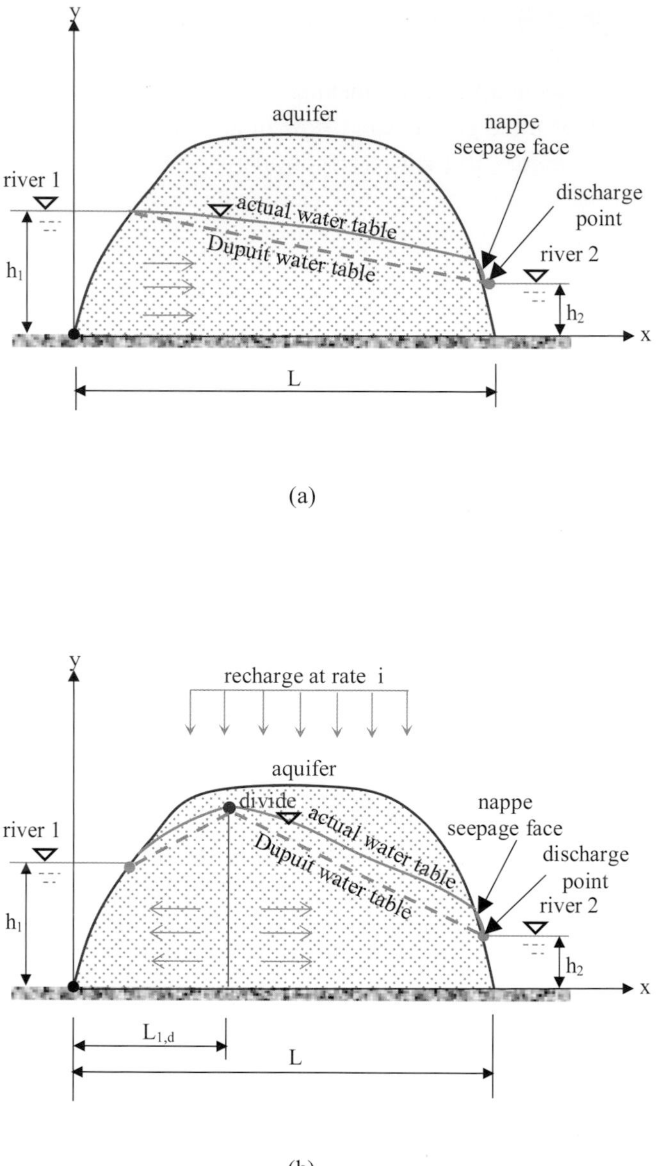

Figure 10.24 **The water table and groundwater flow in an unconfined aquifer connecting two rivers.** The diagrams illustrate transient flow in groundwater. Recharge is: (a) zero; and (b) nonzero. (Reprint of Figure 9.9 for convenience.)

Finally, for a given recharge or withdrawal rate, the sink or source term is computed as:

$$W_{ss,i,j} = \frac{i_{i,j}}{h_{i,j}} \tag{10.39}$$

sink (+) or source (−) water volumetric rate per unit volume of aquifer at node i and time j

recharge (+) or withdrawal (−) rate per unit area of aquifer at node I and time j

hydraulic head at node i and time j

To apply Eq. (10.36), the aquifer is subdivided into a number of smaller segments Δx apart. For a computation time step Δt, apply the equation to the aquifer segment by segments to determine the nodal hydraulic heads at a current time from those at the previous time. The water levels in the two rivers, which can either be invariant or vary with time, define the upper and lower boundary conditions. The stability condition is defined as:

$$\Delta t \le (\Delta x)\frac{\max\limits_{i=0,1,2,\ldots}\{S_{s,i}\}\,h_{max}}{\min\limits_{i=0,1,2,\ldots}\{K_i\}} \tag{10.40}$$

computational time → Δt; length of aquifer segment → Δx; specific storage at node i → $S_{s,i}$; maximum hydraulic head from Eqs. (9.29) and (9.30) using $\min\{K_i\}$ for entire aquifer → h_{max}; hydraulic conductivity at node i → K_i

Example 10.21 Derive Eq. (10.36).

Solution

Start from Eq. (10.35) at nodes i − 1, i, and i + 1, and at time j − 1 and j.

$$\frac{\partial}{\partial x}\left(K\frac{\partial h}{\partial x}\right)=\frac{\left(K\frac{\partial h}{\partial x}\right)_{i+1,j-1}-\left(K\frac{\partial h}{\partial x}\right)_{i-1,j-1}}{\Delta x}=\frac{\left(K_{i+1/2}\frac{h_{i+1,j-1}-h_{i,j-1}}{\Delta x}\right)-\left(K_{i-1/2}\frac{h_{i,j-1}-h_{i-1,j-1}}{\Delta x}\right)}{\Delta x}$$

$$=\frac{K_{i-1/2}h_{i-1,j-1}-\left(K_{i-1/2}+K_{i+1/2}\right)h_{i,j-1}+K_{i+1/2}h_{i+1,j-1}}{(\Delta x)^2}.$$

$$S_s\frac{\partial h}{\partial t}=S_{s,i}\frac{h_{i,j}-h_{i,j-1}}{\Delta t}=\frac{S_{s,i}}{\Delta t}h_{i,j}-\frac{S_{s,i}}{\Delta t}h_{i,j-1}$$

$$\rightarrow\frac{K_{i-1/2}h_{i-1,j-1}-\left(K_{i-1/2}+K_{i+1/2}\right)h_{i,j-1}+K_{i+1/2}h_{i+1,j-1}}{(\Delta x)^2}+W_{ss,i,j-1}=\frac{S_{s,i}}{\Delta t}h_{i,j}-\frac{S_{s,i}}{\Delta t}h_{i,j-1}$$

$$\rightarrow\frac{(\Delta x)^2 S_{s,i}}{\Delta t}h_{i,j}=K_{i-1/2}h_{i-1,j-1}-\left(K_{i-1/2}+K_{i+1/2}-\frac{(\Delta x)^2 S_{s,i}}{\Delta t}\right)h_{i,j-1}$$

$$+K_{i+1/2}h_{i+1,j-1}+(\Delta x)^2 W_{ss,i,j-1}$$

$$\rightarrow h_{i,j}=\frac{\Delta t}{(\Delta x)^2 S_{s,i}}\left\{K_{i-1/2}h_{i-1,j-1}-\left(K_{i-1/2}+K_{i+1/2}-\frac{(\Delta x)^2 S_{s,i}}{\Delta t}\right)h_{i,j-1}+K_{i+1/2}h_{i+1,j-1}\right.$$

$$\left.+(\Delta x)^2 W_{ss,i,j-1}\right\}$$

→ Eq. (10.36).

Example 10.22 In Figure 10.24b, if $h_l = 5.5\,m$, $h_2 = 5.5\,m$, $K = 10\,m\,d^{-1}$, $S_s = 0.0001\,m^{-1}$, $L = 200\,m$, and $i = 0.01\,m\,d^{-1}$. If the initial water table in the aquifer is 5.5 m, determine the variations of water table and discharge using $\Delta x = 20\,m$.

Solution

$L = 200\,m$, $\Delta x = 20\,m \rightarrow$ nodes $i = 0, 1, 2, \ldots, 10$

Computational time step: cells K1:K2 in Figure 10.25.

$$\min\{K_i\} = 10\,m\,d^{-1},\ \max\{S_{s,i}\} = 0.0001\,m^{-1}$$

Eq. (9.29): $L_{1,d} = 1/2 * \left\{200\,m - \left[\left(10\,m\,d^{-1}\right)/\left(0.01\,m\,d^{-1}\right)\right] * \left[(5.5\,m)^2 - (5.5\,m)^2\right]/(200\,m)\right\} = 100\,m$.

Eq. (9.30): $h_{max}{}^2 = \left[\left(0.01\,m\,d^{-1}\right)/\left(10\,m\,d^{-1}\right)\right] * [(200\,m) - (100\,m)] * (100\,m) - [(5.5\,m)^2 - (5.5\,m)^2]/(200\,m) * (100\,m) + (5.5\,m)^2 = 40.25\,m^2 \rightarrow h_{max} = 6.3443\,m$.

Eq. (10.40): $\Delta t \leq (20\,m) * \left[\left(0.0001\,m^{-1}\right) * (6.3443\,m)/\left(10\,m\,d^{-1}\right)\right] = 0.0013\,d$.

Use a computational time step of $\Delta t = 0.001\,d$.

Initial condition: cells D8:N10 in Figure 10.25, highlighted in red.

- At node $i = 0$:

$$h_{0,0} = 5.5\,m$$

Eq. (10.39) $\rightarrow W_{ss,0,0} = \left(0.01\,m\,d^{-1}\right)/(5.5\,m) = 0.0018\,d^{-1}$.
Darcy's law $\rightarrow q_{0,0} = -\left[2 * \left(10\,m\,d^{-1}\right) * \left(10\,m\,d^{-1}\right)/\left(10\,m\,d^{-1} + 10\,m\,d^{-1}\right)\right] * [(5.5\,m - 5.5\,m)/(20\,m)] * (5.5\,m) = 0.0\,m^2d^{-1}$.

- At node $i = 10$:

$$h_{10,0} = 5.5\,m$$

Eq. (10.39) $\rightarrow W_{ss,10,0} = \left(0.01\,m\,d^{-1}\right)/(5.5\,m) = 0.0018\,d^{-1}$.
Darcy's law $\rightarrow q_{10,0} = -\left[2 * \left(10\,m\,d^{-1}\right) * \left(10\,m\,d^{-1}\right)/\left(10\,m\,d^{-1} + 10\,m\,d^{-1}\right)\right] * [(5.5\,m - 5.5\,m)/(20\,m)] * (5.5\,m) = 0.0\,m^2d^{-1}$.

- At node $1 \leq i \leq 9$:

$$h_{i,0} = 5.5\,m$$

Eq. (10.39) $\rightarrow W_{ss,i,0} = \left(0.01\,m\,d^{-1}\right)/(5.5\,m) = 0.0018\,d^{-1}$.

$$\text{Darcy's law} \rightarrow q_{i,0} = -\frac{2\left[2K_{i-1}K_i/(K_{i-1} + K_i)\right]\left[2K_iK_{i+1}/(K_i + K_{i+1})\right]}{\left[2K_{i-1}K_i/(K_{i-1} + K_i)\right] + \left[2K_iK_{i+1}/(K_i + K_{i+1})\right]}\left(\frac{h_{i+1,0} - h_{i-1,0}}{2(\Delta x)}\right)h_{i,0}.$$

Upper boundary condition ($i = 0$): cells D11:D34 in Figure 10.25, highlighted in blue.

At time $j \geq 1$:

$$h_{0,j} = 5.5\,m$$

	A	B	C	D	E	F	G	H	I	J	K	L	M	N
1	Δx =	20	m	i =	0.01	m d^{-1}	h_1 =	5.5	m	Eq. (9-29): $L_{1,d}$ =	100	m		
2	Δt =	0.001	d	min{K_i} =	10	m d^{-1}	h_2 =	5.5	m	Eq. (9-30): h_{max} =	6.3443	m		
3				max{$S_{s,i}$} =	0.0001		L =	200	m					
4									Node i =					
5	j	t_j (d)	Variables	0	1	2	3	4	5	6	7	8	9	10
6			K_i (m d^{-1})	10	10	10	10	10	10	10	10	10	10	10
7			$S_{s,i}$ (m^{-1})	0.0001	0.0001	0.0001	0.0001	0.0001	0.0001	0.0001	0.0001	0.0001	0.0001	0.0001
8	0	0	$h_{i,0}$ (m)	5.5	5.5	5.5	5.5	5.5	5.5	5.5	5.5	5.5	5.5	5.5
9			$W_{ss,i,0}$ (d^{-1})	0.0018	0.0018	0.0018	0.0018	0.0018	0.0018	0.0018	0.0018	0.0018	0.0018	0.0018
10			$q_{i,0}$ (m^2 d^{-1})	0.0000	0.0000	0.0000	0.0000	0.0000	0.0000	0.0000	0.0000	0.0000	0.0000	0.0000
11	1	0.001	$h_{i,1}$ (m)	5.500	5.518	5.518	5.518	5.518	5.518	5.518	5.518	5.518	5.518	5.500
12			$W_{ss,i,1}$ (d^{-1})	0.0018	0.0018	0.0018	0.0018	0.0018	0.0018	0.0018	0.0018	0.0018	0.0018	0.0018
13			$q_{i,1}$ (m^2 d-1)	-0.0500	-0.0251	0.0000	0.0000	0.0000	0.0000	0.0000	0.0000	0.0000	0.0251	0.0500
14	2	0.002	$h_{i,2}$ (m)	5.500	5.532	5.536	5.536	5.536	5.536	5.536	5.536	5.536	5.532	5.500
15			$W_{ss,i,2}$ (d^{-1})	0.0018	0.0018	0.0018	0.0018	0.0018	0.0018	0.0018	0.0018	0.0018	0.0018	0.0018
16			$q_{i,2}$ (m^2 d^{-1})	-0.0873	-0.0502	-0.0063	0.0000	0.0000	0.0000	0.0000	0.0000	0.0063	0.0502	0.0873
17	3	0.003	$h_{i,2}$ (m)	5.500	5.543	5.553	5.554	5.554	5.554	5.554	5.554	5.553	5.543	5.500
18			$W_{ss,i,2}$ (d^{-1})	0.0018	0.0018	0.0018	0.0018	0.0018	0.0018	0.0018	0.0018	0.0018	0.0018	0.0018
19			$q_{i,3}$ (m^2 d^{-1})	-0.1183	-0.0738	-0.0157	-0.0016	0.0000	0.0000	0.0000	0.0016	0.0157	0.0738	0.1183
20	4	0.004	$h_{i,2}$ (m)	5.500	5.553	5.569	5.572	5.572	5.572	5.572	5.572	5.569	5.553	5.500
21			$W_{ss,i,2}$ (d^{-1})	0.0018	0.0018	0.0018	0.0018	0.0018	0.0018	0.0018	0.0018	0.0018	0.0018	0.0018
22			$q_{i,4}$ (m^2 d^{-1})	-0.1454	-0.0957	-0.0268	-0.0047	-0.0004	0.0000	0.0004	0.0047	0.0268	0.0957	0.1454
23	5	0.005	$h_{i,2}$ (m)	5.500	5.562	5.584	5.589	5.590	5.590	5.590	5.589	5.584	5.562	5.500
24			$W_{ss,i,2}$ (d^{-1})	0.0018	0.0018	0.0018	0.0018	0.0018	0.0018	0.0018	0.0018	0.0018	0.0018	0.0018
25			$q_{i,5}$ (m^2 d^{-1})	-0.1696	-0.1164	-0.0386	-0.0092	-0.0014	0.0000	0.0014	0.0092	0.0386	0.1164	0.1696
26	6	0.006	$h_{i,2}$ (m)	5.500	5.570	5.598	5.606	5.608	5.608	5.608	5.606	5.598	5.570	5.500
27			$W_{ss,i,2}$ (d^{-1})	0.0018	0.0018	0.0018	0.0018	0.0018	0.0018	0.0018	0.0018	0.0018	0.0018	0.0018
28			$q_{i,6}$ (m^2 d^{-1})	-0.1918	-0.1358	-0.0508	-0.0146	-0.0030	0.0000	0.0030	0.0146	0.0508	0.1358	0.1918
29	7	0.007	$h_{i,2}$ (m)	5.500	5.577	5.611	5.622	5.625	5.626	5.625	5.622	5.611	5.577	5.500
30			$W_{ss,i,2}$ (d^{-1})	0.0018	0.0018	0.0018	0.0018	0.0018	0.0018	0.0018	0.0018	0.0018	0.0018	0.0018
31			$q_{i,7}$ (m^2 d^{-1})	-0.2123	-0.1542	-0.0631	-0.0208	-0.0052	0.0000	0.0052	0.0208	0.0631	0.1542	0.2123
32	8	0.008	$h_{i,2}$ (m)	5.500	5.584	5.623	5.638	5.642	5.643	5.642	5.638	5.623	5.584	5.500
33			$W_{ss,i,2}$ (d^{-1})	0.0018	0.0018	0.0018	0.0018	0.0018	0.0018	0.0018	0.0018	0.0018	0.0018	0.0018
34			$q_{i,8}$ (m^2 d^{-1})	-0.2315	-0.1717	-0.0755	-0.0275	-0.0078	0.0000	0.0078	0.0275	0.0755	0.1717	0.2315
35	9	0.009	$h_{i,2}$ (m)	5.500	5.591	5.635	5.653	5.659	5.661	5.659	5.653	5.635	5.591	5.500
36			$W_{ss,i,2}$ (d^{-1})	0.0018	0.0018	0.0018	0.0018	0.0018	0.0018	0.0018	0.0018	0.0018	0.0018	0.0018
37			$q_{i,9}$ (m^2 d^{-1})	-0.2495	-0.1884	-0.0878	-0.0346	-0.0108	0.0000	0.0108	0.0346	0.0878	0.1884	0.2495
38	10	0.01	$h_{i,2}$ (m)	5.500	5.597	5.646	5.668	5.676	5.678	5.676	5.668	5.646	5.597	5.500
39			$W_{ss,i,2}$ (d^{-1})	0.0018	0.0018	0.0018	0.0018	0.0018	0.0018	0.0018	0.0018	0.0018	0.0018	0.0018
40			$q_{i,10}$ (m^2 d^{-1})	-0.2666	-0.2044	-0.0999	-0.0420	-0.0141	0.0000	0.0141	0.0420	0.0999	0.2044	0.2666

Figure 10.25 **Excerpt of the spreadsheet for the calculations for Example 10.22.**

Eq. (10.39) $\rightarrow W_{ss,0,j} = \left(0.01\,\text{m}\,\text{d}^{-1}\right) / (5.5\,\text{m}) = 0.0018\,\text{d}^{-1}$.

Darcy's law $\rightarrow q_{0,j} = -\left[2K_0K_1/\left(K_0 + K_1\right)\right]\left[\left(h_{1,j-1} - h_{0,j-1}\right)/(\Delta x)\right] h_{0,j}$.

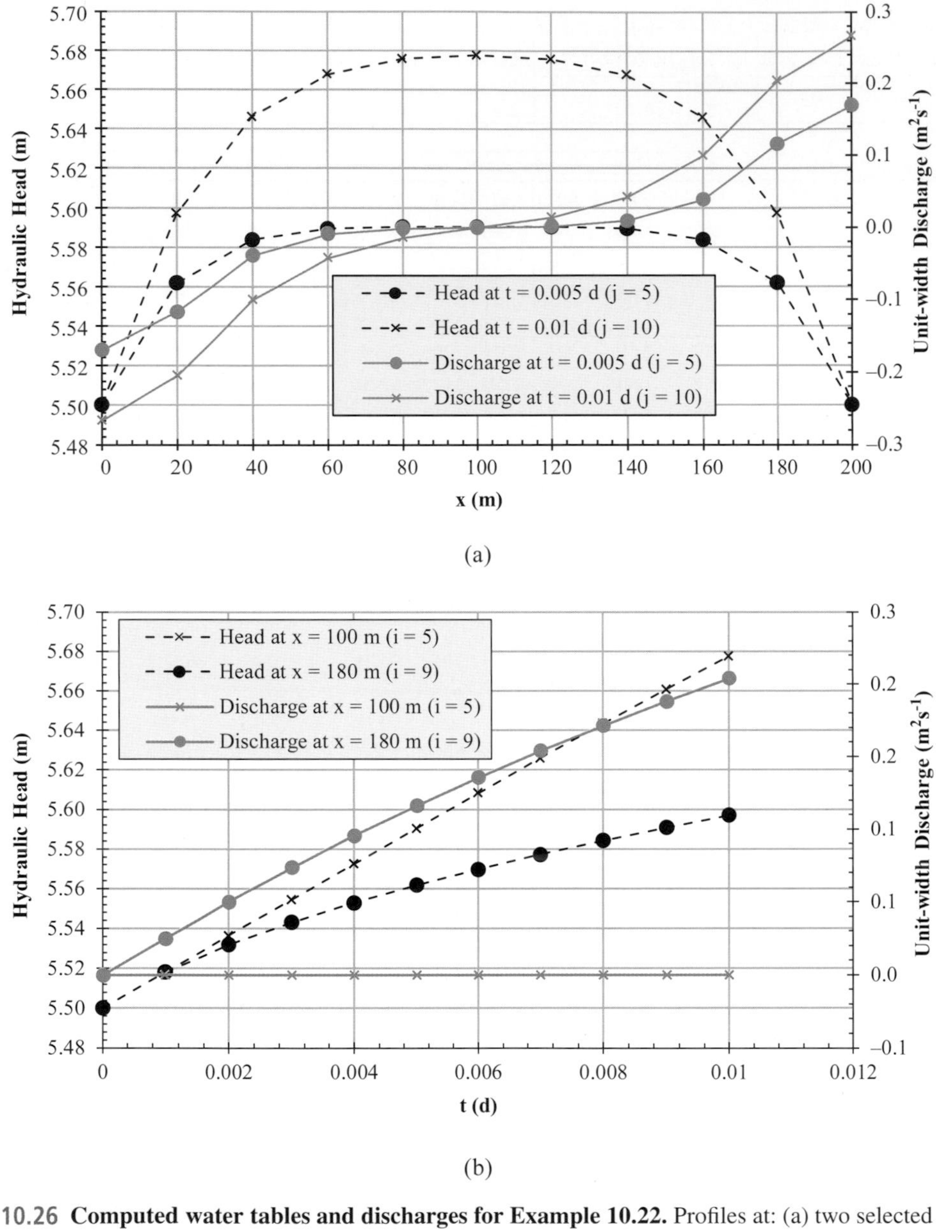

Figure 10.26 Computed water tables and discharges for Example 10.22. Profiles at: (a) two selected times; and (b) Node 5 and 9.

Lower boundary condition (i = 10): cells N11:N34 in Figure 10.25, highlighted in blue

At time $j \geq 1$:

$$h_{10,j} = 5.5\,\text{m}$$

Eq. (10.39) $\rightarrow W_{ss,10,j} = \left(0.01\,\text{m}\,\text{d}^{-1}\right) / (5.5\,\text{m}) = 0.0018\,\text{d}^{-1}$.

Darcy's law $\rightarrow q_{10,j} = -\left[2K_9K_{10}/\left(K_9 + K_{10}\right)\right]\left[\left(h_{10,j-1} - h_{9,j-1}\right)/\left(\Delta x\right)\right]h_{10,j}$.

Interior nodes ($i = 1, 2, \ldots, 9$) at time $j \geq 1$: cells E11:M40 in Figure 10.25

Eq. (10.36) $\rightarrow h_{i,j}$.

Eq. (10.39) $\rightarrow W_{ss,i,j} = i_{i,j}/h_{i,j} = 0.01/h_{i,j}d^{-1}$.

Darcy's law $\rightarrow q_{i,j} = -\dfrac{2\left[2K_{i-1}K_i/\left(K_{i-1}+K_i\right)\right]\left[2K_iK_{i+1}/\left(K_i+K_{i+1}\right)\right]}{\left[2K_{i-1}K_i/\left(K_{i-1}+K_i\right)\right]+\left[2K_iK_{i+1}/\left(K_i+K_{i+1}\right)\right]}\left(\dfrac{h_{i+1,j}-h_{i-1,j}}{2\left(\Delta x\right)}\right)h_{i,j}$.

The calculations can be done in an Excel spreadsheet, as shown in Figure 10.25. The profiles of water table and discharge at 0.005 d ($j = 5$) and 0.01 d ($j = 10$) are shown in Figure 10.26a, while the fluctuations of water table and discharge at Nodes 5 and 9 are shown in Figure 10.26b.

For instance, at time $j = 1$, the upper boundary condition (cells D11:D13) is $h_{0,1} = 5.5\,\text{m}$; $W_{ss,0,1} = 0.0018\,\text{d}^{-1}$; and $q_{0,1} = -\left[2*\left(10\,\text{m}\,\text{d}^{-1}\right)*\left(10\,\text{m}\,\text{d}^{-1}\right)/(10\,\text{m}\,\text{d}^{-1} + 10\,\text{md}^{-1})\right]*$ $[(5.518\,\text{m} - 5.5\,\text{m})/(20\,\text{m})]*(5.5\,\text{m}) = -0.050\,\text{m}^2\text{d}^{-1}$, discharging into River 1. In contrast, the lower boundary condition (cells N11:N13) is $h_{10,1} = 5.5\,\text{m}$; $W_{ss,10,1} = 0.0018\,\text{d}^{-1}$; and $q_{10,1} = -\left[2*\left(10\,\text{m}\,\text{d}^{-1}\right)*\left(10\,\text{m}\,\text{d}^{-1}\right)/\left(10\,\text{m}\,\text{d}^{-1} + 10\,\text{m}\,\text{d}^{-1}\right)\right]*[(5.5\,\text{m} - 5.518\,\text{m})/(20\,\text{m})]*$ $(5.5\,\text{m}) = 0.050\,\text{m}^2\text{d}^{-1}$, discharging into River 2. At node $i = 9$ (cells M11:M13), $K_{9-1/2} = K_{9+1/2} = 2*\left(10\,\text{m}\,\text{d}^{-1}\right)*\left(10\,\text{m}\,\text{d}^{-1}\right)/\left(10\,\text{m}\,\text{d}^{-1} + 10\,\text{m}\,\text{d}^{-1}\right) = 10\,\text{m}\,\text{d}^{-1}$, $S_{s,9} = 0.0001\,\text{m}^{-1}$, $\Delta t/\left[(\Delta x)^2 S_{s,9}\right] = (0.001\,\text{d})/\left[(20\,\text{m})^2*\left(0.0001\,\text{m}^{-1}\right)\right] = 0.025\,\text{d}\,\text{m}^{-1}$, $h_{9,1} = \left(0.025\,\text{d}\,\text{m}^{-1}\right)*$ $\left\{\left(10\,\text{m}\,\text{d}^{-1}\right)*(5.5\,\text{m}) - \left[\left(10\,\text{m}\,\text{d}^{-1}\right) + \left(10\,\text{m}\,\text{d}^{-1}\right) - 1/\left(0.025\,\text{d}\,\text{m}^{-1}\right)\right]*(5.5\,\text{m}) + \left(10\,\text{m}\,\text{d}^{-1}\right)*\right.$ $\left.(5.5\,\text{m}) + (20\,\text{m})^2*\left(0.0018\,\text{d}^{-1}\right)\right\} = 5.518\,\text{m}$; and$q_{9,1} =$ $-\left[2*\left(10\,\text{m}\,\text{d}^{-1}\right)*\left(10\,\text{m}\,\text{d}^{-1}\right)/\left(10\,\text{m}\,\text{d}^{-1} + 10\,\text{m}\,\text{d}^{-1}\right)\right]*[(5.5\,\text{m} - 5.518\,\text{m})/(2*20\,\text{m})]*$ $(5.518\,\text{m}) = 0.0251\,\text{m}^2\text{s}^{-1}$, discharging toward the right. Similar calculations can be done for nodes $i = 8, 7, \ldots, 1$. At other times $j = 2, 3, \ldots$, determine the upper and lower boundary conditions and do the calculations for the interior nodes in similar ways.

Example 10.23 In Figure 10.24b, $h_1 = 5.5\,\text{m}$, $h_2 = 2.5\,\text{m}$, and $L = 200\,\text{m}$. The initial water table in the aquifer varies as $h(x) = 5.5 - 0.015x$. For $\Delta x = 10\,\text{m}$, the values of K_i and $S_{s,i}$, where $i = 0, 1, 2, \ldots, 10$, are given in cells D6:N7 of Figure 10.27. If the aquifer has a recharge rate of $0.5\,\text{m}\,\text{d}^{-1}$ for the first five computation time steps, no sink or source for the successive three time steps, and a withdrawal rate of $0.5\,\text{m}\,\text{d}^{-1}$ for the remaining two time steps, determine the variations of water table and discharge.

Solution

$L = 200\,\text{m}$, $\Delta x = 20\,\text{m} \rightarrow$ nodes $i = 0, 1, 2, \ldots, 10$

Computational time step: cells K1:K2 in Figure 10.27

$$\min\{K_i\} = 5\,\text{m}\,\text{d}^{-1},\ \max\{S_{s,i}\} = 0.0003\,\text{m}^{-1}$$

Eq. (9.29): $L_{1,d} = 1/2 * \left\{200\,\mathrm{m} - \left[\left(5\,\mathrm{m\,d^{-1}}\right) / \left(0.5\,\mathrm{m\,d^{-1}}\right)\right] * \left[(5.5\,\mathrm{m})^2 - (2.5\,\mathrm{m})^2\right] / (200\,\mathrm{m})\right\} =$ 99.4 m.

	A	B	C	D	E	F	G	H	I	J	K	L	M	N
1	Δx =	20	m	i =	0.5	m d^{-1}	h_1 =	5.5	m	Eq. (9-29): $L_{1,d}$ =	99.4	m		
2	Δt =	0.001	d	min{K_i} =	5	m d^{-1}	h_2 =	2.5	m	Eq. (9-30): h_{max} =	31.911	m		
3				max{$S_{s,i}$} =	0.0003		L =	200	m					
4									Node i =					
5	j	t_j (d)	Variables	0	1	2	3	4	5	6	7	8	9	10
6			K_i (m d^{-1})	10	5	15	25	12	20	8	13	9	22	15
7			$S_{s,i}$ (m^{-1})	0.0002	0.0003	0.0003	0.0002	0.0001	0.0002	0.0001	0.0001	0.0002	0.0002	0.0002
8	0	0	$h_{i,0}$ (m)	5.5	5.2	4.9	4.6	4.3	4	3.7	3.4	3.1	2.8	2.5
9			$W_{ss,i,0}$ (d^{-1})	0.0909	0.0962	0.1020	0.1087	0.1163	0.1250	0.1351	0.1471	0.1613	0.1786	0.2000
10			$q_{i,0}$ (m^2 d^{-1})	0.5500	0.5506	0.7875	1.2000	1.0052	0.7784	0.5890	0.5231	0.5398	0.6253	0.6689
11	1	0.001	$h_{i,1}$ (m)	5.500	5.518	5.212	5.153	5.472	4.638	5.063	4.865	3.898	3.674	2.500
12			$W_{ss,i,1}$ (d^{-1})	0.0909	0.0906	0.0959	0.0970	0.0914	0.1078	0.0988	0.1028	0.1283	0.1361	0.2000
13			$q_{i,1}$ (m^2 d-1)	-0.0338	0.2805	0.5102	-0.5823	1.0971	0.6155	-0.3045	1.4526	1.3476	1.9121	2.6174
14	2	0.002	$h_{i,2}$ (m)	5.500	5.800	5.542	5.717	5.944	5.394	5.880	5.685	4.632	4.128	2.500
15			$W_{ss,i,2}$ (d^{-1})	0.0909	0.0862	0.0902	0.0875	0.0841	0.0927	0.0850	0.0880	0.1079	0.1211	0.2000
16			$q_{i,2}$ (m^2 d^{-1})	-0.5505	-0.0427	0.1242	-0.9994	0.7465	0.1113	-0.4531	1.8190	2.0921	3.2765	3.6310
17	3	0.003	$h_{i,2}$ (m)	5.500	6.055	5.886	6.159	6.487	6.030	6.543	6.333	5.232	4.451	2.500
18			$W_{ss,i,2}$ (d^{-1})	0.0909	0.0826	0.0849	0.0812	0.0771	0.0829	0.0764	0.0790	0.0956	0.1123	0.2000
19			$q_{i,3}$ (m^2 d^{-1})	-1.0171	-0.4124	-0.1643	-1.6091	0.3256	-0.1104	-0.5254	2.1303	2.8563	4.5253	4.3510
20	4	0.004	$h_{i,2}$ (m)	5.500	6.289	6.222	6.567	6.953	6.604	7.109	6.882	5.731	4.702	2.500
21			$W_{ss,i,2}$ (d^{-1})	0.0909	0.0795	0.0804	0.0761	0.0719	0.0757	0.0703	0.0727	0.0872	0.1063	0.2000
22			$q_{i,4}$ (m^2 d^{-1})	-1.4459	-0.8016	-0.4645	-2.0877	-0.0984	-0.3325	-0.5242	2.4308	3.6243	5.6552	4.9109
23	5	0.005	$h_{i,2}$ (m)	5.500	6.506	6.548	6.945	7.385	7.120	7.612	7.359	6.156	4.907	2.500
24			$W_{ss,i,2}$ (d^{-1})	0.0909	0.0769	0.0764	0.0720	0.0677	0.0702	0.0657	0.0679	0.0812	0.1019	0.2000
25			$q_{i,5}$ (m^2 d^{-1})	-1.8438	-1.2035	-0.7712	-2.5263	-0.5023	-0.5235	-0.4817	2.7464	4.3791	6.6775	5.3676
26	6	0.006	$h_{i,2}$ (m)	5.500	6.709	6.862	7.301	7.784	7.591	8.065	7.781	6.523	5.079	2.500
27			$W_{ss,i,2}$ (d^{-1})	0.0000	0.0000	0.0000	0.0000	0.0000	0.0000	0.0000	0.0000	0.0000	0.0000	0.0000
28			$q_{i,6}$ (m^2 d^{-1})	-2.2159	-1.6127	-1.0894	-2.9278	-0.8785	-0.6916	-0.4064	3.0781	5.1136	7.6047	5.7513
29	7	0.007	$h_{i,2}$ (m)	5.500	6.651	6.921	7.296	7.516	7.695	7.859	7.517	6.460	4.735	2.500
30			$W_{ss,i,2}$ (d^{-1})	0.0000	0.0000	0.0000	0.0000	0.0000	0.0000	0.0000	0.0000	0.0000	0.0000	0.0000
31			$q_{i,7}$ (m^2 d^{-1})	-2.1104	-1.6681	-1.1961	-1.8871	-1.1675	-0.8569	0.3717	2.6984	5.2150	6.9773	4.9828
32	8	0.008	$h_{i,2}$ (m)	5.500	6.604	6.963	7.253	7.494	7.685	7.728	7.320	6.325	4.512	2.500
33			$W_{ss,i,2}$ (d^{-1})	0.0000	0.0000	0.0000	0.0000	0.0000	0.0000	0.0000	0.0000	0.0000	0.0000	0.0000
34			$q_{i,8}$ (m^2 d^{-1})	-2.0241	-1.7049	-1.2102	-1.6750	-1.2613	-0.5820	0.7472	2.6337	5.1549	6.4225	4.4859
35	9	0.009	$h_{i,2}$ (m)	5.500	6.565	6.986	7.234	7.468	7.655	7.615	7.156	6.168	4.353	2.500
36			$W_{ss,i,2}$ (d^{-1})	-0.0909	-0.0762	-0.0716	-0.0691	-0.0670	-0.0653	-0.0657	-0.0699	-0.0811	-0.1149	-0.2000
37			$q_{i,9}$ (m^2 d^{-1})	-1.9528	-1.7214	-1.2513	-1.5162	-1.2261	-0.3642	1.0076	2.6555	5.0182	5.9415	4.1311
38	10	0.01	$h_{i,2}$ (m)	5.500	6.278	6.760	6.878	6.774	7.288	6.856	6.308	5.604	3.655	2.500
39			$W_{ss,i,2}$ (d^{-1})	-0.0909	-0.0796	-0.0740	-0.0727	-0.0738	-0.0686	-0.0729	-0.0793	-0.0892	-0.1368	-0.2000
40			$q_{i,10}$ (m^2 d^{-1})	-1.4271	-1.3957	-1.0849	-0.0420	-1.0825	-0.1948	1.7816	2.0257	4.3148	4.2224	2.5755

Figure 10.27 **Excerpt of the spreadsheet for the calculations for Example 10.23.**

Eq. (9.30): ${h_{max}}^2 = \left[\left(0.5\,\mathrm{m\,d^{-1}}\right) / \left(5\,\mathrm{m\,d^{-1}}\right)\right] * [(200\,\mathrm{m}) - (99.4\,\mathrm{m})] * (99.4\,\mathrm{m}) - [(5.5\,\mathrm{m})^2 - (2.5\,\mathrm{m})^2] / (200\,\mathrm{m}) * (99.4\,\mathrm{m}) + (5.5\,\mathrm{m})^2 = 1018.286\,\mathrm{m}^2 \rightarrow h_{max} = 31.911\,\mathrm{m}$.

Eq. (10.40): $\Delta t \leq (20\,\mathrm{m}) * \left[\left(0.0003\,\mathrm{m^{-1}}\right) * (31.911\,\mathrm{m}) / \left(5\,\mathrm{m\,d^{-1}}\right)\right] = 0.0039\,\mathrm{d}$.

Use a computation time step of $\Delta t = 0.001$ d.

Initial condition: cells D8:N10 in Figure 10.27, highlighted in red.

- At node $i = 0$:

$$h_{0,0} = 5.5\,\text{m}$$

Eq. (10.39) $\rightarrow W_{ss,0,0} = \left(0.5\,\text{m}\,\text{d}^{-1}\right) / (5.5\,\text{m}) = 0.0909\,\text{d}^{-1}$.
Darcy's law $\rightarrow q_{0,0} = -\left[2 * \left(10\,\text{m}\,\text{d}^{-1}\right) * \left(5\,\text{m}\,\text{d}^{-1}\right) / \left(10\,\text{m}\,\text{d}^{-1} + 5\,\text{m}\,\text{d}^{-1}\right)\right] * [(5.2\,\text{m} - 5.5\,\text{m}) / (20\,\text{m})] * (5.5\,\text{m}) = 0.55\,\text{m}\,\text{d}^{-1}$.

- At node $i = 10$:

$$h_{10,0} = 2.5\,\text{m}$$

Eq. (10.39) $\rightarrow W_{ss,10,0} = \left(0.5\,\text{m}\,\text{d}^{-1}\right) / (2.5\,\text{m}) = 0.2\,\text{d}^{-1}$.
Darcy's law $\rightarrow q_{10,0} = -\left[2 * \left(22\,\text{m}\,\text{d}^{-1}\right) * \left(15\,\text{m}\,\text{d}^{-1}\right) / \left(22\,\text{m}\,\text{d}^{-1} + 15\,\text{m}\,\text{d}^{-1}\right)\right] * [(2.5\,\text{m} - 2.8\,\text{m}) / (20\,\text{m})] * (2.5\,\text{m}) = 0.6689\,\text{m}^2\text{d}^{-1}$.

- At node $0 < i < 10$:

$$h_{i,0} = 5.5 - 0.015\,(i)\,(\Delta x)\ \text{m}$$

Eq. (10.39) $\rightarrow W_{ss,i,0} = \left(0.5\,\text{m}\,\text{d}^{-1}\right) / h_{i,0} = 0.5/h_{i,0}\,\text{d}^{-1}$.

$$\text{Darcy's law} \rightarrow q_{i,0} = -\frac{2\left[2K_{i-1}K_i/(K_{i-1}+K_i)\right]\left[2K_iK_{i+1}/(K_i+K_{i+1})\right]}{\left[2K_{i-1}K_i/(K_{i-1}+K_i)\right]+\left[2K_iK_{i+1}/(K_i+K_{i+1})\right]}\left(\frac{h_{i+1,0}-h_{i-1,0}}{2\,(\Delta x)}\right)h_{i,0}.$$

Upper boundary condition ($i = 0$): cells D11:D40 in Figure 10.27, highlighted in blue.
At time $j \geq 1$:

$$h_{0,j} = 5.5\,\text{m}$$

Eq. (10.39) $\rightarrow W_{ss,0,j} = i_{0,j} / (5.5\,\text{m}) = i_{0,j}/5.5\,\text{d}^{-1}$.
Darcy's law $\rightarrow q_{0,j} = -\left[2K_0K_1/(K_0+K_1)\right]\left[\left(h_{1,j-1} - h_{0,j-1}\right)/(\Delta x)\right] h_{0,j}$.

Lower boundary condition ($i = 10$): cells N11:N34 in Figure 10.27, highlighted in blue.
At time $j \geq 1$:

$$h_{10,j} = 2.5\,\text{m}$$

Eq. (10.39) $\rightarrow W_{ss,10,j} = i_{10,j,} / (2.5\,\text{m}) = i_{10,j}/2.5\,\text{d}^{-1}$.
Darcy's law $\rightarrow q_{10,j} = -\left[2K_9K_{10}/(K_9+K_{10})\right]\left[\left(h_{10,j-1} - h_{9,j-1}\right)/(\Delta x)\right] h_{10,j}$.

Interior nodes ($i = 1, 2, \ldots, 9$) at time $j \geq 1$: cells E11:M40 in Figure 10.27.
Eq. (10.36) $\rightarrow h_{i,j}$
Eq. (10.39) $\rightarrow W_{ss,i,j} = i_{i,j}/h_{i,j}\text{d}^{-1}$.

$$\text{Darcy's law} \rightarrow q_{i,j} = -\frac{2\left[2K_{i-1}K_i/(K_{i-1}+K_i)\right]\left[2K_iK_{i+1}/(K_i+K_{i+1})\right]}{\left[2K_{i-1}K_i/(K_{i-1}+K_i)\right]+\left[2K_iK_{i+1}/(K_i+K_{i+1})\right]}\left(\frac{h_{i+1}-h_{i-1}}{2\,(\Delta x)}\right)h_{i,j}.$$

The calculations can be done in an Excel spreadsheet, as shown in Figure 10.27. The profiles of water table and discharge at 0.005 d (j = 5) and 0.01 d (j = 10) are shown in Figure 10.28a, and the fluctuations of water table and discharge at Nodes 5 and 9 are shown in Figure 10.28b.

For instance, at time j = 1, the upper boundary condition (cells D11:D13) is
$h_{0,1} = 5.5\,\mathrm{m}$; $W_{ss,0,1} = \left(0.5\,\mathrm{m\,d^{-1}}\right) / (5.5\,\mathrm{m}) = 0.0909\,\mathrm{d^{-1}}$; and
$q_{0,1} = -\left[2 * \left(10\,\mathrm{m\,d^{-1}}\right) * \left(5\,\mathrm{m\,d^{-1}}\right) / \left(10\,\mathrm{m\,d^{-1}} + 5\,\mathrm{m\,d^{-1}}\right)\right] * [(5.518\,\mathrm{m} - 5.5\,\mathrm{m}) / (20\,\mathrm{m})] *$

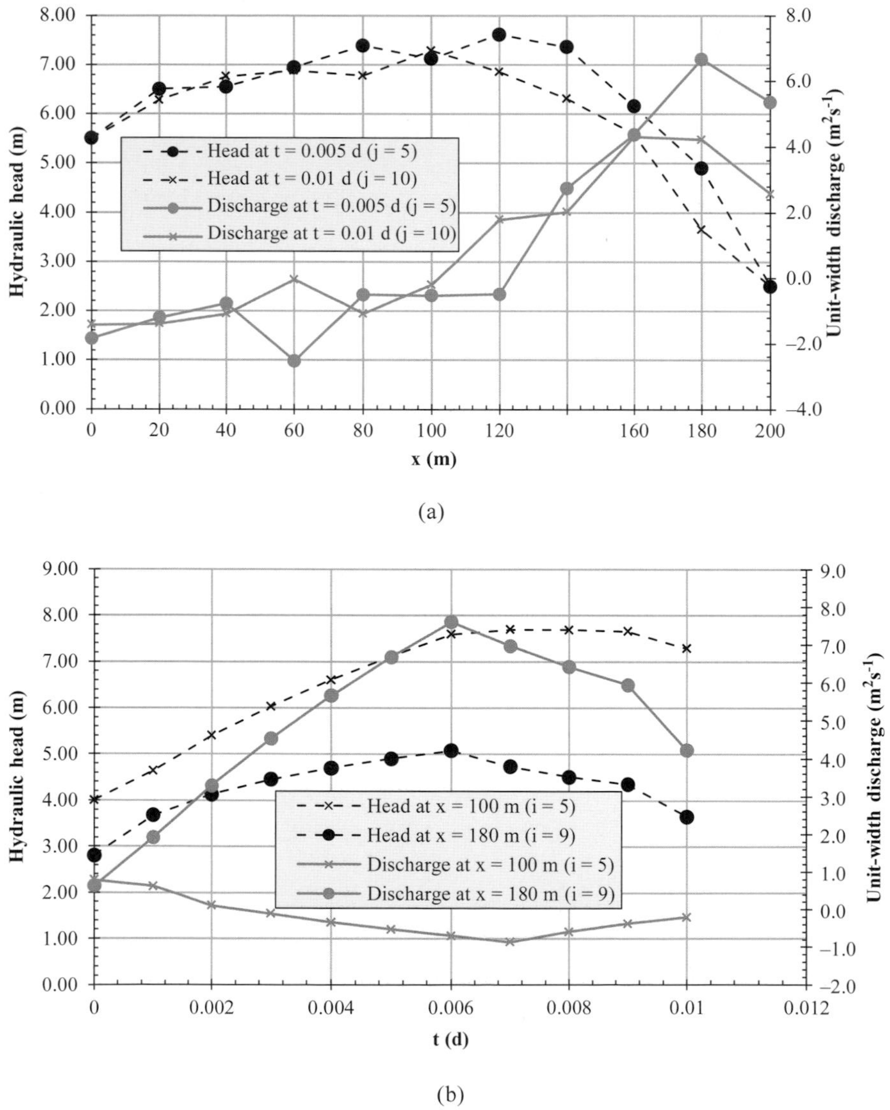

Figure 10.28 Computed water tables and discharges for Example 10.23. Profiles at: (a) two selected times; and (b) Node 5 and 9.

$(5.5\,\text{m}) = -0.0338\,\text{m}^2\text{d}^{-1}$, discharging into River 1. In contrast, the lower boundary condition (cells N11:N13) is $h_{10,1} = 2.5\,\text{m}$; $W_{ss,10,1} = \left(0.5\,\text{m}\,\text{d}^{-1}\right) / (2.5\,\text{m}) = 0.20\,\text{d}^{-1}$; and $q_{10,1} = -\left[2 * \left(22\,\text{m}\,\text{d}^{-1}\right) * \left(15\,\text{m}\,\text{d}^{-1}\right) / \left(22\,\text{m}\,\text{d}^{-1} + 15\,\text{m}\,\text{d}^{-1}\right)\right] * [(2.5\,\text{m} - 3.674\,\text{m}) / (20\,\text{m})] * (2.5\,\text{m}) = 2.6174\,\text{m}^2\text{d}^{-1}$, discharging into River 2. At node $i = 9$ (cells M11:M13), $K_{9-1/2} = 2 * \left(9\,\text{m}\,\text{d}^{-1}\right) * \left(22\,\text{m}\,\text{d}^{-1}\right) / \left(9\,\text{m}\,\text{d}^{-1} + 22\,\text{m}\,\text{d}^{-1}\right) = 12.7742\,\text{m}\,\text{d}^{-1}$, $K_{9+1/2} = 2 * \left(22\,\text{m}\,\text{d}^{-1}\right) * \left(15\,\text{m}\,\text{d}^{-1}\right) / \left(22\,\text{m}\,\text{d}^{-1} + 15\,\text{m}\,\text{d}^{-1}\right) = 17.8378\,\text{m}\,\text{d}^{-1}$, $S_{s,9} = 0.0002$, $\Delta t / \left[(\Delta x)^2 S_{s,9}\right] = (0.001\,\text{d}) / \left[(20\,\text{m})^2 * \left(0.0002\,\text{m}^{-1}\right)\right] = 0.0125\,\text{d}\,\text{m}^{-1}$, $h_{9,1} = \left(0.0125\,\text{d}\,\text{m}^{-1}\right) * \left\{\left(12.7742\,\text{m}\,\text{d}^{-1}\right) * (3.1\,\text{m}) - \left[\left(12.7742\,\text{m}\,\text{d}^{-1}\right) + \left(17.8378\,\text{m}\,\text{d}^{-1}\right) - 1/\left(0.0125\,\text{d}\,\text{m}^{-1}\right)\right] * (2.8\,\text{m}) + \left(17.8378\,\text{m}\,\text{d}^{-1}\right) * (2.5\,\text{m}) + (20\,\text{m})^2 * \left(0.1786\,\text{d}^{-1}\right)\right\} = 3.674\,\text{m}$; and $q_{9,1} = -\left[2 * \left(12.7742\,\text{m}\,\text{d}^{-1}\right) * \left(17.8378\,\text{m}\,\text{d}^{-1}\right) / \left(12.7742\,\text{m}\,\text{d}^{-1} + 17.8378\,\text{m}\,\text{d}^{-1}\right)\right] * [(2.5\,\text{m} - 3.898\,\text{m}) / (2 * 20\,\text{m})] * (3.674\,\text{m}) = 1.912\,\text{m}^2\text{s}^{-1}$, discharging toward the right. Similar calculations can be done for nodes $i = 8, 7, \ldots, 1$. At other times $j = 2, 3, \ldots$, determine the upper and lower boundary conditions and do the calculations for the interior nodes in similar ways.

PROBLEMS

10.1 For the manhole illustrated in Figure 10.1, $D_0 = 8$ ft, $H = 10$ ft, $d_2 = 3$ ft, $P_2 = 0.5$ ft, and $S_2 = 0$. The inlet of the outflow circular concrete pipe has a square edge with headwall. For the inflow hydrographs shown in the following table, determine and plot the variations of water depth in the manhole chamber versus time. Assume that the chamber is initially empty.

t (min)	$Q_{1,t}$ (cfs)	$Q_{1,t}$ (cfs)	$Q_{1,t}$ (cfs)	$Q_{1,t}$ (cfs)
0.0	0.00	0.00	0.00	0.00
0.2	0.61	1.00	7.87	8.94
0.4	1.69	2.00	9.67	15.74
0.6	1.86	5.00	13.84	31.34
0.8	4.90	10.00	15.38	31.7
1.0	5.31	25.00	15.85	32.7
1.2	7.22	13.00	19.87	56.86
1.4	7.98	10.00	22.17	39.74
1.6	8.26	9.00	26.1	40.7
1.8	9.95	8.00	28.43	44.34
2.0	9.49	7.00	26.26	44.48
2.2	8.11	6.00	22.24	52.2
2.4	7.74	5.00	20.35	52.52
2.6	6.88	30.00	16.35	19.34
3.4	0.91	50.00	8.96	17.92
3.6	0.02	50.00	4.47	8.94
3.8	0.00	0.00	0.00	0.00
4.0	0.00	0.00	0.00	0.00

10.2 Repeat Problem 10.1 if the initial water depth in the chamber is 0.3 ft.

10.3 Repeat Problem 10.1 if the outflow pipe has an inlet of groove w/headwall.

10.4 Repeat Problem 10.1 if the outflow pipe has an inlet of groove w/projecting.

10.5 Repeat Problem 10.1 if the outflow pipe is replaced by a 1.5-ft-wide by 1.5-ft-high rectangular box culvert with an inlet of 90° headwall w/45° bevels.

10.6 In the following figure, the exit of the concrete circular pipeline is controlled by a closed valve. The entrance loss coefficient is 0.5. Determine the pressure and velocity profiles using $\Delta x = 100$ ft if the valve is:

(1) Suddenly opened

(2) Gradually opened during four computational time steps. That is, at the first time step, the valve is $3/4$ closed; at the second time step, the valve is $1/2$ closed; at the third time step, the valve is $1/4$ closed; and at the fourth time step, the valve is zero closed (i.e., fully opened). The loss coefficients at $3/4$, $1/2$, $1/4$, and zero closed are 20, 8.5, 2.5, and 0, respectively.

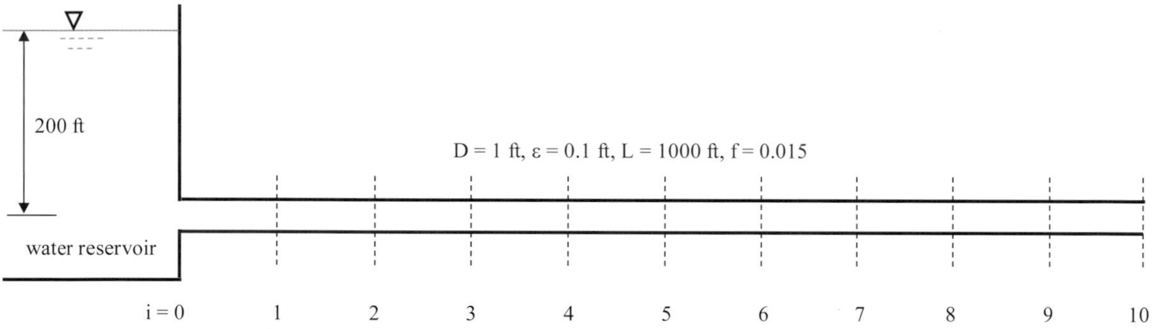

10.7 In the following figure, the exit of the concrete circular pipe is controlled by a valve. The entrance loss coefficient is 0.5. The valve is fully open and a steady-state flow through the pipe has been established. Determine the pressure and velocity profiles using $\Delta x = 50$ ft if the valve is:

(1) Suddenly closed

(2) Gradually closed during four computational time steps. That is, at the first time step, the valve is $1/4$ closed; at the second time step, the valve is $1/2$ closed; at the third time step, the valve is $3/4$ closed; and at the fourth time step, the valve is fully closed. The loss coefficients at zero, $1/4$, $1/2$, and $3/4$ closed are 0, 1.25, 5.5, and 15, respectively.

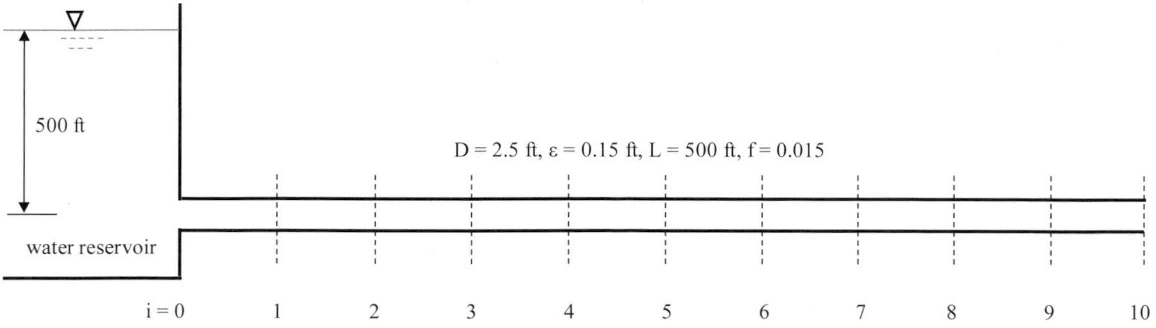

10.8 A 100-m-long, 10-m-wide rectangular channel has a bed slope of 0.001 and a Manning's n of 0.025. The channel, which is designed for a steady-state uniform flow with a normal depth of 1.5 m, is used to divert water from a shallow fresh lake to irrigate a large field. The lake water depth above the channel upstream-end invert is the same as the normal depth. At its downstream end, the channel is controlled by a sluice gate. If the gate is suddenly closed, determine the water depth and velocity profiles using $\Delta x = 10\,m$.

10.9 A 20-m-long, 1-m-wide rectangular channel has a bed slope of 0.001 and a Manning's n of 0.015. The channel, which is designed for a steady-state uniform flow with a normal depth of 0.5 m, is used to divert water from a shallow fresh lake to irrigate a large field. It is controlled by a sluice gate at the downstream end. The lake water depth above the channel upstream-end invert is the same as the normal depth. At the beginning, the gate is fully closed. If the gate is suddenly opened, determine the water depth and velocity profiles using $\Delta x = 2\,m$.

10.10 A 30-m-long, 1.5-m-wide rectangular channel has a bed slope of 0.0025 and a Manning's n of 0.02. The channel, which is designed for a steady-state uniform flow with a normal depth of 1.2 m, is used to divert water from a shallow fresh lake to irrigate a field. The lake water depth above the channel upstream-end invert is the same as the normal depth. At the beginning, the sluice gate at the downstream end of the channel is closed. The gate is gradually opened during the first four time steps, resulting in gate openings of a = 0.1, 0.2, 0.5, and 0.8 m in sequence. At a given opening, the velocity at the channel cross section just upstream of the gate can be computed as $0.35a\sqrt{\frac{2g}{y}}$, where y is the water depth. Determine the water depth and velocity profiles using $\Delta x = 3\,m$.

10.11 A 40-m-long, 2-m-wide rectangular channel has a bed slope of 0.003 and a Manning's n of 0.025. The channel, which is designed for a steady-state uniform flow with a normal depth of 1.35 m, is used to divert water from a shallow fresh lake to irrigate a large field. The lake water depth above the channel upstream-end invert is the same as the normal depth. At the beginning, the sluice gate at the downstream end of the channel is fully open. The gate is gradually closed during the first four time steps, resulting in gate openings of a = 0.8, 0.5, 0.2, and 0.1 m in sequence. At a given opening, the velocity at the channel cross section just upstream of the gate can be computed as $0.38a\sqrt{\frac{2g}{y}}$, where y is the water depth. Once the gate is fully closed, the velocity becomes zero. Determine the water depth and velocity profiles using $\Delta x = 4\,m$.

10.12 In Figure 10.24b, let $h_1 = 3.5\,m$, $h_2 = 3.5\,m$, $K = 8.5\,m\,d^{-1}$, $S_s = 0.0002\,m^{-1}$, $L = 100\,m$, and $i = 0.03\,m\,d^{-1}$. If the initial water table in the aquifer is 3.5 m, determine the variations of water table and discharge using $\Delta x = 10\,m$.

10.13 In Figure 10.24b, let $h_1 = 6.5\,m$, $h_2 = 3.5\,m$, $K = 4.5\,m\,d^{-1}$, $S_s = 0.0001\,m^{-1}$, $L = 50\,m$, and $i = 0.25\,m\,d^{-1}$. The initial water table in the aquifer varies as $h(x) = 6.5 - 0.06x$, where x is the distance from River 1. Determine the variations of water table and discharge using $\Delta x = 5\,m$.

10.14 In Figure 10.24b, let $h_1 = 5.5\,m$, $h_2 = 2.5\,m$, and $L = 150\,m$. The initial water table in the aquifer varies as $h(x) = 5.5 - 0.02x$, where x is the distance from River 1. For

$\Delta x = 15\,m$, the values of K_i and $S_{s,i}$, where $i = 0, 1, 2, \ldots, 10$, are given in the following table. If the aquifer has a recharge rate of $1.0\,m\,d^{-1}$ for the first five computation time steps, no sink or source for the successive three time steps, and a withdrawal rate of $0.25\,m\,d^{-1}$ for the remaining two time steps, determine the variations of water table and discharge.

Node i	0	1	2	3	4	5	6	7	8	9	10
$K_i\ (m\,d^{-1})$	6.07	2.16	1.51	19.74	8.68	14.14	3.65	9.35	3.78	8.56	3.32
$S_{s,i}\ (m^{-1})$	0.0002	0.0001	0.0005	0.0003	0.0002	0.0001	0.0001	0.0002	0.0003	0.0005	0.0002

10.15 In Figure 10.24b, let $h_1 = 5.0\,m$, $h_2 = 8.0\,m$, and $L = 500\,m$. The initial water table in the aquifer varies as $h(x) = 5.0 + 0.006x$, where x is the distance from River 1. For $\Delta x = 50\,m$, the values of K_i and $S_{s,i}$, where $i = 0, 1, 2, \ldots, 10$, are given in the following table. If the aquifer has a recharge rate of $1.5\,m\,d^{-1}$ for the first five computation time steps, a withdrawal rate of $0.5\,m\,d^{-1}$ for the successive three time steps, and no sink or source for the remaining time steps. Determine the variations of water table and discharge.

Node i	0	1	2	3	4	5	6	7	8	9	10
$K_i\ (m\,d^{-1})$	3.88	4.48	5.75	6.93	4.62	16.44	4.05	9.78	3.66	5.35	9.84
$S_{s,i}\ (m^{-1})$	0.0007	0.0002	0.0007	0.0005	0.0003	0.0005	0.0005	0.0016	0.0006	0.0001	0.0002

Appendix I: Notation

AET	actual evapotranspiration
AMC	antecedent moisture condition
ARC	antecedent runoff condition
BEP	best efficiency point
C_H	head coefficient
CN	curve number
C_P	power coefficient
C_Q	discharge coefficient
EGL	energy grade line
ET	evapotranspiration
g	gravitational acceleration
G–A	Green–Ampt
h_c	water depth above centroid
h_f	friction loss
HGL	hydraulic grade line
h_L	total loss
H–W	Hazen–Williams
HW	head water
IDF	intensity–duration–frequency
LULC	land use land cover
MOC	method of characteristics
NRCS	Natural Resources Conservation Service
P	power or precipitation
p	pressure
PET	potential evapotranspiration
PMP	probable maximum precipitation
PMS	probable maximum storm
SCS	Soil Conservation Service
S_f	energy grade line slope or friction slope
TW	tail water
UH	unit hydrograph
USACE	US Army Corps of Engineers
USBR	US Bureau of Reclamation

USDA	US Department of Agriculture
USDOT	US Department of Transportation
USEPA	US Environmental Protection Agency
USGS	US Geological Survey
γ	specific weight
η	efficiency
θ	soil moisture
μ	absolute viscosity
ρ	density
σ	cavitation number
υ	kinematic viscosity or relative velocity
φ_e	peripheral-velocity factor
ψ	capillary suction head

Appendix II: Constants and Units Conversion Factors

Table II.1 Typical constants.

Constant	Symbol	Value	
		BG unit	SI unit
Gravitational acceleration	g	32.2 ft sec^{-2}	9.81 m s^{-2}
Water density	ρ	1.94 slug ft^{-3}	1000 kg m^{-3}
Water specific weight	γ	62.4 lbf ft^{-3}	9.81 kN m^{-3}
Water saturated vapor pressure	p_υ	0.34 psia	2.34 kPa abs
Water absolute viscosity	μ	21.0×10^{-6} lbf sec ft^{-2}	1.0×10^{-3} N s m^{-2}
Water kinematic viscosity	υ	1.09×10^{-5} ft^2 sec^{-1}	1.00×10^{-6} m^2 s^{-1}
Water bulk modulus of elasticity	$E_{m,w}$	0.312×10^6 psi	2.15×10^6 kPa
Standard atmospheric pressure	p_{atm}	14.7 psia	101.3 kPa abs
von Kármán	κ	0.40	

Table II.2 Conversion factors from BG to SI units.

Quantity	BG unit	Multiply by	SI unit
Length	ft	0.3048	m
	in	25.4	mm
	mi	1.609	km
Mass	slug	14.594	kg
	lbm	0.454	kg
	ton	0.907	tonne
Force	lbf	4.448	N
Area	ac	0.405	ha
	mi^2	259	ha

Table II.3 Conversion factors from BG to BG units.

Quantity	BG unit	Multiply by	BG unit
Length	ft	12	in
	mi	5280	ft
Mass	slug	32.174	lbm
	ton	2000	lbm
Force	lbf	4.448	N
Area	ac	43,560	ft^2
	mi^2	640	ac
Power	hp	550	lbf ft sec^{-1}

Table II.4 Conversion factors from SI to SI units.

Quantity	SI unit	Multiply by	SI unit
Length	km	1000	m
	m	100	cm
	cm	10	mm
Mass	kg	1000	g
	tonne	1000	kg
Force	kN	1000	N
Area	ha	10,000	m^2
	ha	0.01	km^2
Power	hp	745.7	W
	kW	1000	W

Appendix III: Microsoft Excel Solver

Solver, a Microsoft Excel add-in program, can be used to quickly find solutions to problems that need to be solved by trial and error. In this regard, the target cell, which stores an objective function, can be maximized, minimized, or equaled to a specific value by adjusting the changing variable cells, which store the decision variables, subject to constraints on the values of other formula cells on a worksheet. Because Solver implements local search engines, if two or more solutions exist, they have to be found one at a time. To find one of the solutions, the initial values of the decision variables that are closer to this solution need to be specified. Herein, any initial values must be within the feasible zone defined by constraints. In practice, for a problem with only one decision variable, the objective function can be plotted to visually determine best initial values; for a problem with multiple decision variables, *a priori* knowledge has to be used to choose best initial values.

Solver provides three local search engines, namely GRG Nonlinear, Simplex LP, and Evolutionary. The GRG Nonlinear engine is for problems that are smooth (i.e., first-orderly differentiable) nonlinear, whereas the Evolutionary engine is for problems that are non-smooth. The simplex LP engine is for linear problems.

To activate Solver, select File → Options. In the Excel Options dialog box, select Add-Ins, click the Go…button, check Solver Add-In, and click the OK button, adding the Solver button under the DATA menu. Clicking the Solver button will open a dialog box, in which the objective function, decision variables, and constraints can be specified. The solution can be obtained by clicking the Solve button in the dialog box.

Example III.1 Find the positive roots of the function $f(x) = 3x^3 + x^2 - 5x\cos(x) - 1$.

Solution

Plotting f(x) reveals that the function has three roots (Figure III.1), one of which is positive. The positive root is close to 1.0. Thus, an initial value of 1.0 is used. For the purpose of setting up the Solver dialog box, the decision variable of x is stored in cell A23 with the initial value of 1.0, while the function f(x) is computed in cell B23 as "= 3*A23^3 + A23^2 – 5*A23* cos(A23) – 1". In the dialog box (Figure III.1), enter *Set Objective* $B$23 *To* 0 *By Changing Variable Cells* $A$23, and check *Make Unconstrained Variables Non-Negative*. Clicking the Solve button, one will get the positive root of $x = 0.975$. Note that the value of cell B23 becomes computationally zero.

The positive constraint is defined by checking *Make Unconstrained Variables Non-Negative*. Given the nonlinear feature of the objective function, the GRG Nonlinear engine should be used.

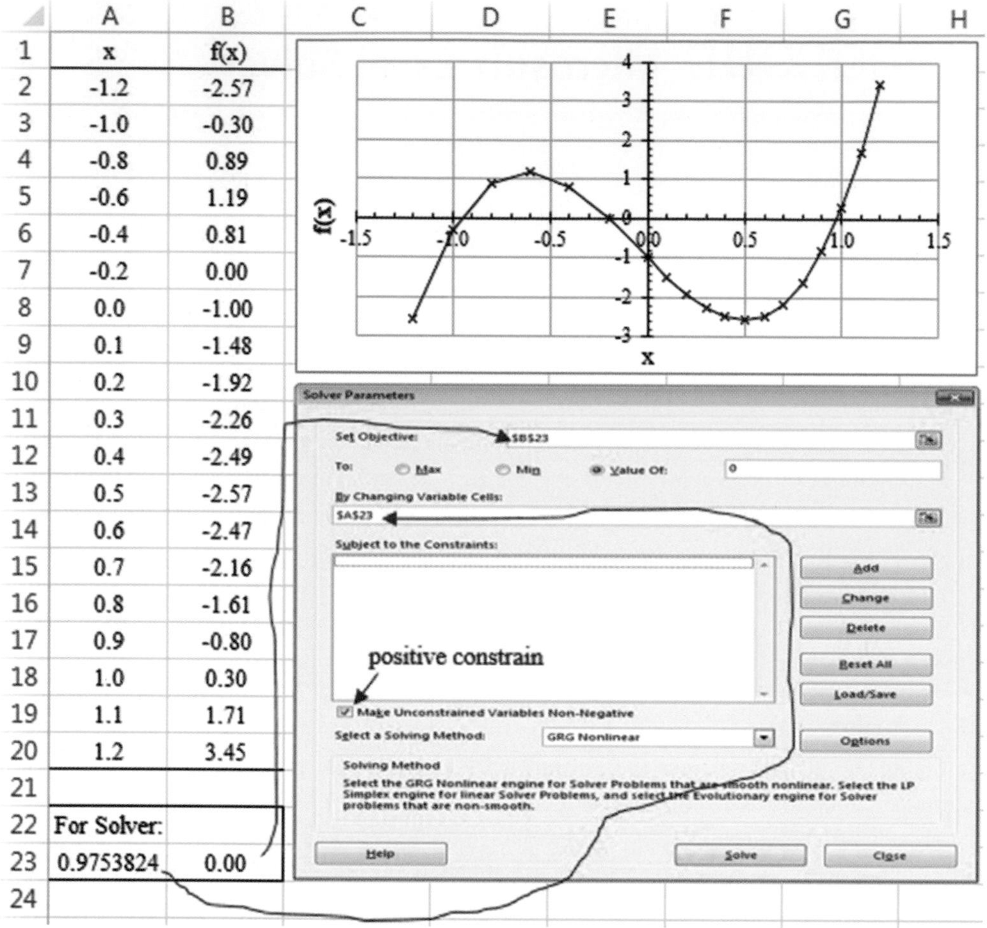

Figure III.1 **Excerpt of the Excel spreadsheet for solving Example III.1.** The plot of the data and the populated Solver dialog box are also shown.

Example III.2

Find all roots of the function in Example III.1.

Solution

As shown by the top-right plot in Figure III.1, the function has three roots. The first root is close to -1.0, the second root close to -0.2, and the third root close to 1.0. Using these values as an initial value one at a time and running Solver as in Example III.1, one will find the three roots to be -0.963, -0.201, and 0.975, respectively. Remember to uncheck *Make Unconstrained Variables Non-Negative* to allow negative roots (Figure III.2).

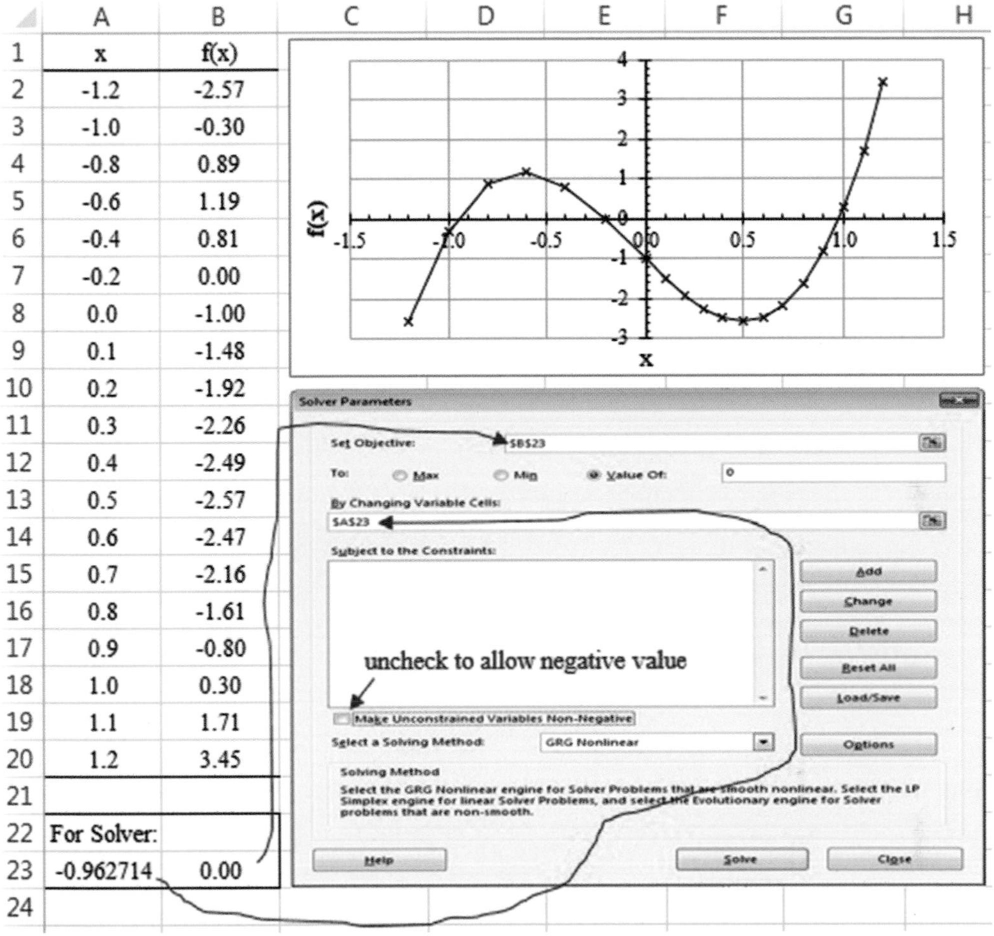

Figure III.2 **Excerpt of the Excel spreadsheet for solving Example III.2.** The plot of the data and a populated Solver dialog box are also shown.

Example III.3 Find x_1, x_2, and x_3 for the following problem.

Objective function: $\max\{f(x_1, x_2, x_3)\} = 10x_1 + 5x_2 + 30x_3$

Subject to: $x_1 + x_2 + x_3 = 100$

$x_1 + x_2 \geq 55$

$x_1 \geq 20$

$x_2 \geq 30$

$x_3 \geq 25.$

Solution

As shown in Figure III.3, for the purpose of setting up the Solver dialog box, the decision variables of x_1, x_2, and x_3 are stored in cells B2:B4, while the function $f(x_1, x_2, x_3)$ is computed in cell B7 as

"= 10*B2 + 5*B3 + 30*B4". The constraints are computed and stored in cells B10:B14: the first is computed in cell B10 as "= B2 + B3 + B4"; the second in cell B11 as "= B2 + B3"; the third in cell B12 as "= B2"; the fourth in cell B13 as "= B3"; and the fifth in cell B14 as "=*B*4". The initial values are empirically specified to be $x_1 = 30$, $x_2 = 30$, and $x_3 = 40$, which satisfy all constraints. In the dialog box, specify *Set Objective* $B$7 *To Max By Changing Variable Cells* $B$2:$B$4. Given the last three constraints, it is not important if *Make Unconstrained Variables Non-Negative* is checked or unchecked. Because both the objective function and the constraints are linear, the simplex LP engine can be used. Of course, the GRG Nonlinear engine should also work. Clicking the Solve button, one will get the solution: $x_1 = 25$, $x_2 = 30$, and $x_3 = 45$, which will result in a maximum objective function of 1750. Excerpt of the Excel spreadsheet for solving Example III-3.

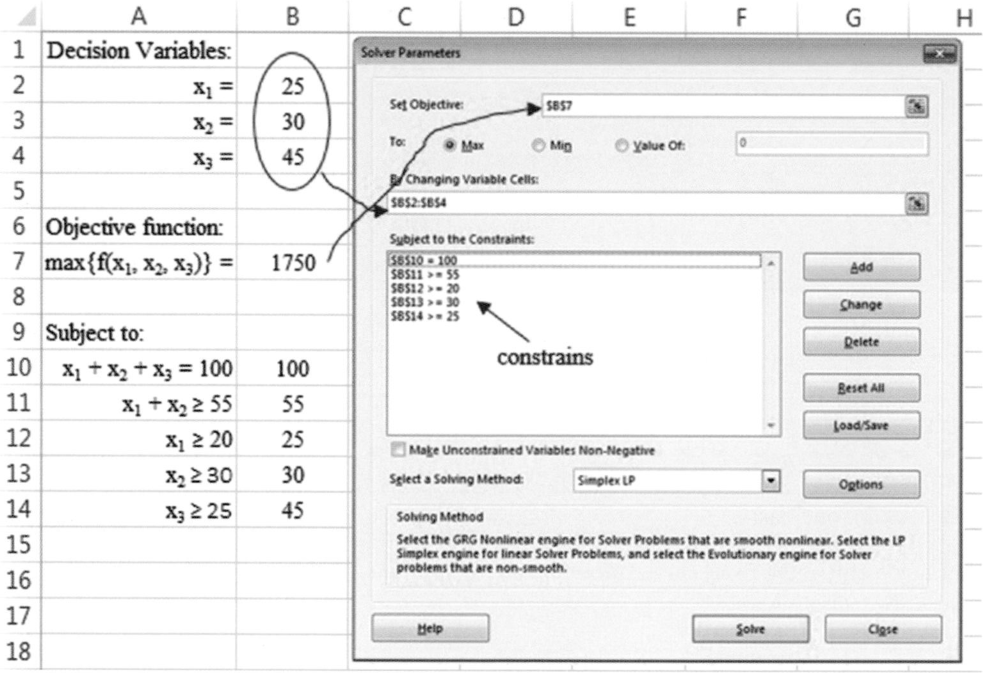

Figure III.3 **Excerpt of the Excel spreadsheet for solving Example III.3.** The plot of the data and the populated Solver dialog box are also shown.

Appendix IV: Derivatives and Integrals

IV.1 Total Derivative

For a multivariable function $y = f(x_1, x_2, \ldots, x_n)$, its total derivative is defined as:

$$dy = \frac{\partial f}{\partial x_1}dx_1 + \frac{\partial f}{\partial x_2}dx_2 + \cdots + \frac{\partial f}{\partial x_n}dx_n = \sum_{i=1}^{n}\left(\frac{\partial f}{\partial x_i}dx_i\right). \tag{IV.1}$$

If the independent variables are functions of a single variable t, Eq. (IV.1) can be rewritten as:

$$\frac{dy}{dt} = \frac{\partial f}{\partial x_1}\frac{dx_1}{dt} + \frac{\partial f}{\partial x_2}\frac{dx_2}{dt} + \cdots + \frac{\partial f}{\partial x_n}\frac{dx_n}{dt} = \sum_{i=1}^{n}\left(\frac{\partial f}{\partial x_i}\frac{dx_i}{dt}\right). \tag{IV.2}$$

Example IV.1 Find the total derivative of $y = 3x_1 + 6x_2$, $x_1 = 2t + 1$, $x_2 = 8t + 5$.

Solution

Using Eq. (IV.2), one has:

$$\frac{dy}{dt} = \frac{\partial f}{\partial x_1}\frac{dx_1}{dt} + \frac{\partial f}{\partial x_2}\frac{dx_2}{dt} = \frac{\partial(3x_1 + 6x_2)}{\partial x_1}\frac{d(2t+1)}{dt} + \frac{\partial(3x_1 + 6x_2)}{\partial x_2}\frac{d(8t+5)}{dt} = 3\times 2 + 6\times 8 = 54.$$

IV.2 Integral by Parts

The integration of the multiplication of two functions $u(x)$ and $v'(x)$ can be computed as:

$$\int_a^b u(x)v'(x)dx = u(x)v(x)\Big|_a^b - \int_a^b u'(x)v(x)dx = u(b)v(b) - u(a)v(a) - \int_a^b u'(x)v(x)dx. \tag{IV.3}$$

Example IV.2 Find the integration of $\int_3^5 \sin(x)\cos(x)\,dx$.

Solution

Using Eq. (IV.3), yields:

$$\int_3^5 \sin(x)\cos(x)\,dx = \int_3^5 \sin(x)\sin'(x)\,dx = \sin^2(5) - \sin^2(3) - \int_3^5 \sin(x)\cos(x)\,dx$$

$$\rightarrow 2\int_3^5 \sin(x)\cos(x)\,dx = \sin^2(5) - \sin^2(3) = 0.8996$$

$$\rightarrow \int_3^5 \sin(x)\cos(x)\,dx = \frac{0.8996}{2} = 0.4498.$$

IV.3 Leibniz's Rule

If the lower and/or upper limits of the integration of function f(x, t) over t are functions of x, the derivative of the integration with respect to x can be computed as:

$$\frac{d}{dx}\int_{a(x)}^{b(x)} f(x,t)dt = f(x,b(x))\frac{d(b(x))}{dx} - f(x,a(x))\frac{d(a(x))}{dx} + \int_{a(x)}^{b(x)}\left[\frac{\partial}{\partial x}f(x,t)\right]dt. \quad \text{(IV.4)}$$

Example IV.3 Compute $\frac{d}{dx}\int_3^5 \sin(t)dt$.

Solution

$a(x) = 3,\ b(x) = 5,\ f(x,t) = \text{six}(t)$

$$\text{Eq. (IV.4)} \rightarrow \frac{d}{dx}\int_3^5 \sin(t)dt = \sin(5)\frac{d(5)}{dx} - \sin(3)\frac{d(3)}{dx} + \int_3^5\left[\frac{\partial}{\partial x}\sin(t)\right]dt = 0 - 0 + 0 = 0.$$

Example IV.4 Compute $\frac{d}{dx}\int_{3x+2}^{5x+6} \sin(x)\cos(t)dt$.

Solution

$a(x) = 3x + 2,\ b(x) = 5x + 6,\ f(x,t) = \text{six}(x)\cos(t)$

Using Eq. (IV.4), yields:

$$\frac{d}{dx}\int_{3x+2}^{5x+6} \sin(x)\cos(t)dt$$

$$= \sin(x)\cos(5x+6)\frac{d(5x+6)}{dx} - \sin(x)\cos(3x+2)\frac{d(3x+2)}{dx} + \int_{3x+2}^{5x+6}\frac{\partial}{\partial x}[\sin(x)\cos(t)]dt$$

$$= 5\sin(x)\cos(5x+6) - 3\sin(x)\cos(3x+2) + \int_{3x+2}^{5x+6} \cos(x)\cos(t)dt$$

$$= 5\sin(x)\cos(5x+6) - 3\sin(x)\cos(3x+2) + \cos(x)\sin(t)\Big|_{3x+2}^{5x+6}$$

$$= 5\sin(x)\cos(5x+6) - 3\sin(x)\cos(3x+2) + \cos(x)\sin(5x+6) - \cos(x)\sin(3x+2)\,.$$

IV.4 Method of Characteristics

The method of characteristics (MOC) is a technique for solving partial differential equations. Typically, it applies to first-order equations, although more generally the method of characteristics is valid for any hyperbolic partial differential equation. The method is to reduce a partial differential equation to a family of ordinary differential equations along which the solution can be integrated from some initial data given on a suitable hypersurface. (Source: https://en.wikipedia.org/wiki/Method_of_characteristics)

The first-order partial differential equation is expressed as:

$$a(x,y,z)\frac{\partial z}{\partial x} + b(x,y,z)\frac{\partial z}{\partial y} = c(x,y,z)\,. \tag{IV.5}$$

In terms of the MOC, Eq. (IV.5) can be reduced to the ordinary differential equations expressed as:

$$\frac{dx}{a(x,y,z)} = \frac{dy}{b(x,y,z)} = \frac{dz}{c(x,y,z)}\,. \tag{IV.6}$$

Example IV.5 Use the MOC to solve $x\frac{\partial 0}{\partial x} + y\frac{\partial z}{\partial y} = 18$.

Solution

$a(x,y,z) = x,\ b(x,y,z) = y,\ c(x,y,z) = 18$

Eq. (IV.6) $\rightarrow \frac{dx}{x} = \frac{dy}{y} = \frac{dz}{18}$

$$\frac{dx}{x} = \frac{dz}{18} \rightarrow z = 18\ln|x| + C_1$$

$$\frac{dy}{y} = \frac{dz}{18} \rightarrow z = 18\ln|y| + C_2.$$

Adding the above two equations and dividing by 2 on both sides, yields:

$$z = 9\ln|x| + 9\ln|y| + \frac{C_1 + C_2}{2} = 9(\ln|x| + \ln|y|) + C.$$

Example IV.6 Use the MOC to solve $x^2\frac{\partial z}{\partial x}+y^2\frac{\partial z}{\partial y}=z$.

Solution

$a(x,y,z)=x^2$, $b(x,y,z)=y^2$, $c(x,y,z)=z$

Eq. (IV.6) $\rightarrow \frac{dx}{x^2}=\frac{dy}{y^2}=\frac{dz}{z}$

$$\frac{dx}{x^2}=\frac{dz}{z}\rightarrow -\frac{1}{x}+C_1=\ln|z|$$
$$\frac{dy}{y^2}=\frac{dz}{z}\rightarrow -\frac{1}{y}+C_2=\ln|z|.$$

Adding the above two equations and dividing by 2 on both sides, one has:

$$\ln|z|=\frac{1}{2}\left(-\frac{1}{x}-\frac{1}{y}+C_1+C_2\right)=-\frac{x+y}{2xy}+C$$
$$\rightarrow z=e^Ce^{-\frac{x+y}{2xy}}=C'e^{-\frac{x+y}{2xy}}.$$

Appendix V: Hydrologic and Hydraulic Models

Over the last few decades, a number of computer simulation models, such as those listed in Table V.1, have been developed for water resources analysis and hydrologic design. These models can be classified in terms of modeling objective, temporal domain, spatial domain, and structure (Figure V.1). For a given model, it might be devised to simulate water quantity, water quality, or both, either for a long-term simulation period or a storm event. The model can be applicable for a spatial domain ranging from fields, watersheds, and basins to streams and rivers. If the model has a distributed structure, the spatial domain needs to be further subdivided into smaller units, each

Table V.1 List of selected computer simulation models.

Developer	Model	Subject	Website
USACE	HEC-HMS	Hydrologic	www.hec.usace.army.mil/software
	HEC-RAS	Hydraulic	
	HEC-FFA	Frequency	
USDA	TR-55	Hydrologic	www.nrcs.usda.gov/wps/portal/nrcs/detailfull/national/water/manage/hydrology/?cid=stelprdb1042922
	RZWQM	Hydrologic and water quality	www.ars.usda.gov/plains-area/fort-collins-co/center-for-agricultural-resources-research/rangeland-resources-systems-research/docs/system/rzwqm
	SWAT	Hydrologic and water quality	https://swat.tamu.edu
	WEPP	Hydrologic and water quality	www.ars.usda.gov/midwest-area/west-lafayette-in/national-soil-erosion-research/docs/wepp/research/
USDOT	HY-8	Hydraulic	www.fhwa.dot.gov/engineering/hydraulics/software.cfm
USEPA	HSPF	Hydrologic and water quality	www.epa.gov/ceam/hydrological-simulation-program-fortran-hspf
	SWMM	Hydrologic and water quality	www.epa.gov/water-research/storm-water-management-model-swmm
USGS	MODFLOW	Groundwater	https://water.usgs.gov/ogw/modflow
	PEAKFQ	Frequency	https://water.usgs.gov/software/PeakFQ

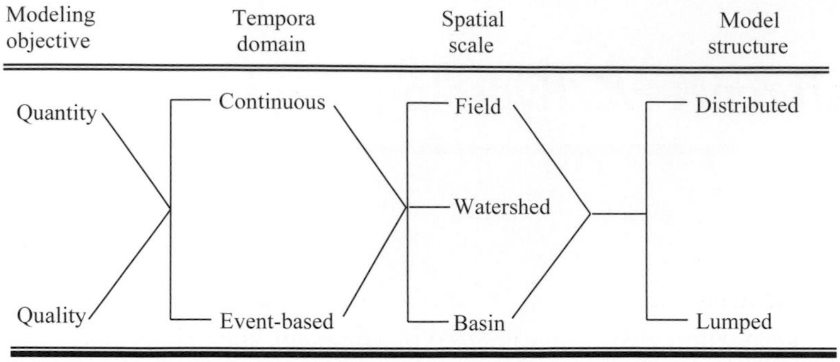

Figure V.1 **Classifications of the existing hydrologic and hydraulic models.**

of which is treated as a basic unit for mass conservation. In contrast, if the model has a lumped structure, the entire spatial domain is treated as a basic unit for mass conservation. Typically, water quantity modeling may be applicable to flood, drought, water balance, and/or water supply, whereas water quality modeling can be related to soil erosion, land management, and/or sediment and contaminant transport.

References

Abbot, J., Armstrong, J.S., Bolt, A. *et al.* 2015. *Climate Change: The Facts* (Woodsville, NH: Stockade Books).

Acreman, M.C. 1989. "Extreme historical UK floods and maximum flood estimation," *Water and Environmental Journal*, 3: 404–12.

Akbari, G., Firoozi, B. 2010. "Implicit and explicit numerical solution of Saint-Venent equations for simulating flood wave in natural rivers," in: *5th National Congress on Civil Engineering*. Ferdowsi University of Mashhad, Mashhad, Iran.

Ali, S., Ghosh, N.C., Singh, R. 2010. "Rainfall-runoff simulation using a normalized antecedent precipitation index," *Hydrological Sciences Journal*, 55: 266–74.

Allen, R.G., Pereira, L.S., Raes, D., Smith, M. 1998. *Crop Evapotranspiration Guidelines for Computing Crop Water Requirements* (Washington DC: United Nations Food and Agriculture Organization).

Amer, K.H., Hatfield, J.L. 2004. "Canopy resistance as affected by soil and meteorological factors in potato," *Agronomy Journal*, 96: 978–85.

Arnold, J.G., Srinivasan, R., Muttiah, R.S., Williams, J.R. 1998. "Large-area hydrologic modeling and assessment: Part I. Model development," *Journal of the American Water Resources Association*, 34: 73–89.

Arnold, J.G., Allen, P.M., Morgan, D.S. 2011. "Hydrologic model for design and constructed wetlands," *Wetlands*, 21(2): 1–16.

Aqua, T.G. 2018. "Partially filled pipe flow measurement." www.aquatechnologygroup.com/partially-filled-pipe-flow-measurement/ (accessed August 15, 2018).

Assoúline, S. 2005. "On the relationships between the pore size distribution index and characteristics of the soil hydraulic functions," *Water Resources Research*, 41: W07019.

ASTM. 2003. "D3385-03 Standard test method for infiltration rate of soils in field using double-ring infiltrometer" (Washington DC: American Society Testing Materials).

Averyt, K., Meldrum, J., Caldwell, P., Sun, G., McNulty, S., Huber-Lee, A., Madden, N. 2013. "Sectoral contributions to surface water stress in the coterminous United States," *Environmental Research Letters*, 8: 035046.

Baltas, E.A., Dervos, N.A., Mimikou, M.A. 2007. "Determination of the SCS initial abstraction ratio in an experimental watershed in Greece," *Hydrology and Earth System Sciences*, 11: 1825–9.

Barnes, G.E. 2000. *Soil Mechanics: Principles and Practice* (London, UK: Macmillan Press).

Bedient, P., Huber, W.C., Vieux, B.E. 2012. *Hydrology and Floodplain Analysis* (New York, NY: Pearson).

Bouchet, R.J. 1963. *Evapotranspiration Réelle Evapotranspiration Potentielle, Signification Climatique* (Berkeley, CA: International Association of Hydrological Sciences).

Bouwer, H. 1989. "The Bouwer and Rice slug test: An update," *Ground Water*, 27: 304–9.

Bouwer, H., Rice, R.C. 1976. "A slug test method for determining hydraulic conductivity of unconfined aquifers with completely or partially penetrating wells," *Water Resources Research*, 12: 423–8.

Bras, R.L. 1999. "A brief history of hydrology," *Bulletin of the American Meteorological Society*, 80: 1151–64.

Brutsaert, W. 2005. *Hydrology: An Introduction* (New York, NY: Cambridge University Press).

Buckingham, E. 1914. "On physically similar systems: Illustrations of the use of dimensional equations." *Physical Review* 4(4):345–76.

Buckingham, E. 1915. "The principle of similitude," *Nature* 96(2406): 396–7.

Charbeneau, R .J. 2006. *Groundwater Hydraulics and Pollutant Transport* (Long Grove, IL: Waveland Press, Inc.).

Chin, D. A. 2000. *Water Resources Engineering* (Upper Saddle River, NJ: Prentice-Hall).

Chow, V. T. 1964. *Handbook of Applied Hydrology: A Compendium of Water-Resources Technology* (New York, NY: McGraw-Hill).

Chow, V. T., Maidment, D. R., Mays, L. W. 1988. *Applied Hydrology* (New York, NY: McGraw-Hill).

Collis-George, N. 1977. "Infiltration equations for simple soil systems," *Water Resources Research*, 13: 395–403.

Corbitt, R.A. 1999. *Standard Handbook of Environmental Engineering* (New York, NY: McGraw-Hill).

Cronshey, R. G. 1983. "Discussion of 'Antecedent moisture condition probabilities,' by D.D. Gray and S.M. deMonsabert," *Journal of Irrigation and Drainage Engineering*, 100: 296–8.

Crowe, C. T., Elger, D. F., Roberson, J. A. 2005. *Engineering Fluid Mechanics* (Hoboken, NJ: John Wiley & Sons, Inc.).

Darcy, H. 1856. *Les fontaines publiques de la ville de Dijon* (Paris: Dalmont).

Davis, C. V., Sorensen, K. E. (eds.). 1969. *Handbook of Applied Hydraulics*, 3rd edn. (New York, NY: McGraw Hill).

De Glee, G. J. 1930. *Over grondwaterstromingen bij wateronttrekking door middel van putten* (Delft, The Netherlands: J. Waltman).

De Glee, G. J. 1951. "Berekeningsmethoden voor de winning van grondwater," in: *Drinkwatervoorziening, 3e Vacantiecursus: 38–80 Moorman's periodieke pers* (The Hague).

Debo, T. N., Reese, A. J. 2003. *Municipal Stormwater Management* (Boca Raton, FL: Lewis).

Dupuit, J. 1863. *Etudes Théoriques et Pratiques sur le mouvement des Eaux dans les canaux découverts et à travers les terrains perméables*, 2nd edn. (Paris: Dunod).

Engematic, E. 2018. *Ultrasonic Open Channel Flow Meter* (Engematic Inc.).

ESRI. 2011. "An overview of the hydrology tools," Environmental Systems Research Institute (ESRI), Inc.. webhelo.esri.com/arcgisdesktop/9.3 (accessed February 13, 2019).

Farnsworth, R. K., Thompson, E. S. 1982. *Mean Monthly, Seasonal, and Annual Pan Evaporation for the United States* (Washington DC: US Department of Commerce, National Oceanic and Atmospheric Administration).

Farnsworth, R. K., Thompson, E. S., Peck, E. L. 1982. *Evaporation Atlas for the Contiguous 48 United States* (Washington DC: US Department of Commerce, National Oceanic and Atmospheric Administration).

FHA (Department of Transportation Federal Highway Administration). 2005. "Design of roadside channels with flexible linings," in: *Hydraulic Engineering Circular No. 15*. US FHA, 154.

Finnemore, E. J., Franzini, J. B. 2002. *Fluid Mechanics with Engineering Applications* (New York, NY: McGraw-Hill).

Flint, R. W. 2004. "The sustainable development of water resources," *Water Resources Update*, 127: 48–59.

Forchheimer, P. 1886. "Über die Ergiebigkeit von Brunnen-Anlagen und Sickerschlitzen," *Z. Architekt. Ing.-Ver. Hannover.*, 32: 539–63.

Forterra. 2005. *Pipe and Products Catalog*. Vol. 1.1 (Little Rock: AR: Forterra), 155.

Fréchet, M. 1927. "Sur la loi de probability de l'écart maximum," *Ann. Soc. Polon. Math. (Cracovie)*, 6: 93–116.

Fristam. 2014. *Centrifugal Pump Performance Curves and Technical Information* (Middleton, WI: Fristam Pumps USA).

Garbrecht, J., Martz, L. W. 1997. "The assignment of drainage direction over flat surfaces in raster digital elevation models," *Journal of Hydrology*, 193: 204–13.

Ghanbarian-Alavijeh, B., Liaghat, A., Huang, G., van Genuchten, M. Th. 2010. "Estimation of the van Genuchten soil water retention properties from soil textural data," *Pedosphere*, 20: 456–65.

Gongol, V., Gongol, B. 2008. "Self-Priming Pumps," DJ Gongol & Associates, Inc.. www.slideshare.net/briangongol/a-brief-introduction-to-selfpriming-pumps (accessed March 10, 2018).

Gorse, C., Johnston, D., Pritchard, M. 2013. *A Dictonary of Construction, Surveying and Civil Engineering* (Oxford: Oxford University Press).

Graf, W. L. 1999. "Dam nation: A geographic census of American dams and their large-scale hydrologic impacts," *Water Resources Research*, 35: 1305–11.

Gray, D. D., Katz, P. G., deMonsabert, S. M., Cogo, N. P. 1982. "Antecedent moisture condition probabilities," *Journal of the Irrigation and Drainage Division*, 108: 107–14.

Green, W. H., Ampt, C. A. 1911. "Studies on soil physics, I. Flow of air and water through soils," *Journal of Agricultural Science*, 4: 1–24.

Grigorjev, V. Y., Iritz, L. 1991. "Dynamic simulation model for vertical infiltration of water in soil," *Journal of Hydrological Sciences*, 36: 171–9.

Gumbel, E. J. 1958. *Statistics of Extremes* (New York, NY: Columbia University Press).

Hager, W. H. 1992. *Energy Dissipators and Hydraulic Jump* (Dordrecht, The Netherlands: Kluwer Academic).

Hager, W. H., Schwalt, M. 1994. "Broad-crested weir," *Journal of Irrigation and Drainage Engineering*, 120(1): 13–26.

Hantush, M. S. 1961a. "Aquifer tests on partially penetrating wells," *Journal of the Hydraulics Division*, 87: 171–94.

Hantush, M. S. 1961b. "Drawdown around a partially penetrating well," *Journal of the Hydraulics Division*, 87: 83–98.

Hantush, M. S., Jacob, C. E. 1955. "Non-steady radial flow in an infinite leaky aquifer," *Transactions of American Geophysical Union*, 36: 281–5.

Hapuarachchi, H. A. P., Wang, Q., Pagano, T. C. 2011. "A review of advances in flash flood forecasting," *Hydrological Processes*, 25: 2771–84.

Harbaugh, A. W. 2005. *MODFLOW-2005, The US Geological Survey Modular Groundwater Model – The Groundwater Flow Process: US Geological Survey Techniques and Methods 6-A16* (Washington DC: US Geological Survey).

Hendriks, M. 2010. *Introduction to Physical Hydrology* (London, UK: Oxford University Press).

Herr, L. A., Bossy, H. C. 1965. *HEC-5 Hydraulic Charts for the Selection of Highway Culverts* (US Department of Transportation, Federal Highway Administration).

Hilpert, M., Glantz, R. 2013. "Exploring the parameter space of the Green-Ampt model," *Advances in Water Resources*, 53: 225–30.

Hjelmfelt, A. T. 1987. "Curve numbers in urban hydrology," in: *XXII Congress International Association for Hydraulic Research* (Lausanne, Switzerland: International Association for Hydraulic Research).

Hjelmfelt, A. T. 1991. "Investigation of curve number procedure," *Journal of Hydraulic Engineering*, 117: 725–37.

Hjelmfelt, A. T., Kramer, K. L., Burwell, R. E. 1982. *Curve Numbers as Random Variables: Rainfall-Runoff Relationship* (Littleton, CO: Water Resources Publications).

Hobbins, M. T., Ramírez, J. A., Brown, T. C., Claessens, L. H. J. M. 2001. "The complementary relationship in estimation of regional evapotranspiration: The complementary relationship areal evapotranspiration and advection-aridity models," *Water Resources Research*, 37: 1367–87.

Hoekstra, A. Y., Chapagain, A. K., Aldaya, M. M., Mekonnen, M. M. 2011. *The Water Footprint Assessment Manual: Setting the Global Standard* (Washington DC: Earthscan).

Hoekstra, A. Y., Mekonnen, M. M. 2012. "The water footprint of humanity," *Proceedings of the National Academy of Sciences*, 109: 3232–7.

Holtan, H. N. 1961. *A Concept of Infiltration Estimates in Watershed Engineering. ARS41-51* (Washington DC: US Department of Agriculture Agricultural Research Service).

Horton, R. E. 1940. "Approach toward a physical interpretation of infiltration capacity," *Soil Science Society of America Journal*, 5: 339–417.

Huggins, L. F., Monke, E. J. 1966. *The Mathematical Simulation of the Hydrology of Small*

Watersheds. Technical Report No. 1 (Layfayette, IN: Purdue Water Resources Research Center).

Hwang, N. H. C., Houghtalen, R. J. 1996. *Fundamentals of Hydraulic Engineering Systems* (Upper Saddle River, NJ: Prentice Hall, Inc.).

HyQuest. 2016. "Instructional Manual: Class A Evaporation Pan." In, 11. Warwick Farm, Australia: HyQuest Solutions Pty Ltd.

Isco, Inc. 2018. "Isco 2150 area velocity flow module". www.avensyssolutions.com/data_AS/File/Solutions/2150_Flow_Module.pdf (accessed August 15, 2018).

ISO. 2014. "ISO 14046:2014 (en) Environmental management – Water footprint – Principles, requirements and guidelines," The International Organization for Standardization. www.iso.org/obp/ui/#iso:std:iso:14046:ed-1:v1:en (accessed December 25, 2018).

Johnson, F. L., Chang, F. M. 1984. *Drainage of Highway Pavements*. (McLean, V: Federal Highway Administration).

Kasdin, N. J., Paley, D. A. 2011. *Engineering Dynamics: A Comprehensive Introduction*, (Princeton, NJ: Princeton University Press).

Kerby, W. S. 1959. "Time of concentration for overland flow," *Civil Engineering*, 29: 60.

Kifissia. 2017. "History of rain gauge," Kifissia Meteo. www.kifissiameteo.gr/Lesson08_Instrument_RainGauge.html#Types (accessed December 28, 2017).

Kindsvater, C. E., Carter, R. W. 1959. "Discharge characteristics of rectangular thin-plate weirs," *Transactions of the American Society of Civil Engineers*, 24: 3001.

Kirpich, Z. P. 1940. "Time of concentration of small agricultural watersheds," *Civil Engineering*, 10: 362.

Kostiakov, A. N. 1932. "On the dynamics of the coefficients of water percolation in soils," in: *Transactions of Sixth Commission, International Society of Soil Science Part A* 15–21 (Graz, Austria).

Kuichling, E. 1889. "The relation between rainfall and the discharge of sewers in populous districts," *Transactions of the American Society of Civil Engineers*, 20: 1–60.

Kuritza, J. C., Camponogara, G., Marques, M. G., Sanagiotto, D. G., Battiston, C. 2017. "Dimensionless curves of centrifugal pumps for water supply systems: development and case study," *Brazilian Journal of Water Resources*, 22: e45.

Laprise, R. 2008. "Regional climate modelling," *Journal of Computational Physics*, 227: 3641–66.

Learn Engineering. 2013. "Comparison of Pelton, Francis & Kaplan Turbine," YouTube. www.youtube.com/watch?v=k0BLOKEZ3KU (accessed March 25, 2019).

Linsley, R. K. Jr., Kohler, M. A., Paulhus, J. L. H. 1958. *Hydrology for Engineers*, 1st edn. (New York, NY: McGraw-Hill).

Linsley, R. K. Jr., Kohler, M. A., Paulhus, J. L. H. 1982. *Hydrology for Engineers*, 3rd edn. (New York, NY: McGraw-Hill).

Longin, F. (ed.). 2016. *Extreme Events in Finance: A Handbook of Extreme Value Theory and Its Applications* (Hoboken, NJ: John Wiley & Sons).

Marengo, J. A., Espinoza, J. C. 2016. "Extreme seasonal droughts and floods in Amazonia: Causes, trends and impacts," *International Journal of Climatology*, 36: 1033–50.

Margat, J., van der Gun, J. 2013. *Groundwater around the World: A Geographic Synopsis* (CRC Press).

Masih, I., Maskey, S., Mussá1, F. E. F., Trambauer, P. 2014. "A review of droughts on the African continent: A geospatial and long-term perspective," *Hydrology and Earth System Sciences*, 18: 3635–49.

Mays, L. W. 2005. *Water Resources Engineering* (John Wiley & Sons, Inc.: Danvers, MA).

McCallum, B. E., Wicklein, S. M., Reiser, R. G., *et al*. 2013. "Monitoring storm tide and flooding from Hurricane Sandy along the Atlantic Coast of the united states, October 2012," in: *Open-File Report 2013–1043* (Reston, VA: US Geological Survey).

McMahon, T. A., Peel, M. C., Lowe, L., Srikanthan, R., McVicar, T. R. 2013. "Estimating actual, potential, reference crop and pan evaporation using standard meteorological data: a pragmatic synthesis," *Hydrology and Earth System Sciences*, 17: 1331–63.

McVicar, T. R., Van Niel, T. G., Li, L., Roderick, M. L., Rayner, D. P. Ricciardulli, L., Donohue, R. J. 2008. "Wind speed climatology and trends for Australia, 1975–2006: Capturing the stilling phenomenon and comparison with nearsurface reanalysis output," *Geophysical Research Letters*, 35: L20403.

Mein, R. G., Larson, C. L. 1973. "Modeling infiltration during a steady rain," *Water Resources Research*, 9: 384–94.

Mekonnen, M. M., Hoekstra, A. Y. 2011. "The green, blue and grey water footprint of crops and derived crop products," *Hydrology and Earth System Sciences*, 15: 1577–600.

Michel, C., Andréassian, V., Perrin, C. 2005. "Soil conservation service curve number method: How to mend a wrong soil moisture accounting procedure?," *Water Resources Research*, 41: W02011.

Miller, J. E. 1984. *Basic Concepts of Kinematic-Wave Models* (Washington DC: US Geological Survey).

Miller, R. A., Frederick, R. H., Tracey, R. J. 1973. "Precipitation-frequency atlas of the Western United States," in: *NOAA Atlas 2* (Silver Spring, MD: National Oceanic and Atmospheric Administration National Weather Service).

Mirzaee, S., Zolfaghari, A. A., Gorji, M., Dyck, M., Dashtaki, S. G. 2013. "Evaluation of infiltration models with different numbers of fitting parameters in different soil texture classes," *Archives of Agronomy and Soil Science*, 60.

Mishra, S. K., Tyagi, J. V., Singh, V. P. 2003. "Comparison of infiltration models," *Hydrological Processes*, 17: 2629–52.

Mishra, S. K., Kumar, S. R., Singh, V. P. 1999. "Calibration of a general infiltration model," *Journal of Hydrological Processes*, 13: 1691–718.

Moody, L. F. 1944. "Friction factors for pipe flow," *Transactions of the ASME*, 66: 671–84.

Morton, F. I. 1983a. "Operational estimates of areal evapotranspiration and their significance to the science and practice of hydrology," *Journal of Hydrology*, 66: 1–76.

Morton, F. I. 1983b. "Operational estimates of lake evaporation," *Journal of Hydrology*, 66: 77–100.

Morton, F. I. 1986. "Practical estimates of lake evaporation," *Journal of Climate and Applied Meteorology*, 25: 371–87.

Mulvaney, T. J. 1851. "On the use of self-registering rain and flood gauges in making observations of the relations of rainfall and of flood discharges in a given catchment," *Proceedings of the Institute of Civil Engineers of Ireland*, 4: 18–31.

Murray, F. W. 1967. "On the computation of saturation vapor pressure," *Journal of Applied Meteorology*, 6: 203–4.

Myers, V. A., Zehr, R. M. 1980. "A Methodology for Point-to-Area Rainfall Frequency Ratios." In.: Washington DC: National Oceanic and Atmospheric Administration (NOAA) National Weather Service (NWS).

Nalbantis, I., Efstratiadis, A., Rozos, E., Kopsiafti, M., Koutsoyiannis, D. 2011. "Holistic versus monomeric strategies for hydrological modelling of human-modified hydrosystems," *Hydrology and Earth System Sciences*, 15: 743–58.

Nash, J. E., Sutcliffe, J. V. 1970. "River flow forecasting through conceptual models: Part I. A discussion of principles," *Journal of Hydrology*, 10: 282–90.

Neter, J., Kutner, M. H., Nachtsheim, C. J., Wasserman, W. 1996. *Applied Linear Statistical Models* (New York, NY: McGraw-Hill Companies, Inc.).

Neuman, S. P. 1975. "Analysis of pumping test data from anisotropic unconfined aquifers considering delayed gravity response," *Water Resources Research*, 11: 329–42.

NOAA. 1978. *Probable Maximum Precipitation Estimates, United States East of the 105th Meridian* (Washington DC: US Department of Commerce National Oceanic and Atmospheric Administration and US Department of the Army Corps of Engineers).

Normann, J. M., Houghtalen, R. J., Johnson, W. J. 1985. "Hydraulic design of highway culverts," in: *Hydraulic Design Series No. 5, Report*

No. FHWA-IP-85-15 (US Department of Transportation Federal Highway Administration).

NWS. 2019. "What is meant by the term drought?," National Oceanic and Atmospheric Administration (NOAA) National Weather Service (NWS). www.weather.gov/bmx/kidscorner_drought (accessed April 12, 2019).

O'Callaghan, J. F., Mark, D. M. 1984. "The extraction of drainage networks from digital elevation data," *Computer Vision Graphs Image Processing*, 28: 328–44.

Overton, D. E. 1964. *Mathematical Refinement of an Infiltration Equation for Watershed Engineering. ARS41-99* (Washington DC: US Department of Agriculture Agricultural Research Service).

Parhi, P. K., Mishra, S. K., Singh, R. 2007. "A modification to Kostiakov and modified Kostiakov infiltration models," *Water Resources Management*, 21: 1973–89.

Philip, J. R. 1957. "Theory of infiltration," *Soil Science*, 83: 345–57.

Philip, J. R. 1969. "Theory of infiltration," in: V. T. Chow (ed.), *Advances in Hydroscience* (New York, NY: Academic Press).

Pitt, R., Lantrip, J., Harrison, R. 1999. *Infiltration Through Disturbed Urban Soils and Compost-Amended Soil Effects on Runoff Quality and Quantity* (Washington, DC: United States Environmental Protection Agency Office of Research and Development).

Prodanovic, P., Simonovic, S.P. 2004. *Generation of Synthetic Design Storms for the Upper Thames River Basin – CFCAS Project: Assessment of Water Resources Risk and Vulnerability to Changing Climatic Condition* (Ontario, Canada: The University of Western Ontario Department of Civil and Environmental Engineering).

Rana, G., Katerji, N. 2000. "Measurement and estimation of actual evapotranspiration in the field under Mediterranean climate: a review," *European Journal of Agronomy*, 13: 125–53.

Roberson, J. A., Cassidy, J. J., Chaudhry, M. H. 1998. *Hydraulic Engineering* (Hoboken, NJ: Wiley).

Rossmiller, R. L. 2014. *Stormwater Design for Sustainable Development* (New York, NY: McGraw-Hill).

Rouse, H. 1983. "Highlights: History of hydraulics," in: *Books at Iowa 38* (Iowa City, IA: University of Iowa).

Sahu, R. K., Mishra, S. K., Eldho, T. I. 2010. "An improved AMC-coupled runoff curve number model," *Hydrological Processes*, 24: 2834–39.

Sahu, R. K., Mishra, S. K., Eldho, T. I., Jain, M. K. 2007. "An advanced soil moisture accounting procedure for SCS curve number method," *Hydrological Processes*, 21: 2872–81.

SAIC. 2006. *Life Cycle Assessment: Principles and Practice* (Reston, VA: Scientific Applications International Corporations (SAIC)).

Salmasi, F., Poorescandar, S., Dalir, A. H., Zadeh, D. F. 2011. "Discharge relations for rectangular broad-crested weirs," *Journal of Agricultural Sciences*, 17: 324–36.

Satterfield, Z. 2013. "Reading centrifugal pump curves," *Tech Brief*, 12: 1–5.

Saxton, K. E., Rawls, W. J. 2006. "Soil water characteristic estimates by texture and organic matter for hydrologic solutions," *Soil Science Society of America Journal*, 70: 1569–78.

Saxton, K. E., Rawls, W. J., Romberger, J. S., Papendick, R. I. 1986. "Estimating generalized soil-water characteristics from texture," *Transactions of the ASAE*, 50: 1031–35.

Schlichting, H., Gersten, K. 1999. *Boundary-Layer Theory* (Cham, Switzerland: Springer).

Seibert, J., McGlynn, B. L. 2007. "A new triangular multiple flow direction algorithm for computing upslope areas from gridded digital elevation model," *Water Resources Research*, 43: W04502.

Shenitech, LLC. 2007. "Non-intrusive ultrasonic flowmeter." www.shenitech.com/support/Brochure_STUF-300FxB_v8_11-08.pdf (accessed August 14, 2018).

Sherman, L. K. 1932. "Stream flow from rainfall by the unit graph method," *Engineering News Record*, 108: 501–5.

Shi, Z., Chen, L., Fang, N., Qin, D., Cai, C. 2009. "Research on the SCS-CN initial abstraction ratio using rainfall–runoff event analysis in the Three Gorges Area, China," *Catena*, 77: 1–7.

Shuttleworth, W. J. 2008. "Evapotranspiration measurement methods," *Southwest Hydrology*, January/February: 22–3.

Sihag, P., Tiwari, N. K., Ranjan, S. 2017. "Estimation and inter-comparison of infiltration models," *Water Science*, 31: 34–43.

Silveira, L., Charbonnier, F., Genta, J. L. 2000. "The antecedent soil moisture condition of the curve number procedure," *Hydrological Sciences Journal*, 45: 3–12.

Singh, V. P. 1992. *Elementary Hydrology* (Upper Saddle River, NJ: Prentice-Hall).

Singh, V. P., Yu, F. 1990. "Derivation of infiltration equation using systems approach," *Journal of Irrigation and Drainage Engineering*, 116: 837–57.

Smith, R. E. 1972. "The infiltration envelope: Results from a theoretical infiltrometer," *Journal of Hydrology*, 17: 1–21.

Smith, R. E., Parlange, J. Y. 1978. "A parameter-efficient hydrologic infiltration model," *Water Resources Research*, 14: 533–8.

Snyder, F. F. 1938. "Synthetic unit graphs," *Transactions of American Geophysical Union*, 19: 447–54.

Sohoulande-Djebou, D. C. S., Singh, V. P. 2016. "Impact of climate change on the hydrologic cycle and implications for society," *Environment and Social Psychology*, 1: 36–49.

Stephan, P., Annette, K., Stefanie, H. 2009. "Assessing the environmental impacts of freshwater consumption in LCA," *Environmental Science*, 43: 4008–104.

Strack, O. D. L. 2017. *Analytical Groundwater Mechanics* (New York, NY: Cambridge University Press).

Strelkoff, T. 1970. "Numerical solution of Saint-Venant equations," *Journal of the Hydraulics Division*, 96: 223–52.

Sturm, T. W. 2010. *Open Channel Hydraulics*, 2nd edn. (Oxford, UK: Butterworth-Heinemann Elsevier).

Swartzendruber, D. 1987. "A quasi-solution of Richards equation for the downward infiltration of water into soil," *Water Resources Research*, 23: 809–17.

Tetens, O. 1930. "Uber einige meteorologische Begriffe," *z. Geophys.*, 6: 297–309.

Theis, C. V. 1935. "The relation between the lowering of the piezometric surface and the rate and duration of discharge of a well using groundwater storage," *Transactions of the American Geophysical Union*, 16: 519–24.

Timár, P. 2005. "Dimensionless characteristics of centrifugal pump," in: *32nd International Conference of the Slovak Society of Chemical Engineering* (Slovak: Tatranské Matliare), 500–3.

Timoshenko, S. P., Gere, J. M. 1961. *Theory of Elastic Stability* (New York, NY: McGraw-Hill).

Todd, D. K., Mays, L. W. 2005. *Groundwater Hydrology* (Hoboken, NJ: John Wiley & Sons Inc.).

Turazza, D. 1880. *Trattato di idrometria o di idraulica pratica* (Padova, Italy: F. Sacchetto).

UN. 1987. *Our Common Future* (New York, NY: United Nations World Commission on Environment and Development).

Urroz, G., Hoeft, C. 2009. "Engineering Field Handbook Chapter 3 (650.03) – Hydraulics," in: *Engineering Field Handbook* (Washington DC: US Department of Agriculture Natural Resources Conservation Service).

USACE. 2010. *HEC-RAS River Analysis System: Hydraulic Reference Manual* (Davis, CA: Hydrologic Engineering Center).

USBR. 2003. "Chapter 8: Flumes," in: *Water Measurement Manual* (Washington DC: US Bureau of Reclamation).

USBR. 2014. *Appurtenant Structures for Dams (Spillways and Outlet Works) Design Standard* (Washington DC: US Bureau of Reclamation, Technical Service Center).

USDA-NRCS. 1986. *Urban Hydrology for Small Watersheds: Technical Release 55* (Washington DC: US Department of Agriculture, Natural Resources Conservation Service).

USDA-NRCS. 2004. *National Engineering Handbook* (Washington DC: US Department of Agriculture, Natural Resources Conservation Service).

USDA-NRCS. 2007a. "Chapter 8: Threshold channel design," in: *Part 654 Stream*

Restoration Design National Engineering Handbook (Washington DC: United States Department of Agriculture, Natural Resources Conservation Service).

USDA-NRCS. 2007b. "Chapter 7: Grassed waterways," in: *Part 650 Engineering Field Handbook* (Washington D.C.: United States Department of Agriculture, Natural Resources Conservation Service).

USDA-NRCS. 2010. *National Engineering Handbook: Part 630 Hydrology* (Washington DC: US Department of Agriculture, Natural Resources Conservation Service).

USDA-SCS. 1989. *Hydrology Training Series: Module 104 – Runoff Curve Number Computations* (Washington DC: US Department of Agriculture, Soil Conservation Service).

USDA-SCS. 1990. *Hydrology Training Series: Module 206A – Time of Concentration* (Washington DC: US Department of Agriculture, Soil Conservation Service). 43pp.

USGS. 1982. *Guidelines for Determining Flood Flow Frequency Bulletin #17B* (Washington DC: US Geological Survey).

USGS. 2001. *Estimates of Evapotranspiration from the Ruby Lake National Wildlife Refuge Area, Ruby Valley, Northeastern Nevada, May 1999–October 2000* (Washington DC: US Geological Survey).

USGS. 2016. "Streamflow: The water cycle," US Geological Survey. https://water.usgs.gov/edu/watercyclestreamflow.html (accessed January 8, 2019).

USGS. 2017. "How much water is on Earth," US Geological Survey. https://water.usgs.gov/edu/gallery/global-water-volume.html (accessed April 12, 2019).

van Mullem, J. 1992. "Soil moisture and runoff: another look," in: E.T. Engman (ed.), *Irrigation and Drainage Sessions at Water Forum '92* (Reston, VA: ASCE), 366–71.

VDCR. 1999. *Virginia Stormwater Management Handbook* (Richmond, VA: Virginia Department of Conservation and Recreation, Division of Soil and Water Conservation).

Viessman, W., Lewis, G.L. 2002. *Introduction to Hydrology* (New York, NY: Pearson).

Wang, X., Liu, T., Li, F., *et al.* 2014. "Simulated soil erosion from a semiarid typical steppe watershed using an integrated aeolian and fluvial prediction model," *Hydrological Processes*, 28: 325–40.

Wang, X., Liu, T., Shang, S., Yang, D. 2010b. "Estimation of design discharge for an ungauged overflow-receiving watershed using one-dimensional hydrodynamic model," *International Journal of River Basin Management*, 8: 79–92.

Wang, X., Liu, T., Yang, W. 2012. "Development of a robust runoff-prediction model by fusing the Rational Equation and a modified SCS-CN method," *Hydrological Sciences Journal*, 57: 1118–40.

Wang, X., Melesse, A. M. 2005. "Evaluation of the SWAT model's snowmelt hydrology in a northwestern Minnesota watershed," *Transactions of the ASABE*, 48: 1359–76.

Wang, X., Melesse, A. M. 2006. "Effects of STATSGO and SSURGO as inputs on SWAT model's snowmelt simulation," *Journal of American Water Resources Association*, 42(5): 1217–36.

Wang, X., Melesse, A. M., Yang, W. 2006. "Influences of potential evapotranspiration estimation methods of SWAT's hydrologic simulation in a northwestern Minnesota watershed," *Transactions of the ASABE*, 49: 1755–71.

Wang, X., Sample, D.J., Pedram, S., Zhao, X. 2017. "Performance of two prevalent infiltration models for disturbed urban soils," *Hydrology Research*, 48: 1520–36.

Wang, X., Shang, S., Qu, Z., Liu, T., Melesse, A. M., Yang, W. 2010a. "Simulated wetland conservation-restoration effects on water quantity and quality at watershed scale," *Journal of Environmental Management*, 91: 1511–25.

Wang, X., Shang, S., Yang, W., Melesse, A. M. 2008. "Simulation of an agricultural watershed using an improved curve number method in SWAT," *Transactions of the ASABE*, 51: 1323–39.

Weibull, W. 1951. "A statistical distribution function of wide applicability," *Journal of Applied Mechanics*, 18: 293–7.

WFN. 2014. "Water footprint and virtual water," The Water Footprint Network (WFN). http://waterfootprint.org/en/ (accessed December 25, 2018).

Wikipedia. 2017. "Water use." https://en.wikipedia.org/wiki/Water_use (accessed December 25, 2018).

Wikipedia. 2018. "Dujiangyan." https://en.wikipedia.org/wiki/Dujiangyan (accessed April 30, 2019).

WMO. 2009. *Manual on Estimation of Probable Maximum Precipitation (PMP)* (Geneva, Switzerland: World Meteorological Organization).

Wurbs, R. A., James, W. P. 2002. *Water Resources Engineering* (Upper Saddle River, NJ: Prentice Hall).

Yarnell, D. L. 1934. "High precipitation for short periods in the United States," *Transactions, American Geophysical Union*, 15.

Index